www.wadsworth.com

wadsworth.com is the World Wide Web site for Wadsworth and is your direct source to dozens of online resources.

At *wadsworth.com* you can find out about supplements, demonstration software, and student resources. You can also send email to many of our authors and preview new publications and exciting new technologies.

wadsworth.com
Changing the way the world learns®

SOCIOLOGY OF MARRIAGE & THE FAMILY

Gender, Love, and Property

Fifth Edition

Scott Coltrane
University of California, Riverside

Randall Collins
University of Pennsylvania

Wadsworth
Thomson Learning

Australia • Canada • Mexico Singapore
Spain • United Kingdom • United States

Publisher: Eve Howard
Assistant Editor: Dee Dee Zobian
Editorial Assistant: Stephanie Monzon
Marketing Manager: Matthew Wright
Project Editor: Jerilyn Emori
Print Buyer: Karen Hunt
Permissions Editor: Joohee Lee
Production Editor: Mike Ederer/Graphic World Publishing Services
Text Designer: Andrew Ogus ■ Book Design
Cover Development: Andrew Ogus

Cover Designer: Stephen Rapley
Photo Researcher: Sue Howard
Copy Editor: Lisa Dunlap/Graphic World Publishing Services
Illustrator: Deep Dutta/Graphic World Illustration Studio
Cover Image: Synthia Saint James
Text and Cover Printer: Transcontinental Printing Inc./Louiseville
Compositor: Graphic World, Inc.

Library of Congress Cataloging-in-Publication Data

Coltrane, Scott.
    Sociology of marriage and the family : gender, love, and property / Scott Coltrane, Randall Collins.—[5th ed.]
        p. cm.
    Rev. ed. of: Sociology of marriage and the family / Randall Collins, Scott Coltrane. c1995.
    Includes bibiographical references and index.
    ISBN 0-534-57960-4 (alk. paper)
    1. Family—United States. 2. Family—Cross-cultural studies. 3. Family life education—United States. I. Collins, Randall, 1941-II. Collins, Randall, 1941- Sociology of marriage and the family. III. Title.

HQ536 .C716 2000
306.8—dc21   00-32498

Wadsworth/Thomson Learning
10 Davis Drive
Belmont, CA 94002-3098
USA

For more information about our products, contact us:
Thompson Learning Academic Resource Center
1-800-423-0563
http://www.wadsworth.com

International Headquarters
Thomson Learning
International Division
290 Harbor Drive, 2nd Floor
Stamford, CT 06902-7477
USA

UK/Europe/Middle East/South Africa
Thomson Learning
Berkshire House
168-173 High Holborn
London WC1V 7AA
United Kingdom

Asia
Thomson Learning
60 Albert Street, #15-01
Albert Complex
Singapore 189969

Canada
Nelson Thomson Learning
1120 Birchmount Road
Toronto, Ontario M1K 5G4
Canada

Brief Contents

Contents

Preface

In this edition of *Sociology of Marriage and the Family: Gender, Love, and Property*, we renew and expand on the vision that inspired us to write a uniquely sociological family textbook. Trained in classical traditions emphasizing history, culture, politics, and theory, we had been disappointed by the narrow view of families presented in college textbooks. Most marriage and family texts took an individualistic or clinical approach to the topic, reducing complex institutional and social phenomena to matters of individual choice and interpersonal problem solving. But families are much more than this, as we demonstrate in our current analysis of the family as a social institution. We not only show how families are deeply embedded in larger economic, political, and cultural processes, but we bring our analysis up to date by investigating a range of new family forms and social issues.

Ours was the first family textbook to detail the many ways that marriage and family practices are linked to larger systems of gender, class, and ethnic stratification. Relying on the concept of the family as a property system, we show how families are both producers and products of patterns of inequality in the larger society. In the past decade, family research has moved in exactly this direction. Issues of ethnic diversity, gender relations, economic inequality, and social policy are now at the forefront of the family studies field, and our textbook provides a theoretical framework to help students understand these emergent issues. In this new edition, we incorporate even more of the latest research on families in their social contexts, offering a wide variety of historical and cross-cultural comparisons to demonstrate how families vary across time and place. In addition, we provide a wealth of information about contemporary personal concerns, family dilemmas, and policy issues that are confronting Americans today.

This textbook is intended for use in several ways, and in different types of courses, but is best suited for classes that emphasize the family as a social institution and those that focus on marriages and families in relation to larger social issues and societal processes. It is ideal for use in any sociology course focusing on families and the social processes that influence them (e.g., Sociology of the Family, Marriage and Families, Contemporary Marriages, Family Diversity, Family Issues, Family History, Gender and Families, Changing Families, Families in Society, Families in Cross-Cultural Perspective). It is also suitable for use in courses offered by Departments of Family Studies, Psychology, Family Science, Counseling, Consumer Science, Human Development, Social Work, Public Administration, Nursing, Criminology, Women's Studies, or any department or program wishing to emphasize a sociological, theoretical, or critical approach to family studies. While we emphasize theory and draw on classical scholarly traditions, we also use real world examples and adopt a conversational tone, making the material highly accessible to students.

What does it mean to study the family as a social institution? Our text focuses on why families take particular forms in response to specific societal contexts. We are especially concerned with understanding what social forces are responsible for shaping specific human forms of romance, sexuality, cohabitation, marriage, childrearing, housework, grandparenting, and a host of other family-related ideals and practices. Because we have been going through a period of rapid social change regarding marital and family processes, and because family issues are near the top of the public policy agenda, it is an exciting time to adopt a sociological perspective in the study of marriages and families. In particular, the shifting positions of women and men in society are remaking contemporary families. However we may stand personally on the issues raised by feminism, they are central to developing a comprehensive understanding of what is happening to families today. Similarly, in an increasingly diverse society with increasing poverty and changing patterns of discrimination, it is important to understand how families can play a role in perpetuating or resisting new forms of inequality.

Analyzing the family as a social institution is also important to understanding a range of practical issues that bear on the work of many different professions. Students preparing to become therapists, counselors, social workers, teachers, childcare providers, parent educators, family law specialists, and law enforcement personnel will all benefit from using this textbook. The family is deeply implicated in the problems of poverty, crime and delinquency, school achievement and career success, and, more specifically, in the controversies over child and spouse abuse, marital and extramarital sex, same-sex marriage, abortion, adoption, artificial insemination, divorce, child custody, and many other social issues. Politicians, political activists, and citizens in general need to be concerned with the sociology of the family in order to act insightfully on these issues as they intersect with public policy.

We are troubled by the tendency of family textbooks to ignore recent developments in social theory and treat family issues as individual matters. Sociological theory is not merely an esoteric subject for intellectuals and can be most useful during times of rapid social change. Insofar as sociology is successful in explaining why certain social patterns exist under certain circumstances, it is presenting information that is of both theoretical and practical importance. This book attempts to show both the basic framework of our knowledge and what we can do with it. Without both theory and application, students will be less able to help diverse families overcome the many problems they face today. And without a firm grounding in sociological theories that encourage students to think for themselves, they will be unable to help families face an uncertain future.

NEW TO THIS EDITION

For this edition, we have added many new features and expanded each chapter. The entire text has been updated with reference to the latest vital statistics and demographics from the Census Bureau, National Center for Health Statistics, Bureau of Labor Statistics, and other government agencies. New issues of concern to family professionals have been added, including discussion of recent developments in dating, romance, sexual practices, cohabitation, communication, interracial marriage, fertility decisions, reproductive technologies, adoption, childrearing, housework, fathering, single parenting, stepparenting, grandparenting, elder care, family violence, and other topics. Material on family diversity is highlighted throughout the text, so that readers can see how membership in a particular ethnic group might influence different aspects of family life including love, sex, marriage, childrearing, and aging. Family policy is also given a more prominent place in this edition, to enable students to better understand how debates over issues like health care, child care, abortion, marriage laws, family leave, workfare, flextime, and social security can have profound influence on the organization of everyday life in American families.

New feature boxes in every chapter, called *A Closer Look*, provide readers with easily accessible and provocative examples of the ideas discussed in the text. These features explore a range of relevant issues, including controversies over family textbooks, cross-cultural comparisons of family practices, the relationship between religion and family, commercial aspects of romance, gay and lesbian sexuality, birth control practices, and other important topics. Many of these new boxes familiarize students with contemporary challenges facing American families by excerpting from popular books about poverty and motherhood (Ruth Sidel's *Women and Children Last*), economic restructuring and racial discrimination (Lilian Rubin's *Families on the Fault Line*), immigration and modernization (John Harriss' *The Family*), the process of breaking up (Diane Vaughn's *Uncoupling*), wife abuse and divorce (Demie Kurz's *For Richer, For Poorer*), and the stresses of juggling work and family (Kerry Daly's *Families and Time*). Every chapter contains new features and artwork, along with new tables and figures, so that students can better understand emerging patterns of family life.

Also new to this edition is a systematic listing and description of relevant theories and concepts in chapter 1. We adopt microstructural and social constructionist approaches to studying families informed by conflict theory, but at the same time, we acknowledge other important theoretical contributions to the study of families. In the early chapters of the book, we reflect on the conceptual foundations of the field, refuting outmoded approaches at the same time that we borrow from classic insights about power and social cohesion. In chapters 3 and 4, we incorporate new scholarship about families from other times and places so that students can place their own experiences in historical and comparative perspective. In chapter 5 we present a sweeping synthesis of family life in the late twentieth century, carefully laying the groundwork for students to make their own predictions about future trends. In chapter 6, we engage a central paradox of late 20th and early 21st century America: In spite of economic prosperity, increasing numbers of children and families are living in poverty or barely making ends meet. Ours continues to be one of the few family textbooks that seriously engages students in debates about class differences and encourages them to think about how families reflect and perpetuate economic inequality. In chapter 7 we explore new data on ethnic/racial diversity in America's families, revisiting some older explanations, and offering some new ideas about the ways that ethnic background and racial identity can influence family life. We provide an expanded treatment of ethnicity and race in this edition because of the central importance of understanding and accepting diversity in America today. In chapters 8 and 9, we incorporate new historical material, contemporary examples, and survey data to show how flirting, courting , and sex continue to be governed by romantic cultural ideals and market principles. In chapters 10, 11, and 12, we include emerging research on the realities of making babies, keeping house, and raising children in the modern era. We document various new developments in the social organization of these activities, but also explore how structural constraints and inherited cultural assumptions about gender continue to shape such family practices. We include recent findings about the harsh realities of family violence in chapter 13, again placing the findings in theoretical perspective so that students can understand reasons for current patterns, and come to grips with how we might reduce the incidence of wife abuse, child abuse, and elder abuse. In chapter 14 we look at recent trends in divorce and incorporate new material on single mothers, single fathers, remarriages, and blended families. Chapter 15 includes recent research findings about later life transitions and related family matters of great concern to our aging population. Finally, we encourage students to think for themselves with a provocative final chapter that challenges them to apply what they have learned to make their own projections about the future of families in America.

SUPPLEMENTS

A variety of new pedagogical tools make the book more instructor-friendly. There is a new glossary (at the end of the book) with important terms bolded in the text, and expanded summary points are listed at the end of every chapter. A variety of instructional aids developed specifically for this edition are now available through the Wadsworth Web site, including a new test bank, classroom exercises, individual Internet exercises, and access to InfoTrac College Edition with exercises. The Wadsworth sociology resource center, with links to these instructional aids, may be accessed via http://sociology.wadsworth.com. Also available is a "print" version of the instructors manual/testbank. Finally, three different "editions" or volumes of the CNN Marriage and Family Today Videos are available for classroom use. These compelling clips are ideal for launching lectures, illustrating applications, and sparking discussion.

ACKNOWLEDGMENTS

No project of this magnitude would be possible without the help of many people. Michele Adams conducted invaluable research into an enormous range of topics and offered expert and detailed technical assistance. To make the text easier to use, Michele Adams and Melinda Messineo created many innovative teaching aids and developed an excellent instructors' manual and test bank. Many colleagues shared comprehensive expert reviews of broad subject areas, including Katherine Allen, Terry Arendell, Rosemary Blieszner, Karen Bogenscheider, F. Scott Christopher, Marilyn Coleman, Martha Cox, Ann Crouter, David Demo, Kathleen Ferraro, Mark Fine, Frank Furstenberg, Christine Ward Gailey, Lawrence Ganong, Michael Johnson, Ross D. Parke,

Maureen Perry-Jenkins, Rena Repetti, Karen Roberto, Judith A. Seltzer, and Susan Sprecher. Careful reviews of previous editions of the text by Bradley Jay Buchner, Bloomsberg University; Joan Huber, Ohio State University; S. Philip Morgan, University of Pennsylvania; and Barbara Scott, Northern Illinois University, helped us formulate specific plans for this edition. Also, Miriam Johnson of the University of Oregon and Stephen Sanderson of Indiana University of Pennsylvania wrote helpful reviews. The editorial, production, and marketing staff at Wadsworth/Thomson Learning provided efficient professional guidance through every phase of the last revision. The leadership and support of Publisher Eve Howard at Wadsworth was extremely beneficial, as was the careful attention to detail of assistant editor Dee Dee Zobian, production editor Jerilyn Emori, and Mike Ederer and all the folks at Graphic World, Inc. We definitely value and appreciate the long-standing encouragement and support of colleagues at our home institutions, the University of Pennsylvania and the University of California, Riverside. Finally, and most importantly, the ongoing love and support we have received from our own immediate and extended families during the many years of working on this project have been profound. Without their encouragement, critical reflection, and patience, this book would not exist.

About the Authors

Scott Coltrane completed his undergraduate studies at Yale University and the University of California, Santa Cruz, and received M.A. and Ph.D. degrees in sociology from the University of California, Santa Cruz. He is the recipient of the C.S. Ford Cross-Cultural Research Prize and won the Distinguished Teaching Award from the University of California, Riverside, in 1997. Dr. Coltrane has published articles on family and gender in various scholarly journals, including *American Journal of Sociology, Social Problems, Sociological Perspectives, Journal of Marriage and the Family, Journal of Family Issues, Gender & Society, Sex Roles, and Masculinities*. He is the author of *Gender and Families* (1998), and his book *Family Man* (1996) was selected as a *Choice* Outstanding Academic Book by the American Library Association. Coltrane is Professor and Chair of the Department of Sociology at the University of California, Riverside; Associate Director of the UCR Center for Family Studies; and 2000 President of the Pacific Sociological Association.

Randall Collins graduated from Harvard in 1963 and received an M.A. in psychology at Stanford University the following year. At the University of California, Berkeley, Collins received M.A. and Ph.D. degrees in sociology, and later taught at the University of California at San Diego and Berkeley, and at the University of Wisconsin, Madison. He has contributed to many sociological journals and is the author of several books, including *Conflict Sociology* (1975), *Sociology Since Mid-Century: Essays in Theory Cumulation* (1998), and *Macro-History: Essays in Sociology of the Long Run* (1999). A member of the Crime Writer's Association of Great Britain, Collins also wrote *The Case of the Philosopher's Ring* (1979), a detective novel set in London and published both in the United States and Great Britain. Collins has been a Visiting Member at the Institute for Advanced Study in Princeton, and was awarded the Theory Prize by the Theory Section of the American Sociological Association in 1982. He was president of the Pacific Sociological Association in 1993. Randall Collins is now a Professor of Sociology at the University of Pennsylvania.

1

Sociological Theory: Explaining Families in a Stratified Society

INTRODUCTION

The family is a multisided reality. From one point of view, there is nothing more ordinary. Parent and child, brother and sister, sitting in front of the TV, doing the dishes, taking out the garbage. What could be more mundane, more routine? The family in that sense hardly seems worth paying attention to in a college course. It is merely what happens when you stay at home, away from where more exciting things are going on.

And yet the very same institution, the family, is incredibly dramatic. Most novels and an extraordinary number of dramas have been written about families, certainly more than about any other subject. One of the oldest of all Greek plays, *Oedipus Rex,* is about a family. *War and Peace* is the story of a family; so are *Hamlet, The Grapes of Wrath, The Godfather, Star Wars,* and even *The Lion King.* One can hardly go to the movies without encountering a story about a family, especially if the movie is a drama or comedy rather than merely an action adventure. And even the latter usually has its family overtones, if we count romantic subjects. For in real life, love, romance, and sexual affairs are all part of the orbit of the family; they are either part of the courtship and mating process through which families are created or part of the way they fall apart and become reshaped. So, ironically, even if one leaves the humdrum of home for a night out "where the action is," one is more than likely to encounter some dramatics related to the family, either in the sphere of personal romance or, even more likely, by sitting in a darkened theater watching the story of some other family.

What should we make of this? It is not simply a matter of literature versus life. Every family has these two characteristics. Great literature captures the conflict between them: it is what Flaubert's *Madame Bovary* is about, as well as many modern novels from the Harlequin romances to modern feminist novels. And the contrast is there in everyone's own life. Falling in love, dating, courting, getting married, and having children are among the more exciting things in life. Family life, too, provides a lot of negative excitement: family quarrels, jealousies, parents' and children's struggles with each other, extramarital affairs, the traumas of separation, divorce, and death. Power struggles in the home are a major theme, both in the psychological literature and in the details of our everyday lives. Sigmund Freud was more than anything an analyst of the aftereffects of family life; most clinical psychology and counseling deal with repairing individuals and reorganizing the way they deal with their families. Even more extreme is the family as a favorite location for violence and abuse. Wives, children, and sometimes husbands have a better chance of being battered in their own house than out on the streets, and murders occur most commonly between spouses.

Families then, are mundane and routine. And families are ultradramatic, accounting for both our personal highs and our worst lows. Moreover, these two sides are connected. The dramatics of love and sex, the ideal image of the bride, the joyfulness surrounding a newborn baby—all of these lead to the routine, mundane side of family life. The transition from dramatic to mundane is often a shock; the letdown causes much of the disgruntlement and conflict that take place in families. We would put it more strongly: the family is the subject of powerful ideals—or illusions, even ideologies—which constitute a major part of its attraction. But as family life gets under way, reality asserts itself: material work has to be done, the diapers changed, the bills paid. The mundane world has its structure too. It is not only alliance and support, but property and power, domination and conflict.

We are not saying that family life simply consists of two phases, a naive, idealistic phase before marriage and nothing but mundane realities and conflicts thereafter. The ideals and the emotional high points keep reasserting themselves throughout the months and years; in some families and relationships, this happens quite a lot. But the mundane world of monthly bills, work that has to be done, and power that some people wield and to which others must acquiesce—all this is always present, whether we are conscious of it or not. Families that enjoy emotional success tend to be those that have worked out their mundane family lives satisfactorily. Otherwise, when the mundane world intrudes, as inevitably it must, its hard realities always take priority and burst the bubbles of our ideals.

Defining Families

The U.S. Census Bureau defines a family as two or more persons related by birth, marriage, or adoption who live together as one household. This definition captures something important about living arrangements and legal relationships, but most of us use the word *family* to refer to much more than this. Sometimes it refers to people: long dead ancestors, distant cousins one never sees, siblings or other blood relations, divorced people who live apart, and friends who are so close that they have become honorary family members. Sometimes the word *family* refers to children, as in "Do you have a family?" or "She's in a family way." Other times, family implies a specific feeling, usually a special type of love and caring. The term "the family" is also frequently used in political and religious debates, as we note later, to claim the moral high ground in arguments about who can get married, who is entitled to family resources, and who should receive public benefits.

Multiple meanings of family are evident in America today, but even more meanings are reflected in the historical and cross-cultural record that we review in the following chapters. For example, in ancient Greece, "family" *(oikos)* referred to the household economy, including the land, house, and servants (see *A Closer Look* 2.5 in chapter 2). In medieval Europe, peasants who lived on feudal estates were considered part of the lord's "family," and he was called their "father" *(pater)* even though they were not related to him by blood. In some cultures, nonkin adults are treated as family members and act as coparents toward children; contemporary examples include godparents *(compadres* among Mexicans and Mexican Americans). Similarly, in contemporary Native American (Indian) families, the terms used to describe family relationships are more encompassing than narrow English usage would imply: a "grandmother" may actually be a child's aunt or grandaunt, and "cousin" may have variable meaning not necessarily based on birth and marriage.

[handwritten margin note: meanings in other cultures]

To understand families and the specific relations they represent, we must recognize that the term and the idea are socially constructed; that is, the meaning of family changes in response to a wide variety of social and personal conditions. To reflect this understanding, we often refer to families in the plural (instead of *the* family, as if there was only one type). As many scholars before us, however, we also sometimes make generalizations about families by referring to "the family" as a social institution—a patterned set of expectations and activities embedded in a larger set of social, economic, legal, political, and cultural practices. We examine definitions of families in chapter 2 and show how academics are especially concerned with such things in *A Closer Look* 1.1.

A CLOSER LOOK 1.1
Controversy over Family Sociology Textbooks

In September of 1997, a private nonprofit advocacy organization in New York released a report on undergraduate college textbooks about marriage and families, giving most of them failing grades and labeling them "a national embarrassment." The report, titled "Closed Hearts, Closed Minds: The Textbook Story of Marriage" claimed that "what students are being taught by these textbooks is probably doing them more harm than good" (Glenn 1997, 3).

As part of its larger mission to revitalize marriage, the Institute for American Values' Council on Families commissioned Norval Glenn, then a Professor of Sociology at the University of Texas and Research Director for the Council, to conduct a study of marriage and family textbooks. They funded distribution of the study results to news organizations, publishers, and others across the country. As a result, news stories summarizing the report's findings appeared during the week of September 16–30, 1997, in the *New York Times, Washington Post, Los Angeles Times, Chicago Sun Times, U.S. News and World Report, Chronicle of Higher Education,* and other publications (Howery 1998). Press coverage resembled the opening of the article from the *New York Times:*

College students are being taught a pessimistic and sometimes inaccurate view of marriage, with great emphasis on issues like divorce and domestic violence, and little attention to the benefits of marriage, particularly for child rearing, according to a new study of undergraduate textbooks on marriage and family. "Both by what they say and sometimes even more importantly, by the information they omit, these books repeatedly suggest that marriage is more a problem than a solution," said the report, issued today by the Institute for American Values' Council on Families, a nonpartisan group of family experts. "The potential costs of marriage to adults, particularly women, often receive exaggerated treatment, while the benefits of marriage, both to individuals and society, are frequently downplayed or ignored."

By looking at chapter titles and counting the number of pages mentioning specific topics, Glenn's report criticized the marriage and family textbooks for putting too much emphasis on adults and too little emphasis on children:

These textbooks are characteristically uninterested in the effects of family change on children. . . . Just 24 of 338 total chapters in these textbooks deal primarily with family effects on children. Moreover, even in some of these chapters, up to half of the space is actually devoted to other matters. Far more space—at least three times as much—is devoted to adult relations, without regard to how they affect children. . . . The average amount of space devoted to child abuse is just over seven pages, compared to a mean of over twelve pages devoted to family violence affecting adults.

THE FAMILY IN CURRENT CONTROVERSY

The family has become a topic of public controversies, cutting in all directions. It is not just a source of drama in one's own life. One can hardly pick up a newspaper or turn on the TV without coming across an issue closely connected to the family.

Take crime, for instance. We hear a great deal about "crime in the streets"—conveying an image of youths away from homes, out of control of parents, and scorning the activities that their parents would approve of. The rhetoric is no doubt exaggerated, but nevertheless, there is a connection. Gangs are a kind of substitute for families and flourish where families are weakest. Another dominant issue today is drugs, and the gang violence that goes on to control the drug trade. Here again we find a public controversy tied to the family in a negative way. The issue, however, is not as simple as some politicians put it—that the family is breaking down, traditional values are disappearing, and therefore crime is rampant.

Finally, the report chastises the family sociology textbooks because they "fail to draw the obvious conclusion that the rapid increase in single parent families and stepfamilies has very likely increased the amount of child abuse in the United States" (Glenn 1997, 14–15).

Other reviews of the same family textbooks, published in scholarly journals at about the same time, did not receive any attention from the media (see Risman and Tomaskovic-Devey 1998; Johnson 1997; Mann et al. 1997). This prompted some family sociologists, like Judith Stacey, author of *In the Name of the Family* (Beacon Press, 1996), to comment on the political ramifications of Glenn's report:

> The Institute for American Values and its Council on Families operate in the terrain not of social science, but of virtual social science. This genre of cultural politics disseminates selective, tendentious representations of social science research via mass media technologies. Closed Hearts is but the latest public relations coup that the Institute has achieved. During the past decade the Institute for American Values has waged a vigorous, influential, political campaign for neoconservative "family values" while successfully representing itself as "nonpartisan" and "devoted to research, publication, and public education." . . . The reality is quite different and a matter for professional concern. Far from nonpartisan scholarship, this is stealth partisan politics costumed in social scientific drag. . . . From the complexity of pertinent research, it selectively broadcasts exclusively data that support its drumbeat message that the decline of married-couple families is the nation's key social problem. Privately funded (including right-wing Olin and Heritage Foundation support) and boasting a full-time staff of five, the Institute deploys academic credentials while operating beyond even the modest constraints of peer review. Its publicists enjoy a direct pipeline to the nation's elite opinion-shaping apparatus, securing habitual prime-time appearances on PBS, CNN, commercial broadcast news and talk shows, and continual coverage of their views in the mainstream print media. In short, Glenn and his associates have made an effective end run around the academy, politicizing scholarship in dangerous ways (Stacey 1998).

Sources:
Glenn, Norval. 1997. *Closed Hearts, Closed Minds: The Textbook Story of Marriage.* New York: Institute for American Values.
Glenn, Norval. 1997. "Marriage Is Not a Dirty Word: College Textbooks Teach Otherwise, Presenting a Dated, One-Sided Distortion." *Los Angeles Times* B-7 (Sept. 16).
Howery, Carla. 1998. "Sociologists Differ about Family Textbooks' Message." *Footnotes* 26 no. 1, (Jan.): 7.
Johnson, Miriam. 1997. "Teaching, Research, and Reference." *Contemporary Sociology* 26:395–399.
Mann, Susan, Michael Grimes, Alice Abel Kemp, and Pamela Jenkins. 1997. "Paradigm Shifts in Family Sociology? Evidence from Three Decades of Family Textbooks." *Journal of Family Issues* 18:315–49.
Risman, Barbara, and Donald Tomaskovic-Devey. 1998. "Sociologists Differ about Family Textbooks' Message." *Footnotes,* 26, no. 1 (Jan.): 10.
Stacey, Judith. 1998. "Sociologists Differ about Family Textbooks' Message." *Footnotes: American Sociological Association Newsletter,* 26, no. 1 (Jan.): 10. Used by permission.

To understand the issue better, we could look at the intersection between the family, race, and crime. A major target of the antidrug crusade is the inner-city youth gangs, especially those composed of African Americans. The inner city is precisely the place where conventional families are in crisis. The majority of African American children now are reared in single-parent families, typically by unmarried or divorced women living at poverty level. Why are there so few marriages in this group? Part of the reason is residential segregation and the lack of jobs available to young men living in the inner city. The result is a vicious circle: poverty and crime greatly weaken the family and its ability to control and support children; and the inability of these families to do much for their children has the effect of perpetuating poverty and crime.

Here we have crime occurring outside the family but related to what is happening inside. We are also faced with a good deal of news about crime and controversy among family members. In fact, some of the longest running and biggest news stories of the 1990s were about family crimes: Did O.J. kill his estranged wife? Did President Clinton lie about an extramarital affair? Did the Menendez brothers kill their parents to get their inheritance? Did JonBenet Ramsey's parents kill her? Family dramas and family tragedy have become a common part of the daily news. Newspaper reporters and television anchors are quick to play on our sympathies about the death of loved ones and our outrage at the cruelty of family relationships gone bad.

Then there are the politicians campaigning for election. How do they treat family issues? Whatever their political stripe, they typically worry publicly about the declining state of American families and usually announce their commitment to family traditions in the most visual way. Right on the front page of their campaign literature, we see pictures of the candidates, posing with spouse and children and smiling at the camera. Apparently the way to get elected is to look like part of a conventional, happy family.

Even academics have joined the act recently, with political think tanks publishing findings about how family sociology textbooks like this one don't say enough good things about marriage (see *A Closer Look* 1.1).

Stratification: Inequality Inside and Outside Families

Is there a more general pattern, a sociological insight that helps explain various kinds of family issues? We propose that one thread that weaves them together is **stratification**—the pattern of social equality or inequality. Most sociologists, one way or another, deal with some aspect of stratification. It has several dimensions, which we may find both inside and outside the family; families are part of the larger system of stratification in American society. In addition, we should pay attention to inequalities inside families—in the relationships between husbands and wives, parents and children, and among sisters and brothers. Gender plays an important role in the patterns of stratification within families (see *A Closer Look* 1.2).

Stratification is usually analyzed in three dimensions: (1) **Economic class** is the dimension ranging from poverty to wealth, and includes the occupations and careers by which people get their income. (2) **Power** determines who controls whom. It ranges from the level of political elites at the most macro level of the state to the most intimate powers, such as who decides when to have sex. (3) **Status** is the dimension concerning who is held in most respect. It is manifested in the way people talk about each other and in the way they display themselves, their homes, and their possessions. It is a realm of symbols and feelings, ranging all the way from admiring smiles to not even noticing someone is there.

We shall say a lot more about these aspects of stratification throughout this book. For now, notice what a stratification perspective tells us about the controversies we have already mentioned. We have seen that issues of crime, drugs, and gangs are woven together with family patterns, especially in the poverty level of economic class. In addition, the status dimension is involved when we examine the way in which racism is a part of the chain of causes that makes poverty, as well as gangs, so common in urban African American and Latino communities and why political sentiment against welfare is so strong. We can examine how youth gangs and rap culture are ways in which youths can claim some power, as well as status, in a society in which they have very little advantage on any of the stratification dimensions.

We should also take a close look at the position of African American families in the entire class structure of American society. Because a sizable proportion of African American families are not in the poverty sector, we can avoid stereotyp-

In the United States, the gap between rich and poor households is increasing. Moreover, the life chances of the poor are jeopardized by their lack of financial resources.

ing African Americans in general by taking class into account. By the same token, we should be aware that African American poverty is only the most visible part of the poverty sector. The majority of poor families are white. Once again, we need to be aware of class if we are to understand the range of families in our society (see *A Closer Look* 1.3).

The instances of crime inside the family should make us aware of the ways in which stratification operates. Spouse abuse is a blatant example of the exercise of *power,* typically by men against women. In the same way, child abuse represents the power of parents over children taken to an extreme. Many parents see this use of power as normal, regularly spanking their children to force outward compliance. Others make few distinctions between spanking and hitting, frequently bruising their children or worse. Fortunately, most husbands and most parents do not push stratification to such an extreme inside families. But abuse happens because there is a larger pattern of family stratification, usually taken for granted. Men who abuse their wives typically feel that men have the right to exercise power in the family, and they become angry when they feel their power is challenged. Violent abuse is only the tip of the power

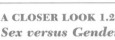

A CLOSER LOOK 1.2
Sex versus Gender: What's the Difference?

The words *sex* and *gender* are used in a special way in this book and in the writings of many contemporary sociologists. Both are ordinary words, but their commonsense usage creates certain confusions that make it preferable to distinguish between them.

In everyday language, the word *sex* has two different meanings. It refers to *(a)* the characteristic of being male or female but also *(b)* erotic behavior. It is in sense *a* that one might refer to "the sexual division of labor," whereas sense *b* comes out in such expressions as, "How's your sex life?"

This double meaning of *sex* can be confusing. *Sexes* (sense *a*) are not necessarily *sexy* (sense *b*).

There is a further problem with an indiscriminate use of the word *sex*. The term has a strong biological connotation, in both sense *a* and sense *b*. Males and females are primarily distinguished by their genitals, and erotic behavior is an act of the physical body. But a good deal of the way in which the sexes (sense *a*) behave is not biologically determined at all, but is *cultural*. Biological males and females learn how to play the *social roles* of men or women. Just what these behaviors are depends on the society they happen to live in and on each person's individual experience.

To distinguish between the *biological equipment* of males and females and the *social roles* built upon them, many sociologists now use two different terms. *Sex* refers to the biological person and his or her physical actions, and *gender* is used to refer to the social roles of being male or female. Thus, female and male styles of dress or talk would be seen as gender differences constructed in daily life; similarly, the fact that women earn less than men is a form of *gender* discrimination. (We should be aware, though, that many sociologists continue to say *sex* discrimination or *sex* differences; it is difficult, if not impossible, to completely standardize everyone's terminology.)

This distinction enables us to use the word *sex* in a clearer way. Its meaning now specifies the biological and erotic dimension. To talk about *sexual* property rights or *sexual* possessiveness, as we will do in this book, is to discuss the rights of erotic access that someone claims over another person's body. How people behave erotically, though, is not just "natural"; it has a biological component, but it, too, is strongly influenced by society.

In sum: **Gender** reminds us that the social roles of being male and female are largely produced by the culture. **Sex** refers us explicitly to biological characteristics and erotic behavior.

iceberg, however, because power is played out in many ways inside and outside families.

Modern Forms of Stratification

One might ask, "What does this have to do with me or with the families I know?" Most families are not involved in murders, serious abuse, or gang crime. Nevertheless, *a family's place in the pattern of stratification is the most important thing we can know about it.* Externally, families are the main way in which each individual becomes placed in the class structure of society. Internally, the degree of stratification between husbands and wives and between parents and children tells us a great deal about what will happen in that family, ranging from housework to sex to family quarrels and personal happiness.

We will say more about these processes later, so let us focus on just one key point there. The family has always been the building block of the class structure. The upper class consists of wealthy families, who usually have managed to pyramid their wealth by inheritance across the generations. The lower class has had a weaker family structure, consisting of many isolated individuals and of families less able to rear children effectively for the competition for economic position.

A CLOSER LOOK 1.3
How to Avoid Talking About Social Class

Most Americans do not like to talk openly about social class. In actuality, social class affects so many aspects of the world around us that we have to invent terms to refer to it. In the 1980s, a favorite newspaper cliché that referred to certain upper-middle-class persons was the word *yuppies.* Other terms that describe the same group are *upscale* and *life in the fast lane.* Back in the 1950s, the term *leadership* was a way of referring to wealthy and powerful people who controlled business organizations, government, and the professions. An earlier generation created the term *high society,* which is now abbreviated to just plain *society,* in the sense of the "society pages" of a newspaper—a peculiar terminology implying that only a few wealthy people count as members of society. For referring to the working class or to the lower class, we have such euphemisms as *low rollers,* or *the streets*—as in the phrases *crime in the streets* or somebody having *street smarts.* Generations ago we spoke of "the other side of the tracks."

Create a game of "sociological trivia." See how many terms you can think of that people use when they talk about different social classes but wish to avoid the words *upper class, middle class, lower class,* and *working class.*

Between these extremes have traditionally been the blue-collar working class and the white-collar middle class. In both of these, the traditional ideal was a stable family, with the husband employed as breadwinner and the wife taking care of home and children.

The main difference between traditional working-class and traditional middle-class families was that in the former, young women usually had to go to work to make ends meet. Traditional middle-class women, on the other hand, did not have to work for pay. The *high-status* thing for them to do was to stay home and take care of the house and children (if possible, with the help of a working-class woman as a servant). For this reason, working-class women have usually wanted to quit their jobs and become housewives because this would have been a sign of success, of living like middle-class women.

Since about 1970, a major shift has occurred in this pattern (Figure 1.1). The *status* system has drastically shifted for middle-class women. It has become high-status for women to pursue careers—not just jobs they hold before they are married, but positions in the professions and in business that are full-time and the equivalent of the best careers of men. A number of things have contributed to this shift—above all, the women's movement, but also the rapid growth in the proportion of women who are college-educated, which now surpasses that of men.

But this is not merely a *status* process—a shift in what is considered respectable and admirable for women to do. It has had a profound effect on *economic class.* As we shall see, *the most important way in which a family gets itself into an upper-middle-class level of income today is by having two middle-class incomes.* The combination of working husband and working wife has become the most important feature differentiating families by economic levels (Figure 1.2).

What we might call the "new upper-middle class" consists of two professionals or business executives married to each other: two incomes of $50,000 to $75,000 or more combining to permit an affluent lifestyle. Further down the class structure, two working-class incomes (e.g., a carpenter and a secretary) will combine to make a decent income at the middle level. In contrast, the traditional single-earner family usually pays an economic price. One

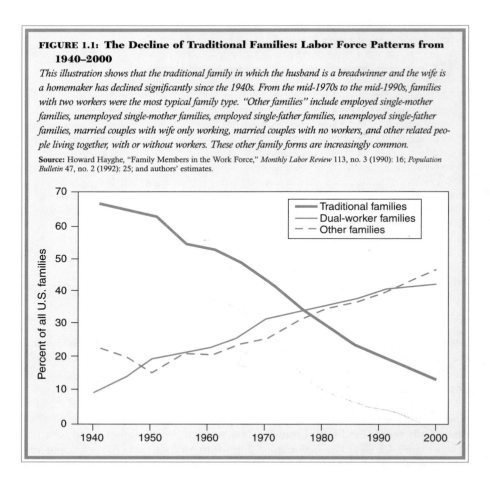

FIGURE 1.1: The Decline of Traditional Families: Labor Force Patterns from 1940–2000

This illustration shows that the traditional family in which the husband is a breadwinner and the wife is a homemaker has declined significantly since the 1940s. From the mid-1970s to the mid-1990s, families with two workers were the most typical family type. "Other families" include employed single-mother families, unemployed single-mother families, employed single-father families, unemployed single-father families, married couples with wife only working, married couples with no workers, and other related people living together, with or without workers. These other family forms are increasingly common.

Source: Howard Hayghe, "Family Members in the Work Force," *Monthly Labor Review* 113, no. 3 (1990): 16; *Population Bulletin* 47, no. 2 (1992): 25; and authors' estimates.

middle-class income today can barely keep a family at an average level of consumption, let alone match the style of life of modern couples with two professional incomes. Further down the scale, one working-class income may be barely enough for survival. Many people in the poverty class are not unemployed or on welfare; they simply are trying to get along on one inadequate income. The upper class stands out as the only group in which the wealth of one individual (usually the husband) is so great that the traditional single-employed pattern still holds. But the upper class is only a small percentage of the population. Everywhere else, the difference between two-income and single-income families is crucial.

The economic class structure puts pressure on every family within it. Whether they want to show off their possessions or not, prices are driven up by the amount of money people have to buy things. This is most obvious in regard to homes. In most major urban areas of the United States, there has been a tremendous inflation in the price of housing. Homes have become so expensive that families without two incomes can scarcely afford to buy anything within many residential areas.

For this reason, American married couples have become locked into the dual-earner pattern. This will no doubt become increasingly true in the future. Older

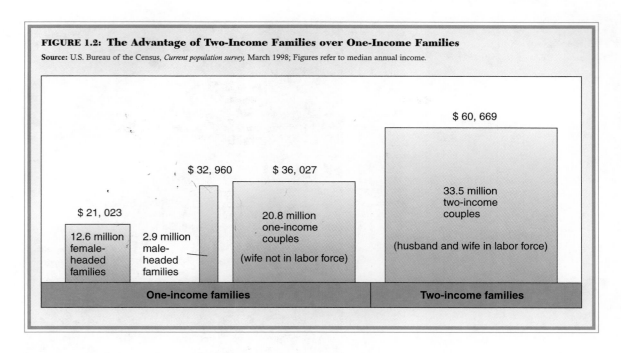

FIGURE 1.2: The Advantage of Two-Income Families over One-Income Families

Source: U.S. Bureau of the Census, *Current population survey,* March 1998; Figures refer to median annual income.

$ 60, 669

$ 32, 960 $ 36, 027

$ 21, 023

33.5 million
two-income
couples

12.6 million
female-
headed
families

2.9 million
male-
headed
families

20.8 million
one-income
couples

(wife not in labor force)

(husband and wife in labor force)

One-income families   **Two-income families**

home owners, who bought at a time when prices were much lower, could afford to live on one income. As these people die or move away, virtually all of the substantial homes, which used to be considered part of a middle-class lifestyle, will be taken over by dual-income couples. One can predict that, except for a small number in the wealthy upper class, families that want to maintain a middle-class lifestyle will have to be headed by dual-income couples.

The Intersection of Gender, Race, and Class

On the other hand, not everyone is going to be successful in this struggle for economic position. A large number of people, for instance, may never be able to afford to own a house. An alarming pattern that emerged in the late twentieth century is the way that the class structure became more polarized. Economic inequality, which declined between 1940 and 1970, has been growing again. Only dual-income families have kept up with inflation in the 1980s and 1990s. The top 20 percent of the population, roughly speaking, the upper-middle and upper classes, has gained an increasing proportion of the income, while everyone else's income has declined.

Several dimensions of inequality intersect. On the privileged side are women who have overcome the barriers of gender stratification and who are pursuing high-paying careers. When these women of the new upper-middle class are married to men of the same class, their families reap double advantages. On the less favored side is a large middle class struggling to hold on. Families in the poverty sector typically suffer from a combination of disadvantages. Race and class tend to reinforce each other at this level. In addition, gender stratification piles on another disadvantage for women who raise their children as single parents on incomes that are already below the level of those of men.

This businesswoman has overcome some of the barriers of gender stratification inherent in the business world.

Equality and Inequality

What will happen to families as we move into the twenty-first century? To get a good grip on this question, we will have to look at today's families at different levels of the class structure. It is apparent that upper-middle class families are moving one way, and lower-class families are moving in the opposite direction. For the families between, there are conflicting tendencies to sort out. Only upper-class families, at the wealthy peak of the society, appear to be holding to the traditional patterns. To determine which way the overall pattern is shifting, we must look at the macrostructure of society and the sociological theories that explain its long-term changes.

In addition, we must look at the micro level of stratification inside the individual family. As sociologists, the most useful thing we can do is to take a good look at how the three dimensions of stratification apply to men and women who are married to each other—or who are thinking about getting married or divorced. (1) *Economic class:* What sources of income does each man and woman have? What effects do their occupations and careers, whether housework or paid jobs, have on how they relate to each other and to their children? (2) *Power:* What kinds of things give a man power over his wife or a wife power over her husband?

The working-class position of these men may affect how they relate to their families.

And what kind of power do they have over their children? To what extent are the children able to evade their parents' power? (3) *Status:* How do people regard each other in the family? Within families, status has a special set of meanings and emotions; love, affection, and admiration are part of it, as are jealousy, disrespect, and hatred.

Our sociological theories are designed to target the causes of these patterns. We want to know what factors will make a husband or a wife more powerful on the micro level and what makes a marriage more happy or more disharmonious. On the macro level, we want to know what changes occur in the pattern of dual-income families in the society and the answers to many other questions about the stratification of the overall structure.

In principle, stratification can range from extreme degrees of inequality on down to zero. Just because we pay stratification a good deal of attention does not mean that inequality is inevitable. Some societies, past and present, and some relationships among individuals are much more unequal than others. Among all these variations, one possible pattern is equality. Men and women could be equal within the family, and conceivably, all families could be equal in terms of economic wealth, power, and status. This is not to say that equality is easy to achieve. Some forms of it, especially class stratification in the larger society, seem to be moving toward greater inequality rather than greater equality.

A particular kind of stratification has traditionally been at the core of the family itself: gender stratification. On this level, there is reason to believe that equality is possible, and indeed it is sometimes achieved in families. We will be especially concerned with the causes of different degrees of gender stratification and will try to pinpoint the conditions under which gender equality exists in families. As we shall see, the degree of gender equality or inequality has a major effect on the amount of conflict and happiness within families.

EXPLAINING GENDER STRATIFICATION IN FAMILIES

A crucial question is how we can explain gender stratification in families. This is a relatively new area in sociology. Before about 1970, few theories raised the question of whether there are differences between men and women in their power, economic position, and social status. A conventional, rather idealized family was taken for granted in the older theories. A husband who goes to work each morning, a wife who stays home and takes care of the kids, a dog named Spot—this was referred to as the sexual division of labor (leaving aside the dog, of course).

In this model, men did the work and supported the family in the economic sphere. Their labor was regarded as **instrumental,** that is, practical, goal-oriented, and unemotional. Women, on the other hand, were regarded as primarily **expressive.** Women took care of the emotions and the personal relationships. They nurtured the children and saw to their husband's emotional needs, providing relaxation and therapy when he came home from the male sphere of practical achievement (Parsons and Bales 1955).

What's wrong with this picture? Listing the defects of this so-called functionalist view has become quite an intellectual industry in the last thirty years. Here are a few of the more important points:

1. Is it really true that husbands worked and housewives did not? This view overlooks what actually has to get done in the home day after day. Cooking food, dusting furniture, cleaning floors, washing and ironing—all this is work too. Even taking care of children is far more than an "expressive" activity. There is a great deal of sheer physical, practical content to it, from changing diapers to getting toddlers dressed, undressed, washed, and fed. When children get older and are able to do some things for themselves, new forms of work appear, such as transporting them from place to place by what seems to many women to be "mom's taxi service." It was a gross misnomer to call men's employed activities "work," but not women's activities at home, even during the 1950s when most wives were not employed. The difference is not whether it is work but the way that it ties into the economic structure. The key point is that men's work was paid, whereas traditionally women's work has been unpaid.

2. Why have men usually worked in the better-paying sector of the economy and women in the unpaid sector? Is this simply a natural arrangement? As we look far enough across different societies and different periods of history, we see that the kinds of work women did varied a great deal. In some hunting-and-gathering or horticultural societies, for instance, women produced the bulk of the food that the whole society needed to survive. It would seem, then, that the "sexual division of labor" is not an inevitable arrangement.

In twentieth-century societies, too, we can see a flaw in the theory that women and men are in different spheres because of purely functional reasons. In all modern industrial societies, including our own, a considerable number of women have worked for pay outside the home. Until the pressure exerted by the women's movement in the last few decades, virtually all of these women worked at jobs that yielded low pay and little power and status. Women were almost totally excluded from positions in the higher levels. So-called "middle-class" or "white-collar" positions for women were largely restricted to secretarial and retail sales positions—jobs with virtually no power and a rate of pay lower than for most manual jobs for male workers. Women in the professions have been heavily concentrated in elementary school teaching or in nursing—both professions in which men tended to dominate the higher-level positions.

There have been plenty of blue-collar working-class women as well. This is especially true in the garment industries, where women operate sewing machines in the so-called "needles trades." We tend to think of a typical factory scene as consisting of armies of men pouring out molten metals in the grim ambiance of the steel mills. A better image today would be long rows of women processing plastic goods or computer chips.

Another sector in which women have always been in the majority is the service sector, for example, working as waitresses and beauticians. This area includes the humblest positions in our society: custodians, hotel maids, and cleaning women who empty the trash and tidy up after the rest of us. When this kind of work is done in a private home by a paid household servant or housekeeper, it is almost always done by women. Here we see a good example of how gender stratification crisscrosses the class structure. Female household servants holding one of the lowest-paying jobs in the working class are typically hired by women who are in the upper and upper-middle classes. These employers used to get their wealth from their husbands, but increasingly they themselves have high-paying professional careers. Class stratification and gender stratification are both hierarchies of privilege in their own right. Sometimes class and gender reinforce each other, as in the case of low-paid housekeepers who are at the bottom on both scales. But sometimes class operates independently of gender, as in the case of the upper-middle-class professional women in a position to hire working-class women to work for them. Even though some women get to the top or near it, other women make up a large proportion of those at the bottom.

In short, women have always worked, either without pay in the home or with pay outside it. In both cases, women have typically received fewer material rewards than men and little power or status from their work. Perhaps this is why women's work has been "invisible" to most men, including the authors of the older sociological theories. The newer theories look for ways in which women are excluded from the higher paying jobs, rather than taking it for granted that these are automatically men's positions. Once we see that men and women have their own economic positions, however high or low, we recognize that the structure of the family is going to be shaped by these economic positions. For economic position outside the family brings resources that give power inside the family. If women have traditionally done household services for men at home, it is at least partly because of the power that has accrued to men who have more economic resources than their wives (Figure 1.3).

3. Is it really true that men are *instrumental*—hard, cold, and practical—and women are *expressive*—emotional, personal, down-to-earth, and uninterested in achievements or in abstract ideas? This is in some ways the most controversial question in the theory of gender. The traditional theories of the sexual division of labor assumed that this distinction was natural. Supposedly, society needs both instrumental and expressive skills; therefore, combining "male" and "female" roles in every family makes a perfect mix. This gave a functionalist justification for gender differences in society.

Newer critical theories are split on this point. On the one hand, theories that might be referred to as "egalitarian" or "socially constructivist" argue that males and females are essentially the same beneath the skin and that the way they behave is socially determined. If men have dominated the social positions in which practical activities were carried out—whether in economics, politics, science, or any other sphere—this has created the image of men as "instrumental," purposeful achievers. By the same token, if women have been channeled into subordinate

FIGURE 1.3: Women Still Earn Less than Men but the Gap is Shrinking

This figure shows that the earnings of women who work full-time, year-round are less than comparable men's earnings, regardless of race or ethnicity. The gap is largest among whites (72%), followed by Asians (80%), blacks (83%), and Hispanics (88%). Earnings vary substantially by race/ethnicity, as well as gender (1997 median earnings for whites = men $35,193, women $25,331; Asians = men $34,682, women $27,781; blacks = men $26,432, women $22,035; Hispanics = men $21,615, women $18,973). Because women are much more likely than men to be unemployed, work part-time, or hold seasonal jobs, the gender gap in pay is even larger than this figure shows.

Source: U.S. Bureau of the Census, *March Current Population Surveys*, Historical Income Tables, p. 33, 1999.

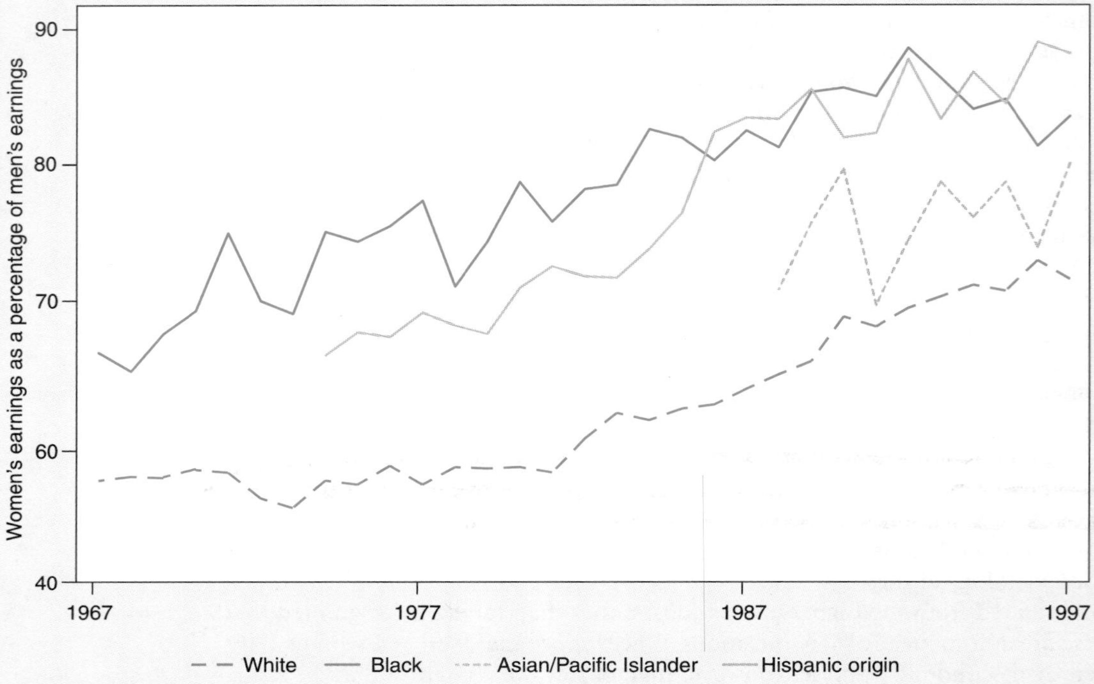

positions and given no opportunity to control business, government, cultural institutions, and so forth, it is not surprising that women develop the image of being nonpractical, uninterested in achievement, and oriented away from the dominant spheres of society. What women were considered to be good at, then, were the virtues of the places where they happened to be. Women's status came from being lovers, mothers, friends, and confidantes; thus, women acquired an image of being subjective and oriented to the personal, the intimate, the immediate, and away from the harsh realities of the world of men. Egalitarian theories hold that these differences are socially constructed and that if men and women are put in the same positions, they behave the same way.

Another line is taken by various theories that have also emerged with the radical criticism of the last few decades. These might be referred to as "essentialist" theories of gender. In some respects, their position agrees with the traditional theories that assert that there are fundamental differences between males and females. However, the radical essentialist theories hold the "male" qualities responsible for most of the evils of society. They posit that the instrumental, impersonal, achievement-oriented approach has been manifested in the pursuit of economic profit, political power, and military power without regard for the value of human beings. In the realm of ideas, the instrumental attitude is linked to the "positivistic" attitude of science, which tries to reduce the world to purely objective elements, without room for the human side of things. This critique links feminism to Marxist and critical theories of society generally. The female qualities, emphasizing personal, intuitive, and humane attitudes, are equated with the ideals that are subordinated and even crushed by the dominance of male institutions: war, violence, the pursuit of profit, and the subjugation of nature.

In this text, we do not attempt to argue all of the wide-ranging issues that are raised by these critiques. In our view, the traditional theories that divided males and females into two different cultures are misleading, to say the least. What shall we put in the place of this reasoning? We can approach this from the point of view of *what kind of theory gives us the most ability to explain social differences.* The traditional, functionalist theory did not convincingly explain the pattern "male = instrumental" and "female = expressive." There is a good deal of evidence that females also engage in many instrumental activities. Much housework is hard, practical labor, as is much of the work that women have put into the paid labor force. Although women have been discriminated against and discouraged from entering positions of political power or economic control, or in the world of science, plenty of evidence shows that when women succeed in attaining these positions, they perform instrumental tasks in the same way as men. (Important studies on this point include Rosabeth Kanter's *Men and Women of the Corporation* [1977], Cynthia Epstein's *Deceptive Distinctions* [1988], and Barbara Reskin and Irene Padavic's *Women and Men at Work* [1994]).

Conversely, there is evidence that expressive, personal, and intuitive activities can also be carried out by men. For example, the realm that has been traditionally considered the sphere of the personal, caring relationship par excellence is mothering. Yet there is evidence (chapter 12) that when men take part in fathering, their personalities take on the qualities of being caring rather than aggressive and impersonal. Similarly, studies of artists and creative scientists—males as well as females—show that their personal style tends to emphasize intuitive judgment rather than the mechanical style of the positivist image. We would conclude that what are traditionally considered to be "male" or "female" styles are really two parts of the range of possibilities within the abilities of all human beings.

In actuality, both men and women tend to have a mix of instrumental and expressive activities. Our task as sociologists is to show the conditions that push the mixture further in one direction or the other, in each particular case.

The question then is: How deep are the conditions that determine what is **gender**, that is, the culturally defined aspect of what is considered to be male or female in each society (see also *A Closer Look* 1.2)? The theories range all the way from conditions that are so primordial that one can doubt they would ever change to conditions that are relatively superficial and can be pushed around with enough personal or political willpower. The continuum would look something like this:

■ Gender roles are determined biologically and cannot change, except as genetic endowments might change over the long course of evolution.

■ Gender and the categories of male and female are deeply embedded in the culture. Everything is seen through the "lens" of gender. This implies we can't get outside of our own gender's field of vision and suggests that the whole culture must be transformed before changes in gender can occur.

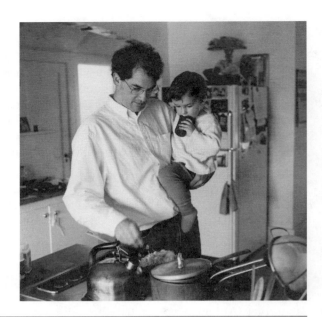

By taking on nontraditional roles, fathers may change their children's perceptions of appropriate gender activity.

■ Gender reflects personality traits developed early in childhood. Relationships between small children and their mothers are crucial in setting boys and girls on the paths that will constitute their gender for the rest of their lives. This position allows a social determinism, but only rather deep in each person's life.

■ Gender is a style of behavior determined by the social structure of each society. As the social structure changes, so too does the content of gender.

Obviously, as sociologists we are likely to be partial to the last line of analysis. We put a lot of attention into showing the social conditions that make males and females act in particular ways and that define the images of men and women. Because we think that child rearing is also a social process, we focus on how much of gender can be explained by what happens in early childhood, and we put this in the larger social setting that affects how partners and parents behave. These changes, in turn, reflect changes in cultural gender lenses that we tend to take for granted.

We also want to be alert to the ways in which individuals do *not* live up to the prevailing social and cultural ideals. Not every male is an unfeeling macho guy and not every woman is a sensitive caring person. We want to be able to see more clearly not only how society puts people into slots, but also how people squirm in these slots and struggle to change the shape of the society into other forms.

Explanation by Comparison

The method of comparison is central to our explanations. This applies to virtually every topic we will consider. If we want to understand how men and women act inside and outside families, we need to make comparisons to see how much things can vary and under what circumstances. For this reason, we need to take a longtime perspective, looking across the centuries and from one society to another. For example, *A Closer Look* 1.4 compares different

A CLOSER LOOK 1.4
Historical Views of Gender and Families

Tribal Marriage Exchange

On reaching puberty, the Konyak Naga boys begin to look for girls from the clan complementary to their own, and they exchange little gifts, the value and nature of which are strictly fixed by custom. These gifts are of such importance that a boy's first question to the young girl whose favours he seeks is as follows; "Will you take my gifts or not?" The answer being, perhaps, "I will take them," or "I have taken the gifts of another man. I don't want to exchange with you." Even the wording of these overtures is fixed by tradition. This exchange of gifts initiates a whole series of reciprocal protestations which lead to marriage, or, rather, constitute the initial transactions of marriage, viz, work in the fields, meals, cakes, and so on.

Claude Lévi-Strauss, *The Elementary Structure of Kinship,* 1949

The Patriarchal Family

By a girl, by a young woman, or even by an aged one, nothing must be done independently, even in her own house.

In childhood a female must be subject to her father, in youth to her husband, when her lord is dead to her sons; a woman must never be independent.

She must never separate herself from her father, husband, or sons; by leaving them she would make both families contemptible.

Him to whom her father may give her, or her brother with the father's permission, she shall obey as long as he lives, and when he is dead, she must not insult his memory.

Laws of Manu, India, ca. a.d. 100

Men have authority over women because Allah has made the one superior to the other, and because they spend their wealth to maintain them. Good women are obedient. They guard their unseen parts because Allah has guarded them. As for those from whom you fear disobedience, admonish them and send them to beds apart and beat them. Then if they obey you, take no further action against them.

Mohammed, *The Koran,* ca. a.d. 630

By marriage, the husband and wife are one person in law; that is, the very being or legal existence of the woman is suspended during the marriage, or at least is incorporated and consolidated into that of the husband; under whose wing, protection, and cover, she performs every thing. . . . Upon this principle, of a union of person in husband and wife, depend almost all the legal rights, duties, and disabilities that either of them acquire by the marriage. . . . A man cannot grant any thing to his wife, or enter into covenant with her, for the grant would be to suppose her separate existence.

Sir William Blackstone, *Commentaries on the Laws of England,* 1765

The Victorian Family

It is really a stillborn thought to send women into the struggle for existence exactly as men. If, for instance, I imagined my gentle sweet girl as a competitor it would only end in my telling her, as I did seventeen months ago, that I am fond of her and that I implore her to withdraw from the strife into the calm uncompetitive activity of

my home. It is possible that changes in upbringing may suppress all a woman's tender attributes, needful of protection and yet so victorious, and that she can then earn a livelihood like men. It is also possible that in such an event one would not be justified in mourning the passing away of the most delightful thing the world can offer us—our ideal of womanhood. I believe that all reforming action in law and education would break down in front of the fact that, long before the age at which a man can earn a position in society, Nature has determined woman's destiny through beauty, charm, and sweetness. Law and custom have much to give women that has been withheld from them, but the position of women will surely be what it is: in youth an adored darling and in mature years a loved wife.

Sigmund Freud, youthful letter to his fianceé, ca. 1885

Egalitarian Ideals

The couple should not be regarded as a unit, a closed cell; rather each individual should be integrated as such in society at large, where each (whether male or female) could flourish without aid; then attachments could be formed in pure generosity with another individual equally adapted to the group, attachments that would be founded on the acknowledgment that both are free. This balanced couple is not a utopian fantasy: such couples do exist, sometimes even within the frame of marriage, most often outside it.

Simone de Beauvoir, *The Second Sex,* 1953

views of gender and families from different societies in different historical periods. In many chapters we examine families from different historical times. Many of these family forms are irrelevant today, for they will never come back again. But comparing the conditions under which the different types of family systems exist gives us the key to the processes that produce variations and thus enables us to understand what is producing the family structures of today. The view from someplace else lets us know when we are taking some feature of our own society as more permanent and fundamental than it really is. This is true all across the board. Every aspect of the family has at least some variations, and these help to reveal the social conditions that are causing them.

We pay special attention to relations between men and women, and comparisons among different types of families in the United States today can help us see how different institutions contribute to differing views of gender in marriage. Table 1.1 shows how different religious denominations define what a marriage is and what the roles of men and women should be within it. Although we tend to think of such things as fixed and stable, even these religious teachings are subject to change. It was not until 1998 that the Southern Baptist Church leaders voted to emphasize their view that the natural role of a woman was to "graciously submit" to her husband (see Table 1.1). Whatever we look at—changing gender relations, how much divorce is taking place today, how much nonmarital sex, how much controversy over abortion, or any other topic—a study of comparisons will help us better understand the issues involved.

Comparison is such an important method for explanation that we often seek out unusual comparisons to make. In chapter 3 we compare male circumcision rituals with female genital mutilation to show how the control of sex and reproduction are political and economic concerns. In chapter 13 we compare wife battery with child abuse to see common features that might promote or inhibit their occurrence. In chapter 12 we compare single-parent families with two-parent families, and in chapter 6 we compare working-class families with middle-class families. These comparisons will help us to identify how the structure of families and the intimate details of life in families are influenced by social forces.

At some points, we will also compare heterosexual couples with homosexual couples on issues like dating, sex, and who does the housework. Again, whether or not we are interested in the specific facts about gay or lesbian couples, the comparisons will show us what processes are quite general and cut across all types of couples. Such comparisons are also useful because they can help us decide which theories best explain what is going on. Finally, using comparisons and employing theories will also give us some hints about the future.

FAMILY THEORIES Table 1.2 presents what we see as the most important theoretical conceptions for understanding families. As sociologists, we draw on those ideas common to our discipline, such as the concept of stratification introduced on the previous pages. As you will see, these themes can also be found in other social sciences (e.g., anthropology, economics, political science, psychology) and the humanities (e.g., history, philosophy, literature), although they often go by different names and have slightly different emphases. And though related, the theoretical ideas in Table 1.2 are not the same as those used historically by sociologists, psychologists, and therapists who attempted to discover a set of universal laws governing families under

TABLE 1.1 How Some Modern U.S. Religions View Marriage and Gender

	Baptist — Southern Baptist Convention	Pentecostal — Church of God in Christ	Mormon — Church of Jesus Christ of Latter-Day Saints	Catholic — Roman Catholic	Episcopal — Episcopal Church	Unitarian — Unitarian Universalist	Islam	Judaism — Orthodox	Judaism — Reformed	Buddhism — American Zen
What is a marriage?	The uniting of one man and one woman in a covenant commitment for a lifetime.	A lifetime union before God, between a man and a woman.	A sacred institution uniting a man and woman for life. (If the ceremony is performed in the temple, marriage will continue through eternity.)	A covenant relationship between a man and a woman in which they establish family life.	A lifelong union between a man and a woman for their mutual joy that may or may not hold within it the procreation of children as a goal.	A union between equal partners.	The union of a man and woman joined together under God of their own free will in a lifetime relationship. The woman receives dowry.	The joining of two individuals with the goal of living a constructive, harmonious life and creating a good environment for bringing up children.	The sacred union of a man and woman in a loving, faithful, monogamous relationship.	The lifetime union of a man and a woman in holy matrimony.
What is a man's role in marriage?	The husband and wife are of equal worth before God. Both bear God's image but in differing ways. A husband is to love his wife as Christ loved the church. He has a God-given responsibility to provide for, protect, and lead his family.	To be the moral and spiritual leader, a protector for the household.	The husband presides over the family in love and righteousness. He is responsible to provide the necessities of life for his family.	He is in a partnership with his wife. Husbands should be self-giving in their love for the family, as Christ was prepared to give his life out of love.	Men and women are absolutely equal in marriage.	He is an equal partner.	A man should provide for the family. The money he earns belongs to him and the family.	A man's role is to oversee teaching and religious ritual for the family. He plays the dominant role in the marriage but defers to his wife concerning domestic matters.	It is the shared role of husband and wife to create a small sanctuary in the home, with shared responsibilities. Both partners are equal.	To be an equal partner with his wife. If there are children, it is incumbent on the man to share in the maintenance of the household and family, as equally as possible.
What is a woman's role in marriage?	A woman is to submit graciously to the servant leadership of her husband even as the church willingly submits to the headship of Christ.	She is an equal partner. The role of both wife and husband is to be mutually submissive.	Women are primarily responsible for raising and nurturing children.	Women and men are equal but not the same. Wives should try to make the will of the husband be the will of the family.	In the marriage rite, women and men make the same promises and accept the same responsibilities. Both husband and wife are submissive to Christ.	She is an equal partner.	She oversees the health of the household which includes physical, spiritual, and material well-being.	A woman's role is to oversee the nurturing of the family. She defers to her husband concerning decisions of religious ritual and study.	There are not particular assigned roles.	To be an equal partner with her husband. Neither wife nor husband has a particular job within the household. Division of labor depends on the couple's circumstances.

Source: Mary Rourke, "A Woman's Place: What the Denominations Think," *Los Angeles Times* E-2 (June 16, 1998). Reprinted with permission.

TABLE 1.2 Major Themes in Explaining Families

Themes	Basic Concepts	Some Applications
Exchange networks and markets	Individual resources Market opportunities Permanent exchanges and "going off the market" Sexual and emotional property	Marriage markets Love Divorce Family power
Situational negotiation and reality construction	Definition of the situation Subjective interpretation Routinization and normalization	Adaptation to abuse "Emotion work" by women or men Child development
Interaction ritual	Group membership "Sacred objects" symbolizing social ties	Love Jealousy Abuse
Moral ideals	Sexual intercourse Moral outrage High- and low-density interaction Conformity Cosmopolitanism	Class differences Incest Child rearing
Social structure and stratification	Classes Wealth Authority Career identification Racism	Lifestyles Early or late marriage and child rearing Teenage strains Neighborhood quality
Gender relations	Gender ideologies Patriarchy Household labor Property rights	Housework Family power Marital conflicts and marital happiness Divorce and remarriage Sexual behavior
Population and child rearing	Ecological approaches Immigration Birth and death rate Infant mortality Early and late marriage Life course issues	Mothering and fathering Economics of the household Women's independence Child development
Politics and institutions	Kinship alliances Fortified households Bureaucratic state Social movements and countermovements Resource mobilization Cultural contexts Nested systems	Historical changes and future trends Feminist and antifeminist movements Abortion Employment discrimination Child care Family values debates

the names of functionalism, family development, human ecology, systems theory, or sociobiology. Rather, our approach and the themes we stress can be classified under the general categories of social constructionism and conflict theory.

The concepts outlined in Table 1.2 constitute a kind of theoretical "toolbox" for analyzing families. We will use them repeatedly in different sections of the book to explain how families operate and how they change. These themes allow

Many traditional religious weddings symbolize marriage as a binding relationship.

us to see beyond the surface of everyday life and identify regularized patterns of social behavior. Ultimately, we hope they will help you to understand your own and others' families and to better understand their relationships to the other parts of society.

1. *Exchange networks and market processes.* Exchange theory is widely used in sociology and the other social sciences, although we need to broaden it to apply to families. **Exchange theory** is, first of all, a social theory of motivation. It says that individuals try to move toward those social relationships in which they receive the best payoff—the greatest rewards for the fewest costs. But there is a structural side to this theory too; not everyone has equal opportunities for payoffs and may have only limited possibilities for exchange. In short, people may be stuck with a very limited or unequal situation in which the best deal they can make is not a very good one. Hence, a theory of exchange must also include the structure of the market, the *opportunities* available to each individual to exchange with others, as well as the lineup of resources that each person on this market has available to exchange with others.

For instance, the market model has often been applied to the process by which persons get married. The "opportunity pool" consists of the women known by a particular man or the men that a particular woman knows. The *resources* that each person has consist of all things that make him or her attractive to someone: personality, good looks, money, social status, connections into which he or she can introduce someone, emotional energy, and so forth. The market theory predicts that people will end up with someone who is "the best deal" they can get in relation to the number of resources they themselves have to exchange. It is important to bear in mind that this process takes place over time. At the outset, a person is likely to be attracted to someone who is "hard to get." One starts out being infatuated with the best looking woman or the most flamboyant or successful man. But then one finds out that there is a great deal of competition for the attention of these "stars" and that some of the rivals have more resources to make themselves attractive. Thus, there are a lot of psychological ups and downs in a courtship market as one finds out what one's own "market position" actually is in relation to who is available and who one's rivals are.

This view of personal relationships no doubt seems like a harsh and cold-blooded one, and it is not the way people ordinarily see their lives. Nevertheless, differences in market structures have a profound effect on people's lives, and if we wish to explain what happens to people, the market theory is one analytical tool that we need to use. The very fact that people don't like to see the world in this way is a revealing one: it tells us that people idealize their relationships. One might say there is a conscious level and an unconscious one. The latter contains these structural conditions—the opportunity networks that surround each individual and the lineup of social resources among these persons—that are generative, and out of which the "surface" of individual experience is constructed.

These market conditions are only one of a set of influences acting on the individual, but they are powerful. We will see some of their effects throughout the book: how they influence what one's social life is like before marriage and whom one will marry (chapter 8); how much power husbands and wives have within their marriages and when conflict is likely to take place (chapters 11 and 13); and how they are one of the pressures involved in divorce and remarriage (chapter 14). Exchange relationships take place between parents and children, too. The fact that children's market positions shift radically as they go through various ages is one of the factors that affects parents' efforts to control their children and the problems that ensue, especially in the teen years (chapters 12 and 15). Even the "midlife crisis," which happens in later years, is a kind of market process, as people become aware of what the limits of their opportunities are going to be for the rest of their lives (chapter 15).

In applying the market/exchange theory to families, it is important to add two special points. One is that individuals can go "off the market" rather than stay in the bargaining process. This is in fact what marriage is: two individuals who have agreed to stop looking for the best partner they can find and agree on a permanent exchange with each other. "Going off the market" turns out to be a matter of degree, in that people do sometimes divorce, and even when they are married, their shifting balance of resources will affect their relationships. But the distinction is still important. We will suggest (chapter 8) that being in love is fundamentally connected with this recognition that one's days on the courtship market are ended.

The second point is that markets imply *property.* Resources can be exchanged with someone else only because they are appropriated in the first place. This leads us to some discoveries that are far from the surface of our consciousness: that rights of sexual access over one's body are a kind of property and so are feelings of emotional possessiveness. This kind of social or symbolic property is the source of some of people's strongest attachments and gives rise to jealousy; it has been a major source of conflict, including violent abuse and murder (chapters 8 and 13). Here again, the idealized surface of our consciousness is tied to profound and potentially very difficult structures in the depths.

Yet another form of property is the straightforward material wealth in the form of incomes and household goods that go into making up a marriage. This is one of the sets of resources that individuals have. Usually we ignore it in the idealization that takes place before marriage, but it plays a crucial part in the marriage routine itself and in what happens if there is a divorce (chapters 6 and 14).

Finally, we should recognize that market structures are not fixed; they have changed through different periods of history. There are quite different kinds of property systems and sexual and emotional exchanges in tribal societies, in the patriarchal households of the Middle Ages, and in modern family systems (chapters 2 to 4). Men and women have typically had very different resources and market opportunities. The series of changes in the market position of women, as we will attempt to show, was crucial in creating the modern system of courtship and marriage. It is the changes in these markets that are going on right now—and that are the object of today's struggles—that will produce the families of the future.

2. *Situational negotiation and reality construction.* People also live in a world of consciousness, and their ideas and interpretations are important. Social relationships are constructed; persons negotiate how they will treat one another, as well as what they will treat as real and important. Hence, all couples and all families live in a world of consciousness that is in some ways uniquely their own. Furthermore, definitions of what is real can change suddenly: a husband deciding that his wife doesn't love him, for example, can have dramatic and far-reaching consequences.

Perhaps even more important than these sudden changes are the effects that cognitive beliefs can have on maintaining a situation. For example, situations of wife or child abuse frequently go on for a long time without the wife deciding to leave, and many women return to their abusive husbands (chapter 13). One of the factors operating here is that the situation is defined as routine, as a normal form of trouble that calls for no special treatment. This is an instance of what sociological theorists in the field of ethnomethodology see as a fundamental process of routinization in ordinary social life.

Social theories that emphasize the situational construction of reality (symbolic interactionism, ethnomethodology, phenomenology, and postmodernism) are often construed as opponents of exchange theories, which emphasize structural determinants. This theoretical opposition, in our opinion, is not necessary, provided that we are willing to be flexible about the use of these theories. Situations can suddenly be changed; under other circumstances, people routinize even very serious troubles and refuse to change their definitions of what is going on. A wide-ranging exchange theory, of the type outlined earlier, helps us to predict when a particular lineup of individuals in a situation will produce just the mixture of cognitive, emotional, and other resources in which someone will suddenly be able to redefine the situation, and when, conversely, he·or she will resist new understandings. A good example of how the two kinds of theories mesh is in Arlie Hochschild's explanation of why men and women have very different outlooks on the world, especially in regard to talking about their emotions and to falling in love. Women do a lot of situational interpretation, what Hochschild calls "emotion work," but the reason they do this more than men is because of the different position they occupy in the marriage market. Further detail on this theory is in chapter 8.

3. *Interaction rituals.* Another theory explains how people's emotions and ideas are influenced by the kinds of interaction they go through in everyday life. Following the theories of Emile Durkheim and Erving Goffman, we call these **interaction rituals.** The most important point is that a group of people who interact intensively create moral pressures about what they believe is right and wrong, and they attach these emotions to symbols that represent membership in their group. These groups can be as small as two people. For instance, a couple in love produces a little private cult in which they idealize one another and create their own symbols or "sacred objects" that represent their bond (chapter 8). Sexual intercourse itself can be analyzed as a naturally occurring form of interaction ritual (chapter 9). But there is a negative side of symbolic ties as well, for when symbols are violated, group members feel righteous anger and tend to feel vindicated in punishing the offender. This is one of the sources of jealous spousal violence (chapter 13). The ritual aspect of social ties is one important explanation of why so often spousal abuse is accepted as "normal" even by the victim.

An important variation is between life situations of "high ritual density" and those of "low ritual density." **High-density situations** are those in which persons are almost always in the presence of the same other persons. This results in strong pressures for conformity, low tolerance for deviation, and rather localistic concrete objects of loyalty. As we shall see in chapter 6, this high density of interaction is most typical of the lives of working-class persons, and it affects both their cultural outlook and their traditionalism in conventional family and gender roles. At the opposite extreme are relatively **low-density situations,** in which individuals are influenced by crisscrossing, cosmopolitan social networks. Here, there is less emphasis on external conformity and more on inner, psychological processes; therefore, there is less respect for traditional, taken-for-granted ideas and more orientation toward abstract and relativistic ways of thinking. This is more typically found in the cosmopolitan social networks of the upper middle class (chapter 6), with its corresponding differences in family life.

We will see, too, that these conditions affect social class differences in how parents bring up their children (chapter 12). In addition, we should recognize that every child begins life in a more restricted, high-density local environment consisting of his or her immediate family and gradually broadens this environment to larger networks. For this reason, every child's emotional and cognitive development is to some extent a movement from localistic and unthinking conformity to conventional symbols toward a more abstract and inwardly reflective consciousness. Because some social classes have more cosmopolitan networks than others, children of different classes tend to develop differently and to end up in different places.

Individuals, as well as social classes, can vary in the density of the social networks in their own lives, and hence, the principles of interaction rituals help explain individual idiosyncrasies. For example (chapter 13), we will see that the type of parent who is involved in incest typically holds extremely authoritarian and traditionalistic beliefs about both morality and family roles.

4. *Social structure and stratification.* Social class conditions affect family life in several ways. There are cultural differences of the sort just mentioned in regard to interaction rituals, and these affect moral ideas and, hence, child rearing, family authority, spousal abuse, and so on. As noted earlier, other cultural differences arise from the kinds of organizations in which people work and how their work lives are structured (chapter 6). Members of the higher social classes have more authority in their jobs; hence, they tend to identify strongly with their careers and to subordinate their family lives to their work. In the lower social classes, people are subjects of someone else's authority; hence, they tend to withdraw from their jobs into their private lives. This indirectly affects such things as how early people will marry and have children. Indeed, the whole process of family aging seems to be affected by class differences; one might say that working-class persons age more quickly in that they go through the phases of the life cycle earlier. The high point of their lives might be the teen culture. In the career-oriented upper-middle class, however, "young" is relatively later because people strive for accomplishments that happen much later. Life chances are also structured according to race and ethnicity, so when considering stratification, we must simultaneously consider the effects of race and class (chapters 6 and 7). We see, for example, that restricted access to jobs and housing have caused special problems for African Americans and Latinos and that immigration policies have influenced the family forms of Asian Americans and Latinos in special ways (chapter 7).

5. *Gender relations.* A particularly important issue of stratification today is one that cuts right through families: the degree of inequality between men and women. This is a complex issue because men and women grow up in the same families and are tied together by the rituals of love and marriage. Thus, a symbolic realm covers over the heterosexual relationships and gives them an idealized definition. This is one reason why the terms *sex* and *gender* are distinguished by sociologists: to emphasize that what is *considered* to be male and female is socially constructed and does not necessarily represent the biological bedrock of sexual difference (see *A Closer Look* 1.2).

Beneath the surface of conventional consciousness, however, men and women have had different ties to the larger system of power and property. In the different family systems that have existed, from tribal societies through the medieval agrarian households and into the various family revolutions of modern centuries (chapters 2 to 4), there have been different forms of gender stratification. Especially in more recent times, the class stratification of family members (discussed earlier in theme 4) has interacted with this gender stratification so that women married to men in the upper classes have faced a different form of gender inequality than women in the lower classes. What is happening in our own lifetimes, as middle-class women have become mobilized into higher education and careers, is another major shift in the relation between these two forms of stratification. This shift will undoubtedly affect the form of families in the future (chapter 16).

The different power and economic resources available to men and women are important determinants of what goes on in many aspects of family life right now. They are a major source of marital conflict (chapters 11 and 13) and divorce (chapter 14) and also seem to be affecting the changing patterns of premarital, marital, and extramarital sex (chapter 9). The different advantages of men and women in the stratification system explain why men remarry after a divorce more frequently and quickly than women (chapter 14). In short, gender stratification and conflict are crucial determinants of many aspects of family life. The happiest marriages are those that have come to terms with this issue (chapter 11).

6. *Population and childrearing.* The only crucial biological difference between men and women is that females can bear children and males cannot. But even here, social conditions intervene.

How many children should women bear? There are social pressures from men in certain societies to produce many children, either as workers for a family farm, as warriors, or to be traded between families in marriage alliances (chapters 2 and 3). Conversely, in bureaucratic industrial societies like our own, the economic costs of bringing up children are no longer balanced by these kinds of economic payoffs; this is one reason the birthrate has gone down in almost every modern society. Birthrates are also affected by the age at which people marry and by health conditions, which determine how many babies survive infancy. In some past societies, women were almost constantly pregnant, going through a large number of births to produce a moderate number of healthy children. Technological advances such as birth-control devices, medical abortion, and the sterilized baby bottle as a substitute for breast-feeding have also affected how many children have been born (chapter 10).

All of these factors have powerful social causes and effects. We can see some of the social causes behind such issues as whether birth control or abortion are encouraged, tolerated, or prohibited by a society. Also, we can perceive very different social effects in the family structure of societies where women are usually pregnant, nursing, or bringing up large numbers of children and societies where these take up only a small portion of women's time. Both gender stratification and the overall shape of society are influenced by these demographic and fertility factors.

Although women bear children, how much responsibility do they have for rearing them? In some societies, women handle all the child care, but there is some variation. As we shall see in chapter 12, this variation has an important connection to the degree of gender stratification in that society. There are also many possible social arrangements, such as every mother rearing her own children; mothers putting their children under the care of a wet nurse (as upper-class women did a few centuries ago); an extended kinship group caring for all the children together (as is found in some tribal societies and in some poverty sectors today); and a collective child-care organization, such as is found in the Israeli Kibbutz. The social conditions that affect these arrangements obviously have an effect both on children and on how much independence mothers have to pursue their own lives.

7. *Politics and institutions.* We do not yet have a good theory that predicts how the family will change under various conditions. To a large extent, we are still navigating "by the seat of our pants," trying to pick out trends and extrapolating them into the future. This is a dangerous form of prediction because it doesn't come from a well-grounded theory. Trends can sometimes reverse themselves rather abruptly; for instance, the baby boom and the wave of traditional familism that happened for more than a decade after World War II was totally unexpected (chapter 5). We would suggest, though, that one of the crucial features moving the family system to newer forms is gender stratification and its interaction with social class and ethnic stratification. But what, in turn, determines how these stratification systems will shift?

One clue, we suggest, is that gender stratification in the past has been heavily influenced by the form of politics and institutions in that society (chapters 2 to 4). The feminist movement in the 1970s, which suddenly rose to prominence (at least within the American upper-middle class), was part of many other political movements of that time. The political shifts of the 1980s and 1990s were in a more conservative direction. The struggle between opposing political positions is far from over, and we anticipate still further conflicts in coming decades.

Our theories as yet are not very good for specific predictions; however, we know enough to point out that family issues are often defined by political processes and shaped by institutional contexts. Abortion, for example (chapter 10), was not a significant public issue until the late 1960s, when it became part of a conscious movement for women's reproductive freedom; later in the next decade, it became the rallying point for the "right-to-life" movement, which attempted to reverse the recent liberal victories. We need to detach ourselves sociologically, at least for a moment, to note several points. One is that politics operates by mobilizing people's moral commitments. It generates symbolic ideas—"reproductive freedom," "right to life"—and it divides up the world into righteousness and evil, according to whether other people support one's ideal or not. Political movements, in other words, are something like large-scale interaction rituals of the kind referred to earlier (theme 3). The process of conflict itself mobilizes people into a more intense network of social interactions and more intense emotions, which in turn create idealized symbols. These symbols are used to mobilize support for government actions with long-term impacts on families, such as making certain sexual or contraceptive practices illegal, making divorce harder or easier, or sanctioning only one type of marriage.

We don't intend to advocate that we try to stay permanently detached, observing the emotional side of such conflicts and staying above them. It is hardly possible to do this. But we can be involved realistically, by recognizing that emotion-laden ideals are being generated by the conflict itself and that they will continue until the underlying conditions change that mobilized the movement in the first place. The "problem" may not actually disappear or become solved, but often the political resources shift so that a movement stops being organized around a particular issue. Politics tends to operate on the rebound; it is a series of movements and countermovements that are mobilized in response. Abortion and contraception issues are just two of many that are affecting the shape of the family. Employment discrimination against women and the costs and availability of child-care provision, as well as issues about public schooling and job opportunities, all affect the kind of families that will be formed in the years to come.

The twentieth century has seen important shifts in virtually every generation (chapter 5). There has been the sexual revolution of the 1920s, followed by the Depression of the 1930s, the war years of the 1940s, the baby boom of the 1950s, the radical family alternatives of the 1960s and 1970s, and the conservatizing moods of the 1980s and 1990s. Indications are that families in the twenty-first century will be increasingly affected by the growing polarization of economic classes and racial groups. Historical precedent suggests very strongly that further upheavals are on the way in the decades to come.

THE USES OF FAMILY SOCIOLOGY

There are a great many issues concerning families to consider. Our aim is to obtain an overall picture of what causes what. Why have families taken particular forms in the past? What is happening to them today? What can we expect for families in the future? These issues are important, for men and women as groups, for the larger society as a whole, and for each of us as individuals.

Social and Political Issues

"The family" has become an object of politics and public controversy. This has happened even though families are private and most people want them to remain so. Issues of employment discrimination against women affect families, whichever way they are decided. The effect is indirect but powerful nevertheless, because the outcome determines how much financial independence women have and thus their desire to get married and to stay married, as well as their power within a marriage. Laws about marriage and regulations on welfare support for dependent children affect families too, especially at the poverty level, where many households are headed by single women. Laws regarding contraceptive information for teenagers or the controversy over abortion can have a dramatic, and sometimes tragic, effect on the flow of many family lives. The politics of the future may deal with such issues as public child-care facilities; these affect the careers of parents and make a link between economics and the shape of families. We live in a time when "the family" is not taken for granted any longer but is constantly being reshaped by public, political, and private decisions.

Many of the current trends in families were being manifested before politicians and other commentators got around to advocating their proposals: the long-term increase in employment among married women, for instance, or the steady rise in divorce and nonmarital births. The world has been changing, and politicians to a certain extent only tinker with it. Nevertheless, the politics of the family is going to be a live issue for the foreseeable future. One aim of a textbook on the sociology of the family ought to be to make us more aware of these realities: what effects different government policies are likely to have on families, and also what aspects of families have been moving along on their own momentum apart from political control.

Our Personal Life Experience

The theories and facts presented from this academic viewpoint have another relevance, too. We all experience families. Virtually everyone has been brought up in one, and most of us will have a family of our own. The family has, to a considerable degree, shaped what we are, and it will be a major part of our life experience in the future. Sociological theories about families have a personal and practical relevance that virtually no other part of sociology shares. This is not to say that sociology is capable of drawing faultless guidelines on how to live a happy family life. Many of the issues are too powerfully embedded in the social world we live in for us to manipulate them easily. However, the sociological perspective gives us some insight into how and why families are different and helps us see the pressures they will face in the future. If it cannot let us evade many of the conflicts, it can at least alert us to what they are and present some ways of dealing with them.

Careers in Family-Related Professions

The sociology of families is also relevant from a third point of view. Increasingly, families are of concern to professional practitioners. Most of the work of psychiatrists, clinical psychologists, counselors, and social workers deals with families and their effects on individuals. We would argue that although some of the problems that one witnesses are *psychological,* in the sense that we can see them in individuals, they are not necessarily psychological in their origins. We need to understand the sociology of families to see how many issues arise. For example, there was a time when women struggling for some autonomous roles in a male-dominated system were diagnosed as merely having psychiatric problems; today, a sociological perspective gives better insight into the causes of their situation and possible courses of action. Families also have a major effect on how children succeed in school and indeed on important aspects of their subsequent careers. Professional educators, career counselors, and person-

---

A CLOSER LOOK 1.5
Some Professional Careers Using Family Sociology

Professional Family Specialists

Family planning counselors

Family life educators

Marriage and family counselors

Psychiatrists and clinical psychologists

Social workers

Sex therapists

Rape counselors

Professions Often Dealing with Family Issues	*Examples*
Lawyers and judges	Divorces, child custody, juvenile crime
Teachers	Learning problems, family motivation and cultural background
Medical doctors and nurses	Child and spouse abuse, family stress
Law enforcement	Spouse abuse; family background of crime, murder investigations
Politicians and social-change activists	Gender equality, public education, child care, abortion, population policy

---

nel managers would also benefit from some lessons in the sociology of families. The growing importance of political issues around the family makes it of practical relevance for all of us to know what aspects of the surrounding world affect the inside of the family, and vice versa. We list some of the professional careers using family sociology in *A Closer Look* 1.5.

SOCIOLOGICAL VERSUS PSYCHOLOGICAL APPROACHES TO FAMILIES

The approach taken in this book is sociological, that is, it analyzes families as structures of relationships among persons and as a part of larger societies. This approach differs from a psychological or individually based approach which would deal with the family from the point of view of a particular person within it. This does not mean that we will pay no attention to individuals' experiences. The sum total of individual experiences and actions, after all, are what make up families. From a sociological viewpoint, we do not merely stop with the individual experiences or motivations but try to show how they fit into a larger pattern.

The **sociological approach** to families, then, is from the outside in, rather than the inside out. The goal is not merely to acquaint the reader of this book with what family life *feels* like but to offer some insight into why people feel as they do and to explain why certain kinds of situations keep turning up. Being in love, for example, is a feeling of warmth, tenderness or passion. As sociologists, we respect that. But as sociologists, we go further and attempt to get "behind the scenes" and explain why these feelings arise at certain times, with certain persons rather than others. Sociology looks for the general process that produces individual experiences.

Sociology does not deny that individuals are unique. Each of us, in certain respects at least, lives a life that is different from other people. But most of what happens to us has happened somewhere else, too—to at least some other persons. We, like other people, are embedded in the larger networks of social structure, and this has a powerful influence on our lives. What makes us unique as individuals is the fact that each of us is embedded in the larger society in different ways; we inhabit different social networks and are subject to a different combination of social influences than are other people. One might say that there is a crisscrossing field of social forces and that each of us is moving around in that field in a different location than other persons.

Thus, each of us has somewhat unique experiences. Falling in love or bringing up a child is never exactly the same from one person to the next. Nevertheless, these experiences are made up of certain common social ingredients. What sociology attempts to do is to analyze how these social ingredients, these general social processes, operate and how they vary systematically. Although these general sociological principles never completely explain everything that happens in one's own individual experience, nevertheless, they provide a crucial background for understanding what is going on. We stand out as individuals against the background of society; but without society, we would not be what we are.

SUMMARY

1. Families are routine and taken for granted, but they are also the source of major emotional experiences—both positive and negative. They have sexual, emotional, and economic sides, all entwined.

2. Many issues surrounding families are connected to patterns of stratification, including the dimensions of economic class, power, and status. These affect the positions of men and women both inside and outside families.

3. Deriving income from two wage earners is now the most important way families acquire a middle-class or upper-middle-class standard of living. Only a small proportion of families reach high income levels with only the man working; if only the woman works, the income tends to be below average or even at poverty level.

4. There is no natural "sexual division of labor" between men and women because all arrangements are socially determined. Women do a great deal of practical, "instrumental" work, and men do carry out "expressive" emotional activities. Gender stratification processes have made women more likely than men to be in the lower-paying and less powerful jobs throughout the different sectors of the labor force, and women do the bulk of the work within the home.

5. Family theories focus on general mechanisms that explain why different aspects of families take various forms under different social conditions. Comparisons among different groups and different societies are an important method of establishing such theories.

6. The family has an important link to social issues because it provides a major basis for perpetuating economic inequalities and ethnic/racial distinctions, as well as the domination of men over women. Families are also central to individual life experience. They have a powerful effect in shaping children's personalities and beliefs, and they are a major determinant of individual happiness at different points during one's life. In addition, family sociology is relevant to careers in many professions.

Key Terms

economic class	high-density situation	sex
exchange theory	instrumental	social class
explanation by comparison	interaction rituals	sociological approach
expressive	low-density situation	status
gender	power	stratification

Sociology Web Site

See the Wadsworth Sociology Resource Center, "Virtual Society," for additional links, quizzes, and learning tools:

http://www.sociology.wadsworth.com

Also on this web site you'll find InfoTrac College Edition, an online library of journals. Here you can search for electronic articles about central topics in sociology.

2

The Social Ingredients of Families: Gender, Love, and Property

INTRODUCTION

What are the basic features of human families. We all know a family when we see one, but what are its essential parts? What makes it tick? The question of defining a family may seem like a simple one. What automatically comes to mind is a husband, wife, and children, living together in a house. But families often are both more and less than this. If we bring other relatives into the household, they too are part of the family. Moreover, there is a sense in which we say people comprise a family even if they don't live together. This is what we mean when we say, "The family all gathered at Grandma's for Thanksgiving."

A family, then, can be more than husband, wife, and children in a home. Can it be less than this? Apparently, it can. The common household does not seem to be necessary. A married sailor in the navy still has a family when he is away for a year on his ship. In some tribal societies of the matrilocal type, a wife lives with her mother, while her husband lives with his mother and only visits his wife in the evenings.

Are children necessary for a family? This is a matter of debate. Some sociologists confine the term *family* to a couple with offspring, using the word *marriage* to refer to the husband-wife relationship by itself. According to this definition, a woman living with children born out of wedlock has a family but not a marriage. Notice, though, that the children do not have to be biological offspring to constitute a family. Adopted children are treated as full members of a family in our own society. In some traditional societies, such as medieval Japan, adoption was very common, especially in families that lacked sons of their own to inherit the family property and carry on the family name. The crucial matter that defines parents and their children, then, is not the sheer biological fact of procreation, but a social consensus of relationship between them. This can be seen in the historical distinction between "legitimate" and "illegitimate" children. Those born without the proper social ceremonies of marriage between their parents were often not considered members of the family (at least not in the full sense), even though the biological ties were there.

Is it then the male-female relationship of a husband and a wife that is the core of the family? There is good reason to argue that this is so, but still we need to be careful. A number of practices found here and there in the world violate this definition. Among some African tribes, for example, there are cases of marriages between two women (*A Closer Look* 2.1). In our own society today, there are communities in which the prevailing form is **polygynous** marriage—one man with a number of wives (*A Closer Look* 3.2), as well as growing numbers of same-sex couples.

What definition we choose, of course, is in a certain sense arbitrary. We can just decide on what word we are going to attach to what observable thing in the world. The word *automobile* was picked out around 1900 when motor cars were invented, but it would work just as well to call them *velocipedes* or whatever we wished—as long as everybody was clear on what the word referred to. So if we liked, we could arbitrarily define the "family" to mean husband, wife, and children living together. The price we would pay for this definition would be that all sorts of arrangements among men, women, children, and households would fall outside the definition. We might overlook these because we wouldn't know what to call them.

In chapter 1, we introduced the U.S. Census Bureau's definition of a **family** as "two or more persons who are related by birth, marriage, or adoption who live together as one household." This definition allows them to decide which households to classify as families when they perform population counts, but we notice

A CLOSER LOOK 2.1
Two Forms of Nonstandard Marriage

An African Marriage Between Women

In the Nuer tribe in the Sudan, it was possible for two women to marry each other. A wealthy woman who believed she was sterile could marry a younger woman by going through the same ceremony with her that was ordinarily used to unite a woman with a man. The older woman would find a man to make her "wife" pregnant. The older woman was called the husband in this marriage, and their children referred to her as "father." In this society, women were important economic producers and were able to accumulate property of their own, thus allowing some of them to forego marriage to a man. (Evans-Pritchard 1951, 108–9).

Polyandry in the Himalayas

In villages of the Himalayan Ladakh region, it is common practice for a woman to have several husbands—usually brothers. A shortage of women contributed to the practice of **polyandry** in this remote mountainous region where the ancient boundaries of Nepal, Tibet, China, India, and Pakistan intersect. This marriage system offers several advantages to women. Since men can be gone for long periods, when one husband is away, the women are not alone, and their children enjoy the benefit of multiple fathers. Women enjoy comparatively high status under polyandry and can end intolerable marriages. Divorce is relatively easy, and when a woman changes husbands, she is likely to command a larger bride-price because of her increased experience and proven fertility (Berreman 1993; Lips 1993, 294; Schuler 1987).

that it leaves out various kinds of relationships that people call families. This definition includes both single-parent and two-parent households, but it excludes such arrangements as married couples living apart, gay or lesbian couples, any nonadoptive and nonbiological "parent," and various relatives separated by divorce. It also excludes those people normally called "family" that don't reside with us—from aunts, uncles, grandparents, and cousins to children who are either off on their own or living in another household. Do we need to include all these people in our definition of the family?

We should try to define the family as broadly as possible, so that we can study as many phenomena of this general type as are of interest to us. There is even a philosophical position, taken by such thinkers as Ludwig Wittgenstein (1953), that definitions of words are always shadowy and elusive at their borders. Any word covers a variety of instances that are connected together like a net but do not actually have any single characteristic in common to them all. According to this approach, we all have a gut-level feeling for what we mean by "family." Instead of trying to arrive at a precise definition that covers all instances, we should take our commonsense understanding as a basis for exploration.

It is still worthwhile, however, to push our investigation far enough to find out what makes a family system work. Exploring the basic concept of the family is a way of explaining why families take the forms that they do.

It is sometimes asserted that the family is universal, found in all societies. It has to be so, according to this line of argument. For without the family, societies would not be able to reproduce themselves, physically or culturally. On the other hand, utopian thinkers from time to time have tried to envision worlds without the family. Attempts have been made to put some of these plans into practice, for instance, in American communes or in the Israeli kibbutz. Other thinkers have predicted, on the basis of current trends, that the family may disappear in the future. This doesn't seem to be happening, although it is true that the forms of the

Not all families are made up of a husband, wife, and children. These gay men, shown at their "commitment ceremony," have committed themselves to each other and made a vow to create their own family unit.

modern family are shifting. But the question remains open: Is it really *possible* that a society could exist without *some* kind of family form? This is one of the questions that a basic theory of families should help us answer. We develop such a theory by considering the family as a property system.

THE FAMILY AS A PROPERTY SYSTEM

Three kinds of property are involved in the family. **Rights of sexual possession** include the rights of sexual intercourse and prohibitions on intercourse with outsiders. Sometimes the rights also extend to claims over a person's emotions or affection, although this is mainly found in modern societies like our own. **Economic property rights** include the material household itself, the income that supports the family, and the labor that different family members put into making the household a living concern. **Intergenerational property rights** include the rights that children have to inherit the family's economic property and also the rights that parents have over their own children, economically and otherwise.

We should bear in mind that not all societies have the same particular rights under each of these headings. In some societies, these rights are heavily concentrated in the power of men over women, parents over children, or a single head of the family over all the other members. In other societies, there are more rights for women and for children. In some, fathers are all-important, whereas in others, a woman's brother has far more power. However, all societies have some way in which these three kinds of property are arranged. If we look at every society, including our own, in terms of these three kinds of property rights, we will gain insight into the basic dynamics that determine what goes on in families. Moreover, the comparisons help tell us under what conditions these property relations vary: when there is the greatest amount of sexist oppression, when we find equality among men and women, or even what conditions produce women's control of such property rights.

Rights of Sexual Possession

The first kind of possession in any marriage consists of erotic rights over human bodies. In all societies, a marriage establishes the right to sexual inter-

A CLOSER LOOK 2.2
What Counts as a Wedding?

Wedding ceremonies vary a great deal from society to society. American wedding customs include many ritual elements: the church ceremony itself, the bride's white dress, throwing rice on the couple as they emerge from the church, tying tin cans to their car, and the bride's throwing her wedding bouquet to the maids of honor. In other societies, the rituals are quite different.

In medieval Europe a couple was not married within the church but outside the church doors. This was because the medieval Church put its highest value upon a life of celibacy, and marriage was regarded as a concession to the weakness of the flesh. The bride wore a red gown, the color of sensual love; the white wedding dress did not come in until the 1800s (Ariès 1962, 357–58).

In the modern Israeli kibbutz, meals are eaten in a dining hall, and children are brought up in a collective nursery. All that is involved in getting married is for a couple to put in an application to live in the same dormitory room (Spiro 1956).

In the matrilineal society of the Trobriand Islands, premarital sexual intercourse is perfectly normal, and nobody comments on it. What the Trobrianders find shocking, though, is the modern Western custom of dating, in which an unmarried man and woman casually go out to eat. For among the Trobrianders, it is sharing a meal that constitutes the wedding ceremony (Malinowski 1929).

In ancient Greek society, the wedding consisted of transferring the bride from her father's domestic religion to her husband's. First her father offered a sacrifice on his domestic altar and declared to his family gods that he was giving his daughter into another family. Then she was veiled, dressed in robes reserved for religious observances, and taken to her husband's house in a procession. Those around her carried a torch and sang a religious chant, with the refrain *O Hymen, O Hymenaie* (Hymen being the Greek god of marriage). When they arrived at their destination, the husband pretended to seize the bride by force and carried her over the threshold, taking care that her feet did not touch the sill. Finally, she touched the sacred fire in her husband's house and was initiated into her new family worship, under the protection of her husband's family gods (Fustel de Coulanges 1973, 44–46).

What all these ceremonies have in common is simply the fact that the couple publicly show the community that they are now going to live together. The wedding is basically a public announcement; all the special ritual is to let everyone know that an important change is taking place.

course between a particular man and woman (see *A Closer Look* 2.2 for a discussion of cross-cultural variation in wedding ceremonies). Sometimes, although not always, this is an exclusive right. Often marriages have given the man exclusive sexual possession over his wife while leaving him relatively free to have intercourse with others. In our modern family system, the ideal is bilateral sexual possession, with both husband and wife confining their erotic activities to each other.

To call this arrangement "property" may sound crass. After all, we usually think of property as a physical thing that someone owns. To speak of someone's body as being the property of another makes us think of slavery, degrading the human person to the status of an inanimate object. Nevertheless, there are good reasons in sociological theory for extending the concept of property to erotic rights in marriage. Kingsley Davis (1936, 1949), who originated the term *sexual property,* pointed out that property is not the thing itself that someone owns, but a *social relationship;* that is, the concept of property is a kind of agreement among people about how they will act toward particular things. If you own a car, for example, that does not mean there is a bond between you and your car. What it means is that (1) you have the right to use the car; (2) other people do not have the right to use your car, and (3) you can call on the rest of society to enforce your rights; for example, you can call the police if someone takes your car without your permission.

Erotic rights over human bodies are property rights in almost exactly the same sense. A marriage may be socially defined in our own society as follows: (1) the husband and wife have the right to sexual intercourse with each other; (2) other people do not have the right to sexual intercourse with either of them while they are married; and (3) if violations occur, the aggrieved party can go to court and demand damages, as in the form of a divorce. These property rights are different in certain respects than the rights people have over their cars. Husbands and wives in our society cannot buy or sell their sexual possessions, for instance, and they do not have the right to destroy them. The exact rules of sexual possession vary from one society to another. Some societies (such as ancient Rome) have allowed the husband to sell or kill his wife and children. The fact that we do not allow this does not mean that there is no property relationship; rather, the state does not allow people to do certain things with their sexual possessions, more or less the same way that the state does not allow people to do just anything they wish with cars, such as drive them without a license or faster than the speed limit.

If it seems crass to define married rights to sexual intercourse as a form of property, the concept nevertheless helps us to see certain things that we otherwise might miss. In many traditional societies, marriages were arranged quite impersonally as a collection of property rights and duties and involved little or no affection between husband and wife. But even in today's society, when there have been major changes in the kinds of property rights that married people have over each other, we still have a hard, tough core of demands that married people make on each other. Sexual rights remain a basic aspect of family possessiveness, and even our ideals of love and affection are actually treated as exclusive rights, as we shall see on the following pages.

How can we show that marriage always involves sexual possession? For one thing, in practice we take marriage as being established by sexual intercourse. Suppose a couple is married in a religious or legal ceremony. Are they really married? Our laws actually say no, in that the marriage license and the ceremony do not have full force until marriage is sexually "consummated," as the expression goes. If the couple never have intercourse, say because of sexual impotence, then this is grounds for annulment. An **annulment** is not a divorce; it simply declares that the marriage itself never came into existence because an essential ingredient was missing. Even in very traditional societies that do not allow divorce (except under extreme circumstances that we will examine), lack of sexual consummation is grounds for annulment.

An act of officially approved sexual intercourse, then, is what constitutes marriage in our society and in all others. We say "officially approved" because intercourse may occur among unmarried people, sometimes quite openly, without implying any consequences. The official sanction takes the form of some kind of public announcement, which brings the rest of society into the picture. Through its representatives, the courts, and/or religious leaders, society thus takes a hand in recognizing that rights to sexual intercourse have been established and thus is ready to back up these rights by appropriate punishments when they are violated. In all societies, a marriage takes the form of some kind of public ceremony that makes this sort of announcement to the community. In early medieval Europe, for instance, the marriage was not religious at all but took the form of a large feast at which relatives and friends were entertained on the night the couple had their first (at least "official") intercourse (*A Closer Look* 2.3).

The fact that sexual possession is a key to marriage is brought home to us by the way in which people react to its violation. The violation of sexual possession is called **adultery.** This is considered a crime in most societies, although some punish it much more severely than others. In conservative, Catholic-dominated countries such as Italy, divorces were not allowed up until just a few years ago, although an exception was made in the case of adultery. This is because adultery was considered such a major violation of the central features of marriage that a marriage could not survive it. The importance placed on sexual rights in many societies is illustrated by the unwritten law that a man would not be convicted of murder for killing his wife's lover. (This unwritten law was usually rather sexist in operation, because it less frequently condoned a wife's killing her husband if she caught him in adultery.) The degree of such sentiments has varied at different times. In ancient Greece and Rome, a husband could kill his adulterous wife and her lover, and the law took no notice of it. Today, a court action would almost certainly be instituted, although juries are inclined to give relatively light sentences to murders committed out of this kind of jealousy.

The modern law also recognizes sexual rights in another sense. In most states, the husband has had the legal right to sexual intercourse with his wife (and vice versa); hence, there was no crime of rape between married people, according to the law. This implies that the marriage contract itself gives away one's right to control one's own sexuality, once and for all; until a divorce takes place, the law has nothing to say over what happens sexually between the couple. Since about 1980, many states, but not all, have passed laws prohibiting marital rape. These laws, and the cases that have arisen since their passage, are controversial, which indicates how strongly held are our beliefs about this kind of "property." Marriages are changing in the direction of sexual rights by mutual consent but are still recognizing married couples as sexual partners.

There are certain practices that might seem to violate this conception of marriage as sexual possession. Some societies allow institutionalized "wife swapping" or extramarital affairs, and others practice **polygamy**, allowing someone to have multiple marriages (see *A Closer Look 2.1*). But here is where our basic theory of property is very revealing. If you own a car, it does not violate your right of property if you let somebody else drive it. In fact, a loan or a gift actually emphasizes whose property it is. To borrow something, one person has explicitly to ask the owner and be granted permission; so the very act of borrowing involves a social acknowledgment of who actually owns the property. The same is true in societies that practice "wife swapping." This was once fairly common, for instance, among Eskimos (Hoebel 1954). Eskimo men took long hunting trips, and when one stopped in for hospitality at a distant house, the host would often lend the visitor his wife. In reciprocity, the host could expect the same courtesy when he traveled far from home. That does not mean that Eskimos had no concern for sexual possession. They certainly did, and Eskimos often quarreled when a man took someone's else's woman without asking. The Eskimo murder rate, in fact, has been one of the highest in the world.

The Incest Taboo and Exogamy

Property rights, as we have seen, have both a positive side—who has the rights to something—and a negative side—who is excluded from it. One of the most important forms of sexual property is the negative side: who is *excluded* from sexual intercourse with whom.

The **incest taboo** is a restriction on sexual intercourse, and hence on marriage, among close relatives. But what has counted as "close" has varied historically. Medieval Christian societies prohibited marriage with cousins, as well as aunts

A CLOSER LOOK 2.3
Brunhild's Wedding Night: A Medieval Battle over Sexual Possession

The *Nibelungenlied* was the most popular German poem of the Middle Ages. In this excerpt from the story, the German King Gunther has brought back his new bride, Brunhild, from Iceland, where he won her favors in a contest to the death. But Gunther's friend Siegfried had secretly helped him win the contest, and Brunhild is suspicious. After a great celebration in his castle with all the knights of the realm, Gunther goes to his room to consummate the marriage.

Listen to how gallant Gunther lay with lady Brunhild—he had lain more pleasantly with other women many a time.

His attendants, both man and woman, had left him. The chamber was quickly barred, and he imagined that he was soon to enjoy her lovely body; but the time when Brunhild would become his wife was certainly not at hand! She went to the bed in a shift of fine white linen, and the noble knight thought to himself: "Now I have everything here that I ever wished for." And indeed there was great cause why her beauty should gratify him deeply. He dimmed the lights one after another with his own royal hands, and then, dauntless warrior, he went to the lady. He laid himself close beside her, and with a great rush of joy took the adorable woman in his arms.

He would have lavished caresses and endearments, had the Queen suffered him to do so, but she flew into a rage that deeply shocked him—he had hoped to meet with "friend," but what he met was "foe"!

"Sir," she said, "you must give up the thing you have set your hopes on, for it will not come to pass. Take good note of this: I intend to stay a maiden till I have learned the truth about Siegfried."

Gunther grew very angry with her. He tried to win her by force, and tumbled her shift for her, at which the haughty girl reached for the girdle of stout silk cord that she wore about her waist, and subjected him to great suffering and shame: for in return for being baulked of her sleep, she bound him hand and foot, carried him to a nail, and hung him on the wall. She had put a stop to his lovemaking! As to him, he all but died, such strength had she exerted.

And now he who had thought to be master began to entreat her. "Loose my bonds, most noble Queen, I do not fancy I shall ever subdue you, lovely woman, and I shall never again lie so close to you."

She did not care at all how he fared, since she was lying very snug. He had to stay hanging there the whole night through till dawn, when the bright morning shone through the window. If Gunther had ever

been possessed of any strength, it had dwindled to nothing now.

Gunther confides in his friend Siegfried, a knight who has magic powers. Siegfried agrees to make himself invisible and come to help Gunther that night.

Siegfried has gone to Gunther's chamber where many attendants were standing with lights. These he extinguished in their hands, and Gunther knew that Siegfried was there. He was aware what Siegfried wanted, and dismissed the ladies and maids. This done, the King quickly thrust two stout bolts across, barred the door himself, and hid the lights behind the bed curtains. And now mighty Siegfried and the fair maiden began a game there was no avoiding and one that gladdened yet saddened the king.

Siegfried laid himself close to the young lady's side. "Keep away, Gunther, unless you want a taste of the same medicine!" But Siegfried held his tongue and said not a word. And although Gunther could not see him, he could plainly hear that no intimacies passed between them, for to tell the truth they had very little ease in that bed. Siegfried comported himself as if he were the great King Gunther and clasped the illustrious maiden in his arms—but she flung him out of the bed against a stool nearby so that his head struck it with a mighty crack! Yet the brave man rebounded pow-

and uncles, nephews, and nieces, out to the seventh degree (Goody 1983, 44); whereas in many tribal societies first cousins are the preferred marriage partners. However, mother-son and father-daughter intercourse has been condemned everywhere as incestuous; brother-sister marriage, although allowed in a few places and circumstances, has also generally been prohibited.

The incest taboo is usually thought of as a way to prevent inbreeding. It is not really useful to think of this as a biological phenomenon; clearly incest prohibitions are not instinctual because they are violated often enough,

erfully, determined to have another try, though when he set about subduing her it cost him very dearly—I am sure no woman will ever again so defend herself.

Seeing that he would not desist, the maiden leapt to her feet. "Stop rumpling my beautiful white shift!" said the handsome girl. "You are a very vulgar fellow and you shall pay for it dearly—I'll show you!" She locked the rare warrior in her arms and would had laid him in bonds, like the King, so that she might have the comfort of her bed. She took a tremendous revenge on him for having ruffled her clothes. What could his huge strength avail him? She showed him that her might was the greater, for she carried him with irresistible force and rammed him between the wall and a coffer.

"Alas," thought the hero, "if I now lose my life to a girl, the whole sex will grow uppish with their husbands for ever after, though they would otherwise never behave so."

The King heard it all and was afraid for the man; but Siegfried was deeply ashamed and began to lose his temper, so that he fought back with huge strength and closed with Brunhild desperately. To the King it seemed an age before Siegfried overcame her. She gripped his hands so powerfully that the blood spurted from his nails and he was in agony; but it was not long before he forced the arrogant girl to recant the monstrous resolve which she had voiced the night before. Meanwhile nothing was lost on the

King, although Siegfried spoke no word. The latter now crushed her on to the bed so violently that she shrieked aloud, such pain did his might inflict on her. Then she groped for the girdle of silk round her waist with intent to bind him, but his hands fought off her attempt so fiercely that her joints cracked all over her body! This settled the issue, and she submitted to Gunther.

"Let me live, noble King!" said she. "I shall make ample amends for all that I have done to you and shall never again repel your noble advances, since I have found to my cost that you know well how to master a woman."

Siegfried left the maiden lying there and stepped aside as though to remove his clothes. And now Gunther and the lovely girl lay together, and he took his pleasure with her as was his due, so that she had to resign her maiden shame and anger. But from his intimacy she grew somewhat pale, for at love's coming her vast strength fled so that now she was no stronger than any other woman. Gunther had his delight of her lovely body, and had she renewed her resistance what good could it have done her? His loving had reduced her to this.

And now how very tenderly and amorously Brunhild lay beside him till the bright dawn!

A sociological comment: The story recalls the violent Viking age several centuries before *The Nibelungenlied* was written. Brunhild is something like the

Valkyrie, the mythical armed women who accompanied the Norse god Odin. There must have been some reality behind these stories. Iceland, where Brunhild had reigned as a maiden queen, was the quasi-mythical Viking frontier where, according to the poet, there was a matrilineal system and female warriors of extraordinary strength. We notice that marriage is conceived of rather crudely, especially from the perspective of the poet's own age, when the Vikings had given way to a polite courtly society of knights and ladies. There is no church ceremony, and the wedding consists only of a large feast and then the first sexual intercourse itself. Brunhild defends herself, in a kind of temporary triumph of the old matrilineal ideal. Then Siegfried shows that his masculine powers are even greater, and she is subdued. Notice that after she has finally submitted to intercourse, she suddenly loses all her strength; the Viking warrioress is turned into a soft and amorous lady of the new courtly style. One can interpret this as a fight between two different systems of sexual property, with a male-dominated system overcoming a female-dominated one.

whereas instincts are incapable of being violated. Nor should we expect that people throughout history and back into primitive times would have had a theory about the genetic effects of inbreeding, since genetics was discovered only within the last 100 years. The incest taboo has been so powerful because it is so central *socially*. As many theorists have pointed out, without the incest taboo, human society would scarcely be possible. The incest taboo forces individuals to go outside of the families in which they were born to find sexual partners. The result is called **exogamy,** meaning going outside one's own family group to marry. Without this pressure, each family could stay isolated, continually inbreeding among its own children. The world would consist of little isolated families, with no contact or exchange among them. The incest taboo is so powerful socially because it lays down the rule that *no family can exist alone.*

How did the incest taboo come about, if it is not instinctual? One cannot simply invoke a psychological explanation. Some theories have suggested that there is a natural lack of sexual attraction among persons who are brought up together. But this certainly isn't true among many animal species, and Sigmund Freud claimed that the unconscious wishes of his patients showed they frequently had incestuous desires. Freud thought that some crucial event occurred early in human history that enforced the incest taboo. Freud's own speculation about what happened is not widely accepted, but he is probably right that the incest taboo did not emerge from inner psychological processes but had to be *enforced from the outside.*

The French anthropologist Claude Lévi-Strauss (1969) proposed a method of reconstructing what happened in unwritten family history. Lévi-Strauss argues that the incest taboo does not stand by itself but is part of a larger system of family structure. The incest taboo can be regarded as a kind of treaty between different families: each one promises that it will send its own children out to find marriage partners if other families will do the same. There are different ways in which this can be done. In some systems, the women leave home, and other women come in as wives for the men who remain there. (This is called a **patrilocal system.**) In other systems, the men live with their sisters, and husbands are temporary visitors in the homes of their wives. (This is called a **matrilocal system.**) In both cases, there is an exchange. Lévi-Strauss argues that families exchange women because the incest taboo requires that they cannot be sexual partners for those who stay at home.

Viewed in this way, the incest taboo is a political treaty among families. Every time a marriage is made, one family has given away a daughter or a niece. In most cases, there isn't a corresponding man or woman from the other family to be given back right away in a compensating marriage. If the Jones family (or the Black Eagle clan), marries a daughter to the Smith family (or the White Eagle clan), it doesn't usually happen that the Joneses (Black Eagle) have a son ready to marry a Smith (White Eagle) daughter at exactly the same time. But a bond is created between the two families because they have intermarried once and expect that when the opportunity arises there will be an intermarriage in the opposite direction. Lévi-Strauss points out that it is even useful that the exchanges do not happen in both directions at the same time. It is better that one family should be waiting in expectation, so that the alliance between the families always has something to look forward to. It is these *overlapping* exchanges that make for the strongest bonds.

Lévi-Strauss goes on to make a technical analysis of many of the different types of marriage exchange systems that exist in tribal societies. In some of them, a man is expected to marry his cousin on his mother's side, and the woman marries her cousin on her father's side; in other systems, the preferred cousins are chosen in a different way. Each type of marriage system links together different sets of families. As a result, there are different types of social structures. For in tribal societies, there is no state, no police force, no stores or factories, and no economic or political organizations other than what the family provides. The network of intermarrying families *is* the entire social structure. By working out these rules of intermarriage, larger units beyond each individual family are created. Moreover, these families do not merely exchange men and women as marriage partners; they also exchange food, help in building houses or clearing land for crops, and cooperate in hunting together. The group of families that has intermarried constitutes a political unit, to defend one another against attack and to mount expeditions attacking their enemies. Intermarriage is thus not just a matter of getting a sexual partner for each individual but creates the entire alliance structure of the society. It is not only a sexual activity but also an economic and political one.

We can see now why the incest taboo was so important in getting human society started. The taboo is part of a system of sexual exchange that makes possible all the other economic and political exchanges. The incest prohibition is the negative rule in this system; it says that families must not use their own members for sexual partners. Corresponding to it is a positive rule, which says that families must go outside to find new sexual partners. By going outside, alliances are made, and the larger society becomes possible.

The incest taboo, then, did not emerge from some kind of repulsion inside the family. It appears to have been negotiated between groups, as part of a package deal that they would begin to intermarry in a particular way. Lévi-Strauss implies that those families that made these treaties and began to prohibit incest became stronger because they gained allies. They were more assured of food and military power than families who did not prohibit incest. Hence, the families with the alliance system would have survived better than the isolated, incestuous families that failed to make the transition. Eventually all families that remained had to abide by the incest taboo and make their children available only to outsiders. When they failed to do so, they were punished from outside. The surrounding society came to expect that children would be reserved as sexual partners for someone outside the family and reacted angrily against violators who failed to keep their part in the social alliance system.

As we shall see in the next chapter, there are quite a few different ways that sexual exchange systems can be organized. In some societies the older generation does most of the exchanging, using their offspring as pawns in a game of marriage politics. In other societies, the younger generation is able to negotiate its own individual exchanges. The shifts among these types of exchange systems make up a major part of the history of the family.

There are also differences in how large a network is involved in exchanges. Some societies have enforced exchanges that go very far from home, whereas others exchange only among a sort of extended family. Continual interchanges among a group in which the cousins are constantly intermarrying would be regarded by us as rather incestuous, even if they did go outside the immediate nuclear family. Like everything else in human societies, the incest taboo is a variable, and its extent is a matter of degree. In one period in ancient Egypt, marriage between brothers and sisters was fairly common. It took special economic and political circumstances for this to happen, and the coming of Christianity soon took family policies to the opposite extreme (*A Closer Look 2.4*).

Unilateral and Bilateral Sexual Possession

Sexual possession, like any other kind of property, can be distributed more equally or unequally. In many traditional societies, it constitutes what may be called one-sided or **unilateral sexual possession:** the woman is the sexual

---

A CLOSER LOOK 2.4
Incest in Ancient Egypt

Marriage or sexual intercourse between brother and sister has almost everywhere been forbidden as incestuous. Nevertheless, there are a few instances historically where such marriages were allowed or even encouraged. In the royal family of native Hawaii, for example, brothers and sisters regularly married, and so did siblings in the succession of the Egyptian pharaohs. Cleopatra was married to her brother (in fact, to two of them, Ptolemy XII and Ptolemy XIII), and so were many other Egyptian queens. Both of these patterns can be explained by kinship politics. Because intermarriage is a way of making alliances between families, royalty attempts to marry only the children of royalty. Both Hawaii and the Kingdom of the ancient Nile Valley were quite isolated from other societies; hence, there were no other kingdoms nearby with which their royalty might intermarry. Hence, brother-sister incest became a solution for keeping inheritance strictly within royal lines.

In ancient Egypt around the time of the Roman Empire, however, brother-sister incest was much more common even than this. Among the landowning classes in general, a very high proportion of marriages were between brothers and sisters. Jack Goody (1983, 43–44) points out that this was a strategy for keeping family inheritances intact. Under Egyptian law of this period, women had considerable rights of inheritance. At the same time, land holdings along the narrow irrigated strip of the Nile were extremely valuable, and the elite of wealthy families who owned them were very reluctant to break them up. Hence, brother-sister incest, which had already been legitimized by the marriages within the family of the pharaohs, was adopted by the landowning classes in general as a way to keep the family property together.

After the Christian church began to become powerful, however, it strongly condemned these marriages as incestuous. Elsewhere in the Roman Empire and in northern Europe, too, the church fought against marriages among cousins, men marrying the wives of their deceased brothers, and other marriages among close relatives. Goody (1983) interprets this as an economic strategy on the part of the church. For the older marriage customs were all ways of keeping inheritance within the extended family, by ensuring that there would usually be an heir available. The church, though, benefited from bequests made to it by people who died without heirs, and this was one factor in its efforts to eliminate these "incestuous" practices. By the late Middle Ages, the definition of incest had been extended so that it included relatives no more closely related than seventh cousins.

---

possession of the male but not vice versa. In medieval Arab societies, for example, women were strongly restricted as the sexual possessions of men. They were kept secluded in harems, and virginity at marriage was considered a prime mark of a woman's value. A man's honor depended on being the only sexual possessor of a woman, and her marriage value was zero if she was not a virgin. However, Arab men were not at all restricted. They could have intercourse with prostitutes and slaves and keep concubines without the wife's having any say in the matter. In such a society, we would say that there is one-sided sexual possession: of men over women but not vice versa.

Sexual possession is not necessarily organized in this extreme male-dominated form. In many societies women have sexual rights over their husbands. This, in fact, is often true in certain kinds of polygamous societies. In Madagascar (off the coast of Africa), wealthier men usually had several wives (Linton 1936). Each wife had her own hut as part of the family compound. The husband was supposed to rotate regularly among them, spending a night with each in her turn. If the husband violated the rotation by spending an extra night with a particular wife, the other wives considered that he had committed adultery. This was a type of horticultural society in which women had considerable economic importance as food

The paramount chief of the Trobriand Islands, off eastern New Guinea in the southwestern Pacific, visits with relatives, who sit respectfully at his feet, at the home of one of his wives. In the Trobriands, polygamy is practiced by a few men of high rank.

producers, and hence women had a certain amount of power to assert some controls over men.

Our own society today is officially closer to a form that can be called **bilateral sexual possession.** This means that in principle, both husband and wife are sexual possessions of each other. Neither one has the right to extramarital affairs, and both have sexual rights over each other's body. This has been the official ideal in Western societies for at least the past few hundred years, although there has been some tendency for it to be violated in practice, especially by men. In recent years, there is some evidence that women are now committing adultery as much as men in some sectors of society (Blumstein and Schwartz 1983), although men still remain ahead on that score (Smith 1991; also Laumann et al. 1994).

The whole issue of sexual possession, and especially its unilateral or bilateral nature, remains a matter of controversy among sociologists and feminist theorists. Using a combination of anthropological evidence and Freudian psychoanalytical theory, some argue that the basic pattern in all societies is for men to exchange women (Rubin 1975); that is, there is a basic taboo, the incest taboo, that requires that individuals must leave their own families to find sexual partners. Hence, one can say that families are actually exchanging sexual partners. In tribal societies, as many anthropologists have pointed out, these exchanges are deliberate political maneuvers by which families establish alliances with one another. In many such tribal societies (as well as medieval societies like the Arab example given earlier), the exchange is explicitly controlled by men: women are treated as sexual property whom men use to make exchanges for the purpose of political alliances. Although the family political alliances have disappeared, the basic psychological structure of *men possessing and exchanging women* continues to exist in our own society.

Thus, Rubin and others argue that one-sided sexual property still exists and is the basis of gender inequalities throughout our society. The deep structure of the problem goes back to a second taboo that is implicit in the incest taboo: a rule of "compulsory heterosexuality" that requires male and female children to give up any erotic drives for members of their own sex. Freud argued that small children have no sexual preference until it is forced on them by the conventions of culture. If males and females took equal part in caring for babies and small children, the infant's original bisexual orientation would not be overturned by the imposition of heterosexuality. Gender distinctions would disappear, as it would make no difference whom one chose as a love object. Males would no longer be regarded as dominant and superior, and men would no longer control women and exchange them in a system of sexual property. The result, in Rubin's theory, would finally be to overturn the culturally based system of gender differences and gender inequality in all its aspects—economic, familial, and sexual.

From a sociological viewpoint, such psychoanalytic theories, though helpful in understanding some of the underlying emotions involved, are incomplete explanations for the emergence and maintenance of current sexual and gender relations. One problem with such theories is the assumption that the system of sexual property exchange in our modern society is basically the same as in tribal and traditional societies. Empirical support for the theory comes from studies of small tribal societies in which the family organizes the entire political and economic system, but large historical shifts occurred in the family, first as the state emerged and broke down these family networks and later as families lost control over the sexual bargaining and marriages of their children. (These historical changes are described in chapters 3 and 4.) It is no longer the case that a few powerful men directly exchange women, although this arrangement was true in some societies at some times. There continues to be an exchange in the form of the modern marriage market, but this is a system in which *individuals,* both males and females, are doing their own exchanging. Gender inequalities enter into this, insofar as males often have more income or higher-prestige occupations than females, but this does not seem to be inevitable. Nor is there any basic cultural rule in our society that says that *men* exchange *women.* Although it is probably not necessary for everyone to become bisexual for gender equality to come about, the assumption of universal heterosexuality is important to the maintenance of the gender hierarchy. It is also quite plausible that gender equality could be accomplished by greater equality in economic opportunities for women and men and that this, in turn, would change the traditional gender roles in the family and the marriage market.

Love as Emotional Possession

The reason that this kind of analysis of sexual possession sounds somewhat improper to our modern ears (although it hardly would have in medieval society) is that we have added another element to the modern marriage. That is the ideal of love. In our society, unlike those of the past, marriages are supposed to be founded on love. Because we commonly distinguish between love and sex, defining marriage in terms of sexual possession sounds unduly cynical: a violation of the spirit of mutual affection and caring that makes up the bond of love.

Nevertheless, our ideal of love does not so much replace sexual possession as add another dimension to it. For one thing, love and sex are not really so separate in our own minds. Sex can take place without love, but our ideal says that they should take place together. Love without sex, between a husband and wife, would be considered rather strange by most of us, and a partner might very well doubt the other's love if she or he refused to "make love," in the telling phrase.

What helps clinch the argument is the fact that we tend to treat love as a form of property too. The very language of love reveals a distinct undercurrent of possessiveness. Probably the most common expressions of love are phrases like: "Will you be mine?" "I'm yours forever." "I've lost my heart." Popular love songs have more of this kind of terminology in them than the ritualistic "I love you." Words like *mine* and *yours* are probably more common in love talk than the word *love* itself.

In our modern system of marriage, we have simply created a new form of property, **emotional possession.** Love is a right that people acquire over each other's affections. According to the modern marriage ideal, this exchange of love vows should take place between every married couple, and traditionally this love bond was supposed to last for a lifetime and be inviolable to any outsider. This is even recognized in the law. The phrase "alienation of affection" describes the violation of emotional property rights that has been recognized as a basis for divorce and damage suits. Ideally, love-property is supposed to be bilateral between both partners; both love each other equally. In practice, of course, this is not always the case. In the love affairs of courtship, men and women often play a psychological game over who is more in love with the other. Like other forms of property, possession of the emotional rights of love can lead to a good deal of maneuvering for domination. In fact, this is one of the traditional points made by most writers on the subject of love for centuries. Love itself is not at all a cynical sentiment, but it is nevertheless implicated in the struggles of a property system. Loving someone and wanting their love is usually a form of possessiveness that works in much the same way as traditional sexual rights.

Sexual Possession or Legitimacy of Childbearing?

The anthropologist Bronislaw Malinowski (1929, 1964) pointed out that sexual intercourse is widely available outside of marriage in many societies, but that bearing "illegitimate" children is prohibited or disapproved of everywhere. Hence, the purpose of the family is to regulate not sex but childbearing. Malinowski's principle of legitimacy states that all societies demand that every child have a sociological father as an official male link between the child and the community. When Malinowski formulated his principle of legitimacy, one of his main arguments was that marriage is not a license for intercourse. This seems to flatly contradict the theory that marriage is sexual possession. Malinowski observed that the Trobriand Islanders, a tribal horticultural society, were shocked by illegitimacy but allowed wide-open premarital intercourse. He pointed out that many other tribal societies put no restrictions on premarital sex. Hence, sex is not what constitutes the family, because that can be procured outside of it. This left Malinowski's candidate theory, legitimacy, as the explanation of the core of the family.

Several points can be raised against Malinowski's argument. Malinowski happened to have studied the Trobriand Islanders, whose culture is similar to many other Melanesian, Polynesian, and Oceanic peoples. These cultures are among the most sexually permissive ever known. Most societies, on a world scale, have been a good deal more restrictive of individual sexual rights, at least as far as women have been concerned. The great agrarian empires and states that have existed in the last two or three thousand years in China, India, and Eurasia have comprised the great bulk of the world's population. There, women were usually very restricted, whereas men were allowed nonmarital carousing. The segment of tribal societies that practiced wide-open premarital intercourse by both sexes includes only a tiny proportion of all the people who have ever lived on the earth. Malinowski was basing his argument on a fairly narrow base. It is true that our own society also has recently moved into a rather widespread pattern of premarital intercourse by both sexes. Does this mean that sexual possessiveness is disappearing? The evidence that we examine in later chapters indicates that it is not.

Malinowski is confusing sex per se with sexual possession. Just because sex occurs outside of marriage doesn't mean that marriage isn't a contract of sexual possession. Possession means that somebody has the perpetual right to intercourse until this right is terminated by death or dissolution of the marriage. It also means, almost always, the *exclusive right* to sexual intercourse. As we have seen, this right has usually been exercised by men over women more than by women over men. Men have almost always wanted their women to be "faithful" and "chaste" at least while they were married to them, without applying the same standard to themselves. This is what we have called one-sided or unilateral sexual possession. Some kind of sexual possession is always present in a marriage. What varies is how widely premarital intercourse is allowed. But this is just part of a larger family system in which sexual possession is a strong feature. Event the Trobriand Islanders believed a married woman should reserve herself for her husband and regarded her violation as adultery.

The comparative evidence is on the side of the property theory. Marriage is an exclusive right to sexual intercourse on the part of a man, a woman, or both. The universal existence of the incest taboo supports this; a mother cannot marry her son, for example, because of the taboo on intercourse between them. Where sexual possession is not allowed, marriage is not allowed. The mother is the exclusive sexual possession of the father; the incest taboo on the son is simply the flip side of the father's property right. Similarly, the incest taboo on father-daughter intercourse upholds the mother's sexual right over her husband, at least from rivals among her own dependents.

The property argument is wider and more inclusive than the legitimacy argument, for the legitimacy argument can be better handled within it. Sexual possession automatically carries with it legitimacy of any children that are born out of the relationship. In effect, the parents announce to the community that they are establishing a long-term sexual relationship. In return, the community regards their child as legitimate. Sexual property is more basic and comes first. Once there is sexual property, you automatically get legitimacy.

From the point of view of the property theory, Malinowski's legitimacy theory is too narrow. It only deals with the third kind of property, intergenerational property. A legitimate child is one who is entitled to inherit the family property and its social status in the outside world. Legitimacy, then, is not merely society's way of connecting children with fathers and hence providing them with a link to the social structure; it is also another facet of the family property system. It is not surprising that the highest nonmarital birth rates in the world occur in the poorest societies and in the lowest economic classes. Those living in slums or shanty towns have little concern for legitimacy because they have no property, and—a related fact—women have more autonomy from men.

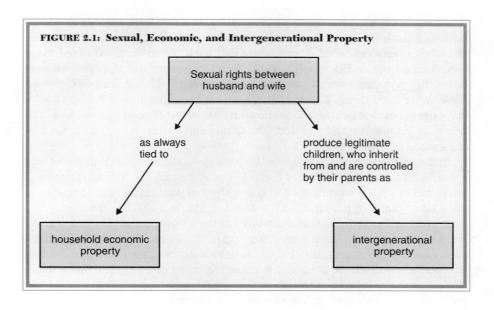

FIGURE 2.1: Sexual, Economic, and Intergenerational Property

Sexual possession links together the other two aspects of the family, intergenerational property and economic property (Figure 2.1). Intergenerational property (children and their inheritance) is a result of sexual possession. Economic property is always implicated in any permanent form of sexual possession, that is to say, in any marriage. A marriage always has some provision for family housekeeping, and it almost always transfers some economic rights among the man, the woman, and their families. For this reason, the sexual rights of women have followed the ups and downs of their economic position. Marriages are an arrangement in which sexual rights and economic rights are always involved. Other things may be involved too, but these form the core that makes the institution work. The sexual rights of each partner depend on how much power they have, and economic resources have been a key to the balance of family power. When women have had proportionately more control over their own livelihood, and even over that of the man, they have had correspondingly greater sexual rights. When men have held all the economic resources, one-sided or unilateral sexual possession has prevailed.

Economic Property and Household Labor

The modern family is not just a sexual arrangement. In day-to-day life, the most obvious aspect is its economic activities. In fact, these are so obvious that they are often taken completely for granted. The family is a kind of business, engaged in the enterprise of running a household. It is an enterprise for cooking food, cleaning clothes, providing lodging, and taking care of children. We might say that the modern family is a mixture of hotel, restaurant, laundry, and babysitting agency, especially from the point of view of the men who tend to be the beneficiaries, rather than providers of these services.

This aspect has always been true. Just how the household economy was arranged has varied among different societies in history, but all of them have taken care of these basic economic needs.

In agrarian societies, like medieval Europe, China, or the Arab world, or in ancient Greece, Rome, and Egypt, households pretty much made up the entire economy. Food was produced by family farms, whether large manors or peasant holdings; cloth was woven by women in their quarters; and most of the necessities of life were produced in the household itself. Even the specialized trades were carried out in households. A medieval merchant did not have business hours, the way that stores today are open between 9:00 A.M. and 5:00 P.M. (or some other fixed time). The medieval merchant (almost always a man) was always open, because he lived in his shop—or, to put it differently, he worked in his home, as did his wife and children. The family and the business had not yet been separated.

In the tribal societies studied by anthropologists, the situation is different once again, but the family is still very central as an economic institution. Tribal societies have much less stratification into wealthy and poor classes than agrarian societies do; what they have instead is a widespread horizontal linkage among families. The family network is the main basis of the tribal economy, its politics, war, and religion. In fact, virtually no other organizations besides the family exist in such societies. Thus, a family that has taken a man or woman in marriage from another family might be obligated to deliver yams or other kinds of food at certain times of the year. In return, the other family might have to help in joint fishing expeditions or other collective economic activities. The family and kinship network *is* the economic system, comprising much of what we would consider the market.

There are three things we should notice from these comparisons:

1. In all societies, the family is economic. In fact, our word family comes from the ancient Roman word for "household," including property as well as people *(A Closer Look 2.5)*. In our modern society, the family is much less important in the total economy than it once was, but it is still very important for the family members themselves. Children, of course, demand economic support from their family, but so do adults. Legally, the husband and wife hold their household property in common. If the husband alone works, the wife has the right to economic support. If she works, her husband shares her income. In both cases, there is also the nonmonetary side of the family economy: the labor put into running the household. Usually, most of this labor is put in by the woman, formerly in the role of housewife, but even today now that most wives are employed.

2. The family economy is always connected to sexual possession. In tribal societies, this marriage market is recognized fairly explicitly as part of the system of economic exchange. Hence, many tribes have specific rules about whom children should marry. For example, some tribes want cousins on their father's side to intermarry (patrilateral cross-cousin marriage), whereas others prefer intermarriage of cousins on the mother's side (matrilateral cross-cousin marriage). These rules keep certain branches of the family tied together by repeated exchanges generation after generation.

Many agrarian societies had the institution of the dowry, wealth that a bride brought to her marriage. Without a good dowry, a woman (at least of the higher social classes) was not very marriageable, and families went shopping rather coldly for a woman with a proper dowry attached. Some tribal societies went to the opposite extreme with the institution of the bride price. A woman was so valuable economically that a man paid to marry her, trading such goods as cattle.

In our own society, the economic trade is almost entirely between the individual husband and wife, not between their families. But the trade remains impor-

> **A CLOSER LOOK 2.5**
> ## *Where Does the Word Family Come from?*
>
> Our modern term *family* comes from the Latin. But the ancient Romans did not use *familia* to mean blood filiation or kinship. It meant rather the household property—the fields, house, money, and slaves. The Latin word *famulus* means "servant." In Rome, the plebian form of marriage consisted of a man buying his wife, and she became recognized by the law as part of his property—his *familia*.
>
> The ancient Greeks used the word *oikos*, which is translated "family." But again, it meant "property" or "domicile." *Oikos*, in fact, is the Greek root for the words *economy* or *economic:* the first economy that the ancient civilizations knew about was the family economy, the property that was produced and consumed in the household. Aristotle said that the family *(oikos)* was composed of three elements: the male, the female, and the servant.
>
> The Latin word *pater,* which is related to our word *father,* also originally had an economic and political rather than a sexual connotation. The term *paterfamilias* could be applied to a man who had no children and was not even married—for instance, if a youth inherited the family property because of his father's early death. The *pater* was essentially the domestic authority. Slaves and clients applied the term to their masters. A man might address a god as *pater,* and poets used the word for anyone they wished to honor. It was synonymous with *rex* or *basileus,* which we translated as "king" or "chief."
>
> In Arabia at the time of Mohammed, the word for marriage was *Nikah,* which literally meant sexual intercourse. In the Koran it was also used to mean a contract. Marriage thus was conceived of as a contract for sexual intercourse (Fustel de Coulanges 1973, 89–90, 107; Bullough 1974, 67, 141; Snodgrass 1980, 79).

tant. Some theorists have argued that the basis of gender inequality in our society is economic inequality: men have higher incomes, and women have to compensate by providing the menial household labor. In all societies, the marital deal seems to be some version of material property traded in return for an arrangement of sexual possession.

3. Sexual possession and economic property thus always go together to make up the family system, but in varying combinations. As we've seen, there are different kinds and degrees of sexual possession, and the same is true of family economic property relationships (Figure 2.2). There are many different combinations possible, and hence we find a variety of family types around the world and throughout history. Our own typical family arrangement is only one among many. It is not always necessary to divide the bundle of property rights among family members in the same way. In matrilineal/matrilocal societies, for example, the husband has sexual rights but little or no economic property in the marital arrangement. His wife keeps her property in her own lineage, often under the control of her brother. Economic property and sexual possession are thus split in this kind of system. Intergenerational property rights are also split off from the husband's control, as the matrilineal principle dictates that children inherit from the female line, not from their father. But even in this case, to comprehend the family institution of the matrilineal society, we have to know how *both* sexual possession and economic property are organized.

Cohabitors, Single Parents, Gay, and Lesbian Couples

In our own society today, the various ingredients of the family seem to be coming apart again. It is as if we were experimenting with all sorts of ways

FIGURE 2.2: Some Patterns of Family Descent and Inheritance

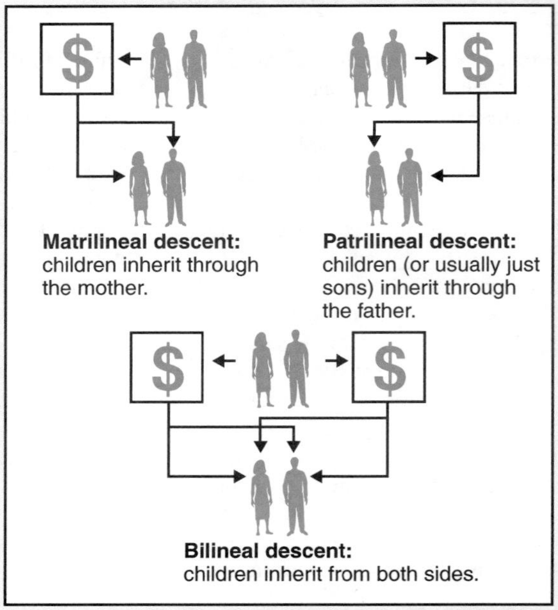

Matrilineal descent: children inherit through the mother.

Patrilineal descent: children (or usually just sons) inherit through the father.

Bilineal descent: children inherit from both sides.

A Note on Ideology:
Interestingly enough, there has been a tendency for beliefs about impregnation to reflect the lineage structure. Many matrilineal societies deny that sexual intercourse with the father has anything to do with pregnancy; the child is viewed as produced entirely by the mother (Malinowski 1964, 7). In patrilineal societies, there is a tendency for men to assert that the child is entirely formed by the father's sperm, whereas the mother's womb provides only the receptacle (Bullough 1974, 62). Sociologically, we would regard these beliefs as ideologies, attempts to justify the prevailing kinship system.

Table 2.1

Some Combinations of Family Ingredients

	Sexual possession	Economic property unit	Childbearing and rearing
Monogamous marriage	Yes	Yes	Usually yes
Cohabiting couple	Yes	Yes	Usually no
Single-parent family	No	Yes	Yes
Gay or lesbian couple	Yes	Partly	Usually no

of combining different pieces of family structures (Table 2.1). Many men and women live together as **cohabiting** couples without legally getting married. For all intents and purposes, **cohabitation** is like a marriage as far as the sexual aspect goes: the couple has sex regularly (chapter 9) and excludes other persons sexually, just like a marital rule of fidelity. They are also sharing a joint household and in that sense have set up something like the common eco-

A lesbian couple at a birth preparation class.

nomic property that constitutes a family. The legal standing of this joint property is still rather iffy. Some courts are treating cohabitation as if it implied the same kind of property contract as a regular marriage, so there are property rights to be divided if the couple breaks up (see *A Closer Look 14.1* in chapter 14). What cohabiting couples usually lack that a marriage has is not property relations but children.

On the other hand, there are many families now in which there is only one adult, but there are children. Here there is no sexual possession (sexual relations between parents and children are prohibited by the incest taboo, of course), and there continues to be economic property. The children have the right to inherit from their mother or father. If this is a mother living in poverty (rather typical of this type of family), with nothing to inherit, there is still an economic relationship implicit in the requirement that parents must support their children. This right is one of the most important "property rights" of children today.

For yet another arrangement, consider gay and lesbian couples. If they live together, there is usually some degree of sexual fidelity or possession (see *A Closer Look 9.2* in chapter 9.) Although they do not usually share custody of children in the same way that heterosexual couples do, gays and lesbians sometimes have children from previous marriages, adopt, or conceive through artificial insemination (chapter 10). Some estimate that as many as one in three lesbians are mothers and one in ten gays are fathers. Because our society has strong taboos against homosexuality, the presence of children in such households is controversial, and most adoption agencies are reluctant to place children with lesbian or gay couples. The public seems to be beginning to accept an ideal-type of gay or lesbian couple, which has some sexual possession and no childbearing or child rearing. Would they have joint economic property? In practice, such couples share the economy of a household, as much as cohabiting heterosexual couples do. So far, economic rights of inheritance among homosexual partners have not been legally sanctioned by the courts. This is an area in which gay rights activists have struggled to change the law, so that when one partner dies, for instance, the survivor would have a right to inherit the deceased one's pension fund. The Hawaii same-sex marriage court decisions and challenges in the 1990s prompted a flurry of legislative activity at the state and federal level to define marriage as only between one man and one woman. Fundamentalist religious movements, some with considerable political influence, have tried to block any attempts by gays and lesbians to realize the economic and practical advantages of marriage.

Looking at the various kinds of property, then, turns out to be a good way of understanding the ingredients that make up family relationships. If the borderlines of what constitutes a family are fuzzy, that is because there are several different kinds of property. Sexual, economic, and intergenerational property do not all neatly overlap; therefore it is necessary for us to look at all three.

HOW TO EXPLAIN FAMILY DIFFERENCES

A theory is useful to us when it explains why things sometimes take one form, sometimes another. In the following chapters, we are going to trace the implications of these theoretical principles. If the family involves three kinds of property relationships, we should ask: What determines the kind of sexual possession there is in a family? What determines how a family's economic property is structured? And what determines parents' relationships with their children? The rest of this book addresses these questions in detail, but let us briefly anticipate what kinds of answers we will encounter.

One type of theory argues that economic property is the key. First, we would want to know what kind of economy characterizes the society in which we find the family. Is it a hunting-and-gathering band, a society of peasant agriculture, an industrial capitalist society, or any one of several other types? Then, within this society, we would look at its economic stratification. Are all families economically equal, as is the case in some tribal societies? Or do some families own most of the property, and other families work for them, as in agrarian societies dominated by landholding aristocracies? Is there a range of social classes through upper, middle, and the lower levels of property?

We should expect that the shape of families will vary with their economic position. Moreover, because a crucial feature of families is the relationship between the men and women within them, we want to examine the ways in which the two genders are connected to property. A theory of this type attempts to show that the power of men and women within families, and hence the kind of family roles they play, is determined by their economic positions.

A second type of theory argues that variations in sexual property are the key. In this line of theory, we would examine: How exclusively is sexual intercourse confined to marriage? Who is sexually eligible for a marriage, and who is tabooed? How permanent are the rights of sexual possession? Does one sex dominate in the degree of sexual possessiveness, or is the relationship egalitarian? Does sexual possession involve emotional possessiveness—feelings of love, including rights over possessing someone's love—or is the emotional side regarded

The Medicis, a powerful Italian family, dominated Florence during the Renaissance and, through kinship politics, extended their influence throughout Europe. Catherine de Medici, pictured here with her son Charles IX by her marriage to Henry II, became virtual ruler of France in the latter half of the sixteenth century.

as unimportant? We especially want to know what causes these different kinds of sexual property arrangements.

This kind of theory proposes that sexual possessiveness is really a question of power. Hence, the variations in sexual property that make families so different at different periods of history are determined by the political structure of the society. As we will see, a key feature of politics is the extent to which the state is organized through the family or independently of it. Before the rise of modern bureaucracy, politics consisted of alliances among families. Sometimes particular families of the aristocracy dominated the rest of the society, and their households—their palaces, fortresses, or country estates—were the centers of government. There were a variety of arrangements of this type. In all of them, politics was essentially "kinship politics." Marriages were controlled by political motives, and sexual possession was a matter of social honor and represented the standing of the whole family in its political networks. Individuals did not control their own sexual lives as private possessions until the point in history when a political shift severed the connection between the family and the state.

Kinship systems are ways of organizing sexual possession so that their members are bound together into alliances. These variations mesh with the historical varieties of political organization. Even in our own society, in which the state is no longer connected to the family, there remain some important indirect connections. For the state always upholds the existing marital property system. It is the state that issues marriage licenses and also authorizes the breaking up of families through divorce. Although family politics no longer determine who is supposed to marry whom, there are still a great many legal rules controlling marriage. These include restrictions on who is too young to marry or to have intercourse and penalties for violations of sexual property, such as adultery. There is also the willingness of the courts to allow spouses to enforce their own sexual property rights through personal violence. State controls also affect childbearing, ranging from prohibition of contraceptives or abortion to their promotion. Under some governments there are subsidies for having children, such as welfare payments or tax deductions; elsewhere we find the opposite, such as tax penalties for children in countries trying to limit the birthrate. We need to be constantly aware of the power of the state, then, in analyzing why one or another of these policies is in effect and how it affects the shape of families. In our own times, political movements for and against the legal rights of women are among the most important influences on the way families have been changing.

These two lines of theory are not mutually exclusive. Both of them focus on some aspect of the property system. Focusing on economic property leads us to consider the predominant technology and its connections to such things as the kind of work that men and women do—their activities in the household and in the labor force. It also implies that the third kind of property—intergenerational property, involving relationships between parents and children—is mainly determined by economic conditions. How many children there are depends largely on economic conditions affecting the birth and death rates. Key factors are whether children can do economically useful work, and on the other hand, how expensive it is to raise children.

The theory focusing on sexual property is a little more exotic. It leads us into the area of highly charged emotions, including pride, jealousy, and violence. But these, too, are socially structured. These emotions are connected with the status dimension of stratification, as family members show off their position in the networks of society, and they are connected to the power struggles of politics. To modern ears, "sex and politics" sounds like the all-too-familiar scandalous headline in a newspaper. But the connection is deeper than that, for both sex and politics are fundamentally concerned with making alliances; also, both sex and politics are overlaid with attitudes about morality, claims about rights that, if necessary, people will act quite violently to defend. Relations of parents and children are political and emotional as well; children were part of the political maneuvers in societies based on family alliances. As the family-state connection has become more remote, children, like their parents' sex lives, have become a more private issue, yet they are still a force in political efforts at control.

In chapters, 3, 4, and 5, we will trace these processes as they have shaped families across history, from the most primitive societies through the family changes of the twentieth century.

SUMMARY

1. Families reflect a set of social rather than merely biological relationships.

2. The property theory sees families as depending on three kinds of property: rights of sexual possession, economic property, and intergenerational property. The amounts of each of these kinds of property held by men and women vary among societies; these different combinations make up the various types of families.

3. Kingsley Davis's theory of sexual property points out that property is not a thing but a social relationship among an owner, a possession, nonowners, and the rest of society, which the owner can call on to back up his or her rights over possession and to keep nonowners away. Marriage is sexual possession involving the exclusive right to have sexual intercourse with a person.

4. The incest taboo is found universally in some form in all societies. The incest taboo is regarded as cultural rather than biological because it establishes a system of exchange between families. Each family must give up its isolation by sending its offspring to find sexual partners outside. In tribal societies, various forms of marriage exchanges made up economic and political networks and thus established the larger society.

5. In some societies, sexual possession is unilateral: the husband has exclusive sexual rights over his wife but not vice versa. In our society, the ideal of sexual possession is bilateral: both husband and wife are supposed to be sexually faithful to each other. The degree to which sexual possession is unilateral or bilateral depends on the relative power and economic resources of men and women.

6. The modern ideal is that marriages take place not merely for sex but because of a bond of mutual love. However, the prevalence of jealousy indicates that love is also a form of socially sanctioned possession. It can be analyzed as an emotional possession that goes along with sexual possession.

7. Families also consist of economic property. This includes both the family's income and the household labor involved in running an unpaid domestic business of cooking, cleaning, lodging, and child care. Our word *family* itself comes from the ancient Roman word denoting the physical household.

8. The family economy is always connected to sexual possession. Marriage systems in different societies have varied depending on how men and women have traded both economic and sexual rights. In our own society, critical theorists have argued that the weaker economic position of women outside families has brought about both sexual subordination and the custom that the wife provides most of the unpaid household labor.

9. The third aspect of families, intergenerational property, is the result of the first two kinds of property. It includes both economic and sexual rights between parents and children. Economic rights include inheritance and the parents' control over children's labor. Parents traditionally have had the rights to control their children sexually, especially in deciding who they marry. This control has declined, but a negative sexual rule continues to apply: the incest taboo.

10.   In the late twentieth century, the different forms of family property are being combined in various ways. Cohabiting couples are similar to marriage partners in sexual possession and in maintaining a joint household, but they usually lack intergenerational property. Single-parent families lack sexual possession; gay and lesbian couples may have sexual possession and joint household property.

11.   To explain variations in families across historical periods, as well as differences within our own society, two main background factors are important: the economic structure of the society, especially the economic property of men and women, and the system of political alliances, which determines the patterns of sexual possession.

Key Terms

adultery

annulment

bilateral sexual possession

cohabitation

economic property rights

emotional possession

exogamy

family

incest taboo

intergenerational property
 rights

matrilocal system

patrilocal system

polyandry

polygamy

polygyny

rights of sexual possession

unilateral sexual possession

Sociology Web Site

See the Wadsworth Sociology Resource Center, "Virtual Society," for additional links, quizzes, and learning tools:

http://www.sociology.wadsworth.com

Also on this web site you'll find InfoTrac College Edition, an online library of journals. Here you can search for electronic articles about central topics in sociology.

3

History of the Family I: From Kinship Politics to Patriarchal Households

INTRODUCTION

In this chapter we look at families through the long sweep of history. Families vary a great deal across these societies. There are tribal systems with patrilineal, matrilineal, and many other forms and complicated rules for marriage exchanges. Some of these forms are among the most egalitarian between men and women that have ever existed, whereas others are characterized by high levels of gender antagonism. In technologically more advanced societies, the so-called ancient and medieval civilizations, the family form changes again, typically into the patriarchal household. Here society is rigidly stratified, and male domination over women is pushed to an extreme. Yet even here women have some resources for gaining power and status.

SOCIETIES: AN OVERVIEW

Let us look briefly at all the major types of societies existing in different parts of the world and historical periods. Figure 3.1 categorizes the main types of societies according to their principal economic base. Sociologists and anthropologists have shown that these economic structures, and the resulting systems of power, affect the overall degree of stratification in the society (Lenski 1991). Because economic and political structures vary widely among these societies, we should not be surprised to find that their family structures are also quite different.

Hunting and Gathering

Hunting-and-gathering societies are small wandering groups of a few hundred people or less. This type of society prevailed during the hundreds of thousands of years of the Stone Age. It has survived almost until today in remote areas like the deserts of Australia, in the Kalahari in Africa, in the Arctic among the Eskimos, in the jungles of the South American Amazon, and on some South Pacific Islands.

In hunting-and-gathering societies, the family tends to be a simple unit, in many cases consisting of no more than the monogamous nuclear family made up of a single couple and their ungrown children. These families are often grouped together into larger bands of people who intermarry, but connections are loose. People wander in and out of the various campsites, so different people may be sleeping there at different nights. There is little property to tie anyone down. Divorce is common, even more so than in our own society.

Primitive Horticulture

Primitive horticultural societies rely on the earliest form of agriculture. Usually the women cultivate the plants by hand and make baskets and pottery, and the men engage in fighting or clearing the land. The society may be as large as a few thousand people, linked together by exchanges, periodic councils, and religious ceremonials. Here we are especially likely to find complex marriage rules of the kind described in Lévi-Strauss's theory. The whole society is connected together by marriage exchanges. Smaller and larger kinship networks are nested inside each other like concentric rings. At the smallest level, homes may be matrilocal

FIGURE 3.1 Typical Family Structure Varies with the Economic and Political Organization of the Society

Inequalities of wealth and power are lowest in hunting-and-gathering societies, highest in state-organized agrarian societies, and intermediate in industrial societies.

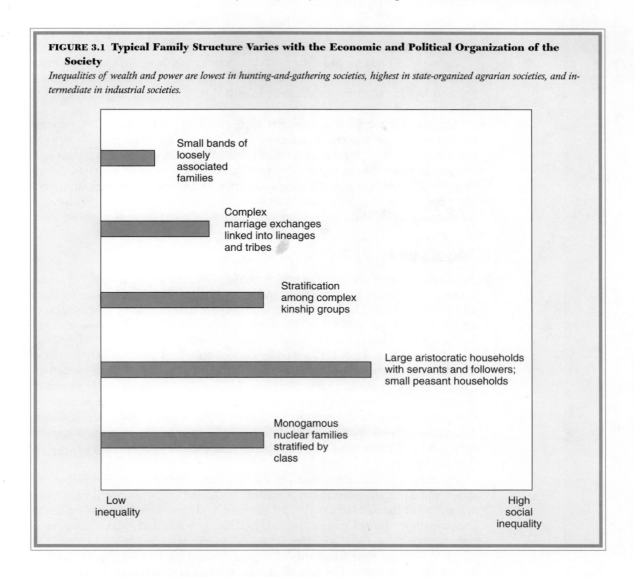

Small bands of loosely associated families

Complex marriage exchanges linked into lineages and tribes

Stratification among complex kinship groups

Large aristocratic households with servants and followers; small peasant households

Monogamous nuclear families stratified by class

Low inequality

High social inequality

or patrilocal. Beyond these are lineages that may descend on the mother's or father's side. Often there are larger clans or tribes, which in turn may be divided into various sections. Men and women may have their own ceremonial groups and centers, often taking the form of secret religious societies. With all these options to choose from, horticultural societies have tremendous variety in their forms of kinship.

These types of societies have been extensively studied by anthropologists in the Americas, Africa, Southeast Asia, and the islands of the Pacific. Somewhat similar forms of organization are found in **pastoral societies,** which live by herding cattle, sheep, goats, or other animals; and in **fishing societies,** which formed in abundant coastal areas like the Pacific Northwest.

Advanced Horticulture and Plow Culture

Through the use of metal tools, **advanced horticultural societies** were able to support larger and denser populations, ranging from 10,000 up to a few million.

Here stratification began to become pronounced, with a hierarchy of chiefs, priests, and military aristocrats exacting tribute from the lower classes.

Examples of advanced horticultural societies include the Mayans and other societies of early Mexico, the Incas in Peru, and some of the societies of sub-Saharan Africa such as the Azande in east Africa or other large societies in Dahomey and Nigeria in west Africa.

Agrarian plow culture is a more advanced form of technology. These societies tend to emphasize property in terms of land, plowing the same ground over and over again for many years, whereas **horticultural societies** usually shift to new land fairly often when the old soil is exhausted. In both these types of societies, the nuclear family form is rare, and a large proportion of them have complex family systems. Because there is more stratification than in primitive horticulture, social class is beginning to become a rival to family structure as the key basis of social position.

Agrarian States

Agrarian societies are those in which the large-scale state emerged. This is because plow culture used animal power rather than exclusively human labor and was able to produce a considerable surplus that enabled other aspects of civilization to be built. The great historic civilizations of Egypt and the Middle East, China, Japan, India, ancient Rome, and medieval Europe fall into this category. They had a variety of technological developments, including the use of irrigation for extensive agriculture, wheeled carts pulled by oxen or other animals, metal weapons, large-scale buildings, and the use of writing. These states might include millions of people over huge territories, organized into a feudal system or a centralized empire. Here stratification was extreme, with more wealth concentrated in the hands of a small aristocracy (about 2 percent of the population) than in any other type of society (Lenski 1991).

We should bear in mind that not all agrarian societies developed large-scale states. Plow farmers lived in the so-called "barbarian" tribes in northern Europe at the time of the Roman Empire, for instance, and there have been plow cultures and irrigation cultures in Asia and in sub-Saharan Africa that did not have strong state structures. In these societies, complex kinship systems tended to remain in effect. Where the state structure did appear, however, it had an important effect on the family system. The households of the military aristocracy became the most important family units. In these societies, the complex marriage rules that tied together horticultural kinship systems tended to fade away. Instead, we find the aristocrats bringing large numbers of nonkin into their households. This was the era of palaces, castles, and fortified buildings, where powerful families employed large numbers of servants or sometimes slaves and surrounded themselves with military followers and political dependents. In these societies, the most important influence on the family was its class position; large aristocratic families were at the center and dictated most of the conditions under which peasants and servants could have families of their own.

Industrial Societies

The use of machinery run on inanimate energy sources such as coal, electricity, and oil characterizes **industrial societies.** Although cities are found in agrarian societies, urban population outnumbers rural residents for the first time in industrial societies. The family system simplifies dramatically as we move toward the wealthier industrial economies. Today, less-developed societies that have taken steps toward industrialization show a drop in the prevalence of complex family systems, whereas among heavily industrialized societies, monogamy and the nuclear family have become almost universal.

This comparison has its limitations, and not all societies fit clearly on the chart. Some have been mixtures of one type with another, such as pastoral societies that borrowed some of the techniques of agrarian civilizations (e.g., Genghis Khan's Mongol empire). Nor do societies necessarily go through all these "stages"; they may skip one or more. We have concentrated only on a crude measure of family complexity that leaves out the variations in details of kinship systems found among societies of the same general type. However, the chart does give us some idea of the typical forms of society that have existed, and we can now look more closely at some of their family structures and at the conditions that caused them.

SOME THEORIES OF GENDER INEQUALITY

How can we explain why family systems have taken different forms? In recent times, a number of sociological theories have addressed this question by analyzing the family as a system of stratification. The most striking differences among types of families lie in the degree of equality or inequality found among their members. Some of this inequality may be between members of the same gender: the rights of the oldest brother against the younger brother; of father versus son; of the mother over her son's new wife who moves into their household; or of the first wife in a polygamous system versus lower-ranking wives or concubines. It is most convenient to begin by analyzing the sources of stratification between men and women. Theorists look for the sources of power that make a society male dominated in various respects, female dominated in certain spheres, or egalitarian. A successful theory of gender stratification should not only be able to explain the varying degrees of inequality between the genders, but go on from there to show how these same resources lead to other features of the family.

Notice that theories must always be comparative. To highlight the conditions for inequality, we must compare them with the conditions in which relative gender equality occurs. More precisely, we look at the variations across a continuum, comparing situations in which gender inequality is high with those in which it is moderate and those in which it is low.

There are two main approaches to the theory of gender inequality. One version, **economic theory,** emphasizes the forms of work that men and women do and the environments in which their societies attempt to make a living. A second version, **political theory,** emphasizes the ways the kinship system produces alliances by marriage exchanges and the various kinds of military situations and state organization that modify the power of men and women over one another. The two theories are not mutually exclusive. In fact, each fills gaps left by the other in explaining the family's historical changes.

Economic Theories

Some version of **economic determinism** has been implied in many of the oldest theories of the family. It was once held (and still is by some sociologists, as seen in the previous chapter) that there is a natural division of labor between the sexes: males evolved as the hunter and later the breadwinner, whereas females specialized in having babies, child rearing, and caring for the home. We know now that this is not necessarily true. In hunting-and-gathering societies, women typically produce more of the food by gathering than men do by hunting. In fact, women may produce as much as 60 to 80 percent of the family's diet. Moreover, in primitive horticultural societies, women not only grow almost all the crops but also do most of the manufacturing of their ceramics and basketry. The idea that men alone provide the income and women merely care for the home is fallacious because it is based on the experience of the middle and upper classes in comparatively modern times. But even in Victorian England, the anthropologists who theorized about "man the hunter" (or breadwinner) ignored the millions of working-class women who had always worked outside their homes.

Economic theories have become more sophisticated. They recognize that the kind of work that women do has varied a great deal from one type of society to another and that the form of *ownership* in that society is crucial for how much benefit women reap from their work. For working does not automatically pay off in control, as we can easily see by comparing the incomes of factory workers today with those of persons who do nothing but invest in the stock market. The hypothesis is that the status of men and women depends on how they fit into the larger economic system rather than simply on how much work they do. This form of analysis goes back to Marx and Engels. Friedrich Engels deserves to be called the originator of the theory of gender stratification, although his precise hypothesis is not much followed today. Engel's book *The Origins of the Family, Private Property, and the State* (1884/1972) proposed that the family originally was a form of "primitive communism," in which equality prevailed and sexuality was promiscuous. After this came a stage of matriarchy and finally monogamy and patriarchy. The last was truly the fall out of the Garden of Eden, because it brought about the subjugation of women and was caused by an economic change, the institution of private property. Along with this, Engels argued, women were guarded as exclusive sexual property, whereas men reserved sexual freedom for themselves in the form of sexual access to slaves, concubines, or prostitutes.

Engel's theory is mistaken in various respects. There was no stage of "primitive communism" and sexual promiscuity, but he was right in the general sense that hunting-and-gathering societies are much more egalitarian than most subsequent forms of economy. There was also no stage of matriarchy; like other early anthropologists, Engels mistook evidence of **matrilineal descent** for a system of female property and power, which did not necessarily follow. Also, the origins of private property cannot be neatly correlated with the rise of patrilineal and male-dominated families. Engels inspired later theories by his emphasis on how an economic system can determine the power of men and women.

There are a number of modern versions of economic theory of gender stratification. Some early models, such as Boserup (1970), Friedl (1975), and Sacks (1979) concentrated on the relationships between women's work, kinship structure, and gender stratification in various kinds of preindustrial societies. More comprehensive models covering the whole range of societies were formulated by Rae Lesser Blumberg (1978, 1984), Joan Huber and Glenna Spitze (1983), and Janet Chafetz (1984, 1990). Figure 3.2 depicts some of the main variables in these models.

The theory centers on the fact that a woman's power depends on how much control she has economically. Working is not enough; it must be supplemented by control over the means of production. Blumberg elaborates this across the en-

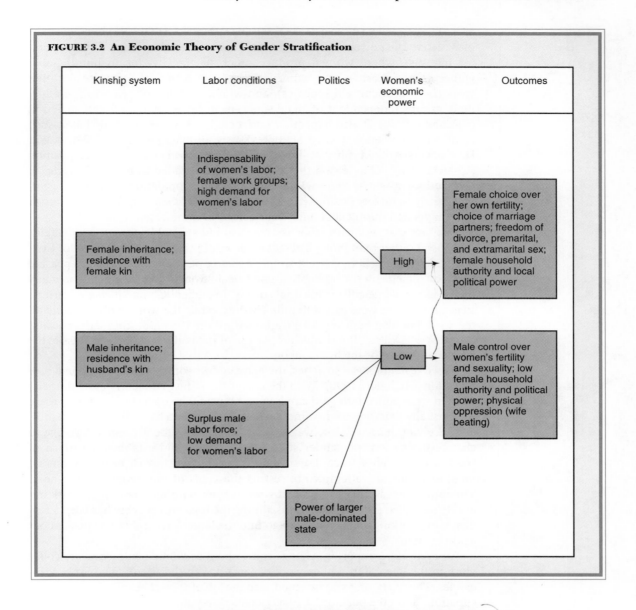

FIGURE 3.2 An Economic Theory of Gender Stratification

tire range of human societies. In many hunting-and-gathering and horticultural societies, women are not only important economic producers but often are able to translate production into economic control and hence into power and status. In agrarian societies with a powerful state, women tend to be excluded from the major areas of work and from ownership as well (although there are exceptions in the upper classes), and hence their status generally becomes very low. In industrial societies, a woman's relative status depends on her income or other property. This is even true in the supposedly egalitarian Israeli kibbutz, where women's status fell when they were excluded from the productive heavy agriculture and manufacturing upon which the kibbutz economy rested (Agassi 1991).

The question then arises: What determines women's economic control? Blumberg (1984) points to three sets of factors: the strategic indispensability of women's labor, the effects of the kinship system on female property, and the stratification of larger societies.

STRATEGIC INDISPENSABILITY OF WOMEN'S LABOR Women have the most economic power when they represent a crucial labor force. This happens, for instance, where women produce much of the diet (as in hunting-and-gathering societies); where women control the technical expertise for their work (as in horticultural societies); and where women do their work in all-female groups independent of male supervision (as in many African societies) (Blumberg 1984). It also happens when there is a shortage of male labor, such as in wartime; hence women held previously all-male jobs during World War II. Conversely, a surplus of labor undermines women's economic position (Chafetz 1990). One reason that women's status tended to be low in agrarian civilizations was that there was always a surplus population of landless peasants, ready to fill any position, and hence forcing women out. Similar processes have happened during economic depressions in more recent times.

The fact that women's labor is *sometimes* indispensable shows, incidentally, that social rather than biological factors determine what work women do. Contrary to modern beliefs, it is not simply a matter of strength. Women in peasant or tribal societies were probably a good deal stronger than women—or many men—in our white-collar industrial society. In ancient China, during a wartime siege all the men were called to military duty, while the women not only did all the work but also built the fortifications (Griffith 1963, 38). Similarly, because World War II killed off a sizable proportion of the male labor force in Russia, the women did most of the heavy labor.

As we reach modern societies, the demand for women's labor takes on a new importance (Chafetz 1990). With the rise of the industrial factory system, many women were pulled into paid employment outside the household. At first, these were usually working-class women, laboring under harsh conditions in the mills and receiving relatively low wages. This gave them little economic leverage to change the system of gender stratification, but it did establish a pattern of women in the labor force. Labor unions became organized, primarily among men, and generally attempted to restrict the kinds of jobs women could hold. The rationale was partly to protect women from hard and dangerous work but also to save their jobs for men. Thus, during the last century there has been a tendency for working-class women to go into the labor force only to be pushed out again by men.

An even more important change came about as women entered white-collar work in the late 1800s. Office work expanded with the rise of modern bureaucracies. The typewriter was invented, and the job of typist became a woman's occupation. Soon the position of office secretary became mostly a woman's job as well. With the growth of commercial business, large numbers of women became clerks and cashiers. Again, these tended to be relatively low-paying jobs, but the pattern was now established in which middle-class women often worked for part of their lives—usually when they were young and unmarried but perhaps longer if their family needed to supplement its income. At first, the highest level job for women was schoolteacher, an occupation that expanded rapidly in the late 1800s and early 1900s with the spread of mass education. Thus, the expanding demand for female labor gave women at least a fingerhold on some of the economic resources that give women more freedom and power in their lives.

Throughout the last century, the movement of women into the paid labor force has been upward, but women's jobs have usually been badly paid compared with men's jobs. Men control the economic balance of power, which has reinforced a male-oriented family structure. In modern societies, women's paid labor has never been the crucial, indispensable economic factor. Nevertheless,

A CLOSER LOOK 3.1
Esan Women and Inheritance

Among the precolonial Esan of Nigeria, women played an important role in agriculture and in trade. Onaiwa Ogbomo describes how women's influence was constrained, but not eliminated, through a patrilineal inheritance system:

> The laws of Esan society excluded a woman from inheriting anything substantial from either her patrilineage or from that of her husband. This exclusion is well expressed in two proverbs: "Okhuo ila aghada bhu uku (A woman never inherits the sword!)" and "Ei bio omokhuo heole iriogbe" (literally, "You do not have a daughter and name her the family keeper!"). The patrilineal and patrilocal nature of Esan society made it impossible for wives and daughters to inherit either their husbands' or their fathers' property. The implication of the second proverb was that the Esan people felt any property handed over to a daughter would ultimately end up in her husband's patrilineage. As a result, families were not prepared to transfer to their daughters' inheritances that would enrich in-laws. On the other hand, wives were considered strangers among their husbands' kin, which again does not qualify them for any form of inheritance. Another curious notion is the belief that women were themselves inheritable. As Okojie noted, in Esan customary laws of inheritance "the woman had

no place, . . . she was one of the inheritable properties!" Thus, to ensure that women were prevented from inheriting any property, a number of cultural norms were established. As a case in point, a woman was not allowed to perform the burial ceremonies of her father. It was a custom that whoever performed the burial ceremonies inherited a man's property. Knowing full well that some women might have been wealthy enough to perform funeral rites, it was deliberately stated that women were prohibited from such rites. Tradition also claims that because women were not permitted to handle an ukhure (a family staff), it was inconceivable for women to want to perform burial ceremonies for their dead fathers. Hence they could not inherit the family shrine where the ukhure was placed. It was therefore logical for a woman not to dream of owning family property. These constraints no doubt prompted women to search for individual ways of amassing wealth that they could hand over to their daughters in the form of either bridewealth or inheritance.

Source: Onaiwa W. Ogbomo. 1995. "Esan Women Traders and Precolonial Economic Power." Pp 7–8 in Bessie House-Midamba and Felix K. Ekechi (eds.), *African Market Women and Economic Power,* Westport, Conn.: Greenwood Publishing Group. Reproduced with permission.

the gradual expansion of women's jobs is one of the conditions that has slowly exerted some pressure for family change.

EFFECTS OF THE KINSHIP SYSTEM ON FEMALE PROPERTY When economic goods are passed along by inheritance from women and to women, female economic power is maximized. Also contributing is the **residence pattern:** where a woman resides with her mother's relatives (a form of matrilocal residence), the female group is most likely to control its own property. This power is lessened when she resides with her brothers and is lowest of all when the woman leaves home to live with her husband's relatives (patrilocal residence). Compared with these, the official **descent rule** (matrilineal, patrilineal, or bilineal descent) is relatively unimportant. In other words, whether children officially belong to a lineage traced through their mother or their father (or both) may not have very much effect on actual economic power. As Alice Schlegel (1972) has shown, in many matrilineal systems, property is passed from mother's brother to sister's son, and husbands or brothers end up dominating their wives or sisters. In some patrilineal societies in West Africa, women still have de facto control of the land, which they work by horticulture (Amadiume 1987). Descent rules thus appear to be kind of ideology that can sometimes mask the actual economic situation between the genders (*A Closer Look* 3.1).

THE STRATIFICATION OF LARGER SOCIETIES In general, women do best in their own local communities and in societies that do not extend far beyond the community. Larger societies tend to have male-dominated networks of warfare and politics, and the state tends to intervene to limit or eliminate local female power. In the great agrarian civilizations, women in the cities close to the upper classes were veiled and secluded, and peasant women in the countryside had much more freedom. In modern times, as anthropologists have pointed out, the British colonial administration in Africa destroyed the local political institutions of native women and replaced them with an all-male native hierarchy. For this reason, women have had greater economic power in smaller and more isolated tribal societies.

The economic model can be visualized as a kind of chain. Labor indispensability comes first, modified by the kind of kinship system. These in turn can be overruled by an unfavorable *stratification system in the larger society.* All these affect economic power, which in turn determines women's status. In real-life terms, this includes the degree to which women make their own decisions about when to have children, contract their own marriages or divorce, have premarital or extramarital sex, have authority within the household, and wield political power in the local community. When female economic power is low, women have few of these rights; men control their fertility and their sexual lives; household authority and community participation are low. Under these conditions, women are most likely to be subject to male physical abuse (wife beating, rape), to be second-class subjects politically, and even to be viewed as unholy according to the prevailing religious beliefs (Coltrane 1996, Johnson 1988, Sanday 1981).

There is a self-reinforcing loop here. When women have low power, they can be forced to bear and raise many children, which in turn makes them less able to participate in productive labor and to acquire economic power. This tends to happen in agrarian societies with a strong state, where women's power is lowest. The men who head these families usually want large numbers of children because they are useful as farmworkers or sometimes as fighters. In addition, a high birthrate tends to produce a surplus population, which further undermines a woman's possibility of wielding economic power. The situation thus becomes a vicious cycle.

These economic theories capture an important part of the explanation for gender inequality, but they do not tell the whole story. They leave unanswered the question as to how men or women gain control of inheritance systems. Certain kinship systems seem to favor women's control over resources, but how those kinship systems arise is not explained by most economic theories. In addition, many economic theories suggest that a population surplus will push women out of desirable jobs, but they do not explain why men should be able to get those jobs. Why do men usually end up controlling the networks of long-distance politics and stratification? These problems are better answered with reference to political theories of gender inequality. Such theories look at kinship as a form of politics and see the larger political system as something that emerges out of shifting alliances of different family systems.

Political Theories of the Family

Political theories of the family go back to Max Weber (1922/1968, 356–84; 1923/1961, 38–53), Collins (1986a), Claude Lévi-Strauss (1949/1969), and subsequent **alliance theories** of kinship. Figure 3.3 illustrates another version of a political conflict theory (Collins 1971, 1975).

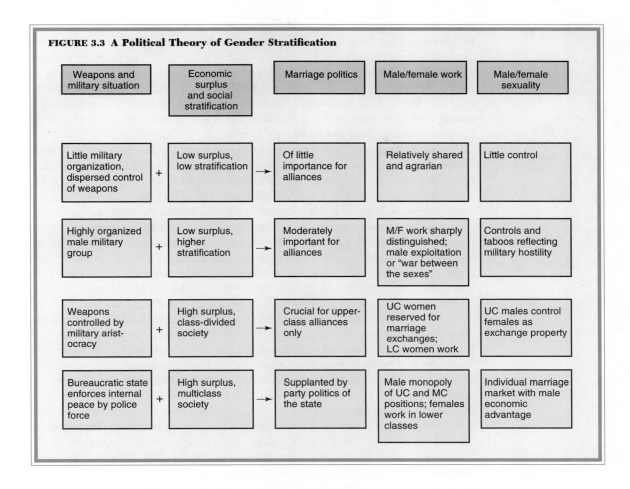

FIGURE 3.3 A Political Theory of Gender Stratification

Weapons and military situation	Economic surplus and social stratification	Marriage politics	Male/female work	Male/female sexuality
Little military organization, dispersed control of weapons	+ Low surplus, low stratification	→ Of little importance for alliances	Relatively shared and agrarian	Little control
Highly organized male military group	+ Low surplus, higher stratification	→ Moderately important for alliances	M/F work sharply distinguished; male exploitation or "war between the sexes"	Controls and taboos reflecting military hostility
Weapons controlled by military aristocracy	+ High surplus, class-divided society	→ Crucial for upper-class alliances only	UC women reserved for marriage exchanges; LC women work	UC males control females as exchange property
Bureaucratic state enforces internal peace by police force	+ High surplus, multiclass society	→ Supplanted by party politics of the state	Male monopoly of UC and MC positions; females work in lower classes	Individual marriage market with male economic advantage

It is convenient to begin with the question, What are distinctive traits of males and females? What is there about men and women that could explain differential advantages they have in family systems but at the same time could account for *variations* in gender stratification in different systems? We are looking for some factor that is not exactly gender neutral—as that would make it impossible to explain why men and women occupy any different positions at all—but that translates into different outcomes in different social situations.

There are three possible candidates: differences in sexual characteristics, childbearing, and differences in size and strength. The first of these, differences in genitals and secondary sexual characteristics, is obvious enough, but it does not seem to make much of a social difference. Male and female genitals are complementary to each other, so to speak, which is one main reason why the family exists in the first place; but in themselves they explain nothing about patterns of inequality or domination and why men should so often dominate the property or political system. Male genitals are not a qualification for work or political positions. As contemporary critics of gender-segregated labor markets have noted, the only job that requires a penis as a necessary qualification is that of male prostitute.

Another candidate is the uniquely female capacity for childbearing and breast-feeding. Early theories made this the crucial basis for gender stratification, but even some modern feminist theories rely on childbearing as the major constraining factor for women. For example, Shulamith Firestone (1970) argued that women have been subordinated since the beginning of the human race because pregnancy made them unable to fight and hunt with men, and childbirth and prolonged breast-feeding extended this vulnerability throughout most of their adult lives. For Firestone, the greatest shift in the status of women was the modern invention of birth-control devices, which finally freed women from the tyranny of birth and made it possible for them to seek careers outside the home—although in an uphill struggle because of the late start.

In the discussion of economic theories, we have already seen some reasons why childbearing need not be associated with disadvantage for women. Women have done quite a bit of the nonhousehold work in most societies throughout history. Moreover, when they have had enough social power, they have been able to control their own fertility; the image of women from less developed countries as continually pregnant or nursing is simply inaccurate. It is also an exaggeration to regard pregnancy as a physical disability; we know that women in most hunting-and-gathering societies continued their normal activities right up to the time of birth, and peasant women in more developed societies certainly did not have the luxury of taking time off from their work when they became pregnant.

After babies are born, moreover, it does not necessarily follow that mothers become specialized primarily in child care. Long before modern experiments with taking babies to work, tribal mothers were carrying infants with them in slings and baskets while they produced what was often the bulk of their family's food supply. Detailed anthropological studies have shown us that rather than limiting women, bearing children provided mothers with power and authority in most kin societies (Johnson 1988). In more stratified societies, children were often cared for collectively, in some fashion, by nurses (including wet nurses for infants), relatives, or others; and this was done not only by the upper classes but often out of necessity by peasant women who could not interrupt their work. There are instances of *some* women specializing in child care to free other women from it, but that it be done by women is not mandatory. The extent to which men have shared parenting tasks varies enormously from society to society. In some egalitarian tribal societies (Sanday 1981, Coltrane 1996), male parenting is considerable. Child care is also sometimes done by older siblings, including brothers. The point is that the female's biological specialization in childbirth is not necessarily a social institution but is made so only under particular circumstances.

This brings us down to the third difference: the fact that on the average, males are somewhat larger than females and have a heavier musculature. The difference is not great compared with some of the primates; it is only about 10 percent, and because these are averages, some proportion of women are bigger and stronger than some proportion of men. Why should this make a difference socially? Despite some common assumptions, it does not make much difference in the kinds of work men and women can do. As shown in the discussion regarding Blumberg's theory of the strategic indispensability of women's labor, women have been able to do all kinds of heavy work when necessary. Perhaps stronger men might be able to do it more easily, but that should not necessarily make a difference unless some kind of competition is going on. If it is just a question of what women can accomplish, there is no reason why women should drop out of heavy work (like plowing in agrarian societies) unless some other social pressure exists, which indeed is found in such societies. Moreover, we should not assume that male work automatically requires more strength. Metalworking and certain other high-prestige crafts have usually been monopolized by men, such as by the secret societies of iron forgers found in tribal societies; but this is a case of social control of a skill rather than of biological fitness. This is even more true of male success in monopolizing positions involving writing after it was invented in the ancient civilizations; this was certainly light labor in terms of muscular effort, but it was the pathway to power and privilege, so males defined it as their own province.

We seem to be running out of alternative explanations for male advantage, but one possibility is left. Size and muscles are useful for fighting, and this is a sphere that is inherently competitive. Here it is not just a matter of being able to do the job adequately but of being able to beat someone else. Fighting is in this sense unlike any productive human activity because there is no intrinsic standard about when the job is well done: it is always a matter of who is stronger relative to someone else. This may be the factor we are looking for; although differences in size between men and women are relatively constant, the importance of fighting is a *variable across societies.*

Fighting brings us into the realm of politics. Politics originally concerned little more than the organization of coalitions for warfare; later, after the rise of the bureaucratic state, violence was monopolized at the state level (this is in fact Weber's definition of the state), and hence peaceful politics emerged. Males gained power primarily by monopolizing the political realm, but again, this is a variable. In some societies (such as our own), women have some political power, and potentially they could have quite a lot; however, this depends on particular kinds of conditions. Partly, this is a matter of how closely politics is connected with warfare; secondarily, it is matter of how the military is actually organized. In general, the closer that fighting comes to the immediate community and to the family, the more male power is enhanced. Conversely, peaceful societies and those in which fighting is done by specialized troops and in distant places are usually favorable to female power.

A lot of variations are noted here, and the different shape of the family across human history is affected by the amount of military threat and by political organization. Let us look briefly at some of the ingredients of these situations.

MEN AND WEAPONS Although not all kinship societies engaged in warfare, among those that did, women often played an important part in its ritual aspects and sometimes had pivotal roles in determining the onset of hostilities and their duration (Brown 1975; Muller 1985). Men typically monopolized the weapons with which fighting was done, but there are some exceptions. Although male-biased reporting by missionaries and early anthropologists tends to deny women's participation in warfare (even while describing it), we do have records of women taking an active role in combat, as among the Kapauku (Gailey 1987; Poposil 1963). In the West African Kingdom of Dahomey, the king had a bodyguard of 5,000 armed women leading to later stories about "amazons" (Sanday 1981). This part of Africa in general has particularly strong political and economic participation by women (Amadiume 1987; Diamond 1996). There are also ancient stories (as in Homer) of societies consisting entirely of women warriors, which some suggest reflect specific historical situations in which the male population was decimated by war (Blumberg 1984). In more differentiated societies, it sometimes happened that all the able-bodied men went off on a foreign military expedition and were then wiped out; under these circumstances, women not only took over the economy but also may have armed themselves and taken control. Similar stories are known among the Vikings, especially in such frontier areas as Iceland. There the god of war, Odin, is depicted as surrounded by female warriors, the Valkyrie, who hover over the battlefield; and Icelandic history contains factual accounts of ferocious queens who ruled by wielding their battle axe. (The story of Brunhild, recounted in chapter 2, depicts some carryovers from these events.)

In more recent times, Israeli women fought in the 1948 war, when the crucial question of establishing the state of Israel hung in the balance. With modern machine guns and other technological weapons, the factor of strength becomes less significant in warfare. In many modern militaries, women now receive approximately the same training as the men, although their role in combat remains controversial. It is clear that physical differences in strength are no longer the crux of the matter. The social beliefs determine whether women should be engaged in warfare.

Sociologically, the significant idea is that men have usually gone to great lengths to monopolize weapons. Weapons offset advantages in sheer strength, especially as they become technologically more advanced. Whatever physical strength one starts out with, one's capability for threatening others goes up rapidly if one is better armed. Men have not only been somewhat bigger than women (on the average) but have made great efforts to accentuate that advantage by making sure that they have been armed and the women have not.

There is a very revealing myth among the Ona tribe, a group of Indians living on Tierra del Fuego at the tip of South America (Cooper 1946; Lothrop 1928). The Ona were hunters who warred among themselves. Bows and arrows, the principal weapons, were forbidden to women, and it was taboo for women to enter the house of men's secret society where the weapons were kept. Periodically the men would carry out a ceremony in which they emerged from the men's houses armed and wearing terrifying masks, representing male spirit powers, to bully the women and children. Adolescent boys, when they were ready to be admitted to the secret society, were first whipped and then finally shown the chamber where the masks were put on. They learned the secret that the terrifying spirits were really the men, the group they were now part of themselves. They were then told this myth: women were witches, responsible for all misfortune and death. At one time in the past, the witches had ruled the Ona, but the men banded together and killed all the women, except for some small girls, who were brought up henceforth to respect men forever. That is why only the men could enter the men's house, don the masks, or handle the weapons. Myths and rituals do not necessarily reflect real events, but the story suggests that the tribe changed from a form in which women had more power. Whatever the historical truth behind the myth, it shows that these men realized that their domination over women depended on their organization and monopoly of the means of intimidation.

It is not only weapons that are power multipliers, but organization as well. Weapons used in a group are much more effective than those used alone. Hence, when men have been organized as a group, their domination over women has been most extreme. The same principle works the other way, of course. The general hypothesis, then, is that military factors (the kinds of threats from outside, the kinds of weapons) determine how males and females are organized and hence help shape the degree of domination in the family system. There are many possible outcomes.

Where there are few weapons and not much military threat, men have little incentive to organize in military groups, and relationships between men and women are more **egalitarian.** This occurs among some gathering societies in environments where there is an abundance of food. For example, the most recently

The legends and myths of many cultures include references to female warriors. The Valkyrie, female warriors in Icelandic mythology, are depicted in a production of Richard Wagner's opera Die Walküre.

discovered tribe in the world is the Tasaday, first noticed in the 1970s in a remote part of the Philippines (Nance 1975). The men did not hunt, and both sexes took part in gathering food. The group was unstratified, but the person with the greatest influence was an older woman. Other very gentle and egalitarian societies existed earlier in the tropical forests of the Malay peninsula, among the pygmies in the forests of central Africa, and in the tribes of the Kalahari desert (Coltrane 1996; Draper 1975; Hewlett 1991; Sanday 1981). Similar gender equality is found at a more advanced level of technology, for example, in some peaceful rice-growing societies in Southeast Asia (Bacdayan 1977; Boserup 1970). Under these circumstances, male or female lineages tend to be relatively unimportant.

Where there is constant fighting and all the males are armed, men tend to be organized very strongly and females are subordinated. This particularly takes the form of **patrilineal** family (inheritance is in the male line) and the patrilocal household or community (the woman leaves home to live with her husband's kin). This family form is predominant in agrarian state societies, especially in the military classes, but it also occurs in many horticultural and pastoral societies.

On the other hand, even in relatively militarized societies the women may be in a stronger position. This occurs whenever the kinship system is **matrilineal** (inheritance passed through the mother's rather than the father's line) and especially matrilocal (wives stay at home with their kin and are joined by their husbands). Such matrilineal and matrilocal systems are not very common (Table 3.1), but where they exist, they split up men's resources: their kin and property are in one place and their home is in another (Murphy 1957). (In a matrilineal/patrilocal system, women go to live with their husbands, but men inherit through

Table 3.1 Matrilineal and Matrilocal Societies at Four Levels of Economic Development

Percent of societies with matrilineal kin groups

Hunting-and-gathering societies	10%
Primitive horticultural societies	26
Advanced horticultural societies	27
Agrarian societies	4

Percent of societies with matrilocal residence

Hunting-and-gathering societies	3
Primitive horticultural societies	15
Advanced horticultural societies	5
Agrarian societies	1

Source: Lenski and Lenski, 1974, 190–91.

their mother's house, which also splits up men's resources.) These forms seem to be good at knitting local groups together, so not too much domestic fighting takes place. The result is that women often hold the balance of power. Matrilineal and matrilocal forms also seem to be correlated with prevalence of foreign wars, in which the men of fighting age are away from home a good deal (Ember and Ember 1971; Divale 1984).

Neither matrilineality nor matrilocality is very common, even at the primitive horticultural level (where we find 26 percent of societies to be matrilineal and 15 percent to be matrilocal; see Table 3.1). However, these dimensions still capture only part of the complexity of family organization. Tribal societies commonly have a larger kin group beyond the individual family; there can be secret societies, clans, totemic groups, and other forms of organization in which both men and women are variously enmeshed. Even patrilineal/patrilocal societies may be offset by other kin-group organizations outside the household, so in some of these societies women may end up being fairly well organized compared with men (House-Midamba and Ekechi 1995). For instance, in the Igbo society of Nigeria, the women had their own political and religious hierarchy separate from the men, culminating in a female monarch who lived in her own palace and who adjudicated all affairs relating to women. In parts of the Igbo realm, each village had a woman's council attended by all married women. Persons, including men, who offended village rules were punished by a mob of women who "trashed" the offender's property and sometimes administered corporal punishment (Sanday 1981, 88–89, 136–40). In general, the better women are organized and the more men are split up, the more favorable the position of women in that family system (see *A Closer Look* 3.2).

Finally, there are situations in which both males and females are relatively well organized in separate groups, but in which men monopolize the use of force while women control the economy. For instance, in some parts of New Guinea or the South American jungle, most of the food is produced by women's horticulture, whereas the men engage in head-hunting or cannibalistic expeditions against rival tribes. Both sexes are well organized: the men are in secret societies or other groups and the women are left alone in their work by the men's incessant preoccupation with warfare. The result is that the men actually depend on the women for livelihood but give nothing substantial (in the form of livelihood) in return. Hence, the men attempt to terrorize the women, through violence as well

A CLOSER LOOK 3.2
Polygyny in West Africa

Many West African societies have a system of polygyny. Most men, if they have any wealth at all, will have several wives. Usually each wife lives in her own hut, inside a walled compound, taking turns feeding the husband and having sex with him. But this is not exactly a masculine paradise. The women have a good deal of economic power, because they produce and market the crops. Women are often richer than their husbands, and some women today own whole fleets of pickup trucks, known in West Africa as "mammy wagons." The women are in a sense better organized than the men, because the men are split up. Even sexual rights are carefully regulated, so the situation is not so much a despotic harem as a group of women sharing a man among themselves. And eventually, because the men tend to be much older than their wives, the women outlive their husbands and remarry; so there is a kind of polyandry that occurs as well, but spread out in time. Divorce is easy, and there is no dual sexual standard, with women having a great deal of leeway in the area of premarital and extramarital sex.

The senior wife has a great deal of power in this situation. She dominates the junior wives, as well as carrying much weight in the community, and she can get considerable deference from the children of the polygynous compound, including grown sons. She takes an active part in the politics of marriage bargaining that goes on in acquiring junior wives. For one thing, any additional wives will have to live with her, in a sense even more closely than with her husband, and their compatibility or suitability as a group of women is to be considered. Also, contrary to our usual notions of jealousy, the senior wife may actually push quite strongly to acquire some junior wives. Her prestige, as well as her domestic authority, depends on how many wives there are ranking below her. And the prestige of her husband in the community, which she shares, depends on how many wives he has. To be married to a man with only one wife is somewhat demeaning, a sign of social failure. It is something that one woman might taunt another with (Lloyd 1965; Clignet 1970).

Note: *polygyny* comes from *poly-gyny,* meaning "many women" (or wives). It should not be confused with *polygamy—poly-gamos,* meaning "many marriages." *Polygamy* is a broader term, including both *polygyny* and *polyandry* ("many men" or "many husbands") (see *A Closer Look* 2.1).

---

as ceremony (Coltrane 1992). At the same time, they are afraid of the women; these societies hold beliefs that women are polluting, place taboos on contact with them, and fear that they are dangerous witches. Divale and Harris (1976) and Murphy (1957, 1959) refer to such societies as exhibiting the "male supremacist complex," but this does not imply that the men felt secure in their manhood nor that the women did not fight back or organize their control over resources. In societies in which men have consolidated their power over women, such as in most agrarian state societies, there is relatively little use of violence. Because power and property are so overwhelmingly in male hands, there is little reason to resort to force. Because women in many hunting-and-gathering and horticultural societies have relatively greater leverage, the men are afraid of them and resort so often to force and vituperation to gain control.

It is also important to note that there are cultural forces shaping gender cooperation or antagonism that operate semiautonomously once they are set in motion in a particular society during a specific historical period. For example, symbolic anthropologists suggest that we can only understand gender inequality with reference to the ways that a particular cultural group constructs meanings about manhood and womanhood. Although cultural practices tend to be initiated in response to specific economic and political situations, once a culture generates a unique system of gender symbols and rituals, the meanings associated with them tend to take on a life of their own (e.g., Ortner and Whitehead 1981; Strathern 1987).

WORK AND CLASS STRATIFICATION   The kind of work that men or women do tends to be an offshoot of the military situation. Where the men are primarily fighting, they will leave the work of producing food to women, as is the case in many horticultural societies. (In some horticultural societies, though, both men and women grow and harvest plants; just as in some hunting-and-gathering societies, both men and women gather.) Early research suggested that men tend to monopolize work that is closely connected to their fighting groups, for example, hunting or herding horses or other large animals. Nevertheless, because hunting in most of human history consisted of "drive hunts" (in which every available man, woman, and child participated), we should not assume that hunting and male control of weaponry are necessarily linked (Gailey 1987).

When plow agriculture and irrigation were invented, the men usually ended up taking over farming. This was not, as is often claimed, because male strength is needed for these tasks. Rather, it took place because such societies produced a considerable surplus and became very socially stratified. This did not happen overnight, and some agrarian societies—people living off plow culture—remained tribal and relatively unstratified. Even these societies usually developed military groups, out of which came a new form of stratification. The stratification involved a very important shift in the military structure; the lower class of men was disarmed, and only a small aristocracy of warriors (knights, samurai, or men with some other honorific title) carried weapons. This aristocracy forced everyone else, female *and male,* to work for them. Nonaristocratic males were transformed into peasants and set to work in the fields, plowing or digging irrigation ditches. Women, too, continued to do a great deal of work, especially women in the peasant classes, who often worked in the fields alongside the men. Women in the lower classes were also called on to perform domestic labor for the upper classes, as well as being encouraged to have many babies who would provide future labor. In contrast, in the higher social classes, women became erotic objects and pawns in the game of marriage politics.

KINSHIP POLITICS

We have already been introduced to Lévi-Strauss's theory that marriage is a form of exchange among families that creates alliances. In chapter 2, we saw how the incest taboo established a link between each isolated family and the outside world and thus made human society possible. There are many different forms of exchange; some are local, whereas others are far-reaching. There may also be formal rules of exchange, such as the rule that a man should always marry his mother's brother's daughter if possible (what is called matrilateral cross-cousin marriage), or exchanges may be negotiated informally. The formal rules tend to be most elaborate in stateless horticultural societies, and informal exchanges are more likely in societies based on both less advanced and more advanced technologies.

Lévi-Strauss's theory has been criticized for depicting the world of kinship in an overly male-oriented way. He describes how men exchange women (their daughters or sisters) to form advantageous political alliances. However, it is also possible to turn the model around; instead of putting the male in the center, we could start from the woman and ask whom she should seek to marry. The rule quoted previously then would state that a woman should always marry her father's sister's son (providing her father has a sister and she has a son), and we could call this patrilateral cross-cousin marriage (which it is from the female viewpoint). This is just a matter of anthropological terminology, but there is a real difference that is often overlooked. We should not assume that women have no initiative or powers in these marriage exchanges. Karen Sacks (1979) points out that women in matrilineal societies often have power as *sisters,* just as even in ostensibly male-oriented patrilineal/patrilocal societies, women may exert considerable influence over marriage negotiations and other family matters. In the latter case, both men and women are moved around as pawns, but primarily because they are members of the younger generation while the older women (as well as men) may have considerable influence in this political game (see *A Closer Look* 3.2). In the Viking period, which included the invasion of Normandy in northern France around A.D. 850–1050, women were very active in pursuing marriage politics to advance their families politically.

Just how much initiative and power men *or* women have in the way this game is played depends on the power situation sketched out earlier. In general, we might say that the more that military factors favor male power, the more that men will control the marriage exchange system. Conversely, where those factors favor female power, women will have a corresponding influence in exchanges. Lévi-Strauss's male-centered view may be more appropriate for certain situations, whereas a female-centered view is more accurate for others. There are also mixed cases, in which both genders are taking part in the maneuvering, and yet another type, in which neither men nor women as a group are particularly influential, but individuals choose their own partners by personal preference.

In the more egalitarian and peaceful gathering and primitive horticultural societies, for example, there is less need for political alliances, and individuals find their own mates. Sexual and even romantic love is not found only in technologically modern societies but is common, for example, in some unmilitarized societies in the Polynesian islands. Again, in our own society, families are no longer politically very important, so they have little choice in the matter of whom their members marry (*A Closer Look* 3.3 presents an exception). It is still true that we fall in love within the constraints of a kind of invisible exchange system, but it is based more on social class than on explicit political negotiations.

The more advanced horticultural societies are where formal rules of the "marry-your-father's-brother's-son" type are most likely to be found. Rules like this, according to Lévi Strauss, are strategies for linking different families together. Some strategies produce what Lévi Strauss (1949/1969) calls a "short cycle," in which the same two families constantly intermarry, like a braid of hair across the generations. Other strategies produce what he calls a "long cycle," in which a long set of families is linked together, like a chain. Lévi-Strauss (1949/1969) theorized that it is riskier for families to follow the strategy of the "long cycle," because one family might always pull out and break the chain. But if the chain is successful, it is like a chain letter in which every person sends a dollar to the ten persons on a list of names, crossing out one name at the top and putting his or her own at the bottom; the end result is that whoever started the chain becomes rich. Some families thus end up with lots of alliances; they become "marriage rich," whereas other families who do not succeed at this strategy end up "marriage poor." Lévi-Strauss argues that at one time in history this could have produced a "family revolution," creating unequal societies with upper and lower classes and thereby bringing to an end the form of marriage politics played by these formal rules.

Whether or not Lévi-Strauss's theory of the "family revolution" is correct, it is clear that more stratified societies emerged at some point. In these, the game of marriage politics was still played, but no longer in the same way. The rules no longer specified which cousins, or kinship groups, were preferred as

A CLOSER LOOK 3.3
Polygyny in an Arizona Town

Many of the marriage and family practices described in this chapter are assumed to happen only in ancient times or in faraway places. On the contrary, because many of the sociological forces leading to such practices still exist, we can find modern examples here in the United States today.

In the town of Colorado City, Arizona, polygyny is the dominant form of marriage. In virtually all families, one man is married to three or more women. Some men have as many as sixty children, and households of thirty or more are not uncommon. Consequently, the houses in this town tend to be huge, rambling structures; they are continually growing as the occupants add on new rooms to accommodate their growing families.

The town is just across the border from Utah and is inhabited by members of a dissident Mormon sect. When the Mormon church (the Church of Jesus Christ of the Latter-Day Saints) was founded in the mid-1800s, some members practiced polygyny, although there was controversy over its religious basis. According to the nineteenth-century teachings of the church, plural or "celestial" marriage was revealed by God to Mormon founder Joseph Smith. Plural marriage was supposed to lead its adherents to a higher level in heaven, where men become gods and can re-create their large families as kingdoms to rule for eternity. In 1890, the Mormon church officially rejected polygyny because the U.S. Supreme Court had declared it illegal; but dissident groups broke away to continue the practice. One of these separatist, fundamentalist groups survives in Colorado City, while there are rival groups in a few nearby towns. All the land in Colorado City is held in common trust. When a young man wishes to marry, he must receive permission from the trustees

to build a house. Because most of the wives go to a few of the men, not all males can marry. Community members say social pressure exists toward boys who rebel against this system. They are urged or even forced to leave town. Men who conform to the system, though, are given extra wives as rewards. The wealthiest men possess the greatest number of wives and derive status from the size of the family they can exhibit.

Girls, on the other hand, are under pressure to marry before they are eighteen, and partners are selected for them. Marriage is described as the only sure route to heaven. Women are subordinate to men, and their primary task is to bear as many children as possible; the fertility rate is almost three times higher than the state average. The males wear more modern clothing styles, and the females wear their hair in the styles of the 1930s and dress in prairie-style dresses, high collars, and long sleeves. Makeup, earrings, or other jewelry are not allowed. Even though the Colorado City High School is public, most of the teachers and the principal enforce the community's strict social prohibitions against boys and girls flirting, seeing each other socially, or even talking privately together.

Although there is some dissension within the system, it also has strong advocates. One man recalled that his father had three wives. "I didn't feel one bit backward about the way I lived," he said. "The kids from monogamous families were the ones that were funny to me." Children generally refer to all their father's wives as "mothers," although some of them refer to them as "aunts." Said the informant: "How can it do anything but strengthen a home if you have two mothers?" (*Los Angeles Times,* April 13, 1986; Williams, 1998).

marriage partners and which were prohibited; such rules did not give individual families much freedom for action. In the more stratified agrarian states, aristocracies and wealthier merchants or landowners wanted to be able to marry off their sons and daughters to form alliances with whoever served their purposes. Permanent linkages between certain families, repeated over and over again across the generations as in tribal systems, gave way to a more flexible and in many ways more cynical use of marriages for immediate political advantage.

Comparing all these different kinds of exchange systems, from the most individualized to the most family dominated, the generalization seems to be that the more militarized the family and the larger the economic surplus and resulting social stratification, the more emphasis there is on marriage politics.

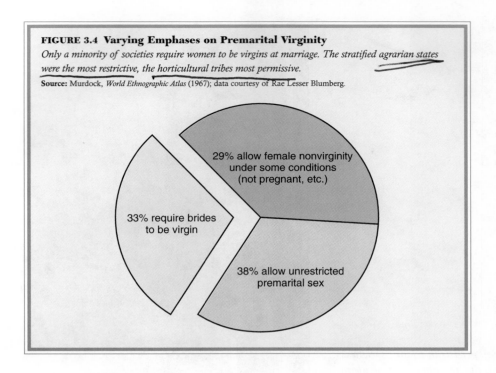

FIGURE 3.4 Varying Emphases on Premarital Virginity

Only a minority of societies require women to be virgins at marriage. The stratified agrarian states were the most restrictive, the horticultural tribes most permissive.

Source: Murdock, *World Ethnographic Atlas* (1967); data courtesy of Rae Lesser Blumberg.

29% allow female nonvirginity under some conditions (not pregnant, etc.)

33% require brides to be virgin

38% allow unrestricted premarital sex

The tendency, then, is for marriage politics to become more explicit as societies become more technologically advanced, from hunting-and-gathering societies up through the agrarian empires. Beyond that point, there is an important reversal. As we shall see in chapter 5, marriage politics suddenly begin to fall away as we get into industrial societies because there *the family is no longer the center of politics.* The modern bureaucratic state takes power away from the maneuvering families of aristocrats and their retainers, and that suddenly puts the family in a different context: the private family, a newcomer in world history, is born. This shift away from marriage politics is the subject of the next chapter.

The Politics of Virginity

One pattern that can be observed is that the greater the emphasis on marriage politics, the more sex becomes treated as property. In unstratified gathering societies and simple horticultural societies, there is very little control over sexuality (Figure 3.4). Premarital intercourse is frequently allowed with any choice of partners, and sometimes marriage is a trial arrangement that can be easily broken up if unsatisfactory. Not all primitive societies are this sexually libertarian, but few of them require that a woman be virginal at marriage; at most it may be expected that she not be pregnant by someone other than her husband-to-be. As we shift toward societies with greater degrees of stratification, the proportion requiring brides to be virgins goes up.

When we reach the highly stratified agrarian societies, restrictions on sexuality become much more stringent. Not only is virginity usually required, but there also tends to be a practice of early marriage, such as found in medieval India. A girl may be betrothed long before the age of puberty, so that she is the exclusive sexual property of her husband-to-be. Any sexual experience she

The wearing of the veil persists in Islamic societies. This custom represents the traditional way of life in which women are secluded from contact with men other than their own husbands. It is also a mark of high status in the society.

may have thus becomes a grave insult to him and a violation of the alliance between the families contracting the marriage. Under these circumstances, women are often secluded, veiled, or locked up in their homes. Any violation makes her completely dishonored and useless for exchange politics. Premarital sex, like adultery, is often punished by death of both man and woman or by the sale of the woman into slavery or prostitution (Vieille 1978). To prevent such immorality, many societies encourage the practice of female genital mutilation (*A Closer Look* 3.4).

This extreme form of sexual property in stratified agrarian societies is very one-sided, however. Women are expected to be virgins at marriage, but men are not. Men may even gain glory by their sexual exploits, and prostitution in such societies is usually quite open and legitimate. There are certain sexual rituals that may be enforced on men, however (*A Closer Look* 3.5).

Another limitation is that sexual property rights are most strongly enforced in the upper classes and in respectable urban society. Peasant women cannot be locked up in women's quarters because they are needed to do the work, outdoors as well as indoors, and they may not even wear veils, although upper-class women in the same society would be dishonored to have a man see their face. The wearing of veils grew up particularly in the Islamic societies of the Middle East. At first it was a mark of social distinction because it meant that a woman could remain secluded from contact with men other than her own husband—which is to say, it implied that she did not have to work in the open like the common people (Hodgson 1974, 342).

In some societies, widows in the upper classes may be prohibited from remarrying (because they are not virgins); in a very status-conscious caste society like India, they were even encouraged (or forced) to commit suicide on their husband's funeral pyre. But a widow of lower rank, especially if she had some property or could make a living, would be more valuable to another husband. Thus in highly stratified societies, lower-class women had more freedom than women of the higher classes. As peasant or village families gained some wealth and respectability, they usually attempted to raise their status by imposing upper-class restrictions on their women. In some agrarian societies modernizing in the twentieth century, change has occurred in opposite directions at once as the wealth of the society has grown. For instance, in contemporary Iran, traditional puritanical restrictions were spreading in the peasant villages, even at the same time that women in the cities were discarding the veil. After the revolution led by Muslim fundamentalists in 1979, the wearing of veils was imposed on all women. Strictly speaking, this custom is not a law of the Koran, but it has become part of the traditional way of life that the fundamentalists were attempting to restore (see *A Closer Look* 5.4).

PATRIARCHAL DOMINATION IN THE AGRARIAN STATES

In the world's great ancient and medieval civilizations, family organization took a new form. No longer were complex kinship groupings as important as they were in tribal societies. Society was starting to emerge beyond the level of kinship; however, it only emerged halfway. For the agrarian state societies were the era of the *household.* Both economically and politically, almost everything was organized around households. The places where people lived were the same places as they worked, and these were simultaneously the places where people had their political loyalty.

If one had visited the bazaar (market) in an Arab city not long ago (or still today in some places), one would have an idea of what the patrimonial household economy meant. There are the merchants, with their goods lined up for display on the street in front of their shops. Step inside a shop, and you find more goods. But the back of the shop is not like an ordinary store. There are children playing around the goods, or more likely helping with the work, coming forward to sell the visitor something or running to do the parent's bidding. If one gets the idea that the father's word is law, that is not far from right. Somewhere in the back, too, may be the merchant's wife, helping with the store but also doing the usual domestic duties of cleaning, carrying water, and cooking. For the store is a home, and the family lives there in the back or perhaps upstairs. That is one reason why the bazaar keeps such long hours, as opposed to our conventional nine to five; the store is almost always open, because the proprietor never goes home: he is already home, available whenever anyone comes in.

Multiply this example by thousands or millions and spread it backwards over the history of medieval Europe, the Middle East, India, China, and Japan, and you will have a fair idea of what life was like in a society of **patrimonial households.** Most such households were agricultural and rural, rather than mercantile like the bazaar example given previously; the key factor is that there is no difference between the place of work and the residence. What is different from a tribal form of kinship is that the households are much more stratified; some are big, as large as palaces, whereas others are very small, mere

> ### A CLOSER LOOK 3.4
> *Female Genital Mutilation*
>
> Female circumcision, also known as clitoridectomy, infibulation, or female genital mutilation, is a traditional practice that is widespread in more than twenty African nations and is also practiced in the Middle East, Malaysia, and Indonesia. The practice also occurs in the United States and Europe, where immigrants from Africa and the Middle East continue to circumcise their daughters, often surreptitiously. It is difficult to estimate the number of girls and women worldwide who have undergone some form of female circumcision because of the secrecy that surrounds the practice. The United Nations reports that the practice affects up to 70 million women, and the Women's International Network estimates that 110 million girls and women in continental Africa alone have been circumcised. The practice of circumcising women
>
> dates back to the Phoenicians of the fifth century.
>
> Female circumcision is usually performed on girls aged seven or eight, although some African tribes perform it on infants and other societies on young adult women. It is usually performed by midwives or elderly women of the community, who often do not have formal medical training. Circumcision is often performed under poor hygienic conditions, using an unsterile knife, razor, or sharpened stone, without anesthetic.
>
> There are four basic types of female circumcision:
>
> *Type 1 is analogous to male circumcision and consists of cutting the clitoral prepuce circumferally to remove it. This is the least drastic type. Type 2 involves removing the glans clitoris or even the entire clitoris; part or even all the adjacent tissues (the labia minora) may be removed as*
>
> *well. Type 3, infibulation or "pharanoic circumcision," involves removing not only the clitoris and adjacent tissues (labia minora), but the external labia as well; the raw edges of the wounds are then sewn together leaving only a tiny opening for urination and menstruation. Type 4, which is rarely practiced, is referred to as introcision and involves enlarging the vaginal opening by cutting the perineum (Rushwan 1990: 24).*
>
> Short-term medical consequences of female circumcision include bleeding, shock, infection, urine retention, and damage to surrounding tissues, all of which can result in death. Long-term complications include cysts, menstrual problems, recurring urinary tract infections, incontinence, infertility, complications during pregnancy and childbirth, sexual problems, and psychological problems such as anxiety, depression, neuroses, and

one-room hovels. They are also much less organized around kinship. There is almost always a family at the center of every household, but there are now other people as well. Servants are found in every household that can afford them; large households might have dozens or even hundreds, dressed in a livery (uniform) that shows off the status of the master, but even the modest middle-level households try to have at least an old woman or a young servant girl, quite possibly a slave.

Slaves, in fact, tend to be very much in evidence. These are rather callous societies, among the cruelest known. Ancient Rome lived off the work of slaves, and they existed in medieval, Christian Europe as well, and in the American colonies. Households of the aristocracy often preferred to own eunuchs, men who had been castrated. Eunuchs were even trusted with considerable authority. This was partly because they could have no families of their own and hence did not pose a threat to the family succession, and partly because they could be used to guard and attend the women without fear of adultery. Eunuchs were especially popular in the Byzantine and Ottoman (Turkish) empires but also elsewhere in the Middle East, the Roman empire, and in China, especially where there were harems.

Somewhat overlapping with the category of servants were apprentices, pages, and retainers. It was quite common, especially in England and northern Europe, for children in their early teens to live in the household of someone of

psychoses. Many women undergo a painful series of de-circumcisions and re-circumcisions after each child is born. Scars from circumcisions are cut open before delivery and stitched together after. In addition, there is evidence that the unsanitary conditions that surround the practice of female circumcision have contributed to the spread of AIDS.

Considering these circumstances, why is female circumcision practiced? The crucial point to bear in mind is that it is a ritual. Rituals are tied in to the social structure of the group and usually are connected to maintaining the pattern of stratification. Rituals reaffirm a traditional sense of identity; for that reason, the admonitions of outsiders—including the host society that surrounds migrants from an area with this tradition—often are regarded as attacks on the integrity of the group. Rituals are founded on the dynamics of group emotions and are surrounded by ideologies. Thus, female circumcision is sometimes defended as a requirement of Islamic law. Strictly speaking, this is untrue; in this respect the ritual is similar to the custom of wearing veils, which is not enjoined in the Koran but has become a tradition over the centuries as it has become a mark of high status in the group. Older women often feel pressured to have their daughters undergo the practice; they fear that their daughters will be ostracized or will not be able to marry if they are not circumcised. Many women accept the procedure because they are taught to believe that sexual pleasure is the exclusive right of men.

Efforts are being made to eradicate these long-held beliefs, including educating midwives, enlisting the support of religious leaders and women's groups, and enacting laws to punish those who practice female circumcision. Strategies to abolish the practice must take into account local customs and belief systems. Efforts to publicize the medical risks are generally more effective than those that emphasize sexual issues, which are highly controversial. The World Health Organization has implemented several programs at the grassroots level that have been somewhat successful in decreasing the incidence of female genital mutilation.

Sources: Efua Dorkenoo. 1994. "Cutting the Rose: Female Genital Mutilation: The Practice and Its Prevention." London: Minority Rights Publications; C.A. Baker, G.J. Gilson, M.D. Vill, and L.B. Cureto. 1993. "Female Circumcision: Obstetric Issues." *American Journal of Obstetrics and Gynecology* 3 (Dec.): 1616–18; "Men Control Female Sexuality with Circumcision." 1993. *AIDS Weekly* (March 29): 10; "Puberty Rite for Girls Is Bitter Issue Across Africa." 1990. *New York Times International* (Jan. 15); Hamid Rushwan. 1995. "Female Circumcision." *World Health* 48:16–17; Women's International Network. 1993. "Progress Report: Campaign to Stop FGM." *WIN News* 19 (Autumn): 29.

higher rank and wealth. A peasant girl might be the housemaid for a yeoman farmer; a boy might go off to the town to apprentice with a baker or some other craftsman. This was partly a way of making a living and partly a form of education of learning a trade, with the hope of amassing enough money to set oneself up in business later on. We have a description of a typical bakery in London in 1619, just after Shakespeare's death; it shows that the household included the baker and his wife, four journeyman employees, two apprentices, two maidservants, and three or four children of the master himself. The total is thirteen or fourteen persons, the vagueness indicating how little attention was paid to small children (Laslett 1971, 1–3). It was not a sentimental society; if everyone was treated as part of some big family, it was an authoritarian and business-like one, not to be confused with today's private families with their emphasis on warm domestic togetherness.

One more important characteristic of these households should be mentioned: they were essentially **fortified households.** Generally speaking, there was no police force; everyone kept arms who could, although this generally meant that the aristocrats had most of the weapons. Shakespeare's *Romeo and Juliet* gives a good idea of what the situation was like. Romeo and Juliet belong to two wealthy patrician families of the Italian city of Verona. Each house is a kind of fortress, and Romeo has to scale the walls to get to Juliet's balcony to see her. Each family is

A CLOSER LOOK 3.5
Politics of Male Circumcision Rituals

Circumcision, the cutting back of the foreskin from the glans of the penis, is carried out unceremoniously in hospitals today as a hygienic measure and taken-for-granted custom. Historically, circumcision had religious significance; for example, it divided the believers from the nonbelievers in both the Jewish and the Moslem faiths. Karen Ericksen Paige and Jeffery M. Paige (1981, 122–66) carried out a cross-cultural analysis of circumcision and found it widespread in a certain type of tribal society.

In these societies, circumcision was a ritual, carried out in public and in the presence of a large crowd. Often the subject was not an infant but a larger boy or even a teenager, who had to undergo an ordeal something like a hazing. There were feasts and elaborate ceremonies, which might include camel races or dancers brandishing swords to drive away evil spirits. It was especially important for all the male relatives to attend, although in some cases females attended as well.

The types of societies in which circumcision rituals of this sort were carried out consisted especially of warlike tribes that ranged widely over grazing land with their camels, goats, sheep, or cattle. They were most likely to be found in the Middle East and North Africa, which are the classic homelands, of course, of ancient Judaism and of Islam. Karen and Jeffery Paige analyze the circumcision rituals as a device used in the political maneuvering of these tribes. They have no regular state but only coalitions of warriors from different families, who band together to defend their grazing territories and to conquer others. Feuds are common, and the whole family lineage is bound to avenge the death or injury of one of its members. Hence, the extended family coalition is very important. These are extremely male-dominated societies, and the main way that a man can become powerful is to have many sons, who in turn might have many grandsons. Thus, by the time he reaches old age, a patriarch might turn out quite an extensive army of descendants and relatives. To do this, men who can afford it acquire as many wives as possible (polygyny), usually by paying a bride price in the form of so many camels, cattle, or other animals (which are the principal form of wealth).

How does the circumcision ritual fit into this situation? Paige and Paige argue that it is a way of monitoring the strength of a coalition. Not only are these societies violent, but they are also unstable. Family coalitions often fall apart, because it is always possible for a son to gather his own wives, descendants, and followers, and move off with their herds of animals to graze somewhere else or to raid someone's grazing lands. There are no fixed political boundaries to hold things together. Moreover, it is often brothers or cousins who get involved in these feuds and splits. The Old Testament gives many examples from early Hebrew history, when the Jewish tribes wandered with their cattle throughout the Middle East. Abraham's followers fought with those of his relative Lot (Genesis 13:6–7); Isaac and Ishmael, who were half-brothers (their father Abraham being polygamous), became rivals, and Ishmael was driven away into the desert like the Arabian Bedouin (Genesis 21:9–21); Jacob and Esau fought over their inheritance and split the tribe once again (Genesis 28:1–5).

The purpose of the circumcision is to bring together all the male kinsmen and to try to bind their loyalty to the military coalition. The ceremony is an occasion on which all who consider themselves a member of the extended family show up to be counted and to stay away is a mark of disloyalty. Moreover, the ceremony centers on the penis of the son, which is the instrument through which a patriarch will acquire more grandchildren and hence a larger coalition. But the penis, is, so to speak, a two-edged weapon: it can increase the coalition, but it also raises the danger of splits between the sons when they gather their own followers—hence, the undertone of hostility and the infliction of pain in the ceremony. Moreover, given the low level of sanitation and of medical practice in these societies, circumcision brings the danger of infection and even of castration. The ceremony shows the willingness of the father to symbolically sacrifice something—the foreskin—and even more to risk his son's penis in the presence of the entire military coalition. It is, in short, a kind of display of bravado, simultaneously boasting about the potential reproductive power of the patriarch's lineage and reminding the group that he is indebted to them and will make sacrifices to hold the coalition together.

This Seigneur's dwelling room of the fourteenth century combines a "living room," bedroom, hearth, and shrine. It exemplifies the medieval Great Chamber, with no connecting hallways between rooms and little privacy.

full of armed men—sons, cousins, retainers, servants—and the two families are feuding because the two groups of armed men have battled on the street and Juliet's cousin has been killed in the fighting. That is the reason why the love affair between Romeo and Juliet is so illicit: the two families not only are not allies, but actually are something like opposing armies.

We have seen that the family, with its servants and apprentices, was the economic unit. At a higher level, the aristocratic family was a political unit. How much power it had depended, first of all, on how much armed force it had within its own walls, and second, on what kinds of alliances it had made with other households. The ruler, originally, was the person with the largest household. Louis XIV of France overawed Europe with his palaces, which might contain ten thousand people besides his own family: courtiers, ladies-in-waiting, guards, and servants. The royal household was the state; below it were lesser households of the nobility and officials, each of these in turn with its own retinue of servants and followers. Just as a peasant boy might be sent to an urban bakery to learn the trade and make his fortune, the son—and also the daughter—of a minor noble-man would usually be sent to be a page or maid-in-waiting at the house of some more powerful lord, with the hope of establishing a connection that would make *this* young person's fortune.

The society was thus held together by a kind of household-to-household link (Girouard 1980; Mertes 1988). We pointed out that this was halfway beyond the kinship system because most of the people in any one household were probably not relatives. In a strange way, there was no clear dividing line between servants, retainers, and guests. A powerful lord built up his reputation and his following by keeping what we might call "open house": people could come and visit him, eat

A CLOSER LOOK 3.6
Medieval Versus Modern Houses

The patrimonial household of the medieval civilizations was a kind of fortress. Some houses of great lords were literally castles, though even urban dwellings tended to be built for protection. Thus, an old European city street typically has stone outer walls, with only a few high windows and great barred gates. The house does not face outward but inward toward the courtyard, where people can be outdoors but under their own protection. Similarly, if one walks down an old street in Japan, one sees nothing but blank wooden walls; only if you are admitted inside can you see the light and airy verandas opening onto an attractive garden. Unlike our modern houses, these medieval houses were not built for external display; the display was for insiders only.

Another difference one would notice is that medieval houses did not have hallways. We tend to take for granted that one of our homes will have a central hallway, a corridor with bedrooms, bathrooms, and other kinds of rooms leading off it. But a large medieval house did not have these corridors or private rooms; instead one large room simply led onto the next. There was no way to get to one room except by passing through all the other rooms leading up to it. This was part of the lack of privacy in medieval society. There were virtually no special-purpose rooms, such as bedrooms or bathrooms. The "Great Chamber" meant a very large room, where the master of the house could entertain large numbers of guests for dinner. The composition of the household shifted, with different guests and retainers there at various times. Beds were trundles that were put away in the daytime and were shared by several people, or sometimes merely cushions or places to sleep on the floor. The high-ranking lords and ladies would have beds, but their servants probably slept in the same room with them. As for bathrooms, there were chamberpots tucked away here and there, which is one reason why people who had servants kept them nearby. Excretion or sexual matters could not be kept very private, and people rarely tried to make them so. In the humbler ranks of society, houses were just as crowded and smelly as among their superiors, but much smaller. A peasant family usually had a hut with a single room, in which all the business of the household was conducted.

at his table, and stay with him for weeks or even years. In return, the guests were expected to be loyal followers, perhaps carry out political duties, fight in his battles, or in the meantime just act as servants. The nearer one could be to the lord, the more favors one could expect from him; hence, it was much better to sit at his own table, though popular lords might entertain so many guests that one was lucky to get a bite at the far end of the hall. In the court of Louis XIV, high lords vied for the honor of doing humble services that might bring them near His Majesty; one was honored with the title of Gentleman of the Bedchamber, which meant that he had the honor of holding the king's robe while he stepped out of bed; another gentleman held a shovel with coals of burning incense to sweeten the air while His Majesty sat on the royal chamberpot (Lewis 1957, 1–61). As one can see, these were societies without much privacy in the modern sense (*A Closer Look* 3.6).

We have not yet arrived at the modern type of workplace, where people live in a private home in one place and go somewhere else to work with people who are not relatives at all. As we shall see in the next chapter, this patrimonial household structure eventually began to break down. Political and military circumstances made it impossible to keep an entire army and an entire government bureaucracy in the household and feed them at the lord's personal table. Organizations began to emerge outside the household, and these giant castles and palaces began to give way to a more private dwelling even for the upper classes. The king's servants gradually became bureaucrats. Sometimes they retained their old names, though their functions changed. *Chancellor,* once the title of the butler in charge of the wine cellar, became the name for a high government official. The secretary took care of the lord's private correspondence (and at this time was always a man); eventually the title was applied to such elevated personages as the secretary of state.

The lower classes, on the other hand, had always lived in much smaller groups, unless they happened to be servants in some great man's household. Because the poor by far outnumbered the rich, one could say that the fortified patrimonial household was for a small but dominant minority, whereas the majority of the populace lived in small nuclear households. In some senses the families of the peasant majority were not too different from families today. But in other ways their lives were quite different: like the upper classes, they had little privacy; they tended to treat their children unsentimentally, as economic tools; where possible, the men were patriarchal and authoritarian; and even poor families emulated the style of the families above them on the rungs of the social ladder, with whom they were often connected by ties of feudal duties. Any family that could afford it tried to acquire servants and to copy the manners of their superiors. They also tried to establish ties with the larger households by going to work in them or becoming their retainers and followers. In these ways, the patrimonial households of the aristocrats dominated the surrounding society, even though this society consisted of a much larger number of small households inhabited by something more like the modern nuclear family.

The Era of Maximal Male Domination

The agrarian state societies had a wider gap between top and bottom of the social structure than any other type of society that has ever existed. It has been estimated that the top 1 or 2 percent of the population owned over half the wealth and that the king or emperor alone usually owned one-quarter of the wealth all by himself (Lenski and Lenski 1991). There was little in the way of a middle class, and some 90 percent of the population might have been peasants in tiny huts, living barely on the edge of subsistence. Stratification between males and females also reached an extreme.

In these societies, women were most callously treated as mere property. Wealthier men, at least in many parts of the world, had harems or concubines. Even where men were restricted to a single wife, they tended to keep their wives and daughters secluded under lock and key or veiled so that strangers could not look at them. The lower classes could not afford to do this; a man could have only one wife, and she was freer because she had to work. But the ideal of the patriarch was widespread, and peasant men lorded it over their women to the extent that they could. Legally and politically, women tended to be treated as minors. Because of the practice of giving a dowry with a woman when she was married, girls were considered an economic imposition and female infanticide was often practiced. Even the religions prevalent in agrarian societies (Islam, Hinduism, Confucianism, Buddhism, Christianity) generally treated women as second-class citizens and as temptations in the way of men's religious duties.

Why did men in these societies want to dominate women? It cannot be explained by invoking universal psychological or biological drives, because we see that this form of extreme domination occurs only in particular historical circumstances. Moreover, even in these medieval agrarian societies, men would not typically have said that they were dominating women just because they wanted to. These were indeed rather callous societies; servants, slaves, and members of the lower classes were treated more or less as dirt beneath the feet of the aristocracy. Upper-class ranking was so ritualized in everyday life that people felt they were divided into two distinct species of humans. Lower-class women were treated badly because they were members of the lower class.

Within their own families, men no doubt believed that they were honoring and protecting their wives and daughters by the tight controls they placed on them. As we have seen, wearing a veil was considered a mark of high status in Muslim societies. Placing strict controls on women's sexual freedom was considered in agrarian societies as a way of protecting the honor of the family and of the woman as part of it. Even the Chinese practice of binding girl's feet, which made them unable to walk freely, was justified as displaying the high status of the family that could afford to have its women in this condition. The ideology of these societies assumed that women were being controlled for their own good, for whatever raised the status of the family raised the status of its women members too. Like all ideologies, this covered up the real power resources that enabled persons in the most favored circumstances to control those below them. Political and economic conditions put upper-class men in positions of almost tyrannical control, and this influenced gender stratification throughout the society.

Countervailing Forces for Women's Status

Nevertheless, there were some conditions that favored women in these societies. The class stratification system itself meant that some women were of very high rank, even though their rank depended on a man. Wives and daughters of important aristocrats or wealthy patricians had an important status in the society. Men of lower rank might try to make their approach to a high-ranking lady, in hopes that she could wield some influence on their behalf. The politics of family alliances produced a good deal of legal maneuvering about inheritance, which became more rigged to favor the female line (Collins 1986b). This was perhaps not done to give more rights to women but because families wanted to receive or give away inheritances through their women. Nevertheless, it had the consequence that upper-class women, at any rate, acquired some property rights. Sometimes, through the death or absence of the males in the family line, a woman might actually wield power as a queen or noblewoman. Interestingly enough, this did not usually change the status of women generally in the society; the reigns of some of the most famous queens, such as Queen Elizabeth I in England or Catherine the Great in Russia, were conservative times when the majority of women were held under tight restraints. But even in an extremely male-dominated world like Ottoman Turkey, women acquired some powers of their own (Dengler 1978). Turkish women were supposed to be secluded from the eyes of men, but one effect of this was that they could move about in their veils more freely and anonymously than anyone else in this rather authoritarian society. The harems of the wealthy had large staffs of female servants because the effort to exclude men meant that women for the most part were left to themselves. Wives and daughters of wealthy men might supervise huge numbers of servants and dispose of considerable economic property, even acquiring property in their own names.

Especially in Europe, the aristocratic women were pioneers. They exerted some pressure, perhaps very gently, on the property system. More significantly, as we shall see in the next chapter, they began to change the culture of male-female relations. In medieval France and then elsewhere, aristocratic women began to create the ideal of courtly love. In the midst of a society in which women were largely treated as pawns in the game of marriage politics or ruthlessly exploited if they were members of the lower classes, an ideal began to emerge that demanded deference toward women on the part of men. It was only one step, but it was to have important consequences.

SUMMARY

1. Family systems rise and fall in a bell-shaped curve as societies become technologically more complex. Relatively simple families are found in hunting-and-gathering societies, while complex kinship systems reach their peak in advanced horticultural and stateless agrarian societies. The kinship system begins to simplify again in the agrarian civilizations, then rapidly loses complexity to form the nuclear family unit in industrial societies.

2. Gender inequality is explained by both economic and political theories. Economic theories propose that women have the greatest power inside and outside the household, and the greatest freedom to arrange their own marriages, divorces, and sex lives, where women make the greatest economic contribution *and* are able to control economic property. This happens when women's labor is indispensable because of female expertise, work organization, and sex ratios in the population. Also contributing is a kinship system that allows women to control inheritance, and it is also important that stratification of the larger society does not interfere on behalf of men.

3. Male strength and female breast-feeding are not very important in determining what work women can do. Where women have the social power and their labor is desirable, they have been able to control their own fertility and work out satisfactory child-care arrangements.

4. Political theories of gender inequality argue that the male size advantage has been used mainly to monopolize fighting and hence politics and that men have accentuated this advantage by monopolizing weapons. Male dominance thus depends on how much fighting there is and how the male military group is organized.

5. Where societies are relatively peaceful and men are not organized in all-male groups, the family system tends to be egalitarian. Societies with patrilineal and patrilocal families isolate women amid strongly organized male groups and are very male dominated. In tribal societies where fighting takes place far from home, however, there is a tendency for organization to be matrilineal or matrilocal and to give women a somewhat better position. In some societies in which both genders are well organized, there is a great deal of antagonism between the sexes and frequent use of male violence.

6. Men tend to monopolize work (such as hunting and herding) that can most easily be transformed into military activity, leaving gathering and horticulture to women. In class-stratified societies, aristocratic males shift the burden of agriculture to male peasants, and women are forced into a more specialized role as baby producers.

7. Tribal societies often practice complex forms of kinship politics, in which families exchange marriages according to certain rules (especially rules emphasizing cousin marriages). In stratified agrarian societies, these formal rules tend to disappear, so that families can maneuver more flexibly to make marriage alliances for their political convenience.

8. In societies that emphasize marriage politics, women are treated as sexual property, and their virginity is closely guarded. Premarital and extramarital sex, however, is allowed in many societies with less political pressure.

9. The ancient and medieval agrarian civilizations were highly stratified between an aristocracy and the peasants, servants, and slaves that they dominated. The upper classes lived in large patriarchal households, in which the family was surrounded by servants and armed followers. The lower classes lived in much smaller units, but they tried to emulate the large upper-class household whenever their income rose.

10. These agrarian civilizations represent the peak of male domination in world history. Especially in the upper classes and in the cities, women were kept secluded or veiled and were legally treated as minors. Female infanticide was often practiced, and women had low religions status. However, countervailing forces existed in some aspects of property systems and marriage politics, so some women acquired economic rights and, occasionally, political power.

Key Terms

advanced horticultural
  societies
agrarian plow culture
alliance theories
descent rule
economic determinism
economic theory

egalitarian
fishing societies
fortified households
horticultural societies
hunting-and-gathering
  societies
industrial societies

matrilineal
pastoral societies
patrimonial households
patrilineal
political theory
primitive horticultural societies
residence pattern

Sociology Web Site

See the Wadsworth Sociology Resource Center, "Virtual Society," for additional links, quizzes, and learning tools:

http://www.sociology.wadsworth.com

Also on this web site you'll find InfoTrac College Edition, an online library of journals. Here you can search for electronic articles about central topics in sociology.

4

History of the Family II: The Love Revolution and the Rise of Feminism

INTRODUCTION

The modern family structure was thought to have emerged quite recently. A shift was supposed to have occurred from the **extended family** to the **nuclear family** within the last 100 years or less. Traditional families were supposed to have been large, with several generations living together in the same household. Along with this went a lack of individualism. All the members subordinated themselves to the good of the family as a whole, contributed their earnings when they could work, and were cared for and respected when they were old. The counterpart of this lack of individualism was that men and women did not get to pick their own spouses; instead, marriages were arranged by the family, often at a very early age. The family encompassed the individual from the cradle to the grave.

Then, according to this analysis, along came industrialization, urbanization, and (in some versions) capitalism. The self-sufficient rural household gave way to the small urban home. Each man had to go out and seek his own work in factories or offices. The craft worker was alienated from the means of production and subjected to the impersonal labor market of the city. No longer did each contribute to the family kitty; no longer were the old people sheltered by their loved ones. With this, family controls over marriage broke down, and each person chose his or her own marriage partner. For better or worse, the modern family was born.

This picture of the modern family transition hangs together nicely, but it has one main flaw: it simply isn't true. Until recently, sociologists, anthropologists, and historians had not looked closely at the families of traditional agrarian Europe. Instead they had relied on a comparison between the nuclear family system that exists now and an idealized approximation of the families of our ancestors from literature and religious writings. Since industrialization happened quite recently, it was assumed that the modern nuclear family emerged recently, too.

WHEN DID MODERN FAMILIES BEGIN?

In the 1960s, a number of British and French historians began to search through local village records, church documents, and other surviving information about medieval and early modern communities. Their aim was to reconstruct what family life was actually like—how many people were living in houses at what time, how often children were born, and so on. The results were quite a surprise, for these western European families did not resemble the "traditional" picture at all. The method has been extended to the early American colonies and to a few non-Western societies such as Turkey and Japan. The picture that emerges shows our former beliefs about the rise of the modern family to include two myths.

Myth 1: Extended Families

When we think of the traditional family, we imagine several generations gathered around the same hearth: father and mother and their children, some of them perhaps grown-up sons who have brought their wives to live in the family home, and grandfather dozing contentedly by the fireplace. But in fact, this type of family was nowhere to be found, at least not in sixteenth- and seventeenth-century England, northern France, or America (Laslett 1971, 1977). Instead, when we look back, we find something that looks very much like the modern nuclear family: father and mother and their young children. When the children were big enough to work, they left home; when they were married, they acquired their own house. Also, old people were not taken care of by their children; one rarely finds evidence of a grandmother or grandfather living in these homes, but one does find evidence of old people living by themselves in their own cottages.

How could we have been so mistaken about the family? One fact that was overlooked was sheer demography. Most people in these traditional societies did not live very long; life expectancy was about 35 to 45 years. (In Plymouth Colony, for instance, it was 45.5 years for men, who lived longer than women [Laslett 1971, 98].) These figures, however, are for life expectancy at birth, which was reduced because many infants and children died. Once individuals reached adulthood, their chances of reaching old age were improved, but it is clear that the proportion of old people was much lower than in our times. Because people did not marry early, it was not so common for men or women to live to see their grandchildren or even to live much beyond the age at which their own children would marry.

Myth 2: Early Marriage

We also have the image of a society in which people married very young. We can think of Romeo and Juliet in Shakespeare's play. Juliet was thirteen, and her mother says to her:

> *Well, think of marriage now; younger than you,*
> *Here in Verona, ladies of esteem,*
> *Are made already mothers. By my count*
> *I was your mother much upon these years*
> *That you are now a maid.*

But this, too, is a myth. Laslett (1971, 86; 1977, 40) found that in Shakespeare's own day (the early 1600s in England) the typical bride was about twenty-four to twenty-seven years old, and the typical groom was about twenty-seven to thirty. In our own century, during the 1950s, the median age of marriage was much lower (twenty for women and twenty-two for men), and teenage brides were much more common than in any other period for which records were kept. (Because the median is the point at which half the marriages are above that age and half below, in the 1950s half of American brides were teenagers.)

Again, what is going on here? For one thing, our image of medieval child marriages is based on a skewed sample: until historians started looking at records of the common people out in the villages, all we had to go on were literary records (like Shakespeare), which are concerned only with the aristocracy. It is true that at the very highest level of society, children might have been betrothed at an extremely early age, with negotiations beginning as soon as a child was born. A study of all the English princesses born between 1035 and 1482 (Boulding 1976, 429–31) found that half of them were married by age fifteen, with some married off by the time they were nine. But these were dynastic marriages, carried on for the sake of political alliances with the royalty of other states. Lower down in the aristocracy, nobles' daughters married a little earlier than peasants' daughters, but not a lot earlier; Laslett (1971, 86) found a mean age of nineteen to twenty-one for noblewomen. Noblemen, on the other hand, tended to marry about as late as other men, around age twenty-six.

Why did most people wait so long to marry, especially considering that their life span did not leave them much time afterward? Primarily, the reasons look economic. The household was an economic enterprise, such as a farm, a bakery, or a mill. The work was done right in the home, and the entire family (together with its servants) constituted the work force. To become the legitimate head of his own business, a man had to be married (and, conversely, a woman could become the head of a business, with very few exceptions, only by taking it over after the death of her husband). So the decision to marry was an economic one. A man could not run such a business without being married, and in fact, he was not even legally allowed to do so in England (Laslett 1971, 12).

On the other hand, a couple could not marry unless there was a house or a cottage they could acquire, that is, a business opening that could be taken over. This operated as an automatic safety valve against overpopulation because the age of marriage adjusted to economic conditions. In good times, people could marry earlier; in bad times, they had to wait, and some of them might never marry. Thus, the number of children born would go down during bad times, decreasing the pressure to find farms and household establishments for them when they became of marrying age. Under these conditions, children tended to be born much later than they are now; in an English town in 1700, for instance, the average age of mothers at the birth of their children was thirty, and the age of fathers was thirty-five (Laslett 1977, 41). Not until an industrial labor force began to emerge after 1750 were many people able to marry and at a relatively younger age (Goldstone, 1986).

What typically happened was that children left home when they were teenagers to become servants or apprentices in someone else's house. This provided labor power to run these household economies: workers had to be gotten from somewhere because it was likely that children in most families were too young to help out seriously. Servants were kept under family discipline, and they could save up their wages until an opportunity came, in their twenties, to leave and acquire a household of their own. There was a typical life phase of working as a servant, at least in England and northwest Europe, experienced by about a third to a half of the population (Laslett 1977, 43). It was not a lifelong position but rather something that people commonly went through in their teens and early twenties.

Western European Family Patterns Compared with the Rest of the World

It looks as if there was (and is today) a distinctive western European family pattern, characteristic of England, the Netherlands, and northern France, which was found already in the late Middle Ages. It was this family structure that was transplanted to the North American colonies that became the United States. Just exactly when it began no one yet knows. Researchers have found it as far back as the 1200s in England (Macfarlane 1979); possibly, it goes back even farther. This "modern" family pattern did not have to wait for industrialization and a modern capitalism to develop; it was there in these rural farming societies. (Conversely, there is evidence now that industrialism did not cause the breakdown of the family when it became ascendent in the late nineteenth and early twentieth centuries. Tamara Hareven [1981] shows that the same New England families had, for several generations, sent their members into the factories at a certain age, a practice not unlike the older system of apprenticeship. This was especially true of single women.)

The western European style family did not exist everywhere, however. It had some distinctive characteristics not found elsewhere in the world (Hajnal 1965; Laslett 1977, 12–49). The household was nuclear, consisting only of parents with their young children. In eastern Europe, southern France, and many other places, by contrast, a complex household was much more likely, with several generations living together. There, married brothers might share the family property, which could not be divided or sold off in shares. Or one brother (usually the oldest) might take over a head of the so-called stem family, living with his parents while they were alive (Berkner 1972). This situation is very much the same today.

In the western European family, there was a relatively small age gap between husbands and wives. Typically the husband was about two or three years older, although in as many as one-fifth of all marriages the wife was older. (This was probably due to the strong tendency for a widow who inherited her husband's business to be a prime target for economically motivated marriage.) In eastern Europe, or Asian societies like China and Japan, husbands were likely to be much older than their wives. This made the relations of spouses especially remote and put women at a special disadvantage because they were young and had acquired few economic and cultural resources compared with their husbands. In western European families, spouses were more likely to have a closer relationship.

In the western European family, women in particular put off marrying and childbearing until they were older. This is one reason that the complex, extended family is more often found in non-Western societies. In Russia, for instance, three-generational depth was found in most families. Women married young and bore children early, thus making it possible for old people to see their own grandchildren grow up. This practice also provided a greater source of labor right within the family, eliminating the need to go outside the family for servants. (The sons' wives who came into the family, though, were the equivalent of such servants; especially before she had borne a child—preferably a son—the bride ranked the lowest in the household and was given all the menial tasks.)

Finally, the western European family was unique in having a high proportion of servants and in typically sending its members out to work as servants during their early adult years. In non-Western European societies, servants were much less common. Typically only upper-class households would have servants, whereas in England and western Europe, even small peasant proprietors were likely to have them. Where servants did exist in eastern Europe or Asia, they were more likely to be lifetime servants who were devoted (or confined) to someone else's family until they died. In western Europe, being a servant was not a permanent condition but a life stage confined largely to the young and unmarried.

If by "the modern family," then, we mean the nuclear household with relatively close and companionate ties between husband and wife, it is certainly not a recent development. Why it should have appeared in western Europe already by the Middle Ages but not elsewhere in the world is a puzzle that has not yet been solved. But this is not to say that *no* important changes have occurred in the Western family in the last few centuries. The most notable changes are those that have affected sexual behavior, family sentiments, and the position of women. The ideals surrounding marriage shifted drastically; marriage became viewed much less as an economic arrangement (though it still was) and more as the expression of mutual love. Sexual puritanism was enunciated very strongly, especially in the 1800s during what might be called the "Victorian revolution." And connected with both of these, the position of women was changed, although in somewhat contradictory directions. On the one hand, women gained a much more exalted position, at least in the popular ideology and in polite society. On the other hand, a doctrine of "separate spheres" grew up that confined women (at least above the working classes) to a role of housewife and mother much more exclusively than before.

This combination of changes is best seen as the result of a conflict. Women were struggling to get out from under the legal and other disadvantages that oppressed them, and the new form of marriage market that was emerging gave them some new opportunities. The sentimental revolution of love and the Victorian revolution of prudery are part of this, but they are also part of a counterattack that gave women a very confined role along with their new gains.

THE LOVE REVOLUTION

The basic principle of the modern love revolution was to connect love with marriage. Love existed in previous societies, to be sure, but it simply was not expected to be the reason for marrying someone. In tribal societies, kinship rules that specified, for instance, that one should marry one's cousin on the father's side but not on the mother's side obviously excluded love as a motive for marriage. Similarly, where marriage was a matter of family politics or economics, the sentiments of the individual counted for very little. By the mid–twentieth century, though, marrying for love became *the* dominant ideal, so much so that one is embarrassed to admit to marrying for any other reason. This change, which came about gradually in the 1700s and 1800s, may be referred to as the **love revolution.**

Ancient Ideals of Love

It is not that love did not exist in other societies. It did, and ancient societies treated it clearly in their mythologies. In ancient Rome, Venus was depicted as the goddess of love, and her son Cupid was depicted as a cherubic figure whose arrows made men and women into helpless lovers. In Greece, love's name was Aphrodite (from which comes the term *aphrodisiac),* and her son's, Eros (the origin of our word *erotic).* Especially in the later phase of Greek civilization called the Hellenistic period (ca. 300 B.C.–A.D. 300), popular literature was full of romantic plots about orphan boys who ran away with girls whom their fathers had betrothed to someone else (Hadas 1950). (In the end it usually turned out that the boy had been exchanged for another when he was a baby, and hence was really the son of a good family, so he could marry the girl.) In Asia, too, love poems from that period speak of yearning and passion.

Love was never depicted as the normal path to marriage. On the contrary, marriage was something quite separate. The Greek and Roman goddess of marriage was not Venus but Juno (Hera), the wife of Jupiter (Zeus), and she was always depicted as a rival and enemy of Venus. Zeus had many legendary love affairs, but his adulteries were an embarrassment to Hera and the object of her revenge. In real life, too, this seems to have been the case. Greek men kept their wives locked up inside their houses while entertaining themselves with cultivated prostitutes called *hetairai* (Pomeroy 1975). The Greek philosophers, including

Plato, wrote a good deal about love, but for them it was an almost exclusively masculine passion. The prevailing form of love-infatuation in the classical period (600–400 B.C.) was homosexual. Typically this involved an older man falling in love with a beautiful adolescent boy (Dover 1978). Obviously it could not be the basis for having children, nor did it even involve the two-sided, mutual love that is the modern ideal of heterosexual love (and of modern *homosexual* love as well). Homosexual love affairs apparently also occurred among women; *lesbians* are named for the island of Lesbos, where the poetess Sappho lived and wrote of her affection for younger women. However, women (at least in the respectable class) were usually severely restricted in this society, and it appears that they had little chance for love affairs, heterosexual or homosexual (Dover 1978, 171–84; Pomeroy 1975).

Such Greek philosophers as Plato spoke much about love, but this kind of love was a combination of the purely physical and the extremely spiritual; it lacked the modern elements of mutual closeness and sympathy between lovers. Plato's image of love was passionate yearning and admiration for someone (usually a young boy, although the object could also be a woman, as in the myths of Venus, Zeus, and the other gods and heroes). Plato spiritualized this emotion by declaring that erotic love was the worship of beauty, which was a reflection of the ideal Forms upon which Plato believed the earthly world was patterned. Thus, one loves somebody for his or her physical beauty, which in turn represents the unearthly beauty of a higher, philosophical realm. Nowhere is the love reciprocal and emotionally intimate. Later Roman writers such as Ovid (43 B.C.–A.D. 18) wrote on *The Art of Love* but confined it to an amusing game of chases and adventures. The goals were sex and beauty, never intimacy—and certainly never *marriage*.

With the arrival of Christianity, love was placed even more in the background. This seems a bit incongruous, in that Christianity began as a religion of love, but the love it exalted was completely spiritual: not the love of ideal Forms as in Plato and later Neoplatonic mystery religions, but the love of Jesus Christ for the world he was to save, and reciprocally the love that was expected of Christians for God and their Savior. Christianity was hostile to the Greek and Roman myths of Zeus and his affairs, which they considered immoral. So, too, were regarded love stories like those of Ovid or the risqué Hellenistic novelists. The eroticism of ancient pagan culture was one of its features that Christianity regarded as most abominable and that it tried hard to suppress.

Early Christianity placed a very strong emphasis on asceticism (Queen and Habenstein 1967, 181–201). Its ideal person was an anchorite or monk, who denied all the desires of the flesh in order to attain holiness. Accordingly, marriage was not looked upon as a desirable state at all, although the church leaders made concessions in the form of accepting marriage as second best, lest one who could not contain the sexual appetites commit the far worse sin of premarital fornication. Saint Paul wrote to the Corinthians: "For I would that all men were even as myself [i.e., a bachelor]. . . . But if they cannot contain, let them marry; for it is better to marry than to burn" (1 Corinthians 7:7–9). This doctrine was to last far into the Middle Ages. Priests and monks were celibate (or at least ought to have been) and regarded anything less than complete celibacy as a condition of semimitigated sin. Marriages had to take place outside the door of the church because they were not holy enough to occur inside. The only kind of love that the Church advocated was purely spiritual and religious.

Economic and Political Restrictions on Love

In all social classes in medieval society, marriages were contracted for reasons other than love. Among the kings and the higher nobility, marriages were a form of diplomacy. Sons and daughters were married off, sight unseen, to the heirs of other states with whom one wished to form an alliance. Diplomats might be sent to negotiate as soon as a prince or princess was born; the same child might be "marketed around" to various kingdoms to find the best deal that could be arranged. Given that kingdoms and other realms like duchies, baronies, and the like were hereditary, the sorts of arrangements to be made were extremely important. A queen might bring as her dowry the right to inherit an entire kingdom. Because not every marriage produced a son, it happened quite often that a throne could pass from one royal family to another via the female line. For instance, as late as 1714, the throne of England was inherited by the House of Hanover, in Germany (which brought to England the first series of kings named George, which lasted through the time of the American Revolution). This was due to a complicated political agreement decades earlier that declared that, in the absence of male offspring to the current king, the throne would pass to the descendants of Princess Elizabeth of the Palatine. The agreement was due to the religious politics of the period, in that the Protestant faction in power in England wanted to make sure that the throne descended to a Protestant family and not back into the hands of the Catholic kings whom they had just overthrown.

Similarly, in the lower ranks of the aristocracy, families arranged marriages to ensure the prestige of their line and the growth of their property holdings. In the middle class, too, marriages did not happen for love. In England (although not necessarily elsewhere in Europe) individuals were generally able to choose their own marriage partners rather than have them chosen by their parents. Such families were too unimportant to make alliances, and the nuclear household structure, as we have seen, tended to put children out on their own by their late teen years. Even so, this did not lead to love marriages. Marriages were just too important economically for other considerations to enter in very much. The man who wanted to be a property owner running his own farm or small business had to be married, and he chose a wife who could help him with the work, oversee the servants and apprentices, and, if possible, bring in some money or property to get started. The woman had even less choice in the matter. She needed a husband to have an economic establishment in which she could share; if she wanted any kind of career or social status, she would achieve it by helping in her husband's trade. After his death, she might even be admitted to his guild (as a weaver, butcher, chandler, smith, or in many other trades [Power 1975, 53-75]) and run his business.

The fact that people did not live long contributed to a somewhat mercenary attitude toward marriage. Because men were usually older than their wives (though not necessarily by much), many women were left widowed. If a widow inherited a substantial business or farm and her children were still too young to run it, she would very likely remarry in order to have a man to help work the enterprise. On the other hand, the life expectancy of women was shorter than that of men, mainly because many women died in childbirth. When this happened, a widower married again as fast as he could, because a wife was needed not for sentimental reasons but to help run the household business.

The same kinds of economic pressures affected marriage far down into the peasant and working classes. Peasant children were especially likely to be hired as servants during their young adulthood and only married when they could afford to acquire a cottage and a little plot of land of their own. Getting married depended on finding a woman and a man who could coordinate this economic arrangement as the opportunities arose. This pattern was especially found in England. In countries where extended families dominated, family pressures regarding the marriages of children were so strong as to exclude even the English pattern of individual choice. In England and northwest Europe and in the American colonies, there was individual freedom to choose, but one had to choose primarily on an economic basis. In eastern and southern Europe and Asia, the choice was in the hands of the family rather than the individual. In both places, when economic times were bad, many people at the bottom of the class structure were condemned to live their lives single.

The Cult of Courtly Love

The so-called High Middle Ages (1100s) saw the rise of a new ideal of male-female relations. This was the cult of **courtly love.** It did not exist across all ranks of medieval society but was confined at first to a small section of the aristocracy, especially in France.

The main carriers of this new ideal of love were the troubadours (Hauser 1951, 202–31; de Rougemont 1956). These were minstrels who composed poems and accompanied their own singing, but unlike other entertainers at the courts of the nobles, the troubadours were not lower class but were themselves knights. The first famous troubadour was William IX, Count of Poitiers in western France. The stylized songs were always about the same subject: the poet's love for an aristocratic lady. The lady is highly idealized, superior, and inaccessible; the lover continually complains and suffers, while the lady always says no.

This manner of love is very different from that depicted in ancient Mediterranean and Asian literature, in which the lover's demands are pure and simple passion for physical possession, which is soon enough accomplished. Here, for the first time, we find love that exalts the woman in a spiritual sense, and a main part of her charm is the fact that she is unattainable.

The troubadour's love ideal was very much a part of the system of feudalism. The lover swears fealty to his lady, in exactly the same way in which a knight becomes the bound vassal of his military lord. In real life, too, knights at this time began the custom of wearing a lady's token (a scarf or other garment) when they fought in a tournament or battle. Love was made a part of the political hierarchy of the day, but in fact, this jump was not so very large. "Love" in medieval society originally meant not a male-female relationship but the fealty between lord and follower. Kissing was originally an emblem of loyalty: a new knight knelt to kiss his master's ring, and we are a little surprised to read in medieval chronicles that a king would kiss his knights on the cheeks when they met after a harrowing escape or a victorious episode in battle. Some of the flavor of that custom still exists in European countries, where the official who bestows a medal will give the recipient, male or not, the ritual kisses on both cheeks. Similarly, Shakespeare's sonnets sometimes seem to avow his love for a man, although this did not indicate a homosexual relationship but simply the polite way of addressing one's patron. We ourselves continue to speak of love in this nonsexual sense in referring to the love that we feel (or are required by custom to feel) toward our parents and children.

The medieval knight was thus doing something unprecedented when he vowed subordination and fealty to a woman. Such love was vowed only to aristocratic women, preferably of the highest rank. The same authors who speak of the rules of courtly love make it clear that low-ranking people cannot feel or be the object of love; if a knight feels a physical passion for a peasant woman, he is advised simply to rape her because she cannot possibly appreciate his poetic sentiments. The cult of courtly love was part of the status of the aristocratic knights. It was more than a literary convention, for it was just at this time that the crude medieval warriors were beginning to live at a higher standard of comfort, and codes for polite conduct were being established. Knights had originally been recruited from the peasant class based on their qualities as brutal fighters, but during the 1100s, the knighthood began to establish itself as a hereditary aristocracy. The requirements of courtly love and the courtesies of chivalry that went along with it were ways that these newly established aristocrats tried to distance themselves culturally from the crude soldiers they had so recently been. Bowing and deferring politely to ladies was one of the marks of their new refinement. Whatever its motive, courtly love began a new era in Western society and marked the first time that women had deference paid to them precisely because they were women.

Courtly love was not entirely platonic, however. In the romances that became popular at this time, the predominant theme is adultery. The most popular romance, which set the pattern for the others, is the story of Tristan and Iseult. Sir Tristan is a knight who is sent by his king to negotiate for the hand of a neighboring princess and to bring her home to be his queen. On the way they fall in love and consummate their passion. The rest of the story concerns their further adventures as the king finds out about their affair, while Tristan and Iseult try simultaneously to keep their loyal vows both to their lord and to each other. (A similar theme, you may remember, can be found in the German tale of Brunhild and Siegfried, quoted in chapter 2.)

The adultery could be real enough. The troubadours and other knights were often attending ladies in their castles while their husbands were away fighting. The noble lady might be left in charge at home, which meant in military charge as well, because the home was fortified and garrisoned by knights. So the oaths of fealty in the courtly poetry might well be serious. Moreover, the high-ranking lords and their ladies had not married for love and had not developed a personal attachment to each other; often enough, they had been negotiated for by diplomatic missions exactly like that which begins the story of Sir Tristan and Queen Iseult. Poetic love affairs were a game to pass the time at these courts, nicely fitted inside the status hierarchy of the current military and domestic situation (*A Closer Look* 4.1).

The more open avowals of the cult of love were rather daring and only existed where the political situation permitted. They depended on a decentralized feudal system in which military power rested on the armed knights. Wherever possible, the kings attempted to gain personal control by replacing the knights with paid armies of nonnoble soldiers. This became increasingly possible as guns were introduced into warfare in the 1400s and the following centuries. Military chivalry gradually disappeared, but the courtly practices of politeness and idealization of upper-class women remained. In fact, the rise of professional armies meant that the nobility often had less to do and therefore spent their time in idleness at the courts (Dickens 1977). Hundreds or even thousands of people might live in a king's palace, like the one Louis XIV built at Versailles. To pass the time, card games had been invented in the 1400s, along with gambling, hunting for the men, lawn games like croquet and tennis for the ladies—and love affairs, as often as not adulterous. Courts acquired a reputation for dissoluteness, which sometimes caused their political downfall. The complaints of the English Puritans against the sexual practices of the courts of James I and Charles I helped fuel the revolution that cut off Charles' head in 1649; and the court of the French kings,

The cult of courtly love, which appeared in Europe in the twelfth century, exalted women in the spiritual sense, but love still had little to do with marriage. For aristocrats, marriages were arranged to cement alliances and to insure prestige and wealth. For peasants, marriage was a matter of economic necessity. (Scene of Courtly Life, anonymous. France, ca. 1490. Philadelphia Museum of Art. Purchased: Subscription and Museum Funds.)

with their official mistresses and hedonistic atmosphere, was one of the factors leading up to the execution of Louis XVI on the guillotine in 1793 during the French Revolution.

This type of courtly love was very much confined to the upper classes. It was neither an ideal nor a reality for the bulk of the population. The peasants

A CLOSER LOOK 4.1
Medieval Courts of Love

In the courts of medieval France, England, and Spain, love was a game with elaborate rules. Courts were convened, modeled on law courts, to decide such questions as: Can true love exist between married persons? The answer, handed down by the comtesse of Champagne (near Paris) at a Court of Love held in 1174, was no; true love can only exist when the lovers are under no constraint, whereas married people are bound by duty. Another favorite question was: Is marriage a sufficient excuse to refuse a lover? The answer again was no; the rule was set out with concurrence of the assembled ladies of the court, with the admonition that to disobey the law of love was to incur disgrace before every woman of gentle birth.

The laws of love, published by André, the chaplain to the King of France, went on to detail such judgments and to lay out a series of laws. They included the following points:

1. Love could only take place in the aristocratic class.

2. Multiple affairs could be carried on at once, and jealousy was part of the game.

3. Affairs should be kept secret.

4. Lovers were to be emotionally obsessed with seeing each other, adoring tokens of their clothing, and so forth.

5. No pleasures were to be taken by force.

6. A woman should refuse nothing to her lover, including the "most intimate embraces."

7. Most emphatically, love did not lead to marriage and was not to be constrained by any economic motives. There was, however, a kind of pseudomarriage, in that if a lover died the remaining lover must remain faithful to his or her memory for two years.

Courtly love, although openly spoken of in many places, was nevertheless risqué. It was scandalous in the eyes of the serious church (although it should be kept in mind that religion for many people at this time, including many of the clergy themselves, like André, the king's chaplain, was more of a formality than a personal moral code). Of course, the real or even the literary adulteries undermined the sexual property rights of powerful lords. This is what makes the medieval courts of love unique in world history, for in other agrarian states, upper-class men enforced a very rigid dual sexual standard and kept their women locked up in harems and women's quarters.

What made this cult possible in western Europe at this time was the political situation. Kings were not very powerful because they depended on the military forces raised by their noblemen. France itself was split into a network of little feudal states, which shifted alliances constantly. The king of England at various times in the 1100s held more than half of France, although never very firmly. This situation gave the local nobles room to maneuver, especially through dynastic marriages. This in turn gave noblewomen a great deal of importance. Eleanor of Aquitaine (1122–1204) was the heiress to the territory of southwest France; she was married first (at age fifteen) to the king of France but later divorced him and took Aquitaine into alliance with England by marrying King Henry II. It was her own daughter, Mary Comtesse of Champagne, who held the famous Court of Love in 1174. Another daughter became queen of Spain, while a granddaughter became queen of France. This reliance on intense marriage politics gave aristocratic women comparatively great freedom and set the stage for the cult of courtly love.

and the small craftsmen not only continued to base their marriages on practical economic necessities but were also often forced to maintain sexual continence. A woman simply could not afford to become pregnant until she was married (or just before), and illegitimacy in the bulk of the population was kept under strong controls until the rural family structure began to break down in the 1700s and 1800s (Laslett 1971, 1977). Morals were fairly rigidly enforced in small rural communities, where church attendance was not only compulsory but easy to check up on, unlike in the cities and the courts. Throughout the medieval period, then, there were two sharply distinguished sexual cultures: one for the aristocracy, and another for the rest of the popu-

A CLOSER LOOK 4.2
A Victorian Love Poem

How do I love thee? Let me count the ways.
I love thee to the depth and breadth and height
My soul can reach, when feeling out of sight
For the ends of Being and ideal Grace.
I love thee to the level of everyday's
Most quiet need, by sun and candle-light.
I love thee freely, as men strive for Right;
I love thee purely, as they turn from Praise.
I love thee with the passion put to use
In my old griefs, and with my childhood's faith.
I love thee with a love I seemed to lose
With my lost saints—I love thee with the breath,
Smiles, tears, of all my life!—and, if God choose,
I shall but love thee better after death.

Elizabeth Barrett Browning, *Sonnets from
the Portuguese*, ca. 1830

lation. Only later would the two cultures come together in the modern ideal of love.

The End of Patrimonial Households

In the 1600s and 1700s, the structure of medieval society began to change. The patrimonial household began to give way to the modern home (Girouard 1980). The major difference between the two was the emergence of privacy. The medieval house, we may recall, was a public place as well as a dwelling. The house of a king or high lord was also the seat of government. The hundreds of knights and courtiers congregated there because it was their place of employment: the place where law courts were held, orders given, taxes collected, and all the other business of government carried out. It was also a fortified castle, serving among other things as a military barracks for its defenders.

Two of the main reasons for the decline of the patrimonial household were the rise of the state bureaucracy and the advent of the paid professional army (Stone 1977). Medieval lords had a few hundred or at most thousands of troops, whereas the kings of the 1600s had armies of foot soldiers numbering in the tens or hundreds of thousands and large numbers of government officials spread out over their domains. Government was separated from the aristocrat's home. Similarly, at the lower level of small businesses (although happening a hundred or so years later than the rise of the bureaucratic state), enterprises began to move into separate factories or office buildings. For the first time, the private home, reserved for the family alone, began to appear.

The Rise of Private Middle-Class Marriage Markets

Along with this development came a similar change in the nature of marriage. Marriages became more of a private affair between the couple themselves, less of a political or business arrangement (Macfarlane 1986). At the aristocratic top of society, dynastic marriages became less important with the rise of bureaucratic government and the gradual loss of royal power. The great national revolutions in particular, such as the English revolutions of 1642 and 1688 and the French Revolution of 1789, transferred power to parliaments or other strictly national bodies of politicians and officials. It was no longer possible that a whole state might pass into the possession of a foreign power merely by marriage. (This is so even though royalty today still marries, for the most part, foreign royalty. Thus, the present royal family in England is descended from the royal houses of Greece and Germany; but because the English monarchs are figureheads, this has no effect on international relations.)

At the lower level, too, marriages became less a matter of practical economic necessity. As business moved out of the home, the work of wives (at least in the middle class) became less crucial economically. As the private domestic sphere emerged, women found themselves confined within it. At least on men's part, the incentive for getting married became less an economic one and more a matter of acquiring a sexual and domestic partner. The modern marriage market began to emerge, and with it came the ideal of love. But this was a new and different ideal of love than that found in previous societies. It was not simply the erotic passion of ancient times, which went along with the dual standard of male-dominated patrimonial households. Nor was it the courtly love of the aristocracy, which was a pastime with adulterous overtones. For the first time, the ideal emerged that love was a mutual sentimental bond between a man and a woman, a relationship of caring that was supposed to last a lifetime and was connected to marriage. Love was no longer extramarital but at the very core of the marital relationship (Flandrin 1979; Shorter 1975).

This **sentimental ideal of love** as a marital bond arose as part of the new marriage market. Marriages were no longer held together by the larger political and economic structure; therefore, whatever bond there might be had to come from within. The basis for this bond, as suggested elsewhere (Collins 1971), comes above all from the motives of women in their marriage-market situation. Although men now needed little from marriage besides sex and domestic help, women still found themselves economically dependent on a husband. To ensure their own economic well-being, it was important to attach men to themselves personally and with a strong and lifetime tie if possible. This is what the new love ideal did. It lifted love from the status of a game, a sideline amusement that the wealthy aristocracy could afford to play, and made it the emotional insurance that kept a woman and her loyal breadwinner tied together "until death do us part." To do so, it was necessary to evolve a new attitude toward sex, to keep it highly connected with sentimental love and confined to marriage.

THE VICTORIAN REVOLUTION AND THE SEXUAL DOUBLE STANDARD

Besides the new emphasis on marriages for love, the transition to the modern family involved a sharp conflict over sexual practices. The older **sexual double standard** prevalent in patrimonial societies was challenged by a new, more equal but puritanical sexual standard, which strongly confined sex to the married couple. The earlier practice enforced ideals of premarital virginity and chastity upon women but left men free to have concubines, slave women, and courtesans (if they could afford them) and even to practice rape against lower-class women. After considerable struggle, the ideal was established that both

A CLOSER LOOK 4.3
Two Famous Fictional Affairs

Samuel Richardson's *Clarissa* (1749) was the most popular novel of the eighteenth century. It tells the story of Clarissa Harlowe, the daughter of a family attempting to climb into the ranks of the nobility. Her father wants her to marry a wealthy but unattractive boor named Soames, while she prefers a rakish young nobleman, Lovelace. In defiance of her family's wishes, she runs off with Lovelace, hoping to reform him by her own example of sexual purity. Lovelace tries all his wiles to make her yield to him and finally has to resort to raping her while she is drugged with opiates. Having won his conquest, he offers to marry her, but Clarissa declares she has been eternally dishonored, falls sick, and dies. Lovelace thereafter repents and seeks his own death in a duel.

Richardson expresses an early version of the idealization of women and the fight against the double standard, themes that were to become dominant in the Victorian era. The novel is not yet full-fledged Victorian; although Richardson is very moralistic, he nevertheless is willing to talk openly about an explicit sexual theme. In the next century, the Victorians would present the sentiments but censor any reference to sexual issues.

In contrast to the middle-class center of moral gravity in the English novel, Pierre-Ambroise-François Choderlos de Laclos's *Les Liaisons Dangereuses* (1782) reflects the courtly love games of the French aristocracy at its most decadent. It tells of a plot by two sophisticated and amoral aristocrats, the Vicomte de Valmont and his female friend the Marquise de Merteuil, to arrange the seduction of a naive young girl just out of convent school, Cecile Volanges. Their motives are partly revenge against the girl's mother for a previous slight and partly sheer diabolical interest in entertaining themselves through their leisure hours. The plot goes off badly, and the characters all come to a bad end; but unlike Richardson, Choderlos de

De Laclos's novel, Les Liaisons Dangereuses, *reflects the courtly love games of the French aristocracy at its most decadent. In this still photo from the 1988 film version,* Dangerous Liaisons, *the scheming Marquise de Merteuil (Glenn Close) and her co-conspirator, the Vicomte de Valmont (John Malkovich), plot a virtuous woman's seduction.*

Laclos presents a cynical and completely idealized portrait of both villains and their victims. The novel caused a sensation in France. It was condemned as immoral by official opinion, but the first edition was sold out within days, and subsequent editions were eagerly snapped up. Parisian ladies retired behind locked doors to read it. After the French Revolution a few years later, a copy was found in the library of the executed Marie Antoinette herself. Choderlos de Laclos, an army officer with time on his hands, wrote no more novels. He ended up as a general in Napoleon's army. In the nineteenth century, when Victorianism hit France, his book was banned and condemned to be destroyed as "dangerous." In 1988, it became a popular film, *Dangerous Liaisons.*

men and women ought to adhere to the same sexual standard—and that the female ideal of marital fidelity should be the standard. Needless to say, the official ideal was often violated. The Victorian period, during which this ideal was most strongly upheld in public, was also a time when men frequently consorted with prostitutes and mistresses. Some Victorian women pursued sexual adventures as well. It would be more realistic to say that the official sexual standard did not come in without a struggle. Various forces opposed it or favored it at various times. No sooner did official puritanism reach its height than it began to give way to a sharp reaction, culminating in the sexual permissiveness of the twentieth century.

This whole process may be referred to as the **Victorian revolution,** because it was during the reign of Queen Victoria in England (1837–1901) that sexual prudishness reached its height. Absolute propriety was the imperative for all respectable people. The sight of a woman's ankle was considered extremely risqué; it was even considered improper to use the word *leg* in reference to a woman. Nevertheless, this "Victorianism" goes back much further into European history. The rules of sexual behavior that Victorianism strove to enforce were reflected in ancient Christianity and were especially heavily touted during the High Middle Ages (A.D. 1000–1400). Even during Victoria's own lifetime, the high point of prudery was the earlier part of the century (around 1800–1870), while in her old age, even London high society was rent by numerous scandals of adultery and homosexuality, and her own son the Crown Prince Edward was the center of a notorious "fast crowd."

The "Victorian revolution," then, is only a metaphor for something larger and more complex. Just as love has a history going back before the "love revolution" of early modern times, the battle over sexual standards seen in "Victorianism" goes back to the 1700s, with roots even earlier. It was with the rise of the private household and the romanticized marriage market that sexual prudery became prominent. Already in the 1700s, Englishmen of the middle and upper classes were referring to women as "the delicate sex" and attempting to keep "improper" topics from their ears (Watt 1957). In 1818, the English physician Thomas Bowdler brought out a censored edition of Shakespeare because the bawdy talk of the Elizabethan era was now considered too obscene for women and children to read. By the mid-1800s, as the new privatized family and the sentimentalized conception of women had spread widely throughout society, sexual prudery had reached an extreme. Even references to pregnancy and childbirth were considered obscene, to be avoided especially by women themselves.

> *A few days before a baby calf had been born and I had seen it . . . but then my father and mother forced me to keep out of sight of the field where the mother and calf were, and where I had been a few moments before. The thing I had seen I dared not talk about or ask about without "deservin' to have my ears boxed."*
>
> *Even when my little brother was about to be born, we children were hurried off to another farmhouse, and secrecy and shame settled like a clammy rag over everything. At sunset, a woman, speaking with much forced joy and a tone of mystery, asked us if we wanted a little brother. It seems a stork had brought him (Wertz and Wertz 1977, 79).*

Battles Against the Double Standard

The extreme prudery of this period capped off the separation that had been emerging between male and female spheres. Men and women were to make no suggestive allusions in each other's company, nor were women, as the "purer sex," to discuss such subjects among themselves. Sex was left in a male backstage of private clubs (for the upper class), saloons (for the middle and lower), smoking parties, hunting trips, and the "sporting world" of the theater and prostitutes (Chesney 1970).

Prostitution had existed, of course, since ancient times. Medieval European lords did not have official concubines or practice polygamy like their counterparts in Moslem, Hindu, and Chinese societies (there are exceptions: Charlemagne, who reigned early in the Middle Ages, ca. A.D. 800, had four wives). However, they did allow themselves considerable license to have mistresses. Some of these, such as Louis XIV's favorite at Versailles, Madame de Maintenon, or Louis XV's Madame de Pompadour, had official titles and residences and were persons of greatest importance in court politics. Prostitution existed quite openly in ports and military towns for the sailors and soldiers. The acting profession was considered very close to professional prostitution. In India and China, female entertainers were identical with prostitutes and often were admired for their cultivation and admitted into high society—among men only, of course. All this continued in the Victorian period, only now it went underground. Cities like London and Paris had huge districts where prostitutes walked the streets, as well as hotels and rooming houses ranging from squalid dives to luxurious houses of prostitution for the upper classes (Chesney 1970; Marcus 1964).

A battle went on between the male and the female spheres or, rather, between the cult of domestic purity on the one hand and the male backstage with its "bad girls" and "fallen women" on the other. Some of the moves in this battle were legal (see the following discussion on the feminist movement), but much of it was waged in the moral tones of everyday life. Women, at least in the middle class, worked hard at making men behave "decently," and a central tactic was to desexualize proper conversation entirely. Nancy Cott (1978) calls this the tactic of "passionlessness," which women used to remove themselves as much as possible from being subject to men's sexual desires. "The belief that women lacked carnal motivation was the cornerstone of the argument for women's moral superiority, used to enhance women's status and widen their opportunities in the nineteenth century" (Cott 1978, 233) . Barbara Welter (1966, 152) refers to this as "the **Cult of True Womanhood,**" characterized by "piety, purity, submissiveness, and domesticity. Put them all together and they spelled mother, daughter, sister, wife—woman." The one thing that they did not spell was sex object.

One can also say that women now had to be especially careful to control sex in the new individual marriage market. They could no longer count on being married off by their parents or making an easy match based on their economic potential as workers. Instead, they expected to have someone fall in love with them and that because of purely personal attractiveness. Sexual prudishness had the effect of confining sex within proper marriage bonds. At the same time, there was the widespread sentiment that love led to marriage. One can easily interpret the feelings of love that large numbers of people were beginning to experience for the first time in history as the form that sexual passion took when it had to pass through a filter of refinement and idealization. Because true love is forever (another point in contradiction to courtly love), it neatly adds up to a lifetime marriage vow.

By the early 1800s, this doctrine was strongly in place. It had become the formal moral code of society, even though the old courtly love game might still be played in the salons of the wealthy (Stendhal 1967), and the workers and peasants still lived for the most part in the traditional hard-fisted business-enterprise family. The history of the next hundred years was to include the gradual spreading out of the middle-class ideal until it encompassed virtually all of society. The "Victorian revolution" that it represented was the result of a conflict over male and female status. In one respect, it was an important historical victory for women because it raised their status (at least officially) to a very idealized level. At the same time, it guaranteed most middle-class women the economic support of a marriage at a time when they were excluded from independent careers of their own. The price that was paid was to confine women very strictly to the domestic realm and to make this realm almost the polar opposite of the male world outside.

THE FIRST WAVE OF FEMINISM

Until the end of the nineteenth century, women were distinctly second-class citizens. In England, the United States, and elsewhere, women had virtually no legal rights. The husband was entitled to collect his wife's wages, if she worked, and to dispose of her property, if she had any when they were married. He was the head of the household and represented his wife in all public and legal matters. He could decide how their children were to be educated and in what religion they were to be brought up. A dying husband could will his children, even unborn, to other guardians; in the case of divorce (not easy to obtain in those days) he had control of the children. A wife was the humble subordinate of her husband; if she disobeyed him, he had the legal right to chastise her physically and could even hand her over to the law for punishment.

Similar restrictions held in public life. Women were not allowed to preach in church (except in radical sects like the Quakers). They could not vote (although in England it is true that men of the lower social classes could not vote either until 1884, and in the early days of the United States, various states restricted the vote to property-owning males). Even speaking in public was a scandal for a woman.

It was this segregation that first gave rise to the organized women's movement (Chafetz and Dworkin 1986; Flexner 1959; Sinclair 1965; O'Neill 1970). The mid–nineteenth century was a period of liberal social reform, and the most important of all the moral crusades was the campaign to abolish slavery. In the 1830s there were hundreds of antislavery societies in the United States, a large number of them separate organizations of women. Women made up more than half the signatories of the huge petitions periodically sent to Congress on the slavery question. In the 1830s, Angelina and Sarah Grimke, two wealthy South Carolina sisters, were among the first women to speak in public. Their theme was the abolitionist issue, but the furor roused against their public participation (especially among the conservative New England preachers who supported abolition) led them to defend the rights of women.

The Grimke sisters were eventually persuaded to drop the feminist issue lest it compromise antislavery. It was a type of political bind that was to plague the feminist movement continuously. However, not all women reformers were willing to give in. In 1840, when the World Anti-Slavery Convention was held in London, the two leading American women abolitionists, Lucretia Mott and Elizabeth Cady Stanton, were excluded from the meeting and had to sit in the spectators' gallery behind a screen. To his credit, the great abolitionist orator William Lloyd Garrison furiously withdrew and sat with the women. Out of this incident was organized the first women's rights organization, which held its opening convention in Seneca Falls, New York (the home of Elizabeth Cady Stanton), in 1848. Its members passed a resolution calling for women's suffrage, though only narrowly: votes for women were regarded as so extreme a demand that it might compromise everything else.

Nevertheless, the movement began to spread. At first the women's movement largely relied on the fervor of larger reform campaigns. In America, most of the early feminists were abolitionists. Others, like Susan B. Anthony, came to feminism from the temperance movement, where women at first were arrogantly subordinated to male crusaders against alcohol (mostly ministers). Even more women were involved in missionary associations, drives to organize philanthropy for "the deserving poor," and the very popular crusade to suppress vice and reform prostitutes. These were all middle-class movements; a smaller number of women were involved in efforts to improve the conditions of the working class, by outlawing child labor and establishing protective legislation for women to ensure minimum wages and maximum working hours.

These different reform crusades had varying fates. Slavery was abolished in the United States in 1863 (having been stopped in England in 1807). The temperance movement, fueled by women's votes in some states where local suffrage existed, succeeded in having a constitutional amendment passed in 1919, although Prohibition was repealed again in 1933. Protective labor legislation for women and children was finally achieved early in the twentieth century, although surrounded by controversy (some of the more radical feminists declared that it restricted the chances of women to work). Charity organizations, which were relatively uncontroversial, became firmly established.

The politically most popular crusades were those for the suppression of vice. In the 1870s and 1880s in particular, widespread campaigns cracked down on prostitution, which had been officially condoned especially in large cities. In the United States, these movements were connected with what later became called Progressivism, which was especially concerned with overturning "boss rule" in the urban immigrant communities. A disproportionate number of the immigrants were men, and they tended to come from countries with patriarchal traditions and a definite double sexual standard. The political and social centers of the immigrant communities were usually the saloons, which was one reason why these were special targets for the Anglo-American women of the temperance crusade. The various movements tended to overlap; the WCTU (Women's Christian Temperance Union) had its own section for "Social Purity" and another devoted to eradicating obscene literature, while all of these movements generally supported votes for women as a means of implementing their political programs.

In England, similar movements developed, although with some differences: temperance was never a very strong movement in England, and working-class reforms were much more strongly backed. Slowly some of the women's legal disabilities were removed. In 1857, a bill gave women some rights of divorce; in 1884 Parliament gave them the right to their own earnings and abolished the penalty of imprisonment for women who denied their husband his conjugal sexual rights. In 1869–1870, women property owners were allowed to vote in municipal elections and to serve on school boards.

In the United States, the territory of Wyoming in 1869 gave the vote to women, partly as an effort to attract them to this sparsely settled territory; when it became a state in 1890, it was the first state whose women could vote in national elections. A few years later, Utah, Colorado, and Idaho followed suit. But progress was hard and slow, and many men worried that women's increasing independence would undermine society (see *A Closer Look* 4.4). More than 480 campaigns were waged before 1910 in different American states to have the issue of the female franchise brought up, but only a tiny proportion were successful. Just after the Civil War, when the vote was guaranteed for Negro men by a constitutional amendment, the demand of women to be included was turned down by the abolitionist leaders, who argued it would jeopardize the passage of the amendment.

In England, too, there was very slow progress. In the 1860s, liberals like John Stuart Mill unsuccessfully introduced a women's suffrage bill into Parliament. In 1869, Lady Amberly (who was to be the mother of the philosopher and peace-movement leader Bertrand Russell) caused a furor by breaking the taboo on women speaking in public when she addressed a suffrage meeting. Queen Victoria was so outraged by this breach of decorum that she declared that Lady Amberly ought to be horsewhipped (O'Neill 1970, 31). Although some prominent political

A CLOSER LOOK 4.4
Women's Independence and the Death of Romantic Love, Circa 1899

A hundred years ago, social commentators worried that women's increasing independence might signal the end of romantic love.

[T]he greater securing of divorces, the growing independence of women and their disinclination to domesticity—are undermining that family life which civilization has so slowly and laboriously built up. . . . In recent years the notion that family life is not good enough for women, and that they should be brought up in a spirit of manly independence, has come over society like a noxious epidemic. It is quite proper that there should be avenues of employment for women who have no one to support them; but it is a grievous error to extend this to women in general, to give them the education, tastes, habits, sports, and politics of the men. It antagonizes that sexual differentiation of the more refined sort on which romantic love depends and tempts men to seek amusement in ephemeral, shallow amours. In plain English, while there are many charming exceptions, the growing masculinity of girls is the main reason why so many of them remain unmarried; thus fulfilling the prediction: "Could we make her as the man, sweet love were slain." Let girls return to their domestic sphere, make themselves as delightfully feminine as possible, not trying to be gnarled oaks but lovely vines clinging around them, and the sturdy oaks will joyously extend their love and protection to them amid all the storms of life (820–21).

Cultural ideals at the turn of the century made sharp distinctions between men and women and placed women on a pedestal. According to authors like Finck, separate gender spheres and romantic love ideals of "civilized" societies in Europe and America are beneficial for both women and men: *"man's adoration of woman as a superior being . . . creates an ideal which has improved women by making them ambitious to live up to it. No one who has read . . . [about] the treatment of women before romantic love existed, can fail to recognize the wonderful transformation brought about by gallantry and self-sacrifice—altruistic habits which have changed men from ruffians to gentlemen"* (822).

How similar are these arguments to gender and family values debates of today?

Source: Henry T. Finck. 1899. *Primitive Love and Love Stories.* New York: Charles Scribner's Sons. pp. 820–22.

leaders and intellectuals supported women's suffrage, the political parties were either opposed to it or regarded it as an issue that could be sacrificed to more important things. Benjamin Disraeli, the Conservative prime minister, favored suffrage, perhaps out of his perception that women would likely support the Conservative party. The Liberal party, on the other hand, although it tended philosophically to believe in equal rights, dragged its feet because it felt women's votes would go to the Conservatives.

At the turn of the twentieth century, the pressure finally began to build. New Zealand in 1893 became the first country to give women the vote, and Australia, Sweden, Norway, and Finland soon followed suit. In the United States, the growing wave of belief that non-Anglo immigrants were swamping the country gave support to the idea that native women's votes could help turn back the tide. World War I brought the goal of women's suffrage in sight at last. The process here was much more peaceful than in England (*A Closer Look* 4.5), although in 1913 a group of women protesting at the inauguration of Woodrow Wilson was attacked by a mob, and in 1917 women pickets at the White House were jailed and abused. In fact, the movement had been building momentum just before the war. In 1910 the state of Washington passed a women's suffrage bill, followed by narrow victories in referendums in California, Arizona, Kansas, and Oregon. Suffrage was winning the West, although this was followed by a reaction in the East as Ohio, Michigan, and Wisconsin defeated suffrage bills; further defeats followed in 1915 in New York, Pennsylvania, Massachusetts, and New Jersey. These were the states with large immi-

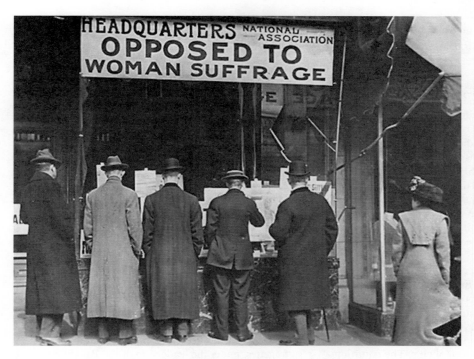

Because the women's suffrage and anti-immigration movements became aligned, eastern states with large immigrant populations tended to be antisuffrage. Their organized resistance to women receiving the vote helped defeat many state suffrage bills. Western states, however, were pro-immigration and also hoped to attract women settlers.

grant populations, which tended to be antisuffrage. But the suffrage movement was large and well financed, and it kept up its pressure. The confrontations at the White House in 1917 brought the matter to an emotional crisis point. In 1918 President Wilson called for women's suffrage as a war measure, and by 1920 the thirty-sixth state legislature had approved the constitutional amendment.

Elsewhere around the world, progress was much slower. Although German women received the vote in 1919 with the new Weimar constitution that followed the overthrow of the Kaiser, the civil code continued to give husbands control over family place of residence, behavior of children, and most economic issues in the home. In France, women did not receive the vote until the late 1940s (Tilly 1981). This was despite the very notable participation of French women in the war effort in World War I; it is apparent that the war provided an emotional opportunity to get the franchise through in England and the United States, but that it could not have been done without the strong pressure of an organized movement. In Switzerland, women could not vote until the 1960s.

After the winning of the franchise in England and the United States in 1918–1920, the feminist movement in both countries collapsed. Despite great expectations, few significant changes followed. As it turned out, women voted more or less the same way as their husbands, and neither the predicted era of

A CLOSER LOOK 4.5
Radical Suffragettes in England

In England, the movement for **women's suffrage** seemed to have come almost to a standstill in the early twentieth century. After decades of legislative rebuffs at the hands of both the Conservative and Liberal parties, a radical group of suffragettes was formed in 1903. This was the Women's Social and Political Union (WSPU), led by Mrs. Emmeline Pankhurst. She was described as a strikingly beautiful woman; perhaps this asset, together with the fact that she was the widow of a former member of parliament, gave her the courage to challenge the male establishment as it had never been challenged before.

Together with her daughters Christabel, Sylvia, and Adela, Emmeline Pankhurst advocated a militant approach. In 1905 Christabel and other women were arrested for interrupting a political meeting; by 1907, the suffragettes' marches were being broken up by the police and by violent attacks by male opponents. The confrontations attracted widespread support and sympathy for the WSPU, including huge financial contributions from wealthy supporters and thousands of new demonstrators. As the persecution became more severe, the women escalated their tactics. They were being arrested and sentenced to months in jail for speaking at illegal rallies; in response they picketed and chained themselves to public buildings to keep from being hauled away. The Liberal government, which had taken over from the Conservatives in 1906, was embarrassed, because its own principles called for women's suffrage, but its reaction was primarily to impose yet longer jail sentences. The confrontation reached its climax in 1912–1914, when hundreds of militant suffragettes were arrested for smashing windows in a demonstration in downtown London, during which the house of the Liberal cabinet minister Lloyd George was burned down. Christabel Pankhurst escaped to Paris, where she established an underground command post. Emmeline Pankhurst, in and out of jail, continued to generate support for her endurance of brutality.

The outbreak of war in 1914 brought the drama to its conclusion. The Liberal government had already been looking for an excuse to grant women's suffrage without appearing to back down. The widespread participation of women in the British war effort would challenge the idea that women's place was only in the home, and the suffragettes, under the leadership of the Pankhursts, called a truce and threw themselves into the war effort. In 1918, British women were finally given the franchise, although in a graduated package that at first gave votes only to women over age thirty, with full equality deferred to 1928. The military tactics of the British feminists were widely criticized, but they set an example that was later copied by the nonviolent movement for Indian independence led by Gandhi in the 1930s and '40s and later by the black civil rights movement in the United States and by peace demonstrators, antinuclear activists, and others.

peace nor the hoped-for nonpartisan reform happened, nor did women's legal conditions improve greatly. Some modest victories had been achieved decades earlier, when women won rights to divorce and to keep their own earnings. However, as late as the 1960s, women in some states were still required by law to take their husband's name and to live in the residence that he chose (which implied that his job legally took precedence over hers); she also had no legal rights over his income, nor a legal voice in spending it, nor a right to independent financial credit (Kanowitz 1969).

Instead, the family as it emerged after the suffrage movement looked very much like a renewed version of the traditional model with its sharp separation of male/female spheres. After some initial inroads into the professions and the opening up of new clerical positions, women made no further occupational progress; in the early 1960s the degree of occupational segregation was about the same as it had been in 1900 (O'Neill 1970, 93). In education, women actually lost ground; in 1920 they accounted for 47 percent of all college students, but by 1950 their proportion had fallen to 30 percent. In advanced degrees, women fell from one of six doctorates to one of every ten. As educational degrees became more important for jobs, men rushed in to snap them up, especially with the aid of the G. I. Bill after World War II. Not until 1970 did women get back to where they had been before the war *(Historical Statistics of the United States* 1965, 384–85).

In the larger view, there were two main reasons for the failure of the first wave of feminism to achieve its goals.

The Victorian Strategy

One weakness was the extent to which the women's movement relied on moralistic arguments to generate support. The movement began as an offshoot of various reform groups, and it gained a good deal of strength from the shame it could impose on Victorian men for the vices they condoned. The campaigns against prostitution were among the most successful, and they held up the image of women as morally superior—"purer" than men. This attitude carried over into sexual matters generally. Thus, although feminists often recognized that unrestricted childbearing was one of the things that kept them tied to the home, nevertheless they generally opposed birth control and joined in the campaign against abortion that swept the United States in the late nineteenth century (chapter 10; also Gordon 1982). The obscenity laws passed in the 1870s cut off information on birth control, besides silencing the few militant feminists who spoke out against the restrictions inherent in puritanism and who advocated free love or other utopian experiments (*A Closer Look* 4.6).

The result was to draw the boundaries even more sharply between men's and women's spheres. The women's campaigns against alcohol, prostitution, and obscene language did not make any of these things go away but rather removed them into an even more sharply delimited "man's world" that was not even supposed to be heard of by their wives at home. Some feminists wanted to limit sex as much as possible not only outside of marriage but in it; in the 1890s they caused a sensation by advocating separate bedrooms for husbands and wives; one even called for separate residences (O'Neill 1970, 41). On the other hand, these women were fighting against a situation of rather extreme sexual exploitation; in the absence of other weapons, one of their stronger tactics was to place restrictions on sex, in conjunction with the love revolution that led to the first great rise in women's status. If it also resulted in an idealization of women in a **separate sphere,** confined to a decorous courtship and then a restricted home life, that was part of the price in this battle of unequal forces.

Neglect of Working-Class Women

The other weakness was that the feminist movement was essentially a movement of the middle class (Chafetz and Dworkin 1986). Working-class women were not much involved. Such women had no need to fight for rights to seek work outside of the home because most of them were forced to work anyway (Roberts 1986). In the countryside, peasant and farm women were used to doing heavy work in the fields. Throughout the early industrial period, they helped their husbands at their trades. Coal miners' wives would drag carts of ore through the mine shafts, and craft workers' wives picked up and delivered heavy loads as intermediaries between their husbands and their masters (Scott and Tilly 1975). With the progress of industrialism, women workers were concentrated in textile factories. Many women worked for other women in private sweatshops as piece

A CLOSER LOOK 4.6
A Free Love Scandal

The most flamboyant of the early American feminists was Victoria Woodhull. A stylish and intellectual woman, she made friends with the robber-baron millionaire Cornelius Vanderbilt and broke into the male business world as a stockbroker. Together with her sister, Tennessee Claflin, she established a magazine in New York City in the late 1860s; it advocated radical causes, including not only women's suffrage but *free love.* Her argument was that marriage was the basis of women's second-class citizenship and that exclusive sexual possession was its main prop. This prop she proposed to overthrow. Her argument should be seen against the background of the various utopian communities that had sprung up in the 1840s and 1850s, including the Oneida Community, which actually practiced group marriage.

For a time, Victoria Woodhull acquired a leading position among New York suffragists, and was the first woman to run for president of the United States. In 1871, she was allowed to address the House Judiciary Committee, the first time that Congress actually recognized female suffrage as a legislative issue. Stirred by her success, in November of the same year she announced from the stage of a public hall in New York City that she believed in free love. She was immediately denounced from all sides, not only by clergymen and the press but by other feminists. Landlords turned her out of lodgings.

Not to be easily disposed of, Woodhull struck back by making public an instance of free love among her respectable compatriots, an affair between Henry Ward Beecher, a famous abolitionist preacher, and Mrs. Elizabeth Tilton, the wife of a liberal editor. Woodhull declared that they, too, were believers in free love but were hypocritical about coming out and supporting her on the subject. The result was a lawsuit by Theodore Tilton, Elizabeth's husband, against Beecher, who was acquitted by a sympathetic jury, although probably guilty of the charge. In the ensuing scandal, Theodore Tilton, Victoria Woodhull, and her sister all had to flee the country. The American women's suffrage movement was smeared by its opponents and for decades thereafter had to live down the taint of free love. The movement's reaction was to emphasize sexual puritanism even more strongly.

workers in the needle trades, in return for miserable wages and a confined place to stay. A large proportion of women workers were servants, living under conditions of patriarchal discipline. Virtually all these women either worked directly under the control of their parents or husbands or else had to give up their wages to their family. The fate of women workers was to be exploited, whether by men or other women.

Where working-class women had any political consciousness, it was generally in connection with a male-dominated labor union or (in Europe) a radical socialist movement. The unions were generally unfavorable to organizing in the women's occupations themselves; instead they joined in the push for protective legislation that would keep women from competing with men. "Some socialist newspapers described the ideal society as one in which 'good socialist wives' would stay at home and care for the health and education of 'good socialist children'" (Scott and Tilly 1975, 63). Middle-class feminists generally did not appreciate the position of working-class women, although one wing of the feminist movement campaigned for social legislation to eliminate child labor and restrict women's hours and regulate their working conditions. The main effect of such laws, where they were actually applied, was to separate women even more into segregated occupational spheres or to take them out of the labor force entirely.

While all this was happening, though, an occupational revolution was coming about that had been anticipated by no one. After 1880, women began to appear in the labor force in large numbers as clerical workers, such as secretaries, typists,

Women clerical workers in a typical pre–World War I office: a large, American mail-order house.

file clerks, bookkeepers, retail clerks, as well as schoolteachers and nurses. These were the first important inroads women had ever made into working in non-manual labor. Previously, the working woman was always a member of the lower classes, working as a domestic, farm laborer, or industrial employee. Now, at last, there were career opportunities open for middle-class women and for working-class women attempting to break into the middle class on their own. There was a price to be paid: occupations that had formerly been held by men and that led upward into higher careers now became sex-typed and segregated. Secretaries and clerks had formerly been men, and their jobs could lead on up into the higher management hierarchy. The new women secretaries had no place to go up; they became an occupational caste restricted by gender and separated from male white-collar workers whose careers could still take them higher. Moreover, those jobs were confined almost exclusively to single women; married women were expected to quit their jobs.

The growth of the female clerical "ghettos" was an ambiguous legacy of the nineteenth century. On the one hand, it institutionalized the Victorian doctrine of **separate spheres** and cut off any further progress that women might make in gaining alternative sources of support for themselves so as to become less reliant on the patriarchal home. On the other hand, it did get middle-class women out of the house, at least for part of their lives, and opened the way to changes in cultural and sexual matters that gradually transformed the ethos of male/female relations. Out of this would eventually come the family changes of the twentieth century.

SUMMARY

1. The traditional family in western Europe and North America never was an extended, multigenerational household, nor did it practice early marriage (except in the highest aristocracy). Men and women tended to leave home in their teens to become apprentices or servants in other households and to marry only in their late twenties when they could afford to start their own homes. By the time their children were grown, the parents were likely to be dead because people on the average did not live much past their forties.

2. In eastern and southern Europe and in Asia, however, traditional families tended not to have servants and had a pattern of older men marrying younger women. These families were more likely to have an extended, multigenerational structure, whereas families in England, northern Europe, and the American colonies tended to be more individualistic.

3. In ancient societies, love was generally regarded as purely spiritual or physical and not at all associated with marriage. Early and medieval Christianity was not favorable even to marriage because its ideal was the celibate priest or monk. In traditional Europe, marriages were arranged either as diplomatic alliances (among the aristocracy) or to procure work partners to run the household business or farm (among the middle class and peasants). In neither case was love a consideration.

4. The first period of idealization of women came in the 1100s in France, when the cult of courtly love developed among the knights and ladies of the feudal castles. Only women of high rank were subjects of this love game, which was always adulterous. The chivalric code had the long-term effect of introducing politeness toward women as part of upper-class culture.

5. In the 1600s and 1700s, the professional army and the bureaucratic state began to displace the armed patrimonial household and to separate work from the private home. Along with this came a new, individually based marriage market, based on personal attractions alone. The new ideal, for the first time in Western history at least, was that marriages should be based on a lifetime bond of personal, mutually sympathetic love. The new love ideal reflected the individual freedom of the marriage market but also the fact that the middle-class woman was not cut off from the economic world that had moved outside the home and had to attach herself permanently to a husband who could support her.

6. During the Victorian era of the nineteenth century (actually it began a little earlier in England and America), society became extremely puritanical about sex, at least in its official standards. This may be interpreted as a struggle by women against the sexual double standard, which had given comparative sexual freedom to men, both in the old patrimonial household and continuing in a Victorian "underground" of widespread prostitution. It also reinforced the new private family by attempting to confine sex strictly to marriage, for both males and females.

7. The first organized feminist movement began as an offshoot of the movement to abolish slavery in the 1840s. It made slow progress but did eventually succeed in giving women certain legal rights, such as the ability to sue for divorce and to keep their own property after marriage, and it gained some protections for women factory workers. The right to vote was not fully won in the United States and England until the political upheavals of World War I, eighty years later. Nevertheless, the winning of suffrage had little effect on the family, which continued to feature a very sharp separation between male and female spheres. The most important source of change at the end of this period was the opening up of white-collar employment to middle-class women for the first time, although in newly sex-segregated positions.

Key Terms

courtly love	love revolution	sexual double standard
cult of true womanhood	nuclear family	Victorian revolution
extended family	sentimental ideal of love	women's suffrage
first wave of feminism	separate spheres	

Sociology Web Site

See the Wadsworth Sociology Resource Center, "Virtual Society," for additional links, quizzes, and learning tools:

http://www.sociology.wadsworth.com

Also on this web site you'll find InfoTrac College Edition, an online library of journals. Here you can search for electronic articles about central topics in sociology.

5 Family Trends: The Twentieth Century and Beyond

INTRODUCTION

Amerian families have recently been going through a period of major change. It has been suggested in some quarters that even more significant changes will occur in the future. Some have gone so far as to predict that the family as we now know it is in the process of disappearing entirely. Others lament recent family changes and talk about recreating older patterns of family life that were popular in the 1950s. The rise of diverse and increasingly fluid family forms has spawned intense controversies over what a family is and what it should be (see *A Closer Look* 1.1). Only by moving beyond the nostalgic political rhetoric of such controversies will we be able to comprehend the sociological forces shaping modern families.

In this chapter, we explore some of the major changes that families in the United States have experienced during the twentieth century. We document shifting patterns of sexuality, marriage, fertility, and divorce, and identify some of the causes of long-run changes. Rapid and widespread modifications in family structure and process occurred during the century, but such changes were largely unanticipated. Few predicted the baby boom of the 1950s, the rise of divorce in the 1970s, or the nonmarital childbirth explosion of the 1980s and 1990s. As we move into a new century, an increasingly diverse range of family types has replaced the breadwinner-homemaker household that once predominated. By analyzing some of the underlying forces behind the rapid rise and fall of isolated suburban nuclear family households in the twentieth century, we can better understand what is happening to American families today.

WHAT IS HAPPENING TO MODERN FAMILIES?

Sex, Marriage, and Divorce

Some important shifts have been taking place, and one in particular has been a big change in sexual behavior. Since the early part of the century, premarital sex has greatly increased (Figure 5.1; chapter 9). Once considered largely taboo, especially for women, sex before marriage is now experienced by a rather considerable majority. One recent national study found that most women born before World War II (55%) were still virgins when they got married, compared with only about one in four men (26%) (Laumann et al. 1994, 214). Since then, fewer than one in three women and about one in five men have been virgins at marriage. Living together before marriage, which not so long ago would have been considered scandalous, at least in polite middle-class society, is now widely taken for granted. In the 1990s, about a third of women of ages twenty-five through thirty-nine lived with a man (cohabited) before their marriage, and about half had an unmarried cohabitation at some time in their lives (National Center for Health Statistics 1997). An interesting note about these trends is that men's and women's social practices and living arrangements have become more similar.

At the same time, marriage rates have fallen, especially for young people. Once it was common for working-class women to marry in their teens and for middle-class women to marry about the time they graduated from college. Now both groups are waiting longer before marrying. Most people now delay marriage until their mid-twenties, and one of four adults is still unmarried past age thirty. As the **marriage rate** has gone down, the **divorce rate** has gone up, although it peaked in the early 1980s (Figures 5.2 and 5.3; chapters 8 and 14).

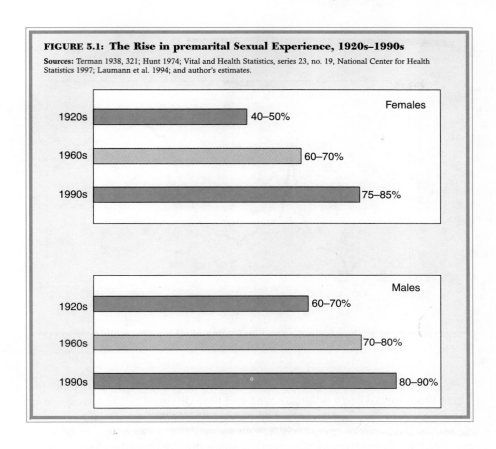

FIGURE 5.1: The Rise in premarital Sexual Experience, 1920s–1990s

Sources: Terman 1938, 321; Hunt 1974; Vital and Health Statistics, series 23, no. 19, National Center for Health Statistics 1997; Laumann et al. 1994; and author's estimates.

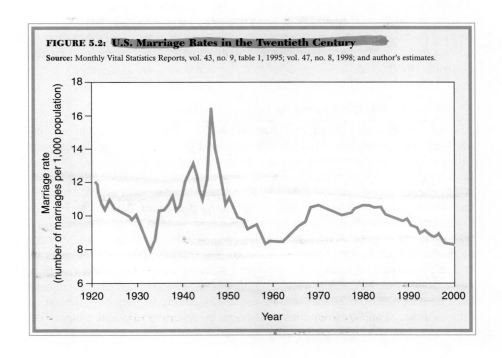

FIGURE 5.2: U.S. Marriage Rates in the Twentieth Century

Source: Monthly Vital Statistics Reports, vol. 43, no. 9, table 1, 1995; vol. 47, no. 8, 1998; and author's estimates.

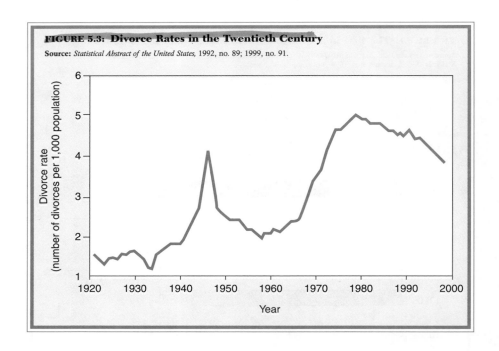

FIGURE 5.3: Divorce Rates in the Twentieth Century

Source: *Statistical Abstract of the United States,* 1992, no. 89; 1999, no. 91.

A generation ago divorces were considered scandalous by many people. There was a time when a politician who had been divorced would have given up hope of public office. Legal restrictions on divorce required people to go through painful public trials to prove extreme cruelty, neglect, or adultery. In the 1960s and 1970s the laws changed, and in many states divorces became much easier to obtain on a "no-fault" basis. However, it would be inaccurate to say that easier divorce laws are the major reason for the increase in divorces. The rising trend in divorces was going on for a long time, and now we are seeing a slow drop-off. During the 1980s and early 1990s, five of ten marriages ended in divorce. In the coming decade, we can expect continued high levels of divorce, perhaps dipping down a bit so that four of ten marriages will dissolve (National Center for Health Statistics 1999).

Nonmarital Childbirth and Fluctuating Birthrates

The combination of a falling marriage rate and a rising divorce rate is one reason why some observers have predicted the end of the family. Such observers can also point to the massive increase in unmarried childbirth rates in the last decades (Figure 5.4, chapter 10). Where once **nonmarital birth** was a relatively minor phenomenon, it has now reached considerable proportions. One of three births now occurs to a woman who is not married. Among African Americans it accounts for two-thirds of all births. Bearing a child without being married was once the most scandalous behavior of all, more so than divorce or premarital intercourse. Now, it is on its way to being accepted as normal, and this is not just happening in the poverty-stricken lower class. Among whites, one of every four births is now a nonmarital birth, and rates for this group have been increasing faster than for others, especially among women over the age of thirty. Although still frowned on by many, an increasing number of women feel that it is a woman's right to bear her own child without having to undergo a con-

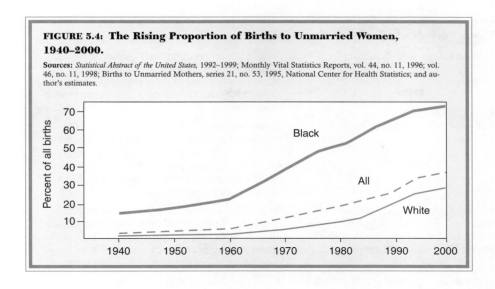

FIGURE 5.4: The Rising Proportion of Births to Unmarried Women, 1940–2000.

Sources: *Statistical Abstract of the United States,* 1992–1999; Monthly Vital Statistics Reports, vol. 44, no. 11, 1996; vol. 46, no. 11, 1998; Births to Unmarried Mothers, series 21, no. 53, 1995, National Center for Health Statistics; and author's estimates.

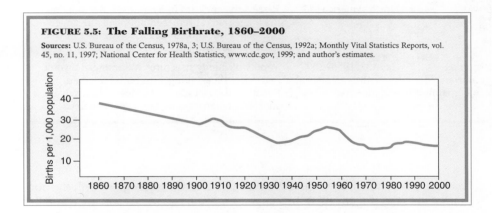

FIGURE 5.5: The Falling Birthrate, 1860–2000

Sources: U.S. Bureau of the Census, 1978a, 3; U.S. Bureau of the Census, 1992a; Monthly Vital Statistics Reports, vol. 45, no. 11, 1997; National Center for Health Statistics, www.cdc.gov, 1999; and author's estimates.

ventional marriage with a man. This is reflected in the behavior of famous people, such as Madonna, who decided to have a baby without having a husband and without planning to marry.

Another important trend has been the shift in the overall birthrate. A few decades ago there was a lot of concern with the problem of overpopulation. At the rate people were reproducing, it looked as if we were heading for a twenty-first century in which the United States would have to support a huge population. This image of wall-to-wall crowds of people has faded, at least for the advanced industrial countries. Birthrates began falling in the late 1950s and by the mid-1970s reached a level at which Americans were not quite replacing themselves. By the early 1990s, the **birthrate** had crept back up to near replacement level (Figure 5.5; chapter 10). Since then, the birthrate has declined slightly.

This trend, in a sense, is the opposite of the trend in nonmarital childbirth. That is, on one side we see more sexual activity spilling over outside of marriage, driving up the nonmarital birthrate. On the other hand, we have fewer children being born in general and a declining birthrate (Figure 5.5). Actually the two trends are not incompatible. One reason we have such a high nonmarital birth ratio (i.e., the ratio between nonmarital births and marital births) is that the marital birthrate

Table 5.1 Some Major Periods of Change in the Family

Human nuclear family emerges.	1–5 million B.C.?	Strong sexual bonding ties together males and females to cooperate in raising children. Necessary because of long period of immaturity among human offspring as compared with other species.
Complex kinship exchange systems in tribal societies.	8000 B.C.	Extended family networks with complex marriage and economic exchange rules. Some family systems are matrilineal and/or matrilocal, although most are patrilineal and patrilocal or take still other forms. Family networks make up all of social organization.
Emergence of nonkinship social organization. Decline of complex exchange networks. Rise of the patrimonial household.	Ca. 3000 B.C. in Mesopotamia and Egypt; ca. 600 B.C. in Greece and China	Rise of the state as an organization outside family networks. Tribal family systems decline and are replaced by the patrimonial household, dominated by property-owning males and containing many servants and armed warriors.
Rise of the bureaucratic state and capitalist economy. Emergence of the private household.	A.D. 1700s/1800s in Western Europe and North America	Work and defense shift out of the household. Dominant patrimonial households disappear. Rise of the nuclear family as ideal. An individual marriage market replaced politically arranged marriages. Ideal of male breadwinner and female housewife.
Diversified marriage market. Tendency toward egalitarian male/female marriage bargaining.	Ca. A.D. 1950–2000 in wealthy industrial societies	Variety of family structures. Widespread divorce and remarriage, along with both male and female careers, produces a series of relatively short-term marriages and informal cohabitations.

has been going down, so there are fewer marital births to compare with the nonmarital ones (see Figure 10.3). In a larger sense, the two trends add up to another challenge to the traditional family. The old-fashioned ideal of the large family, with Mom and Dad surrounded by a brood of four or five children, has clearly become a subject of nostalgia. New families are mostly small, and many of the children are born into families that do not include a father at all.

Change, Not Disappearance

The prognosis that the family is going to disappear is overdrawn. In our opinion and in that of many sociological experts on the family, what is happening now is a change in typical family patterns. However, a change does not mean the whole family institution is disappearing. Important changes have occurred in the past (Table 5.1), and at each period of change, conservatives have claimed that the family itself was disappearing and that society was being undermined. Instead, what had disappeared was simply a particular form of the family, one that people had been used to for perhaps hundreds of years, but the newer forms that developed were also versions of the family. Hundreds of years later, when these "new" forms themselves were giving way to something more recent, social commentators again complained that the family was disappearing and that society was in jeopardy. In the late 1990s, some conservative foundations even blamed sociology of the family textbooks like this one for a supposed breakdown in family values among the younger generation (see *A Closer Look* 1.1). (For contemporary debates on this issue, see Cherlin 1997; Coltrane 1997; Coontz 1997; Glenn 1997; Johnson 1997; Mann et al. 1997; Popenoe 1996; Scanzoni 1997; Skolnick 1997; Stacey 1996.)

New Variants of Family Structure

The new family system we are now seeing differs from the older one in several ways. With the increase in premarital sex, cohabitation, and nonmarital birth, a good deal of what used to be reserved for legal marriage is now taking place outside of it. These nonlegalized sexual and parental activities are not chaotic but exhibit a pattern that has social significance. They may even be called new versions of family structure itself. Persons who have premarital intercourse, for example, do not simply sleep with everyone who comes along. There is a fairly limited number of sexual partners, and premarital pairings are put together and come apart in definite ways. Living together, if not seen through the old scandal-tinted eyeglasses from former days, actually looks a good deal like a conventional marriage. It lacks the formal ceremony, but most people in our large urban society do not personally know about other people's marriage ceremonies anyway. Without being told, it is often hard to tell the difference between an unmarried or married couple living together.

Even in the case of unmarried women with children, a certain family pattern is involved. These relationships are at least one part of the traditional family structure: parent and child. Often there is a network among single mothers that is very much a family situation, perhaps one that makes us think of the tribal kinship networks studied by anthropologists rather than the conventional American nuclear family dependent on a single male breadwinner.

The latest forms of sexual and parental arrangements, then, should be seen as new variants of family structure rather than phenomena that have nothing to do with a family system. At the same time, we should bear in mind that the conventional nuclear family is still here. Many households still have a father, mother, and minor children living together, although this makes up only about one-fourth of all households—a significant drop from 1970 (Figure 5.6). The most common type of household now consists of a married couple with no children. The family structure in the United States is becoming more complicated because there are different kinds of families and **household types** existing at the same time. The family is becoming more diverse, and it is harder for advocates of just one form to say it is the ideal or norm.

For instance, a steadily increasing percentage of Americans are living past the time when their children have grown up and left home, and a steadily increasing percentage of married couples are waiting longer to have children. Therefore, many households consist only of husband and wife without children. After children leave home and as couples grow older, one spouse or the other (typically the male) dies first, leaving a household of a single person. When we add these households together with those of young people who have not yet married and are living alone or with roommates or who are cohabiting (both gay and straight), then we find quite a substantial proportion of households that are not the conventional father-mother-children nuclear family. When we add in the people who are divorced but have not yet remarried and never married women with children, the total of nonstandard households goes up still further. Also, the number of smaller households is increasing, whereas large households are becoming rather scarce (see Figure 5.6).

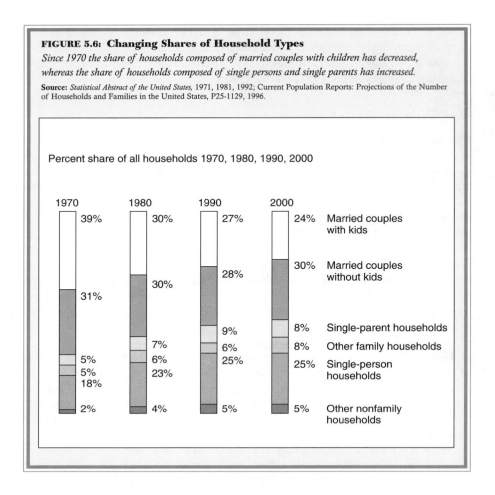

FIGURE 5.6: Changing Shares of Household Types

Since 1970 the share of households composed of married couples with children has decreased, whereas the share of households composed of single persons and single parents has increased.

Source: *Statistical Abstract of the United States,* 1971, 1981, 1992; Current Population Reports: Projections of the Number of Households and Families in the United States, P25-1129, 1996.

Percent share of all households 1970, 1980, 1990, 2000

	1970	1980	1990	2000	
Married couples with kids	39%	30%	27%	24%	
Married couples without kids	31%	30%	28%	30%	
Single-parent households	5%	7%	9%	8%	
Other family households	5%	6%	6%	8%	
Single-person households	18%	23%	25%	25%	
Other nonfamily households	2%	4%	5%	5%	

Diversity in Families

The stress we ought to place here is on the permanent *diversity* of types of families rather than on one form taking over entirely from the others (Figure 5.7). The nuclear family is not disappearing either, because it keeps on being re-formed. People put off getting married longer these days; hence, there are more of them to count as single-person households, shared-with-roommate households, or cohabiting arrangements. The rate at which people eventually marry is still high. Similarly, although there is a high divorce rate, there is also a high rate of remarriage. During the time when people are divorced, they add to the alternative-family total. Because of a constant flow in and out between divorce and marriage, at any given time there is going to be a fair number of people in both kinds of living arrangements.

One interesting result of all this divorce and remarriage is that the kinship structure is becoming more complicated. In the tradition of the **nuclear family,** each child has one father and one mother and lives in a single household until he or she is old enough to leave home and establish another household. With a high rate of divorce and remarriage, though, many children have two sets of parents and two households (or sometimes even more). Instead of merely having one set of brothers and sisters, they have two or more sets, living in different places. It is said that there is a whole new problem of etiquette at marriage receptions today; for when children of divorced people get married, a quite complicated set of

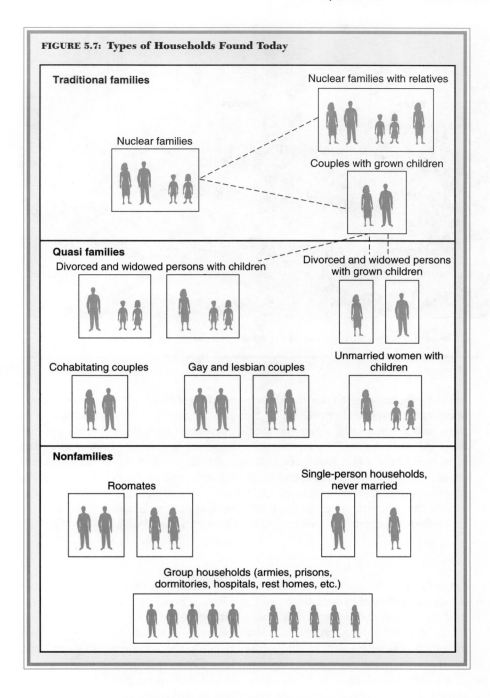

FIGURE 5.7: Types of Households Found Today

Traditional families

Nuclear families

Nuclear families with relatives

Couples with grown children

Quasi families

Divorced and widowed persons with children

Divorced and widowed persons with grown children

Cohabitating couples

Gay and lesbian couples

Unmarried women with children

Nonfamilies

Roomates

Single-person households, never married

Group households (armies, prisons, dormitories, hospitals, rest homes, etc.)

families may have to be invited to the reception. In this sense we may actually have *more* family than before.

CAUSES OF LONG-RUN CHANGES

If we look at the way families have changed throughout the twentieth century, one pattern stands out. The period around 1950 is different from every other time. This is the time of the so-called **baby boom,** when suddenly the long-standing decline

The baby boom was definitely underway outside this Boston clinic in 1948.

in the birthrate reversed itself and the population shot up. Along with the baby boom came several other reversals in trends that had stood for decades: the marriage and divorce rates. After the 1950s, the baby boom disappeared, and the other trends in the family also reversed themselves once again. In retrospect, the time around the 1950s looks like a strange aberration, a bump in the long-term curve. Andrew Cherlin (1992) has claimed that the real problem is not to explain why there was a liberalization of the family in the 1960s and 1970s (that was merely the continuation of a long-term trend), but to explain why there was this anomalous countertrend, this revival of tradition, in the 1950s.

Why the Anomalous 1950s?

Let us examine the charts again. Figure 5.5 shows us that the birthrate had slowly been declining ever since the Civil War, a period of well over one hundred years (possibly longer, if it was declining in the early 1800s, too, which seems likely). The only exception is in the years from about 1940 to 1955, when the rate briefly rose again. In the 1920s and 1930s, the decline had been so striking that statisticians at the U.S. Bureau of the Census predicted that U.S. population would level off in the 1960s at about 140 million people. The baby boom caught them by surprise, and their prediction turned out to be about 60 million people short.

Along with the rising birthrate came a rise in the marriage rate (see Figure 5.2). Actually, this began a little earlier than the baby boom. The rate of marriage was the lowest in the early 1930s, but it climbed after the depression, fluctuated during the war years, and then rapidly shot up in the postwar 1940s. It fell again thereafter but rose again in the late 1960s, and it has been falling since about 1980. Divorce rates also dipped down during the depression and rose dramatically in the 1940s. There was a peak immediately after the end of the war in

FIGURE 5.8: Women in Their Twenties Who Have Never Married

Source: U.S. Bureau of the Census, 1992h, P-23, no. 181, *Current Population Reports;* 1992–1998; 1996, P25-1129.

1945, which is usually interpreted to mean that the disruption of relationships in the war years was being made official. Then the divorce rate dropped sharply during the 1950s. Divorces began to go up again in the 1960s and increased dramatically during the 1970s, beginning to fall again in the 1980s and 1990s.

Finally, Figure 5.8 shows us one more piece of the picture. American women were marrying earlier during the baby boom years. Early in the century, women typically married in their mid-twenties, and the level held steady for three decades. However, in the 1940s, the age of marriages took a rapid drop, and half of all women were married by about the age of twenty-one. In the 1950s, it dropped even lower, and almost half were married before they turned twenty. Only in the 1970s did this pattern reverse itself, and the age of marriage began to rise again toward the pre–World War II pattern. Now the median age at first marriage is twenty-seven for men and twenty-five for women.

The different figures hold together. In the 1950s, marriage was especially popular. That generation had the highest percentage of married individuals on record: 96 percent of females, 94 percent of males (Cherlin 1992). They married younger, divorced less, and had more children. No wonder the 1950s acquired a special reputation as the era of the "familistic generation," an era of tradition and conformity. Actually, "the fifties" is a little inaccurate; the phenomenon began in the 1940s, with some trends beginning even before World War II. The rates of marriage and birth, for instance, were already inching up during that time, although they started from the low point of the previous downtrend, so the full "baby boom" and "marriage boom" were not apparent until the late 1940s. Then in the late 1950s, most of the trends reversed themselves (although again some of them hung on longer, into the 1960s, and did not really shift until the 1970s).

How do we explain this fifties (really 1940s–1950s) bump in the trends? One argument often advanced is that the return to the traditional family was a reaction to the postwar situation. According to this theory, after the end of wartime disruption with its temporary surge of divorces, people naturally turned toward

The years between 1920 and 1950 were overshadowed by two great totalitarian systems. The Communist party came to power in Russia after the Revolution in 1917. Despite its idealistic beginnings, by the mid-1920s, following Lenin's death, the Soviet Socialist Republic degenerated into a dictatorship under Joseph Stalin. At about the same time in Germany, Adolf Hitler organized his Nazi movement around a militant core of disgruntled World War I veterans. After a long campaign of street demonstrations and battles, mostly aimed against the socialists but also against the Jews, Hitler's party was elected in 1933. Soon after, Hitler abolished the German democracy and set up his fascist dictatorship.

The Nazi program was committed to, among other programs, overthrowing the gains recently made by the German feminist movement. Women had won the right to vote in 1918; the Nazis officially excluded women from public life. The purpose of the German woman was "to minister in the home," devoting herself to "the care of man, soul, body and mind," from "the first to the last moment of man's existence." In *Mein Kampf,* Hitler declared that "the aim of feminine education is invariably to be the future mother." Contraception and abor-

tion were made illegal, and the sole purpose of sex was to procreate new members for the German nation. Marriage was encouraged by taxing bachelors and spinsters, while taxes were rebated and interest-free loans given for each child born under a state-supported contract. (A woman took out the loan, but it was paid to her husband.) The measures were successful, at least in producing babies: the birthrate was raised 30 percent in the first two years after the Nazis took power.

Although the Nazi ideal was to keep the woman in the home, this policy failed. For one thing, the huge military losses of World War I had left Germany with two million more women than men, and these women had to support themselves at work. When military preparations began for World War II, the Nazis made both men and women liable for state labor and called out women in increasing numbers. At the height of the war effort in the 1940s, when the German army had conscripted virtually all the able-bodied men, most of the civilian work was done by women. As many as 80 percent of all adult women were working for the Third Reich. The state took a contradictory stance toward females: the police attempted to prevent women from smoking or wearing cosmetics, while at the same time houses

of prostitution were maintained for the military and the Nazi leadership.

In the Soviet Union, the Revolution had begun with a principled effort to liberate women from the patriarchal Russian family. Women were given the right to choose their own domicile, name, and citizenship; to marry and divorce independently of the wishes of their parents or spouses; to pursue economic independence; and to control their own sexuality, with contraception and abortion on demand. Illegitimacy, adultery, and homosexuality were eliminated from the code of criminal offenses. In order that women should be able to participate economically and politically on a par with men, the plan was to establish public nurseries and a collective housekeeping service, along with maternity leaves at work.

But economic and then political forces intervened to prevent the carrying out of this liberal plan. During its early years the revolutionary government witnessed civil war and economic depression. In the early 1920s, there was widespread unemployment, which hit women especially hard in their campaign for economic equality. The public child-care services could not be built, and mothers had to con-

reestablishing normalcy and the family. There are several problems with this explanation. One is that some of the trends (such as the rising rates of birth and marriage and the falling age of marriage) actually began before the war. Moreover, postwar situations do not invariably have this result (see *A Closer Look* 5.1). We can see this from Figures 5.2 and 5.5, which show that the rates of marriage and birth did not go up in the 1920s, after World War I, nor was there a rise in the birthrate after the Civil War (which ended in 1865). World War II is thus

tinue as before, often with job responsibilities on top of their other tasks. When the Soviet economy recovered, all resources were channeled into building heavy industrial equipment and armaments. By the 1930s, Soviet policy had shifted to support for the traditional family. The early revolutionary feminists like Alexandra Killontai were publicly censured. The purpose of sex, it was now declared, was solely to produce children. As in Germany, there was a great deal of rhetoric about "preserving the race," and an effort was made to build up the population. Women who had six or more children were given bonuses; mothers of seven were given medals. Abortion was again made a criminal offense. The old czarist legislation against homosexuality was reintroduced, providing for sentences of three to eight years, enforced by mass arrests. Illegitimacy again became a legal category, and both mother and child (but not father) were stigmatized. In 1936, couples who divorced were fined 30 to 50 rubles for "mistaking infatuation with love," and in 1944 the fine was raised to 500 to 2,000 rubles (which made divorce virtually impossible).

The rearing of children, which originally was planned on a libertarian basis, also took an authoritarian turn. An authoritarian youth organization was formed, with compulsory partici-pation, under the leadership of Makarenko, a Secret Political Police official who had been in charge of delinquent boys. Progressive schools founded in the 1920s were abolished and replaced by traditional authoritarian ones. By 1943, coeducation was eliminated, and the schools preached a puritanical doctrine of sexual abstinence in the name of commitment to the state. At home, parents were held responsible for teaching their children the correct ideological line.

Do the Soviet and Nazi cases show the impossibility of changing the traditional family? Although this conclusion has often been drawn from a superficial acquaintance with the facts, the lesson actually is somewhat different. The Nazis, on political grounds, attempted to reestablish the most sexist form of the traditional family. But their effort was undermined by economic realities: they needed women in the labor force and ended up undermining the traditional family more than supporting it. The Soviets started out with an extremely liberal plan but were unable to put it into effect for lack of economic resources. The key to equal rights for women was in the provision of state-supported child care and housekeeping, but these quickly went by the board because of the early economic crisis and then stayed low priority because of the subsequent military buildup. By the 1930s and 1940s, at the height of the Stalin dictatorship, the Soviet policy on the family and sexual behavior had become as authoritarian as the Nazis'. This provides a second lesson: there is a tendency for politically authoritarian regimes to invoke sexual puritanism and the authoritarian family as part of their general ideology of control over the individual.

These extreme instances of the authoritarian family are now in the past. The Nazi regime fell in 1945. After Stalin's death in 1953, Soviet policy gradually eased. The right to abortion was reestablished in 1955, and illegitimacy ceased to be registered in 1965. Women have moved closer to equality in educational access, although they are still concentrated in the lower-paid occupations. Women do make up a majority of Russian doctors, but even these are ranked lower than men in the bureaucratic medical system. Child-care facilities have expanded, but women continue to be responsible for most of the traditional housework and domestic child care. After the downfall of the Soviet Union in 1991, the situation of social services for women has gone into chaos. Feminist interests have been pushed into the background. Women there have a long way to go to catch up with their sisters in many parts of the West (Millett 1970, 157–76; Neumann 1944; Geiger 1968; Fisher 1980).

more of a coincidence on the chart than a full explanation of what actually happened in the 1940s and 1950s.

Part of the answer is that the war came on the heels of an economic depression. The war itself created considerable employment (the military boom), and after the war came a huge expansion of the American economy, which lasted into the 1950s. In general, people tend to marry when they feel they can afford it. The better their economic prospects are, the younger they are when they decide they can get married. The converse of this is that during the Great Depression of the 1930s, there were strong reasons why many people did *not* get married, and why those who married felt they could not afford to have many children.

When the economic situation improved—culminating in the postwar boom, although there were smaller improvements earlier—it was as if a dam had burst. Suddenly there were two groups of people who had a chance to get married or to go ahead and have children if they were already married. There were those who put off getting married or put off having children in the 1930s and soldiers returning from the war. Along with this, the generation just coming of age began to marry extremely early and to have children quickly. **Familism** had become something of a new cultural movement, and everybody was jumping on the bandwagon.

Working-Class Affluence and the Growth of Suburbs

We can understand this atmosphere better if we look at the way the 1950s were perceived at the time. It was the so-called age of suburbia, when mass housing developments sprang up for the first time outside the major cities. For a hundred years previously, the tendency had been for the population of the countryside to decline, while large city populations grew continuously larger. In the 1950s a countermovement developed: people were leaving the apartment houses of the city for single-family dwellings in the suburbs.

Culture critics of the time generally regarded the suburbs with horror. The houses stood in mass-produced developments like the famous Levittown; this was taken as a sign of uniformity invading our culture. The plethora of clubs, school teams, teenage gangs, and other organizations that went along with suburban life was also seen as evidence of growing conformity. David Riesman wrote in his famous sociological best-seller, *The Lonely Crowd,* that the traditional American inner-directed personality was on the decline and the other-directed conformist was proliferating. William H. Whyte described the impersonality of the suburbs as the habitat of *The Organization Man,* who was moved from place to place by gigantic bureaucracies. One folk song satirized the suburbs as "little boxes, on the hillside, little boxes made of ticky-tacky" that "all look just the same."

The critics, however, missed one crucial fact: the newly developed suburbs were not, in general, full of "organization men" working their way up the management ladder, nor did many doctors and lawyers live in the folksinger's "little boxes." The new suburbs were to a considerable extent working class (along with young professionals bound for better things). Houses were cheap, standardized, and prefabricated because that was the only kind of house the working class could afford. From their point of view, at least they had a house rather than an apartment, and it was in the countryside rather than in the tenement district of a city. Americans in general have trouble seeing social class, and the upper-middle-class critics fell into the trap of assuming that the suburbs were built for people like themselves. Used to the luxury of old and grander houses, with more space between them and more investment in individualistic touches, they regarded the new suburbs with horror, as a decline in American taste. For most of the working class, moving out to own their own houses for the first time felt like a tremendous step up.

The familism and "conformity" that were such popular topics of discussion in the 1950s, then, were apparently really to a large extent a movement of the working class (Berger 1960, 1971; Gans 1967). The immediate postwar decades have been described as an "economic miracle": for the first time in history, a middle-class standard of living seemed to be in sight for most or many of the

Mass-produced developments like Whitestone, in the New York City borough of Queens, aroused fears that the uniformity of suburbia was invading our culture.

working class. To own their own home, fill it with labor-saving appliances, own a car or even two—all these had been reserved for the rich, until this boom time in America. For a working-class married woman to stay home, like a respectable middle-class housewife, instead of going off to work with the masses, this was a new frontier of social status.

The 1950s was one of the wealthiest periods in the history of the country. The United States, with its industrial plant running at full blast on the rich natural resources of North America, was producing the highest standard of living in the history of the world. The middle and upper classes got richer, but they already had their single-family dwellings. For the working class, this was something new, and it threw itself into the suburban lifestyle with enthusiasm. The "do-it-yourself" craze of putting in one's own room additions, plumbing, and wiring and doing all sorts of other jobs around the house made a splash in the 1950s. The traditional homeowner had been a white-collar worker who hired a handyman for these kinds of repairs; the new working-class suburbanites, though, could only afford to make these kinds of improvements by putting in their own manual labor.

This is one reason for the "conformity" charges made by the critics. For as we will see (chapter 6), working-class culture has traditionally been more group oriented and less individualistic and achievement oriented than upper-middle-class culture. The new suburbs looked roughly like middle-class communities on the outside, but culturally they continued to be working class and hence emphasized conformity (as Herbert Gans showed in his book *The Levittowners,* 1967). Moreover, working-class culture also tends to be more **familistic** than middle-class culture. Working-class people marry younger, have children earlier, and confine more of their socializing to their immediate relatives and neighbors.

The rising rates of marriage and birth and the declining age of marriage, then, were to a considerable extent economic in origin. In that era of affluence, working-class people and younger, less affluent members of the middle class were more able to do what they preferred to do: to marry as soon as possible and to raise a large family. Early economic constraints had kept them from doing these things; early in the century, the age of marriage was much higher, simply because most people couldn't afford to set up housekeeping on their own until they had worked for a number of years. In the economic boom that followed World War II, workers were finally able to indulge their familistic concerns more than ever before. It is probably this, more than anything else, that contributed to the unusual family pattern of the 1950s.

Why did this pattern come to an end, beginning in the late 1950s and the 1960s? In Cherlin's view (1992) the long-run trend just reasserted itself. The new **suburban working class** had a chance to emulate what it thought was a middle-class lifestyle, but actually it only moved the working-class style out to the better material conditions of the suburbs. The wave of familistic sentiment that went along with this transition began to be undercut by forces that had been operating in the middle classes for some time.

A Comparison: The Roaring Twenties and the Rebellious Sixties

If our analysis is right, the 1950s was a period when the working class captured the center of cultural attention in America. Because working-class culture is typically familistic and traditional, the family life of the 1950s looked like a trend back toward tradition. During most of this century, though, the cultural initiative that has been most influential has come largely from the middle and upper classes. Its thrust has been in a very different direction, especially regarding sexual liberalism. This is apparent in two of the more spectacular eras of change, the so-called **roaring twenties** and the **rebellious sixties.**

THE 1920s The 1920s was the era of our first great popular sexual revolution (*A Closer Look* 5.2). Before this, the respectable middle and upper classes had attempted to live with Victorian propriety. Gradually, among the upper-class youths before World War I, a hedonistic culture was appearing. Drinking and sexual flirtation became the vogue, but this was confined to a rather small group and received little publicity until after the war. Then middle-class youth culture as a whole caught on to the new style, and it invaded the mass media.

Some of the changes were highly visible. One was the shift in clothing styles. Women's skirts suddenly went up, to the knee and even above. This does not seem like much to us now, but at the time it was a shock and a revolution. For the Victorian style had been long dresses, touching the ground and not revealing so much as an ankle. Legs had not even been a fit subject for proper talk, and now they were in full view. This was probably the single largest change in women's clothing styles in several thousand years of Western history. Women had worn a long robe or dress ever since Greek and Roman times, and the custom of covering the female body had scarcely varied throughout the centuries: style changes had been confined to the cut of the garment and to hair and head covering. Suddenly we were in a new era, and the style definitely announced a shift toward a sexier appearance.

In the 1920s, an independent youth culture sprang into being. Previously, middle-class parents had controlled their offsprings' socializing, which featured carefully chaperoned formal dances and gatherings of whole families with their

In the 1920s, the youth culture began to experiment with new forms of music and dancing, more revealing clothing and the use of alcohol, which was at that time illegal. Here, young New York City women begin a Charleston endurance contest at the Parody Club in 1926.

guests at home. Now the youths went off and had their own parties, driving their own cars and frequenting roadhouses and dance halls. Prohibition of alcohol had just been enacted by constitutional amendment, but this seemed only to make the use of illicit alcohol more exciting. Dancing changed from the graceful ballroom traditions of the waltz to the new "hot" styles: the fox-trot, the Charleston, and many more. Jazz bands and their syncopated rhythms became the vogue; modern popular music was born. Jazz was considered somewhat immoral by traditional elders; as F. Scott Fitzgerald pointed out, the terms *jazz* or *jazzy* themselves first had the connotation of sex, and then wild dancing, before they finally came to mean a kind of music.

The upper-class youth started this new sexually oriented popular culture, but it soon spread. By the early twenties, according to Fitzgerald, it had infected the youths of the middle class in the smaller cities across the country. Within a few years, their elders had begun to catch on, and drinking parties, jazz, and the new styles had spread far and wide. The new dating style had set in, and the traditional family as the center for social life was now becoming a thing of the past, found only among the elderly and in rural sectors of America.

With the fall of the stock market in 1929 and the coming of the Depression in the 1930s, the wild and hedonistic Jazz Age was over. However, the changes that it made in the family, as well as in sexual styles and mass culture, were not to be undone. The modern age was firmly launched.

THE 1960s "The sixties" did not quite fit that decade chronologically. The era of political and cultural ferment that is usually called "the sixties" started just after the assassination of President John F. Kennedy in 1963 and went on until about the resignation of President Richard M. Nixon in 1974. It was a time of the civil rights movement, with its demonstrations, sit-ins, mass arrests, and even murders of civil rights workers in the Deep South, which finally culminated in the overturning of legal segregation. (A vivid account is *The Sixties* by Todd

A CLOSER LOOK 5.2
F. Scott Fitzgerald on the Jazz Age

The first social revelation created a sensation out of all proportion to its novelty. As far back as 1915 the unchaperoned young people of the smaller cities had discovered the mobile privacy of that automobile given to young Bill at sixteen to make him "self-reliant." At first petting was a desperate adventure even under such favorable conditions, but presently confidences were exchanged and the old commandment broke down. As early as 1917 there were references to such sweet and casual dalliance in any number of the *Yale Record* or the *Princeton Tiger*.

But petting in its more audacious manifestations was confined to the wealthier classes—among other young people, the old standard prevailed until after the War, and a kiss meant that a proposal was expected, as young officers in strange cities sometimes discovered to their dismay. Only in 1920 did the veil finally fall—the Jazz Age was in flower.

Scarcely had the staider citizens of the republic caught their breaths when the wildest of all generations, the generation which had been adolescent during the confusion of the War, brusquely shouldered my contemporaries out of the way and danced into the limelight. This was the generation whose girls dramatized themselves as flappers, the generation that corrupted its elders and eventually overreached itself less through lack of morals than through lack of taste. May one offer in exhibit the year 1922!

That was the peak of the younger generation, for though the Jazz Age continued, it became less and less an affair of youth.

The sequel was like a children's party taken over by the elders, leaving the children puzzled and rather neglected and rather taken aback. By 1923 their elders, tired of watching the carnival with ill-concealed envy, had discovered that young liquor will take the place of young blood, and with a whoop the orgy began. The younger generation was starred no longer.

A whole race going hedonistic, deciding on pleasure. The precocious intimacies of the younger generation would have come about with or without prohibition—they were implicit in the attempt to adapt English customs to American conditions. (Our South, for example, is tropical and early maturing—it has never been part of the wisdom of France and Spain to let young girls go unchaperoned at sixteen and seventeen.) But the general decision to be amused that began with the cocktail parties of 1921 had more complicated origins.

The word *jazz* in its progress toward respectability has meant first sex, then dancing, then music. It is associated with a state of nervous stimulation, not unlike that of big cities behind the liens of a war. To many English the War still goes on because all the forces that menace them are still active—Wherefore eat, drink and be merry, for tomorrow we die. But different causes had now

brought about a corresponding state in America—though there were entire classes (people over fifty, for example) who spent a whole decade denying its existence even when its puckish face peered into the family circle. Never did they dream that they had contributed to it. The honest citizens of every class, who believed in a strict public morality and were powerful enough to enforce the necessary legislation, did not know that they would necessarily be served by criminals and quacks, and do not really believe it today. Rich righteousness had always been able to buy honest and intelligent servants to free the slaves or the Cubans, so when this attempt collapsed our elders stood firm with all the stubbornness of people involved in a weak case, preserving their righteousness and losing their children. Silver-haired women and men with fine old faces, people who never did a consciously dishonest thing in their lives, still assure each other in the apartment hotels of New York and Boston and Washington that "there's a whole generation growing up that will never know the taste of liquor." Meanwhile their granddaughters pass the well-thumbed copy of *Lady Chatterly's Lover* around the boarding school and, if they get about at all, know the taste of gin or corn at sixteen. But the generation who reached maturity between 1875 and 1895 continue to believe what they want to believe.

Even the intervening generations were incredulous. In 1920

Gitlin, an activist who became a sociologist.) It was the time of the antiwar movement, centered on Untied States involvement of half a million troops in support of a military dictatorship in Vietnam, which eventually resulted in U.S. withdrawal in 1973. Both the civil rights and the antiwar movements were extremely popular on college campuses, where they led to numerous demonstra-

Heywood Broun announced that all this hubbub was nonsense, that young men didn't kiss but told anyhow. But very shortly people over twenty-five came in for an intensive education. Let me trace some of the revelations vouchsafed them by reference to a dozen works written for various types of mentality during the decade. We begin with the suggestion that Don Juan leads an interesting life (*Furgen,* 1919); then we learn that there's a lot of sex around if we only knew it (*Winesburg, Ohio,* 1920), that adolescents lead very amorous lives (*This Side of Paradise,* 1920), that there are a lot of neglected Anglo-Saxon words (*Ulysses,* 1921), that older people don't always resist sudden temptations (*Cytherea,* 1922), that girls are sometimes seduced without being ruined (*Flaming Youth,* 1922), that even rape often turns out well (*The Sheik,* 1922), that glamorous English ladies are often promiscuous (*The Green Hat,* 1924), that in fact they devote most of their time to it (*The Vortex,* 1926), that it's a damn good thing too (*Lady Chatterley's Lover,* 1928), and finally that there are abnormal variations (*The Well of Loneliness,* 1928, and *Sodom and Gomorrah,* 1929).

The Jazz Age had had a wild youth and a heady middle age. There was a phase of the necking parties, the Leopold-Loeb murder (I remember the time my wife was arrested on Queensborough Bridge on the suspicion of being the "Bob-haired Bandit") and the John Held Clothes. In the second phase such phenomena as sex and murder became more mature, if much more conventional. Middle age must be served and pajamas came to the beach to save fat thighs and flabby calves from competition with the one-piece bathing-suit. Finally skirts came down and everything was concealed. Everybody was at scratch now. Let's go—

But it was not to be. Somebody had blundered and the most expensive orgy in history was over.

It ended two years ago [1929], because the utter confidence which was its essential prop received an enormous jolt, and it didn't take long for the flimsy structure to settle earthward. And after two years the Jazz Age seems as far away as the days before the War. It was borrowed time anyhow—the whole upper tenth of a nation living with the insouciance of grand ducs and the casualness of chorus girls. But moralizing is easy now and it was pleasant to be in one's twenties in such a certain and unworried time. Even when you were broke you didn't worry about money, because it was in such profusion around you. Toward the end one had a struggle to pay one's share; it was almost a favor to accept hospitality that required any travelling. Charm, notoriety, mere good manners, weighed more than money as a social asset. This was rather splendid, but things were getting thinner and thinner as the eternal necessary human values tried to spread over all that expansion. Writers were geniuses on the strength of one respectable book or play; just as during the War officers of four months' experience commanded hundreds of men, so there were now many little fish lording it over great big bowls. In the theatrical world extravagant productions were carried by a few second-rate stars, and so on up the scale into politics, where it was difficult to interest good men in positions of the highest importance and responsibility, importance and responsibility far exceeding that of business executives but which paid only five or six thousand a year.

Now once more the belt is tight and we summon the proper expression of horror as we look back at our wasted youth. Sometimes, though, there is a ghostly rumble among the drums, an asthmatic whisper in the trombones that swings me back into the early twenties when we drank wood alcohol and every day in every way grew better and better, and there was a first abortive shortening of the skirts, and girls all looked alike in sweater dresses, and people you didn't want to know said "Yes, we have no bananas," and it seemed only a question of a few years before the older people would step aside and let the world be run by those who saw things as they were—and it all seems rosy and romantic to us who were young then, because we will never feel quite so intensely about our surroundings any more.

F. Scott Fitzgerald, "Echoes of the Jazz Age," in *The Crack-Up.* Copyrighted 1945 by New Directions Publishing Corp. Reprinted by permission of New Directions.

tions, sit-ins, and strikes, some of which resulted in massive confrontations with police. Though the student demonstrations generally adhered to the nonviolent-protest philosophy espoused by the civil rights movement, the authorities sometimes put them down with force (such as at Kent State University in Ohio in 1970, when the National Guard opened fire on a demonstration and killed four

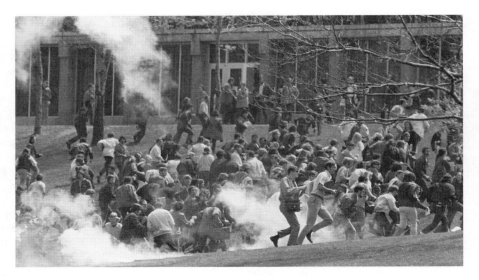

On May 4, 1970, the National Guard, called in to put down antiwar protests at Kent State University, fired tear gas and bullets into a crowd of about 500 students, killing four and wounding several others.

students). By the mid-1960s, over half of America's population was under the age of thirty.

Not surprisingly, the prevailing mood of American youth during this period was one of intense alienation from traditional "straight" society. Conventional society was regarded not only as racist and militarist but also as sexually and culturally restrictive. A **counterculture** emerged. Many young people rebelled against the Protestant work ethic and traditional patterns of deference to authority. The established churches that had been a bulwark of conservative social values, especially in the United States, began to lose both their political and cultural influence. Official piety and church attendance declined, even though religious beliefs remained strong (*A Closer Look 5.3*). There was an anti-authoritarian spirituality to the counterculture, as young people rejected the rationalism of the scientific age and embraced romantic and utopian dreams.

For fifty years, the American style of grooming had been for men to be close shaven; now beards and mustaches became the vogue. The conservatism of the 1950s had been manifested in military-style crew cuts for men, and women also had short, conservative coiffures. In the 1960s, naturalism was expressed in long hair for both sexes; the eye was jolted when men's hair began to be as long as women's. Part of the youth culture developed a "hippie" style that included beads, loose-fitting clothes, and bright colors. Some went to live in tepees in the woods; others established communes. LSD and other "psychedelic" drugs were popular for inducing visions and altered states of consciousness, and many people became devotees of Eastern mystical religions.

The counterculture and the political rebellion of the time were the most apparent public manifestations of the changes that were going on, but they were only part of a deeper change. As we have already seen, the 1960s and early 1970s formed a period during which the conservative family trends of the 1950s re-

A CLOSER LOOK 5.3
Churches and Society in the Twentieth Century

The British historian John Harriss notes that the influence of churches over culture and politics has declined since midcentury. Do you agree?

After World War II, the long process of **secularizaiton**—*the decline of religious belief—accelerated. Although the great majority of people in western societies continued to express their belief in some kind of deity, active church membership dropped. By the early 1960s only about 10 percent of the British people—mostly the elderly, women, and young children—went to church at all regularly. Religious observance was not much more active in Catholic France and Italy. Amongst the industrialized countries only the United States did not conform to the trend. A survey carried out in 60 countries showed that the United States had a higher level of religious commitment than any other, except for India. Outside America the cultural role of churches largely disappeared. Outside Ireland churchmen lost their political influence.*

Throughout the earlier twentieth century the Roman Catholic Church, in particular, remained a profoundly conservative influence. There were glimmerings of change after the war when . . . "worker priests" . . . took up manual jobs and, identifying with workers, opposed injustice of the existing social order. . . . Meanwhile, Protestant theological reform exposed the Church, seeming to suggest that its message had become irrelevant. In this context the papacy of Pope John XXIII after 1958 brought dramatic change. The Ecumenical Council which he summoned (known as Vatican II) and which met from 1962 to 1965 reformed much of the Church's social teaching as well as rewriting liturgy and lowering old regulations. The Mass, for example, was no longer said in Latin. After the Council the Church made strides toward ecumenism and became more open even on issues such as mixed marriage. Yet it still remained resistant to the practice of birth control, to abortion and to homosexuality, all condemned by John's successor Pope Paul VI, under whose leadership the pace of reform slowed down.

Later in the twentieth century the Christian churches were to become more divided, between liberal theology and a return to "fundamentals," and between the pursuit of an active social role (as in "liberation theology," for example) or a conservative one.

What do you think the role of the Church should be in shaping family policy? Should religious organizations promote laws regulating sexual practices, birth control, abortion, marriage, and divorce? Should they participate in political campaigns about these issues?

Source: John Harriss (ed.). 1991. *The Family: A Social History of the Twentieth Century.* New York: Oxford University Press, p. 196. Copyright © 1991 by Oxford University Press, Inc. Used by permission.

versed themselves. The baby boom was over, marriage was being put off to a later age, and the divorce rate was rocketing up (see Figure 5.3). The nonmarital birthrate was taking off (see Figure 5.4), and spokespersons of the counterculture argued that it was discriminatory to punish either mother or child for an illegitimate birth. Sexual relations had traditionally been closely confined to marriage or else to the courtship period that quickly led to marriage, but now sex was being explicitly severed from family legalities. The most striking form of this was the upsurge in the practice of unmarried couples living together, or cohabiting. In every generation up through the 1950s, this would have been considered scandalous. Now it was widely practiced in the middle-class youth culture, openly and shamelessly avowed, and the traditional public was forced to back down in its judgment.

What was responsible for these changes of the 1960s? The "revolution" was partly a revolution in the family, comparable with that of the 1920s. In both cases, there was a popular movement, with the most attention going to changes in clothing style, dancing, music, and illicit "trips" (prohibition alcohol versus psychedelic drugs). The 1960s did have an idealistic political aspect, the movements for civil rights and against the Vietnam War, which the 1920s did not share. In both cases there was a spectacular public movement that lasted for about ten years and then died away as economic conditions changed. The "roaring twenties" ended with the stock market crash in 1929, which ushered in the Great Depression. "The sixties" movement died partly because it won some important victories: the Vietnam War did end, and laws were passed overturning racial segregation. In both the 1920s and the 1960s, the period of upheaval left a more permanent monument in the shape of a changed family system and a changed public culture.

The Second Wave of the Women's Movement

A politically vocal women's movement appeared on the scene late in the sixties period, that is, especially in the years just after 1971. The earlier political movements of the sixties had been male-dominated movements; and the hippies, too—though they broke out of the conventional family structure and espoused sexual liberation—tended to have a traditional setup of male leadership and female subordination. The women's movement challenged all that. More significantly, it challenged male supremacist practices throughout ordinary society: the exclusion of women from important managerial and professional jobs, their segregation into female "work ghettos" of secretarial and service work, and their domestic segregation into the traditional careers of housework and child care.

The **women's movement** acted politically to overcome legal discrimination against women through such organizations as the National Organization for Women (NOW) and the National Women's Political Caucus (NWPC). This public part of the movement constitutes a second wave of feminism, following the long first wave that spent the years from the 1840s to 1920 getting basic legal rights and political suffrage for women. If we look at the trends in the family structure, we can see that the political aspect of the women's movement in the 1970s was just a manifestation of forces that had already been in motion, some for as long as twenty years (Chafetz and Dworkin 1986).

One crucial shift of this sort has been the trend for women to work outside the home. In the mid-1960s, only about 38 percent of women worked at paid jobs, and most of these were unmarried women, who often quit their jobs after they were married and almost certainly after they had children. Now 60 percent of all women aged sixteen or older work. This trend is strong among married women with children; over three-fourths of women with school-age children now work, two-thirds of those with preschool-age children are employed, and even among women with infants less than one year old, over half are in the paid labor force (Figure 5.9). The number of working women has doubled since 1970—from 30 million to over 60 million, with three of four of these employed full-time.

Moreover, the reasons why women work have changed. Formerly the bulk of working women were from working-class families. They worked at relatively low-paying and low-prestige jobs as waitresses, store clerks, domestic servants, and factory operatives. These jobs did not attract them; they worked out of economic necessity. Now we see large numbers of middle-class wives working to have independent careers of their own. This is happening despite the fact that women still have acquired only a small proportion of the desirable managerial, professional, and technical positions, although there has been some improvement in this direction.

One of the main manifestations of the feminist movement at home appears to be an increase in the amount of dissatisfaction that women are showing with traditional domestic roles. This is one of the reasons the divorce rate has gone up, especially at the height of the recent wave of feminist consciousness in the 1970s. Women are putting off marriage longer, partly to pursue their own careers and partly because married life of the traditional sort is less appealing to them.

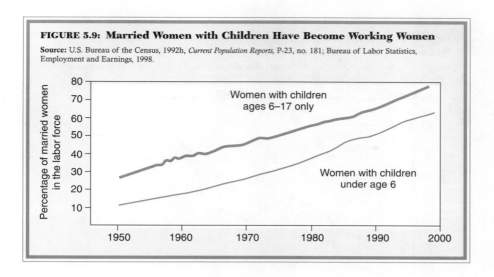

FIGURE 5.9: **Married Women with Children Have Become Working Women**

Source: U.S. Bureau of the Census, 1992h, *Current Population Reports*, P-23, no. 181; Bureau of Labor Statistics, Employment and Earnings, 1998.

The women's movement has also taken a stand against many other areas of gender discrimination. Some of these have changed relatively rapidly once the pressure was turned on. In 1965 it was difficult or impossible for women to obtain credit cards in their own names or to get a loan from a bank. Now women have access to consumer credit. It used to be common for restaurants, clubs, and bars to publicly announce a "men only" policy. This kind of segregation, too, has changed, although it hangs on in some quarters. For example, the U.S. Junior Chamber of Commerce, which started admitting women in the late 1970s, reversed itself and attempted to become sex-segregated again in the 1980s. Cases of this sort remain in litigation.

Other areas of gender discrimination remain. Women still make up less than 10 percent of high-level government and business officials. At somewhat lower levels, the number of women holding elected office has gone up, from only one in twenty state and local officials in the 1970s to about one in four or five today. The number of women who are U.S. senators or congressional representatives has increased significantly since the 1970s, but as of 1999, women still held only about twelve percent of those seats (9 of 100 senators and 54 of 435 representatives). Women are paid a good deal less than men for similar kinds of work; across the board, working women make about seventy-four cents for every dollar a man makes (see Figure 1.3). One might predict that if the feminist movement is successful in raising female wages up to the male level, women will have even less incentive for marrying, and current trends in the family toward fewer and later marriages will become even stronger.

The women's movement obviously has a long way to go to reach its goals. One reason to expect that it is going to continue having an effect over the long run is that many men support it. In fact, on some issues men are even stronger supporters of feminist goals than women are. For example, national polls conducted in the 1990s show that men are even more likely than women to approve of wives earning money when they have husbands capable of supporting them. As we will discuss in future chapters, women's entry into the labor force has not necessarily been accompanied by a lowering of expectations for wives and mothers, and some women do not support feminist political goals. As we can see from Figure 5.10, about six percent of women even recently would not have

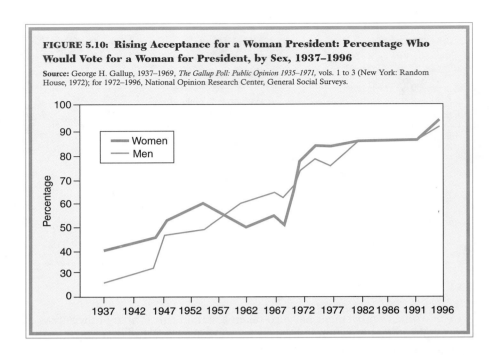

FIGURE 5.10: Rising Acceptance for a Woman President: Percentage Who Would Vote for a Woman for President, by Sex, 1937–1996

Source: George H. Gallup, 1937–1969, *The Gallup Poll: Public Opinion 1935–1971,* vols. 1 to 3 (New York: Random House, 1972); for 1972–1996, National Opinion Research Center, General Social Surveys.

voted for a woman for president. However, the opposition has been going down since the 1930s, when 60 percent of women and 75 percent of men were opposed to having a woman president.

Conflict over Family Change

Some of today's population not only does not support feminist goals but is actively opposed to them. This group is most heavily represented among the older, less educated, and poorer people and in the rural areas of the country. Recent estimates place it at about 25 to 33 percent of the population (Grigsby 1992; Liebman and Wuthnow 1983; Washington Post 1998). This group, too, has become politically organized in recent years, sometimes in religious form, such as the Reverend Jerry Falwell's Moral Majority and the popular Promise Keepers movement led by former football coach Bill McCartney. These groups were formed largely in reaction to the political and legal victories of feminists and to protest against changing trends in family structure. Just as feminists have used politics to press for antidiscrimination laws, the traditional "profamily" movement has fought for legislation that would get women back in the home, reestablish the old ideal family form, make divorce more difficult, and reinstitute some of the traditional taboos.

Social movements often operate in opposing pairs of this sort. The women's movement of the 1970s gave rise to a countermovement that favors an ultratraditional family (Stacey 1996). In the 1920s, the Prohibition movement had its strength in rural and small-town America (Gusfield 1963), against which the Jazz Age culture rose in rebellion. Society is never uniform and homogeneous, and different classes and social sectors often go in different directions or openly conflict with one another. It is not surprising that today there are both feminist and antifeminist movements.

Chafetz and Dworkin (1987, 1989), comparing data on where feminist and antifeminist movements have emerged around the world, propose a theory that accounts for these movements. Feminist movements emerge when there are favorable economic opportunities for women. (This is in keeping with the economic theory of gender stratification presented in chapter 3.) Women's educational level has been rising, and so has the demand for white-collar labor. As women move increasingly into the labor force, more middle-class women have the resources to organize a social movement attempting to break down gender disadvantages and discrimination.

STATUS CONFLICT AMONG WOMEN Not all women are equally well situated to take advantage of these new opportunities. Some women lack the education or the job chances to move into nontraditional occupations and to earn higher incomes. Other women have already organized their lives around the traditional pattern of wife, homemaker, and mother. To the extent that the feminist movement succeeds in redefining women's roles, this traditional group comes under pressure. Their roles are devalued by the new standard that defines a modern, successful woman as someone who has a professional or management-level job that was formerly reserved for upper-middle-class men. The entire status structure of stratification has changed, and traditional women are losing the idealized status that used to surround them as homemakers for their husbands and mothers for their children. This group is the source of an antifeminist backlash (Klatch 1992).

Antifeminist feelings thus are really a form of status conflict, arising from the changes in the structure of gender stratification. In the traditional structure, women's class position was seen by almost everyone as attached to their husband's position (or their father's, when they were younger). Now independent career women are pushing to define their own class positions in addition to those of men they might be related to. For some women, especially the group from whom the feminist activists are most likely to come, this means a push upward, into the upper-middle class. For other women who are unable to keep up, their class positions are reduced in the harsh light of the new standard. Hence, conflict breaks out over the standards by which women are to be judged: on their own, according to their personal occupational success, or by the older tradition of making a home for a man. The antifeminist movement that recruits from the traditional group has become active on various political and legal issues, but its real concern is to defend the traditional status ideal; it fits the model of what Gusfield (1963) calls a "symbolic crusade."

UNIVERSALISTIC VERSUS TRADITIONALISTIC IDEALS There is another way of looking at the kinds of ideals that are defended on both sides of this conflict. In each case, we should see the social movement as like a ship, floating on a much larger tide. The movement is the organized, self-conscious part of the female population; underneath are much larger numbers of women who are simply responding to the economic and social realities around them. On one hand, there are the women responding to the new economic opportunities, especially in the elite professions and in higher education. These women tend to be situated in social networks that are **cosmopolitan,** open to diverse experiences. As Chafetz and Dworkin (1987) put it, they are in closer contact with men's occupations and take these as their reference group. Using the theory of interaction rituals (described in more detail in chapter 8), we would expect these women to have abstract and universalistic ideals. This is reflected in their belief in universal rights and abstract philosophies to justify the improvements they are seeking.

A CLOSER LOOK 5.4 *Men's Reactions to Women's Independence: Islam, Immigration, and Family Change at the End of the Twentieth Century*

The late twentieth century saw the resurgence of Fundamentalist religious movements throughout the world. In 1979, a revolution overthrew the Shah of Iran and created an Islamic fundamentalist state under the Ayatollah Khomeini. Under the sacred law of Islam *(Sharia),* there is no clear distinction between the state and religion. Throughout its history, there have been disputes within Islam about the correct basis for the law *(Quran* or *sunna),* and tensions routinely surface between more orthodox and more pragmatic approaches to Islam. One area of major contention in the recent past has been the proper role of women.

Women Under Islam

In Islam, women are regarded with a mixture of fear—as a source of evil—and of paternalism—because they are vulnerable and in need of protection. They are seen as being dominated by "unruly passion," in contrast with the "calm and orderly nature of men." Men have thus been given a status above women and authority over them. Women are thought of as being threatening to the stability and good judgement of men. For this reason it is extremely important that their sexuality should be under the control of men. They must be modest, their "adornments" concealed by a veil except in the intimacy of their own bedrooms. They should be married as soon as possible and "give themselves unquestioningly to their husbands."

There is justification in the Quran for these ideas. Yet it also contains what amounts to a charter for marriage as a flexible contract between two consenting adults. In spite of this, after the revolution in Iran, Quaranic justification was used for the reversal of rights which Iranian women had won—rights to education; to leave off the veil; to vote; to contest the custody of children in the case of divorce; to abortion on demand; and a ban on polygamy. Women were required again to wear the veil; their rights in marriage were annulled while men were allowed up to four permanent wives and were given exclusive rights to divorce at will. Some religious leaders equated unmarried women with terrorists; their approval of polygamy was in spite of a Quranic injunction "to marry only one wife." (Harriss 1991, 219). Used by permission.

Gender Struggles After Immigration

The largest concentration of Iranian immigrants in the Western World is in Southern California, where an estimated 600,000 Iranians resided in 1999. Gender relations in this community are a source of conflict, as women's increasing independence clashes with the teachings of Islamic fundamentalism. Between 1980 and 1990, the employment rate of Iranian women in the Los Angeles area almost doubled, from 27% to 48%. Such extrafamilial resources provided leverage for many Iranian women to resist restrictive customs, and divorce rates rose quickly (with 66% reportedly initiated by women). Some Iranian women began to enter into marriages that were virtually indistinguishable from those in the gen-

Antifeminists, on the other hand, are on board a ship floating on a very different tide. Their movement consists of women who are largely encapsulated in traditional female roles and spend most of their time in situations where they have little contact with more cosmopolitan occupations. Here we can say that the social structure is one of **high ritual density;** instead of being in cosmopolitan networks, these women are confined largely to the local routine of family, neighborhood, or gender-segregated job. The theory of interaction rituals (Collins 1988, 187–226) predicts that there is a high emphasis on conformity to traditional, particularistic symbols. Instead of dreaming of ideals of universal equality or the ascendance of feminist values, these women give their loyalty to their family or their community tradition. Because religion is one of the strongest examples of a set of traditional symbols and rituals, it is not surprising that antifeminist movements are often attached to the most traditional churches. In fact, such churches draw on the same social base as do antifeminist movements.

eral population, but others struggled against men who tried to control them according to the old customs.

One Iranian man living in the United States complained that his imported Iranian wife had become too Americanized, so he flew back to Iran for his second arranged marriage. After he brought his new wife Nili (a pseudonym) back to Los Angeles, he ordered her to stay in his apartment and refused to let her learn English for a year. When she questioned his tyranny, he quoted religious doctrine calling for women to "submit" to their husbands.

Nili loved Iran, but she found her country's strict laws controlling women—reimposed after the revolution—a cruel and capricious torment. It depressed her to see pictures in the newspaper of adulterous wives being stoned. Her working-class family married her at 16 to a man she had glimpsed from behind a veil. When she bore no children, he took a second young bride. Eventually, he told Nili he could not afford two wives, and he divorced her—knowing full well he was condemning her, at 28, to live

in a stigmatized social limbo. . . . Taboos encircled Nili like an electric fence. When she went back to school, she was forced to walk about without a chaperone, and her neighbors gossiped that she was really going to meet men. Such talk is no small matter in a country where unmarried daughters who become pregnant—or anyone reputed to be a "loose woman"—are sometimes executed by relatives in "honor killings."

So when the Los Angeles Iranian came looking for a wife, her brother, a conservative Muslim army officer, told her it would be best if she married him. Her new husband promised she could go to school in America, but once she got to the United States, he made it clear she was there to cook and clean and take care of two children from his previous wife. He didn't introduce her to anyone. He began to disappear for days to the house of a girlfriend, and when she protested, he quoted religious scripture on the proper role of women. And, she said, he began to physically abuse her. "I told him, 'Where in the Koran does it say you can abuse me all day long?'" Nili said. Without fathers or brothers to speak up for them, unable to speak

English, drive or get a job, such women are worse off than if they had stayed at home. . . .

Eventually, Nili's husband moved in with his girlfriend, taking his children. Nili moved in with her brother-in-law, but her husband kept showing up to rage against her. And her brother-in-law told her something that chilled her blood: Her husband had taken photos of her as she bathed and dressed that he could use to defame her back in Iran. Men who refuse to file for the wives' immigration papers can use the legal limbo to exact obedience. The threat to have the women deported is more intimidating if the husbands have such photos. One woman went back to Iran with her husband, not knowing he had sent such photos home, and was given a public beating by a religious committee when she stepped off the plane. . . . Today Nili, 32, works in a store and is finally learning English. She keeps her story to herself. . . . Women who come forward and speak out against their husbands are seen as violating the sanctity of the family.

Source: Anne-Marie O'Connor, "New Lives for Women from Iran," *Los Angeles Times* (Dec. 10, 1998) A1, A16–18. Reprinted with permission.

WHO'S WINNING? The 1980s were a period in which the women's movement, after its spectacular growth in the previous decade, appeared to slow down. Antifeminist movements mobilized around particular issues, such as opposing the Equal Rights Amendment (successfully) and becoming involved in the crusade against abortion. On these political and legal issues, particular events shifted from year to year, and undoubtedly they will go on shifting somewhat unpredictably through the next few decades. It is important to bear in mind the image of two ships floating on two different tides. The female population, like modern society in general, is stratified in complex ways. Neither of the two big "waves," or segments, is going to disappear; however, the long-term flow of the tide has been in the direction of expanding the well-educated, middle-class population of women and decreasing the segment that consists of full-time housewives and perhaps even the portion in traditional gender-segregated jobs.

Although gender discrimination still exists, fewer young people are willing to tolerate outright gender segregation.

Between the feminist and antifeminist movements, we would have to say that the initiative has always come from the feminist side, whereas the opponents tend to react against them. Chafetz (1990) points out that the strongest determinant of an antifeminist movement anywhere in the world is the fact that a feminist movement already exists. Since the feminist movement itself emerges because the underlying economic tide has already started to flow, the existence of these movements means that changes are already under way. As Chafetz argues, women's movements act primarily to expedite a process already in motion.

What about the role of men in all this? On the overt level, men react in various ways. Some attempt to protect their existing advantages, holding on to male-dominated jobs. As we will see in subsequent chapters, men continue to use their balance of resources when they can to keep power in the home (chapter 11). Men often rely on so-called "traditional" family practices to maintain their dominance over women. As *A Closer Look* 5.4 shows, however, when women gain economic resources, they are able to exercise more freedom at home. Many men feel threatened by women's assertion of independence, and when violence is their only resource, some men use it to reassert symbolic dominance in ugly outbursts of spouse abuse (chapter 13). Because men are the dominant members of traditional organizations, including religious and political ones, it is not surprising that men should be prominent in the leadership of organized antifeminist movements and other conservative groups (Chafetz and Dworkin 1987).

However, even this is subject to social forces for change. Men are also very prominent in the kinds of cosmopolitan networks, in the professions and educational institutions, that promote universalistic values. Therefore, many men believe in the values of gender equality (to what extent they actually put them into effect in their own families, however, we shall soon see). Perhaps even more importantly, these egalitarian ideals are also reinforced by the way the class structure has been changing. The most important determinant of stratification now has become whether a family has one or two income earners. Large numbers of middle-class men are strong supporters of women's rights because these rights make a huge difference to their own lifestyle. The most important way that a family gets into the upper-middle class is to have two middle-class careers in it (as we saw in chapter 1). Hence, many men now have a stake in gender equality too. This is one more reason why we may expect that the move toward the modern family structure will continue.

SUMMARY

1. The present is a time of considerable changes in the American family. Rates of premarital intercourse, nonmarital childbirth, and divorce have gone up. People are putting off marriage to later ages, and the birthrate has declined. Some observers have suggested that the family is in the process of disappearing.

2. Periods of major change in the form of the family have occurred before in history. We seem to be going through another important period of family change rather than seeing its disappearance.

3. Some new variants on family structure have become important, including the temporarily cohabiting couple and the string of marriages and divorces sometimes referred to as "serial monogamy."

4. The main trend is toward diversification of family forms rather than a single standard form. The nuclear family of father, mother, and minor children now makes up only about a quarter of all households. The most common form consists of a married couple living with no children.

5. The 1950s, characterized by early marriages, the "baby boom" in birthrates, and a declining divorce rate, were an anomaly in the long-term trends of the twentieth century. One possible cause may have been the movement of working-class families, for the first time, into the *physically middle-class* single-family dwellings of suburbia, while importing working-class family styles.

6. The 1920s and the 1960s were comparable periods of liberalization in sexual manners and family formation. In both cases, the new style began in the youth culture and spread rapidly to older age groups.

7. Family changes of the last two decades are also connected with the "second wave" of the social movement for women's economic and occupational rights. The increasing mobilization of women through higher education and their shift into full-time careers are the major factors producing changes in the structure of the modern family.

8. Women in middle- and upper-middle-class careers have redefined women's status as depending on occupational success. Traditionalist and antifeminist movements have emerged in reaction, to defend the status of traditional female roles. Women in cosmopolitan networks favor universalistic ideals of equality, while women encapsulated in gender-segregated roles conform to symbols of loyalty to family and community.

9. Feminist movements appear when economic conditions have already begun to change and attempt to accelerate the change. The weakness of antifeminist movements is that they are reacting against changes already occurring. Men take part on both sides of these conflicts. Because contemporary stratification now depends on the economic advantages of two-income families, many men now have a stake in the gender-egalitarian family as well.

Key Terms

baby boom
birthrate
cosmopolitan
counterculture
divorce rate
familism

high ritual density
household types
marriage rate
nonmarital birth
nuclear family

rebellious sixties
roaring twenties
secularization
suburban working class
women's movement

Sociology Web Site

See the Wadsworth Sociology Resource Center, "Virtual Society," for additional links, quizzes, and learning tools:

http://www.sociology.wadsworth.com

Also on this web site you'll find InfoTrac College Edition, an online library of journals. Here you can search for electronic articles about central topics in sociology.

Families and Work: The Economy, Social Class, and Inequality

6

INTRODUCTION

One of the most important facts about families is that they are not all alike. There are considerable differences in family lifestyles, and among the most important of these are differences resulting from social class.

What do we mean by *class?* Classes are different categories of persons, usually thought to be socially unequal. It should be admitted right away that sociologists are not in agreement about what measures should be used as the bases for class ranking. Should they include income, education, social prestige or status, occupation, ownership or nonownership of property? Our belief is that all of these should be taken into account; in other words, we should deal with a multidimensional theory of social class. For purposes of introduction, it is easy enough to reduce most of these to a single dimension, because persons possessed of one indicator of a certain status tend to have the other characteristics as well.

The basic idea of this chapter is that family life and family structure are profoundly affected by the economy. This is most evident when we look at economic disruptions, such as the industrial plant closings in the American "steel belt" during the 1970s, the farm foreclosures in the American midwest during the 1980s, and the layoffs in the defense and aerospace industries in the 1990s. Extreme changes in the national or local economy place enormous burdens on families to survive on limited resources, but the economy also influences family structure and family interaction when it is running smoothly. How families react to localized economic conditions and how they are stratified by their positions in the larger economy are the subjects of this chapter. We will first take a look at the simplest means of ranking families—in terms of income. Then we will examine the lifestyles of two very different kinds of families: working class and upper-middle class. Finally, we will discuss labor market trends that affect women's job opportunities and begin to explore how families are responding to a changing economy.

STRATIFICATION AND WEALTH

The most obvious basis of **social stratification** is how much money people have. This is surprisingly little analyzed by sociologists, who tend to regard money as too vulgar a measure. It is often pointed out, for instance, that money is not equivalent to status. The truck driver who earns $50,000 per year will still have lower status than the schoolteacher who earns $30,000 and will have a different lifestyle as well. Similarly, radical Marxian theories emphasize that the quantity of money is less important than how one earns it: wage earners, however much they make, have a different class interest and different class conflicts than business owners. These points are valid enough, but they are often overstated. The plain fact is that one of the most obvious differences among people is how much money they have, and money itself has major effects on lifestyle, family crises, politics, and almost everything else.

Even though psychologists and other scientists try to identify generic patterns of family life and consistent stages in human development, all these processes are influenced by the availability of money. For example, how does a child's capacity for thinking develop? The amount of money a family has affects things like the quality of housing, the types of toys in the home, and the number of books available to a child. Because of money—or the lack of it—a child's perceptual, cognitive, and motivational development will be different, even when

A CLOSER LOOK 6.1
What Is an "Old Family"?

Members of the upper class like to pride themselves on belonging to *old families*. But stop and think about it: all families are equally old, because everyone has grandparents, great-grandparents, and so forth as far back as you wish to go. What they are really saying is that their families have had a lot of money for several generations. This is never stated openly, because claiming status on the basis of their money would be considered the sign of an upstart, a *nouveau riche*. Part of upper-class culture is the belief that talking about one's money is vulgar. Instead, they like to claim that their superiority is based on their better taste, their contributions to charity, their "public service," and above all on their breeding and their families—as if they were the only ones who had families more than a generation back. Translated, this is a covert form of boasting because families who have a great deal of money can invest it in education, travel, acquiring a taste for the arts, and other learning experiences that set them apart from people who have merely money and not "background." Of course, such an investment sets them apart even more from people who have neither money nor the kind of culture that one can buy with it.

two children start out the same or when their parents use the same child-rearing techniques (Elder and Caspi 1988). As we will see in this and later chapters, however, parents from different social classes also tend to act differently, primarily as a result of environmental factors linked to the availability of money.

Most of what we know about the influence of money on families comes from studying the poor. Researchers have consistently found that the life chances of the poor are jeopardized by their lack of financial resources. People with fewer economic resources have the highest infant mortality rates; they are more likely to have premature or mentally retarded babies from prenatal malnutrition; they have a shorter life expectancy; and they are more likely to die from accidents, tuberculosis, influenza, and pneumonia. They are also more likely to get sick and stay sick longer because of deficiencies in diet, sanitary facilities, shelter, and/or medical care. In addition, people with limited incomes are less likely to go to college and are more likely to be arrested, found guilty, and given longer sentences for a particular violation. Finally, families with limited incomes are more likely to experience child abuse, spouse abuse, divorce, and desertion (Children's Defense Fund 1994; Council of Economic Advisors 1998; Eitzen 1983; Lichter 1997; United Nations Development Programme 1998).

Lack of resources in poor families clearly limits one's life chances, but an abundance of resources is equally consequential. It is an oversimplification to claim that the upper and middle classes derive their status more from culture than from wealth, for the marks of an upper- or middle-class lifestyle—taste in clothes, housing, and entertainment; educational attainment; familiarity with the arts; and savoir faire (social know-how and tact)—can largely be acquired with money, especially over generations. The upper classes usually have a high level of culture because their families have had money long enough to spend it on education, art, the opera, and so forth (*A Closer Look* 6.1). The truck driver who earns $50,000 a year has lower prestige partly because his (or her—but mostly his) income has been earned only recently, and he has not had time to translate it into a genteel lifestyle. Also, the income earned by the working class is generally much less secure; the truck driver may have a few good years, but he often has a short earning span that is vulnerable to sudden disruptions. Middle- and upper-class occupations have more financial security built into them. One of the main differences among social classes, then, is money, except that we should bear in mind that *long-term wealth has a much stronger social effect than short-term wealth*. For this reason, it is important to look at accumulated wealth in stocks, bonds, real estate, and other investments, as well as looking at a family's income or earnings.

Has Inequality Been Increasing?

For the most part, the world's wealth is concentrated in the hands of very few families. Recent estimates show that the world's 225 richest people have a combined wealth of over $1 trillion, equal to the annual income of the poorest 47 percent of the world's people (approximately 2.5 billion) (United Nations Development Programme 1998). The United States ranks first in the world in the number of ultrarich people, with 60 of the 225 (combined wealth $311 billion), followed by Germany with 21 ($111 billion), and Japan with 14 ($41 billion). Popular stereotypes of the ultrarich suggest that sheiks, sultans, kings, and dictators would be in the majority, but most of the world's rich come from modern industrial democracies. Industrial countries have 147 of the richest 225 people ($645 billion combined), developing countries have 78 ($370 billion combined), Arab states have 11 ($78 billion), and African states have just two (under $4 billion—both from South Africa) (United Nations Development Programme 1998). The concentration of wealth in the hands of relatively few families is related to the dominance of industrial nations in the world economy and the role of corporations and governments in regulating global markets. For families, the important thing to remember is that class position determines resources and shapes consumption patterns. On a global scale, the 20 percent of people who live in the richest countries consume 11 times as much meat, 7 times as much fish, 17 times as much energy, 49 times as many telephones, 77 times as much paper, and 145 times as many cars as the 20 percent of people who live in the poorest countries (United Nations Development Programme 1998). These global disparities in income and consumption patterns, like the differences between rich and poor within any particular country, are important considerations when analyzing the forces that shape what goes on inside of families.

In the United States, 1 percent of the population owns over one-fourth of the combined market worth of everything owned by every American. Contrary to popular belief, the proportion the wealthy own has remained about the same since the 1940s, and for some forms of wealth, the concentration at the top has increased. For instance, even though the use of 401(k) plans has widened the extent of stockholding, roughly 60 percent of U.S. families holds no stock whatsoever, and half of all stock held by U.S. families is held by the best-off 5 percent (Mishel 1997). We know something about the most visible of these families—the Rockefellers, DuPonts, and Mellons—but a look at the *Social Register* of every major American city would yield additional less affluent, but fantastically wealthy families. By controlling corporations and influencing the policy process, these families are able to maintain their economic and social position (Domhoff 1990; Parenti 1983; Zarsky and Bowles 1986). (See *A Closer Look* 6.5 for a brief look at some aspects of elite upper-class families.)

Despite the fact that so much wealth is concentrated in the hands of a few, many assume that Americans are coming to share a bigger portion of the economic "pie." That seemed especially true when the size of the total pie was increasing. From 1940 to 1974, real earnings of Americans improved steadily, but since then, real earnings have stagnated. Figure 6.1 shows that the income

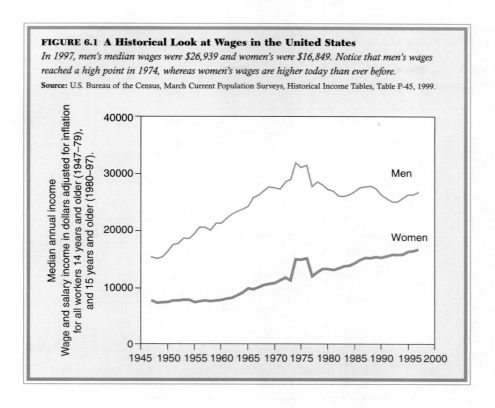

FIGURE 6.1 A Historical Look at Wages in the United States

In 1997, men's median wages were $26,939 and women's were $16,849. Notice that men's wages reached a high point in 1974, whereas women's wages are higher today than ever before.

Source: U.S. Bureau of the Census, March Current Population Surveys, Historical Income Tables, Table P-45, 1999.

of male workers peaked in 1974 and has fluctuated since. The wage and salary income of women still lags well behind that of men, but the trend for women has been more steadily upward (see Figure 1.3). Because of the drop in men's wages, most Americans are less well off than their counterparts born earlier. For the first time since the Great Depression of the 1930s, American men face the prospect of lower lifetime real earnings than men born ten years before them.

The economic situation for working-class families seems to be diverging more and more from that of upper-middle-class families. Although most reports reveal that the overall economy is strong, it is not equally strong for all workers. Starting in the late 1980s, hourly wages fell among the bottom 80 percent of the workforce. Several government data sources show that the wages of high-wage workers grew whereas those of the rest of the workforce flattened or declined (Bernstein and Mishel 1997). Most observers agree that the life chances of working-class men have deteriorated over the past thirty years, even though working class women have relatively better job opportunities. As noted later, the new class divisions increasingly revolve around the number of workers in each family.

How Families Make Up the Income Hierarchy

In spite of popular rhetoric to the contrary, it seems that the rich keep getting richer and the poor keep getting poorer. During the 1980s and 1990s, income inequality actually increased. The changes were not primarily the result of increasing unemployment but stemmed from stagnant or declining incomes

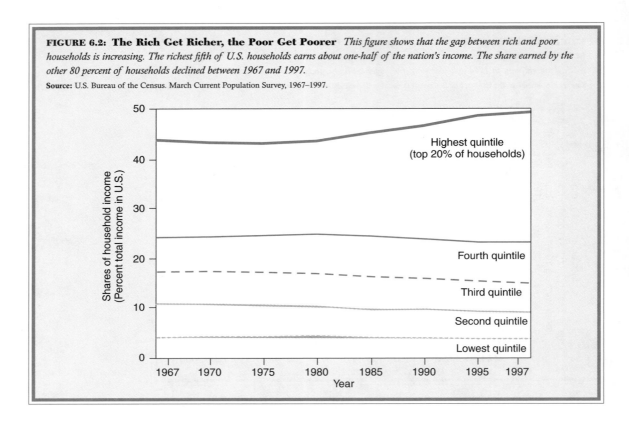

FIGURE 6.2: The Rich Get Richer, the Poor Get Poorer *This figure shows that the gap between rich and poor households is increasing. The richest fifth of U.S. households earns about one-half of the nation's income. The share earned by the other 80 percent of households declined between 1967 and 1997.*

Source: U.S. Bureau of the Census. March Current Population Survey, 1967–1997.

among employed people. The number of full-time, year-round workers with low earnings has been rising, and the real mean income of the majority of households has been going down or staying about the same. The share of aggregate income earned by the top fifth of all American households rose between 1980 and 1997, whereas the share earned by the lower four-fifths stayed about the same or dropped (figure 6.2). Figure 6.3 shows the way income is distributed among families in the United States and gives the occupations of both husband and wife. It also shows people who are not in the labor force (housewives and househusbands, the unemployed, and retired persons), and it indicates whether families consist of married couples or single-family heads (widows and widowers, divorced persons, unwed mothers, etc.).

Several features should strike the eye. First, notice that occupations are distributed across the various income categories. Managers and professionals dominate the higher income brackets (and go way up to the top: this part of the income distribution goes up so high, relative to where most people are, that it would go far off the top of the page); but some managers and professionals are much further down, some even below the poverty line. This is partly due to the fact that people at the beginning of their professional careers may barely eke out a living; it also reflects the fact that some business managers (like the owner of a small cigar stand) may not be employed in a very lucrative establishment. Other occupations follow similar patterns. Each occupation clusters around some income level, with white-collar jobs higher on the average than blue-collar jobs. Some blue-collar families, however, especially those of skilled workers (plumbers, electricians, heavy-equipment operators), are in the higher income brackets.

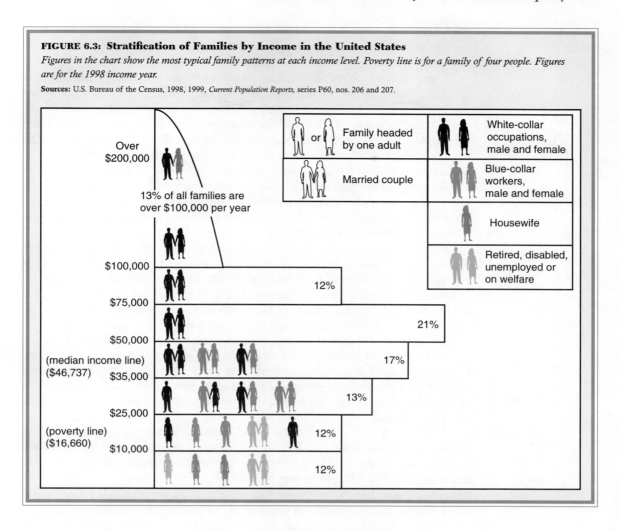

FIGURE 6.3: Stratification of Families by Income in the United States

Figures in the chart show the most typical family patterns at each income level. Poverty line is for a family of four people. Figures are for the 1998 income year.

Sources: U.S. Bureau of the Census, 1998, 1999, *Current Population Reports*, series P60, nos. 206 and 207.

A second feature to notice is that the income structure is not a triangle, widest at the bottom and narrowing toward the top. Instead, it is narrower at the bottom, bulges out toward the middle, then shrinks inward until it reaches the higher income brackets, where it suddenly becomes a tiny spike (it would look narrower if it could be drawn to scale). One fact that this demonstrates is that the poorest people in the United States are not the majority. That is one reason—though there are others—why they have not been a revolutionary force or even able to claim very much support in the form of welfare and other social reforms.

Third, notice how important it is for people to be in the labor force. The two poorest segments of the population are heavily made up of people who are retired, unemployed, or on welfare. Social security and welfare payments do not amount to much, despite some well-publicized cases that make it seem as though welfare recipients are driving Cadillacs and living in luxury. Whatever welfare fraud there may be, most people are better off working.

A fourth point worth noticing is that marriage can make a tremendous difference in economic standing, especially when both the husband and wife are employed The poorest group on the chart consists almost entirely of families headed by a single adult (mostly women). Many of these people are also outside the labor force, which makes them doubly poor. The next category up also contains many households with only one adult. This person is working, but there is only one income.

When we look above the median income line on the chart, a striking fact emerges: most middle- to upper-middle-income families in the United States have two incomes. Notice, for instance, that quite a few blue-collar workers (especially skilled workers) fall above the medium budget line. Virtually all of them are there because both husband and wife are employed, usually with the woman in a clerical job. The same is true for white-collar families. Male professionals, managers, and clerical/sales workers are most likely to be in the middle to upper-middle range of income if their wives are employed. If both husband and wife are working in high-income white-collar jobs—managers and professionals—their combined incomes put them not merely into the middle class but into the affluent reaches of the upper-middle class.

Some families maintain the traditional pattern, with the man as sole bread-winner and the woman as housewife. Such arrangements are found in both white-collar and blue-collar families, and both of them pay a price. The blue-collar family in which only the husband has an income is almost always doomed to stay in the bottom half of the income distribution; the white-collar family has a good chance of not making it out of the lower-middle-class income bracket. Of course, unemployed wives do make economic contributions that do not show up on income statements, and some husbands and wives place great value on at-home wives' emotional contributions to a family, especially during children's early years. One might say that traditional men who do not want their wives to work or wives who prefer the traditional role are paying a serious economic price for their tradition.

There is only one section of the labor force where this is not true. That is at the top. Up on the needle of income distribution, where it rises into the salary range from an affluent $100,000 per year on into the stratosphere of salaries in the range of $200,000 and higher, families revert to the traditional gender-role pattern. Most of the top earners in this income range are men in the managerial and professional categories, but they are not in ordinary managerial and professional positions. These managers are the corporate executives, the managers of the few hundred giant corporations that dominate the American economy. The professionals are the most successful practitioners in a few of the most prestigious specialties: medical doctors, lawyers in the elite law firms (which work especially for corporate clients), and a few others. In this group, the men earn so much by themselves that it makes little difference whether their wives add to the family income. They can afford a traditional family lifestyle; as we shall see, quite often they demand it (see *A Closer Look* 6.5 for a description of the activities of upper-class women). The difference between the top of the income distribution and the great bulk of the families that make it into the affluent upper-middle levels by having two family incomes is one of the most important splits within the structure of stratification in America.

Fifth, the stratification system, to a considerable degree, is based on gender. This is true in the direct sense that women are much more apt than men to occupy the low-ranking and poorly paid jobs. It is also true that if women who head households are not employed—because they are on welfare, retired, or unwed mothers—they almost always end up in the poverty class. Men are less apt to head households if they cannot provide—one might say the women are left "holding the bag." In retirement, men are much more likely to have reasonably adequate pension benefits.

Families headed by women represent the fastest growing poverty category. By 1990, more than half of the families below the official poverty level in the United States were female-headed households with no husbands present.

Families in Poverty

The number of Americans living in poverty has grown substantially since the early 1970s. Official poverty income thresholds were set in the early 1960s at three times the income necessary to maintain an economy food plan devised by the U.S. Department of Agriculture for families of various sizes. These rates are adjusted each year for inflation using the Consumer Price Index (CPI) and used by governmental agencies to establish criteria for various social insurance programs and public assistance. In 1998 the poverty threshold for a family of four was $16,660. Using the official guidelines for each year, about 23 million Americans, or 11 percent of the population, lived below the poverty level in 1974. By 1998 there were about 35 million people below the poverty line, representing 13 percent of the population. The poverty rates for African Americans and Hispanics are twice the rates for whites (chapter 2). When welfare and other government money transfers are excluded from the calculations, over 50 million Americans, or over one-fifth of the population, are below the official poverty level (U.S. Bureau of the Census 1998b).

Families headed by women represent the fastest growing poverty category. By the late 1990s, more than half of the families below the official poverty level in the United States were female-headed households with no husband present. The growth of female-headed households is the primary reason why greater numbers of people are susceptible to poverty than in the recent past. Also, because female-headed households are likely to include children, four of every ten poor people in the United States are now under the age of eighteen (Lichter 1997; U.S. Bureau of the Census 1998b). In fact, the poverty rate for most age groups has declined or leveled off since the early 1970s, but it has increased for children. In 1997, one of five American children under eighteen was being raised in a family below the poverty level. Children under the age of six were particularly vulnerable. In 1997, the overall poverty rate for children under six was 22 percent. Those living in single-mother households had a poverty rate (59 percent) that was over five times the rate for their counterparts in married-couple families (11 percent) (U.S. Bureau of the Census 1998b). The rate of child poverty is at a thirty-year high, and the income gap between rich and poor children is greater than at any time in recent memory (Fisher et al. 1996; Lichter 1997).

The increasing likelihood that women and children will live below the poverty level has been labeled the **feminization of poverty.** Because about half of all U.S. children are projected to spend some time living in a single-parent household, the proportion of the nation's poor who are children is projected to rise. As we discuss in more detail in the next chapter, a disproportionate share of families living in poverty are racial minorities. Over one in four African Americans and Hispanics live below the official poverty level. The situation is even worse for children, with 40 percent of African American and Hispanic children living in poverty, as compared with fewer than one in five white or Asian American children (Council of Economic Advisors 1998). The rates of child poverty have been falling for African Americans and increasing for Hispanics, but public perceptions about welfare mothers continue to be shaped by prejudicial attitudes (see *A Closer Look* 7.1 in chapter 7). Even though people of other races are more likely to be poor than whites, about two-thirds of all people below the poverty level in 1997 were whites (U.S. Bureau of the Census 1998b). Prospects for a reversal in the feminization of poverty—or more accurately, the "childhoodization" of poverty—are slim because American politicians have been reluctant to commit resources to poor children. As Table 6.1 shows, the United States ranks high in initial child poverty compared with other industrialized nations, but because we do so little to help them, we top the list in the proportion of children who are poor after government assistance.

Family Events Cause Income Fluctuation

Researchers have found that changes in income are fairly common for households in the United States. For instance, only about half of people in the bottom fifth of the income pyramid remain there a decade later, and every year about as many families fall into poverty as climb out of it. For this reason, researchers have begun to measure income on a month-to-month basis, computing episodic and chronic poverty rates. In the mid-1990s, for example, most people who experienced poverty were poor for a few months, with the median spell lasting 4.5 months (U.S. Bureau of the Census 1998c, 3). Figure 6.4 shows how episodic and chronic poverty rates differ by race/ethnicity and family types. Like the annual poverty rates noted earlier, both of these measures of poverty are higher for nonwhites and for single mothers. The differences are largest for chronic poverty (this rate is for people who were continuously below the poverty rate for two years). The rates for African Americans and Hispanics were over three times higher than for whites, and the rate for single mothers was over eight times higher than for married couples. For episodic poverty, on the other hand (which measures just those who were poor for two or more months in a year), the differences by race/ethnicity and family type were smaller—though still quite large. Whites were half as likely as African Americans or Hispanics to have

Table 6.1 Child Poverty in Seventeen Developed Countries

The United States ranks high in child poverty before public assistance and does less than other developed countries to lift poor children out of poverty.

Country	Percent of Children in Poverty		Percent of Children Lifted Out of Poverty by Government Programs
	Before Assistance	After Assistance	
United States	25.9	21.5	17
Australia	19.6	14.0	29
Canada	22.5	13.5	40
Ireland	30.2	12.0	60
Israel	23.9	11.1	54
United Kingdom	29.6	9.9	67
Italy	11.5	9.6	17
Germany	9.0	6.8	24
France	25.4	6.5	74
Netherlands	13.7	6.2	55
Norway	12.9	4.6	64
Luxembourg	11.7	4.1	65
Belgium	16.2	3.8	77
Denmark	16.0	3.3	79
Switzerland	5.1	3.3	35
Sweden	19.1	2.7	86
Finland	11.5	2.5	78

Source: Adapted from William P. O'Hare. 1996. "A New Look at Poverty in America." Population Bulletin, vol. 51, no. 2, Washington DC: Population Reference Bureau, p. 36; Lee Rainwater and Timothy M. Smeeding. 1995. "Doing Poorly: The Real Income of American Children in a Comparative Perspective." Working paper No. 127, Luxembourg Income Study, Maxwell School of Citizenship and Public Affairs. Syracuse, NY: Syracuse University.

episodic poverty, and married couples were one-third as likely to experience such short-term poverty spells as single mothers (see Figure 6.4).

The single most important factor accounting for changes in poverty or family income is a change in family structure: divorce, death, marriage, birth, or a child leaving home. In other words, changes in economic position of families are highly dependent on changes in the composition of families. When someone gets married, he or she often changes from living in a one-earner household to living in a two-earner household. When couples have children, parents often don't have time to work as many hours at their jobs. When people divorce or become widowed, they typically lose the benefit of having the other person's income. Although family composition changes provide the best explanation of changing economic position, other events are also important and reflect a certain fluidity in income levels. For instance, movement into or out of the labor force, such as being laid off from a job or taking a new one, has substantial effects on family income. Because women's hourly pay rates and employment hours are lower than those of men, women's movement into and out of the labor force produces an average change in annual income that is only about a quarter of that produced by men's movement into or out of the labor force (Duncan et al. 1984). For those who remain in the labor force continuously, variations in economic situation reflect changes in work hours brought about by second jobs, overtime, job changes, and brief spells of unemployment. However, these changes have less impact on family earnings than the sorts of family structure changes noted earlier.

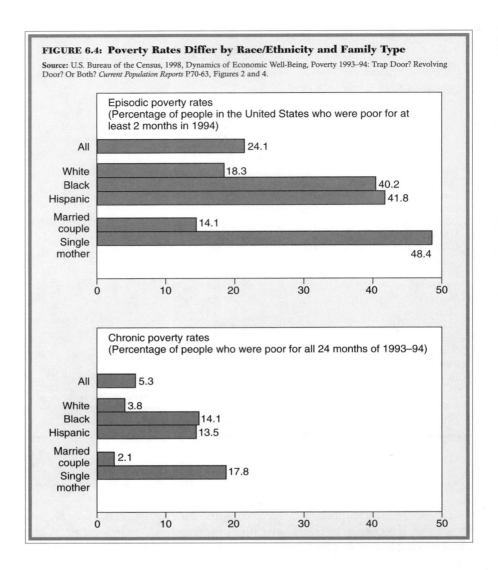

FIGURE 6.4: **Poverty Rates Differ by Race/Ethnicity and Family Type**

Source: U.S. Bureau of the Census, 1998, Dynamics of Economic Well-Being, Poverty 1993–94: Trap Door? Revolving Door? Or Both? *Current Population Reports* P70-63, Figures 2 and 4.

Episodic poverty rates
(Percentage of people in the United States who were poor for at least 2 months in 1994)

- All — 24.1
- White — 18.3
- Black — 40.2
- Hispanic — 41.8
- Married couple — 14.1
- Single mother — 48.4

Chronic poverty rates
(Percentage of people who were poor for all 24 months of 1993–94)

- All — 5.3
- White — 3.8
- Black — 14.1
- Hispanic — 13.5
- Married couple — 2.1
- Single mother — 17.8

Even though income changes occur frequently, most people and families maintain the same overall class position. That is, those who hold working-class jobs may go through bouts of unemployment or overtime work, but they are unlikely to move to upper-middle-class jobs. Similarly, entry into the elite upper class is severely restricted, and the wealthiest families control about as much of the country's wealth as they did a half-century ago. For that reason, sociologists talk about stratification by social class as a relatively stable phenomenon. We will see how families play a big part in the maintenance of this stratification system by preparing children to occupy different positions in the class structure. More immediate changes in a family's economic situation are the result of having one earner or two earners in the family, with the majority of families now composed of two earners. Changes in family composition, particularly through marriage or divorce, are now the most important precursors to changes in the economic position of families. As we shall see in later chapters, the downward mobility experienced by women after divorce, particularly if they have not been previously employed, reflects a profound change in class position. We will also see how the economic disadvantage of being in a mother-headed family has made more children poor than ever before.

THE MIXTURE OF SOCIAL CLASS STYLES IN AMERICA

We must face one important difficulty in dealing with social class in American society. Except for the very rich and the very poor, social class differences are not highly visible. Between the extremes, there are no sharp barriers between different occupations or income levels. The neighborhoods or suburbs in which people live are not usually divided into "working-class areas" and "middle-class areas," as was true earlier in our history and as is still likely to be the case in Europe. A majority of American manual workers own their own homes, and the areas in which they live include a proportion of white-collar workers as well, and very likely some business owners (perhaps of small businesses), and professionals and technical people of various levels. Only the upper-middle-class and upper-class neighborhoods are likely to be class segregated, consisting of expensive homes of the most successful lawyers, doctors, administrators, business managers, and others. Because neighborhoods are likely to be divided by race, Americans tend to be much more race conscious than class conscious.

Similarly, today's Americans tend to dress in much the same styles of clothes, no matter what their social class. At least this is so when they are not working. Factory workers, delivery truck drivers, or repair persons may wear special work clothes, not at all like the suits worn by business executives, but their leisure outfits are much the same. Moreover, some high-level personnel (scientists, professors, engineers, mass media personnel) tend to wear casual clothes at work, as do most college students. Though there are subtle differences in styles, it is no longer possible to tell people's social class at a glance, except when they are actually in their work situation. Furthermore, our leisure-centered society tends to have much the same culture shared by most people. The same professional spectator sports are followed by most fans; the same popular music, television shows, and movie stars are widely admired across social class lines.

It is true that there is also a "high culture" style of reading "serious" books and going to the opera, the theater, art museums, and "art" movies (see *A Closer Look* 6.5). However, even in the highly educated section of the upper and middle classes, only a *minority* of persons participate much in this "high culture"; the popular cultural style of TV, professional sports, and pop music has been widespread even in these social classes (Halle 1984; Gans 1974). At most, we can say that there is a distinctive elite culture only within a highly educated fraction of the upper-middle class, consisting especially of people who are themselves cultural specialists (teachers, musicians, aspiring artists, and the like). For the rest of the population, blue collar, white collar, and managerial alike, there seems to be little in the way of class difference in cultural lifestyle. This is one reason why Americans are not very conscious of social class, except at the extremes.

Does this mean that social class is no longer of any importance in explaining how people will behave and think and what kinds of family lives they have? That conclusion would be too hasty. Social class factors continue to operate, even if people are not aware of them as such. The economic pressures of different kinds of career trajectories have a strong effect on such things as whether one marries, the timing of marriage or children, and the kinds of family crises that occur. Also, there are subtle cultural patterns that affect child-rearing styles, sociability, and the status concerns of the family members. We shall see, for example, that the higher social classes tend to be career oriented and to subordinate everything else to their careers, whereas the working classes are more detached or alienated from their jobs and stress the leisure and personal sides of their lives. The higher social classes tend to be involved in cosmopolitan social networks, whereas the working classes are more localistic and kin centered; different styles of thinking tend to follow from those social patterns. We shall examine these differences in more detail in this chapter.

Are White-Collar Jobs Middle Class?

There have been many debates among sociologists about how "working class" and "middle class" ought to be defined. Our ordinary terminology distinguishes "blue-collar" and "white-collar" jobs: on one side are people who do manual work, usually wearing special work clothes because they get dirty on the job; on the other side are people who work in offices at clean jobs in their ordinary (if somewhat dressed-up) clothes. Nevertheless, the distinction may no longer be important. Skilled manual workers (electricians, mechanics, etc.) or those with strong labor unions often earn more than many white-collar workers. This is one reason why these manual workers can live in the same neighborhoods, wear the same leisure clothes, drive the same cars, and afford the same entertainment as white-collar workers. Moreover, the distinction between manual and nonmanual work may be more a matter of image than reality; many "white-collar" workers spend their time operating copying machines, typewriters, and computer keyboards and cash registers, whereas "manual" workers may be checking dials and switches and writing tallies. In short, the jobs may be different in their prestige and public image, but the actual work of both may be much the same: repetitive, boring, and subject to the authority of supervisors. Hence, sociologists such as Erik Olin Wright (1985, 1997) have argued that the majority of lower white-collar jobs are really working class, and other jobs entail a mixture of class conditions.

If this is so, we should not be surprised that neighborhoods tend to mix white-collar and blue-collar workers, or that both of these occupations tend to wear the same styles of clothes and have the same culture. In other words, the "working class" in America is bigger than we would believe if we used the traditional criterion of manual occupations. Perhaps we need some new terms such as "blue-collar working class" and "white-collar working class." In fact, public consciousness has already arrived at this conclusion, except that it refers to all these people as "middle class" rather than "working class." This is an example of how status consciousness in the United States is skewed in an upward direction, allowing most people to inflate their relative position in their own eyes and in the public impression they make on others.

Husbands' and Wives' Occupations and Mixed-Class Marriages

Finally, we have the question of how we are to define the social class level of a marriage in which both husband and wife have jobs. The traditional method was to classify by the husband's occupation, as if the wife's position did not count in society's class rankings (Sørenson 1994). This introduces a considerable bias: for example, if a male factory worker is married to a schoolteacher, it would be a mistake to consider the family as purely working class. In fact, this sort of combination is quite common. In the working class, wives typically have more education than their husbands, and many wives hold jobs in the lower professions such as teaching, social work, or nursing. Even more common, wives of blue-

Table 6.2 Most Women Hold Working-Class Jobs

White-Collar Working-Class Women

Clerical employees	25%
Retail sales	13%
Technical support	4%
Total white-collar working-class women	42%

Blue-Collar Working-Class Women

Blue-collar workers	10%
Service workers	17%
Farming, forestry, fishing	1%
Total blue-collar working-class women	28%

Middle- and Upper-Middle-Class Women

Professional and technical	17%
Managers and administrators	13%
Total middle and upper-middle class	30%
Total employed women	100%

Source: U.S. Bureau of Labor Statistics, 1998. Employment and Earnings. Current Population Surveys.

collar men have clerical jobs as secretaries, typists, or sales clerks. Hence, the blue-collar/white-collar distinction is crossed in a great many families; a majority of what used to be considered "working class" families on the basis of the husband may really be only partly "working class."

Nevertheless, we have to consider this discussion in conjunction with the points raised in the previous paragraphs; that is, we must determine whether "white collar" occupations really ought to be analyzed as "middle class." We would suggest the following: that secretaries, typists, and most clerks are actually performing working-class types of jobs in office settings and that we should consider them "white-collar working class" rather than "middle class." Federal data on the occupations of employed women show that 70 percent of women workers could thus be classified as working class (Table 6.2).

If we count this way, it turns out that most working-class husbands do have working-class wives, and there are not so many mixed-class marriages after all. Where mixed-class marriages do exist, we should recognize them as a separate category. For instance, a substantial number of working-class men are married to women who hold genuinely middle-class jobs, such as teaching or social work. To be honest, it should be admitted that sociologists do not have much precise information about these kinds of marriages because most of the research has classified the family by the social class of the husband and has not systematically separated out wives who are in genuinely higher occupations. (As we have said, we do not think a bias was actually introduced by subsuming wives who were secretaries and clerks as working class.) A related problem, with which research has not specifically dealt, is *husbands* who are in what we have called the "white-collar working class"; they are more typically referred to as "lower middle class" and have usually been ignored in sociological studies of family styles. We suspect, though, that their lifestyle is similar to that of the manual working class.

Because clerical jobs are still the single most common form of female employment, we should be aware of a rather different form of mixed-class marriage: middle-class husbands who are married to "white-collar working-class" wives. Although recently there has been a push for women to enter the higher-paid professions and administrative and management positions, the occupational

Table 6.3 Common Class Types of Marriages

Husband	Wife	Class Consistency
Poverty class	Poverty class	Consistent
Absent	Poverty class	Consistent
Working class (blue collar)	Working class (blue or white collar)	Consistent
Working class (blue or white collar)	Middle class	Inconsistent
Working class (blue or white collar)	Housewife	Consistent
Middle class	Working class (white collar)	Inconsistent
Upper-middle class	Middle class	Semiconsistent
Upper-middle class	Housewife	Consistent
Upper-middle class	Upper-middle class	Consistent
Upper class	Housewife	Consistent

structure has been slow in opening up to women. Among employed women who are married to middle- and upper-middle-class husbands, clerical positions are still the most common work. Thus, there is a sense in which the class dividing line is most often crossed *downward* from husbands to wives, rather than vice versa. As we shall see, this is one reason why there tend to be strong power differences at home between upper-middle-class husbands and their wives and an underlying source of conflict within many families. Even when both husband and wife hold genuinely middle-class jobs, there is a strong tendency for the husband to outrank his wife because of the general pattern of employment discrimination against women.

The structure of social class in the United States, then, is probably rather different than it appears on the surface. Several class levels and many possible occupational combinations exist for husbands and wives. There are poverty-level persons, those who are either out of the labor force or only sporadically employed. Then there are both blue-collar working-class and white-collar working-class, as well as middle-class (mixed authority positions), upper-middle-class, and upper-class occupations. Husbands and wives could be consistent—in the same class level—or inconsistent. Moreover, we should add cases where women are housewives and cases of single-adult families, where the household has only one source of class position.

Instead of diagramming all these possibilities, we use Table 6.3, which shows only some of the most common class types of marriage.

WORKING-CLASS FAMILIES

It is not possible to describe all the different family types in this chapter. Instead, we have picked out two types from different parts of the spectrum: the working-class family and the upper-middle-class family. Moreover, we concentrate on the class-consistent types. These are working-class men (mostly blue collar, in the studies we have) married to women who hold blue-collar or white-collar working-class jobs or who are housewives. Furthermore, we should emphasize that the most vivid studies of family life focus on the lower part of the working class, where there is more economic stress. For the other type, we deal with upper-middle-class families of the most traditional sort, where the man alone is employed. Because this group is similar to the upper class, we will add some materials about upper-class wives, who are one of the most culturally distinctive groups in the

Idealized images of weddings have encouraged many women to marry early.

stratification of American families. We will also briefly discuss an increasingly popular upper-middle-class family type—the dual-career professional couple. The family types we have chosen are not at the extremes of the class system; they are all substantially employed, but they show the contrasting tendencies that exist beneath the superficial uniformity of cultural styles in America.

Early Marriages

What is family life like in the **working class?** One notable fact is that working-class people tend to marry at an earlier age than their counterparts in the upper-middle class. In the 1970s, working-class women married around age seventeen to nineteen, and working-class men married around age nineteen to twenty-one (Rubin 1976). As women's labor force participation has increased, everyone is marrying later. Children of working-class parents now marry in their early to mid-twenties, compared with the children of middle-class parents who now marry in their mid-twenties to late twenties.

Why do working-class people marry at a younger age? For one thing, working-class youths are less likely to go on to college and therefore have less incentive for putting off marriage. Moreover, they are usually not looking forward to an all-encompassing career. When they finish or drop out of high school, long-term

A CLOSER LOOK 6.2
A Working-Class Woman's Image of Marriage, Circa 1970

> When I got married, I suppose I must have loved him, but at the time, I was busy planning the wedding and I wasn't thinking about anything else. I was just thinking about this big white wedding and all the trimmings, and how I was going to be a beautiful bride, and how I would finally have my own house. I never thought about problems we might have or anything like that. I don't know even if I ever thought much about him. Oh, I wanted to make a nice home for Glen, but I wasn't thinking about how anybody did that or whether I loved him enough to live with him the rest of my life. I was too busy with my dreams and thinking about how they were finally coming true (Rubin 1976, 69).

career goals are often secondary, and going to work is not something they eagerly plan for (Rubin 1986, 1994; Sennett and Cobb 1973). As compared with middle-class youths, working-class teenagers tend to name fantasy careers rather than realistic ones: boys want to be professional athletic stars, and girls want to be singers or movie stars. When those dreams fade, and the excitement of the teen years is over, they are already full-fledged adults, and it is time to settle down.

Another reason for early marriage is an economic one. Working-class youths are eager to get out of their parents' homes. Because their parents don't have much money, their homes are usually small and cramped; there is little privacy, and working-class parents often try to maintain strict control over their children. If working-class teenagers work, as many of them do, they are often asked to contribute their earnings to the household. Working-class parents face unique struggles in trying to keep their children safe and maintain respectability. In her studies of working-class families, Lillian Rubin documents how patterns of parenting go beyond the economics of the situation: "The need for the kind of iron control working-class parents so often exhibit has another, more psychological dimension as well. For only if their children behave properly by their standards and only if they look and act in ways that reflect honor on the family, can these parents be assured that they will be distinguished from those below. This is their ticket to respectability—the neat, well-dressed, well-behaved, respectful child who can be worn as a badge, the public certification of the family's social position" (Rubin 1994, 60). As a response to this situation, Rubin observed rebellion among young adult children who lived at home and described how working-class fathers try to assert their authority over their children, especially daughters. Working-class parents, worried that their children might get into trouble in the rough neighborhoods that surround them, often attempt to control their activities with strict rules. The children, now grown and wanting independence, look for a way out. That was true when Rubin began her studies in the 1970s, but it was also true in the 1990s: "Every one of the working-class young adults I interviewed was itching to find a way out of the parental home" (Rubin 1994, 61). Until recently, the solution for most working-class youth was to get married. In past decades, this was one of the only legitimate reasons for leaving home and spending one's income on oneself. It was also a reason for leaving school and eliminating other adult controls over oneself. In short, it marked the beginning of independence. Among working-class women, in particular, marriage has been a tremendous ideal; it is hardly an exaggeration to say that the single most important thing in a working-class high school girl's life was her image of herself as a bride (*A Closer Look* 6.2). In Rubin's first study of working-class families, the women had married at an average age of eighteen, and the men at the age of twenty. Two decades later, when she reinterviewed those families, she discovered that only one of the daughters married before eighteen and none of the sons before age twenty (Rubin 1994). Rubin suggests that young working-class women have become more ambivalent about early wedding bells as a way to escape the family home, even if they still hold onto romantic images of fulfillment through marriage and parenthood.

Economic Realities in Working-Class Marriages

After a working-class couple marries, there is usually a rude awakening, especially if they followed the early marriage pattern so prevalent in past decades. The couple may move into their own house or apartment, but the early married years are almost always extremely hard. The couple experiences a great deal of conflict, problems they never thought they would have to deal with. Middle-class marriage counselors have usually explained this as the result of marrying too young: the couple did not know each other well enough and were not realistic or mature enough to cope with marriage. However, such an analysis ignores the economic facts, for these are troubles especially found in working-class marriages. (It is fairly typical of middle-class psychologists to think that all problems can be resolved by adopting a middle-class lifestyle, without considering whether this is economically feasible.) The most serious source of the fights and unhappiness that are common at the beginning of working-class marriages is simply economic pressure.

This pressure comes mainly from two sources. One is that the husbands usually have low-paying jobs. The young man living with his parents who earns money at the auto shop or factory feels that he has more than enough pocket money and would be able to support himself if he didn't have to give it to his parents. He finds that when he has to pay all the household bills (many of which are invisible to teenagers—insurance, loan payments, unexpected car and home repairs, and so forth), there is much less money than he thought. Added to this is the fact that working-class jobs are notoriously vulnerable to disruption by ups and downs in the economy; the young married man may find that his source of income is disrupted when he gets his hours cut back or loses his job.

A second source of economic pressure is that the wife's income is often reduced. Most working-class young women have jobs before they marry, and the couple may expect that their two incomes together will make it possible for them to maintain their own home. The problem is that young working-class wives tend to have children very early in their marriages. In Rubin's sample of working-class women in the 1970s, most were pregnant at the time they were married, and half of these had given birth by the time they had been married seven months. (In contrast, the group of middle-class women she studied at the same time had their first child three years after marriage.) Many of these young working-class mothers then left their jobs because they hardly earned enough to cover the costs of child care and other work-related expenses.

This means that economic issues come to a head very soon. There is a tremendous feeling of letdown. Instead of living the happy, independent life they had imagined, the young couple feel trapped and may even fear that they are going under. It is this economic situation, not psychological problems, that sets off most of the fights and personality conflicts. Both husbands and wives experience a great deal of anger. Women attack their husbands for being poor breadwinners. The men are stung by this attack, because most accept the

A CLOSER LOOK 6.3
The Link Between Economic Stress and Racial Prejudice

Two decades ago, when I began the research for Worlds of Pain, we were living in the immediate aftermath of the civil rights revolution that had convulsed the nation since the mid-1950s. Significant gains had been won. And despite the tenacity with which this headway had been resisted by some, most white Americans were feeling good about themselves. No one expected the nation's racial problems and conflicts to dissolve easily or quickly. But there was also a sense that we were moving in the right direction, that there was a national commitment to redressing at least some of the worst aspects of black-white inequality.

In the intervening years, however, the national economy buckled under the weight of three recessions, while the nation's industrial base was undergoing a massive restructuring. At the same time, government policies requiring preferential treatment were enabling African-Americans and other minorities to make small but visible inroads into what had been, until then, largely white terrain. The sense of scarcity, always a part of American life but intensified sharply by the history of these economic upheavals, made minority gains seem particularly threatening to white working-class families.

It isn't, of course, just working-class whites who feel threatened by minority progress. Wherever racial minorities make inroads into formerly all-white territory, tensions increase. But it's working-class families who feel the fluctuations in the economy most quickly and most keenly. For them, these last decades have been like a bumpy roller coaster ride. "Every time we think we might be able to get ahead, it seems like we get knocked down again," declares Tom Ahmundsen, a forty-two-year-old white construction worker. "Things look a little better; there's little more work; then all of a sudden, boom, the economy falls apart and it's gone. You can't count on anything; it really gets you down."

This is the story I heard repeatedly: Each small climb was followed by a fall, each glimmer of hope replaced by despair. As the economic vise tightened, despair turned into anger. But partly because we have so little concept of class resentment and conflict in America, this anger isn't directed so much at those above as at those below. And when whites at or near the bottom of the ladder look down in this nation, they generally see blacks and other minorities.

traditional role and feel demeaned that they cannot support their wives. Rubin's studies show how this sense of inadequacy as a breadwinner often is redirected as anger toward the race and ethnic groups that are just below them in the economic hierarchy (*A Closer Look* 6.3). Sometimes marginalized working-class men respond by pulling "male prerogatives"—telling their wives to shut up, beating them, or going off to drink with their old friends of unmarried days. Such emotional escapes only reinforce the wife's viewpoint that her husband is irresponsible; because he doesn't have the economic resources to back up his claim to male privilege, she continues to criticize him. Both sides feel done in, their dreams crumbled.

A baby's demands usually make matters worse. There are unexpected chores and pressures: getting up in the middle of the night to feed or quiet the baby,

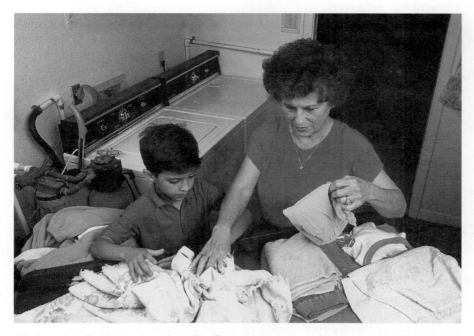

Working-class families tend to rely on kin for economic and social support. A grandmother, for example, may help out with tasks at home while the mother works at her daytime job.

changing diapers, arranging the daily schedule around the new arrival. The parents' old independence is eroded even more. The baby also means added expenses, just at the time when the working-class couple is least able to afford them. To top if off, the wife spends much of her time and attention on the baby and tends to ignore her husband. In particular, their sex life tends to suffer, and the resulting sexual frustration creates more anger on the husband's part. Often, he is jealous of the baby. Numerous cases of child abuse arise out of such a scenario. This pattern of early troubles in the marriage is not limited to working-class couples or to those who marry young, but the economic troubles that exacerbate these struggles are more common lower down on the income pyramid.

Fortunately, the hard times at the beginning of the marriage do not last forever. One solution is that the marriage simply comes apart. As we shall see in chapter 14, most divorces and separations occur soon after the marriage; the curve peaks right away and then gradually declines. Many working-class couples simply do not make a go of it.

The other common solution is for the couple to call on their families to bail them out. One of the most consistent findings about working-class families is that they rely on kin for economic and social support, a practice often labeled "pooling resources" (Bott 1957; Rosen 1987; Stack 1974). Young working-class couples often live near parents and other kin and spend considerable time with them. This means that mothers, mothers-in-law, aunts, sisters, and other family members typically help out with tasks such as child care or shopping. In addition, they frequently will offer advice on the "best" methods of doing housework, disciplining children, or solving marital problems.

When a young working-class couple is forced to move back in with their parents, it is often a bitter step to take, because many married in the first place to gain independence. With a husband out of work, or a pregnant wife, or a young mother caring for a small baby, often the couple finds they cannot afford to live in their own home or apartment. The young couple are a strain on their parents' household too, and carping over their money problems leads to many quarrels. Even without economic problems, the strain of depending on in-laws can lead to a great deal of bitterness. The American folklore about the bad feelings between a man and his mother-in-law probably has a lot to do with this kind of situation, typical in many working-class marriages.

Eventually, the economic crisis passes. The husband finds work again or moves to a better paying or more secure job. When the baby is older, the wife too returns to work and once more adds her income to the family's support. At some point, they are able to maintain their own home, and their sojourn with relatives is over. Nevertheless, the couple is still likely to rely on kin for ongoing help. Even though conventional wisdom suggests that this informal reliance on kin for economic support and exchange is an important survival strategy (especially among African-Americans and Latinos), recent research paints a less sanguine picture. Quantitative studies show that white children from nonpoor households receive much more material support from kin (Lichter 1997). Although money, food, housing, cars, phones, and other resources are shared by working-class grandparents, parents, and other relatives, the amount that can be shared is so limited that it provides only temporary relief from economic hardship.

What we have just described, of course, is only a tendency. Not all working-class families go through this kind of economic crisis, though it is especially common in lower-working-class families and immigrant families. In the more highly skilled part of the working class (Halle 1984), there are fewer of these economically induced strains. As times goes on, many such families achieve relative affluence.

After about five years of marriage, a working-class couple has usually settled down. The family and community pattern that emerges is a very localistic one. Working-class couples tend not to go out very much. Their social life together consists largely of visiting relatives. This is one very noticeable difference between the traditional working class and the middle and upper classes (Fischer 1982; Bott 1957). One reason is, of course, economic: working-class people simply have less money to spend on entertainment. Young married couples often look back nostalgically to their teenage days, when they went to movies or fast-food restaurants or listened to music more often. Once married, they spend most of their time together at home, watching television.

Nor do they regularly invite other couples over to have dinner with them; the dinner party is a middle-class, not a working-class phenomenon. According to custom, there is little male/female mixing. Both sexes tend to socialize predominantly with their own friends and in different places. The house is usually the woman's sphere. She may drink coffee or soda there during the daytime with neighbors or, more likely, female relatives. The men, on the other hand, tend to congregate outside the home. Sometimes they go to bars with their friends, or, even more commonly, spend their leisure time outside the house with other men, often fixing their cars. Sports consume much of the leisure time of working-class men. The men may be involved as participants in softball, bowling, hunting, and fishing or as spectators of professional teams. Often they are involved in gambling (Halle 1984).

Many working-class families, at least on the surface, are rather strongly dominated by men. In the families studied by Rubin in the 1970s, the men insisted on making the major family decisions themselves: what car to buy, when and where to go on vacation, whether to buy a house, and how to spend money. They often turned over the family checkbook to the women, but only because paying bills was a chore and there usually wasn't any discretionary money to be spent. Women were expected to care for the children and do the housework, without much help from their husbands, even when they were working at paying jobs. In fact, most such working-class wives did have jobs, usually part time. Some men, however, would not allow their wives to go to work. They felt that as long as a man could support his wife, it would reflect poorly on him if he let her work. Some of these men would not even let their wives go back to school.

These working-class couples felt a great deal of dissatisfaction, especially the women. There was often a feeling of being cramped. Women said they were trapped in the house with their continuous housework. They complained that their husbands did not want to go out, even now that they were more secure and could afford it. They only wanted to get home from work, putter with their cars before dinner, and then watch TV before going to sleep.

In other cases the men did not come home after work but went out to drink with their friends. For the women in Rubin's study, this was worse yet than feeling trapped at home. Many of these women were fearful of alcoholism. They remembered their own fathers coming home drunk and abusing their mothers. They constantly repeated the gossip and folklore about men who took to drinking and "running around," who stopped supporting their families and ran off with another woman. Much of this may well have been true, although such behavior likely could have been a reaction to the economic pressures of working-class life, such as the shame of being unable to support one's family properly.

Most of the women in Rubin's study were economically secure, although not prosperous. There was little reality to fears of their husbands' being unable to support them, at least at that time. But the "action crowd" at the bars was a constant image before their eyes. The women wanted to get out of the home, but they feared the milieu that their husbands went into when they got out. For them, the family circle was the only secure place. We find the same thing in other studies. Herbert Gans (1962) described working-class life as sharply divided between the routine of the self-enclosed family setting and an "action crowd," mostly male, that frequented bars and sporting events and went in for a culture of gambling, fights, and sexual prowess. Similarly, Halle (1984) found that about half of the oil refinery workers he studied (a rather well-paid working-class group) spent time at their favorite bars as part of their daily routine, instead of going directly home from work.

To an extent, working-class people have traditionally passed through both milieus: the typical teenage experience is the "action scene," which is abruptly broken up by marriage. The women, in particular, get out of that environment and try to keep their husbands from ever returning to it, even though they themselves often look back nostalgically on a time when their lives were more interesting. For the men, the "action scene" is an escape from their wives and home life and the routine of their jobs. It is a place where they gather with their old friends and remember what it was like when they were free (Le Masters 1975).

Conflicts Between Men's and Women's Cultures

The settled life of the working class, then, involves a rather severe split between men's and women's cultures. In her first study, Rubin found there was little sharing of housework and only limited communication between husbands and wives, especially on personal and emotional matters. Even in the 1990s, she found that the men and women experienced family life differently. The men gained their sense of identity from their paid work, and even though they were more willing than men in their fathers' generation to help around the house, they rarely took responsibility for anticipating or initiating parenting or housekeeping chores. One of her informants, a thirty-five-year-old office worker, commented that even the smallest request for help was met with resistance: "I really don't ask for that much. I just want him to help out a little more. . . . It isn't like I'm asking him to cook the meals or anything like that. I know he can't do that, and I don't expect him to. But every time I try to talk to him, you know, to ask him if I couldn't get a little more help around here, there's a fight" (Rubin 1994, p. 88). Even though contemporary working-class couples say men ought to help with the business of raising children, cooking, and keeping the house running, responsibility for these things continues to fall to the women. Rubin found that only among the African-Americans in her 1990s sample did the men do a substantial amount of the cooking, cleaning, and child care. The Latino and Asian men she interviewed found support for a traditional male breadwinner role in their intergenerational ethnic communities and so rarely felt obligated to help with housework or child care. The working-class women (of all ethnicities) were left to juggle competing visions of equality and service to their husbands: Maria Acosta, a white, twenty-eight-year-old secretary commented, "to tell you the truth, I think [feminists] go too far. I'm a firm believer in making your man feel like a king: you know, a man's home is his castle and all that. But I don't mean I want to be like my mom. She waits on my father all the time, and he never does anything for her. It's like everybody's supposed to bow down to him, just because he's a man, and she does. I won't do that for anybody" (Rubin 1994, 74). The contradictions and ambivalence in working-class women's lives are difficult to ignore.

Halle (1984), too, found a split between working-class men and working-class women (see also Locksley 1982). He found that 40 percent of the men were unhappy with their marriages and another 25 percent said their marriages were somewhat happy; only a third said they had happy marriages. These figures should make us aware that the often-cited public opinion surveys tend to be superficial. At the time when Halle was making his study, about 70 percent of Americans were reported as saying that their marragies were "very happy," and only 3 percent would admit they were "not too happy" (*Public Opinion,* 8 [1985] Aug./Sept.: 24, and Oct./Nov. 23). It appears that when asked a brief question by a stranger, most people tend to give the conventional answer, whereas Halle, who spent seven years getting to know his male subjects well, found much higher levels of unhappiness (as did Rubin with her working-class wives.) Moreover, the accounts of marriages by field researchers often have not picked up the depths of the split between husbands and wives, because they have usually interviewed the husbands at home, on the women's "turf." Hence, these husbands did not reveal much about their own lives in the masculine sphere, which often had an element of disreputable or even illicit activities such as gambling or sexual entertainment, activities they did not want their wives to know about.

Several typical sources of strain exist between working-class husbands and wives. One is the sharp split between the male and female social spheres. Working-class men spend most of their leisure time with other men participating in sports, hunting, and fishing or watching spectator sports; when they are home, they are likely to be absorbed in sports on TV. Husbands and wives share few activities except eating. To some extent, of course, this is also true of middle- and upper-class families, but it is especially severe in the working class. Moreover, what is distinctive about working-class marriages is that wives tend to be suspicious or even hostile to the male companionship groups. They view the male bar scene, male drinking and carousing, as a potential threat to themselves and their marriages. There is, no doubt, some reality to this, but we should understand that it has a symbolic dimension, for studies show that consumption of alcohol and the rate of alcoholism or "problem drinking" is approximately equal in all social classes. Yet, it is working-class wives who feel particularly threatened by it; drinking is an emblem of the split between women and men. Thus, Halle (1984) found that the closer and happier marriages were those in which the men had rejected participation in the male bar culture, the marriages in which the husbands came home directly from work and spent more time with their wives.

Halle (1984) also reports that working-class marriages change over time. As blue-collar men grow older, they enjoy more economic security, their children grow up and move out, their wives often take jobs, and their home mortgages are paid off (or mostly paid off). Seniority on the job protects them from cutbacks, gives them choice vacation times, and ensures them a pension on retirement. At the same time, working-class men often begin to develop physical ailments that slow their mobility, and most of them know someone who has died suddenly. A shift in leisure time activities often occurs at this point, with men who once spent most of their free time with male friends now spending most of their time with their wives. The wife becomes more of a companion, and couples travel together, particularly after retirement. As we shall see in later chapters, aging and maturity can also place special strain on families.

Earlier in the marriage, however, Halle (1984) reports that wives tend to be more socially ambitious than their husbands. Wives put on pressure to rise in the social class system, and they were sometimes critical of their husbands for the low status of their jobs. The husbands, on the other hand, were more likely to withdraw into their all-male sphere of sports, drinking, and the like and to ignore the differences between themselves and the higher classes. Perhaps these women were status conscious because so many of them worked in white-collar positions, as secretaries and clerks. Even though their jobs tended to be relatively low-paying order-taking positions (and hence could be classified as "white-collar working class"), they had one distinctive feature that set them off from the men's blue-collar jobs: the women were in offices with a middle-class atmosphere and were usually in close contact with middle-class or upper-middle-class managers or professionals. Thus, they picked up a standard of status respectability from their bosses (who were usually men) by which they compared their own working-class husbands.

Halle found that many family disputes broke out because wives criticized their husbands for their crude manners or for embarrassing them in public. He quotes a working man:

> [My wife and daughter] wanted to get something from the store. So I gave them a lift, and Deirdre met a friend of hers and started chattering. You know, I was tired—I'd been working since 8:00 the night before. So I said, "Deirdre, get the fucking hell in here!"
>
> . . . [My wife] kept yammering on. I called her just now [on the telephone], and she still says I shouldn't have used that language in front of Deirdre's friend.
>
> So I'm in a bind. I know what's waiting for me back home. She'll start to yell at me and turn the children against me. (Halle 1984, 60)

This is an example of the split between the standards of the male culture of the working-class group and the middle-class standards of public respectability upheld by the wife. Actually, we find when we turn to upper-middle and upper-class marriages (discussed later) that wives often seem to specialize in keeping up the cultural image of the family. The main class difference is that in the higher social classes, the husbands go along with this status seeking, whereas in the working class, it tends to be a point of contention between the marital partners.

Finally, we should notice that working-class families experience a distinctive strain that comes from conditions that make them age faster psychologically. This results, in part, from marrying earlier and reaching their peak earning capacity and on-the-job responsibility much sooner than professionals of the middle class. For working-class people, there is often a feeling that life is already over at the age of thirty—a time when most upper-middle-class people are feeling that their lives have scarcely begun. The working class idealize their earlier years, and probably with good reason. It was then that they had the most sense of excitement and expectation, even if they were pursuing independence and expectations that turned out to be illusory. Among the women in Rubin's first study, there was almost unanimous agreement that the fault lay in early marriage. They wanted to keep their daughters from making the same mistake. One woman said: "I sure hope she doesn't get married until she's at least twenty-two or three" (Rubin 1976, 113). In fact, when Rubin interviewed some of the same families twenty years later, most of the daughters had waited longer before marrying (Rubin 1994). As we will see in chapter 14, however, delaying marriage does not eliminate the sorts of problems that working-class couples face, nor ensure that they will stay together.

Working-Class Culture and the Family

Why does working-class family life have these unique features? Part of the reason, as we have seen, is economic, but not all of it is. For instance, although working-class wives are generally fearful and worried about their husbands' potential drinking, violence, and sexual infidelity, these problems are not unique to the working class. Studies of marital violence find that it happens surprisingly often in the middle class as well, as does marital infidelity (chapters 9 and 13). Why then do working-class women react more strongly than middle- and upper-class women?

Part of the reason is that the working class tends to have a particular culture that might be called "localistic" and "group conformist." The British sociologist Basil Bernstein calls it the "restricted" code, as compared with the "elaborated" code of the middle class. The working-class culture, or code, sets a very strong line between insiders, whom one can trust, and outsiders, who are to be feared. Further, it has a strict sense of the moral rules incumbent on insiders and makes heavy use of authority to enforce those rules.

The fear of outsiders manifests itself in numerous ways. When working-class mothers have to go back to work, they often leave their children with relatives instead of taking them to a day-care center. When asked the reason, they say they do not want to leave their children with strangers. The same fear is shown when wives complain about their husbands going out to bars and getting in with "the wrong crowd." They are more comfortable when their husbands stay at home, even if they feel bored and restricted there.

It is the same with working-class sociability—there is a sharp barrier between those whom one will invite into one's home and the rest of the world. The former consists almost entirely of relatives and a few old friends whom one has known since childhood. The contrast to middle-class sociability, which consists so centrally of dinner parties, is striking. But then, middle-class culture is different precisely because it does not have this same insider/outsider barrier. (We will discuss some of the contradictions and hypocrisy of the middle-class patterns in the next section.)

The other feature typical of the working-class culture is its moral code. People with modest income and limited opportunities who grow up, marry, and stay in the same neighborhood tend to see a clear, black-and-white dividing line between right and wrong and strictly punish any deviations. The working-class parents in Rubin's study tended to be critical of the public schools and did not like having their children there because they felt that the schools did not enforce strict enough discipline. The psychological techniques used by middle-class parents in controlling their kids are not as evident in the working-class home. Most studies show that working-class parents are more likely to demand strict obedience from their children and to use physical punishment to enforce it. Some studies question whether working-class parents are any more likely to use physical punishment than middle-class parents, but studies consistently show that parental values and discipline practices differ according to social class. Research in industrialized countries from the 1950s to the 1990s confirms that working-class jobs tend to promote conformity and obedience, whereas middle-class jobs tend to promote individualism and problem solving. Parents from all social classes discipline children for misbehavior, but working-class parents emphasize conformity to external authority, whereas middle-class parents emphasize self-direction. Working-class parents are more likely than middle-class parents to value characteristics in their children such as manners, neatness, cleanliness, being a good student, honesty, and obeying orders. These traits are similar to those needed for success in working-class jobs. In contrast, middle-class parents tend to emphasize consideration, curiosity, responsibility, and self-control; traits that are required for many managerial and white-collar jobs (Kohn 1977, 1979; Kohn et al. 1990; Schooler 1996). Research thus shows that differences in life experiences, job demands, and economic stress lead to class differences in parenting styles: working-class parents are most concerned with outward conformity and discipline, whereas middle-class parents are most concerned with positive emotional communication and individuality. In addition, the simple fact of having more financial resources allows upper-middle-class parents to provide many more activities and experiences for their children, preparing them for a life that revolves around more cosmopolitan social networks (*A Closer Look* 6.4).

The same kind of differences between middle-class and working-class cultures has been found in many surveys of attitudes. The working class is much more likely to be moralistic and to adhere to traditional standards. This is found especially in attitudes about sexual behavior (there is much more intolerance of homosexuality, for example, or of premarital sex on the part of women), political dissent, and religious conformity. For violations, the working class is more likely to favor strong and violent punishments. This is one reason why working-class people sometimes support more conservative law-and-order or "family values" politicians instead of the liberal or progressive candidates who purport to serve their interests.

UPPER-MIDDLE-CLASS FAMILIES

The most important basis of social class differences is occupation. We have seen some of the reasons why it makes a difference in families' cultures: it affects the sheer amount of money people have, as well as their financial security or

A CLOSER LOOK 6.4
Class Differences in Mothering

There are clearly noteworthy differences between the beliefs and practices of working-class and poor mothers, on the one side, and those of middle-class and upper-middle-class mothers, on the other. Given differences in their financial resources, their reference group, and their cultural milieux, this is not surprising. The women in these two groups have different baseline standards for what "good" mothers should provide for their children, as well as differential means and differing images of how to achieve what is best for them. For instance, although all these mothers want to spare their children future financial hardship, for working-class and poor mothers this leads to a tendency to stress their children's formal education, whereas middle-class mothers are more likely to be engaged in promoting their children's "self-esteem." In a connected way, working-class and poor mothers tend to emphasize giving children rules, while middle-class mothers are busy providing their children with choices. And working-class and poor mothers are somewhat more likely to demand obedience from their children while their well-to-do counterparts are more likely to negotiate with theirs. But all these mothers, I will argue, share a set of fundamental assumptions about the importance of putting their children's needs first and dedicating themselves to providing what is best for their kids, as they understand it.

Wealthier mothers have the money and the access to transportation to send their children to tumbling classes, piano lessons, judo classes, swimming lessons, child psychologists, and, perhaps most crucially, to day-care centers, preschools, and private schools where the facilities are clean and bright, where the toys and games and instruction are clearly chosen, and where the teachers and caregivers are well trained. Laura, for instance, takes her ten-year-old son to soccer practice and her five-year-old daughter to a therapist. As a homeroom mother, she also arranges school parties, sets up fund-raisers, and provides gifts for her daughter's teacher. She has always hired someone to clean her home, and she currently has a live-in au pair; in the past she has had nannies. Her younger children attend an elite preschool; the older children go to private schools.

Lupe, unlike Laura, considers a trip to the grocery store a special outing for the kids; piano lessons would not even cross her mind. Even when free activities are available for the kids, there is always the problem of transportation. (Lupe does not have a car, and the public transportation system in her area is grossly inadequate.) And Lupe felt compelled to leave her last job when she discovered that the family caregiver she used was not only neglecting the children but physically abusing them as well.

Source: From Sharon Hays. 1996. *The Cultural Contradictions of Motherhood*, pp 86, 89. Used with permission of Yale University Press.

insecurity, and hence has an important effect on their lifestyles. It also helps produce the cultural codes that derive from a high-social-density/localistic or a low-social-density/cosmopolitan life experience. A further dimension of occupations is the way in which people experience power (see Collins 1975, 61–79).

Working-class people are generally **order takers:** they have little power on their jobs and take orders from others. Because taking orders is a psychologically demeaning experience, working-class people try to distance themselves from it by not identifying very strongly with their jobs. They tend to be cynical about the official ideals that they hear coming out of the mouths of their bosses because these are used mainly to justify ordering them around. Working-class persons emphasize the leisure side of their lives and support the "cool," rather alienated themes of popular culture. Child-rearing practices tend to reinforce these values and prepare working-class children to assume jobs in which they must take orders from others.

People who are **order givers,** on the other hand, have a very different psychological perspective on their work. High-ranking managers and professionals tend to be proud of themselves and to identify strongly with their work roles. The experience of telling other people what to do gives them these qualities, both because it is more psychologically gratifying to give orders and because they have to take their work roles seriously if they are going to make other people respect their authority. Upper-middle-class child-rearing values tend to emphasize taking responsibility for others' actions and feelings, as well as one's own, thereby preparing children to occupy positions of authority. Members of the higher social classes have more incentive to identify with the ideals of their organizations and of the official, "front-stage" part of society in general. Although the working class tends to be alienated from the larger society and to find refuge in their private lives, the upper classes live more in the larger world and are the main carriers of its various ideologies.

Order givers and order takers are not two mutually exclusive classes, and they do not exhaust all the levels of occupational classes. In between the higher social classes and the working classes are various intermediate groups (e.g., middle-level bureaucrats and officials, supervisors, and foremen) that take orders from people above them and pass them along to others below. These kinds of workers usually share parts of both the upper- and the lower-class cultures; they identify with rules and regulations in some ways but are not as committed to the larger ideals as high-ranking persons. There are also many white-collar workers who tend not to experience much order giving *or* order taking but who work in an office or professional setting surrounded by people more or less of the same rank. This is quite common among engineers, many scientific researchers, and other technical workers. In these white-collar groups, there is a typically "middle-class" style of being casual and informal on the job. This style is characteristic of the United States rather than of other countries, where the middle class is much more likely to be rather formal and concerned about the rituals of deference and power.

All these aspects of class cultures influence family life. Here, we will pick out just one type of class culture, that of the upper-middle-class, which contrasts most sharply with the family life of the working class. (See *A Closer Look 6.5* on the very highest social class.)

Upper-Middle-Class Cultural Traits

We already know some of the cultural traits found in the upper-middle class. It is the class of high-ranking order givers: high-level managers in large corporations, the most important professionals (successful lawyers, professors in major universities, doctors, notable scientists, renowned architects, and so on), and higher government officials. Their culture has a number of things in common. As we might expect, these are the individuals who identify most strongly with the official ideals of their jobs. It is the successful scientist who cares most about the intellectual accomplishments of science; the doctor who goes on at greatest length about the need for proper medical care; the business executive who most extols the values of free enterprise in general and the merits of his (or her, though this is less common) corporation in particular. These people get the most out of their jobs in terms of prestige, power, and income, and hence they have a very favorable attitude toward their work. Many of them put in quite long hours, more than the workers below them. This is partly because they enjoy their jobs and tend to be highly satisfied with them, and partly because the higher levels of big organizations and the professions are very competitive places, and people need to put in long hours to stay ahead of their peers.

Members of the upper-middle class also are more concerned about general and abstract ideas than other social classes. They are more likely to have taken a position

A CLOSER LOOK 6.5
Women of the Upper Class

The family lifestyle of the wealthiest class can best be seen by looking at the wives of those men who control the major businesses. This is what Susan Ostrander (1984) did in her book *Women of the Upper Class* and Arlene Daniels did in her book *Invisible Careers* (1988). The husbands of these women are the presidents or chairmen of the board of major corporations, or they own old family firms, or they are partners in the law firms that handle corporate business, or they have major roles in stocks and finance. The **upper class** is not difficult to pick out; they announce their own membership in the *Social Register,* a volume published for most big eastern cities (and a few western ones). They belong to exclusive clubs. They usually attend exclusive private secondary schools, and their colleges tend to be the traditional Ivy League ones.

The daily routine of these women is not quite what the popular image of the rich would have us believe. "The life of the upper-class woman is not all champagne and roses, trips to Paris and Palm Beach. The upper-class woman is also not very interested in high fashion, nor is she a jet-set party goer. Her days are not spent lying around the country club pool or attending elegant ladies' luncheons" (Ostrander 1984, 3). They are much more conservative. They believe that they are morally superior to other people, that they are more responsible parents, that they contribute in essential ways to the community, and that they uphold the standards for the rest of society. They also have a distinct sense of being different from other people, and have a preference for associating with people like themselves. In the generally un-class-conscious U.S. society, upper-class women are among the most class conscious, and they have the sharpest sense of class boundaries.

"The club is a place to go where you can have lunch and discuss business without having to wade through the mass of people. You want to have people who are congenial, all of the same social group," said one woman (Ostrander 1984, 97). There is an emphasis on belonging to clubs in which membership is by invitation only; one must be known and acceptable to the existing members, and new members must be sponsored by one of them. Many upper-class women are members because their families have been members. Women whose husbands have made it to the top of the corporate hierarchies can be invited to join. One of the main activities of established upper-class women is socializing these wives into the standards of the upper class and of picking out those who are to be sponsored into club membership. "Whenever a new executive wife comes along, I give her quite a lecture on ceasing to be an emotional person. She should never be nervous or cross. These men can't take that. She should always be prompt and ready to go" (Ostrander 1984, 47).

The same applies to the women's activities as mothers. They take their family roles seriously and believe they should protect their children from exposure to gangs, drugs, blatant sex, and the rest of what they perceive as the popular culture of the day. This is one reason why they want their children to attend private schools. But they are also concerned with maintaining upper-class homogeneity. Upper-class mothers are expected to prepare a new generation of the elite for its future responsibilities. These women generally consider it important for their children to marry someone of the same social class. For this reason, they want them to participate in dancing schools, debutante balls, and other sociable activities that are carefully arranged and controlled. These activities are supposed to be pleasant, but they are also viewed as a duty. One woman from a long-standing upper-class family, for example, commented: "My mother was one of the women who started [the debutante ball] so I felt the children should do it whether they wanted to or not" (Ostrander 1984, 89). These social events serve as rites of passage into membership within upper-class circles, and hence they are a barrier that status-seeking individuals would like to overcome from below. Upper-class women are aware of this and regard it as their duty to carry out the "dirty work" of separating the "acceptables" whom they will sponsor from the "unacceptables" whom they regard as mere "social climbers" (Ostrander 1984, 93–94; Daniels 1987, 224).

Another aspect of upper-class women's lives is participating as a community volunteer in charitable organizations. This does not so much mean the front-line activities of actually sitting at tables or ministering directly to the needy, but of serving on boards that plan functions and raise

Debutante balls signify wealthy young women's social availability and readiness to assume formalized roles as upper-class wives.

money. They themselves are expected to be big contributors. This can be time consuming, and these women sometimes will describe it as a burden; but their charity work is part of their self-image as responsible persons and community leaders, and they regard volunteer work as superior to working for pay. Because virtually none of these upper-class women has a career of her own, charity work is a kind of substitute activity. For women whose husbands are in the upper-middle class, charity work is a way of becoming known and associating with the upper-class network. On the other side, upper-class women see these relationships as yet another aspect of their "sponsorship" tasks and are suspicious of "newcomers" who they feel are pushing too hard. A moderate proportion of upper-class women regard their charity work as an obligation that they took on reluctantly. Some complained that the real power to make deci-

sions was still held by the men of these charity boards; others suggested that participation was only symbolic and that it accomplished little to alleviate real problems. One woman said she had "quit the Junior League because it seemed to her that 'the most important thing was how neat your hat was'" (Ostrander 1984, 125). In short, charity participation is essentially a ritualistic activity, felt to be an obligation ("noblesse oblige") but nevertheless an important way of displaying one's class position. Although upper-class women feel obligated to participate in such activities, they generally do not stress the importance of maintaining their class position. In general, they find "class" an "ungenteel matter for discussion" (Daniels 1987, 227).

The lives of upper-class women are both formalized and constrained. Despite their wealth, they have little freedom. At home, they have servants to

take care of the housework. However, their husbands are the dominant figures in their lives; these are the men who control the major organizations of society, and their business activities take priority over everything else. "My husband never asks me what I think. He just tells me how it's going to be," said one woman (Ostrander 1984, 37). Wives make social arrangements for their husbands; they travel with them, when requested, even at short notice; they put up with their husbands' frequent trips and meetings away from home, making sure the home the men come back to is well ordered. Their task is to shield their husbands from having to be distracted by any domestic problems or issues.

Although upper-class women's volunteer activities help to support their husbands' careers, they typically minimize the importance of their work for this purpose. Because the women hold positions on powerful and prestigious governing boards, their husbands have the opportunity to make professional contacts and acquire "inside" information about future changes in the community that can affect their business or professional practice (Daniels 1987, 228).

Many of these women have inherited considerable money of their own, but this typically gives them little power. They usually put the money into their husband's care. These men, after all, are in the financial power centers of the world. It is their social networks at work that are the core of all this privilege. Living in intimate relationships with the sources of upper-class power makes these women aware of how dependent they are on the power structure.

on most issues and less likely to answer "don't know" to public opinion polls. Of all social classes, they are most likely to be active in politics—not only to vote but to run for office, campaign for other candidates, or contribute money to campaign funds. It is not much of an exaggeration to say that the upper-middle class (together with the upper class, which has been much more widely studied) runs politics in America. That is to say, the upper-middle class runs both the liberal and the conservative political parties. It is true, of course, that the business upper class and its higher-ranking managers tend to be the bulwark of the Republican party and to support conservative policies. However, the Democratic party as well, including its liberal wing, is largely run by members of the upper-middle class: not so much the business upper-middle class but professionals such as lawyers, academics, and media professionals. Political campaigns in American are almost always between rival factions of the upper-middle class. The reason for this is partly that this class has many more resources (money, contacts) to put into politics, but it is also that upper-middle-class culture is more concerned than working-class culture with general, abstract ideas (e.g., Does the environment need to be saved from pollution? Is racial inequality an important problem? Does free enterprise need to be defended?). They admire more abstract forms of art, whereas the working class prefers more sentimental and concrete forms of artistic "prettiness" (Bourdieu 1984).

The upper-middle class is the cosmopolitan class par excellence. It is a class of social joiners and group members. The upper-middle class also participates more in churches, professional associations, and clubs of various sorts. It is the wives of upper-middle-class and upper-class men who support the various women's clubs and organize charity exhibitions, luncheons, and balls; it is the upper-middle class, not the working class, that makes a point of belonging to the country club and the tennis club and takes part in the golf tournaments. (The same is true of more recent athletic fads such as long-distance running; the people who turn out for marathons and ten-kilometer races are largely upper-middle class, almost never working class.)

One big contrast between the upper-middle class and the working class, then, is the way their social lives are organized. The working class is organized very locally, around the home, relatives (for women), and nearby groups of male friends (for men). The upper-middle class belongs to special-purpose organizations, with members who come together only occasionally and for one special activity only. Upper-middle-class people thus encounter many others in the course of their activities, although they interact with them much less closely. This is what is meant by the terms **high social density** (in the working class) and **low social density** (in the upper-middle class). The same is true on the level of purely sociable relationships. The higher social classes tend to have many more friends than the lower classes, but they see them less often and are less close to them. Business executives, for instance, often send out Christmas cards to hundreds of their "friends"—persons they have socialized with at clubs, over business dinners, on trips, and so forth. Some higher executives even keep a secretary busy full time just taking care of files, so that the executives can be sure of who their "friends" actually are.

All this, of course, has an important effect on upper-middle-class marriages. Upper-middle-class people tend to get married a good deal later than working-class people. Their lives tend to be oriented toward the future, not the present, and they are trained to wait for things. Not only do they marry later, but they put off having children longer. When children do arrive, they are more likely to have been planned through the careful use of birth control, so that they do not interfere with the husband's (and sometimes the wife's) career.

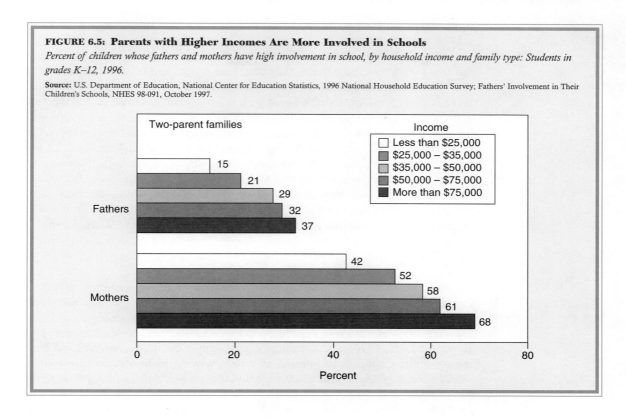

FIGURE 6.5: Parents with Higher Incomes Are More Involved in Schools

Percent of children whose fathers and mothers have high involvement in school, by household income and family type: Students in grades K–12, 1996.

Source: U.S. Department of Education, National Center for Education Statistics, 1996 National Household Education Survey; Fathers' Involvement in Their Children's Schools, NHES 98-091, October 1997.

Social Networks

The upper-middle class is less family oriented than the working class. Working-class people don't see many other people besides their own family members and relatives, whereas the upper-middle class see families, relatives, *plus* lots of other people. Their lives are less exclusively oriented toward the family network (Collins 1975, 79–80; Fischer 1982).

Upper-middle-class careers, for reasons given earlier, are both more attractive and more demanding than working-class jobs. The husband in particular is likely to work long hours or be away for business trips or professional meetings rather frequently. Also, if he is not gone physically, he is often absent psychologically: his body may be present at home, but his mind is elsewhere, preoccupied with his work.

At the same time, because members of the upper middle class are strongly oriented toward the more intellectual media and keep up with ideas coming out of the technical professions (science, medicine, law, psychology), they tend to have well-defined opinions about family life and the resources to pay for various lessons and classes for their children (see *A Closer Look 6.4*). Because they have gone further in school themselves and have strong opinions about the right way to teach their children, they tend to get involved in their children's schools. As Figure 6.5 shows, mothers and fathers with higher incomes are the most likely to attend school meetings, come to school or class events, participate in parent-teacher conferences, and serve as school or classroom volunteers. Such activities, along with reading to children and talking about homework, are predictors of later success in school (U.S. Department of Education 1997). For most upper-middle-class parents, it is assumed that children will do well enough in school to go to college.

Although upper-middle-class parents may not spend as much time with their family as working-class parents, they have well-articulated beliefs about it, just as they do on everything else. In the conservative part of the upper-middle class, there is likely to be strong agreement with religious and political leaders who assert that the traditional family is the backbone of America; such persons will have strong opinions about the necessity for women to be full-time mothers, the need to protect the sanctity of family life, and so forth. On the liberal side of the upper-middle class, there is a strong concern with the psychological dimension of relationships between husband and wife, parents and children. Books are read, specialists sometimes consulted, family discussions are deliberately held, and avant-garde styles of relating are explicitly practiced. Yet, all of this tends to be crammed into lives that are also busy with many things outside the family.

Dual-Career Families

As more women seek education and training to enter professional careers as corporate managers, medical doctors, lawyers, and scientists, researchers have turned their attention to **dual-career couples** (e.g., Barnett and Rivers 1998; Deutsch 1999; Hertz 1986; Pepitone-Rockwell 1980; Peters 1997; Rapoport and Rapoport 1971, 1976). Most studies have described the unique stresses and coping strategies common to families in which wives and husbands have professional careers. Although dual-career couples represent only a fraction of families in which both wives and husbands are employed, the tendency has been to assume that what goes on in these "upscale" families is equally applicable to other middle- and working-class families (Ferree 1987). Using dual-career couples as models suggests that paid employment is the major source of personal reward for most people and that family work, like cooking, cleaning, and child care, is burdensome.

Although dual-career marriages may be "more equal than others" (Hertz 1986), most researchers find that there is significantly more sharing of egalitarian *ideals* than sharing of domestic *tasks* such as child care and housework (chapter 11). Most studies also show that having two careers provides professional couples with the resources to hire someone else to perform housework and child care and that two relatively high incomes allow the couple to enjoy a high standard of living that includes frequent entertainment and dining out, as well as expensive homes, cars, and clothes. Many researchers focus on the egalitarian attitudes of such couples, but Hertz (1986, 415) suggests that "it is not an ideology of marital equality that determines careers; instead it is because of the two careers that equality becomes an issue." What appears to happen is that as the woman's career becomes as demanding as her husband's, there is no time for her to be a typical wife. Consequently, the "invisible" work of the home comes into plain view for some dual-career husbands.

Hertz (1986) claims that the dual-career couple should be characterized as a situation with two husbands and no wife. She suggests that in this situation, both people feel pressure to limit their career aspirations and goals. If men and women in demanding professional careers cut back on their employment hours, however, they are often perceived as unserious about their careers. This tendency is reflected in an informal two-tiered system of job tracking in the corporate world: those who limit their time commitments to career because of family obligations are placed in a so-called "mommy track" or more recently "daddy track." Such unofficial categorization often leads to slower and fewer promotions and puts dual-career professionals who spend time with their families at a disadvantage in relation to their "fast-track colleagues" (Coltrane 1996; Schacter 1989). It appears that major adjustments in the requirements of professional careers are yet to come.

GENDERED JOB MARKETS AND FAMILY CHANGE

As we have seen in this chapter, patterns of family life are shaped by family members' incomes and their place in the economy. It follows that many family changes are precipitated by changes in the overall economy. As a result of restructuring in our national and global economies in the latter part of the twentieth century, we have witnessed a steady increase in women's labor force participation. To assess how women's employment has influenced families, we need to understand how their job opportunities have evolved and how they might change in the future.

Since the 1970s, dramatic growth in the service sector of the economy and a decrease in the relative importance of agriculture and manufacturing have increased the demand for secretaries, typists, and clerks, and related support services. Over the years, these positions have gradually become stereotyped as "women's jobs." Today, occupations with especially high concentrations of women include secretaries and other office workers, retail clerks, maids, electronics assembly-line workers, school teachers, nurses, real estate agents, and social workers. Despite some entry by women into traditionally "male" occupations in recent years, men continue to predominate in management, higher-status professions (e.g., doctors and lawyers), skilled crafts (e.g., carpentry or plumbing), manufacturing, and jobs involving outdoor labor (England and Browne 1992; Reskin and Padavic 1994). Although men's and women's jobs require roughly equivalent amounts of formal education before entry, women's occupations have less on-the-job training, offer fewer opportunities for advancement, and are less likely to entail supervising other workers (Miller-Loessi 1992).

Occupational segregation by gender has weakened somewhat over the past several decades as more women have become professionals and managers. Jobs like reporter, bus driver, bartender, pharmacist, and insurance adjuster have opened up to women, but the jobs women hold are still typically less prestigious than corresponding men's jobs. For example, school buses (the old yellow ones) tend to be driven by women, whereas metropolitan transit buses (the big silver or blue ones) tend to be driven by men. Transit or long-distance bus drivers make much more money than school bus drivers. Although there are more female doctors and lawyers than ever before, the more prestigious specialties, like surgery and corporate law, continue to be dominated by men (Reskin and Padavic 1994). Decreasing job segregation is therefore both real and illusory. Approximately 60 percent of men or women would have to change occupations for the workforce in the United States to become truly gender balanced (England and Browne 1992).

Although most Americans say they believe in gender equality, most also recognize that labor markets are still biased against women. According to a recent Washington Post poll, about three-quarters of Americans agree that women are discriminated against on the job market. Both men and women believe that women do not receive equal salaries with men for doing equal work, with women significantly more likely to agree that this type of job discrimination exists (80 percent of women and 66 percent of men). Most Americans also blame men for workplace discrimination against women. Sixty percent of women and 43 percent of men say that a major reason that very few top level executive and professional positions are filled by women is that men don't want women to get ahead in the workplace (Grimsley 1998; Morin and Rosenfeld 1998). These figures show that women are most likely to notice sex discrimination, but both men and women tend to see the disadvantages that women face on the job. On the positive side, as shown in Figure 6.1, the gender wage gap appears to be narrowing. Women's wages in the United States have shown steady increases since the 1970s, at the same time that men's wages have remained stagnant or declined.

Why does the labor market remain biased against women? The most frequent justification is that it is natural for women to care for children and tend homes (chapters 11 and 12). This "separate spheres" assumption is usually accompanied by claims that women are paid less or assigned to "female" jobs because they are not temperamentally suited for the competitive business world or that they are poor risks for job training because they exert less effort on the job and are likely to quit work to have babies. In fact, most empirical studies find that if there are differences between men's and women's efforts on the job, it is women who expend more effort than men, in part to overcome stereotyped notions of female incapacity (Bielby and Bielby 1988; England and Browne 1992).

Others argue that occupations are segregated because of self-selection; that is, women prefer female-typed jobs because they are less demanding and allow more time off for family responsibilities. There is more support for this hypothesis than the first, especially regarding the management of child care, but only for some women and only under certain conditions (Coltrane 1998; Moen 1985). Although women do tend to work part-time more often than men, the general trend is toward more women working full-time, and evidence suggests that most women's jobs have closer supervision and less schedule flexibility than men's jobs (Miller-Loessi 1992). The claim that women earn less because they frequently quit work to have babies is contradicted by the finding that women do not have higher turnover rates than men when the wage level of the job is considered. Other commentators suggest that women favor lower-paying gender-typed jobs because they like the kind of work involved. Self-selection into "helping" occupations such as nursing or teaching may indeed occur, because the work bolsters one's self-image as a caring woman (Cancian and Oliker 1999). As we will see in later chapters, whatever the cause of labor market inequities, job segregation and unequal pay, along with women's unemployment, are associated with marriage bargains that include wives' obligation to perform domestic labor and husbands' sense of entitlement to receive unpaid domestic services.

Links Between Jobs and Family Life

Sociologist Rosabeth Kanter (1977) was one of the first scholars to point out that it is misleading to treat jobs and families as separate spheres. Rejecting popular beliefs about the separation of work and family that were popular from the late nineteenth century to the late twentieth century, she called for examining the interconnections between the two. Today, most social scientists recognize and appreciate the numerous mutual influences between work and family for both men and women (for reviews, see Aldous 1982; Bielby 1992; Ferree 1990; Menaghan and Parcel 1990; Spitze 1988; Thompson and Walker 1989; Voydanoff 1987). In this chapter, we have focused on how different forms of income-producing work are linked to specific class cultures and different family patterns. Most of the research on work-family linkages focuses on the one-way effects of jobs on family life, with long work hours, harsh working conditions, low wages, and periods of unemployment associated with various family and health problems (e.g., Catalano 1991; Conger et al. 1990, 1994; Mason 1996; Vinokur, Price, and Caplan, 1996). Also noteworthy are the findings, summarized earlier in this chapter, showing that harsh living conditions and closely regulated work environments are associated with restrictive parenting practices, whereas workplace managerial responsibilities are associated with parenting practices that promote individuality and self-expression.

Our review also shows that work-family linkages are shaped by cultural ideals about gender. Despite recent changes in labor markets, men continue to be identified most strongly with paid work, and men's authority continues to be determined by success on the job. Even though male-breadwinner families with stay-at-home wives are now relatively rare, husbands tend to retain symbolic responsibility for earning the family's money. Studies of working-class men show that they resent women's entry into formerly all-male occupational enclaves, but men from the middle and upper classes are also reluctant to accept wives as equal providers. Even when both spouses work full-time, husbands tend to define their jobs as more important than their wives' and use employment as a reason to avoid taking responsibility for family work (Bernard 1981b; Coltrane 1996; Hochschild 1989; Hood 1983; Lamphere, Zavella, and Gonzales 1993; Pyke and Coltrane 1996; chapter 11).

In working-class couples, husbands and wives typically work out of financial necessity. It is here that the wage gap between spouses is smallest, so women tend to see themselves as shouldering a significant part of the provider responsibility, and many of their husbands acknowledge it (though some reluctantly, as Rubin's studies show). Because upper-middle-class men's salaries are usually considerably higher than their wives' incomes, women in this class have an even harder time receiving recognition as providers (Ferree 1987). Among upwardly mobile dual-career professional couples, there is also more pressure to acknowledge the abstract ideal of women's equality. This results in some contradictory findings in survey studies about family activities, with couples often reporting that middle-class husbands share more family tasks than working-class husbands (chapter 11). Nevertheless, it is more likely that middle-class couples are more concerned with talking about the importance of sharing than they are with instituting actual sharing of household tasks (chapters 11 and 12).

Balancing Commitments to Job and Family

Today most people say they hold jobs to make money, but the majority also report that they achieve personal satisfaction from their paid work. Although upper-middle-class jobs are more likely to define one's character, some working-class jobs also provide an important source for building one's identity. One recent study by sociologist Arlie Hochschild (1997) even suggests that people may now get more satisfaction from their paid work than from their family life. Most studies show that satisfying, well-paid work is related to enhanced well-being for both men and women (Barnett and Rivers 1998; Thompson and Walker 1989). The only exception seems to be when people believe that they should *not* be working, but circumstances force them to take jobs or work longer hours than they want. Things are especially troubling for women who want to spend more time with their children, but whose families cannot survive without their incomes. The dilemma is most acute for single mothers, who are usually both sole caregiver and major financial provider for their children. Many must work long hours to earn a living wage and then feel guilty that they cannot give more time to their children (Kurz 1995). In general, however, as men's and women's jobs and work histories begin to look more alike, they share similar family concerns. Polls find that over 70 percent of American men and over 80 percent of American women feel torn between the demands of their job and wanting to spend more time with their family (Gerson 1993). However, women feel the most stress about balancing work and family obligations. A recent Washington Post poll found that 69 percent of American women, compared with 50 percent of American men, said there is too much pressure to have it all—marriage, family, and a successful career (Merida and Vobejda 1998).

Researchers have attempted to measure people's work and family attachments by asking questions about how committed they are to each. Although somewhat superficial, answers to these survey questions indicate whether someone gains special meaning from family and work activities and how willing they might be to cut back on one or the other. Both men and women say they are strongly committed to both work and family, but men tend to be slightly more identified with work than with family, and women tend to be slightly more identified with family than work (Bielby 1992). As noted earlier, it is middle-class people—both men and women—who tend to establish the strongest commitment to their work and base their identity on it. Working-class people, especially women, tend to derive a greater sense of self from their families. If women have work statuses and experiences similar to men's and have the opportunity to identify as strongly with the work as men, however, gender differences in commitment to work and family are negligible (Bielby 1992; Bielby and Bielby 1989; Rosen 1987). As women's labor force participation has increased, their commitment to work has become more similar to men's. This has created some new challenges for married-couple families, as we will see in later chapters, but it has also provided the economic resources for maintaining families when no husband is present. Thus, at the same time that gender differences in commitments to work and family are diminishing, social class differences may be increasing. And as noted earlier, gender and family structure are also linked to social class, with single mothers and their children now the most likely subgroups of the American population to be poor. We investigate these linkages further in the next chapter as we explore diversity among families according to race and ethnicity.

SUMMARY

1. The class structure in the United States is relatively stable, with most of the nation's wealth controlled by very few families. Individuals tend to move up or down the income pyramid because of changes in family composition—things like getting married, having a baby, or getting a divorce. Since 1973, real earnings for the average American male worker have declined, and most families now have two wage earners.

2. The poorest group in the United States consists of households of people who are out of the labor force because they are retired or unemployed. Households headed by women are especially likely to be at the bottom.

3. Families are most likely to be in the prosperous upper parts of the income distribution if both husband and wife are working full time, especially at white-collar jobs. Only in the very highest income groups is there a preponderance of families with the traditional structure of an employed husband and a nonemployed housewife.

4. Working-class people are especially likely to marry young, although this often results in severe economic strains early in the marriage, which are intensified by the early arrival of children. If the marriage survives this crisis, it tends to settle down into a routine in which the family sticks close to home for most of its nonworking hours. The woman tends to socialize with her relatives, the man with a local group of friends. There is usually a sharp split between male and female cultures.

5. Working-class culture tends to be highly localistic, with strong pressures for conformity within the group and distrust of outsiders. These conditions make working-class people typically moralistic and traditional in their beliefs, favoring strong discipline to control children and punish deviancy.

6. Upper-middle-class people emphasize delayed gratification and future planning and tend to both marry and have children later than lower-class persons, so as to pursue their careers. They are involved in numerous networks of professional and business contacts and do much more nonfamily socializing than working-class people.

7. Upper-class families generally separate themselves from other classes by attending private schools, belonging to exclusive clubs, and acting as patrons of the arts. These families socialize their children to accept the responsibilities and obligations of the elite and ensure their class position by passing on their enormous wealth.

8. Dual-career couples have received much attention from family researchers, even though they are less common than dual-earner couples who have working-class jobs. Dual-career couples can usually hire other people to perform child care and cleaning, and men and women professionals who limit their work hours to take care of children are still considered less serious about their careers than others.

9. Economic restructuring and the growth of the service sector have created more job opportunities for women. Women earn less than men, are concentrated in service occupations, and are less likely to be promoted to management positions than men.

Key Terms

class	occupational segregation	social stratification
dual-career couples	order givers	upper class
feminization of poverty	order takers	upper-middle class
high social density	poverty level	working class
low social density		

Sociology Web Site

See the Wadsworth Sociology Resource Center, "Virtual Society," for additional links, quizzes, and learning tools:

http://www.sociology.wadsworth.com

Also on this web site you'll find Info Trac College Edition, an outline library of journals. Here you can search for electronic articles about central topics in sociology.

7 Race and Ethnicity: Understanding Structural and Cultural Differences

INTRODUCTION

In this chapter we look at some of the ways in which family life is affected by membership in different racial and ethnic groups. Except for the descendants of the indigenous tribes of North America, all families in the United States came here from somewhere else. All Americans are members of at least one ethnic group, insofar as everyone's ancestors came from a region of the world with unique language, culture, and customs. Even white Americans whose ancestors came from England or Europe centuries ago and who like to think of themselves as the "real" American families are members of immigrant families who carried unique cultural traditions from somewhere else. In fact, the United States is often referred to as a nation of immigrants. As we will see in this chapter, different ethnic groups moved to the United States at different times, and both their cultural heritage and their experiences in this country helped shape unique family styles among various groups.

Is There an American "Melting Pot"?

The idea of the great American **melting pot** is that diverse racial and ethnic groups will blend into a standard prototype and that all families will eventually become similar. This idea stems, in part, from watching the experiences of repeated waves of European immigrants to North America. Founded by people from different lands, the United States witnessed massive waves of **immigration** during the nineteenth and twentieth centuries. Although they brought unique cultural customs and practices from their native countries, many of the early immigrants became assimilated into the emerging American culture and slowly lost many of the unique characteristics that identified them as different from others.

Between 1830 and about 1880, over 10 million immigrants came to America as the land west of the Mississippi was opened for settlement. New roads, canals, and railroads were built; new areas were settled by farmers; and growing industrialization provided many jobs for those seeking a better life in the New World. Most of the new immigrants during this period were from Northern Europe, principally England, Ireland, Germany, and the Scandinavian countries. From about 1880 to 1930, an even larger number of Europeans immigrated to the United States, with both northern Europeans and southern Europeans contributing an influx of over 22 million new American citizens. Chief among the latter group were Italians, Poles, Greeks, Russians, Austro-Hungarians, and other Slavic groups. Industrialization uprooted millions of Europeans, but it generated many jobs in America for unskilled workers in mining, construction, and manufacturing. This tremendous influx of people into the United States—the largest in recorded history—was mostly composed of peasants and others of the lower social classes but also contained some families from the middle and upper classes (Gordon 1964; Zinn and Eitzen 1987).

In major cities like New York, Chicago, and Boston, ethnic enclaves developed with neighborhood and community patterns that were distinct from other groups. Like the Asian and Latino immigrants of recent decades, these European groups carried Old World traditions with them but transformed them by establishing unique patterns of family and community support. *A Closer Look* 7.1 describes how Italian-American immigrants living in urban ethnic enclaves fared after the turn of the century. Because immigrant families came from different

A CLOSER LOOK 7.1
Life in Italian–American Neighborhoods in the Early Twentieth Century

By 1910 one-third of the population of the twelve largest American cities was foreign born. Different groups of immigrants tended to be concentrated in particular industries, for example, the Italians in construction and textiles. Whatever the industry, they undertook the more disagreeable jobs and worked long hour in poor, often hazardous conditions. To find work, non–English speakers tended to rely on compatriots who had preceded them. These established men who acted as intermediaries became "work bosses"—padrones among the Italians; they were paid commissions by the men to whom they gave jobs. Not all padrones cheated, but enough did to give the system a bad name.

By the beginning of the century, peasants from southern and eastern Europe formed the overwhelming majority of immigrants. Their adjustment to urban, industrial life was eased by the tendency of different groups to occupy particular neighborhoods within the congested slums in which they mostly lived. Here, despite appearances, they did not so much recreate the communities from which they had come as establish them afresh. An Italian immigrant and community leader, Constantine Panunzio, said of the colony in Boston in which he lived that though it was "in no way a typical community neither did it resemble Italy." The sociologist Robert E. Park wrote in 1920 that "In America the peasant discards his habits and acquires ideas. In America, above all, the immigrant organizes." Characteristic of Italian communities were mutual-benefit societies, Italian-language newspapers, Italian theater, opera and marionette shows, the celebration of religious festivals involving elaborate processions, and an allegiance to Catholicism that was combined, however, with the strong resistance to Irish domination of the Church in America. Mutual-benefit societies, of which there were, for example, 150 in Manhattan alone in 1900, required small but regular financial contributions from their members and provided notably for medical and funeral expenses. But the importance of the societies went beyond

When a large number of Italians immigrated to the United States near the turn of the 20th century, they adapted peasant customs to urban neighborhoods.

the merely instrumental, for the fraternities they established filled a social and psychological void for those uprooted from familiar surroundings and ways of life. One society described itself as an organization that "unites us and gives us strength, and will make us more acceptable in the eyes of the American people." The ethnic press also served as a bridge between life in the European village and that in the American city, providing leadership and voicing group demands and complaints. The Italian newspapers were not like those of Europe but were "addressed to the common man" (Park) and much more popular in language. The Italian theater, too, acquired a distinctive "Italian-American" style and idiom.

Exploited though they may have been, the immigrants enjoyed more abundant and varied food than they had known before. Part of what has become known internationally as "Italian food"—notably pizza—is rather an Italian-American creation.

Source: Excerpted from John Harriss (ed). *The Family: A Social History of the Twentieth Century.* Copyright © 1991 by Oxford University Press, Inc. Used by permission of the publisher.

cultures, had different resources at their disposal, settled in different geographical regions, and had varying patterns of employment for different family members, it is impossible to characterize one typical immigrant family type—even among the Europeans who constituted the bulk of immigration during the early part of the 1900s. In general, immigrant families relied on traditional customs from the old country when they were compatible with life in the new country, and ethnic groups adopted the language, customs, and beliefs of the host country with varying speed. Although many of the immigrant groups maintained a unique ethnic identity, they were generally able to find work at a variety of jobs, could accumulate wealth, and in time experienced about the same social mobility as those who came before them. This gradual incorporation into mainstream society, and the gradual loss of a unique cultural identity, reflect the idea of the "great American melting pot."

Traditionally, studies of the experiences of new immigrants in America were guided by what has become known as **straight-line assimilation theory** (Landale 1997). This view suggests that immigrants become increasingly similar to the existing population as they spend more time in the country. Eventually, and typically after several generations, immigrants were supposed to lose their cultural and socioeconomic distinctiveness, blending into dominant American society (Gordon 1964; Park 1950). Early proponents of this theory recognized that the assimilation process was complex and varied, but they assumed that the process would be essentially similar for all groups.

The assumption of straight-line assimilation for all ethnic groups in America was challenged over a century ago (e.g., DuBois 1899) but has experienced renewed attacks recently. Other race and ethnic groups, notably African-Americans, Latinos, and Asians, were not integrated into American society in the same fashion or at the same rate as European-Americans, principally because of economic, legal, and social discrimination. Analysis of the experiences of these groups and the changing global and national contexts for immigration thus have led many scholars to challenge the assumptions of straight-line assimilation theory. Because there are multiple pathways of assimilation and conforming to the dominant culture is not necessarily an inevitable nor unidirectional process, melting-pot assumptions are not as widely accepted as they once were (Landale 1997). By focusing on different groups, their reasons for immigration, the varying resources they have under their control, and regional variations, modern scholars have developed more complex theories of segmented assimilation to help us understand the diverse experiences of ethnic families in the United States (e.g., Buriel and De Ment 1997; Portes and Zhou 1993; Rumbaut 1997). Later in the chapter, we will explore how immigration, ethnic identity, and economic opportunity are related as we focus on the different historical and contemporary experiences of African-American, Latino, and Asian-American families.

Race and Ethnic Categories

Most people think of race and ethnicity as things with fixed and stable meanings, but these terms and the meanings associated with them are socially constructed and constantly subject to change. Although people used various racial terms in everyday speech, Federal government agencies in the United States did not even regularly collect data on race until they attempted to reduce discrimination through the 1964 Civil Rights Act and the 1965 Voting Rights Act. There are now four official U.S. Census categories for race (white, black, Asian/Pacific Islander, and American Indian/Alaska Native) and two for ethnicity (Hispanic, non-Hispanic), but this scheme was not developed until 1977 (DeVita 1996). In the current Census categorization of "minorities," **racial identity** is distinct from **ethnic identity,** so Hispanics can be of any race. As we will also see, the rapidly growing number of people with mixed racial/ethnic identity makes categorization even more difficult for demographers and other record keepers.

Although people tend to identify with a specific race and/or ethnic group, it is important to note that these are categories based on cultural or social factors rather than on scientific evidence (in fact, biologists remind us that distinct races do not actually exist). In addition, people tend to change the way they classify themselves or others depending on social context and the meanings of specific terms used for categorization. For example, many Americans of African descent used to be identified as "Negro," and in the 1970s, many government agencies adopted the term *Afro-American*. The term *black* has gone in and out of favor and is still widely used today. The more recent term *African American* has gained widespread acceptance, though arguments continue over whether it should include a hyphen. Such debates signal the fluid nature of racial categories and show how terminology is affected by cultural and bureaucratic concerns.

Similar debates occur over use of terms for other racial and ethnic groups. The current Census term *Hispanic* implies a Spanish-speaking heritage, whereas the designation includes many subgroups of the population with widely different origins. People in this general category in the United States tend to identify themselves more specifically in everyday speech, often depending on the country or region from which their ancestors came (e.g., Mexico, Puerto Rico, Cuba, Central America). Others use the term *Latino* to refer to people who come from Spanish-speaking countries, and some use *Chicano* (which has taken on contemporary political significance, but originally referred to the people of Spanish/Mexican and sometimes mixed Native American descent who lived in the Southwest before it was part of the United States). Similarly, diversity among Pacific Islanders and Native Americans is quite extensive, and Americans of Asian descent are more likely to use cultural and regional labels like Chinese, Japanese, Korean, Vietnamese, Indian, Pakistani, or Persian. This diversity and flexibility in labeling are confusing for survey takers who would like to have simple and fixed racial/ethnic categories, but it reminds us that such categories are fluid and that people have some control over the ways they see and talk about themselves and others. We explore some of this diversity in this chapter and typically employ the aggregate labels of *African American, Asian American, Native American,* and *Latino* to talk about some general marriage and family patterns. Because we report demographic data from the U.S. Census Bureau, note that these terms also correspond to the labels of *blacks, Asian* and *Pacific Islanders, American Indians/Native Alaskans,* and *Hispanics.*

Increasing Diversity Among American Families

Because the population of the United States is becoming more diverse all the time, it is especially important to understand and appreciate differences and similarities among different American families according to race and ethnicity. In 1970, African Americans, Asian Americans, Native Americans, and Latinos represented only 16 percent of the total U.S. population. By 1998, this share had increased to 27 percent, including over 70 million people. African Americans constituted the largest segment, about 34 million people or over 12 percent of the total U.S. population. Hispanics numbered over 30 million, or 11 percent of the population. Over 10 million Asians and Pacific Islanders constituted almost 4 percent of the total U.S. population, and over 2 million Native Americans represented almost 1 percent (*Statistical Abstract* 1999). Because of higher than average birthrates and increasing levels of intermarriage, the Census Bureau projects that by the year 2050, these ethnic/racial groups will account for nearly half of the total population of the nation (Council of Economic Advisors 1998). In many areas today, these so-called minority groups are now in the majority, especially when we focus on families with young children.

What difference does ethnicity make to household form or the experience of family life? We know, for example, that the household composition of people from different race/ethnic groups tends to vary. Only about a sixth of black households and a quarter of white households are married couples with children, whereas over a third of both Asian and Latino households are of this type. Only one in six Latino or Asian households are single people living alone, compared with one in four African American or white households. Also, fewer than one in sixteen white or Asian households are single parents, whereas almost one in four African American households are of this type (DeVita 1996).

If we exclude single people and other nonfamily households from the calculations, we can see how families differ according to race and ethnicity. (Remember that the Census Bureau defines a **family** as "two or more persons who are related by birth, marriage, or adoption who live together as one household".) We can see in Figure 7.1 that Asian American families are the most likely to contain married couples, followed closely by white households. Hispanic and Native American families are next most likely to include married couples, followed by African American families. Conversely, African American families are more likely than others to be "female headed" according to the Census, or in other words, to include a single mother and at least one child. The rates for mother-headed families among Native Americans and Hispanics are in the middle and are lowest for whites and Asians. Finally, single-father households are relatively rare for all groups, with Native Americans and Hispanics showing the highest rates.

AFRICAN AMERICAN FAMILIES

Although in many respects African American families reflect the general culture, certain unique pressures on them, coupled with their own individual cultural heritage, have caused these families to devise some distinctive structures for surviving and getting ahead. Yet despite such commonalities, there are also important economic and social differences among African American families. For example, blacks whose families came from different regions of Africa or the Caribbean have very different linguistic and cultural backgrounds. Blacks raised in the American South have different experiences and traditions from those raised in the North, and those living in urban areas differ from those living in rural areas or those residing in the suburbs. One thing African Americans typically are not, however, is a reflection of the stereotypes commonly applied to them.

The Myth of the Black Family

It is widely believed that the black family deviates greatly from white family structure in the United States. The general image is of marital violence, broken homes, large numbers of children, and a resulting cycle of poverty, illegitimacy, delinquency, welfare, and unemployment. During the racial upheavals of the 1960s, Daniel Patrick Moynihan wrote a report for the U.S. Department of Labor, the so-called Moynihan Report, which argued that the economic problems of blacks were basically due to the weakness of their family structure and a **culture of poverty.** He described it as a "tangled pathology . . . the principal source of most of the aberrant inadequate, or antisocial behavior that . . . perpetuates the cycle of poverty and deprivation" (Moynihan, Barton, and Broderick 1965, 47). Moynihan, later a U.S. senator from New York, argued that

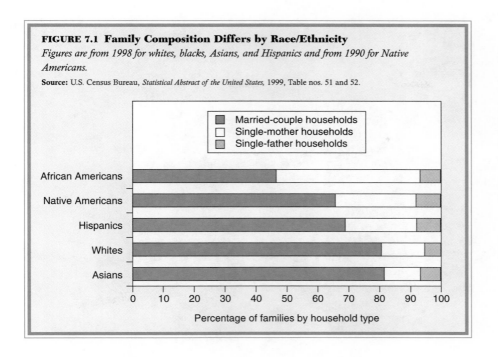

FIGURE 7.1 Family Composition Differs by Race/Ethnicity

Figures are from 1998 for whites, blacks, Asians, and Hispanics and from 1990 for Native Americans.

Source: U.S. Census Bureau, *Statistical Abstract of the United States,* 1999, Table nos. 51 and 52.

the source of this family structure was the historical experience of slavery, which separated fathers from their children and left the black family with a heritage of dependence upon mothers. "The Negro community has been forced into a matriarchal structure which, because it is so out of line with the rest of American society, seriously retards the progress of the group as a whole" (Moynihan, Barton, and Broderick 1965, 29).

The key attribute of the so-called broken family structure seen by Moynihan was the absence of the father and the unusual amount of power wielded by women. Allegedly, the psychological results of this include a low self-image on the part of male children; lower IQ scores with resulting low school achievement; high dropout rates; delinquency; and unemployment (Moynihan, Barton, and Broderick 1965). Like E. Franklin Frazier (1950) before him, Moynihan blamed African American families for many of the troubles of blacks, but unlike Moynihan, Frazier was keenly aware of the ways that American society created insurmountable problems for black families (Zuberi 1998). Moynihan's image of the black family has several faults. It ignores social class and especially overlooks the large number of black families that do not fit the stereotype. Less commonly noticed but quite apparent once one thinks of it, Moynihan's indictment assumes a rather sexist and privatized model of the "normal" family.

Social Class Differences

One important fact missed by the broken/matriarchal image of the black family is that a large number of African American families do not have this structure. Almost half of all African American family households are married-couple households, and the men in these families are likely to have dependable incomes. More than a fourth of all black families have incomes above the median for the United States, and in this group, the majority have two parents present. Still, African Americans are much more likely than members of other racial/ethnic groups to be unmarried, to live in single-parent families, and to live below the poverty line.

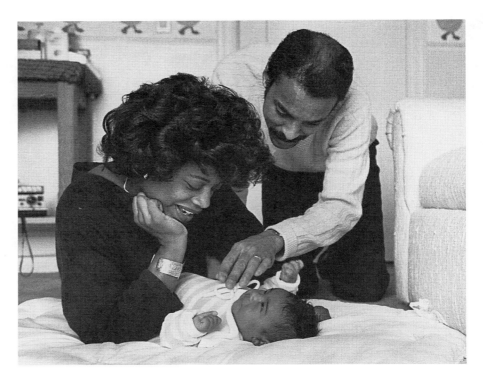

Two parents are present in just under half of all black families.

In 1998, fewer than half of all black adults were married, and fewer than half of all black children were living with two parents (*Statistical Abstract* 1999).

Racial Discrimination and Its Paradoxical Consequences

Until the 1950s, the United States was a thoroughly and even openly racist society. Segregation laws were enforced in southern states, while de facto discrimination prevailed in the North, and racial stereotypes were an accepted part of popular culture. The civil rights movement of the 1950s and 1960s, after both nonviolent demonstrations and violent uprisings, brought a rapid change in the legal structure, although informal racial discrimination has proved more resistant to eradication. One effect of this discrimination has been a serious financial penalty paid for being black.

Because of racial discrimination, the employment patterns of African Americans have taken a peculiar form. They have been able to acquire decent white-collar jobs, or even stable blue-collar positions, mainly by working in government rather than private business. Many black women have been able to pull their families up to the middle-class level by working as schoolteachers. This has given a special importance to education in the black community, where women have often had more education than their husbands. It has been more difficult for black men to find comparable middle-class jobs. Where they have found them, it has often been through government employment, such as the post office. Cities with large-scale government employment have thus tended to attract a large black population; one of the largest percentages of blacks in any city (about two-thirds) is in Washington, D.C.

Such economic pressures have also affected forms of family organization typically adopted by both the African American middle class and the African American working class (see *A Closer Look* 7.2). Neither fully resembles its white counterpart.

Among affluent blacks, distortions in employment patterns owing to discrimination are played out in family relationships. White upper-middle-class families tend to subordinate everything to the career of the father (as we have seen in the previous chapter on social class). Among blacks, however, this is seldom the path to success. The father alone usually cannot attain upper-middle-class status for his family with only their moral support. He needs that, but in addition, the whole family works together as an economic unit to get ahead. Women's achievements are important. In fact, the pioneering African American upper-middle-class, seeing itself as part of an embattled minority making its way against heavy odds, places special emphasis on self-discipline and achievement by *all* its members. There is great pressure on all the children to go to college and have successful careers. Willie (1981, 176–79) has argued that, as a result, the affluent black family has held together more firmly than the comparable white family—perhaps too much so because, he suggests, both adults and children of such families pay a price in subordinating their individualism and spontaneity to their achievement.

For the African American working-class family, discrimination has had a paradoxical effect. It has kept some blacks in the working class who might otherwise have risen higher. By putting a lid on *all* black achievement, discrimination has in effect transformed the black working class into a middle class. Compared with their chief reference group, other blacks, their achievements *are* in the middle of the spectrum. It was already noticeable at the time of the Second World War that black working-class families were much more like middle-class families than their working-class counterparts in the white community (Drake and Cayton 1945). They belonged to more organizations, attended church more, and held a higher status ranking within their own community than white workers did in theirs. This may still be so. Although the latter, as we have seen, are alienated and withdrawn from a world run by affluent whites, working-class blacks are more of the respected middle stratum in a delimited social context. Perhaps for this reason, some studies have shown that blacks at the working-class level actually have similar or lower rates of mental illness and higher levels of social participation and self-esteem than comparable whites (Farley and Hermalin 1971; Olsen 1970; Regier et al. 1993; Surgeon General 1999; Willie 1985).

The Question of Black Matriarchy

Another argument concerns the supposed deleterious effects of female domination in African-American families. Earlier, we have argued that this actually does not occur as much as some writers would have us believe: many black families are in fact rather like the conventional two-parent white family. At the same time, a substantial number of black families, usually in the urban underclass, are headed by women, and black women also play an important role within middle-class families.

The question is: Are the effects of this female power good or bad? The assumption has been made by Moynihan and many others that the psychological effects of this situation have led to delinquency and nonachievement in the black lower class. Others have also focused on "restoring the dignity" of the African American male, giving him a dominant place in the family and implicitly

A CLOSER LOOK 7.2
Sharing Networks Among Lower-Class African American Women

Lower-class black women have evolved special structures of their own for dealing with economic insecurity. African American extended kin patterns and social families have been viewed as a vestige of West African cultures, as well as an adaptive response to situational constraints in America (Billingsley 1968; Hatchett and Jackson 1993; McAdoo 1997; Rivers and Scanzoni 1997; Sudarkasa 1997). The anthropologist Carol Stack (1974) found that, instead of conventional marriages, African American mothers use a network of mostly female relatives and friends who help each other out regularly in emergencies. They provide each other with places to stay, take care of each other's children, share the use of a refrigerator, and use whatever money is available to pay the most pressing bills.

This "fictive kin" network makes it possible for these women to survive at the very depths of the class structure, but it also tends to keep individuals from breaking out of this position. Stack found that one young woman wanted to get married, but the other members of her network discouraged her because they needed her economic contributions. An older couple in a network received a windfall inheritance of $1,500, which they decided to use as a down payment on a house. But other members of the network put in their requests too: one needed $25 to keep her phone, which they all used, from being cut off; another was about to be evicted if she couldn't come up with the rent; still others asked for train fare to attend their mother's funeral; then the couple's own children were dropped from the welfare rolls. Within a few weeks the money was gone. The network was still intact, and so were the bonds of poverty.

advocating that black women should be in a subordinate position (see Ransford and Miller 1983). However, the evidence does not support the contention that equality for women is responsible for psychological problems and low achievement in children.

What *is* true is that black women have more family power than their white counterparts. This is especially seen in the lower class, where most families are headed by women, but it is also true in the intact families of the working class and middle class. Because of economic discrimination, black men have had unusual difficulties in getting steady or well-paying jobs, and black families have tended to rely on women's incomes (as well as children's incomes). Black women rate themselves as having more power than white women, and black husbands attribute more power to their wives than white husbands do. Over the years, researchers have concluded that black families tend to be more egalitarian than white families at the same class level (Broman 1988; Creighton-Zollar and Williams 1992; Scanzoni 1977; Staples 1997; Taylor et al. 1991).

From the point of view of gender equality, this is a positive development. However, there are signs that it may also be a source of continuing, or even increasing, conflict. For example, much African-American rap music in the 1990s was criticized for using hostile and degrading language toward women, treating black females as sex objects and antagonists. Even among the black middle class, men tend to hold fairly traditional notions of appropriate gender roles, perhaps because black men have been denied access to the jobs and earnings that have enabled their white counterparts to enjoy dominance over women. Ransford and Miller (1983) found that black men were much more likely than whites to agree with statements such as "Women should take care of running their homes and leave running the country to men," and "Most men are better suited emotionally to politics than are most women." Furthermore, black middle-class wives often tend to agree with their husbands, expressing much less support for gender equality than white middle-class wives. This fits Willie's (1981) picture of the black middle-class family, with its strong internal pressures subordinating everything to the task of getting ahead economically. Willie even refers to such families as "affluent conformists." The result is something of a paradox. We see the black middle-class family fighting its way to some success against racial discrimination but losing some of the female equality it had previously held. Even though middle-class black men tend to believe in male domination, they have a realistic attitude about the importance of their wives' incomes, which have always been so crucial to the success of black families.

At the same time, there are countertrends and signs of impending strain. Working-class and lower-class black women do not share this conservatism about gender equality. On the contrary, many of them are rather militant in their attitudes toward men, not in the larger political sense of taking a feminist position on issues in the larger society but in the immediate personal realm. Black women have been found to be much more likely than white women to regard men as untrustworthy and economically unreliable. Black women are much less likely to feel they need a husband to have children, contributing to more black babies being born outside of marriage than within it. With black men's economic prospects declining and black women's increasing (though still at the bottom), many African American women have given up on the white middle-class ideal of the isolated and self-sufficient nuclear family.

The Case of the Black Underclass

The female-headed black family that makes up such a large proportion of the urban ghettos has given most of us our stereotyped image of the African American family. Although only a minority of black families fits this description, this group nevertheless is a crucial focus for the problems of poverty, crime, and other social problems of our day.

However, the majority of poor people in the United States are actually *not* black. Their situations no doubt differ from those of African Americans in many important ways. By studying black poor people, then, we can by no means learn everything significant about *all* poor people. As it happens, however, sociologists have studied black poor people much more than white poor people; much of what we know about poverty refers to black poverty. Keeping these limitations in mind, we will see what we can learn about major social problems through the study of the black underclass.

One conclusion, we would suggest, is that although this lower-class family is female headed, that is not necessarily the source of its problems. In fact, one might say that black women have been particularly strong here, not only in holding the family together as much as they have but also in carving out a protected sphere for themselves as women in a world of chronic poverty and under the threat of considerable male violence (Sudarkasa 1993).

Let us consider the historical trends. It has often been asserted (Moynihan, Barton, and Broderick 1965; Billingsley 1968) that the black family broke down and became female headed because of the experience of slavery. Some later historical studies (Gutman 1976) suggested that most slaves grew up in two-parent families and that the family began to break down only after 1925. In contrast, more recent demographic studies conclude that African American households were less likely to be nuclear and more likely to be headed by women than white families during the late nineteenth and early twentieth centuries (McDaniel 1994; Morgan et al. 1993; Ruggles 1994). Hence it is most appropriate to view African American deviation from white family patterns as persistent rather than of recent origin. At the same time, the tendency for black families to be headed by a single parent accelerated after 1960 (McDaniel 1994; Wilson and Neckerman 1986). Just under half of black families are married-couple households, but those with children are more likely to be headed by women than others. In 1970 over half of all black children lived with two parents (59 percent), but by 1998 only 38 percent lived with two parents. At the end of the 1990s, a majority of black families with children were single-mother households (57 percent).

There are other trends, too, that indicate that the female-headed lower-class black family of today represents a relatively recent type of family structure. We know, for instance, that although blacks used to marry at younger ages than whites, ever since about 1950, blacks have been waiting longer to marry than whites (Cherlin 1992; Koball 1998; Tucker and Mitchell-Kernan 1995). Blacks have a higher divorce and separation rate than whites at all class levels. They wait shorter intervals after marriage to divorce and take longer to remarry; more blacks never marry (O'Hare et al. 1991; *Statistical Abstract* 1999; Taylor et al. 1991). Although most blacks have been married, the proportion has dropped significantly since the 1970s. In 1970, once of five black adults had never been married, but by 1998, more than one of three were in this category (Figure 7.2). The trend toward nonmarriage has been especially strong in the black lower class, whose difference in this respect from whites has been accelerating in the last few decades. In the late 1990s, the proportion of never married adults was almost twice as high for blacks as it was for whites (*Statistical Abstract* 1999). Nonmarital birthrates, too, show a sharp disparity between blacks and whites, with blacks twice as likely to have a child outside of marriage. Nevertheless, rates of nonmarital birth for African American women have been dropping, and the gap between blacks and whites has been shrinking. In 1970, the birthrate for unmarried black women was nearly seven times as high as the rate for white women, but by the mid-1990s it was only about twice as high (Moore 1995). Because of trends in marriage, divorce, and nonmarital birth, the percentage of children living with two parents is lowest for African Americans (Figure 7.3). Among black teenagers, birthrates have declined in recent years, but because of lower rates of contraception and abortion, the rates are still much higher than for white teens (Taylor et al. 1991).

Unmarried women are only slightly more likely to give birth to a child, but because both blacks and whites are waiting longer to get married and sometimes not marrying at all, the number of nonmarital births is increasing. Also, married women are having fewer babies than they used to, so a larger percentage of all babies are born to unmarried women. All of these factors contribute to a much greater likelihood, especially among blacks, for children to be born outside of marriage. There is now little or no stigma attached to having a child outside of marriage, especially among poorer blacks. About two-thirds of all black births occur to women who are not married (O'Hare 1992).

Explanations for Marital Instability

What underlying factors account for the greater instability of marriage and higher rate of nonmarital births for blacks as compared with whites? The most important explanations include (1) differences in social class or economic position, (2) the impact of family assistance benefits (welfare), (3) the relative economic status of men and women, and (4) a culture of poverty. We review these explanations in turn; a fifth explanation, scarcity of black men, is discussed later, in *A Closer Look* 7.4.

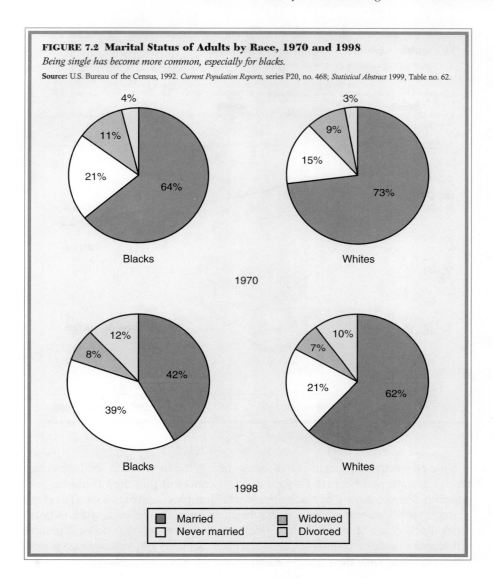

FIGURE 7.2 Marital Status of Adults by Race, 1970 and 1998

Being single has become more common, especially for blacks.

Source: U.S. Bureau of the Census, 1992. *Current Population Reports,* series P20, no. 468; *Statistical Abstract* 1999, Table no. 62.

1970

Blacks: 4%, 11%, 21%, 64%

Whites: 3%, 9%, 15%, 73%

1998

Blacks: 12%, 8%, 39%, 42%

Whites: 10%, 7%, 21%, 62%

Legend: ■ Married   ■ Widowed   □ Never married   □ Divorced

DIFFERENCES IN SOCIAL CLASS OR ECONOMIC POSITION Researchers have found that high rates of birth to unmarried women are regularly associated with the following conditions: low socioeconomic status, urban residence, little education, times of economic depression, prior prevalence of divorce and separation, home background of unwed parenthood, weak parental controls over children, and lack of severe censure or social sanctions for premarital sexual relations and pregnancy. Because blacks lag far behind whites on most indicators of economic and educational status, they are much more likely to be exposed to the conditions that produce nonmarital births. The aforementioned background conditions are typically found in clusters, so families with few economic and social resources are least able to control youths or to reward them for confining childbirth to marriage (see *A Closer Look* 7.3 for a discussion of cultural stereotypes about families in such conditions).

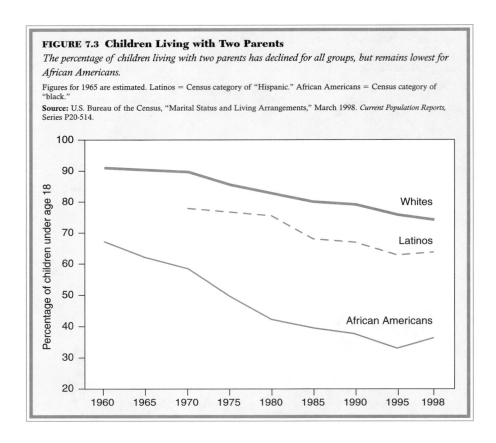

FIGURE 7.3 Children Living with Two Parents

The percentage of children living with two parents has declined for all groups, but remains lowest for African Americans.

Figures for 1965 are estimated. Latinos = Census category of "Hispanic." African Americans = Census category of "black."

Source: U.S. Bureau of the Census, "Marital Status and Living Arrangements," March 1998. *Current Population Reports,* Series P20-514.

One persuasive argument is that black men at the lower-class level are reluctant to get married because they do not feel confident that they could support children. Studies show that black men in stable employment are twice as likely to marry as black men who are not in school, in the military, or at work (Mason 1996; Tucker and Mitchell-Kernan 1995). Those marriages that do exist tend to fall apart because of economic pressure. There is plenty of evidence that people are dissatisfied with their marriages and families when economic pressures are severe (McLoyd 1997; Voydanoff 1991), and these pressures hit especially hard in the black lower class.

Studies show that black marriages are more likely to end in disruption than white marriages. White women are more likely to be divorced than separated from their husbands, whereas black women are more likely to be separated than divorced (Wilson and Neckerman 1986). These differences in marital disruption and marital status for blacks versus whites primarily stem from the underlying effects of social class and racial segregation. During the 1970s and 1980s, when rates of marriage for blacks were declining, the black underclass was growing. Economic restructuring and inflation drove up poverty rates for blacks, but underclass communities were mainly created where there was also a high degree of spatial segregation. Patterns of isolation and segregation in housing encouraged physical decay, crime, drug use, joblessness, welfare dependency, and nonmarital birth among the black underclass living in urban ghettos. According to Massey and Denton (1993), spatial segregation of blacks in this country is a relatively recent phenomenon. "American Apartheid," as they call it, is encouraged by discrimination and prejudice, but once in place, such racial segregation tends to

A CLOSER LOOK 7.3
Stereotyping Poor African American Mothers

Over the past several years, but particularly since the 1994 congressional election, we have witnessed the systematic stereotyping, stigmatizing, and demonizing of the poor, particularly of poor women. They have been pictured as the embodiment of the characteristics Americans revile—laziness, willful dependence on government, wanton sexuality, and imprudent, excessive reproduction. They are frequently described as transmitters of negative values, or, even worse, of no values at all; their family structure and child rearing have been blamed for fostering violence, crime, school failure, out-of-wedlock births, and above all, for passing their poverty on to the next generation. In the eagerness of many in positions of power to deny the structural causes of poverty and the other ills that beset American society, politicians and policy makers have revived the "culture of poverty" analysis and laid the responsibility for poor people's problems and many of the problems of the wider society at the feet of the most impoverished and powerless among us. "They" don't want to work, it is said. "They" want something for nothing. "They" are like animals who have been given too much, conditioned to be dependent, and consequently can no longer make it on their own. "They" have too much sex, have too many babies, and all too often care for them miserably. "They" breed violent children who, when they reach adolescence, will rape, pillage, murder, and burn.

    "They" as a group have been portrayed as the "underclass," a group with multiple, severe problems, increasingly considered impossible to reach, a breed apart from mainstream Americans. The image of the underclass has expanded to include virtually all of the poor, particularly single mothers and welfare recipients, and is envisioned by many to consist largely of African Americans. And it is the race factor, the stereotype that most poor people are black, that holds the entire image together. This mythology is in part embedded in our history. Large elements within the United States have since colonial times despised and feared the poor and have particularly despised and feared African Americans. A critical consequence of dividing people into "them" and "us" is that it clouds our perceptions of the reality that millions of people face daily and therefore propels society toward solutions that will harm rather than help those in need.

Source: Excerpted from Ruth Sidel, *Keeping Women and Children Last* (New York: Penguin, 1996), 166–68. Copyright © 1996 by Ruth Sidel. Used by permission of Viking Penguin, a division of Penguin Putnam, Inc.

produce even more economic and social disparities between blacks and whites, leading to further discrimination and more segregation. In this view, black marriage patterns are a major outcome of segregation and economic inequities, rather than their cause.

FAMILY ASSISTANCE BENEFITS Some conservative observers have suggested that it is primarily welfare laws that have caused nonmarital birth among blacks. Because mothers have received aid according to how many dependents they have and whether the father was absent, antigovernment politicians blamed welfare for "out-of-wedlock" births. Some have argued that welfare payments (formerly Aid to Families with Dependent Children [AFDC] and, more recently, Temporary Aid to Needy Families [TANF]) encourage unmarried pregnant women to have more children and set up their own households and for married mothers to divorce their husbands. The assumption is that in the absence of such payments, women and men would be forced to rely on other options, such as avoiding pregnancy, having an abortion, living together, getting married, sharing a residence with parents, or staying in an unhappy marriage. If this was the case, then those areas with higher levels of benefit payments might be expected to have higher rates of births to unmarried women and higher divorce rates. A number of studies have tested this hypothesis with statistically detectable but modest associations between these factors (Abrahamson 1998; Duncan, Hill, and Hoffman 1988; Ellwood and Bane 1985; Garfinkel and McLanahan 1986;

Lichter 1995). Ellwood and Summers (1986) found no significant relationship between the percentage of black children in single-parent households and the level of welfare benefits in different states. In a major review of studies on the relationship between welfare payments and nonmarital births, Moffitt (1995) concluded that the more adjustments a study makes for differences among states and the women in those states, the lower the estimated effect of welfare on non-marital childbearing. Some studies show no relationship between welfare and nonmarital birth, whereas others show a negative relationship, and still others a positive relationship. Of particular note is that higher welfare payments are linked to higher nonmarital birth rates for whites more often than for blacks (Moffitt 1995). Perhaps the most important evidence on this question compares overall trends in welfare and nonmarital birth. Because nonmarital birthrates increased at the same time that welfare payments were decreasing, it is unlikely that such benefits were a major causal force. Similarly, African American marriage rates are generally unrelated to receipt of public assistance (Lichter 1995; McLanahan and Casper 1995). Cherlin (1992) notes that even if assistance payments accounted for some part of recent nonmarital birth trends, this does not explain why divorce rates and births to unwed mothers grew to be so high in the period between 1925 and 1960, before significant welfare benefits were available. In sum, the evidence seems to suggest that family assistance benefits play a relatively small role in the overall tend toward greater divorce and nonmarital birth among blacks.

Research does seem to suggest, however, that public assistance payments may reduce the likelihood that an unmarried pregnant woman marries in haste or obtains an abortion. Higher family assistance benefits are also sometimes associated with mothers marrying less rapidly after divorce. Perhaps the strongest and most consistent finding is that the availability of welfare encourages single mothers to form their own households rather than living with their parents (Ellwood and Bane 1985). Even before recent reforms, most welfare recipients tended to stay on welfare for a limited amount of time, so we must look elsewhere to discover why marriage and birth patterns have been changing.

THE RELATIVE ECONOMIC STATUS OF MEN AND WOMEN Many researchers suggest that high rates of nonmarital birth and disruption among African Americans result from the relative economic status of black men and women. Wilson (1987) claimed that low rates of employment for black men make marriage less attractive for both men and women. In the 1990s, a black male was about twice as likely as a white male to be unemployed or out of the labor force. About 20 percent of young African American men are neither in school nor working, compared with 9 percent of young white men (Council of Economic Advisors 1998). The percentage of young men who are in this situation increased during the 1980s and did not fall substantially during the 1990s. In general, black marriage rates have declined most in those regions of the country where the ratio of employed black men to employed white men has declined most. The ratio of employed black men to women in younger age groups has fallen, and these changes parallel changes in marriage rates. Although joblessness is an important concern, even earnings of fully employed black men have declined since 1973, with blacks in all income categories earning much less, on average, than their white counterparts.

As the employment and earnings of black men have decreased, employment and relative earnings of black women have risen. In the 1940s, the wage and salary earnings of black women were about a third of those of black men; by the

1950s black women's earnings were over half of black men's; in the 1960s they reached two-thirds; and by the 1970s, black women's income was three-fourths of black men's income. By the 1980s, black women were earning over 80 percent of what black men did, and during the 1990s, they earned about 86 percent of what black men did. For whites, a different pattern holds: in the 1940s, women's median earnings were about half of men's, but from the 1950s to the late 1980s, white women's earnings had only increased to about 64 percent of white men's earnings. It was not until the late 1990s that white women's earnings reached over 70 percent of white men's earnings (and note that these comparisons use only full-time year-round workers, overestimating women's actual incomes). To be sure, black women are disadvantaged relative to both white women and black men, but the relative gap between their earnings and those of black men is much less than for whites. The question thus arises: Does increasing income independence for black women make marriage less attractive to them? Some researchers find that increased economic independence for women leads to increased divorce and separation, as well as lower marriage rates (Garfinkel and McLanahan 1986; Lichter 1995; McLanahan and Casper 1995), but declines in marriage have been more pronounced among less-educated women. In addition, employment tends to delay marriage for African American women, but education and better employment tend to increase their chances for eventual marriage, provided they can find a partner with similar earnings (Tucker and Mitchell-Kernan 1995). With black women earning more and with the scarcity of black men who are able to support families, many black mothers are likely to remain unmarried. However, the economic interpretation may not be the only one, for it is apparent that black women, *as women,* have been fighting for power and independence vis-à-vis men longer and more successfully than white women. The circumstances of racial discrimination and poverty are largely responsible for this, but black women have taken the opportunity and created a type of family system that gives them greater power and economic leverage, while reducing their subordination to men.

Poverty and Black Family Form

Although changes in family form have often been thought of as *causing* poverty among blacks, it is important to remember that family arrangements are more typically the *result* of economic conditions. Research indicates that for whites, the formation of female-headed households from divorce or separation can be seen as a cause of poverty. For blacks, however, poverty is often the result of an already poor two-parent household splitting up. Because the long-term income of children in two-parent black families over the decades has been lower than the income of children in female-headed nonblack households, we should be careful not to blame family structure for all forms of poverty. The economic disadvantage faced by African American children in single-mother households is profound, but because of limited earnings for black men, having a father around does not necessarily lift the family out of disadvantaged economic situations (Zuberi 1998). We can thus conclude that it is not family structure per se that causes poverty among blacks.

The important trends in the marital status and living arrangements of African Americans closely parallel those of whites and most other groups. Since 1960 these trends include the following:

- Lower marriage rates
- Delayed age at first marriage

- Lower birthrates
- Higher divorce rates
- Earlier and increased sexual activity among adolescents
- Higher proportion of births to unmarried mothers
- Higher percentage of children living in female-headed households
- Higher proportion of women working outside the home
- Higher percentage of children living in poverty

With so many similarities between trends for all groups, it is misleading to place sole reliance on race-specific differences to account for recent trends. Black men, like other men, are much more likely to get married, stay married, and be involved fathers when they are employed in secure jobs that earn decent wages (Mason 1996; Taylor et al. 1991; Zuberi 1998).

All of the factors discussed previously undoubtedly contribute something to higher rates of nonmarital birth and marital disruption for blacks, but the most convincing explanation is economic. A much larger proportion of blacks than whites live in poverty.

The crucial group of black females that accounts for the growing proportion of female-headed households and nonmarital childbirths is the poverty class. Not all female-headed households are of this type; many consist of women who have been married but who are now divorced or separated from their husbands. Although teenage birthrates have been falling for all groups, African American women under the age of twenty are almost three times more likely to give birth than white teens. Even though the birthrate for unmarried African American teens fell by 20 percent during the 1990s, almost a third of all births to unmarried African American women were to those under the age of twenty (Ventura et al. 1998). Staples (1985, 1997) points out that these women have particularly poor prospects of marriage. Not only are there relatively few men available, but the ones with whom they are in close contact tend to have poor economic prospects. Unemployment rates are over 50 percent for black teenagers in inner-city neighborhoods. If we count dropping out of the labor force entirely—that is, giving up on seeking a job and registering for employment—the figure rises to 75 percent. It is estimated that one-third of these black inner-city teenagers have drug or alcohol problems. Given these conditions, desirable marriages for young African American women are not very available (see *A Closer Look* 7.4). When marriages do take place, the chances of divorce are high, especially because of the volatile economic positions of the men.

Added to this picture is the fact that poverty-level African American women tend to have traditional attitudes about motherhood. They are much less likely than white teenagers to use contraceptives, and only a small minority terminate premarital pregnancies by abortion. Staples (1985) suggests that for these young women, bearing a child is the most important thing they can do with their lives. Their own career prospects are low; their chances of a traditional, stable marriage to a man with a respectable job are even lower. Rather than live without having children and a husband, they choose to have children and rear them themselves with the help of other women (see *A Closer Look* 7.2).

The effects of extreme poverty conditions on many young black mothers and their children are reflected in recent data on infant health. Overall infant mortality rates are significantly higher in the United States than in most other industrialized countries, but the rates for U.S. blacks are more than double the rates for whites. Nearly 28,000 of the 3.9 million babies born each year in the United

A CLOSER LOOK 7.4
The Scarcity of "Marriageable" Black Men

Staples (1985) has argued that there are structural causes for the decline in the proportion of traditional two-parent households and the continuing increase in female-headed families. The most important point is: *there are not enough men for black women to marry, especially black men who can make economically viable marriages* (see also Bennett, Bloom, and Craig 1989; Guttentag and Secord 1983; Kiecolt and Fossett 1995; Spanier and Glick 1986). Of the black women who do marry, 98 percent marry black men, but there are almost 1.5 million more black women than men over the age of fourteen.

Because women tend to marry men who are two to three years older than themselves, when the population is growing, there are more women than men in the same "marriage pool." For instance, when the black population growth rate is 2 percent per year, as it was for over a decade after World War II, the number of women born in any year will be 6 percent greater than the number of men born three years earlier. What this means is that black women born during the post–World War II baby boom must "compete" for relatively few men. The sex ratio for young unmarried whites in 1985 was 102 men per 100 women, but only 85 men per 100 women for blacks (Jaynes and Williams 1989, 537–38). This does not take into account other factors, such as the exceptionally high mortality rates among black males or their lower earning potentials.

Although there are slightly more males than females born, there is a much higher mortality rate among young black males. One of every 21 black males is murdered before age twenty-four, whereas the white ratio is 1 in 131 (*Science* 1985, 30: 1257). Rate of suicide, accidental death, and drug overdose are also much higher among young black males than females. Blacks make up a disproportionate number of the armed forces and are especially likely to be assigned front-line duty and to be killed in combat. A substantial number of black men are in prison. Recent studies estimate that one in four black men between the ages of twenty

and twenty-nine is either in jail or otherwise under the control of criminal courts through parole or probation (Savage 1990). Others, with little chance of employment, have dropped out of the labor force and live a transient existence.

Marital "availability ratios" take into account the fact that women looking for marriage partners may exclude those too old or too young, those who are incarcerated, those with educational attainment deemed too little or too much, or those with too little income (Goldman, Westoff, and Hammerslough 1984). Using these techniques, Jaynes and Williams (1989, 539) calculated that black unmarried women above the age of twenty-two are in marriage pools with relatively few men. For example, at age twenty-six, black women with less than a high school education are in a marriage pool that has 651 men per 1,000 women. For twenty-six-year-old black women with some college education, the ratio is 772 men per 1,000 women. Conversely, black men can select from a large pool of unmarried women. Black men who are twenty-five years of age or older are in a marriage market containing between 1,100–1,200 unmarried women per 1,000 unmarried men.

Of the black males who are actually available for marriage, those who are in the proportion who do well economically tend to have quite conventional family lives. Approximately 90 percent of black male college graduates are married and living with their spouse, but this group of successful black males is not enough to provide marriages for most black females. For one thing, twice as many black men as women marry outside their race. For another, black women have higher average levels of education than black men. They are more likely to finish high school, and they make up 57 percent of black college students and an even larger proportion of those earning degrees. In the past, black women tended to marry down in educational level. However, now there is increasing resistance to doing so, and hence, about one-third of college-educated black women are remaining unmarried into their thirties.

States die before they reach the age of one year. The infant mortality rates in cities with large African American populations, such as Washington, D.C., and Philadelphia, are higher than those in some Third World countries, such as Jamaica and Costa Rica. In the United States, not only are black infants more than twice as likely to die in their first year as white infants, but black babies are

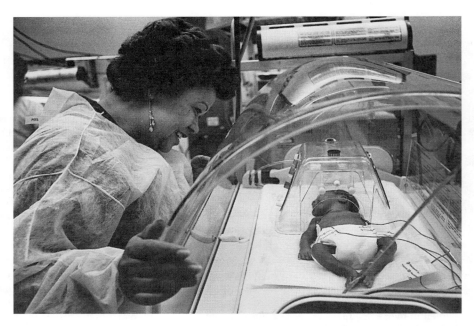

Low birth weight is a significant factor in infant mortality. Black infants are more than three times as likely as white infants to be born at less than 3.25 pounds in the United States.

more than four times as likely to be born at less than 2,500 grams (MacDorman and Atkinson 1999). Low-birth-weight babies are forty times more likely than other babies to die during the first months of life and two to three times as likely to be disabled by conditions such as blindness, deafness, and mental retardation (Hale 1990). For other complications leading to death, such as congenital anomalies, sudden infant death syndrome (SIDS), respiratory distress syndrome (RDS), and maternal complications, African Americans had significantly higher rates than whites (MacDorman and Atkinson 1999). These trends are directly related to poverty and the attendant lack of adequate nutrition and medical care. For this group of black families, things appear to be getting worse.

LATINO FAMILIES

One of the fastest growing segments of the American population is composed of Latinos, members of ethnic groups that migrated to the United States from Spanish-speaking areas. There were over 30 million Hispanic Americans in 1998. Latinos, or Hispanic Americans, comprised less than 3 percent of the total U.S. population in 1950, but by the late 1990s, they were over 11 percent of U.S. residents (Pinal and Singer 1997; *Statistical Abstract* 1999). According to census projections, Latinos will constitute the largest "minority" group around the year 2005 (Council of Economic Advisors 1998).

Latino Subgroups

"Hispanic origin" is considered an ethnic identity rather than a race, which presents problems for census takers who must classify people into categories. People

Table 7.1 Profile of Latino Subgroups

	Mexican American	Puerto Rican	Cuban	Central South American	Other Hispanic
Age					
Under fifteen years old	33%	29%	18%	24%	28%
Sixty-five years and older	4	6	20	4	7
Social Class					
Education					
High school graduate or higher	48	64	68	65	72
Four or more years of college	8	12	22	17	16
Median income	$27,000	$24,000	$38,000	$32,000	$30,000
Percent home ownership	49	34	56	32	48
Families below poverty level	26	32	16	18	25
Family Type					
Female-headed households	20	38	15	23	34
Married couple	72	54	81	68	60
Fertility					
Birthrate per 1,000 persons	27	20	11	25	NA
Percent of births to teen mothers	18	22	8	9	NA
Percent of births to unmarried mothers	33	56	18	41	NA

Figures are for 1998, except for birthrates, which are from 1997.

Source: *Statistical Abstract* 1999, table nos. 55 and 92.

of any race can be classified as Hispanic (or Latino), on the basis of tracing their ancestry to countries where Spanish is the primary language. There are several rather different Latino groups in the United States: Mexican Americans (or **Chicanos**) make up 64 percent of the Latino population; Puerto Ricans make up 11 percent; Cubans make up about 4 percent; persons from Central and South American countries (primarily El Salvador, Nicaragua, Colombia, Guatemala, and Ecuador) constitute 23 percent, and "other Hispanics" (primarily Dominicans) make up the final 7 percent (Pinal and Singer 1997). Mexican Americans are largely located in the Southwest, from Texas to California. Puerto Ricans are located primarily in New York and surrounding states; Cubans are mainly located in Florida. Most Cubans migrated as refugees from the Castro regime between 1960 and 1973, though there have been more recent waves of immigrants in the 1980s and 1990s. The fastest growing group is composed of immigrants from Central and South America. Most Cuban Americans came from the upper and middle classes; hence, they are quite different from the other two large Latino groups. Cubans on the whole tend to be much older than the average of the American population, whereas Mexican Americans and Puerto Ricans tend to be much younger. Cubans in most respects are much like the majority white population: relatively high levels of education, relatively few families below the poverty line, a high incidence of marriage, low birthrates, and low rates of teen or unmarried childbearing (Table 7.1).

The other two large Latino groups are quite different from the American mainstream. Not only are there more young persons and far fewer old persons among Mexican Americans and Puerto Ricans, but both groups have relatively low levels of education and lower incomes. Puerto Rican families, in particular, are likely to be below the poverty line and to be female headed with no husband present. Mexican American families have a poverty rate that is higher than that of the mainstream but lower than that of Puerto Ricans. Mexican American families have a rather low level of female-headed households. An overall impression is that Puerto Rican families are somewhat like the black families in the poverty sector, where there is a predominance of female-headed families.

The proportion of adults who divorce is also higher for Puerto Ricans than whites. The Puerto Rican family reflects a more independent female-centered pattern, whereas the Mexican American family tends to be viewed, at least in theory, as a conservative, father-centered structure (Chilman 1993; Mirandé 1997). Mexican Americans are less likely to be divorced than the majority population and are much less likely to divorce than blacks or Puerto Ricans. When separation is included with divorce, Mexican-Americans look much more like non-Hispanic white households in their rates of marital breakup. This means, incidentally, that we must look elsewhere for causes of poverty among Mexican American families. Although there are no conclusive studies of this issue, which has received much less attention than the issue of black poverty, it is quite likely that a combination of low educational credentials and employment discrimination is responsible.

Mexican American Families

Recent research on Mexican American families parallels the larger body of research on African American families. As a corrective to earlier studies that held minority family structure accountable for poverty and other problems, recent scholarship has focused on the strengths and adaptability of Chicano families. Although some of this research tends to idealize or romanticize certain aspects of Chicano family life, it avoids the previous tendency to blame the victim (Mirandé 1997; Vega 1991). Social scientists, many of them Chicanos or Chicanas, have tended to focus on three general topics related to Chicano family form and functions: (1) historical trends, (2) family solidarity, and (3) gender relations. We will briefly review some of the major findings in each of these areas.

HISTORICAL TRENDS Spain ruled much of what is now the southwestern United States until the middle of the nineteenth century. Mexico won independence from Spain in 1821, and Texas won its independence from Mexico in 1836, joining the United States in 1845. In 1848, after the Mexican-American War, the United States annexed territory that later became the states of Arizona, California, New Mexico, Nevada, Utah, and parts of Wyoming and Colorado. In general, land ownership was transferred from Mexican to American hands, despite treaty provisions that guaranteed the original Chicanos continued ownership of their lands. In the late 1800s, indigenous Chicanos who had worked for themselves as subsistence farmers began to be employed as subordinate wage laborers in agriculture, ranching, railroads, and mining (Barrera 1979; Pinal and Signer 1997). According to Bean and Tienda (1987, 18), Mexican Americans in the Southwest "lost their land, their social mobility became blocked, and . . . their brown skin and indigenous features encouraged racism and discrimination by the Anglo majority." The population of the American Southwest grew rapidly, supplemented by large-scale immigration from Mexico. People tended to move from Mexico to the United States in "immigration chains," whereby family units once living in Mexico were gradually relocated over time in the United States.

Divisions of labor in traditional Mexican families tended to be rigidly defined by gender. This was also true of Mexican American families living in the southwestern United States in the mid-nineteenth century. Men performed virtually all of the productive work outside the household, and women did virtually all the housework, child care, and—in rural areas—tended gardens. After about 1870, Chicanas and their children were increasingly recruited into the labor force as domestics, as well as in the textile industry, in canning and packing houses, and in other agricultural endeavors, especially seasonal harvesting. Wages tended to be extremely low in all of these jobs (Zinn and Eitzen 1987).

One of the most distinctive historical features of Mexican American family life was the reliance on extended networks of kinfolk for emotional and material support. Not only was household size larger for Mexican American families throughout the Southwest, but the family also consisted of an extended network of relatives living outside the household. "La familia" included grandparents, aunts, uncles, cousins, married brothers and sisters, and their children. A unique *compadrazgo* system also linked godparents and children in a system of mutual exchange and support. *Padrinos,* or godfathers, and *madrinos,* or godmothers, were nonbiologically related individuals who became members of the extended family and participated in the major religious ceremonies of the children's lives, including baptism, confirmation, first communion, and marriage. Godparents acted as *compadres,* or coparents, disciplining children, offering companionship to parents and children, providing emotional support, and offering financial aid (Griswold del Castillo 1984).

Because of high fertility, Chicano household size tended to be large. Many households were made even larger with the addition of compadres, boarders, adopted and visiting friends, and servants. Although almost all Mexican American families had kin who lived in the same town during the nineteenth century, research shows that most did not share their houses with relatives. Griswold del Castillo (1984) notes that extended family households were a temporary and impermanent creation of circumstance arising out of old age, sickness, death, or economic misfortune.

Maxine Baca Zinn (Zinn and Eitzen 1987, 92) describes Mexican American women of the late nineteenth and early twentieth century as doubly oppressed: they held the most subordinate jobs outside the home and in their private lives were subject to "a distinctive system of Mexican patriarchy." Mexican-American women were instructed to be obedient and subservient to their husbands and parents and were obliged to shoulder full responsibility for all housework and child care. Chicanas were viewed primarily as wives and mothers; however, they often performed wage work to make ends meet, although such work was considered much less important than that of the men. On the whole, the women guarded Mexican cultural traditions through daily family life and with customs and rituals including folklore, songs, birthday celebrations, saints' days, baptisms, weddings, and funerals (Garcia 1980; Zinn and Eitzen 1987).

FAMILY SOLIDARITY The concept of **familism** has often been invoked to describe both traditional and contemporary Mexican-American families with their higher than average birthrates and household size. Familism refers to "a constellation of values which give overriding importance to the family and the needs of the collective as opposed to individual and personal needs" (Bean, Curtis, and Marcum 1977, 760). Embeddedness in close-knit extended kin networks and valuing family solidarity over individuality or personal achievement have typically been thought to be Mexican cultural patterns handed down through the generations. Recent research contends, however, that reliance on actual and **fictive kin** (non-relatives defined as family members) is common among lower-class blacks and whites, as well as Mexican Americans (Zavella 1987). For this reason, dependence on close-knit kin networks might more appropriately be viewed as an adaptation to a marginal existence. Nevertheless, Mexican Americans are typically found to have the highest levels of extended familism when compared with blacks and Anglos of various class levels. One of the benefits of strong patterns of kin interaction among Chicano families appears to be enhanced mental health.

Acculturation theories based on a melting-pot ideal would predict that Mexican Americans' patterns of extended family interaction might become more like Anglo American family patterns over time. Recent research reveals that acculturation is not a simple linear process, however, particularly when applied to Mexican Americans. Keefe and Padilla (1987) found that long-term residents of Mexican descent did not rely on compadres as much as earlier generations had. Nevertheless, they did rely heavily on primary and secondary kin for emotional and material support. Counter to the acculturation theories, local extended families were more *typical* of long-term residents than new immigrants, and the local extended family network grew stronger in conjunction with many processes assumed to promote individualism and sole reliance on the nuclear family—including English language acquisition, urbanization, and social and economic assimilation. Similarly, researchers report that many Chicano families are simultaneously modern *and* traditional. In light of findings like these, many contemporary scholars conclude that there is no *one* type of ethnic family structure and that Mexican Americans construct unique family patterns based on cultural inheritance, economic necessity, religious beliefs, and a host of other factors (Keefe and Padilla 1987; Lamphere, Zavella, and Gonzales 1993; Zavella 1987; Williams 1988; Zinn 1980).

GENDER RELATIONS Reflecting the biases of the times, early social research characterized Mexican American families as rigidly patriarchal (e.g., Clark 1959; Heller 1966; Rubel 1966). The father was seen as having full authority over the mother and children, and wives were described as passive, submissive, and dependent. For example, William Madsen (1973, 22) focused on the destructive aspects of Mexican-American **machismo** by comparing the men with roosters: "The better man is the one who can drink more, defend himself best, have more sex relations, and have more sons borne by his wife." According to such depictions, authoritarian child rearing and wife beating were assumed to be common in Mexican American families.

Some researchers acknowledge male dominance as a persistent feature of Mexican American families but see it as benevolent rather than malevolent. In a review of literature on Mexican American fathers, Mirandé (1988, 1997) suggests that genuine machismo is characterized by bravery or valor, courage, generosity, and ferocity. Sociologically, what should be recognized here is that we are describing a family structure imported into the United States from a largely traditional and rural society. Thus, Mexican American and many other Hispanic families come from a background that is much like the agrarian societies described in chapter 3. We saw there that these societies are highly stratified and organized around patriarchal households, with the upper class dominating the population of small landholding peasants and farm workers. This structure has changed rather slowly in Latin America, and many of the Latino migrants to the United States have come to escape from the unfavorable conditions of this form of stratification.

The culture of machismo thus looks very much like the culture of an agrarian society of patriarchal households. Where coercion and even violent threat are prominent features of class relations, men value courage and loyalty to their group. Moreover, as we should recall from the theory of gender stratification (chapter 3), male dominance over females is maximal where there is a high level of violent conflict and where there are weapons in the household. As we recall, gender stratification is at its height in agrarian societies. Men's and women's roles are sharply distinct, especially in regard to sexual behavior; men act both to protect and control the women of their own families.

The ethos of the original agrarian society is slowly adapting to new conditions in the largely urban and industrial United States. The conditions that supported a patriarchal family structure are gradually eroding. The machismo style is apparently weakening, and Chicano spokespersons such as those quoted previously are attempting to redefine it to fit a less coercive modern setting. Although it may be true that youth gangs in working-class Latino neighborhoods maintain some aspects of the traditional conflicts and the machismo style, the realities of adult life are now quite different than in a traditional community dominated by the power of caudillos and rural landlords. The status of Latino families in the United States now depends on their individual economic success rather than on the power coalitions to which they belong. The power resources of Latino men are declining as well. Simultaneously, the movement of women into the labor force is creating a new balance of gender resources and reshaping the Hispanic family. It appears that gender expectations in Hispanic families will change as social and economic conditions require (Lamphere, Zavella, and Gonzales 1993; Vega 1991).

Some recent research on marital interaction seems to support the notion that gender relations in Mexican American families are more egalitarian than the traditional model assumes. Studies of marital decision making have found that Chicano couples tend to regard their decision making (but not necessarily their division of household tasks) as relatively shared and equal. Mexican American women, like their Anglo American counterparts, exercise more marital decision-making power if they are employed outside the home than if they are housewives (Coltrane and Valdez 1993; Lamphere, Zavella, and Gonzales 1993; Ybarra 1982; Zinn 1980).

Mirandé (1988, 1997) suggests that contemporary Chicano fathers take more responsibility for child care than in the past, and Zavella (1987) notes that Chicano husbands sometimes respond to demands from employed wives to do more around the house. Nevertheless, most researchers reject the notion that marital roles are truly egalitarian in Mexican American households (Williams 1988; Hartzler and Franco 1985; Zavella 1987). (Neither are marital roles truly equal in Anglo-American couples; see chapter 11.) Because Mexican cultural ideals require the male to be honored and respected as the head of the family, Zinn (1982) contends that Chicano families maintain a facade of patriarchy. At the same time, Chicano families can be characterized as mother centered because women are responsible for child rearing and domestic chores and thus assume authority over day-to-day household activities. It may be that many minority families grant symbolic power to the male head of household as a gesture of appreciation for trying to earn a family wage in a discriminatory labor market. A similar pattern of symbolic deference often occurs in white working-class families. Power in such families, in terms of everyday interaction, may in fact reside more securely in the women's domain.

Mexican American women, as a group, have been concentrated in the lowest-paying jobs and have made limited progress over the past decade in terms of educational attainment, occupational status, or income (Reyes, Kobus, and Gillock 1999; Segura 1984). Limited employment opportunities for Chicanas coincide with a strong sense of obligation to the family and primary self-esteem being derived from being wives and mothers. Garcia-Bahne (1977) notes that adopting the individualism common to many white middle-class families runs counter to the collectivity of the Chicago family but that idolizing familism can be limiting for Chicanas. She describes how various myths about Chicano families serve both positive and negative functions, with family loyalty, self-sacrifice, and modesty often contributing to Chicanas' social and economic oppression.

Pressures for Change in Chicano Families

In a study of Chicana cannery workers in central California, Zavella (1987) reports on how some cultural ideals about the Chicano family directly affect women's lives. Tensions and conflicts in the families she studied tended to revolve around the uses of women's time and labor. "Because women were considered primarily as homemakers who happened to work, husbands (and even children) expected the women to continue deferring to them and maintaining the needs of family members. When women contested these assumptions, conflicts emerged" (Zavella 1987, 159). As the Chicana cannery workers began to view their jobs as essential to the family's well-being, they challenged taken-for-granted assumptions about Chicano family structure. One of the cannery workers talked about how she automatically enacted traditional gender expectations but then began to question obligatory deference to her husband:

> *I found myself doing things that I didn't agree with. I'd tell the kids: "Be quiet because your father's asleep, and he's tired and he needs his rest." I'd make a meal and leave something for him to warm up. I always had this feeling, "Well he's the daddy, so I'd better make the kids be quiet, and I'd better cook his meals, or whatever." He's a grown man; he could do things for himself. But you're taught to do these things without thinking, because he's the man of the house. It's inbred in you. (Zavella 1987, 87)*

Reflecting conflicting pressures for change in such families, recent research has produced contradictory findings about the sharing of family work in Latino families. Some studies suggest that there is slightly more sharing than among white families (Mirandé, 1997; Shelton and John 1993), but others suggest there is less sharing in Latino families (Golding 1990). Most research does reveal that the same things that encourage Anglo families to share family work also encourage Latino families to do so. As discussed in chapter 11, these include longer employment hours and better pay for women, along with lower levels of employment for men and more liberal attitudes about gender on the part of both men and women (Coltrane and Valdez 1993; Golding 1990; Herrerra and Del Campo 1995; John, Shelton, and Luschen 1995). Reflecting generally high levels of gender traditionalism among Latinos, DeMaris and Longmore (1996) also report that Latino men and women are less likely than Anglos to view current divisions of housework as unfair to the wife.

Because many immigrant Latino women are employed as domestic laborers—cleaning wealthier families' houses, cooking for them, and caring for their children—it is sometimes difficult for them to challenge assumptions that they should be responsible for all the domestic labor in their own homes (Hodagneu-Sotelo 1992; Romero 1992). Because Latino women have lower levels of labor force participation and lower average wage rates than any other group, their bargaining position is often constrained (U.S. Department of Labor 1998). Nevertheless, researchers have noted how life in both working-class and middle-class

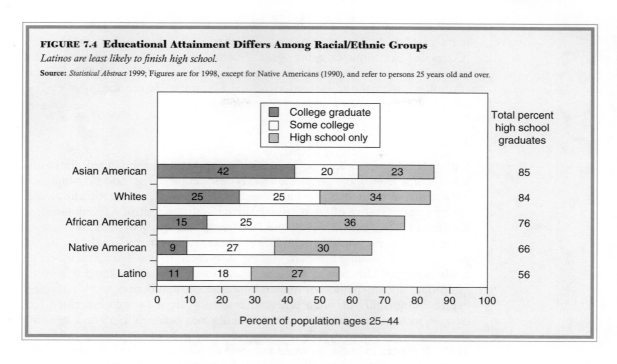

FIGURE 7.4 Educational Attainment Differs Among Racial/Ethnic Groups
Latinos are least likely to finish high school.
Source: *Statistical Abstract* 1999; Figures are for 1998, except for Native Americans (1990), and refer to persons 25 years old and over.

Chicano families begins to change when employed wives demand more participation from husbands and children (Coltrane and Valdez 1993; Lamphere, Zavella, and Gonzales 1993; Zinn 1980). One Mexican American woman interviewed by Maxine Baca Zinn commented:

> *Since I've gone to work, the kids have more responsibilities. Sometimes I have to tell them what to do, but you know, they're more independent now. They don't rely on me so much. Believe it, even Juan [husband] helps, but this has been a long time in coming. At first, he reacted by going against me; he wouldn't get after the kids, but things are different now. And it's good for the whole family. Just this morning when I was getting ready for work, there was Christopher sewing his pants. (Zinn 1980, 53)*

Zinn also reports that wives' employment outside the home and attempts to share some of the household labor did not weaken the families' efforts to maintain and develop their Mexican heritage. All of the families were Catholic, and many of their activities with kin tended to revolve around religious celebrations. Although many were "much like non-Chicano families in day-to-day activities" (p. 57), they spoke Spanish at home, preferred Mexican and Spanish food, and listened to Spanish or Mexican music. In these ways, many Chicano families combined some modern family patterns with more traditional ethnic practices.

As with African American families, there is no one typical form of Latino family. Even within Latino subgroups, there is great variation depending on social class position, recency of immigration, place of residence, education, and other factors. Although Latinos are less segregated than African Americans, where urban barrios exist, the Chicanos or Puerto Ricans who live within them are faced with many problems that shape the nature of family life.

Among adults, Latinos have the lowest educational levels of any race or ethnic group. This is partly due to the large number of Latino immigrants who had little schooling in their home countries, but it is also true for American-born Latinos (Figure 7.4). Educational levels for all racial ethnic groups have been improving but only for some members of each group, so there is increasing educational disparity within minority groups. Educational levels for Latinos have only gone up slightly in the past two decades, with four of ten Latinos still lacking a high school degree (Council of Economic Advisors 1998). If Latinos graduate from high school, they are about as likely as whites to go on to college (Pinal and Singer 1997). Because education is often the key to a good job and a promising future, the lower levels of schooling among many Latinos is of major concern.

Latino Families in the Future

Latino families have become increasingly important in the United States and will become even more so in the twenty-first century. The rate of population growth of the white majority has slowed, whereas the Latino populations constitute a major source of what growth there is. Almost half of all immigrants to the United States between 1960 and 1999 were Latinos. Like other immigrants, some come to escape harsh conditions at home, and many are drawn by the prospects of economic prosperity. Many also come to join family members already here, and although most come legally, many slip across the border without proper documentation or remain long after their student or work visas have expired. One study of undocumented migration from Mexico estimated that over 36 million undocumented entries took place between 1965 and 1990 but that over 30 million of these immigrants returned to Mexico (Massey and Singer 1995). It is hard to know the exact numbers because of the large amount of illegal migration; hence, a sizable population, especially of Mexican Americans, is not reported by the Census.

We do know that the Latino population in the United States has been growing because of its high birthrate. Even if there were no more immigration (whether legal or illegal) from now on, the Latino population would still reach a sizable proportion of the U.S. population in the early part of the twenty-first century. Hispanic Americans have a birthrate that is over 50 percent higher than the average of non-Hispanic women. Leaving aside Cubans, who are more similar to the majority population, we can see (see Table 7.1) that Mexican Americans have a birthrate of 27 per 1,000, and Puerto Ricans of 20, which compares with the majority white birthrate of 14 per 1,000. Latino women tend to bear their children earlier, too. In addition, considering that the Mexican American and Puerto Rican populations are a good deal younger than the majority population, we can expect that proportionately more Hispanic persons are going to reach childbearing age in the coming decades. These are some of the reasons for the present Hispanic population bulge, which will continue for some years to come. By the year 2050, projections indicate that Hispanics will make up over 20 percent of the total U.S. population. Around the year 2005, Hispanics are projected to be more numerous than blacks, whose percentage is also increasing (Council of Economic Advisors 1998).

What difference will it make to the United States as the population becomes proportionately more Hispanic? In part, this depends on how much these families are assimilated into the majority culture (and to which culture they assimilate). The question has been little investigated, but there is some interesting information about the use of the Spanish versus the English language. Use of English depends, in part, on whether Latinos are foreign or native born, and if native born, how long they have lived here. Preference for English or Spanish depends on one's job, neighborhood, and family lifestyle. In urban centers with large Latino populations (New York, Los Angeles, Miami), primary Spanish speakers have access to newspapers, radio, and television stations, and many businesses cater to Spanish speakers. Elsewhere, pressures to adopt English are greater. Nearly two-fifths of all Hispanics reported that they did not speak English well or at all in 1990. However, recent studies show that Spanish speakers in the United States want to learn English and feel it is important for their advancement. As Pinal and Singer (1997, 35) note, "Hispanic immigrants, and especially their children, do learn English—just as Italian, Polish, and German immigrants did, even though they also lived in ethnic communities where little English was spoken" (see also Porter and Schauffler 1996). About three of four Hispanic immigrants speak English on a daily basis by the time they have been in the country for fifteen years. Younger immigrants and the children of immigrants usually become fluent English speakers, even though many retain some knowledge of Spanish and use it in the home. Most observers predict that bilingualism will persist longer for Hispanics than for other immigrant groups, though it is likely to decline if immigration drops significantly (Valdivieso and Davis 1988).

ASIAN AMERICAN FAMILIES

Asians are the fastest growing minority in the United States. Their numbers nearly doubled between 1980 and 1990, and they are likely to double again by 2010 (Lee 1998). In 1990, there were more than 10 million Asian-Pacific Americans in the United States, representing about 4 percent of the total U.S. population (Hooper and Bennett 1998). The Census Bureau enumerates and identifies a total of seventeen Asian American groups and nine groups of Pacific Islanders, the most numerous being Chinese (24 percent), Filipino (21 percent), Asian-Indian (13 percent), Vietnamese (11 percent), Japanese (10 percent), Korean (10 percent), other Asian (e.g., Pakistan, Bangladesh; 7 percent) and other Southeast Asian (e.g., Cambodian, Hmong, Laotian; 5 percent) (Lee 1998, table 3). Although these groups are often lumped together in reporting demographic statistics, the groups are quite different from one another. Many Chinese and Japanese families have lived in the United States for several generations, whereas groups like the Vietnamese and Cambodians tend to be recent immigrants.

Historical Precedents

In the mid-nineteenth century, many Chinese men came to western United States to work on the railroads and in the mines. Most were unable to establish families because of limits on immigration and laws that prohibited Asians from becoming citizens. Prejudice, violence, and legal and extralegal discrimination against the Chinese in the late 1800s and early 1900s kept them segregated, poor, and undereducated. As a result, Chinese Americans tended to rely on their own resources and developed their own institutions. They created their own businesses, ran their own hospitals, formed their own insurance companies, and established their own social clubs. This self-reliance led in turn to accusations that Asians were separatists who were not assimilable into the dominant community (Kitano and Daniels 1995).

The stereotypes applied to Asians in this country have changed dramatically in the latter half of the twentieth century. Asian Americans, especially Japanese and Chinese, have increasingly come to be seen as **model minorities** who work hard and easily conform to American ways of life. Although Asian men were previously stereotyped as wily and devious and Asian women as exotic and mysterious, these images have given way to the stereotype of Asian families as cohesive, conformist, and highly successful (Lee 1998; O'Hare 1992; Sue and Kitano 1973). As we will see, however, the stereotype does not fit all Asian American families.

Modern Demographics

The majority of Asian Americans live in western United States, including Hawaii, with only Asian-Indians, and to a lesser extent Koreans, spread relatively evenly across the country. Asians make up 48 percent of Hawaii's population and 9 percent of California's population, with a full 40 percent of all Asian Americans—about 4 million—living in California (Lee 1998). Other states with over 200,000 Asians include New York, Texas, New Jersey, Illinois, Washington, Florida, Virginia, and Massachusetts.

Most Asian Americans live in large metropolitan areas, though their neighborhoods of residence are about evenly divided between central cities and suburbs. About half of all Asian Americans live in just five large urban centers: Honolulu, Los Angeles, New York, Chicago, and the San Francisco Bay area. This pattern contrasts with general residence patterns among whites, two-thirds of whom live in the suburbs (O'Hare and Felt 1991). Forty-five percent of Asian Americans live in central city areas, as compared with 22 percent of whites and 55 percent of blacks. Although immigration patterns have a tendency to concentrate Asian populations, they tend to live in less segregated areas than other minorities (Edmonson 1997; Lee 1998).

Population gains among Asian Americans have resulted primarily from changes in U.S. immigration policy. Beginning in the 1960s, a new wave of Asian-American immigrants began coming to the country in record numbers. This trend continued and even accelerated during the 1970s and 1980s as the U.S. government established a program of resettlement for refugees from Southeast Asia (Vietnam, Cambodia, Laos, etc.). Although the number of Asians coming to this country is less than the number of Hispanics, the rate of growth for Asians is higher than for any other group, including whites. This phenomenal growth is likely to continue because recent immigration policies favor family reunification. Newer immigration policies also favor admitting more highly educated and well-trained workers. Because Asians are more likely than others to have advanced degrees and job skills, continued high levels of Asian immigration are expected in the future.

Although birthrates for Asian American women have been below those of whites in the past, they now exceed slightly those of whites. Hispanic and black women tend to have their first child at a younger age than do whites, whereas Asian American women tend to have first births later than white women. Because Asian American women are more likely than others to stay married, however, they tend to have slightly more children than their white counterparts. Newly immigrating Asian groups tend to have the highest birthrates. For example, Vietnamese Americans have birthrates approaching those of Hispanics. In contrast, Chinese American and Japanese American women have fertility rates well below the replacement level of 2.1 children per woman—the number of children needed for a generation to replace itself. These two groups have other characteristics associated with low fertility: high education levels, high average incomes, and later age at first birth. Without immigration, the U.S. Chinese and Japanese populations eventually would de-

Asian Americans set a premium on their children's educational achievement and obedience to authority.

cline. Despite such internal differences among Asian Americans, it is predicted that their overall numbers will continue to increase to over 30 million by 2050, when they will represent about 8 percent of the U.S. population (Lee 1998).

Cultural Patterns

Research on Asian families has typically focused on the importance of unique cultural influences. The cultural approach to studying Asian family life focuses on traditional child socialization practices and shows how individual needs are sacrificed to the overall welfare of the family unit. Asian child-rearing patterns are characterized as instilling obedience, loyalty, self-control, and a desire for educational achievement. Traditional Asian family relations also include rigid task and role divisions, with wives expected to stay home and assume major responsibility for household tasks and child care, while husbands fill the role of breadwinner. In the traditional Asian family model, the father has final authority and wields power sufficient to enforce his decisions. Children and wives are expected to be obedient and submissive, and children are supposed to defer to both parents' wishes (Kitano and Daniels 1995). Only recently has research on Asian families moved beyond this cultural approach, which tends to homogenize Asian families and ignore the many differences that exist between Asian families living in the United States (Chao 1994; Espiritu 1997).

Autobiographical accounts reveal some of the stresses and strains that result from traditional Asian family patterns. For instance, Jean Wakatsuki Houston (1985) discusses how her mother, a first-generation immigrant from Japan, subordinated herself to her husband and sought ways to elevate his position in the family:

He was always served first at meals. She cooked special things for him and sat next to him at the table, vigilantly aware of his needs, handing him the condiments and pouring his tea before he could ask. She drew his bath and massaged him and laid his clothes out when he dressed up. As I was growing up I accepted these rituals to be the natural expressions of a wife's love for her husband. . . . This attitude, that to serve meant to love, became an integral part of my psychological make-up and a source for confusion when I later began to relate to men. (1985, 13)

Houston describes an issue common to many Japanese American families: the tension between first-generation immigrants *(issei)* and their children *(nisei)* that stems from the younger generation adopting the dominant culture's language and customs:

Whenever I succeeded in the hakujin [Caucasian] world, my brothers were supportive, whereas Papa would be disdainful, undermined by my obvious capitulation to the ways of the West. I wanted to be like my Caucasian friends. Not only did I want to look like them, I wanted to act like them. I tried hard to be outgoing and socially aggressive, and to act confidently, like my girl friends. At home I was careful not to show these traits to my father. For him it was bad enough that I did not even look very Japanese; I was too big, and I walked too assertively. My breasts were large, and besides that I showed them off with those sweaters the hakujin girls wore! My behavior at home was never calm and serene, but around my father I still tried to be as Japanese as I could. (Houston 1985, 17)

Like many nisei, Jean Wakatsuki eventually married a Caucasian (the novelist John Houston). Younger Japanese Americans and Chinese Americans have extremely high rates of **out-marriage,** that is, marriage to people of a different race (Chapter 8). For example, a majority of third-generation Japanese Americans marry non-Japanese, with Japanese women much more likely than men to out-marry. Scholars have suggested that more acculturated Asian women are motivated to out-marry because of dissatisfaction with more traditional Asian men's limited attitude toward women (Kikumura and Kitano 1973; Staples and Mirandé 1980). As the population of Asians in the United States grows and becomes more diverse, it is likely that patterns of intermarriage will also change. For example, patterns of out-marriage are much higher for U.S.-born Asians than for foreign-born Asians, especially women (Figure 7.5). This gender gap in out-marriage was influenced by patterns of marriage between Asian women and U.S. military personnel stationed in Japan, South Korea, Vietnam, and the Philippines but is also present for U.S.-born Asian women (Lee 1998; Lee and Fernandez 1998). Inter-ethnic marriages—those between Asians of different national origins—have been increasing in recent years. Scholars suggest that inter-ethnic marriage among Asians signals the development of a pan-Asian consciousness and the breakdown of social and physical distance among Asian groups in the United States (Espiritu 1992; Lee and Fernandez 1998; Shinigawa and Pang 1996).

Stresses and Strains

The stereotype of Asian Americans as model minorities can be harmful in that it masks some of the unique problems and strains they face. For instance, the common perception that Asian families are extremely cohesive is supported by the relatively low number of divorces among Asians, especially first-generation immigrants. Nevertheless, divorce rates are increasing for Asians, as they are for

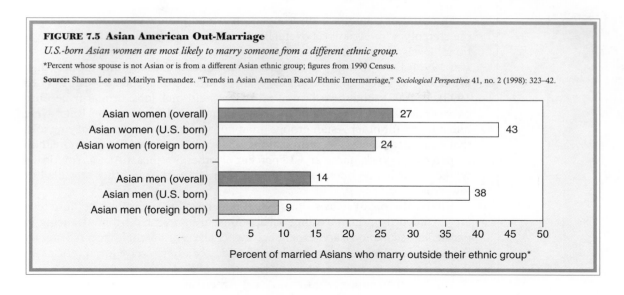

FIGURE 7.5 Asian American Out-Marriage

U.S.-born Asian women are most likely to marry someone from a different ethnic group.

*Percent whose spouse is not Asian or is from a different Asian ethnic group; figures from 1990 Census.

Source: Sharon Lee and Marilyn Fernandez. "Trends in Asian American Racal/Ethnic Intermarriage," *Sociological Perspectives* 41, no. 2 (1998): 323–42.

other races. Acculturation and generational changes have produced different mo-tives and expectations for marriage among Asians, and these differences have contributed to increases in rates of marital disruption. Because most Asians come from societies in which women rarely work for wages outside the home, the mod-ern necessity of having two wage earners can lead to special conflicts and ten-sions in Asian American families. As noted previously, intergenerational conflicts between old-fashioned parents and their Americanized children have been com-mon for most immigrant groups. Because Asian American families tend to ob-serve strict privacy norms, they rarely seek counseling and thereby reinforce the notion that they are "problem free" (Kitano and Daniels 1995).

Another area in which Asian Americans are assumed to be an ideal minority is educational attainment. For instance, in 1998, 42 percent of Asian Americans over 25 years old had graduated from college (see Figure 7.4). With the exception of recent immigrant groups from Southeast Asia, such as the Vietnamese, Asian groups have higher rates of completion for high school, college, and doctoral de-grees than other minorities or majority whites. Selective migration of better edu-cated people from Asian countries partially accounts for the higher levels of ed-ucation, but native-born Asian Americans also have extremely high levels of educational attainment (Lee 1998). The stereotype of Asians doing well in school is also supported by higher-than-average Scholastic Achievement Test (SAT) scores in mathematics (Kitano and Daniels 1995).

Nevertheless, not all Asians do equally well in school, and one of five Asian-American adults has less than a high school education, about the same percentage as for white adults. The stereotype of Asians as overachievers can have deleterious effects on good students as well as on below-average students. Some researchers report a "pushy parent" syndrome, in which children are forced to study continu-ally and are channeled into focusing only on the natural sciences. Instead of praise for a report card with all As and one B, the pushy parent criticizes the student for receiving the one B grade. Young Asian Americans can also receive mixed mes-sages from their parents. On the one hand, they are expected to stay in close con-tact with their parents and remain dependent on them. On the other hand, they are expected to work very hard, be independent, and earn money while in school. The result can be high levels of stress (Kitano and Daniels 1988, 165).

Once out of school, job discrimination can add to the pressures inspired by the stereotype of Asians as overachievers. In all age groups and at all educational levels, white Americans earn more than Asian Americans (O'Hare and Felt 1991). Even though Asians have more education than other groups, they are about as likely to hold managerial or professional jobs as whites (Lee 1998). In 1997, about one-third of all Asians held such jobs, primarily because of exceptionally high rates for Chinese, Japanese, and East Indians. Rates were much lower for other Asian groups, many of whom work in lower paying jobs that require few skills. According to the last available census figures, more than 20 percent of Vietnamese and 37 percent of other Southeast Asian Americans worked as machine operators, fabricators, or laborers or in other unskilled or low-skilled jobs.

Kitano (1976, 93) suggests that Asian men have lower earnings because they are excluded from high-earning occupations, are prevented from advancing in particular occupations, and experience higher levels of unemployment. Thus, although Asian Americans tend to be well-educated, they receive lower returns on their education than European Americans.

Unlike other minority groups, Asian Americans tend to have higher-than-average household incomes. Similar to blacks and Hispanics, however, the differences among Asian families are substantial, and recent trends indicate ever-widening gaps between families in this large and diverse group. For example, the poverty of Asian Americans has increased over the past decade and is nearly twice as high as that of whites (though it is less than half the poverty rate for Latinos or African Americans). At the same time, the average household income of Asian Americans is higher than that of whites (*Statistical Abstract* 1999).

What might account for these differences? For one thing, Asian Americans tend to live in households with more workers than any other group. Although official government figures do not even include many Asian Americans who work in family-owned businesses, almost one of every five Asian American families includes at least three wage-earning workers. The relatively high number of workers is related to typical living arrangements. Three of four Asian Americans live in married-couple families, slightly more than the percentage for whites. However, Asian Americans are only half as likely to live alone, and they are twice as likely to live in extended family situations or to double-up in one household. More workers means more earnings, and the average income of Asian Americans reflects this. In 1998, a greater percentage of Asian Americans than white Americans lived in households with incomes of over $50,000 per year (*Statistical Abstract* 1999).

Another reason for the wide variation among Asian American families is the composition of recent immigrant pools. Less advantaged migrants from Southeast Asia have swelled the ranks of the poor at the same time that more highly educated people with special skills have come here with immediate high earning potential. It makes a big difference whether one considers the social and economic situation of recent Laotian or Cambodian immigrants—who tend to be poor and to utilize government welfare programs—or the situation of Japanese or Taiwanese immigrants—who tend to have higher earnings than virtually any other foreign-born group in the United States today. Second- and third-generation Asian Americans also move into the ranks of the middle and upper-middle class at higher than average rates.

Although researchers who study African American and Latino families have long supplemented attention to culture with an emphasis on institutional and economic constraints, those studying Asian families have only recently moved away from an almost exclusive emphasis on cultural explanations. Evelyn Nakano Glenn (1983) criticized the cultural approach for portraying Asian American families as essentially static. By tracing historical shifts in the economic, legal, and political constraints faced by Chinese American families, she illustrates how Chinese family structure has changed drastically: from split-households, to small-producers, to dual-wage earners. Rather than seeing traditional Chinese culture as determining Chinese American family form, she showed that individual families devised strategies that varied according to the structural conditions that prevailed during a given historical period. Split-households, prevalent until about 1920, adopted the strategy of sending married men to the United States to make money, while the wife and children maintained a separate household in China. The small-producer household, prevalent from about 1920 to the 1960s, relied on the labor of all family members in small family-run businesses such as laundries or restaurants. The dual-wage pattern, common among immigrants arriving after 1965, is based on a strategy of both husbands and wives working in low-wage jobs outside the home. Such analyses remind us that culture alone does not determine family form and that families continually adapt to changing economic and institutional circumstances (see also Espiritu 1997; Kibria 1997; Min 1995; Zhou 1997). This is as true for Asian families as it is for African American, Latino, and white families.

NATIVE AMERICAN FAMILIES

The term *Native American* (sometimes called "American Indian") refers to a diverse assortment of tribes and ethnic groups living in North America before Europeans and others settled the continent. Although estimates vary, there were probably between 10 and 18 million Native Americans living north of Mexico when Europeans arrived in the late fifteenth century (John 1988). Native American family and kinship structures varied enormously among different cultural groups, as did social customs, lifestyle, subsistence practices, religion, and political organization. North American Indians, including Eskimos and Aleuts, spoke over 300 different languages and represented over 100 different major cultural groups before contact with white settlers. With such diversity, it is impossible to identify one dominant historical type of Native American family.

European settlement almost wiped out the native population through disease, destruction, and displacement. By the end of the nineteenth century there were only about 250,000 Native Americans in the entire United States (Dobyns 1983; Martin 1978; Taylor 1994). Despite high levels of infant mortality and a relatively short life span, the North American Indian population began to grow in the twentieth century. The 1990 Census counted 1.9 million American Indians, including 81,000 Eskimos. Part of this dramatic increase is due to the fact that Native Americans have the highest birthrate of any ethnic group (2.9 children per woman). Part of the increase is also due to better census reporting, though there is still probably substantial undercounting, especially on reservations. The remarkable increase in the number of people who identify as American Indian is also likely the result of a resurgence of ethnic pride among Native Americans (O'Hare 1992). In 1990, American Indians claimed membership in over 500 tribes. Most, however, identified with one of the eight largest tribes: Cherokee, Navajo, Chippewa, Sioux, Choctaw, Apache, Iroquois, and Pueblo (includes Hopi, Zuni, Acoma, and Laguna).

Most Native Americans traditionally have derived their identity from their family and local kindred group, with wide variation in tribal customs. Today, about half of all Native Americans live in suburbs or cities.

Nearly half of all Native Americans live in the western part of the United States. Very few tribes have been able to retain their ancestral homelands, though some native peoples were never forced to leave their native geographical areas (mostly in Alaska, Arizona, and New Mexico). About half of Native Americans lived in rural areas in 1990, with about half of these living on one of the 314 reservations left in the United States. Still, over a fourth of Native Americans now live in suburban areas, and another fourth live in larger cities such as Los Angeles, Phoenix, Tulsa, and Milwaukee (O'Hare 1992; Snipp 1989; Taylor 1994). Not surprisingly, Native American families look different depending on tribal affiliation, geographic location, and socioeconomic status.

Diversity Among Native American Families

Research on Native American families is still relatively sparse compared with the large number of studies on other racial/ethnic groups in the United States. We are beginning to identify some of the important features of American Indian families, including extended family structure, respect for elders, and distinctive socialization practices. Nevertheless, diversity among different tribal groups and differences between rural and urban residence make generalization about American Indian families difficult.

In the past, nearly all the principal variants of marriage and family found among nonindustrial peoples were commonly practiced by Native Americans. Looking across societies, one can find arranged marriages; bride price and dowry; adoptive and interfamily exchange marriages; trial marriages, monogamy, polygamy, and temporary polyandry; premarital and extramarital sexual relations; divorce; patrilocal, matrilocal, and neolocal residence patterns, patrilineal, matrilineal, and bilineal descent systems; and other kinship variations (Driver 1969; John 1988; Taylor 1994). Differences among indigenous North

American families were undoubtedly as numerous as the similarities among them.

Historically, most Native Americans derived their identity from the family and local kindred group, so the externally imposed label of "Indian" made little sense to one who lived daily life as a Navajo or Cherokee. Even today, there is a strong sense of tribalism and pride among specific groups. Native American families continue to speak different languages; observe different customs; and maintain different kinship, political, and social practices. For example, most Navajos tend to socialize with and marry other Navajos, and most Sioux tend to socialize with and marry other Sioux, in spite of residence on reservations or in cities. Some Native American groups, such as the Pueblo people of the American Southwest, maintain especially strong tribal and family identities. The Hopis, Zunis, and other descendants of the ancient Anasazi remain loyal to their matrilineal clan systems, native tongues, and religious ceremonies, emphasizing sobriety, self-control, and inoffensiveness (Benokraitis 1993; Olson and Wilson 1984).

Despite tribal differences, a common history of oppression and the recent upsurge of mobilization of protest movements have helped to foster a shared sense of American Indian culture among diverse peoples in the United States and Canada and to promote a modern "pan-Indian identity" (Jarvenpa 1988). Continuing high levels of spatial segregation on reservations have also promoted a sense of unique American Indian culture and have allowed many traditional family practices to be retained (Red Horse 1980; Taylor 1994). Over the past several decades, however, the range and diversity of Native American family patterns have been reduced, as more American Indians take up urban residence, attend majority schools, and marry non–American Indians.

Common Features of Native American Families

Some researchers claim that "the extended family network" is a universal cultural feature among Native Americans of both the past and the present. American Indian extended families or family networks have been described as structurally open, including several households of significant relatives and nonkin as family members. Extended family members engage in obligatory mutual aid and actively participate in important ceremonial events (Red Horse 1980). Some scholars suggest that the extended family form was a fair description of Native American households in the past; however, today it is more of an ideal than a reality (John 1988).

It is hard to determine whether extended family structures and practices are widespread because we have few representative studies of Native American families. Another problem is that modern survey measures may not be sensitive enough to capture unique Native American family structures. The U.S. Census Bureau defines a family as two or more persons, related by birth, marriage, or adoption, living together as one household, but this definition can obscure the existence of extended family relationships among Native Americans. Yellowbird and Snipp (1994, 180) point out that the Census definition considers all related people under a single roof as just one family, treating parents or grandparents living in a mobile home near children or grandchildren as entirely separate households and families. A related problem is that Census terminology does not reflect Native American cultural conceptions of family membership. For example, a Native American "grandmother" may actually be a child's aunt or great-aunt in the English use of the term, and "cousin" may have variable meaning not necessarily based on birth and marriage (Yellowbird and Snipp 1994). The picture is further complicated by the fact that terms, meanings, and family relations can vary significantly from tribe to tribe.

Native Americans are about as likely to live in married-couple families as Latinos, and like Latinos and African Americans, they are more likely than whites to have more and younger children, to have teenage pregnancies, and to be single mothers (*Statistical Abstract* 1999). As noted previously, however, these demographic data don't tell us much about what goes on within and between families. Most researchers have concluded that American Indian families are more firmly based on interdependence than white American families. Although joint residence among Indians may be decreasing, a strong norm of interfamily and intrafamily exchange is still common, especially on the reservations. For many, family and tribal identity are made salient through patterns of reciprocal exchange, ancient cultural customs, and public participation in important life course rituals (Taylor 1994; Yellowbird and Snipp 1994).

Using an acculturation framework, some researchers have identified different Native American family types. For example a three-part longitudinal study of 120 American Indian families living in the Oakland, California, area found four distinctive styles of adaptation to city living (Miller 1979). A "traditional" group endorsed and retained American Indian values and behaviors, living similarly to ways they had on the reservation. A "transitional" group left traditional American Indian values behind and tried to adopt white values and behaviors. A "bicultural" group held onto American Indian values and behaviors at the same time as they adapted to white expectations for behavior. A "marginal" group abandoned American Indian ways but did not fit into the white culture either, feeling alienated from both worlds. Like studies of other ethnic groups, this one found that the bicultural group, by retaining some native language and customs, was the most successful as judged by standard measures of social and psychological functioning.

American Indian families have also been characterized as affording a special role to elders. Through elders' meetings and councils, older tribal members have directed the spiritual, social, and cultural needs of the family and community. Children are taught to respect their elders, and elders expect family members to offer assistance without being asked. By virtue of their age, older Native Americans are assumed to have pleased their creator and moved into a stage of life that Red Horse (1980, 464) calls "assuming care for," wherein they are seen to be important stabilizing forces in the extended family and in the tribe. Grandfathers take grandsons and nephews for long walks to talk about life, nature, and tribal values. Grandmothers are often the center of family life and assume substantial responsibility for child care and for passing on tribal language and family traditions.

Some of the "core values" of American Indian families are different from those of the dominant culture (Snipp 1989). Besides specific tribal identity, scholars have pointed to Native American conceptions of "time," "leadership," "sharing," "cooperation," and "harmony with nature," as departing from Western ideals. For example, the Pueblo Indians blend past and present in their conception of time and say "time is always with us." Western cultures, in contrast, view time as linear and are often preoccupied with "watching the clock." Leadership among American Indians entails a kind of servitude, and the good leader is one who is more sacred, person oriented, honest, and intuitive. For the dominant culture, in contrast, leadership is often characterized in terms of ambition, aggressiveness, and the ability to dominate situations

(Yellowbird and Snipp 1994, 196). These cultural differences help shape family life and lead to some unique socialization practices in Native American families.

Responsibility for children tends to be shared by a wider range of adults than in the dominant white culture. In general, there is little stigma attached to having children outside of marriage, reflected in rates of nonmarital childbirth for Native Americans that are second only to those of African Americans. Most Native American cultures encourage children to become self-reliant at earlier ages than European American families and place more emphasis on traditional ceremonies and spiritual values. Rather than direct intervention, many Native American families teach by example and attempt to foster good listening skills and respect for authority (Yellowbird and Snipp 1994). These values sometimes conflict with the emphasis on competition and accumulation common to many American families.

Poverty rates are about as high among Native Americans as they are among African Americans, leading to a variety of social problems and need for social services. High levels of infant mortality, suicide, alcoholism, psychological distress, poor health, and shorter life expectancy are strongly associated with poverty and are higher among Native Americans than among most other groups. Although poverty rates are highest for Native Americans living in rural areas, the extended family structure that is more common in those areas helps to mitigate some of the adverse personal impacts of impoverishment. Transition to urban areas often brings increased economic advantages for Native American families but also carries risks associated with social isolation and loss of supportive extended family networks.

Rates of out-marriage are higher for Native Americans than for any other racial/ethnic group in the United States. Over half of all marriages of Native Americans are to people of other races (O'Hare 1992). These high rates of out-marriage are probably due to the small number of Native Americans in the population, the increased migration of American Indians to urban areas, expanding opportunities for education and employment in nonreservation settings, and a shift toward more favorable attitudes about Native American culture. Unlike among Asian Americans, out-marriage is not strongly associated with gender. Intermarriage rates for Asian American women are much higher than for Asian American men, whereas rates for Native Americans are over 50 percent for both men and women.

Interracial marriages are also leading to more multiracial children. Since 1970 the total number of multiracial children in the United States has increased more than fourfold, led by high rates for Native Americans and Asians. In over half the births to a Native American parent in the 1990s, the other parent was of another race, usually white. In addition, American Indians have the highest total fertility rate of any ethnic group, with almost three children born in a lifetime to every Native American woman (O'Hare 1992). Although the dominant culture usually assumes that assimilation is a desirable objective for any minority group, there is some concern among American Indians that they will "marry themselves out of existence" (Yellowbird and Snipp 1994, 239).

As the structural and social conditions affecting Native Americans change, we can expect that their family structures will also change. For example, some American Indian scholars observe that rigid gender roles in the family are loosening as more women are employed outside the home and as more Native American families move to urban or suburban areas. Others point to a history of matrilineal and matrilocal family systems that promoted "loose" marriage ties and afforded women substantial authority in many Native American families to begin with. As with other ethnic minority groups, there is a growing split among Native Americans between the haves and the have-nots. Many of those in rural areas and on reservations live in substandard housing and rely on government subsidies just to get by. Native Americans living in cities are also more likely to be poor than their white counterparts. Urban residence and high rates of single motherhood in the cities tend to compound the problems of some Native Americans because they cannot rely on extended family support networks more typical of rural areas. Nevertheless, some tribes are able to generate substantial income though the operation of casinos, increasing numbers of Native Americans are finishing high school or attending college, and the percentage of American Indians in professional and managerial jobs is increasing (O'Hare 1992). Combined with a greater likelihood of intermarriage and increasing numbers of dual-earner couples among Indians, we can predict that the American Indian middle class will continue to grow and that these families will be most likely to assimilate into the dominant culture. Of course, even modern middle-class families can identify with their Native American ancestry and maintain important aspects of their cultural identity, but without being embedded in dense social networks and participating in routine ritual practices, their ethnic identification is expected to become more diffuse. In the face of changing social and economic circumstances, we can expect American Indian families to flexibly adapt, but because the life circumstances of Native American families will continue to differ, we can also expect to see continuing diversity among Native American family forms and practices.

SUMMARY

1. African Americans (blacks), Latinos (Hispanics), and Asians are the most numerous minorities in the United States. Because of higher-than-average fertility rates for Latinos and African Americans and because of high rates of immigration for Asians and Latinos, the proportion of minority children and families is increasing.

2. Because of slavery, persistent racial discrimination, spatial segregation, and other cultural and economic factors, African American households have historically been less likely to be nuclear and more likely to be headed by women than white families.

3. Almost half of African American families in the United States are not "broken" or matriarchal but include a husband and wife. Another third of the black population, however, has very low levels of income and education, and families here tend to be headed by women. Economic pressures seem to be primarily responsible for the low levels of marriage in this group.

4. As a result of racial discrimination, African American workers earn less than their white counterparts at all class levels. Because of economic discrimination experienced by African American men, black families have relied heavily on women's incomes. Disparities between black women's and black men's real earnings have been lower than for whites. One result has been that black marriages are more likely than white marriages to end by separation or divorce. Another result is that black women have had unusually strong positions within their families compared with white women.

5. Since about 1950, there has been an accelerating tendency for women in the black lower class to remain unmarried and bear children outside of marriage. In fact, two of three African American children are now born to unmarried mothers. One interpretation for this phenomenon is that these women have become increasingly unwilling to enter into a relationship of subordination to men that gives them little economic advantage. Another structural cause is that there are relatively few marriageable men available in relation to the number of women. The availability of welfare payments to single mothers has sometimes been blamed for encouraging divorce and nonmarital births among blacks, but these payments have contributed relatively little to the overall trends in marital disruption and birth.

6. Latino or Hispanic families include Mexican Americans, Puerto Ricans, Cubans, and Central and South Americans. The groups vary in terms of family demographics and social class standing, with Cuban families most likely to be middle class, Mexican American families most likely to have large numbers of children, and Puerto Rican families most likely to be female headed and poor.

7. Mexican American family ideology, with its emphasis on family cohesion and male dominance, derives from its agrarian and patriarchal roots. This ideology is undergoing change as most Mexican American families now live in urban areas and most Mexican American wives and mothers are now employed outside the home.

8. Asian American families, including those of Chinese, Japanese, Korean, Filipino, Vietnamese, and East Indian descent, tend to be concentrated in the western United States. Although these groups differ from one another, most are characterized by high levels of educational attainment and high rates of out-marriage (particularly for women). Traditional Asian families value family well-being over individual needs and grant respect to an authoritarian father. As economic conditions shift, and as native-born Asians become acculturated, Asian American family patterns are coming to resemble those of majority whites.

9. Native American families were so diverse historically that it is impossible to identify one typical American Indian family form. Nevertheless, researchers have suggested that Native Americans tend to have open and extended family networks that provide mutual support, respect elders, collectively raise children, and reinforce tribal identity. The American Indian population is growing because of high birthrates, but more Native Americans are moving to cities, getting an education, and marrying non–American Indians than ever before. American Indian families, like African American families, have higher rates of poverty and nonmarital childbirth than other racial/ethnic groups.

10.   Variations in family form and lifestyle are most often the result of economic and institutional constraints, and minority family patterns respond to changes in economic conditions. For instance, heavy reliance on extended kinship networks or child care, emotional support, and financial assistance is typical of poor white families, as well as of poor African American, Latino, Asian American, and Native American families.

11.   Rates of marriage, divorce, and birth differ between ethnic groups, but the overall trends for all groups parallel those of whites. Since 1960, these include lower marriage rates, later marriage and birth, higher divorce rates, more out-of-wedlock birth, more employed mothers, more single-parent households, and more children living in poverty.

Key Terms

culture of poverty

ethnic identity

familism

fictive kin

immigration

machismo

melting pot

model minorities

out-marriage

racial identity

straight-line assimilation theory

Sociology Web Site

See the Wadsworth Sociology Resource Center, "Virtual Society," for additional links, quizzes, and learning tools:

http://www.sociology.wadsworth.com

Also on this web site you'll find InfoTrac College Edition, an online library of journals. Here you can search for electronic articles about central topics in sociology.

8 | Romantic Markets: Love, Cohabitation, and Marriage

INTRODUCTION

Americans seem to be obsessed with romantic love and intimate sexual relationships. Popular movies almost always have a romantic subplot, steamy romance novels are perennial best-sellers, and soap operas focusing on shifting erotic relationships dominate daytime television. Sex scandals and the love affairs of movie stars continue to be mainstays of the tabloid newspapers; radio call-in shows about love and sex garner huge followings; and countless magazines at supermarket checkout stands offer advice on how to attract, impress, satisfy, keep, or even dump the mate of your choice. Americans did not invent this obsession with love and sex, but Hollywood-style romance featuring beautiful women and daring men has become one of our most influential and profitable cultural exports. Remarkably, each week *Baywatch* is seen by more viewers than any other television program in the world.

Most of the romantic imagery that we see in the popular media is unrealistic, but even when such images are understood as unattainable fantasies or out-of-date nostalgia, we still tend to carry them around in our heads. Turn on the radio and the first song you'll hear is probably a love song—about the joys of falling in love or the sorrows of breaking up. If you happen to tune into a call-in radio talk show, the subject matter is as likely to be about sex and love as it is to be about politics. On the Internet, chat rooms are used for "cyber romance" and "cyber sex." Television talk shows are even more dominated by bizarre testimonials about romantic and sexual desires, practices, and disappointments. Movies and television shows, even when they focus on action, almost always contain a story line that includes heartthrobs or heartaches. Most of these stories are about conventional heterosexual loves, and most include some version of a strong male hero rescuing a beautiful and (at least partially) helpless woman and defeating the evil bad guys. Whether boy meets girl, girl meets boy, or some creative blending of the two, love imagery is rarely absent from the popular media (Coltrane 1998).

This heavy emphasis on romance in popular culture provides us with a unique view of love, but it also sets us up for disappointment by creating unrealistic expectations. As you will see, some critics even suggest that romantic love is a kind of trick because it recruits women into relationships where they are expected to perform free domestic labor for their lovers or husbands and take care of them emotionally. Most of us don't like to think about courtship and marriage in these terms because our romantic ideals tell us that love transcends worldly matters. However, sociological analysis shows how love and marriage have always been mixed up with issues of power and control and have always been shaped by a variety of practical and economic concerns.

In this chapter we investigate how patterns of meeting, matching, and mating have changed. Building on the brief history of love and marriage presented in chapters 3 and 4, we also show how gender differences in dating and courtship have persisted. Drawing on ideas abut exchange, networks, and markets from chapter 1 and applying the theory of the family as a property system outlined in chapter 2, we look at how patterns of attraction, relationship formation, and marriage matches are all strangely predictable. Although modern marriages are no longer arranged by one's family, we investigate how individual choices about love and marriage continue to be structured by age, race, gender, and social class. Finally, we lay the groundwork for understanding the importance of sexuality to men, women, and marriage, a topic explored more fully in the next chapter.

Love and Markets

Love and markets: this seems an incongruous pair of topics. The first has the idealized ring of one of our highest values; the other sounds cold and cynical. Nevertheless, there is good reason to treat them together. The relationships between men and women before marriage involve both. A few years ago a standard text would simply have called this courtship. If that term sounds a little archaic now, it is because romantic relationships are changing. Many people are putting off marriage until they are older, though they are still planning for it eventually. Others are less interested in marriage. The old term **courtship** carried with it the idea of old-fashioned boys and girls sitting on the porch swing under the watchful eyes of their relatives in the parlor, with wedding bells on everyone's mind. Instead, we now have men and women of all ages who are involved in romantic relationships but who do not feel that they are courting; that is, they are not following the old rules of dating with a prescribed set of steps leading ultimately to marriage.

Beneath the surface, however, the relations of unmarried males and females may not have changed so much. They still involve two old themes: love and the kind of bargaining that leads to cohabitation or a more permanent bond like marriage. Love is highly idealized, but we are becoming more realistic about what it really involves, even to the point of seeing how it relates to something as practical as the economic pressures of a marriage market. Feminist theories point out how the traditional ideal of love contributed to a system of gender domination. Love may be a beautiful interpersonal bond, but the family structure built around it can also be a part of a system of gender inequality.

We like to regard love as something ultimate that transcends worldly considerations. Many people see love as a magical and mysterious emotion that strikes suddenly and defies logical explanation. If there are material interests involved in a relationship, or even only sexual ones, we tend to regard it as not really love; in fact, however, our feelings abut love are solidly anchored in the social, material world. For one thing, our very idealization of love is a construct of our culture. To think about romantic love as an ideal is unusual in the history of the world. Most traditional societies, such as ancient or medieval China, the Middle East, and Europe, regarded it as rather dangerous and aberrant. We live in a historical era that is almost unique in making romantic attachment the proper, and indeed the only real, basis for marriage. We don't live up to this ideal so very often, but the fact that almost all of us believe in it shows that cultural forces are at work here (Barich and Bielby 1996; Illouz 1997).

Moreover, *when* people fall in love and *with whom* are patterned in predictable ways. "Love matches" are distributed across social classes, races, and age groups in very standard fashion—within, of course, a certain range of statistical variation. Romance typically affects males and females differently. Men, for instance, tend to fall in love more quickly than women do. This can be explained by the use of sociological theories, especially those informed by a feminist viewpoint. Love, in short, is tied into the market system through which people choose the partners with whom they enter the sexual, economic, and emotional relations that characterize cohabitation and marriage. As we shall see, even if those relations don't turn out to be marriages in the conventional legal sense, they still are shaped by the social structures that we inhabit.

We begin with a discussion of love and its ambiguities, introducing a theory of love as a ritual symbolizing emotional possession. Then, using advice manuals, personal advertisements, and other cultural artifacts, we turn to a sociological analysis of the distribution of love: how often it happens, when and to whom

> **A CLOSER LOOK 8.1**
> *Three Love Poems*
>
> I thought it was snowing
> Flowers. But, no. It was this young lady
> Coming towards me.
>> *From the Japanese of Yori-Kito,*
>> *Nineteenth Century*
>
> Will he be true to me?
> That I do not know.
> But since the dawn
> I have had as much disorder in my thoughts
> As in my black hair.
>> *From the Japanese of Hori-Kawa*
>
> Dew on the bamboos,
> Cooler than dew on the bamboos
> Is putting my cheek against your breasts.
> The pit of green and black snakes,
> I would rather be in the pit of green and black snakes
> Than be in love with you.
>> *From the Sanskrit, Fifth Century*

and with whom, and how long it lasts. To find out why these patterns exist, we examine the second component of our analysis: the marriage market. We look at who marries whom and at explanations of how various personal attributes of men and women are matched up or traded off on the marriage market. Romantic love, we argue, is a feeling that appears at certain moments in this bargaining and matching process.

Some form of love has probably been present in all societies (Jankowiak and Fischer 1992; see *A Closer Look* 8.1 for examples of love poems from Japan and India); however, modern Western culture (European/American society) places unique emphasis on romantic love. As we pointed out in earlier chapters, in past times and in many cultures, marriages were arranged by parents or other community members, not by the prospective bride and groom. Western culture, and especially contemporary American society, is unusual in considering romantic love to be the only proper basis for marriage. Because we are so preoccupied with being "in love" before we marry, we often overlook the fact that selecting a marriage partner is influenced by social and market forces outside of ourselves.

Call it courtship, a romantic market, or a modern sexual friendship, the basic theory is the same, even if we get somewhat different outcomes now than we did in the past. If the market seems a rather harsh environment in which to situate something as tender and beautiful as love, that is largely due to our society. It is what constrains us now and puts the ambivalence into modern love relationships.

WHAT IS LOVE?

According to various theories and literary traditions, love can involve sexual passion, romantic idealization, affection, companionship, altruism, dependence, attachment, shared experiences, and caregiving (Hendrick and Hendrick 1996; Cancian 1987). Social science theories about love focus on how similar (or different) potential mates are, what they get out of the relationship, whether the relationship seems fair or equitable to them, and how ritual interaction promotes intense feelings of love and attachment. One of the most interesting aspects of this research is the attention paid to similarities and differences between men and women in their experience of love, potential causes of these patterns, and how they might change over time.

Gender Differences in Love

Popular writers, psychologists, therapists, and sociologists have focused on the troubles caused by gendered differences in intimate relationships. They describe a typical pattern in which men approach relationships primarily with sex in mind, whereas women downplay the element of sex and emphasize emotional attachment and intimacy. A battle then ensues over who has to give what. Men may become cynical about emotional attachment and regard it as a feminine trick. When they do fall in love, other men regard them as having lost the game—and they may come to feel that way themselves. On the other hand, men may feign love in order to win sex. Then there is a battle over the level of commitment in the relationship: Will there be a brief encounter or a lifetime marriage? (e.g., Gray 1992; Rubin 1983; Safilios-Rothschild 1977).

This is not to say that men and women don't *both* experience sexual desires and feelings of love, but they tend to experience them in culturally prescribed ways. Feminist writers have looked at this issue from various perspectives in an effort to understand the complex relationships between love, sex, and power. Some draw on Freudian theory to suggest that child-rearing arrangements produce men and women with basically different personalities (Chodorow 1978; Gilligan 1982; Rubin 1983; Williams 1993). Briefly, they suggest that exclusive maternal care encourages women to value and depend on close emotional relationships, whereas it produces men who fear close emotional connection and value independence. Cancian (1985, 1993) suggests that men and women define love differently because of differences in power. Men equate love with sex or providing help, whereas women define it as verbal exchange and emotional rapport. By focusing on conflict and exchange theories, these approaches show how men's style of loving reaffirms and perpetuates men's dominance over women.

When men fall in love, they tend to overromanticize and idealize the women they seek. In the typical romantic pattern inherited from former times, she becomes an idol, to be protected from the world. She is someone whose favors have to be striven for mightily; once she is won, she might be displayed but is taken away to a man's private castle. Although these attitudes seems to glorify women, they end up imprisoning them. The protected woman on the pedestal is as confined and dependent as a child, and when the man is finally sure of possessing her, he begins to lose interest. She is no longer a mysterious and lofty goal to be striven for; she turns out to be an ordinary person, and the ideal vanishes into somber reality. What was once idealization can turn into subtle contempt.

Drawing on such analyses, feminist writers have argued that the scenario of romantic love in our culture and the social structures built around it constitute one of the main forces keeping women tied into traditional gender roles and subordinate to men. It is because women are tied by bonds of love and dependence to their husbands (and subsequently to their children) that they have fitted so easily and without protest into the traditional gender-segregated division of labor.

Weddings are emblems of romantic love in our culture.

Traditionally, love meant wedding bells and a happy ending. After the wedding cake was divided up and the white dress put away, there came the reality of "forever and ever, until death do us part"—a young woman retiring to the interior of her new home to become a housekeeper, wife, and mother. Romantic love was a glorious episode, but it exacted a price; it meant the end of whatever independent career she might have started. "Let me take you away from all this," said the prince in the fairy tale. The unfortunate reality is that he does, and that is the end of the story. If perchance the woman wanted to get out again on her own, the traditional answer from her husband invoked the sentimental tie: "Don't you love me anymore?" In the traditional scenario, then, love served to recruit women into subservient positions in a system of exchanging goods and services (Rapp 1992).

Is the feminist position too cynical? Although they consider power to be part of romance, feminist theorists, like other sociologists and psychologists, distinguish among various types of love. Love can involve sexual passion, romantic idealization, affection and companionship, or altruism (extending oneself for the well-being of the other). There are many gradations and combinations that can occur between people in various sorts of relationships (Berscheid 1994; Blumstein and Kollock 1988). In general, researchers conclude that love means different things to different people in different relationships at different points in time (Hendrick and Hendrick 1992). When feminists critique the bondage of love relations in our

A CLOSER LOOK 8.2
Love and Marriage

If someone had all the other qualities you desired, would you marry them if you were not in love with them?

_____ yes   _____ no   _____ undecided

In 1967, William Kephart asked over 1,000 U.S. college students this question. Almost two-thirds of the men (65 percent) but just under one-quarter of the women (24 percent) answered no, that they would not marry someone if they were not "in love" with them. Only 12 percent of the men and 4 percent of the women said yes, they would marry the person. Less than one-quarter (24 percent) of the men were undecided, as compared with 72 percent of the women. One woman commented "I'm undecided. It's rather hard to give a 'yes' or 'no' answer to this question. If a boy had all the other qualities I desired, and I was not in love with him—well, I think I could talk myself into falling in love!" In 1967, Kephart suggested that a young woman has "a greater measure of control over her romantic inclinations" than a young man.

In 1976, after the women's movement had gained more widespread support, researchers again asked college students if they would consider marrying someone even if they were not "in love" with them. Men were still more likely than women to say they would not marry without being in love (86 percent versus 80 percent). In fewer than ten years, the percentage of women saying "No" to the question more than tripled, and the percentage of men increased by over 10 percent. By 1984, responses had changed even more. The college women's opinions had caught up to the men's, with about 85 percent of each reporting

that they would not marry someone without being in love with them. The researchers concluded that the reason for the shrinking gender gap was that women had become more economically independent and were therefore less willing to settle for someone who was only a good provider (Simpson, Campbell, and Berscheid 1986).

In the 1990s, researchers turned their attention to student attitudes in other parts of the world. When researchers asked the romantic love questions in the United States, about nine of ten college men and women continued to answer that being in love was essential to marrying (Allgeier and Wiederman 1991; Levine 1993). In contrast, when students in India and Pakistan were asked the same question, three-quarters reported that they would have no problem marrying someone they did not love (Levine 1993). Although romantic love is found in many societies, the extent to which it provides the primary legitimate basis for marriage varies considerably. In both developing countries and wealthier industrialized nations, a strong women's movement and economic opportunities help create an environment in which women are more willing to consider their own needs and less willing to make themselves subservient to men.

Source: Adapted from William M. Kephart, "Some Correlates of Romantic Love," *Journal of Marriage and the Family* 29 (1967): 470–74; Allgeier, E. R., and Wiederman, M. W., "Love and Mate Selection in the 1990s." *Free Inquiry*, 11 (1991): 25–27; Levine, R. V., "Is Love a Luxury?" *American Demographics* (Feb. 1993): 27–29; Simpson, J. A., Campbell, B., and Berscheid, E., "The Association between Romantic Love and Marriage: Kephart (1967) *Twice Revisited*." *Personality and Social Psychology Bulletin*, 12 (1986): 363–72.

society, they have love of the passionate/romantic type in mind. Their criticism is stated from the point of view of a higher ideal—love that is more truly spontaneous and unmanipulative, love that allows equality. It is what some have called "mature love"—genuine caring for the other person while preserving one's own honesty and respecting one's own rights. A two-way relationship of this sort can only take place between mature persons (of whatever ages), both of whom give support and affection to the other while allowing a breathing space for their own independence (Mills and Clark 1994; Safilios-Rothschild 1977).

Equity and Exchange Processes

Elaine Walster (Hatfield) and G. William Walster (1978) proposed a model of falling in love that ties the psychological level of interaction to a larger sociological tradition of exchange theory. Relationships work best when both individuals feel they are getting a "fair exchange" for what they have to offer. The man and woman do not have to offer the same things, but the total "worth" of what the two are offering must somehow feel approximately equal. This worth may include attractiveness, social status, personality traits, and admiration by other people around them. A man who felt he could easily get other partners, whereas his girlfriend could not, would tend to move out of the relationship; a woman who had more opportunities than her boyfriend because of her attractiveness would similarly be motivated to leave. According to this research, falling in love could be predicted by matching up the traits of the two individuals and finding whether the combination gave them both a sense of an equitable situation.

Equity theory (Berscheid 1985; Walster, Walster, and Berscheid 1978) is a comprehensive framework within which the findings of other social psychologists can be included. Individuals may be paid off somewhat randomly at first, but they gradually gravitate toward those with whom they feel most appropriately matched. This process seems to go through various stages, in which more superficial traits are matched first, then other similarities or complements are worked out. Individuals need not be exactly alike, although similarities do help to establish a sense of a "fair trade"; the two can also offer different traits, which must compensate for each other so that the whole package is balanced. This is essentially a social-psychological view of the market process that we will examine subsequently. It also is a process that can be rather callous and put considerable strain on the individuals involved. One of its typical outcomes, especially in many of those couples studied a decade or more ago, was a trade-off in which traditional gender roles of domination and subordination were negotiated.

Love as High-Intensity Ritual

The foregoing theories may seem somewhat remote and abstract. They bracket love inside a large-scale frame, as if viewed through a telescope. Love, after all, is something dramatic and emotional. Can sociology focus in on the immediate reality of love at all?

It can, but we must shift our level of attention down to the microstructure of interaction and consider love as a ritual. We can make use of the theory of **interaction ritual** inspired by the classic sociologist Émile Durkheim (1893/1947), and developed by Erving Goffman (1967) and Randall Collins (1975, 97–102, 153–54). A ritual has the following ingredients:

1. Brings people together face to face

2. Focuses their attention on some common object or activity

3. Promotes a shared emotional tone among the participants, which grows as the ritual proceeds

4. Produces an emotionally charged symbol, which represents the partial sense of membership in the group

Romantic love can be seen as a ritual in precisely this sense. In fact, it is an extremely high-intensity ritual. Let's look at the list of ritual "ingredients" again.

FACE-TO-FACE INTERACTION The face-to-face quality of love is famous. Lovers want to be in each other's sight; when they are together, they tend to ignore everyone else, "on a cloud by themselves." Lovers prefer to be alone, with no one else around. They constitute a group of two, cut off from the rest of soci-

The intensity of love is correlated with the length of time lovers spend staring at each other.

ety: so much so that sociologists in the past have referred to love as "dyadic withdrawal" (Slater 1963). Sociologists who have actually measured the process find that the intensity of love is indeed correlated with the length of time lovers spend staring at each other.

FOCUS OF ATTENTION Not only are lovers together, but they have a common object of attention: themselves. Passionate love means being obsessed with the person one loves, forgetting everything else. Lovers talk about each other and about their love; mundane subjects apparently do not exist. Or, instead of talking, they are looking, touching, making love: activities that absorb all one's attention. Love, in fact, probably involves a more intense focusing of attention than any other type of interaction. Love may well claim to be the most high-intensity ritual.

SHARED EMOTION Love, of course, is an emotion. It is other things too; without the face-to-face encounters and the strong focus of attention on each other, the emotion of love could not reach its full power. What kind of emotion is it? Initially, it is desire and admiration, perhaps mixed with apprehension; or it may even be a much weaker emotion at first, a vague kind of attraction or interest. But *the process of going through the high-intensity interaction strengthens the emotion* *and turns it into passion,* for the fact that the lovers isolate themselves from the rest of the world, and focus intensely on each other, means that whatever emotions each has are then shared by the other. They settle into a common mood—one that is intensified by their exclusive attention to each other.

The physical acts of making love have exactly this structure. Kissing, stroking, intercourse itself; all these involve a narrowing in of one's attention to just oneself and one other person. Sexual excitement builds up between one partner and the other; each one's arousal makes the other more aroused. We are not saying here that love is simply equivalent to sexual arousal, but the two are connected, above all by the fact that both of them are highly intense forms of ritual interaction between two persons. Sexual arousal, precisely because it is contagious and shared, can generate the sense of oneness that is the hallmark of love.

SYMBOLS AND SACRED OBJECTS Love is known for its symbols: hearts, rings, presents given by one lover to another. If love is a high-intensity ritual, then the ritual must work to attach the shared emotions to some emblem that can represent them. This symbol serves as a reminder of the emotions and also as a touchstone for setting the boundary between insiders and outsiders. Bear in mind, the theory of rituals was originally developed as an explanation of how religious symbols acquired their sacred status: the idol of a pagan cult, the Bible, or cross of Christianity (Durkeim 1912/1954). The ritual of love, then, creates a little private cult with its own object of veneration—the loved person. That is why lovers idealize their loved one to such a high degree: the intense ritual absorption that they are involved in automatically elevates its object into something that is more than human.

Lovers idealize each other most strongly early in a love relationship. Later, they acquire a more realistic view of one another, probably because the stage of highly ritualized interaction cannot last forever. The lovers find they can't spend all their time together, staring at one another and ignoring the rest of the world. Eventually, the claims of ordinary life reassert themselves. When that happens, the idealization diminishes, and so does the intense emotion of love. It is for this reason that lovers, after they are married, inevitably feel some letdown from their emotional peak. They stop carrying out the social conditions that made up a high-intensity ritual, and the effects of the ritual diminish as well.

The ritual-charged symbols, however, can retain their power for some time after the rituals are over. Such symbols represent membership in a group of two, much as respect for the symbol of the cross represents membership in the religious group of Christianity. That is why lovers are so touchy about the emblems of their relationship. The man who carelessly loses a trinket given to him by his lover may precipitate an emotional crisis in their relationship because the gift has become charged with symbolism. In losing it, he appears to be reflecting a loss of devotion. Actions, to—kissing, holding hands, not to mention sexual intercourse—symbolize the emotional bond. Forgetting to perform them—or performing them with the "wrong" person—opens the way to the emotional strains and jealousies that make love affairs so dramatic.

Love and Jealousy

It is because of love's inherent riskiness and the lovers' vulnerability that love and hate are so closely tied together. Passionate love and hate might seem to be polar opposites, but they actually have a great deal in common. Both of them are high-intensity emotions; both of them tie individuals together in a passionate relationship. The opposite of love is not hate but indifference; hate has a different emotional quality, but it is of the same structural nature as love. Hate arises when the emotions of love are threatened or negated; its intensity is related to the intensity of the love it supplants.

Lovers are intensely connected; the whole ritual process works toward that result. Their emotional bond amounts to a form of emotional property, for the group that is formed by the love ritual consists of just two persons, and there is no room for outsiders. Moreover, the ritual has charged up various symbols representing this exclusive relationship, and these symbols are vulnerable to abuse. For example, holding hands is a symbol of their relationship; hence, if the woman so much as touches another man, her lover may become jealous. Because staring at one another is part of their love ritual, if the man glances at another woman a split second too long, his lover may go into a rage.

These types of jealousies and angers belong especially to the most romanticized part of a love affair—a first love or, for some experienced loves, the earliest phases of intense feeling. However, long-term commitment certainly does not preclude jealousy. When the emotional bond incorporates a strong sexual tie, the lovers tend to demand a kind of property right over their partner's body. Any act of physical sex comes to symbolize the whole emotional relationship. Because the emotional relationship is exclusive, so must the sexual relationship be. If this sexual exclusiveness is violated, the excluded partner often feels an uncontrollable rage, perhaps out of all proportion to any consequences. The emotions are parallel to how a worshiper would feel if his religious symbols were desecrated by an unbeliever—for example, if someone spit on a Bible or tore down the cross. The aggrieved lover feels a righteous anger: he has no doubt of being in the right and may not feel inhibited from punishing the offender, even violently.

This, then, is the negative side of the intense and exclusive love bond. As long as all goes well, there is strong devotion and idealization between the lovers. If some symbol of their relationship is violated, the result can be jealousy and uncontrollable and seemingly irrational anger. The outcome, as we will see in later chapters, may be quarrels, violence, or even murder. As with most aspects of families and close relationships, researchers are beginning to uncover interesting differences in patterns of jealousy across cultures and between men and women (Berscheid 1994; Salovey 1991). Most of these differences can be explained with reference to theories of love as an interaction ritual and romantic relationships as part of a system of emotional and material property relationships (chapters 1 and 2).

Love Advice in the Twentieth Century

Americans have long relied on "expert" media advice to learn about love relationships. Radio talk shows such as "Dr. Laura" or "Loveline" receive top ratings, and the most popular television situation comedy of the 1990s, *Frasier*, is about a psychiatrist who offers relationship advice over the airwaves. Print sources for such advice have never been more popular. Teenage girls turn to magazines like *YM, Seventeen*, or *Teen* to find out how to attract boys or solve relationship problems. Women can turn to a huge array of magazines for such advice, from the often steamy *Cosmopolitan, Ebony, Glamour*, and *Elle*, to the more discreet *Redbook, Woman's Day*, or even *Working Woman*. Men's magazines devote fewer pages to romantic tips, but single men can sometimes find out

A CLOSER LOOK 8.3
Advice for the Lovelorn, 1900–1950

Dorothy Dix (whose real name was Elizabeth Meriwether Gilmer) was born in 1861 and married in 1882. After discovering that her husband would be ill for life, she won a prize in a newspaper short story contest and became a journalist. In 1897, she was sent to cover Queen Victoria's Diamond Jubilee by the New Orleans *Picayune.* By 1901, she was writing for the *New York Journal.* She showed a remarkable ability to reply to readers seeking help with relationships, and her advice-to-the-lovelorn column was syndicated and read by millions of people in the United States and throughout the world. When she died in 1951, at the age of 90, she had been counseling readers for more than half a century.

In a collection of her advice columns (appropriately titled *How to Win and Hold a Husband),* Dorothy Dix (1939, 66–69) tells "the girl who wants to catch a man" to go where the fish are plentiful, for "there is no use in throwing a line and praying for luck in waters that have been fished out." Make yourself as good looking as possible, says Dix, pick a likely target, and "dangle before him the charms which he prefers and at which he is most likely to bite." Dix coaches the would-be fisher to vary her technique to suit her subject, appearing "bookish or golfish or musical or domestic as the case calls for." If the man you desire is very bashful and timid, Dix recommends, "your best play is a bold move." Although she suggests taking the initiative, she cautions, "do it so quietly and unobtrusively that he never suspects that you are starting his feet along the path that leads to the altar." So that he does not get scared off, Dix suggests that the young woman "ask favors of him, make him feel that you depend on him, ask his opinion and show him that you regard him as an oracle. No one on earth is so amenable to flattery as the shy man, and any woman who is a deft salve spreader can get him."

If, on the other hand, a young woman has set her affections on a "bold, bad lady-killer," the best way to get your man, according to Dix, is to scorn him: "Be snooty. Show no interest in him whatever. Snub him. Break dates with him. Confide to some little cat who will purr it to him that you don't consider him so much of a much anyway and nothing to write home about." Mocking behavior, Dix opines, will give him the shock of his life, because he is used to having women "kowtow before him." With this technique, coaches Dix, "you will put him on his mettle to make a conquest of you, and the chances are that while he is doing this you will bowl him over." Dix comments, "that type of man always hankers after the peach that hangs highest on the tree. He never wants the one that is ready to fall into his mouth." Besides instructing women to vary their techniques according to their target, Dix tells women that they can "sell themselves" to men by feeding them, flattering them, and listening to them talk about themselves. She also gives young women a pep talk about the importance of persistence: "virtually any woman can get her man if she will just stalk him long enough."

Source: Dorothy Dix (Elizabeth M. Gilmer), *How to Win and Hold a Husband* (New York: Arno/New York Times/Doubleday, 1974). (Originally published in 1939.) See Coltrane *Gender and Families* 1998.

about how to woo (or bed) the woman of their choice in the pages of *GQ, Men's Fitness, Penthouse,* or *Playboy.* The romantic advice is not directed just to singles; magazines like *Family Circle, Good Housekeeping,* and *Parenting* regularly have articles and columns that tell mothers (but rarely fathers) how to keep love alive in their relationships. Love advice is also offered to Americans of all ages through widely read newspaper columns like "Dear Abby," "Ann Landers," and "Helpful Heloise." You may have seen these newspaper advice columns, but you may not know that they have some important predecessors. "Dorothy Dix" was the "Dear Abby" of the early twentieth century, yet her practical advice could easily be mistaken for something right out of the pages of *Cosmo* (*A Closer Look* 8.3).

Dorothy Dix's advice about love and marriage both reflected and shaped Americans' attitude toward romance in the early twentieth century. In *Middle-*

In her love advice column, Dorothy Dix instructed women to get a man's attention by position-ing themselves nearby, wearing special clothes, smiling encouragingly, and playing up to them. Notably, she never questioned whether all this effort was worth it.

town, a classic sociological study of family life in 1920s America, Robert and Helen Lynd wrote that Dorothy Dix's column was "perhaps the most potent sin-gle agency of diffusion from without, shaping the habits of thought of Middle-town in regard to marriage, and possibly represent[ing] Middletown's views more completely than any other available source" (Stein and Baxter 1974, i; Lynd and Lynd 1929/1956). Her advice column was successful because she of-fered down-to-earth advice that resonated with many women's experiences of trying to get the right man to notice and marry them. Dix's analogy of "fishing" and "catching" a man suggests that the successful woman approached courtship and marriage with shrewdness and clever calculation. Note, however, that Dix never questioned whether all this effort was worth it. A product of her time, she assumed that every woman needed a husband.

Dix instructed women to use indirect tactics to get a man's attention, like positioning themselves nearby, wearing special clothes, smiling encourag-ingly, and playing up to them. This type of flirting was more common among young women of past generations because it was one of the only resources they could use to try to win a mate. Being "forward" was unacceptable be-cause it might send the "wrong" message. All this advice about "catching"

A CLOSER LOOK 8.4
Love Advice from the Kama Sutra

Now, a girl always shows her love by outward signs and actions such as the following: She never looks the man in the face, and becomes abashed when she is looked at by him; under some pretext or other she shows her limbs to him; she looks secretly at him, though he has gone away from her side; hangs down her head when she is asked some question by him, and answers in indistinct words and unfinished sentences, delights to be in his company for a long time, speaks to her attendants in a peculiar tone with the hope of attracting his attention toward her when she is at a distance from him, and does not wish to go from the place where he is; under some pretext or other she makes him look at different things, narrates to him tales and stories very slowly so that she may continue conversing with him for a long time; kisses and embraces before him a child sitting in her lap; draws ornamental marks on the foreheads of her female servants, performs sportive and graceful movements when her attendants speak jestingly to her in the presence of her lover; confides in her lover's friends, and respects and obeys them; shows kindness to his servants, converses with them and engages them to do her work as if she were their mistress, and listens attentively to them when they tell stories about her love to somebody else; enters his house when induced to do so by the daughter of her nurse, and by her assistance manages to converse and play with him; avoids being seen by her lover when she is not dressed and decorated; gives him by the hand of her female friend her ear ornament, ring, or garland of flowers that he may have asked to see; always wears anything that he may have presented to her, becomes dejected when any other bridegroom is mentioned by her parents, and does not mix with those who may be of his party, or who may support his claims.

We shall now speak of love quarrels.

A woman who is very much in love with a man cannot bear to hear the name of her rival mentioned, or to have any conversation regarding her, or to be addressed by her name through mistake. If such takes place, a great quarrel arises, and the woman cries, becomes angry, tosses her hair about, strikes her lover, falls from her bed or seat, and, casting aside her garlands and ornaments, throws herself down on the ground.

At this time the lover should attempt to reconcile her with conciliatory works, and should take her up carefully and place her on her bed. But she, not replying to his questions, and with increased anger, should bend down his head by pulling his hair, and having kicked him once, twice, or thrice on his arms, head, bosom, or back, should then proceed to the door of the room. Dattaka says that she should then sit angrily near the door and shed tears, but should not go out, because she would be found fault with for going away. After a time, when she thinks that the conciliatory words and actions of her lover have reached their utmost, she should then embrace him, talking to him with harsh and reproachful words, but at the same time showing a loving desire for congress.

When the woman is in her own house, and has quarreled with her lover, she should go to him and show him how angry she is, and leave him. Afterward the citizen having sent the Vita, the Vidushaka, or the Pithamarda to pacify her, she should accompany them back to the house, and spend the night with her lover.

Thus ends the discourse of the love quarrels.

Attributed to Vatsyayana, India, ca. A.D. 400.

and "keeping" a man seems a bit old fashioned today, but as we will see, there are sociological reasons why the pattern developed and still persists. When we look at male-dominated cultures in other times and places, we can see that women were similarly coached to act coy and childlike to win over their lovers. In India in the fourth century, the Kama Sutra, attributed to Vatsyayana, offered advice to would-be lovers (*A Closer Look* 8.4).

Dorothy Dix counseled women to be cautious and calculating in their approach to romance. Her advice illustrates an interesting pattern of gender difference in initial emotional commitment that might at first appear counter intuitive—that men are quicker to fall in love than women. Dix, along with most contemporary purveyors of relationship advice, coach women to make men fall in love with them before they commit. According to most studies, this is a pattern that still holds in modern courtship. Men are more likely than women to say that they believe in "love at first sight" and to report that they fall in love earlier in the relationship. In addition, as Dix and other love advisors suggest, if women want to "catch" the man of their choice, they would do well to enhance their beauty with the latest clothes, makeup, and hairstyles. Since the 1960s, studies have shown repeatedly that men are more likely than women to choose a partner on the basis of physical attractiveness, and men are more likely than women to say they are "smitten" by a potential mate's good looks (Berscheid 1994; Patzer 1985; Sprecher 1989). Just as in Dix's day, as we discuss next, women tend to place more emphasis than men on a potential marriage partner's ambitiousness, industriousness, and financial prospects (Buss 1989, Feingold 1992).

What's going on here? In a sense, we can say that men can afford the luxury of using "falling in love" or physical beauty to pick their marriage partners, whereas women have had to be more careful about the practical aspects of selecting someone to marry. As noted in our discussion of gender differences in love, women do care about the emotional aspects of marriage and usually spend more time than men thinking and talking about such things (Frazier and Esterly 1990; Thompson 1993). However, they tend to focus on the man's love for them as much as or even more than their own feelings for him. Although both men and women worry about meeting the right person, it is still more typical for women to worry about getting men to notice them and fall in love with them than the other way around (Wood 1996). As the women's magazines suggest, most studies show that women spend more time and effort than men on staying attractive to their love partner and keeping him interested in them (Rubin 1983; Wood 1994). Not only have women generally worked to get men to notice them, but they have also taken the lead in talking about the relationship, taught men how to talk about their feelings, and have sometimes given in to men's sexual advances out of a sense of obligation. These actions relate to the larger gender balance of power in the society and show that women have traditionally been more dependent on marriage than men (Cancian 1993; Rubin, Peplau, and Hill, 1981; Sattel 1992). This dependence has encouraged women to emphasize relationships but to be less impulsive than men about falling in love or wanting to get married.

Studies of college students' attitudes about the need to be "in love" with someone before marrying (described in *A Closer Look* 8.2) suggest that women used to be much more careful than men about falling in love, in part, because they realized that their choice of a marriage partner would have profound implications for their economic futures. Today that is less the case, so it is not surprising that the answers of male and female students have converged dramatically. At the same time, researchers discovered that for other mate selection criteria, gender differences remained. For example, 1990s women ranked physical attractiveness and social skills as more important in a mate than had their sisters from an earlier era. As before, however, 1990s women ranked physical attractiveness much lower than did the men who constituted the pool of their potential mates. Studies continue to show that women are much more concerned than men about the future earning power of their prospective spouse. The gender gap in romantic love is thus shrinking, but there are still some differences in what men and women desire in a mate. In short, as we show in our analysis of personal advertisements that follows, men are still valued more for providing and women more for their beauty (Allgeier and Wiederman 1991; Berscheid 1994).

Gender and Emotion Management

Because most women want to fall in love *and* find a good provider, how do they manage it? In part, women are able to align the two by consciously analyzing and discussing their feelings about their partners. In general, this allows them to fall in love less impulsively than men and makes it more likely that their love will develop incrementally or practically (Hendrick and Hendrick 1989; Wood 1994). Of course, both men and women manage their emotions through countless personal adjustments and internal conversations. Compared with men, however, women more actively manage or regulate their emotions through talk, thought, and interaction with other people, usually ending up with a better fit between what they "want" and what they "feel" (Cancian 1987; Hochschild 1983; Sattel 1992).

Arlie Hochschild (1983) uses the term **emotion work** to describe this process of aligning one's emotions with the dictates of the situation and finds that women tend to do more "emotion work" than men. For example, if a woman thinks a man is a good match, she is more likely to allow herself to be overwhelmed by her emotions. On the other hand, if she is attracted to a man who is unacceptable as a marriage partner, she is likely to deliberately work at falling out of love with him. Men do this kind of emotion work too, but traditionally, they have not been as focused on it as women.

A woman's tendency to "work" on her emotions and to focus on getting the right man to love her is partly a matter of personal choice, but it also reflects women's historically weak bargaining position in the society and on the marriage market. Because women have been dependent on marriage for economic support, they end up paying more attention to initiating and maintaining love relationships and are more likely to adjust their feelings through emotion work, as recommended by the romance advice columns. Men, on the other hand, have been able to follow their impulsive feelings and let women pay attention to building intimacy. Traditionally, they have been quicker to be smitten by Cupid's arrow and to fall head over heels in love with a vision of their ideal mate.

These basic differences in the ways that men and women have approached romantic relationships do not hold for every person because they are just general tendencies that are subject to change (Hendrick and Hendrick 1996). What happens to men sometimes happens to women, and vice versa, and the gender differences seem to be getting smaller as men's and women's lives become more similar. The ideals and practices of romantic love also differ according to ethnicity, religion, and the like, with enormous variation between different groups. Nevertheless, a consistent message that we get from the popular culture and hear from many people around us is that gender differences in courtship are natural, unchanging, and beneficial for both men and women. Most people would concede that there are some general differences in women's and men's approach to romance, but increasingly, people are questioning whether they are necessary or helpful.

THE ROMANTIC MARKET

As noted previously in our brief discussion of equity and exchange theories, it is not far fetched to think in terms of a market for love, sex, or marriage partners. In fact, when close friends are not afraid of being overheard, they often use market language or a point system when discussing what is sometimes called "rating, dating, and mating" (e.g., "she's/he's a perfect 10"). In most cases, the romantic relationship market does not involve a product being exchanged for money but instead entails bartering between two people who have distinct personal needs and attributes. Each prospective romantic partner offers a set of physical characteristics, personality, skills, and so on, which are symbolically exchanged for those of a partner who seems interesting (Laumann et al. 1994). See *A Closer Look* 8.5 for a discussion of the role of physical characteristics in attractiveness, dating, and marriage.

Although it is not the typical case, people in search of romantic partners can advertise their attributes in "personal ads" appearing anywhere from the local weekly recycler newspaper to high-tech computer or video matching services. Consulting advertisements in the "personals" section of most newspapers reveals how people are engaged in some very explicit bargaining about what they have to offer and what they want from someone else. Take a look at the ads in *A Closer Look 8.6* and use the key of abbreviations to figure out the characteristics of the advertiser and the person they are seeking (e.g., *SAF* for "single Asian-female").

In the thousands of newspapers that carry personal ads, advertisers and their respondents run the gamut from heterosexual to homosexual; from single to married; from teenagers to octogenarians; from those advertising for one lifetime marital partner to those seeking candidates for impersonal, commercial sex, "discreet" afternoon trysts, threesomes, foursomes, or one-night stands. Still, Erich Goode (1996) and other social scientists studying such things, suggest that the overwhelming majority of personal ads are written by heterosexual singles between twenty-two and sixty-five claiming to seek long-term, marriage-oriented relationships. The procedure for using personal ads as a means of dating is approximately the same for all periodicals running personal services. The people placing the ads describe their own characteristics or qualities, along with the characteristics they are seeking in a dating partner, sometimes also including information about the nature of the relationship desired. Each ad includes a post office box number and/or a telephone code run by the publication. Voice mail is becoming more common than letters or notes for initial responses, though both forms are common. Responses are forwarded to the individual advertiser, who then makes a decision about which ones to answer.

According to Goode (1996, 143), these "mediated" modes of courtship differ from more conventional forms in at least three important ways: "First, in 'mediated' courtship, the initial selection process takes place prior to a face-to-face meeting. Second, participants have a great deal more deliberate control over the information they allow potential dating partners to have access to than is true of more informal, naturalistic avenues of courtship (Woll and Young 1989, 483). And third, while courtship through traditional channels is overwhelming a product of personal associations (friendship networks, school, place of employment, etc.) mediated courtship transcends this restriction by attempting to establish an intimate relationship with strangers (Laumann et al. 1994, 4)." Although somewhat different from more mainstream forms of courtship, personal ads via print or computer are definitely increasing and provide researchers with an opportunity to examine some of the emerging norms governing "mediated" courtship.

Because personal advertisements usually contain both the characteristics of the advertisers and the qualities they seek in a partner, such ads have been a valuable source of information on types of self-presentation, preferred social roles, gender norms, and judgments of attractiveness in both heterosexual and

A CLOSER LOOK 8.5
The Influence of Physical Attractiveness

We like to believe that we are above merely physical considerations, that we choose our friends and loved ones for their "selves," not their bodies. But there is a lot of evidence to the contrary, summarized in Gordon L. Patzer's book, *The Physical Attractiveness Phenomena,* 1985. (See also Feingold 1992; Goode 1996; Hatfield and Sprecher 1986; and Surra 1991.)

Generally, there is consensus about how attractive someone is among both males and females. The major exceptions are as follows:

1. Individuals tend to be inaccurate in judging their own attractiveness; they think they are better looking than the way in which they are perceived by other people. Interjudge reliabilities are about .87–.89; whereas self- and other ratings correlate .17–.22 (Patzer 1985, 24–26). Better-looking persons tend to be more accurate about their own appearance, especially if their attractiveness is actually quite high. Low- and average-attractiveness persons tend to overestimate themselves. Men are especially likely to overestimate their own looks, whereas unattractive females are more accurate about other's estimations (though still overestimating themselves somewhat). This is because women get more feedback about their attractiveness and are made aware of it. In general, the contrast of self-concept between females of high and low attractiveness is greater than between their male counterparts.

2. Spouses, as well as engaged couples, overestimate the attractiveness of their husband or wife (Patzer 1985: 25–26; Murstein 1972; Murstein and Christy 1976). Wives overestimate their husbands' attractiveness even more than husbands overestimate their wives'. On a 10-point scale, with 1 very unattractive and 10 very attractive, the average values in one study found husbands rated 3.6 by judges, 5.4 by self, and 7.2 by wife; wives were rated 3.6 by judges, 5.1 by self and 6.5 by husband.

This says something about the process of marriage. Not only is there a market in which people select someone of about the same attractiveness level as themselves, but their own egos become involved in their beliefs about the attractiveness of their partners. Spouses who overestimate their partners' attractiveness the most have greater marital satisfaction.

Patzer (1985: 140–63) reviews evidence on what traits people judge are the basis of attractiveness. In men, the face is most important (explaining 50 percent of the variance; the next most important, explaining 10 percent, is weight and its distribution); in women, by far the strongest factor is not being fat (50 percent of variance; this goes up to 75 percent when females judge females); face is second at 10 percent. A woman's bust size makes little difference; medium-sized breasts were judged as more attractive than large ones. People tend to believe a large bust creates an air of sexiness, but the large bust also calls up a stereotype of immorality, lack of intelligence, and immodesty. Fat stomachs are generally considered the worst trait in men.

Females are concerned with their own attractiveness more than males are concerned with their own attractiveness, but males consider their partner's attractiveness more important than females do. For this reason, there is higher correlation between dating popularity and attractiveness for females than for males (Berscheid et al. 1971). For example, college females' attractiveness correlated .61 with the amount of dating; males' attractiveness correlated .25. Better-looking women date more, have more friends, are in love more, and have more sexual experiences than average or less-attractive women.

Before meeting, both males and females choose dating partners largely on the basis of attractiveness. For actual dates, though, males select the most attractive females only if they feel assured of being accepted by them; otherwise they pick partners of moderate attractiveness, assuming more-attractive girls would be less accepting of them as dates. Females do the same. Females also are less likely to pick the better-looking male as a date if their own self-rating of attractiveness is low. In bars, men approach women more often whose attractiveness matches their own. In other words, although physical attractiveness is highly valued, people are usually aware of the realities of competing on the sexual marketplace and adjust their aims accordingly.

Once dating takes place, personality plays a role in the couple's involvement, but attractiveness has a continuing and undiminishing effect on romantic attraction through at least five encounters (Patzer 1985, 87). Persons who considered their partners to be more attractive tended to have greater love for their partners; they were also more submissive to better-looking partners (Critelli and Waid 1980). Better-looking persons worried less about their partners becoming involved with other persons. Hill, Rubin, and Peplau (1976) found that breaking up is seldom mutual; the female partner is more likely to perceive problems and initiate a breakup. Differences in physical attractiveness are recognized as a significant factor in ending relationships.

The process of forging a romantic tie is structured around *similarity* of the partners' attractiveness. This happens because the marriage market brings people together according to their own levels of resources. Feingold (1982) found that the couples who fall in love or become more intimate are those who are more closely matched in looks, and those well-matched individuals

Although we want to believe that we are above merely physical considerations when choosing our romantic partners, there is considerable evidence to the contrary.

progress more rapidly in their romance. Dating couples, like married couples, tend to be similar in attractiveness (Patzer 1985, 92). A marriage is less stable when the attractiveness of husband and wife is not similar. Matching for attractiveness is correlated with marital adjustment even among elderly couples.

The general pattern is that women are more conscious than men of attractiveness ratings, that they compare themselves more with other women than men do with other men, and that women suffer more anxiety over these ratings. Women are particularly severe in judging other women, and themselves, by their weight. This is probably the most important cause of anorexia nervosa, an eating disorder. All these are strains resulting from the competitive marriage market, in which men's main resource is their economic prospects and women's main resource is their personal appearance. This pattern reflects the sexual stratification of the larger society. As women break through into better-paying, independent careers of their own, attractiveness should become somewhat less important because it will no longer be women's only resource on the marriage market.

Look over these personal advertisements and see what patterns you can discern. Do men and women advertise the same things? Can you find any patterns according to race, ethnicity, age, or sexual orientation?

SAF, 24, 5', 100 lbs, n/s, drug-free, no kids, enjoys fine wine and beach. ISO SA/Wm 25-36, Thai-spkg w/similar interests.

SAF, passionate, athletic, educated, high integrity, delicate, feminine, luvs rose gardens, 30s, slim, wants 1 special Gent!

SAF, attractive, honest, simple, seeks feminine, intelligent B/H/AF, with good heart for good times

SBF, attractive, funloving warm, 43, good sense of humor, looking for my hero. ISO SBM, 40-50.

SBF, cosmopolitan, 28, ISO well-edu SB/H/WM, 25-40, 6'+, mature, open-mind, n/s, n/d; friends, dance, books, movies. No airheads please!

SBF, attractive, feminine, articulate, intel. ISO SBF, 26-36 5'5"+ 4 friendship, movies, dining and possibly more.

SJF, slim, strawberry blonde/blue, 53, loves opera, theater, trvl. ISO DJM, affluent, 50-60. Let's explore life's pleasures.

SJF, beautiful, dimpled, vivacious, blonde with knock-out figure ISO tall, fit, witty, adventurous SJM with champion personality, 48-58, n/s.

SHCF, 41, 5'3", 123#, meticulous, stable, no vices. ISO phys fit M, 39-49, avg build, 5'9"-6'2", 170-200#, finan. secure, similar quals, no kids.

SHF, beautiful, full figured, college student, likes dancing, roller skating ISO SM 21-40, financially secure.

DWF, very attractive, prof, brown/green, 5'8" slender, young-looking 45, ISO SWM 40s attractive, honest, sense of humor, outgoing.

SWF, stunning figure, tall, leggy, blonde, sense of humor, n/s, n/d, ISO fin sec entrepreneur, humor, knight in armor. Friendship 1st.

SWF, attractive, fem, fun, erly 40s, gnuine, sxy, crtive, enjys thtr, mvies, rstrnts, ISO fem attr SWF, prefer blnd, hnst, 30s-40s.

GWF, caring, 34, prof, slim, fem, with young son. ISO soft butch, to share fun and love. n/d. No games.

SHM, financially sec, 40, 5'10" 175# jewelry desgnr ISO slim, pretty, well-balncd W/HF, 24-38, n/s, 4mus, dining, romance.

SHM, 27 ISO 50-60, <5'7" AM for gd times, frndshp.

SHM, 27, 5'8" 160# clean, honest, fun, fit. ISO SAF, 22-30, 4 LTR, enjoy life, n/d, n/kids, let's talk.

SHM, 33, attr, prof, ISO SW/HF 4 fun, adventure, and romance.

SAM, 39, 5'6" 160#, computer prof, ISO SWF, <35, tall, full-figured, single mom OK. prefer policewoman.

SWCM, 6'5" handsome, athletic, fun-loving, intelligent, well-educated, strong, tender, sensitive, ISO sim S/DW/A/HCF.

SWM, tall, attr, 35, fun, intel, stable, knight in dull armor. ISO bright creative, very curvaceous, sweet princess w/polish.

SWM, affluent, trim, athletic, 5'8", generous nature, ISO attr. petite curvaceous, n/s lady for mutually rewarding rel.

GWM, 50, 5'10", 180#, vry masc, good-looking, HIV-, ISO masc, pot-bellied WM, 25-45.

GWM, prof, 33, 5'7", 145#, masc, rmtc, down-2-earth ISO HIV-, str8-act, fit Guy, 18-38, safe desrt, 4 fun.

GAM, 28, 5'6", 125#, fit, enjoy conversation, movies, and wlks on the beach, ISO WM, 30-40, for long-term relationship.

BiBM, smart handsome, 39, safe, tall slim, fit, ISO manly HM, 25-45.

SWM, romantic, tall, good-looking, affluent, ISO bright, vry attractive, warm SWF, 40s.

SJM, attorney/fin. prof, cute, ISO exotic lady w/dk hair, 28-39, smart funny, pretty for romance, travel, and more.

SWM, 50, 5'10", 170lbs, n/s creative, rom, hndsm ISO queen-sized Dolly Parton/Anna Nicole Smith-type, vibrnt, affec.

Abbreviations:

A.......	Asian
B.......	Black
C.......	Christian
D.......	Divorced
F.......	Female
G.......	Gay
H.......	Hispanic/Latino
J.......	Jewish
L.......	Lesbian
M.......	Male
S.......	Single
W.......	White
ISO.......	In Search Of

Excerpted and modified from the *Los Angeles Times*

homosexual samples. Most researchers assume that the qualities advertised are those that an individual believes are attractive to a potential partner and further that such preferences reflect stereotypes from the culture at large (Child et al. 1996; Deaux and Hanna 1984). Most researchers studying personal ads use a content analysis of the ads themselves, though a few also look at responses to the ads.

Researchers document how personal advertisements tend to differ according to gender. Ads written by men seeking women tend to emphasize the importance of conventional physical attractiveness, using terms such as *attractive, slender, petite,* and *sexy.* Men's ads are also likely to seek youth, sexual characteristics, and offer financial security (Child et al. 1996; Cicerello and Sheehan 1995; Davis 1990; Deaux and Hanna 1984; Smith, Waldorf, and Trembath 1990). Men's ads are also more likely to offer a casual relationship (Cameron and Collins 1998). Women's ads may also indicate that they are seeking attractive male partners but mention this much less frequently than do men's. Women are more likely than men to seek sincerity and permanent relationships and to offer information about their own personalities, physical appearance, and youth. Regarding qualities sought in a partner, women's ads tend to emphasize status and success with words such as *secure, affluent, professional,* and *successful* (Child et al. 1996; Cicerello and Sheehan 1995; Davis 1990; Deaux and Hanna 1984; Smith, Waldorf, and Trembath 1990). In addition, men and women seeking heterosexual partners tend to emphasize the things that the other party seeks; that is, women tend to say they are pretty, and men tend to say they are successful. The ads thus reflect conventional expectations for men to be providers and for women to be objects of beauty. At the same time, a few studies have found that heterosexual men tend to offer expressive traits (caring, affectionate) and women, instrumental traits (intelligent, ambitious), suggesting that the advertisers are assuming that potential intimate partners will value these traits (Cicerello and Sheehan 1995).

Turning to analyses of homosexual advertisements, gay men—like straight men—emphasize physical attractiveness and mention sexuality more often in their ads than do women (Child et al. 1996; Deaux and Hanna 1984). When compared directly with lesbian women, gay men are significantly more likely to mention physical attractiveness and less likely to express an interest in the personality characteristics of their potential partner (Hatala and Prehodka 1996). In one study, homosexual men mentioned attractiveness more often than male transsexuals, who appeared to be seeking friendship (Hatala and Prehodka 1996). Finally, heterosexual men have been found to seek long-term relationships and mention sincerity more often than gay men (Gonzales and Meyers 1993).

In one of the most innovative studies on this topic, Erich Goode (1996) catalogued responses to four personal ads he placed in each of four East Coast metropolitan newspapers or magazines (*A Closer Look* 8.7). By placing ads that varied according to attractiveness, success, and gender, he was able to analyze the pattern of responses and move beyond the cataloguing of preferred traits. He found that men-seeking-women ads were more plentiful than any other type of ad in the journals he used and that men were most likely to send in multiple responses to the ads. Men were much more influenced to respond by the listing of physical attractiveness in a potential female date, whereas women were substantially more influenced by financial success. In addition, Goode called on working-class and upper-middle-class raters to estimate responders' chances of landing dates with the (fictional) advertisers (including just responders who sent in pictures). His female panelists tended to rate the male responders as unacceptable dating candidates, suggesting that the men were far less fearful of rejection than

A CLOSER LOOK 8.7
A Study of Responses to Personal Advertisements

Erich Goode placed the following ads in four East Coast newspapers or magazines with the listed replies:

	No. Replies
Beautiful, shapely, slender blonde, 5 ft 7 in, waitress, 30. Would like to meet a man for lasting relationship.	668
Handsome, athletic, single male, broad shoulders, 6 ft, 32, cab driver. Seeks woman for lasting relationship.	15
Successful woman lawyer, intelligent, financially secure, 30, average appearance. Seeks man for lasting relationship.	240
Successful, financially secure, intelligent, single male, attorney, 32, average looking. Wants to meet woman for lasting relationship.	64

Which ad generated the most responses?

The four "beautiful waitress" ads attracted a total of 668 responses, or a bit more than 160 per ad. The successful, average-looking female lawyer ads garnered 240 replies, or 60 per ad. The average-looking male lawyer ads generated 64 replies, or 16 per ad, and the handsome taxicab driver ads generated a total of only 15 responses, or less than 4 per ad.

Which gender is most likely to respond to a personal ad?

The ads placed by women attracted 908 replies from men, whereas the ads placed by men attracted 79 replies from women, a ratio of over 11 to 1. In this study at least, men were more likely to respond to a personal advertisement than women.

According to researcher Erich Goode (1996, 169), large numbers of men who responded to the ads "overestimated their worth on the dating marketplace and approached an unrealistically desirable candidate." He concludes, "The fact that men are far likelier to advance themselves as sexual candidates, in spite of their undesirability, even in this fairly unusual and atypical dating venue, reveals the crucial role of entitlement in heterosexual dating relations."

Source: Erich Goode, "Gender and Courtship Entitlement: Responses to Personal Ads," *Sex Roles* 34 (1996): 141–70.

were female responders. Goode (1996) suggests that "men are far more likely to display inappropriate entitlement in their dating choices: Among men, less desirable partners are far more likely to consider themselves acceptable dating choices for those who are more desirable than is true of women." He concludes that equity theory works better for the courtship strategy pursued by women than for that of men. As others have suggested, Goode calls attention to the fact that our cultural ideal of masculinity promotes a sense of entitlement among men and encourages them to expect more from women than women feel the men deserve (see also Pyke and Coltrane 1996).

As a whole, the research on personal advertisements shows how people are figuratively trading valued sexual and emotional commodities in a kind of intimate relationship marketplace. Some of the ads are more about sex, and some are more about companionship, but they all attempt to match people according to valued characteristics. Note from the examples in *A Closer Look* 8.6 or from your own local sources how many of the listings advertise for people similar to themselves in terms of physical characteristics, age, education, race, ethnicity, social class, and so on. There is some variation, especially when we look at same-sex or bisexual relationships, but even here we can see gender stereotypes at play when lesbians describe themselves as "feminine" or when gay men describe themselves as "masculine"—the more valued gender characteristics in our society. As a result of such findings, researchers conclude that the pattern of requests and offers expressed in personal advertisements match the stereotyped expectations of men and women in more typical courting and dating situations.

Market Resources and Opportunities

As the market metaphor suggests, meeting and being attracted to someone take place in a large-scale system of exchange. It is like a market, only what is being exchanged are not economic goods and services but people's own social and personal traits. This characterization seems harsh, as if people were cattle to be bought and sold. In fact, the process has a rather harsh side, particularly when it takes place in situations, like our society, where there is a good deal of status stratification, as well as inequalities between the sexes.

For us, love is supposed to be a highly personal experience. Hence, we do not wish to be conscious of the market forces at work around us that mold our opportunities for falling in love or marrying. In traditional agrarian societies, where the ideal of love was not important, people were much more open about marriage as a market. In Middle Eastern societies even today, or places such as rural Greece or Sicily, men speak quite plainly of the value of women they might acquire or of their own daughters and sisters as property that they might use to make an advantageous deal with some prospective suitor. Traditionally, a man's value was his social status and wealth. A woman's value was measured by her beauty and youth and especially by her virginity. A woman who had lost her virginity before marriage was no longer of value on the marriage market. Her father or brothers felt dishonored and might even take her life; such murders tended not to be punished because they were regarded as crimes of honor. In recent wars in Kosovo, Afghanistan, and Bangladesh, women who were raped by soldiers, through no fault of their own, were nevertheless devalued; men refused to marry them, and some husbands would not take back their own wives.

Such societies show a market system at its most brutal. Even though some of the characteristics of the market have changed, we are still subject to similar pressures. One of the major differences is that families no longer control the marriage market for their children, and men no longer barter away women. Instead, all individuals, male and female, have to enter the marriage market on their own and strike the best deal they can for themselves. A further difference is that the emphasis on virginity as a key attribute of a woman's value has declined precipitously, especially in recent years.

The market still operates in other respects. Social status, earning capacity, and attractiveness are among the resources that each person has to offer. A person's ability to attract others is heavily influenced by the bundle of resources at his or her disposal. The market consists of all the people who might come into contact with each other and then "compare resources." Eventually, they find out not only who is available and whom they want but also whom they can get; the latter depends on how valuable one is compared with the other people on that market.

To put it very simply, everybody has a market value. Whom a person will be able to "trade" with depends on matching up with someone of about the same market value. Thus, when any two people come together, each has the following market position:

1. Your own resources include your social status (class, race, ethnicity; your family background; your current occupation), your wealth and your prospects for making more in the future, your personal attractiveness and health, and your culture. Even personality traits such as your "magnetism" or charisma are social resources (moreover, as we shall see, they are produced by one's social position, including one's position on the market itself). Your market opportunities consist of the people you know who are of the desired sex and sexual orientation and who are not already involved in a relationship that prevents their becoming involved with you. The people in your "opportunity pool," of course, have their own degrees of attractiveness depending on their resources.

2. The other person's resources are also social, economic, and physical. His or her opportunity pool also consists of persons of the desired sex who are known and available, ranked by the attractiveness of their resources.

In every encounter, then, an implicit comparison takes place. You weigh how attracted you are to a person by comparing that person with others whom you know *and* by determining how confident you feel about being able to "strike a deal" with him or her. Most people, it should be stressed, are not consciously very calculating about this. However, they do make such comparisons, *whether they consciously want to or not,* partly because people have been conditioned to do so on the basis of previous success or failure in both dating and other social situations.

Suppose that a woman meets someone she finds extremely attractive. He must have traits that she values—appearance, self-confidence, good humor, style, interesting conversation, politeness, and sympathy—but her evaluations themselves, if we analyze them further, turn out to be influenced by her previous experiences in the social marketplace. If one finds someone physically attractive, that always means that one finds him more attractive than some other people; that, in turn, is affected by how many other attractive, or unattractive, people one has known. A woman who knows a lot of handsome men is going to be affected differently by meeting a moderately handsome fellow than another woman who has associated with mostly unattractive guys.

The same holds true for other traits. One's style comes to a considerable degree from one's social class background; the things one has to talk about, too, depend on one's social experiences, most of which are acquired as a result of social status (education, travel, acquaintances in important positions, etc.). Even one's emotional qualities are affected by social experience. People who appear relaxed, charming, and good humored are that way to a large extent because they have had many favorable experiences in previous social encounters. That is to say, they have had a balance of social resources that has made them attractive to most of the people they have met. In the social world, and especially in romantic encounters, nothing breeds success like previous success.

Even how polite and attentive the other person is to you has been influenced by the market, for a high degree of attentiveness means that you are a relatively attractive person to him or her; it implies that the resources you present rank relatively high compared with what the other person has been used to. There is evidence that men tend to prefer women who are hard for others to get (i.e., who have a high market value) but who are easy *for themselves* to get, instead of uniformly hard-to-get or easy-to-get women.

The romantic market, then, is a vast sorting process. People move from one encounter to the next, attracted or unattracted by the people they meet, depending on whom else they have met or expect to meet. Sometimes one is attracted to another person but finds that the attraction is not reciprocated; one's resources do not match up with what other people have been showing to that person. Sometimes the shoe is on the other foot, and you find someone is much more attracted to you than you are to him or her. Both kinds of mismatches tend to make those encounters relatively brief, as one or the other person will move away from it. Someone may be very persistent, of course, in pursuing a consistently rejecting partner. When this happens, it may not be so much because of that particular person's personality traits (stubbornness, insensitivity, passion) but simply because of the way the market works for him. A man may chase a woman who rejects him simply because the other women he knows are much less attractive or because he does not know very many other women. In fact, as we saw in responses to the personal ads, men tend to feel entitled to relationships with women who have more desirable traits than they do.

Eventually, people discover that they get the fewest rebuffs and find the most rewarding person they can by sticking to someone whose resources most closely match what they themselves have to offer. This is why partners tend so often to be similar in social class background, education, and many other social traits. There is even some evidence that couples remain together most often when they are similarly matched at levels of attractiveness.

The Winepress of Love

We are beginning to see some of the reasons, then, why love can be an emotionally draining, or even brutal, process. As long as a market is operating, people are being shunted around by forces beyond their own personal control. One's attractiveness to other people changes as rivals come on the scene, as new opportunities open up, or others close off, and these things can go on without our conscious awareness. New resources elsewhere in the network of people who make up a community can send rippling effects all down the line. When an attractive man or woman moves next door, takes a new job, or joins a new social group, his or her mere presence may be enough to change the character of several existing romantic liaisons.

Even in the normal course of events, a market of this sort puts a great deal of strain on the people who go through it. One's status goes up and down without one's knowing precisely why or being able to control it, depending on whom one happens to encounter. There is a lot of manipulation, especially when people of different social classes put their unequal resources into the market. Most Americans are not very aware of social-class distinctions, but the resources operate nevertheless. People from higher social-class backgrounds have resources that enable them to be more socially attractive and popular than people from lower ranks. Invitations to attend parties or to join fraternities, sororities, and clubs are a normal part of many youthful settings, and whether one is invited can have a powerful effect on one's self-esteem. Hidden in the background is the fact that all of these are affected by social class. It is the higher social classes that are much more likely to give parties; they have more money to spend on entertainment, and their lifestyle is much more oriented around acquiring the cultural skills for this sort of leisure mingling. Even though no overt class discrimination may be in anyone's mind, nevertheless, the socially popular youths tend to be those with the higher class backgrounds. All of this makes up a major part of the setting for the marketplace of sexual relationships. It is no wonder that young people often feel a good deal of strain in their lives.

Ironically, romantic love is itself produced by the experiences of being on this market. The market is the major reason that love is so dramatic. For many people, it is the single most exciting thing that happens in their lives, and it does seem to happen, at least once, to a large proportion of people today. Why should this be so? Because virtually everyone is on the sexual marketplace at some time in his or her life; in modern societies, one has to find one's own partner and negotiate one's own relationships. Thus, everyone experiences the ups and downs of the market.

There is the uncertainty about whom a person will meet and about whether a given pair will like each other. There is no guarantee that one will hit on a compatible person right away. Even if one does, the pushes and pulls of rivals and rejections in the larger market will usually make for a lot of momentarily raised hopes followed by disappointments. When one finally finds "the right person," there are still uncertainties. Will he call? What does she really think of me? Quarrels, misunderstandings, reconciliations, rivals, and rivals' rivals all come into play, spreading out into a network of men and women whom one has not even seen.

All these dramatic shifts and uncertainties make the experience an emotional one. Negative emotions—anxieties, anger, and the like—are set against the positive emotions that arise when a person finds someone compatible, someone who seems to reciprocate one's positive evaluation. Love, we have argued in the foregoing model, is produced by a type of social ritual. Its ingredients include, besides the privacy that focuses the energies of a small group of just two, the sharing of a common emotion. Where does that emotion come from? It comes from the experience of the market itself. The special feeling about the person whom one loves, which is so strong right at the first, freshest point in the relationship, is the joy that comes from having successfully negotiated a passage through the uncertainties of the market. After the inevitable buffeting about that comes from encounters with persons whose market resources are either too great or too small, a match is generally found. Not only are the two compatible, because having similar resources tends to make them socially similar, but also, the matchup of market resources means that both have the same motivation to stay together. The market bargaining, with its disappointments and blows to the ego, is over. A haven from the market has been found, and the feeling of joy and relief a couple experiences provides the first impetus for the ritual that generates the symbolic attachment of love. Love is a sentiment that arises on escaping from the market into a safe harbor.

LOVE MATCHES

Contrary to the old saying, love is *not* blind. As we have seen, it does not generally appear as a bolt from the blue or a tidal wave sweeping away all in its path. Rather, there is considerable predictability in who gets together with whom and how they go about doing it. Because people are so preoccupied with being "in love," they often ignore the fact that selecting sexual and marriage partners is still strongly influenced by social circumstances. To begin with, one generally must have close contact with another before falling in love with them (or even "in lust" with them). People who end up getting intimately involved generally meet where they spend most of their time: in the neighborhood, at school, on the job, at a friend's house, and so on. So the first thing shaping love matches is **propinquity,** or nearness. Not surprisingly, the places where people spend time and the people one comes into contact with tend to be divided according to social class and ethnicity. Because of residential and occupational segregation, as well as other cultural factors, on average, people get together with people like themselves.

Notice that this sociological explanation for similarity between romantic partners does not depend on people being outwardly prejudiced against those unlike themselves (though, as we saw in chapters 6 and 7, overt racism and class antagonism are still with us). On the contrary, people end up meeting potential romantic partners like themselves because social gathering places tend to be relatively homogeneous in terms of age, income, race, ethnicity, and, to a lesser extent, religion, political views, and cultural tastes. For example, consider the commercial establishments where two people might go out for a date or where two single people might meet. Although few such locations restrict attendance to just one type of person, most end up having a specific type of crowd. Most bars and clubs do not explicitly prohibit certain types of individuals from entering (except those who are under the legal drinking age). Nevertheless, the patrons select themselves on the basis of the reputation of the place, its location, its cost, the type of music it plays, the typical person who frequents it, and so forth. Similarly, anyone who buys a ticket can go to a sports event, but the average profile of those attending a professional football game, a wrestling match, or a tennis tournament would be quite disparate. And the crowds at various music concerts would be just as distinct, as evidenced by the stereotypical fans of country-western, heavy metal, alternative rock, hip-hop, pop, folk, or classical symphonies. The point is that the locations where people might take or meet romantic partners reflect specific social boundaries. Whether at clubs, bars, work, school, church, the gym, or private parties, people are likely to be surrounded by similar others (Laumann et al. 1994).

This tendency toward **homophily,** or association with others of similar and equal status, is one of the most consistent findings from past research on social relationships. People typically initiate and maintain relationships with those possessing comparable social characteristics, such as social class, age, race, education, and religion. Studies not only show that people choose to interact with others who have similar traits, attitudes, and values, but there is also evidence that people tend to adopt the opinions and attitudes of those with whom they interact on a regular basis (Axinn and Barber 1997). Thus, romantic homophily of the sort we have been discussing can be seen as both a cause and a consequence of patterns of segregation among social groups. Not only do "birds of a feather flock together," but "flocking together" also tends to reinforce and perpetuate attitudes about in-group and out-group members. Romantic relationships play an important part in this process. When people enter love relationships, they develop and reinforce a specific view of the world through their talk, engaging in a couple-based construction of reality (Berger and Kellner 1964). Extended family members also play an important role in shaping one's choice of love partners by reacting to potential mates. In fact, one of the most commonly used measures of race/ethnic prejudice assesses social distance by asking if one would object to having someone of a different group date or marry one's son or daughter. The assumption is that someone might feel OK about having a different ethnic group member live in the neighborhood but that allowing such a person to have sex with one's adult children or to marry into the family would be crossing a social boundary.

Even casual romantic encounters are governed by propinquity and homophily, so we should not be surprised to discover that even these temporary relationships tend to occur between similar others. Contrary to popular stereotypes about romantic "flings" with exotic others, homophily is almost as common for those who have short-term sexual relationships (under one month), as for those who have longer-term sexual relationships. Ninety-one percent of short-term sexual partners in a national random sample were found to be of the same race/ethnicity, 87 percent were similar in educational achievement, 83 percent were within five years in age, and 60 percent had a similar religion (Laumann et al. 1994). Although the common pattern is for people of similar race, ethnicity, social class, religion, and so on to get together, the rates of cross-category romantic relationships are on the rise. Even though most people continue to pick someone like themselves for romantic partners, increasing numbers of people ignore such social divisions. For example, it used to be rare for a person of one religion to date or marry someone of another religion. As discussed in the following text, this social barrier has broken down considerably in the past decades.

Cohabitation

One of the most dramatic trends in American courtship and marriage patterns in the past few decades is the increasing likelihood that people will cohabit. **Cohabitation** simply means living together, and though it technically refers to any two people living in the same household, the term is commonly used to refer to romantically involved couples who live in the same household. Such arrangements include heterosexual and homosexual couples, but they can be distinguished from households composed of roommates who are not romantically involved with each other. According to the Census Bureau, approximately half of all American women between the ages of 25 and 40 have cohabited (*Statistical Abstract* 1999, table 66).

As Figure 8.1 shows, fewer than one in ten women born before the 1940s cohabited with a partner, but over half of those who came of age in the 1980s and 1990s have done so. As an element of "mate selection," cohabitation is probably at least as commonplace today as "going steady" was 30 years ago (Nock 1995). What was previously a scandalous and shameful practice—it was called "living in sin"—has become normal. Note also in Figure 8.1 that men used to be much more likely than women to live with a sexual partner outside of marriage, but now the difference between men and women is negligible. This trend, coupled with findings from recent research, suggest that although women used to suffer negative consequences from living with nonmarital sexual partners, they may now actually benefit from living in cohabiting relationships (chapter 10). What we are witnessing is a transformation of the marriage system driven by larger changes in the social contexts for romantic partnerships. Though some pundits claim that this signals a breakdown in the American way of life (see *A Closer Look 1.1*), it is more accurate to see the rise of cohabitation, and the subsequent delay in marriage or remarriage, as a subtle transformation in patterns of heterosexual union formation tied to increasing gender equality. As we note later, nine of ten Americans continue to marry and will marry for the forseeable future—a rate much higher than that of most nations—suggesting that trends in cohabitation are not a threat to the institution of marriage.

Some see cohabitation as a part of the modern courtship process rather than an alternative to marriage, whereas others see cohabitation as an informal type of marriage (Schoen and Weinick 1993). We will consider the legal aspects of marriage and domestic partnerships in chapter 10, but for now, we can summarize some of the similarities and differences between the two forms of unions as described in previous studies. Cohabiting partners resemble single people in

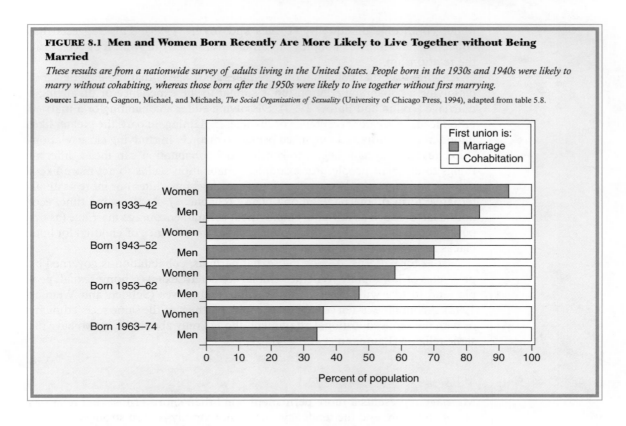

FIGURE 8.1 Men and Women Born Recently Are More Likely to Live Together without Being Married

These results are from a nationwide survey of adults living in the United States. People born in the 1930s and 1940s were likely to marry without cohabiting, whereas those born after the 1950s were likely to live together without first marrying.

Source: Laumann, Gagnon, Michael, and Michaels, *The Social Organization of Sexuality* (University of Chicago Press, 1994), adapted from table 5.8.

some ways and resemble married spouses in others. One of the ways that cohabitation differs from marriage is that it tends to be much shorter in duration. Though many marriages end, and some end abruptly, cohabitation is even more likely to end quickly, usually within two years. An important difference, however, is that many cohabitations end in marriage, rather than in separation or divorce. Nevertheless, most researchers report that cohabiting unions are less stable than marital unions (Nock 1995).

Consistent with the interpretation that cohabitation is more like singlehood, researchers find that young cohabiting couples are less likely than married couples to share finances or to own a home. Cohabitors are also less likely to have children or to want children in the near future (Axinn and Barber 1997; Schoen and Weinick 1993). Some cohabitors are also like some singles in rejecting the idea of marriage (Bumpass, Sweet, and Cherlin 1991). On the other hand, cohabitation, like marriage, involves coresidence between sexual partners and carries with it some of the same everyday demands and consequences. For this reason, cohabitation is sometimes considered to be a sort of trial marriage. In fact, it is now often part of the normal transition to marriage. At the same time, recent research shows that the more young people are exposed to cohabitation, the less enthusiastic they are toward marriage or childbearing (Axinn and Barber 1997). Cohabitation can thus act as a reality check, showing prospective partners what they might be getting into if they married. In part because of such experiences, researchers find that cohabitors are less likely to be committed to long-term relationships and more likely than others to terminate them (Nock 1995). In an interesting recent study, however, researchers discovered that cohabitors, unlike married couples, were more likely to remain together under conditions of equality (Brines and Joyner 1999). This suggests that cohabitors and married couples can differ substantially in what they seek from the relationship.

Another way that cohabitation resembles marriage is that the experience of dissolving a cohabiting relationship is similar to the experience of dissolving a marriage (Axinn and Barber 1997). Dissolving either a cohabiting or a married union leads to greater acceptance of divorce, explaining, in part, the greater likelihood of cohabitors and remarried people to divorce. In studying close relationships, researchers have begun to specify how cohabitation can mean different things to different people. For example, cohabitation seems to act more like a precursor or transitional stage leading to marriage for whites but more as an alternative form of marriage among blacks (chapter 7). At the same time, economic factors such as full-time employment seem to encourage marriage (as opposed to cohabitation) for both groups, as does the presence of children for both blacks and whites (Manning and Smock 1995).

Just as for patterns of selecting sexual partners, cohabitation is governed by principles of homophily. Whether unions are heterosexual or homosexual, people tend to live with others who resemble themselves (Schoen and Weinick 1993). About nine of ten cohabitors are with partners of the same race/ethnicity or with the same educational attainment, though only about one in two have the same religion (Laumann et al. 1994).

Marriage and Homogamy

Marriage represents a more permanent bond than short-term sexual relations or cohabitation, and the tendency toward homophily is even stronger than for sexual relationships or cohabitation. For marriage, the tendency to couple with like others is called **homogamy** (literally "same marriage"). People tend to marry others similar to themselves in terms of age, social class background, race, religion, education, and even in personal traits like body type and personality (Kalmijn 1998; Laumann et al. 1994). This certainly doesn't mean that everyone follows the pattern, because most of us can think of counterexamples: a tall person married to someone short, an African American with a European-American, a Catholic with a Protestant, or an older man with a younger woman. Some of these contrasts are even romanticized in contemporary films, like *Pretty Woman,* where a street prostitute (Julia Roberts) marries a rich businessman (Richard Gere). In real life, such extreme cross-class marriages are very rare, though we do see a gradual increase in the numbers of people marrying across various social categories, as described in the text that follows.

Who Marries Whom?

An old saying goes: "Don't marry for money. Go where the money is and marry for love." Consciously or not, this is the way it tends to work.

CLASS BACKGROUND Most people marry within their own social class. This is true whether one measures both the husband's and wife's class background by that of their parents or by their own occupations (using the wife's father's occupation as a measure of her social class if she does not work.) It is also true using education as a marker of social class. Not everyone marries into exactly the same class, of course, but when people move outside their social class, they usually do

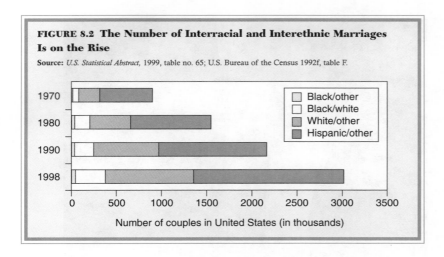

FIGURE 8.2 The Number of Interracial and Interethnic Marriages Is on the Rise
Source: *U.S. Statistical Abstract*, 1999, table no. 65; U.S. Bureau of the Census 1992f, table F.

not move very far; they are most likely to marry into the adjacent social group, up or down. The farther apart social classes are, the fewer people climb or descend that distance through marriage (Kalmijn 1998).

In most industrialized countries, there has been a decline in the importance of social class background for marriage. As young adults have become more independent, parents have exercised less direct or indirect control over the choices their children make. People also spend more time in school settings, where there tends to be a greater mix of class levels than in parental neighborhoods (Kalmijn 1998). Spending more time in school has also increased educational homogamy. Especially for the college educated, Americans are increasingly likely to meet their spouse in school, ensuring that spouses will have approximately equal levels of education, at least in the beginning of the relationship.

When there is a difference in the social class or education level of spouses, the difference tends to be in a predictable direction with regard to gender. That is, women have tended to use marriage to improve their overall social standing, a practice referred to as **hypergamy** or "marrying up." Although the practice is weakening, women still tend to court and marry men with higher social standing and resources than themselves. In other words, some women still use marriage as a path to financial security and upward mobility. It is likely that the ways women and men approach marriage will become even more similar in the future, provided that women's resources and opportunities on the job market continue to improve.

RACE Most marriages occur within the same racial group. Some of this, of course, may not simply be due to the marriage market. Seventeen states still had laws prohibiting **miscegenation** (interracial marriage) when the U.S. Supreme Court declared such laws unconstitutional in 1967. Even today, over 90 percent of all marriages take place within the same race. Nevertheless, the number of interracial and interethnic marriages has increased significantly since 1970 (Figure 8.2), and attitudes toward intermarriage have become more favorable.

Contrary to the common image, marriages between blacks and whites are not the most common type of interracial or interethnic marriage. Less than 1 percent

The majority of black-white intermarriages are between black men and white women, unlike this couple.

of all marriages are black-white intermarriages. The most common type of interracial marriages, in fact, are not very highly publicized. Native American women are the most likely to intermarry (over 50 percent of their marriages are interracial). Next come Asian American women, with over 40% of those born in the United States marrying someone from outside their ethnic group (see figure 7.5). The majority of these interracial marriages are with white men, though intermarriage among Asian subgroups is on the increase (Kitano and Daniels 1988; Shinagawa and Pang 1996).

When intermarriages do occur, there are very distinctive patterns by gender of the out-marrying partners. Sixty-four percent of the black-white marriages in existence in 1998 were between black men and white women. In Asian-white marriages, the situation is almost the reverse, with mostly white men marrying nonwhite women. For Native American–white marriages, the ratios are about even. How can these patterns be explained? As we discussed in the last chapter, some may be due to the experience of American military men abroad, who brought home Japanese, Filipino, or Vietnamese brides. The question remains why they would be particularly likely to marry Asians. It is sometimes suggested that Asian cultures encourage women to be more passive and subservient to males; hence, American men who marry them might be looking for more dominance in their marriages than they could get from other brides. Some have also suggested that out-marriage rates are higher for Asian-American women than for Asian-American men because the women can exercise more independence if married to non-Asians (Kitano and Daniels 1988).

The black-white pattern is more mysterious. We do know that these marriages are most likely to occur among highly educated persons. Black men who have attended graduate school are the most likely to have interracial marriages (Carter and Glick 1976; Kalmijn 1999). But why is there less of a tendency for highly educated black females, or white males, to marry interracially? Moreover, it is not just a matter of availability. There is a fairly severe imbalance of numbers in the black population, with more females than males; hence, the intermarriage ratio is just the opposite of what one would expect. The higher tendency of black males to marry outside their race reduces even further the chances of black females marrying (see *A Closer Look 7.4*).

One explanation may be that our cultural ideals of beauty have been based on white standards. Images of thin young Caucasian models predominate in the media, and although this may be changing somewhat, the stereotype of female beauty tends to be defined in terms of white features and styles.

Another possible explanation may be that there is a particularly strong distinction between male and female cultures in the black community and a particularly vehement battle for gender domination. African-American women, as we have seen (chapter 7), have a tradition of independence; they are not very subservient to men. This may contribute to their own lower rate of intermarriage and to the tendency of black males to seek wives elsewhere.

RELIGION There used to be a strong tendency to marry within the same religious faith. Seventy-nine percent of Catholics, 91 percent of Protestants, and 92 percent of Jews married someone of the same religion in the 1960s (Carter and Glick 1976). This sounds like Catholics were most likely to marry out and that Protestants and Jews were both equally **endogamous;** however, we must take into account the fact that these groups are of different size and that the sheer chances of marrying someone from a large group are much greater than from a small group. Protestants are by far the largest religious group in the United States (about 70 percent of the population), whereas Catholics constitute about 26 percent and Jews approximately 4 percent. Thus, Protestants marry other Protestants slightly more than we would expect just by chance, by a ratio of about 1.7 to 1. Catholics marry within their religion at an even higher rate, about 5 to 1 over what would happen purely randomly; and Jews marry in at a ratio of about 52 to 1.

Today, religious intermarriage rates are much higher for all groups except Fundamentalist Christian groups (Kalmijn 1998). For example, Jews marry non-Jews about 40 percent of the time (Krantowitz and Witherspoon 1987). Some researchers have found that religious intermarriages are less stable (i.e., more likely to end in divorce) (Bumpass and Sweet 1972; Glenn 1982). This is probably because those who marry someone outside of their own faith are also less bound by a variety of traditional norms. However, interreligious marriages are not less happy than others. Recent research indicates that Catholics who marry someone of a different religion are just as likely to report being happily married as Catholics who marry Catholics (Shehan, Bock, and Lee 1990).

PERSONAL TRAITS People tend to marry others who resemble themselves in quite a number of ways. The most prominent feature is the degree of physical attractiveness (see *A Closer Look* 8.5). Married partners tend to be relatively similar physically, correlating in weight (given, of course, the fact that men tend to be heavier), height, hair color, condition of health, and even basal metabolism (Burgess and Wallin 1953; Murstein 1972; Patzer 1985). Some of these features are probably artifacts of the factors mentioned earlier. Hair color, for instance, is partly an index of race and ethnicity; and height, weight, and health are correlated with social class. Homogeneity even extends to psychological characteristics. Husbands and wives tend to correlate in IQ scores at a level of about .50, which is approximately the same as the correlation between the IQs of siblings (Taylor 1980). Because there is evidence that a good deal of IQ is cultural (chapter 12), this probably indicates, more than anything else, that people are attracted to each other because of similarity in what we would call their "cultural capital."

CONCLUDING THOUGHTS ON LOVE AND MARRIAGE

The foregoing analysis shows that people continue to fall in love with, cohabit with, and marry similar others. Our summary of recent trends also shows that some of the norms surrounding *who* one should love, live with, or marry have loosened over the years. Normative restrictions governing *whether* or *when* one should take a lover or cohabit with a romantic partner have also changed— sometimes dramatically. Does this also mean that marriage is weakening or might disappear? As we suggested earlier, our assumption is that marriage is under no threat of disappearing. The timing and content of marriage, as you will see in future chapters, are indeed undergoing some significant shifts. In addition, the laws and customs surrounding marriage and domestic partnerships are similarly undergoing some important changes. At the same time, the basic institution of marriage has remained remarkably stable.

Contrary to popular rhetoric from politicians and religious leaders about the disappearance of marriage, the number of married people in the United States continues to grow. Between 1970 and the late 1990s, the number of married people actually increased by more than 20 percent. At the same time, the proportion of all adults who are married has declined somewhat, primarily because people marry later and divorce more, because many people now live together for awhile before getting married, and because a greater number of people are choosing not to marry at all. Nevertheless, the majority of Americans are eager to try marriage at some point in their lives. Over 90 percent of Americans still get married, which is down from about 95 percent during the middle part of the century.

Marriage rates vary according to many factors, including trends in fertility and population size, average age at marriage, school attendance, labor force participation, economic prospects, social conditions, and cultural traditions. For example, the stagnation of men's wages during the 1970s and 1980s made marriage less affordable for many men, and marriage rates dipped, just as they had during the depression in the early 1930s (Cherlin 1992). Most people are pragmatically taking employment and earnings into account when deciding who and when to marry (DeVita 1996). Because African American men are less likely to find good paying jobs than white men, it is not surprising that rates of marriage for African Americans, especially women, are among the lowest in the nation (chapter 7). When African American men have access to good paying jobs, however, the chances of getting married and staying married increase dramatically (Mason 1996).

Most demographers project that over 90 percent of Americans will continue to marry in the future (DeVita 1996). The overall marriage rate in the United States, although lower than it was in the 1970s and 1980s, is still much higher than those of most other industrialized countries (DeVita 1996; Sorrentino 1990). At the same time, our divorce rates are almost twice as high as other countries' (chapter 14). Americans are thus more likely to marry but also more likely to divorce than people in other industrialized countries. This general pattern has existed ever since reliable cross-national statistics have been recorded.

As we will see in chapter 14, most people who divorce also remarry, with about a third of all Americans expected to marry, divorce, and remarry (Cherlin and Furstenberg 1994). Remarriage rates have exceeded marriage rates for more than 50 years with about half of all U.S. marriages now remarriages. Though remarriage rates are still high, they have been decreasing since the 1960s. Today, about two-thirds of separated and divorced women will remarry, compared with more than four-fifths in the 1960s (Cherlin and Furstenberg 1994). Remarriage rates also differ between ethnic groups: about half of white women are expected to marry within five years of their separation, as compared with one-third of Mexican American women and one-fifth of African American women (Sweet and Bumpass 1987).

The main point to remember from this chapter is that we cannot understand love, romance, courtship, cohabitation, or marriage without looking at how they are shaped by larger social and cultural forces. As much as it goes against romantic sensibilities, we cannot understand how love and marriage are socially organized without invoking some notion of bargaining and markets. Gender is particularly important to our understanding of such things, because men and women learn to feel differently about love and marriage, and because they experience different opportunities on the romantic market. Not only are they exposed to different messages about love from the culture, but they are encouraged to play different roles in scripted courtship rituals. In later chapters, we will see how gender differences are also promoted by expectations about breadwinning, emotional attentiveness, housework, child socialization, and other institutional influences on family roles. First, however, we will explore in more detail how sex plays an important part in the formation and maintenance of human family life.

SUMMARY

1. Feminist theories regard love as a bond that has tied women into domestic roles and kept them from economic independence. Traditionally, women have had to appear sexually attractive but control their emotions to acquire a man of the right social status.

2. Equity theory predicts that people fall in love when they feel they are making a "fair exchange" of resources and traits.

3. Ritual theory analyzes love as based on face-to-face interaction, with an intense focus of attention on each other, excluding all outsiders; this intensifies the lovers' shared emotions and results in symbols representing common membership in the dyad. Love is thus a kind of private cult of mutual worship and idealization.

4. Because love produces strong bonds of attachment, amounting to emotional property, small symbolic violations are likely to provoke righteous anger. For this reason, love easily turns to hate.

5. Men are more likely to fall in love at first sight and act on spontaneous feelings of love. Because more of their social and economic status has depended on marriage, women have done more of what Hochschild calls "emotion work" to shape their emotions toward the appropriate man.

6. People looking for romance sometimes advertise in newspaper or magazine "personal ads" for the characteristics they seek in a romantic partner. Ads written by men are generally different than those written by women. Men's ads are likely to seek youth and particular sexual characteristics, offer financial security, and mention interest in a casual relationship. Women's ads for men, on the other hand, seek sincerity and permanent relationships, while offering information about their own personalities, physical appearance, and youth. Women's ads tend to emphasize status and success in a partner rather than attractiveness, a trait which is more often sought by men in ads for women.

7. In the romantic market, research has shown that "less desirable" men approach women more desirable than themselves more often than "less desirable" women approach men more desirable than they consider themselves to be. This illustrates our cultural ideal of masculinity, which promotes a sense of entitlement among men and encourages them to expect more from women than women feel the men deserve.

8. Cohabitation, once considered a scandalous and shameful practice, has now become normal. As a union formation style, cohabitation has characteristics of both marriage and singlehood. Like marriage, cohabitation involves coresidence between sexual partners and thus entails a number of the same everyday demands and consequences as marriage. On the other hand, similar to single individuals, cohabitors are less likely than married couples to share finances, to own a home, or to have children. As with others forming sexual unions, cohabitators tend to be homophilous, living with partners of the same race/ethnicity and similar educational attainment.

9. The experience of finding the right person on the marriage market can involve much uncertainty and many bruises to one's ego. The emotion of joy and relief at finally finding a match is a prime ingredient that goes into the ritual producing love.

10. Most marriages match persons who are similar in social class background, race, religion, and many personal traits.

11. These similarities in marriage partners can be explained by the workings of the marriage market, which enables people to exchange mainly with others who have social and personal resources similar to their own, in competition with the "opportunity pool" of persons whom they know.

Key Terms

cohabitation	equity theory	interaction ritual
courtship	homogamy	miscegenation
emotion work	homophily	propinquity
endogamous	hypergamy	

Sociology Web Site

See the Wadsworth Sociology Resource Center, "Virtual Society," for additional links, quizzes, and learning tools:

http://www.sociology.wadsworth.com

Also on this web site you'll find InfoTrac College Edition, an online library of journals. Here you can search for electronic articles about central topics in sociology.

9 Erotic Ties: Marital and Nonmarital Sex

INTRODUCTION

What is sex? The answer seems obvious, but once we begin thinking about the subject, the possibilities seem limitless. Sigmund Freud (1924) once commented in a lecture that sex is so hard to define that he finally concluded it is whatever we consider to be shameful or dirty. He was joking, but only partly. For the rather Victorian audiences of the early twentieth century, sex was a uniquely shameful, unmentionable subject. It is no longer unmentionable, but it is still surrounded by a tension of its own that makes it different from anything else in our lives.

Why did Freud say that sex is so hard to define? From a biological viewpoint, sexual behavior can be seen as that which leads to procreation. But what about kissing, or masturbating? These are obviously sexual, so Freud pointed to the need for a wider definition. What about anything having to do with the genitals or erogenous zones? The hand is not an erogenous zone, although holding hands, especially the first time a couple touches one another, can have tremendous erotic significance at the moment, as can stroking someone's hair. Freud pointed out, moreover, that a common theme in many of his neurotic patients was that they had eroticized certain objects in their environment while repressing and displacing their erotic drives for the genitals themselves. For some patients the foot was the prime object of sexual arousal. Others had made a fetish out of an object of clothing or had a neurotic phobia of snakes or small dogs because they unconsciously symbolized a penis. These examples come from the realm of psychiatric symptoms, but Freud held that the borderline between the normal and abnormal in matters of sex was not very wide. Whether or not we accept the details of Freud's theories, it is not hard to think of examples of normally accepted sexual symbols: a woman's spiked-heeled shoes are considered sexy in our culture, although technically speaking, one might call this a shoe fetish.

In short, we must distinguish between the biological and the cultural. Recall the discussion in chapter 1 regarding the terms *sex* and *gender. Sex* refers to the biologically given genitals and secondary characteristics (facial and body hair, pelvic size, musculature, etc.) that make one male or female. Most persons have a distinct sex, although there are occasional hormonal mixtures and even cases where a person's genitals are partly of one sex, partly of the other (hermaphroditism). *Sex* also refers to behavior related to eroticism and procreation. *Gender,* on the other hand, is a social role, a learned way of acting masculine or feminine, which sometimes includes performing erotic or sexual behavior according to normative scripts for men or women.

BIOLOGICAL VERSUS SOCIAL
EXPLANATIONS FOR SEXUAL PRACTICES

Biologists and evolutionary psychologists tend to see men's and women's reproductive systems as sources of inevitable difference in all other aspects of their lives (Buss 1998; Ericksen 1999; Laqueur 1990). In this approach, fundamental sexual differences are assumed to lead males and females to experience intense physical attraction for the "opposite sex," to pursue distinct "mate selection strategies," to form heterosexual unions, and to express a wide range of other "sex-linked traits." Notice that this line of reasoning posits heterosexuality as the only normal or natural form of sexuality. Even Freud, who accepted that everyone had sexual attraction for both sexes, considered it normal and natural (though sometimes difficult)

Affectionate gestures and embraces can be erotic for intimate couples.

for the developing child to transfer erotic desires to the "opposite" sex. As Julia Ericksen (1999) points out, if only a handful of sexual deviants engaged in homosexual practices, this belief in the biological inevitability of heterosexuality would remain unchallenged. If, on the other hand, researchers showed homosexuality (or even bisexuality) to be widespread in the population, it would call into question the assumption that men and women are natural sex partners. These social and political contexts, along with other cultural and religious biases against homosexuality and taboos against acknowledging sex, have limited the amount of research on Americans' sexual practices.

Because most scientific research on sexuality has been conducted by biologists and psychologists, it has focused on sexual behavior purely as an "individual level" phenomenon (Laumann et al. 1994). Such research generally tries to find causes for variation in sexual activity, defined as thoughts, feelings, and physical actions of individuals. According to Laumann and colleagues (1994, 3), a good example of this type of approach is the study of sexual drives or instincts. "In drive theories, people are assumed to experience a buildup of 'sexual tension' or 'sexual need' during periods of deprivation or during particularly erotic environmental stimulation. When sexual activity is experienced, the drive is satiated and the need reduced." Note that humans in such models of behavior are assumed to respond almost automatically to various environmental or internal stimuli based on underlying biological predispositions or drives. The goal of such research has typically been to identify variation in individuals' sexual behaviors and to attribute it to biological or psychological differences in the individuals themselves.

As Laumann and his colleagues (1994) point out, the major shortcoming of these individualistic approaches to studying sexuality is that they can explain only a very small part of the variations they observe. Whatever the biological underpinnings of human sexuality—and obviously there are important physiological processes at work—an individualistic paradigm for studying sexuality simply cannot comprehend the multitude of cultural and social factors influencing human sexual interaction (see *A Closer Look* 9.1). The sexual behavior of some species appears to be determined by biology (e.g., salmon, who are genetically programmed to swim upstream and to spawn during the appropriate season), whereas human sexual behaviors are much too variable and shifting to be considered fixed by biology.

Although the observation that the social environment shapes sexual practices seems fairly obvious to most social scientists today, it is only relatively recently that researchers have seized on this insight and used it to design their studies of sexuality. Only anthropologists have tended to appreciate cultural variation in sexual practices as natural and inevitable. Over the years, anthropologists have documented how many societies combine forms of homosexuality and heterosexuality in their normal sexual practices and how a few societies even recognize more than two sexes (e.g., *berdache* or two spirits). For heterosexual practices, the range of variation among societies is much wider than simple biological explanations would allow. In some societies, women initiate sex, whereas in others they do not. In some, women act intensely aroused and active during sex, and in others they have no concept of orgasm. In some settings, anthropologists found that when they told women about orgasms, they refused to believe they even existed (Schwartz and Rutter 1998). Men's sexual practices are similarly influenced by varying cultural beliefs, with widely different approaches to men's ejaculations. Some societies consider it healthy for men to ejaculate early and often, whereas in others, men are told to conserve semen and ejaculate as rarely as possible or even to ritually ejaculate into another man's mouth so that none of the precious fluid will be lost (Ford and Beach 1951).

In cataloguing the sexual experiences of so-called primitives, many anthropologists and other researchers in the nineteenth and early twentieth centuries adopted and perpetuated biases of the dominant cultures they represented. Often, Europeans characterized colonial subjects as sexually uncontrollable or perverse, and many also took advantage of them sexually (e.g., Kath Weston [1998, 19] speculates, "it's a sure bet that adolescents who tease their friends about a sexual repertoire limited to the 'missionary position' seldom have in mind the centuries of violence and religious/political repression interred in that phrase.") Recent critics have suggested that early anthropological research, in spite of documenting a wide range of cultural sex practices and beliefs, rendered them exotic and downplayed the overall importance of sexuality to the organization of society (Weston 1998).

Much previous sexuality research has also been shaped by the underlying idea that there is an inevitable negative conflict between the biological nature of human beings and the cultures in which they are reared. At least for Western cultures, most researchers have assumed that social factors function solely to inhibit or constrain people's intrinsic sexual desires and urges; for example, biologically mature adolescents will naturally want to have intercourse, so those who do not are simply better at "controlling their urges." According to Laumann (1994, 5), the problem with such an approach is that it takes a narrow view of the role of social processes as merely constraining sexual conduct. In fact, social factors shape what people perceive to be sexual, and the conceptual lenses of one's

A CLOSER LOOK 9.1
A *Famous Literary Love Scene*

Premarital sex has a power in real life that is scarcely hinted at by bare sociological statistics. It is a matter of passion and adventure, for many people perhaps the most dramatic experience of their lives. It is no wonder then, that it is the subject of some of our most famous literary works. The following is a notable literary love scene from James Joyce's *Ulysses*. An Irish woman lies in bed next to her sleeping husband, as thoughts of their courtship go through her head:

so there you are they might as well try to stop the sun from rising tomorrow the sun shines for you he said the day we were lying among the rhododendrons on Howth head in the grey tweed suit and his straw hat the day I got him to propose to me yes first I gave him the bit of seedcake out of my mouth and it was leap-year like now yes 16 years ago my God after that long kiss I near lost my breath yet he said I was a flower of the mountain yes so we are flowers all a woman's body yes that was one true thing he said in his life and the sun shines for you today yes that was why I liked him because I saw he understood or felt what a woman is and I knew I could always get round him and I gave him all the pleasure I could leading him on till he asked me to say yes and I wouldn't answer first only looked out over the sea and the sky I was thinking of so many things he didn't know of Mulvey and Mr Stanhope and Hester and father and old captain Groves and the sailors playing all birds fly and I say stoop and washing up dishes they called it on the pier and the sentry in front of the governors house with the thing round his white helmet poor devil half roasted and the Spanish girls laughing in their shawls and their tall combs and the auctions in the

morning the Greeks and the jews and the Arabs and the devil knows who else from all the ends of Europe and Duke street and the fowl market all clucking outside Larby Sharons and the poor donkeys slipping half asleep and the vague fellows in the cloaks asleep in the shade on the steps and the big wheels of the carts of the bulls and the old castle thousands of years old yes and those handsome Moors all in white and turbans like kings asking you to sit down in their little bit of a shop and Ronda with the old windows of the posadas glancing eyes a lattice hid for her lover to kiss the iron and the wineshops half open at night and the casanets and the night we missed the boat at Algeciras the watchman going about serene with his lamp and O that awful deepdown torrent O and the sea the sea crimson sometimes like fire and the glorious sunsets and the figtrees in the Alameda gardens yes and all the queer little streets and pink and blue and yellow houses and the rosegardens and the jessámine and geraniums and cactuses and Gilbraltar as a girl where I was a Flower of the mountain yes when I put the rose in my hair like the Andalusian girls used or shall I wear a red yes and how he kissed me under the Moorish wall and I thought well as well as him as another and then I asked him with my eyes to ask again yes and then he asked me would I say yes to say yes my mountain flower and first I put my arms around him yes and drew him down to me so he could feel my breasts all perfume yes and his heart was going like mad and yes I said yes I will Yes.

From *Ulysses* by James Joyce. Copyright © 1934 and renewed 1962 by Lucia and George Joyce. Reprinted by permission of Random House, Inc.

culture, upbringing, and social network provide frameworks for perceiving and understanding sexual fantasies and thoughts.

It is misleading to assume that biology "causes" specific behaviors or that natural human "sexual instincts" must be kept in check by societal restrictions. Although human activity is organized, to some extent, by the physical equipment men and women possess, the social world influences biology at least as much as the reverse. Biological factors matter for sexual behavior, but they play only a small role in determining what those specific behaviors will be and how they will be interpreted. As we have noted, even though we tend to think of sex as something personal, private, and natural, it is heavily dependent on prior socialization and current social context. For example, one of the most consistent findings from past research is that the sexual practices of working-class people vary in predictable ways from those of the upper-middle class. Another consistent finding that argues against a simple biological model is the fact that adolescents involved in religious activities delay first intercourse longer than those who are not religious (Laumann et al. 1994).

Just as for love and romance, we can better understand sexuality by focusing on how it is organized according to specific historical and social contexts. The French intellectual Michel Foucault (1978) suggested that when feudalism declined and the bureaucratic state expanded in Europe, sexual identity arose as a mechanism for monitoring individual behavior. As cities grew and society became more impersonal, Foucault argues that people learned to control their own behavior by checking on the normality of their acts. What was previously public became more private, and as science replaced religion as the principal source of knowledge, a wide range of experts began giving advice about normal sexual activities. Citizens in earlier villages and towns policed the intimate details of each others' lives, whereas the growth of privacy and the anonymity of cities afforded people more opportunity for illicit sexual activity. According to Foucault, new ways to regulate sexuality emerged, a new emphasis on normality developed, and guilt replaced shame as a mechanism for controlling sexual practices. Although Foucault wrote little about gender, other scholars influenced by his views have emphasized that ideas about normal sexual behavior were also extremely gendered (D'Emilio and Freedman 1988; Ericksen 1999; Foucault 1978; Laqueur 1990). As noted in chapters 3, 4, and 8, the sexual and romantic scripts that emerged for women in Western cultures differed substantially from those of men.

Sex As an Interaction Ritual

Is there any sociological significance in research on what happens during sex? One pattern that stands out is the underlying similarity of the sexual response in men and women. The sex researchers Masters and Johnson (1966) found that in more intense sexual experiences, there is the same sexual flush over the body and the same postorgasmic reaction. Timing of rhythmic contractions up to and during the orgasm itself is remarkably parallel in men and women. One recent study showed that men and women use nearly identical terms to describe the feeling of orgasm (Hatfield and Rapson 1996). All this suggests that males and females are fundamentally similar beneath the surface, although the similarity can be covered over by the way their lives have been socially organized.

We are reminded that embryonically, both sexes start out the same, and only in the third month does the undifferentiated genital tissue shape into male or female organs. Thus, the penis is shaped from the same embryonic tissue as the clitoris, the scrotum is analogous to the labia minora and labia majora, and there are vestigial breast nipples in the male. Intense sexual arousal strips off the social differences, so to speak, and brings a man and woman closer together in a common pattern they share as human beings.

We can understand sexual intercourse as an intense kind of interaction ritual. We have met this line of sociological theory, deriving from Durkheim and Goffman, at various points earlier in the book. It was used in chapter 6 to examine the emotional and moral consequences of social class differences in the "ritual density" of interaction. Even closer to our concerns here, the theory was spelled out in chapter 8 to analyze love as a powerful interaction ritual creating a symbolic bond between two persons. Sexual intercourse, on a more physical level, also has the ingredients of a naturally occurring interaction ritual.

Let us recall the components that go into generating a ritual. A group is assembled; they focus their attention on the same activity; and they share a common emotional mood, which builds up as they become increasingly aware of each other's focus on the same thing. If all this happens, the result is a feeling of group membership, attached to symbols that are respected because they represent the relationship. Sexual interaction certainly has the elements of this: it brings together a little group of two persons; they focus attention on each other's bodies; and there is a common emotion of erotic arousal.

It is well known, of course, that both partners are not always equally aroused. One may be dominating the lovemaking while the other is passively following along or even just lying there letting the partner "get off." A couple are not always in the same mood or even on good terms with each other at the moment. Though the basic ingredients are there, sexual intercourse does not always build up the intensity that makes it a successful interaction ritual. However, this is true of other interactional situations; sometimes an intense ritual focus is achieved and sometimes not. The key is whether a cumulative process is set in motion, so that the mutual focus gets stronger as the interaction goes along.

Sexual intercourse has a somewhat unique quality, though, as compared with other kinds of potential interaction rituals; if the physical bodies can start getting attuned, there is a kind of physiological escalator that will pull the partners together more and more intensely as they go along. The right emotional mood helps to get it started; so does the right physical arousal—of the genitals or other erotic zones. Often a couple has their own private arousing gestures—not necessarily overtly erotic, but just a touch of the hair or the face, perhaps—that have come to symbolize their own personal connection to each other. (That is to say, they carry over symbolic gestures charged with meanings from previous interaction rituals.) On the other hand, although the lovemaking couple may be psychologically comparative strangers to one another, they may turn each other on enough physically so that the physiological escalator starts taking off of its own accord.

On the sheer physical level, sexual intercourse has a built-in rhythmic mechanism. If it can get started, the rhythm can take over on its own. Furthermore, this rhythmic sexual arousal is contagious from one partner to the other. The intensity of one partner building up to an orgasm spills over to the other partner and builds up his or her intensity in turn. That is why lovemaking couples are usually concerned about each other's sexual arousal; even if they are not in love, and the sex is engaged in fairly casually or even selfishly, each person's own arousal is usually more intense if the other person is aroused. Sophisticated lovers try to turn each other on because there is a kind of self-interested, as well as altruistic, interest in the other's orgasm.

When and if this happens, intercourse truly becomes an interaction ritual, on an unintentional, physical level. The contagious sexual arousal is like the emotional contagion that happens in more ordinary rituals, pulling in the participants' attention and making them more and more sharply focused on their common experience. It is no wonder, then, the sexual intercourse itself becomes a powerful symbol of the social bond and even gets invested with a "sacred" quality. It becomes more than a matter of individual pleasure (though it is that, too); it becomes an emblem of the relationship between those two persons. Sex comes to symbolize love. It comes to symbolize the possessive aspects of a marital (or quasi-marital) relationship, too. Lovemaking is an exclusive possession, and its violation—or even jealous thoughts about it—becomes the most serious of all transgressions.

To be sure, not every act of sexual intercourse between a couple actually builds up to this level of intensity. Much of the time, probably, one or the other partner is not so aroused. Their attention may be wandering from their bodies, and there is little mutual build-up of contagious physical or emotional effects. Total absorption does occur at least occasionally, and people orient themselves toward those high points. However infrequently this level is achieved, sexual intercourse nevertheless is the most potent symbol of an intense social tie. Even comparatively unsuccessful acts of lovemaking become significant representatives of the couple's relationship. Whether they are only maintaining their marital routine or changing their relationship for better or worse, it is likely to be felt symbolically in their lovemaking.

MEASURING TRENDS IN SEXUALITY

It is often claimed that there has been a sexual revolution in the last few decades. Undoubtedly, many things have changed, particularly in the area of premarital sex. College dormitories housing both males and females were introduced only in the 1970s; before that there were numerous regulations concerning hours during which members of the opposite sex could be in one's room, whether the door had to be kept open or others had to be present, or the like. In the 1950s and before, even regulated intersex visiting was prohibited entirely on many campuses. Despite all that, the change has not necessarily been so drastic as it appears. There had been a great deal of change already throughout the earlier part of the century, and our grandparents and great-grandparents may not have been completely restrictive in their sexual behavior, although they were not very open about it. On the other hand, premarital sex is not the *official* standard even today. Very few countries (e.g., Sweden) do not have laws that officially confine sex to within the bonds of marriage. Most states prohibit sex below age eighteen, except for married partners, as "contributing to the delinquency of a minor." These laws are not enforced much, although numerous people believe they are morally correct, and teenage girls (but not boys) are sometimes arrested on charges of delinquency if they are caught committing sexual intercourse.

It would be more accurate to say that the United States today has a variety of sexual standards and practices and that different groups are often in conflict. Surveys of the American public in recent years show that 38 percent believe that premarital sex is not at all wrong, and about 23 percent think it is sometimes right or wrong, depending on the circumstances (Christopher and Sprecher 2000; Smith 1994). In other words, the country is split, with a little over a third at each extreme and the rest in the middle. These percentages have stayed about the same since the mid-1980s, but they indicate much more acceptance of nonmarital sex than was common for most of the century. In 1963, more than 75 percent of the populace stated that premarital intercourse was always wrong (Reiss 1967), although, as we shall see, most of them did not live up to their own statements. Today, the division is sharply along age lines. Only about one in six men in their twenties and one in five women that age believe that premarital sex is always or almost always wrong. In contrast, one in three men over fifty and one in two women over fifty believe that premarital sex is always or almost always wrong (Laumann et al. 1994).

Words Versus Actions

Perhaps the clearest difference between today and earlier times is that sex is talked about more openly. In the Victorian period, anything involving sex was unmentionable, although a surprising amount went on behind the backs of polite society. This was a hundred years ago, but the situation lasted for most of the century in the United States, which has been in many ways an especially puritanical society. Farther back in history, however, there were societies in which sexual matters were much more open; for instance, public rituals of some tribal societies featured sexual interaction.

A notable change that has occurred in the last few years is a major increase in public talk about sex. Popular magazines are full of discussions of it; men's magazines are openly pornographic; movies and some television channels have explicit eroticism; advertisements place tremendous emphasis on the sexy; and formerly taboo four-letter words are commonly used by both men and women (although at the same time, some persons strenuously object to this). However, all of these references are impersonal. People may sometimes allude to their own sexual experience, but it is still not quite an ordinary topic of conversation. One does not often hear people talking about exactly what happens when they make love, even with their own partners; in fact, this lack of explicit communication has been cited by sex therapists as a common cause of sexual difficulties.

Following are some questions worth keeping in mind as we examine the data. Is there more sex today or only more talk? Has the kind of sex changed? Is there still a dual standard for males and females, so that what is allowed to be experienced by one is prohibited for the other? We shall see some trends, but at the same time, we should be aware of the possibility that trends are not all moving one way, that there are strong differences among social groups, and that there are currently ideological battles over conflicting standards for sexual behavior.

Sex Surveys

Getting reliable information on sexual behavior is a problem that has still not been completely solved. Standards of greater openness in talking about sex make it easier to do surveys today, but the fact that these standards have changed recently means that older information may not be strictly comparable; previously, people may have been more inclined to hide what they were doing. Even today, there is considerable disagreement about the appropriateness of discussing sex, so one's sample may be badly biased toward those who are willing to talk about it. In 1974, *Playboy* magazine carried out a national survey of sexual behavior; 80 percent of those contacted refused to cooperate (Hunt 1974). Another *Playboy* poll (Petersen et al. 1983) was done simply by having readers who felt like it mail in questionnaires torn out of the magazine. This resulted in 80,000 responses, which is by far the largest number of people ever surveyed on sex, but they are obviously among the segment of the populace most interested in it. The fact that 60 percent of the *Playboy* respondents almost always engaged in oral-genital sex, for instance, or that 35 percent had had sex with more than one person at the same time probably does not mean that the rest of the population was doing the same.

This problem of biased sampling can be overcome by selecting respondents randomly from the population. Such surveys are attempted most often of women of childbearing age in connection with attempts to estimate the birthrate (e.g., see the figures in Table 9.1, which summarizes some results from the National Survey of Family Growth sponsored by the National Center for Health Statistics). This sample not only leaves out men's sexual practices but also tends to yield relatively superficial data. A stranger asking persons a few questions about their sex life is likely to get perfunctory answers, skewed toward what is the conventionally respectable thing to say in a particular social group. For this reason, really good, in-depth surveys of the sex lives of the general population were comparatively rare until the 1990s. Before then, one of the best sources of data was the Kinsey reports (Kinsey et al. 1948, 1953), carried out by the Institute for Sex Research at Indiana University.

Alfred Kinsey was not a sociologist but a zoologist, who in 1937 realized that he knew more about the sexual behavior of insects than anyone did about the sexual behavior of humans. Between 1938 and 1950, Kinsey and his colleagues

Table 9.1 Federal Government Statistics on Women's Sexual Partners

The National Center for Health Statistics collects data on the number of sexual partners women have to estimate future fertility rates and to assess the potential for spread of sexually transmitted diseases (STDs). This table, excerpted from the 1999 *Statistical Abstract of the United States* shows that almost half (45.8 percent) of women under the age of twenty had not engaged in voluntary sexual intercourse with a male but that more than half of women between the ages of twenty-five and forty had experienced sexual intercourse with four or more partners.

Women	*Number of partners in lifetime by percent distribution*							
	0*	1	2	3	4	5	6–9	10+
Age								
15–19 years old	45.8	19.4	11.0	7.9	3.6	2.9	5.7	3.7
20–24 years old	11.0	21.6	13.4	11.5	9.0	7.7	11.9	13.9
25–29 years old	4.3	22.0	13.5	10.1	9.2	10.1	13.8	16.9
30–34 years old	2.8	23.4	11.8	9.1	11.5	9.7	13.7	18.0
35–39 years old	1.8	24.3	11.7	9.4	8.5	9.3	14.5	20.5
40–44 years old	2.4	29.4	12.6	9.7	8.0	8.4	11.7	17.7
Married	0.5	34.5	13.8	9.9	8.9	8.0	11.2	13.2
Unmarried	20.2	12.8	10.9	9.4	7.9	8.2	12.8	17.7
Education								
No high school diploma	1.3	27.7	15.8	10.1	7.6	10.1	11.7	15.8
High school diploma	2.5	23.5	13.0	10.8	9.8	9.6	13.7	17.1
Some college education	4.6	22.3	10.3	9.6	9.1	10.2	14.1	19.9
BA/BS or higher degree	4.7	28.1	11.7	8.6	9.5	7.1	13.5	18.9

Source: *Statistical Abstract of The United States,* 1999.

*Never had intercourse or never had voluntary intercourse if first intercourse was not voluntary.

interviewed 16,000 persons, making strenuous efforts to ensure reliable responses—for instance, by comparing the answers of husbands and wives. Kinsey attempted to bring in the full range of individuals by contacting groups such as churches and trying to interview 100 percent of their members. By questioning a wide selection of groups, the effort was made to represent the entire population of the United States. Kinsey did not succeed in this, but different social classes, educational levels, and other subgroups were sampled, and it was possible to compare their patterns of sexual behavior. Probably the weakest part of his data is on the working class; for this group, Kinsey relied heavily on interviews with prisoners, although their sex lives had probably been more active than the rest of the working-class population.

Since the late 1980s, because of social awareness of the acquired immunodeficiency syndrome (AIDS) epidemic, there have been new efforts to carry out large-scale surveys that accurately represent the sexual behavior of the whole population. These surveys have generally found that sexual behavior is more conservative than the impression given by previous research, especially by the more popular surveys. Even these modern surveys are not entirely ideal; often, they have a narrow interest only in the question of how a disease like AIDS is transmitted and do not provide information that is relevant to how sexual relations operate to tie people together and how they affect the family in this era of increasing gender equality.

Perhaps the best survey of sexual behavior ever conducted was the National Health and Social Life Survey (NHSLS) conducted by Edward Laumann, John Gagnon, Robert Michael, and Stuart Michaels (1994). For this study, a probability (random) sample of 3,432 Americans, ages eighteen to fifty-nine, were interviewed about a wide range of social behaviors and attitudes, and they completed a brief questionnaire with a variety of sensitive questions about sexuality. Approximately 54 percent of the sample were married, and another 7 percent were in cohabiting relationships. Several other ongoing and first-time, large-scale, random sample surveys provided data about sexuality in the 1990s (e.g., General Social Survey, The National Survey of Men, and The National Study of Adolescent Health). Because of the relevance of sexual information to the AIDS crisis, during the 1990s it became more legitimate to ask about sexual behaviors and attitudes in national studies, and government and private funding for survey research on sexual practices increased substantially (Christopher and Sprecher 2000). Unfortunately, even these newer national probability samples did not include enough homosexual participants to systematically analyze their results separately, leaving the Blumstein and Schwartz study from the 1980s (see *A Closer Look* 9.2) as the most extensive source of information on the sexuality of gay and lesbian couples to date.

NONMARITAL SEX

It is clear that the percentage of persons who are virgins at marriage declined considerably over the twentieth century, but recent surveys show that in the 1990s, that trend may have abated. Four surveys of high school students collected between 1991 and 1997 showed an 11 percent increase in the incidence of virginity, driven by lower rates of intercourse for white and black male youths (Christopher and Sprecher 2000). For slightly older males, no such drop in the incidence of sexual intercourse was evident. Almost nine of ten men of ages twenty to thirty-nine were "coitally experienced" (nonvirgins) in the 1990s. For women of the same age, about eight of ten were nonvirgins (Christopher and Sprecher 2000). Nonvirginity appears to have been fairly widespread all the way back to the beginning of the century, and it gradually increased until the 1990s. Among men in the 1920s, about two-thirds (67 percent) had had premarital intercourse; in the 1940s, Kinsey found the figure was already up to about 85 percent. There has been no sexual revolution for men as far as nonvirginity is concerned.

For women, the shift has been much sharper. Around 1900, about three-fourths (73 percent) of women had no premarital intercourse, and of those who did, it was generally with their fiancé (husband-to-be) (Reiss 1960). The percentage of virgin brides declined in the 1920s and 1930s to around 45 to 50 percent; most of the increase in nonvirginity came about because more women were having intercourse with their prospective husbands. As late as 1960 to 1964, a slight majority (52 percent) of women marrying were nonvirgins; this number rose sharply to 72 percent in 1970 to 1974 and 79 percent in 1975 to 1979 (*Family Planning Perspectives* 1985). By the 1990s, 80 percent of women born after 1953 reported that they were nonvirgins at the time of their marriage (Laumann 1994). In the long run, there has been a continuous decline in virginity for women throughout the twentieth century, although there was a big leap in the late 1960s and the 1970s. This period seemed even more revolutionary at the time, because the nonvirgins became the majority, and the public finally realized that the traditional position enunciated in public all along was now practiced only by a minority.

By comparing the sets of figures for men and women, we see that consistently, a larger share of men have been sexually active. In Kinsey's day, about 50 percent of American women had premarital intercourse, compared with 85 percent of American men. Not only that, but most of the women's premarital sex was with their prospective husbands, whereas at least half the men had intercourse with women other than their prospective wives. How was this possible? Obviously a small proportion of women were having intercourse with a larger number of men. Historically, for the most part, this occurred through prostitution.

A CLOSER LOOK 9.2
Husbands, Wives, Gays, and Lesbians: The Influence of Gender Roles on Sex

Blumstein and Schwartz's study, *American Couples* (1983), provides an unusual opportunity to examine the effects of gender—the cultural expectations and definitions of what it is to be male or female—on sexual behavior. Their study consisted of ordinary married couples and cohabiting couples who were not married. In addition to these heterosexual couples, Blumstein and Schwartz also included two types of homosexual couples: gay males who lived together in a permanent arrangement and lesbians in similar arrangements. This sample gives us the opportunity to see what effect cultural gender is having, because it separates out the effects of one's sexual partner from one's own sexual behavior.

In heterosexual relationships between a man and woman, how much of their behavior is determined by the male or by the female? What we observe is the interaction between the two of them, and this can be affected by both. One side may be more influential than the other, but it is difficult to separate out how much. In *American Couples,* though, we can also see what happens when a relationship has two males in it or two females. By comparing these patterns with the one male/one female pattern, we can see in what direction the cultural role of being male or female is exerting its influence.

For example, consider the following figures on the frequency of sexual activity for couples who have been together for two years or less, shown in the table above.

	Married Couples	Gay Males	Lesbians
Sex 3 or more times a week	45%	67%	33%
1 to 3 times a week	38	27	43
Once a week or less	17	6	24

Source: Calculated from Blumstein and Schwartz, 1983, 196.

It is apparent that gay males are much more sexually active than heterosexual couples and that lesbians are the least sexually active. In other words, when there is a male and a female in a relationship, the amount of sex is moderate. When there are two males, the amount of sex goes up. When there are two females, the amount of sex goes down. Similar figures apply to couples who have been together for longer periods. In all cases, the frequency of sex drops after they have been together for two years, and drops still further after ten years, but the pattern between gay males, heterosexuals, and lesbians remains the same.

It looks as if sexual activity is culturally defined as a male province, and this turns out to be true whether the activity is heterosexual or homosexual! Furthermore, males are more attached to their sex lives. Married men, as well as gay men, tend to be dissatisfied with their whole relationship when there isn't much sex in it, whereas lesbians (like female cohabitors) are no less satisfied with little sex (Blumstein and Schwartz 1983, 201).

We see a similar pattern in regard to whether persons from various groups feel it is important that their partner be sexy looking (Blumstein and Schwartz 1983, 249):

Gay males	59%
Husbands	57%
Wives	31%
Lesbians	35%

Again, the pattern is that males are quite similar, whether they are homosexual or straight; females, too, generally are similar, in a much lower emphasis on the sexiness of their partner.

If we ask what percentage of each group is possessive of their partner, we find another interesting pattern (Blumstein and Schwartz 1983, 254):

Wives	84%
Husbands	79%
Lesbians	74%
Gay males	35%

Sexual possessiveness and jealousy over breach of possession is widespread among wives and almost as widespread among lesbians. Married men are also quite possessive, but less so than their wives. Gay males are generally rather unpossessive. What this pattern

Frequency of sex is highest among gay males.

seems to mean is that in our society, possessiveness is an especially strong part of the female cultural role when it comes to sex. In the ordinary marriage consisting of a heterosexual couple, the male tends to be quite possessive as well. The possessive bond goes both ways, although it seems to be particularly oriented around the female. In a couple in which there is no female—that is, gay males—most of the possessiveness drops out, perhaps because there is no female to be possessive or to be possessed. This isn't the whole story, of course, because about a third of the gay males are possessive. There is at least some influence in any kind of sexual relationship that works in the direction of possessiveness.

The same pattern is apparent in the extent to which people think it is important that they themselves be monogamous (Blumstein and Schwartz 1983, 272):

Wives	84%
Husbands	75%
Lesbians	71%
Gay males	36%

In general, then, it appears that sexual activity remains more a male than a female province in our society. Left to themselves, men have more sex, women less. Each group has corresponding levels of concern about sex as a basis for relationships. Females are more concerned about emotional ties, possessiveness, and their own fidelity, although when men are married to women, the possessiveness of that relationship affects them too.

Because Blumstein and Schwartz collected their data in the early 1980s, does this analysis still tell us anything sociologically significant? For one thing, the AIDS epidemic has dramatically grown in the public consciousness. In the intervening period, one of the original authors, Philip Blumstein, has died of AIDS. In response to the AIDS epidemic, gay men have to some extent reduced their number of sexual partners (Kimmel and Levine 1992; Weinberg, Williams, and Pryor 1993). However, the relative differences remain, including lower levels of sexual activity in general as we go down the table from gay males, through heterosexuals, to lesbian couples; this is what we would expect theoretically, if there are socially constructed roles for sexual behavior (Kleinberg 1992; Kinsman 1992). Gay men still share in the aspect of our culture that makes sexual initiative and adventure more of a male province than a female province. Conversely, the same pattern is found in the opposite direction, in studies of lesbian women, and discussions by lesbian activists about the relatively lower level of lesbian sexual activity (Vance 1989).

Another theme that has developed in recent years is the argument that homosexuality is genetically determined and not social at all. A modest but growing number of studies by geneticists lends support to the argument that at least some proportion of homosexuality is genetically transmitted. However, an important point should be borne in mind; even if there is a genetic mechanism, that does not make social factors irrelevant. For instance, homosexuality in ancient Greece was predominantly between older men and young boys, whereas modern homosexuality is overwhelmingly between adults. In our own society, the number of persons who are bisexual exceeds those who are exclusively homosexual by a ratio of about 8 to 1; and bisexuals switch back and forth at different times in their lives between an emphasis on same-sex and heterosexual partners (Smith 1991; Weinberg, Williams and Pryor, 1993). There is much to be understood about how biology interacts with society and culture, but it is clear that how genetic predispositions are expressed, and whether they get expressed at all, is affected by social factors. For this reason, it is important for us to be able to examine the key comparisons, which show us how various social factors operate, while keeping biological factors constant. It is this type of comparison that makes Blumstein and Schwartz's data so significant and more useful than statistics that merely count how many people are doing what at some point in time. His ongoing contribution to this understanding makes this research a fitting memorial to the life of Phil Blumstein.

Back in the 1940s, Kinsey, Pomeroy, and Martin (1948) found that about 70 percent of American men had visited a prostitute at least once. (This activity was not entirely confined to unmarried men; 10 to 20 percent of married men at various ages also had intercourse with prostitutes.) Intercourse with prostitutes was most common among men born before 1900 and has declined for each generation since then. Thus, the *Playboy* survey of 1974 (Hunt 1974) found that males were much less likely to use prostitutes than were Kinsey's subjects, even though this group was probably above average in sexual activity. Prostitution still exists today, but it is no longer as visible as it was previously, when organized brothels flourished in "red-light" districts of major cities. In its place is a smaller scale of prostitution consisting of streetwalkers, escort services, and brothels disguised as massage parlors, often catering to older rather than young unmarried men.

The decline in male use of prostitutes is probably due to the trends in premarital sex. Prostitution was to a considerable extent the result of a large imbalance between male and female sexual behavior. As female behavior has become more like the male pattern, prostitutes have become less important for unmarried men. At the same time, female sexual behavior continues to place more emphasis on affectionate contacts rather than sheer recreation; in that sense, the shift of male attention from prostitutes to girlfriends probably indicates that in some ways the male pattern is getting closer to the female one as well.

Who Does What How Often?

Figures on nonvirginity are a little misleading in some respects. The loss of virginity requires only a single act of intercourse; how frequently premarital sex takes place is another question. The rise in female nonvirginity, or for that matter, the long-standing, near-universal male nonvirginity, does not mean that either women or men are having a great deal of nonmarital sex.

Figures on frequency of intercourse indicate that the nonvirginity issue does indeed overdramatize what is actually going on. Whether one is a virgin has been a major symbolic issue in the past, and hence, such figures still make headlines; but in fact, the rate is rather low. Among never married, noncohabiting women who had intercourse at all, about a third have sex only a few times per year, and another third have sex a few times per month (Laumann et al. 1994). If we go back to Kinsey's data for the years before 1950, we find that frequencies then were lower. Kinsey and colleagues (1953) found that the average premaritally experienced woman had intercourse once every month or two. Unmarried men with sexual experience in Kinsey's day were likely to have sex more often: about once per week. By the 1990s, the frequency of sex for unmarried men and women had converged, with men reporting only slightly more sexual acts than nonvirgin women (Laumann et al. 1994).

TEENAGERS COMPARED WITH YOUNG ADULTS

The public image is that the teens are the wild years, the stage when all the premarital sex occurs, but in fact, this is a myth. Teenagers tend to be a good deal more conservative than the adults just ahead of them, in sexual behavior as in most other respects—in religious, political, and social attitudes. The young unmarried adults, especially those in their twenties, are most sexually active. What this means, though, is that teenagers do not engage in intercourse as often as young adults.

This is clearly a social pattern, not a biological one. Male youths in their late teens have the maximal amount of sexual drive according to early sex researchers (Kinsey, Pomeroy, and Martin 1948). A surprising proportion of

Although males in their late teens have a high sexual drive, teenagers, in fact, do not engage in intercourse as often as young adults.

very young males are capable of as many as five ejaculations within an hour or less, although capacity for multiple orgasms (at the rate of two or more in short succession) falls off quite rapidly with age. Among females, an equivalent figure is rather hard to estimate. Kinsey and colleagues (1953) did find that a woman's tendency to have multiple orgasms becomes higher than a man's by about age twenty-five and remains higher throughout the rest of their lives. It makes sense biologically that in the human species, as in all others, peak reproductive ability is attained soon after sexual maturity and falls off gradually thereafter. We see this in the declining fertility of women after their twenties, until the age of menopause around fifty.

The lower incidence of sex among teenagers is yet another instance of how sexual behavior is affected by social processes. Sexual relationships among teenagers are more unstable than those among adults; there is a considerable amount of change going on in all aspects of teenagers' lives, and this affects sex, too. Most sex takes place during an affair that is monogamous during the time when it is going on. For instance, according to one study, about four-fifths of male teenagers had some experience of sexual intercourse; however, their frequency of intercourse was only 2.7 times per month, that is, less than once a week (Sonenstein, Pleck, and Ku 1991). One reason the average was so low was that these sexually experienced youths typically spent six months out of the year without any sexual partner; only 21 percent of them had more than one partner in any one month during the year. There are not many teenagers who are having roving sexual adventures day after day; much more typical is two sexual affairs per year, each lasting a few months, separated by several months without a sexual partner. It is just because of this switching between partners that the amount of sex is restricted; so much time is taken up in breaking up and finding a new partner.

How Many Partners?

Of all the different aspects of sex, one of the most difficult to obtain reliable information on is how many sexual partners people have. The NHSLS found that among unmarried female teenagers, the majority (55 percent) of those who had sexual relations at all had them with only one person. At the upper extreme, 12 percent had had intercourse with five or more different partners (Laumann et al. 1994). Interestingly enough, these figures do not seem to have changed much since Kinsey's day (1953); he found that 53 percent had had a single premarital partner only, and 13 percent had had six or more partners. Unmarried teenage men in the 1990s were likely to report having more sexual partners than their female contemporaries. According to the NHSLS, about a third of nonvirgin unmarried male teenagers reported having five or more partners, with over a third having two to four partners, and another third having just one partner (Laumann et al. 1994).

Somewhat surprisingly, the number of partners reported by men declined between the time of the Kinsey surveys and the 1970s (Weinberg and Williams 1980). Kinsey found that high school–educated men (i.e., the lower social classes) had about nineteen premarital partners, and college-educated men (the higher social classes) had about ten. Thirty years later, these figures had *dropped* to below thirteen and eight, respectively. It is true that in some nonrandom sample surveys, we get very high figures; the 1983 *Playboy* survey reported an average number of twenty partners for men, sixteen for women (Petersen et al. 1983), and about 10 percent of each sex said they had had more than fifty. However, this group is probably rather out of the ordinary. In the overall population in the late 1980s, adult men were saying they had had, on the average, twelve sexual partners since age eighteen, and adult women were averaging a little more than three (Smith 1991). Probably, the men were exaggerating upward and the women may have been minimizing somewhat. In general, men seem to be dropping closer to the female pattern, and in some ways women are behaving more like men.

With young men still more likely than young women to lose their virginity and to have multiple sexual partners, we are seeing both an erosion and a continuation of the sexual double standard. In the middle part of the twentieth century, it was common for boys to "prove" their virility by trying to get girls to have sex with them. Girls, on the other hand, were admonished to avoid having sex or at least to maintain their virginity until they got married. As Schwartz and Rutter (1998, 79) point out, this sexual double standard constricted the pool of available sexual partners and encouraged a "good girl/bad girl" dichotomy. Girls who followed the masculine code and experimented with sex were labeled "bad," "loose," or "easy." To maintain one's reputation as a "good" girl, one had to remain chaste or at least appear like it, thus maintaining one's potential worth on the marriage market. The double bind for the young women was that the young men they were dating were simultaneously cast as sexual predators to be fought off and as masculine saviors needed to protect them from the other men.

This difference in sexual scripts for men and women is weakening, but we can still see evidence of it in gender-linked patterns of sexual desires. A common assumption in the United States and many Western countries is that women's sexual desire is more "relational" and men's is more "recreational" (Hatfield and Rapson 1996). In other words, women's sexual desire tends to focus on a specific person and is ratified by love and mutual passion, whereas men's sexual desire is more likely to focus on sex for fun and personal gratification (Schwartz and Rutter 1998). Reflecting a convergence in sexual scripts, recent surveys show us that men, like women, now tend to prefer relational sex; that is, both men and women experience heightened sexual desire and pleasure when they share a mutual and loving personal relationship with their sex partner. Nevertheless, studies also continue to show that emotionally valuing a partner tends to motivate women—more than men—to engage in sexual intercourse (Christopher and Sprecher 2000).

Surveys and studies also continue to show that more men than women seek sex for recreational purposes. The NHSLS shows that one in three men between the ages of eighteen and forty-four find the idea of having sex with a stranger very or somewhat appealing, whereas only one in ten women feel that way. Similarly, almost half of men this age find the idea of group sex appealing, whereas under 10 percent of women do (Laumann et al. 1994, 162–63). Another indication of women's general preference for relational sex and lower interest in recreational sex is revealed in their use of "escort services." Women are much less likely to use prostitutes than men, and when they use escorts, it is usually to have a date or companionship for the evening, rather than solely for having sex with them (Schwartz and Rutter 1998).

Another comparison of relational versus recreational sex can be made between lesbians and gay men. Contrary to some stereotypes, both gay men and lesbian women prefer relationships, but the sexual norms for gay men and lesbians differ from each other and from heterosexuals (see *A Closer Look* 9.2). Particularly before the outbreak of AIDS and recognition of human immunodeficiency virus (HIV), gay male magazines and culture championed recreational and sometimes anonymous sex. There has been a substantial shift toward more relational sex and couplehood among gays, but recreational sex is still much more common than among lesbians and heterosexuals. After pointing out some recent convergences between men and women, Schwartz and Rutter (1998, 47–48) sum up the differences between them this way:

> [T]he evidence suggests that men are more likely to see sex as fun for its own sake and that women are more likely to believe sex can be enjoyable only when it has meaning, affection, and, for some, love. Do women like quickies? Some do. As time marches on, women are at greater liberty to express lust, in such settings as male strip bars like Chippendale's. But the more impersonal the sex is, the more likely it is that men will approve of it and women won't.

Who Initiates Sex?

One interesting finding from at least one study in the 1990s is that college women *think* about initiating sex about as frequently as men (O'Sullivan and Byers 1992). We also know that sex is more often initiated in steady as compared with less-committed dating relationships and that these initiations involve both indirect verbal messages and nonverbal behaviors for both men and women (Christopher and Sprecher 2000). Nevertheless, reflecting a continuing sexual double standard, it is men who more often initiate sex and women who more often refuse. Is this because men have a stronger sex drive than women, or is it because men have been socialized into the job of initiating sex and women have been given the prerogative to accept or reject? As noted

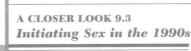

A CLOSER LOOK 9.3
Initiating Sex in the 1990s

In *Hot and Bothered: Sex and Love in the Nineties* (1992) Wendy Dennis reveals how sex is much more out in the open than it used to be. For example, she begins with the observation, "In the sixties, when I discovered sex, there was no such thing as a politically correct blow job" (p. 1). Like other love advice from contemporary magazines, Dennis blends appreciation for the benefits of feminism with a lamentation that women's gains have made romance more complicated. According to Dennis, it is difficult for women in the nineties to perform the delicate tightrope act that forces them to balance "dogma and desire, sexism and sexiness, feminism and femininity" (p. 268).

Wendy Dennis's advice to young women in the 1990s also shows how the sexual double standard has changed little since Dorothy Dix was giving advice to women about how to catch a man in the early part of the century (see *A Closer Look* 8.3). Suggesting that most men still find the notion of a sexually aggressive women distasteful, Dennis coaches women to flirt with men, sending out subtle cues to ignite their passion. "Women draw men to them through flirtatious behavior: glint of eye, seductive posture, tone of voice, pressure of touch. Predictably, men warm to a laugh, a smile or a touch. But when a woman carelessly executes a hair flip, head toss, skirt hike or lip lick, the needle goes right off the attraction-meter and the guys are off and running" (p. 92).

This 1990s advice to women carries some of the same gender assumptions as earlier advice columns, insofar as women are expected to indirectly entice men into paying attention to them.

Dennis tells a story about one man who fondly remembered a woman he had met who was an expert flirt: "Within an hour of meeting we were flirting like mad. She was instigating it, for sure, but I found it utterly charming. We had a mock fight in a subway station, and she touched me and the touching was pregnant with meaning. As she left, she sort of turned over her shoulder and asked in this curious, little-girl voice, 'Do men like anal sex?' God, it was erotic. It was the perfect combination of innocence and sleaze (Dennis 1992, 95).

Like women, Dennis encourages men to master the fine art of flirting: "A great flirt always leaves a women feeling good about herself. He knows how to compliment her on her charms without making her feel as if she is a piece of meat. He looks her in the eye, listens to what she is saying, jokes and spars with her in a delightful, engaging way" (p. 95). Wendy Dennis does not imply that the task of wooing a woman is easy for the 1990s man, for women want contradictory things from him: "He has to be gentle but not weak, malleable but not limp, masterful but not macho, sensitive but not sappy and stylish but not shallow. He has to cook! He has to clean! He has to garden and decorate! If he's married, he has to chauffeur the kids and pontificate knowledgeably on the subject of fatherhood. In his leisure time, however, should he experience a moment of heaving desire, he has to radiate the animal magnetism of a Tom Cruise, do a me-Tarzan, you-Jane routine and (with her lusty consent, of course) pin Jane against the fridge, remove her panties and make the earth move for her" (Dennis 1992, 69).

previously, the biological and social aspects of sexuality are so intertwined that this question cannot be answered one way or the other.

We do know that most people see **sexual initiation** as "men's work" and refusal as "women's work" (see *A Closer Look* 9.3 for an example of Initiating Sex in the 1990s). In fact, the norm against women initiating sex is so strong that women who step out of line and initiate sex "too much" often receive a hostile reaction. Blumstein and Schwartz (1983) found that husbands had very negative reactions when their wives showed more sexual initiative than they did. Such reactions run counter to a simple biological explanation for gender differences in sexual behavior. If men simply had a stronger appetite for sex, they should welcome the sex regardless of who initiates it. Schwartz and Rutter (1998) suggest that we can understand these men's reactions only by understanding that initiation is also a reflection of power. When men are the sexual aggressors, they exercise control. When they are responding to women's advances, it changes the power dynamics of the interaction and of the relationship. Many men become uncomfortable when they lose control, and it is easy to see how this might impede their sexual feelings.

Differences in initiation also occur in lesbian and gay relationships, and as for heterosexual couples, these differences often reflect an unequal distribution of power. Because gay men, like straight men, have been socialized to be in control, they are comfortable initiating sex. When Blumstein and Schwartz (1983) asked gay men, "Who initiates sex?" both partners claimed that they were the initiator. When they asked lesbian couples the same question, both partners claimed that the other partner did more of the initiating. Both straight and gay men try to prove their masculine prowess through initiating sex, whereas lesbian and heterosexual women tend to consider sexual initiation as too aggressive (Blumstein and Schwartz 1983). Thus, whatever sexual differences between males and females exist in the first place, socialization into more passive femininity encourages women to curb or deny sexual initiation, just as socialization into masculine entitlement encourages men to act on their sexual feelings. Schwartz and Rutter (1998, 55) conclude that "initiating sex may not be any more 'natural' for many men than it is for women, but the pressure to perform is different."

Compliance and Coercion

How does the sexual double standard and pressure to be the initiator play itself out in early sexual experiences for young men and women? As you might guess, men more often push, and women more often go along with men's sexual advances, sometimes regretting it later. In the 1996 book *Going All the Way,* teenage girls revealed how they sometimes traded sex when they were promised love, only to find that the love did not follow and the sex was unfulfilling (Thompson 1996). Many girls gave in to repeated requests because it seemed easier than always saying no and because they wanted to be with a boy. Survey research also shows that it is women who are likely to be unwilling to have sex but to comply reluctantly to a partner's sexual wishes to maintain the relationship (O'Sullivan and Gaines 1998; Shotland and Hunter 1995). In later stages of dating, reluctant women report giving into sexual demands because they do not want to disappoint their partners and do not want to risk damaging the relationship. On the other hand, when women try to get reluctant male partners to be sexual, they are more likely than men to be rebuffed (Christopher and Sprecher 2000).

In other recent studies of teenage sexual experiences, young women who were sexually coerced were found to have poorer peer relations than others, to be more susceptible to conformity pressures, to have less parental monitoring, and to have been exposed to more parental abuse—a cluster of characteristics typically associated with low self-esteem (Christopher and Sprecher 2000). Sexually coercive men, in contrast, tend to be manipulative, have few communication skills, and be conflicted and ambivalent about relationships with women. Such men tend to be callous toward women and to ignore their wishes. Studies using videotaped interactions show that sexually coercive men tend to discount the truthfulness of women's rejection messages and misinterpret women's verbal responses to sexual advances. Such men have more dating relationships than others and are often preoccupied with seeking novel and casual sexual encounters (Christopher and Sprecher 2000). This combination of predatory boys and easily influenced girls is a volatile mix.

In the NHSLS, respondents were asked whether their first experience of sexual intercourse was "something you wanted to happen at the time; something you went along with, but did not want to happen; or something that you were forced to do against your will" (Laumann et al. 1994, 328). Over 92 percent of men but only 71 percent of women answered that they wanted their first intercourse experience to happen when it did. This provides evidence for the operation of a sexual entitlement on the part of men, insofar as they were more likely to be in control of the situation.

These data also reveal who is more likely to give in to sexual advances and to comply with demands even when they do not want to have sex. About one-fourth of the women reported that they did not want to have vaginal intercourse, a proportion three times higher than that for men. Finally, just over 4 percent of women reported being forced to have sex their first time, compared with less than half of 1 percent for men (Laumann et al. 1994). We thus see that **date rape** is a fairly common occurrence.

Of the women who went along with vaginal intercourse without wanting it, 39 percent cited affection for their partners as the reason they did it. Another 25 percent reported curiosity about sex, and 25 percent reported peer pressure as reasons for giving in to the man's sexual advances. Among the men who went along with the experience of sexual intercourse reluctantly, over half said they did so out of curiosity about sex and more that 29 percent did so because of peer pressure (Laumann et al. 1994). Even in this intimate first encounter with sexual intercourse, we can see social forces at work. Especially for the women, gender ideals, relationship dependence, and peer pressure encouraged them to perform an act that they did not choose. We should be aware that reporting biases might also be operating here, with young men feeling they should say they wanted sex, and young women feeling that they should say they did not. Even so, these results and those of other studies show clearly that the sexual double standard has not disappeared. Even though more young men are focused on sex as a relational experience that builds intimacy in a relationship, others remain sexual predators. Because of the continuing double standard, many men still feel entitled to demand sex from women or to force it on them, and some women continue to feel obligated to provide it for the sake of maintaining the relationship.

Gay and Lesbian Sexuality

The period since the late 1960s has been a time of gay liberation—more accurately, the movement of gay men and lesbian women to overcome discrimination and gain equal rights in society. Here again we need to distinguish between changes in what people are willing to say in public and changes in what they are actually doing. There has been a huge change in the degree of openness in talking about homosexuality and some changes toward greater tolerance in the attitudes that people express about allowing homosexual preferences. Through all the controversies and conflicts, it is not clear if there is a real change in sexual behavior or only greater attention to the topic. Is a shift taking place from heterosexual to homosexual preference? Does homosexuality coming into the open, along with the greater tolerance of premarital sex in general, indicate that the family is gradually declining as the main center of sexual life?

Surveys of sexual behavior made in the late 1980s have begun to give a picture of what has been happening. Somewhat surprisingly, these figures turn up less homosexual behavior than was suggested by the earlier Kinsey studies, which may have been biased toward sampling the more sexually active respondents. The NHSLS asked people about their sex lives since puberty, with respondents

The gay liberation movement has been successful in reducing some of the most overt forms of discrimination against homosexuals. However, greater tolerance of homosexuality does not necessarily indicate a widespread shift in sexual behavior.

indicating whether they were sexually active and the types of sexual activity in which they engaged (Table 9.2).

About 97 percent of the adult population was sexually active at some time in their lives (2 to 3 percent never had sex); less than 1 percent had exclusively homosexual partners; about 5 percent had both homosexual and heterosexual partners; and the rest had only heterosexual partners. It also appears that gay and lesbian sex is concentrated in particular periods of a person's life, rather than going on constantly throughout. When the survey asked about sexual activities that happened during the past year, 2 percent of men and 1 percent of women said they were exclusively homosexual during that time.

Table 9.2 also shows two other common measures of sexual orientation: attraction and identity. Again, we see that most people fall into the heterosexual category, but over 4 percent of women and over 6 percent of men report that they are attracted to same-sex others either exclusively or in addition to opposite-sex others. The lowest numbers are for people claiming a homosexual or bisexual identity. We can thus see that homosexuality is not a unitary phenomenon and that it is difficult to measure. Because the concentration of gays and lesbians is much higher in cities (typically over 10 percent), they seem more numerous than figures from nationwide surveys.

If one can believe these most recent surveys, gay men and lesbian women are a rather small minority. This of course does not justify any kind of social discrimination against persons on the basis of their sexual preference. In terms of the sociology of sex and the family, there probably never was a threat to the institution of the family from this direction. There are still many questions unanswered. We lack good trend data on the relative amounts of homosexuality and heterosexuality in the past, and we do not know whether the AIDS epidemic that began in the 1980s was responsible for reducing the amount of homosexual sex.

Table 9.2 How Much Homosexuality Do People Report in Sex Surveys?

In the National Health and Social Life Survey (NHSLS), a random sample of 3,432 Americans, aged 18 to 59, answered questions about who they had sex with, who they were attracted to, and what their sexual identity was. Less than ten percent reported having engaged in sex with a same gender partner, with homosexuality higher among men than women.

	Percentage Reporting Sexual Activity (since puberty)	
	Men	**Women**
Sexual Activity—Partners		
Heterosexual activity only	90.3	94.3
Homosexual and heterosexual partners	5.3	3.3
Homosexual activity only	0.6	0.2
No sexual activity	3.3	2.2
Expressed Sexual Identity		
Heterosexual	96.9	98.6
Bisexual/other	1.1	0.6
Homosexual	2.0	0.9
Sexual Attraction for		
Other sex only (heterosexual)	93.8	95.6
Both sexes (bisexual)*	3.9	4.1
Same sex only (homosexual)	2.4	0.3

Source: Edward Laumann, John Gagnon, Robert Michael, and Stuart Michaels. 1994. *The Social Organization of Sexuality.* Chicago: University of Chicago Press.

*Includes "both," "mostly opposite," and "mostly same."

Although these large, random, sample sex surveys probably underestimate the real extent of homosexual activities (Ericksen 1999; Weston 1998), it appears that the majority of sexual relationships have been heterosexual in the past and continue to be so in the present.

Understanding Changes in Sexual Activity

Although changes in sexual behavior in recent years have received a great deal of publicity, they have not always been accurately understood. There has been a revolution, perhaps, in the way we talk publicly about sex. Also, there is less insistence that sex be tied to marriage, although the change partly means less hypocrisy about the realities of premarital sex. Changes in the overall frequency of sexual intercourse and other forms of sexual behavior are not so great. The importance of commercialized sex in the form of prostitution has actually declined, and premarital sex for both men and women is now more likely to take the form of affectionate and quasi-stable, if not permanent, pairing. Cohabitation of unmarried couples has gained wide acceptance as part of this trend toward relationships that fall somewhere between traditional marriage and courtship.

Two major factors appear to be involved in this change. One that is often mentioned is the "contraceptive revolution" that took place with the development of the birth-control pill in the 1960s. Other forms of birth control were available before that time, such as rubber condoms, vaginal foams, and diaphragms. But the birth-control pill made contraception easy, unobtrusive, and reasonably reliable, and this may have sharply reduced women's fears of pregnancy and hence

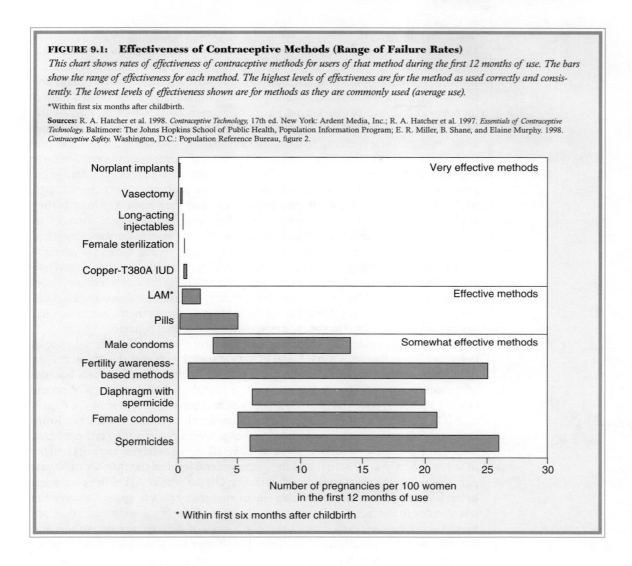

FIGURE 9.1:   Effectiveness of Contraceptive Methods (Range of Failure Rates)

This chart shows rates of effectiveness of contraceptive methods for users of that method during the first 12 months of use. The bars show the range of effectiveness for each method. The highest levels of effectiveness are for the method as used correctly and consistently. The lowest levels of effectiveness shown are for methods as they are commonly used (average use).

*Within first six months after childbirth.

Sources: R. A. Hatcher et al. 1998. *Contraceptive Technology*, 17th ed. New York: Ardent Media, Inc.; R. A. Hatcher et al. 1997. *Essentials of Contraceptive Technology*. Baltimore: The Johns Hopkins School of Public Health, Population Information Program; E. R. Miller, B. Shane, and Elaine Murphy. 1998. *Contraceptive Safety*. Washington, D.C.: Population Reference Bureau, figure 2.

increased their premarital sexual behavior. (See Figure 9.1 for further information on the effectiveness of various forms of contraception.)

The revolution in contraceptives, however, is not likely to be the whole story. Changing social attitudes were necessary for the pill and intrauterine devices (IUDs) to become widely and even openly available. Moreover, it is clear that many women do not use them, as indicated by the rising rate of nonmarital births. Contraceptive styles shift from time to time, as when health hazards reduce the popularity of a method (witness the decline in the use of the IUD). It is likely that contraceptives of various kinds, although contributing to sexual behavior, are only part of a larger story that is basically social rather than a matter of medical technology.

The other major shift that coincides with the sexual revolution is the feminist movement. For some people, this may seem incongruous because feminists have been scapegoated as antimen or antisex. In fact, the structural pattern makes a good deal of sense. The feminist movement is part of a wholesale shift in the way women organize their lives. It involves the tendency for women to finish college and professional training, to pursue full-time careers, and to put off marriage while they are doing such things. All these are trends that became very strong in the 1960s and 1970s. Moreover, they fit together with the sexual revolution, for *that* revolution has consisted mainly of closing the gap between men and women in sexual behavior.

Thus, there has been more premarital intercourse on the part of women, and it is now reaching almost the same level as for men. There has been a slower increase in the number of sexual partners women have, although part of the pattern here may be that men's premarital sex has become less promiscuous and closer to the female pattern. To some extent the double standard, which had prescribed one pattern of sex for men and another for women, has been diminishing. The shift in the way we now publicly talk about sex is part of this change because the old taboo on using sexual language in the presence of women was part of the double standard that allowed men to do what they wished as long as they kept it among themselves (and with a small female minority of prostitutes).

More important structurally has been the fact that as women pursue careers and put off marriage, they are putting off marital sex as well. Hence, the rise in women's premarital sex is partly a way of finding the sexual gratification that they would otherwise find in marriage. Also, as women emphasize having their own incomes, it becomes less important for them to confine sexuality to marriage as a way of acquiring a husband to support them. For all these reasons, the women's movement has undoubtedly contributed to the changing patterns of sexual behavior.

Has the AIDS epidemic, which began in the early 1980s, had an effect on sexual behavior? There has been a good deal of public speculation that the sexual revolution has come to an end and that people are behaving more conservatively because of the deadly disease. Survey data show that a relatively small proportion of people (10 to 15 percent) have changed to less risky sexual behavior because of the danger of AIDS; that is, they have delayed having their first intercourse, shifted to a lower number of sexual partners, or avoided casual sex with persons they did not know well. The same surveys, though, show that overall sexual behavior has not changed much in recent years; there is still the same general level in the number of sexual partners and in the amount of marital fidelity. On the whole, AIDS does not seem to have had much impact. In part, this is because many people ignore dangers, but it is also due to the fact that the majority of people are fairly conservative in their sex lives. As we have seen earlier, they may have a series of different sexual partners during their lives, but usually they have only one partner at a time; also, many unmarried persons go through periods without any sexual partner. About 25 percent of both unmarried and divorced persons are completely abstinent in any one year (Laumann et al. 1994; Smith 1991). For these reasons, most people don't feel much need to change their sexual behavior.

MARITAL SEX

As we have seen in chapter 2, a main ingredient of the family is exclusive sexual possession between husbands and wives. If lovemaking is an interaction ritual establishing relationships, marital sex is the interaction ritual that most centrally symbolizes the ongoing bond. During the marriage, this is supplemented by other ties—especially shared property and children—and the amount of sexual activity goes down. Sex continues for most married couples at least occasionally, even into old age, and serves as a reminder of what their tie is about.

Premarital sex is now quite widespread, but is not generally as frequent or as regular as during cohabitation or marriage. Moreover, premarital sex is often a pathway to marriage in that a considerable percentage of people who engage in

Table 9.3 Frequency of Marital Intercourse over One Year

Women	Men	Frequency
3.0%	1.3%	None
11.9	12.8	A few times per year
46.5	42.5	A few times per month
31.9	36.1	Two to three times a week
6.6	7.3	Four or more times a week

Source: Laumann et al. 1994, table 3.4.

premarital sex end up marrying each other. Because sex and marriage are legally and morally linked, marital sex is generally not viewed as a social problem or likely to lead to negative outcomes. As a result, sex in marriage has not received as much research attention as nonmarital sex (Christopher and Sprecher 2000).

Marital sex is less of a negotiation and an adventure and more of a routine. But what kind of routine is it, and how do married people feel about it? Is it the essence of wedded bliss or a source of domestic frustration? Are we dealing here with couples making love forever and ever until death do them part, or with horny husbands and unsatisfied wives? Marriage in general is a situation of power and conflict in which happiness and unhappiness depend on many conditions. These conditions carry over into a couple's sexual life as well. How a couple gets along sexually is an indication of how their marriage is going in general.

Rates of Sexual Intercourse

How often married couples have intercourse varies a great deal, depending on how old they are, how long they have been married, their social class, and other factors. Overall, the rate is about once or twice a week, but different couples may range all the way from no intercourse at all to every day (Table 9.3). The National Survey of Families and Households (NSFH) reported that married couples had sex an average of 6.3 times per month (Call, Sprecher, and Schwartz 1995). More recently, the NHSLS reported that married men said they had sex an average of 6.9 times per month, whereas married women reported an average of 6.5 times per month (Laumann 1994). These numbers show that husbands say they are having more sex than wives say they are having. In other surveys, researchers have found that men tend to claim that their last episode of sexual activity lasted longer than the women claim it did (Laumann et al. 1994). Schwartz and Rutter (1998, 39–40) say "men boast and women are demure when it comes to reporting sexual experience." As we have noted about survey data in general, self-reports are influenced by social desirability, so men may say they have more sex because they think they should be having more of it. It is also possible that men are having more sex than women outside of their marriages (as discussed later). Like other information about Americans' sexual practices, the survey frequency numbers about marital sex seem to indicate that it is more acceptable for both men and women to do it or at least talk about it, particularly when figures are compared with those collected in the 1950s and early 1960s.

Why has the rate of sex gone up since about 1965? This happened among all age groups; therefore, it was not merely a phenomenon of the youth culture of the 1960s. It seems to be correlated with the increasing use of contraceptive pills, which came into popularity at that time. Also women who were employed had higher rates of intercourse; so the increase in marital sex may also be due to the greater career liberation of women in general during that period. Again, as we've seen in regard to premarital intercourse, the general trend of growing feminism has not been antisexual, as some critics contend, but quite the opposite.

We should also note that although the rates of sex in marriage are higher than rates outside of it, people in cohabiting relationships have the highest rates of all (Laumann 1994; Call, Sprecher, and Schwartz 1995). Rates of sex early in a marriage are also higher than those later in a marriage, so one way to interpret the higher cohabitation sex rates is to see them as "early-in-the-relationship" rates of sexual activity. When people first get together, they tend to have sex more frequently than after they have been together for a few months or a few years. Because cohabiting relationships tend to be short lived and include more people in earlier stages of their relationship, this may account for the higher levels of sexual activity. Another factor to consider is age; people in cohabiting relationships tend to be younger.

Differences by Age and Social Class

Almost all studies examining such things find that the frequency of sexual intercourse tends to decrease with age. Even though rates of activity have gone up for all age groups, this pattern of a relative drop-off as people age has remained. Married couples under the age of twenty-four have sex an average of almost twelve times per month (three times per week). The rate falls off gradually thereafter, until it is about once per week for people in their fifties and less than once per month for people over the age of seventy-five (Call, Sprecher, and Schwartz 1995).

There are also interesting patterns by social class. Contrary to the image that the lower social classes are less inhibited and more direct about sex, earlier researchers found that higher proportions of middle- and upper-middle-class persons have intercourse in total nudity or with the lights on, whereas the working class tends to be more prudish about these matters (Kinsey et al. 1948, 1953; Hunt 1974; Westoff 1974). Later researchers found that the higher social classes were also more likely to use a variety of positions in intercourse and to practice oral and other forms of sex. More recent data indicate that these social class differences have narrowed somewhat but still persist. For example, according to the NHSLS, men and women with more education (a marker of social class) are more likely to masturbate and to "always" or "usually" reach orgasm during masturbation. Although more educated men and women were not likely to have more frequent sex with a partner, the character of their sexual activity differed somewhat. For example, both men and women with college degrees were more likely than others to give and receive oral sex, to have anal sex, and to find a wider range of activities sexually appealing. Finally, both men and women with a high school education or less were much more likely than their college-educated counterparts to report that the last time they had sex, it lasted less than 15 minutes (Laumann et al. 1994).

There seem to be several reasons for these class differences. One is that working-class homes are smaller and more crowded. Especially at the poorest levels, numerous people are likely to be sleeping in the same room, perhaps including children and adults. Under the circumstances, sex is more difficult to arrange and has to be more surreptitious. It simply isn't possible to have a leisurely, totally nude erotic experience with the lights on if there is little privacy.

Another important factor is that generally, the lower the social class, the stronger the division between traditional male and female roles. Studies have long pointed out that poor working-class men tend to regard sex "as a man's pleasure and a woman's duty." "Good girls" are not supposed to know anything about sex, whereas men take pride in whatever sexual pleasure they can get from prostitutes or "loose women." There is a strong belief in the importance of possessing a virgin at marriage. Given these different attitudes about sex, the elaborateness of erotic behavior tends to be kept to a minimum. The honeymoon experience is often traumatic for the women, and it sets the tone for unpleasant sexual relationships during the subsequent marriage (Rainwater 1964).

Masters and Johnson (1966) also found that in their sample only 14 percent of the males whose education was below college level expressed any concern with whether their sexual partners achieved satisfaction, as compared with 82 percent of the college-educated group. As Lillian Rubin (1976) pointed out, working-class sex was not as elaborate as in the middle classes. More recent data show that the traditional sexual attitudes of working-class men have become more accepting of women's rights to sexual expression and fulfillment. Some scholars question whether the class bias of the earlier sex researchers didn't influence their findings and lead to pejorative depictions of working-class men's callous sexual practices (Ericksen 1999).

Extramarital Sex

We turn now to examine **extramarital sex**—sexual activity that occurs outside the marriage bond. This has always been a favorite dramatic topic, but there is a sociological reason for looking at extramarital sex that is more significant than the fact that it entertains us with its scandal. We have seen that marriage involves an element of possessiveness, which can be analyzed as a form of sexual and emotional "property." In traditional societies, this sense of marital property was strong. It was part of one's public honor (especially of a husband and father) to maintain control of one's sexual property. That proprietary ideal is not so strong today, and the very notion of someone being someone else's "property," especially sexually, is generally considered distasteful. Nevertheless, as we have seen, the ritual aspects of love (chapter 8) and of intercourse (earlier in this chapter) tend to make one's spouse's body into a symbol of the couple's special bond. Even among persons who consciously disapprove of jealousy and favor individual freedom and openness, there are powerful social mechanisms in the forms of intimate interaction itself that tend to produce an emotional, symbolically loaded sense of possessiveness about sex.

The prohibition on extramarital sex is a very strong one in our cultural history. Officially, sex is supposed to be confined to marriage. Unofficially, there has been a double standard, in that infidelity has been much more strongly condemned when it is the woman who is unfaithful. Traditional English has the term *cuckold* for the husband whose wife has sex "behind his back" but no equivalent term for the wife whose husband does the same. The implication is that it is much more of a disgrace for a man to fail to keep control of his sexual "property" than it is for the wife. In fact, the opposite is much more likely to be the case because men traditionally have committed "adultery" and condoned it in other men.

Although some groups now have shifted their standards somewhat, there is still a great deal of strong belief that extramarital sex is morally wrong. Recent polls of the entire American population find that about 74 to 77 percent believe that it is always wrong, and another 13 to 15 percent believe it is almost always wrong (Laumann et al. 1994). This level of disapproval has been fairly constant since the early 1970s after a small surge of liberalization in the late 1960s. Only about 1 percent of women and 3 percent of men say there is nothing wrong with extramarital sex.

EXTRAMARITAL AFFAIRS Kinsey and colleagues (1953, 437) found that about half of all men and quarter of all women had at least one extramarital affair. About half the women who had any extramarital sex had only one affair, and about 20 percent of them had more than five extramarital partners. (Kinsey did not collect systematic data on the number of the husbands' extramarital sex partners.) That extramarital affairs happen at all is probably more significant than how much intercourse happens in that fashion; Kinsey found the largest proportion of extramarital sex involves men and women in their forties, when it makes up about 10 percent of their total sex lives. The incidence of extramarital experience generally increased from the beginning to the middle of the century (Kinsey et al. 1953, 422; 1948, 414), although it should be noted that this increase was largely on the part of women rather than men.

Surveys using nonrandom samples in the 1970s and 1980s reported that from 25 to 30 percent of married Americans had engaged in extramarital affairs (Blumstein and Schwartz 1983). Nonrandom surveys conducted by popular magazines like *Playboy* or *Redbook* reported even higher rates, often up to 50 percent (Hunt 1974; Schwartz and Rutter 1998). More recent surveys using random samples have found somewhat lower rates. Approximately 15 percent of women and 25 percent of men report that they have had sex with someone other than their spouse at least once while they were still married (Laumann et al. 1994; Schwartz and Rutter 1998). Under 4 percent of married respondents reported that they had engaged in extramarital sex in the past year. Cohabitors are found to have even higher rates of extrarelationship sex than married couples (Christopher and Sprecher 2000; Laumann 1994).

It is hard to estimate the accuracy of such statistics (Ericksen 1999). The methods used in some of the random sample surveys like the NHSLS may depress the actual numbers because respondents had to pass their questionnaires back into the hands of the interviewers and might have been worried about confidentiality. Also, many individuals may not want to talk about their infidelities because they do not want to admit them, even to themselves, much less to some impersonal interviewer (Schwartz and Rutter 1998). Different people have different definitions of what constitutes an affair: one recent national politician even admitted that he preferred to have oral sex with women other than his wife because it "wouldn't count" as an affair. Even taking such definitional and methodological issues into account, it appears that a large majority of marriages contain sexually faithful spouses.

What kind of people are most likely to be involved in these affairs? For one thing there is a social class difference, at least for males. Men of both higher and lower social classes have extramarital affairs, but for the lower classes, extramarital sex begins very soon after marriage, while for the higher classes it happens later. Kinsey, Pomeroy, and Martin (1948) found that working-class women were most likely to have extramarital sex when they were in their early twenties; the middle- and upper-class men were beginning it past the age of thirty.

Most recent surveys show that rates of affairs are highest among those with the least education (high school dropouts) and with the most education (earned a postgraduate college degree). Rates for the middle groups were in between. In general, very poor people are likely to have higher rates than richer people, but as we have seen the timing of those affairs tends to vary (Schwartz and Rutter 1998). People with more education who have ever had an affair also tend to report a greater number of sexual partners than those with less education. People living in urban areas and those less religious are also more likely to report having sex with someone other than their spouse or cohabiting partner (Laumann et al. 1994). There are many

Most extramarital affairs are transient episodes.

reasons why someone might engage in a love affair outside of marriage, but no single cause is predominant. Pepper Schwartz suggests that there are seven major reasons: emotional incompatibility, boredom, sexual incompatibility, anger, flattery, a way out, and love (for a more complete discussion of these reasons, see Blumstein and Schwartz 1983; Schwartz and Rutter 1998, 151–54).

About half of wives who have extramarital affairs have only one extramarital partner, both in the Kinsey data and in more recent surveys (Laumann et al. 1994). There is also evidence that employed women have higher rates of extramarital sex than traditional housewives. Undoubtedly, this is due partly to greater opportunity and to changing attitudes that women should be able to choose to do what men have traditionally done. Because both the degree of premarital experience and the proportion of women who work have gone up, the rate of extramarital sex on the part of women has also increased, though it has not caught up to the rates for men, except in the youngest age groups (Schwartz and Rutter 1998). Interestingly, there has been little apparent change in the rate of extramarital sex by men, which is understandable enough in that no major change in their premarital experience or occupational status has occurred.

A certain amount of what is technically "extramarital" occurs during the period when a couple has separated but is not yet legally divorced (Hunt 1974). Because the rate of divorce has gone up since the 1960s, the increase in extramarital sex may be partly due to this factor as well. We might imagine, conversely, that extramarital sex is a prime cause of divorce, either because somebody gets caught or because the affair leads a partner to fall in love with a third party or otherwise desire to break up the marriage. However, this does not seem to be the case.

Most affairs are transient episodes, a moment of erotic pleasure rather than any emotional commitment. Spouses seem to know this and may tend to condone affairs for that reason. Kinsey and colleagues (1953) found almost half of the women who had affairs said that their husbands were aware of it, but major difficulties occurred in only about 40 percent of these cases. Blumstein and Schwartz (1983) found that over two-thirds of adulterous spouses said their husband or wife was aware of their affair. There was some tendency for the discovery to lead to divorce, but a majority of these marriages nevertheless survived. Edwards and Booth (1994) reported that only about 5 percent of a national sample of married Americans said that extramarital sex had caused a problem in their marriage.

The major reasons for women having less sex outside of marriage than men are decidedly unromantic. On average, women are simply more economically insecure than men and consequently fear losing monetary support for their children. As we've seen, women's traditional economic dependency on marriage has translated into more conservative attitudes about love and sexuality and increased their concern with their "reputation." In addition, as Schwartz and Rutter (1998) suggest, most women have been trained to be champions of family values, and their sexuality has been geared to that responsibility.

Not all extramarital sex involves male-female intercourse. Researchers have found that there was quite a lot of extramarital kissing and petting, sometimes but not necessarily to the point of orgasm. Because many survey respondents may think of "extramarital sex" only as intercourse itself, we have probably underestimated how often this goes on. In our society, the definition of sexual "fidelity" is so sharply concentrated on intercourse that many people—perhaps even a majority—indulge in these other forms of extramarital sex without thinking of themselves as truly unfaithful.

It is also a mistake to think of extramarital sex as necessarily heterosexual. A considerable proportion of persons who are homosexual—at least part of the time—are married, and their homosexual affairs are thus necessarily adulterous. There is little information about how spouses react to such affairs. Does the spouse get more or less outraged to discover he or she is being "cheated on" for a homosexual partner? (How often does anyone say, "I might have expected I would be jealous of another woman, but of another man?") Perhaps the number of people affected in this way is rather small, or homosexual affairs may be particularly easy to conceal. There is some evidence that women who are struggling with a traditionally sexist marriage may turn to other women for both emotional and sexual intimacy, and sometimes this turns into an exclusive love relationship (Blumstein and Schwartz 1976). Homosexual affairs outside of one's primary relationship are even more common than among married couples. Gay men's likelihood for extrarelationship sex has been calculated as high as 95 percent, and lesbians are more likely to practice nonmonogamy than heterosexual couples. An estimated 38 percent of lesbian pairs together for two to ten years have engaged in extrarelationship affairs (Blumstein and Schwartz 1983; see *A Closer Look* 9.2).

Extramarital affairs sometimes occur because people are dissatisfied with their relationships, but it is not clear if they usually play a causal role in marital breakups. In some studies, marital dissatisfaction seems to have only an indirect influence on the likelihood of extramarital sex, mediated by such factors as premarital sexual permissiveness, a lower value placed on fidelity, and having a reference group or social network that tends to support extramarital sex (Christopher and Sprecher 2000; Greeley 1991). In studies conducted before the 1990s, some researchers found that marital dissatisfaction was related to extramarital sex, particularly for women. The *Redbook* sample of women found that half the unhappily married women had had an affair, but so had a quarter of the happily married ones (Tavris and Sadd 1977). There is also some evidence (Glass and Wright 1977) that after couples have been married longer than about ten years, the rate of extramarital sex is the same whether the marriage is happy or not. Marriages that have held together that long tend not to break up.

In general, the pattern seems to be that adultery is least disruptive to the marriage if it is purely sexual rather than emotional. It is falling in love with the wrong person that is most likely to pull a marriage apart. To be realistic, though, this is not so removed from sexual affairs, because falling in love usually brings sex along with it; however, it is a particular form of sex: the claim to have sex as an exclusive relationship, a kind of token of one's emotional property rights over someone else.

Recent research suggests that men become more upset by the sexual aspect of a partner's infidelity, whereas women become more upset by the emotional aspect (Christopher and Sprecher 2000). Violating either symbol of the marital bond, however, carries the risk of upheaval and threatens the relationship.

COMMUNES AND SEXUALITY A sociologically interesting form of experiment with regularized extramarital sex occurred in certain types of communes. These were group-living arrangements, which became quite famous, although not necessarily very widespread in the 1960s and 1970s. Communes were organized primarily as part of an explicit political and religious program. Their members tended to be young and militantly unconventional, dropouts or rebels against existing society and its corruptions.

One conventional practice that some communes rebelled against was the "normal" standard of sexual possessiveness. Some regarded this as a manifestation of capitalist society or of pathological individualism or the result of a lack of sufficient love and trust to share oneself physically and emotionally with those around one. The conventional form of marriage was seen as a form of exclusive sexual property rights over another human being, a kind of dehumanization that communes set out to replace with a better and more open way of personal life. The communes organized group households to combat selfishness fostered by holding private property; many also tried to overcome the concept of "private property" in the realm of sex. This was a revolutionary aim. In fact, however, communes usually substituted one form of sexual controls for another.

All kinds of different sexual arrangements were found in communes. Contrary to the popular image, however, more communes were on the highly restrictive side sexually than on the libertarian side. Thus, out of 120 communes (including many of the larger ones in the United States) studied by Benjamin Zablocki (1980, 339): 1 percent had **group marriage,** in which everyone was included; 12 percent had partial group marriage (some individuals not participating); 19 percent had shifting sexual relationships, not organized under any particular rule; 47 percent had monogamous couples, in some cases officially approved and enforced by the commune; and 13 percent had **celibacy,** either as a compulsory policy or as a voluntary religious observance. What is surprising here is that the most common form of sexual organization (observed in almost half the communes) was **monogamy.** This was particularly the case in communes under religious auspices, some of which insisted quite strongly on the sanctity of marriage. (These usually involved Christian rather than Eastern religions.) Quite a few of the other types of

Table 9.4 Sexual Activities Declined After Joining a Commune				
	Before		*After*	
Percent who have ever participated in:	**Males**	**Females**	**Males**	**Females**
Sex with more than one person at a time	23%	24%	14%	14%
Homosexuality	15	11	8	11
Open marriage	16	16	6	7
Group marriage	3	1	1	1
Public or group nudity	44	35	32	31
Falling in love	85	83	47	50
Celibacy	23	24	31	30

Source: Zablocki, 1980.

communes (political, psychotherapeutic, hippie counterculture) also advocated monogamy, apparently as a fairly obvious practical solution to the need for regulating sexual life. In the close quarters of a group, the emotions and strains involved with sex are a major cause of problems.

One way to handle sex, then, is for the communes to try to keep traditional couples together. Another method is to prohibit sex entirely. In fact this was done quite often—not so much by prescribed celibacy for everyone (found in only 13 percent in Zablocki's sample), but by the individual practice of celibacy (chosen by about 30 percent of all commune members) (Table 9.4). Most sexual practices tended to become less common after people joined the commune. Beforehand, commune members were more likely to have been involved in group marriage, open marriage, homosexuality, public nudity, and sexual orgies than after they joined. This was even true of falling in love.

The communes seemed to have known what they were doing, in that sex is rather dangerous to the group structure. The primary aim of regulating sexual life is to avoid disruptive attachments. Thus, among the minority of communes that did not have monogamy or celibacy, there was an effort to put sex on the basis of a regular schedule, so that everyone slept with everyone else and individual possessiveness and jealousy were avoided. In the group marriages studied by Larry and Joan Constantine (1973), about half had a rule of fixed rotation among sexual partners. Sex still was mostly private in all these kinds of communes and free-for-all orgies were very rare. (As we see from Table 9.4, about 14 percent of the commune members did at least occasionally have sex with more than one partner at a time.)

All of the communal ways of dealing with sex turned out to have their own strains. Communes tended to disintegrate rather frequently, and this happened most often when there were intense emotional relationships among most of their members. Zablocki found that it made no difference whether these relationships were sexual or not; even nonsexual ties, if they were intense and loving, tended to create crisscrosses of strain and jealousy that pulled the groups apart. Besides, sex tended to be a problem even when it was tightly regulated. Even those communes that enforced celibacy as an official rule often had illicit affairs—rather like a conventional marriage, one might say, except that one is committing adultery not against a particular individual but against a whole group! Some groups attempted to provide an outlet by enforcing celibacy only on the premises, while visits to outside lovers were allowed. The result, though, was that individuals began to leave more and more often, until they finally stayed away entirely (Zablocki 1980). On the whole, Zablocki found that in at least 60 percent of the communes that tried to restrict sex, by either celibacy or strict monogamy, violations had come to the attention of the group. Moreover, these often involved the leader of the commune, the charismatic individual who laid down the moral rules in the first place.

Given the social structure of the commune, this is not surprising. The commune is an intentionally self-enclosed group, isolated from the outside world. It constantly focuses everyone's attention on one another, constituting what we have referred to (chapter 6) *as high ritual density.* The leader, if there is one, is especially likely to be at the center of the group's emotions. Sex is an emotional process, hard to separate entirely from other emotions, and hence it is not surprising that a leader—even a celibate or highly moralistic leader—should be a kind of sexual magnet in the group. The upshot, of course, is to put a good deal of strain on the ideological beliefs of the groups, and scandals or disagreements involving the leader are one of the main ways communes have come apart. For ordinary married couples, too, the commune pressure chamber tends to be fatal. There was a high rate of marital breakup in the communes Zablocki studied, and in the great majority of cases the cause was a dispute over an extramarital affair (1980).

CONCLUDING REFLECTIONS ON THE SIGNIFICANCE OF SEX

Aside from being a real-life version of X-rated entertainment, does sexual behavior have any theoretical significance for the family in general? It does, if we recall that the family consists of economic, sexual, and emotional possession. There is much more sex in marriage than outside it. Young married couples have more sex than unmarried people. Unmarried couples who are cohabiting have a higher rate of intercourse, but they are also involved in a kind of exclusive sexual possession (which is to say, cohabitation is really an informal variation of marriage).

Of course, sex isn't love, and love isn't sex, as people say over and over. But in fact, there is a tendency for *permanent and mutually possessive* sex to become the same thing as the bonds of love (and vice versa). In real life, a great deal of what is implied in "love" is exactly this claim for permanent possession of another, sexually and emotionally. Marriage is not merely sex, but it is a claim to *exclusive* sexual possession theoretically, quite apart from its soap-opera aspects, because it tends to prove that marriage is an institution of sexual possessiveness. Affairs are usually clandestine because people think they are wrong; even those who don't think so usually try to hide them anyway because of pressures from their partners, who want to maintain exclusive possession.

Various modern, liberalized sexual experiments, in which the partners try to allow open extramarital sex, tend to confirm this. They break down precisely because they cannot carefully arrange limited and balanced sexual trades; instead, they put both husbands and wives back on the sexual market. This is devastating to the marriage because marriage is a way of getting *away from* the sexual marketplace, with its constant negotiations, the need to make efforts to find attractive partners whom one is able to attract, and the need to constantly check how one is doing in the eyes of others. Worse yet, open marriages make husbands and wives compete against *each other* over how extramaritally popular they are. Before long, whichever spouse feels most left out or abandoned wants to end the arrangement. This, of course, is a likely reaction to most cases of conventional, clandestine adultery.

Communes teach a similar lesson. They try to avoid jealousy and possessiveness, surprisingly often by banishing sex (through celibacy) or enforcing traditional monogamy. (Not that they do this very successfully, because the opportunities for sexual relations in a living group of this sort are continually tempting.) Or else communes try to enforce nonpossessiveness by having universal and equal sex (e.g., by rotation), as well as mutual group love. But these sexual arrangements are extremely hard to maintain, and only a tiny percentage of communes ever put them formally into effect. Most others that try it only manage a loose version of sexual sharing, in which some people are always left out and some are more sexually popular than others. In other words, the market for sexual relationships rears its ugly head once again. Along with these market negotiations, of course, comes the dramatic and sometimes pleasant side—love—but also attachment, possessiveness, jealousy, strains, and—the equivalent of extramarital scandals for a commune—blowups. Communes are only the most spectacular example of how sexual relationships turn into *emotional* property.

The conclusion seems to be that sexually possessive marriages are here to stay. This isn't to say that monogamous marriages work so smoothly either (or that monogamy is the only way in which sexual possession can be organized; chapter 3). Conventional marriages do tend to strain over the issue of possessiveness, and there is a constant undercurrent of extramarital affairs. Nevertheless, by far the greatest amount of Americans' sex is marital rather than extramarital.

SUMMARY

1. Most sex research has taken an individualistic approach. The goal of such research has typically been to identify variation in individuals' sexual behaviors and to attribute it to biological or psychological differences in the individuals themselves. An individualistic paradigm for studying sexuality cannot comprehend the multitude of cultural and social factors influencing human sexual interaction. The social world influences what people feel, how they experience it, how they express it, what behaviors they choose to take, and how they react to others' behaviors. Sex is social and biological.

2. Sexual intercourse at its most intense is a naturally occurring interaction ritual, in which the mutual physical arousal of the partners makes them strongly feel their personal tie. Though intercourse may not often reach high levels of intensity, it tends to become a symbol of the couple's social relationship.

3. The greatest change in sexual practices during the twentieth century is that they have recently become publicly acknowledged. Traditionally, the populace was always more sexually active than official standards admitted.

4. During the 1920s, about two-thirds of men had at least some premarital intercourse; this figure has risen to around 90 percent in recent years. For women, the percentage of nonvirgins at marriage has risen more sharply, from about 45 percent to about 80 percent.

5. Nonvirginity, however, does not mean that premarital intercourse happens with great frequency; the average rate is a few times per month. Earlier in the century, men resorted to prostitutes a good deal more than in recent generations. This shift is probably due to the growing equalization in sexual behavior between the majority of men and women.

6. Earlier in this century, college-educated males had much lower rates of premarital intercourse than men from lower-status groups. In recent years, this class difference in premarital virginity has disappeared, although the lower social classes still begin sex earlier. Among women, all social classes used to be almost equally restrictive sexually, and all have become more sexually active.

7. The sexual double standard has weakened but not disappeared. We can still see evidence of it in gender-linked patterns of sexual desire. Women's stereotypical sexual script is more "relational" because it focuses on the person or the relationship. Men's stereotypical sexual script is more "recreational" because it focuses on fun, conquest, or "scoring" (seeking multiple short-term partners). Reflecting a convergence in sexual scripts, recent surveys show that men, like women, now tend to prefer relational sex.

8. Most people continue to see sexual initiation as "men's work" and refusal as "women's work." The norm against women initiating sex has weakened but is still used to control women's sexuality. Initiation reflects power, and by being the aggressors, men tend to exercise more control over the sexual relationship, with women sometimes exercising veto power. Men are much more likely to coerce women into having sex or to use force. Women are more likely to have sex when they do not want it for themselves.

9. Marital intercourse takes place much more frequently than nonmarital intercourse. The average rate in the total married population is once or twice a week, although this varies from considerably higher in young married couples to less than once a week for couples beyond age fifty-five.

10. The higher social classes tend to have more elaborate erotic practices than couples of lower social class levels. This is connected with opportunities and a sharper distinction between male and female sexual cultures in the lower classes. However, these class differences have been diminishing in some respects in recent years.

11. A majority of Americans believe that extramarital affairs are morally wrong. Even among those who say they do not approve of extramarital sex, however, clandestine affairs are not uncommon. The rate of extramarital sex has gone up in recent years for women but apparently not for men.

Key Terms

celibacy	group marriage	sexual double standard
date rape	monogamy	sexual initiation
extramarital sex	sex surveys	

Sociology Web Site

See the Wadsworth Sociology Resource Center, "Virtual Society," for additional links, quizzes, and learning tools:

http://www.sociology.wadsworth.com

Also on this web site you'll find InfoTrac College Edition, an online library of journals.

Here you can search for electronic articles about central topics in sociology.

10 Contraception and Reproduction: Making Babies in a Technological Age

FERTILITY CONTROL IN CROSS-CULTURAL AND HISTORICAL CONTEXTS

Before exploring the ways that our own society organizes contraception and birth, it is useful to look at how other societies have regulated birth and population growth. All human populations must strike a delicate balance with nature if they are to survive (chapter 3). Overpopulation has often been avoided because many people die from disease or warfare. Societies have also regulated their birthrates to ensure a sufficient food supply for their members. Intentional population control has typically been accomplished through infanticide (the killing of infants) and contraception (techniques to prevent pregnancy). The most common forms of contraceptive practices have been celibacy, delayed marriage, abortion, and minimizing sexual intercourse through taboos or prolonged breast-feeding. Late weaning of young children and low percentage of body fat also seem to have been effective means of birth control among hunting and gathering societies (Blumberg 1984). Even without scientific knowledge of human reproduction, many societies discovered and used various contraceptive barrier devices (e.g., roots, pods, animal intestines, grasses) and contraceptive or abortive oral concoctions (mostly herbs, roots, or metals). It is unlikely that these contraceptive devices, or the ritual practices that accompanied their use, were very effective, but they probably did reduce the chance of pregnancy to some degree.

High infant mortality rates from disease or malnutrition have naturally limited population growth, but the most common form of *intentional* population control has been infanticide. Although it is comfortable to attribute the killing of newborn babies to some distant and primordial tribes, it may have been most common among those we like to think of as most civilized. For example, Plato and Aristotle advocated infanticide as a means of population and disease control. William Langer (1972) reports that infanticide was especially common in Europe and Britain during the eighteenth and nineteenth centuries and that women were almost never punished for taking the life of their infants:

> *In the eighteenth century it was not an uncommon spectacle to see the corpses of infants lying in the streets or in the dunghills of London and other large cities. . . . In 1862 one of the coroners for Middlesex county stated infanticide had become so common that the police seemed to think no more of finding a dead cat or dog." . . . The Morning Post (September 1863) termed it "this commonest of crimes." (Langer 1972, 96–97, cited in Skolnick 1987)*

Abandoning children, or putting them in foundling homes, was also quite common in eighteenth-century Europe and amounted to almost the same thing as infanticide. Although foundling homes were started by reformers, such as St. Vincent de Paul and Napoleon, who were shocked at infanticide, the conditions were so bad in these homes that between 80 and 90 percent of infants placed there died (Skolnick 1987, 317). In some traditional peasant societies, such as China or India, the killing of girl babies has also been common for many centuries.

These historical practices suggest that there is no universal "maternal instinct" that causes mothers to treat babies a particular way and that belief in the sacredness of infant life is the product of specific cultural and historical influences. Cross-cultural studies show us that when women are given free choice—that is, when children are not needed for old age security, and religious, cultural, and economic pressures are minimal—they choose to have only a few children (Steinmetz, Clavan, and Stein 1990).

Variation in Birthrates

Women in a given society tend to give birth at similar ages and to have a typical number of children. Theoretically, it is possible for women to have thirty or more births, but even ten is rare. Partly, this is a matter of health, because many women used to die in childbirth before the development of modern standards of medical practice, but especially it implies that social factors are more important than biological ones in determining how many children are born.

Our own birthrate has been hovering around the point of **zero population growth** for decades. In part, this is because children are increasingly expensive to have and raise, and they bring in no income (chapter 12). In many traditional societies this was not the case. In the agrarian, or medieval, system of peasants and rural landlords, for example, child labor was useful for the parents. Children could perform many tasks, such as caring for livestock or helping in the largely self-subsistent household economy. Moreover, in these societies the landlords were usually pressuring the peasants hard and taking most of their surplus crops, and hence there was even greater pressure for peasants to have as many children as possible to squeeze out even more agricultural productivity. On top of this, as Rae Blumberg (1984) points out, women in these societies tended to have little power, so economic pressures plus male dominance encouraged multiple births in rapid succession, despite the difficulty of childbirth under poverty conditions and the high maternal death rate.

Conversely, in many traditional societies where women had greater power, they took measures to keep the birthrate more comfortably under their own control. In tribal hunting-and-gathering societies (see more on these in chapter 3), women usually spread their children quite far apart, typically more than four years; so they had fewer small children to deal with. Today, in the agricultural countries of the Third World, women also tend to reduce their own birthrates when they acquire more economic resources of their own. This is sometimes done in opposition to their husbands, who cling to the more traditional, machismo ideal in which a man's status is measured by the number of his children (which ensures that his wife is too busy to challenge his authority). Even in traditional parts of Latin America, women have increasingly resorted to illegal abortions to gain some control of their own lives (Remez 1995). When women in Puerto Rico entered the labor market on their own, for instance, the first thing they did was to initiate contraception. Because their husbands often depend on their income, they tend to get their way regarding family size.

These same causal conditions seem to operate in our own society. We are at an advanced stage of this process because the movement of women into full-time work has proceeded quite far here. Thus, all the general factors point toward a decline in our birthrate. First, children are not an economic asset but a sheer expense; hence, the economic motivation for having children has disappeared. Second, women have entered the labor force in large numbers and are now committed to independent careers. This both reduces their motivation for having children (especially early in their lives) and increases their power to make their own decisions about childbearing. The result is a long-term slowing of the population growth.

We are currently very close to zero population growth, with a birthrate of about two births per woman. Our population is not quite stable yet for two

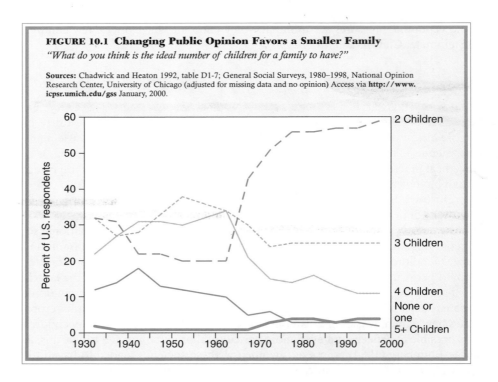

FIGURE 10.1 Changing Public Opinion Favors a Smaller Family

"What do you think is the ideal number of children for a family to have?"

Sources: Chadwick and Heaton 1992, table D1-7; General Social Surveys, 1980–1998, National Opinion Research Center, University of Chicago (adjusted for missing data and no opinion) Access via **http://www. icpsr.umich.edu/gss** January, 2000.

reasons. One is that immigrants continue to come in, primarily from Latin America and from Asia. The other reason is that the actual numbers of children born depend on how many women are of childbearing age. Because there was a "baby boom" at midcentury (roughly 1946–1964), an exceptionally large number of women reached childbearing age during the 1970s and 1980s. This resulted in what is sometimes called the demographic "echo boom." However, in the 2000s, the number of mothers will be rather low, and the number of children being born will decline.

There are a number of factors that enter into this low birthrate. On the whole, people are older when they marry. Although many births occur outside of marriage, marital births account for the majority of births. More and more women are waiting longer to have their children, many delaying into their thirties. The result is that as women reach childbearing age, they do not add their children to the population so quickly, which also reduces the rate of population growth.

Most people still want to have children. They just do not, for the most part, want to have large families any more. As Figure 10.1 shows, a majority of Americans (59 percent) regard two children as the ideal family size, a proportion that has been growing ever since the early 1960s. In the 1940s and 1950s, over 40 percent of Americans thought that four children or more was the ideal number for a family. Today, it is the trendsetting sections of the population who are most strongly in favor of having only two children: the college educated more than the less educated, urban dwellers more than rural, and professionals more than service workers or manual laborers. Women of childbearing age—the persons who are most directly affected by having to care for children—are especially likely to favor two children, although a majority of men also favor that number. All this suggests that the two-child family will become even more strongly entrenched as time goes on.

Most indicators suggest that the two-child family will become increasingly entrenched as the standard of family size.

NONMARITAL CHILDBIRTH

"There was something mysterious about that house," related a women interviewed in a midwestern town, speaking about a time fifty years ago. "There was somebody in there who was never seen, never spoken about. Everybody knew she was there, but nobody would ever let on that she existed. It must have been terrible for her. She used to come out at night sometimes, just to drive around in a car, but never stopping anywhere, just to get out." She was a woman who had had an "illegitimate" child. Abortions were not easily available; she had given birth, put the child up for adoption, and gone back to live with her parents. It was all done in great secrecy, but everyone in the town knew about it, speaking in hushed tones with accents of gravest scandal. Her parents took her in, and she was rarely heard from again.

This kind of story, with all its variants, must have happened millions of times throughout the last centuries. A woman's life was circumscribed between social pressures: men pressing her toward sexual adventure (perhaps along with her own rebelliousness, independence, or sexual desires), while social condemnation awaited her if she did not avoid any appearance of sex outside of marriage. In the 1500s and 1600s in England, when most of the wealthier households had female servants, it often happened that a young woman was made pregnant by her employer, then dismissed from service with no chance of getting another job; the result was that she would often drift into poverty or prostitution (Stone 1977). Though men were often the exploiters in these everyday dramas, the worst condemnation often came from other women, who scorned unmercifully any woman who fell from the narrow path of socially defined virtue.

Times have changed, but only relatively recently. Nonmarital childbirth and various ways of dealing with it are no longer hidden in the shadows or subject to traditional persecutions. Nevertheless, major problems and controversies remain, and various movements are pressing either for new solutions or for ways to turn back the clock. We now examine some of these trends.

Pregnant Brides

Pregnancy can be, among other things, a route to marriage. In colonial America, as well as in England during the late 1700s, for example, almost half of all brides were pregnant at the time of the wedding ceremony (Stone 1977). It has been suggested that a "shotgun marriage" was one way that a woman could maneuver a less-than-willing boyfriend into getting married—a rather dangerous game, in view of the dire social sanctions if she failed. For that very reason, however, pregnancy was likely to put tremendous moral pressure on the man "to make an honest woman out of her." There may actually have been a fair amount of complicity on the part of the man, however. Such couples, especially in the lower classes, seem to have often deliberately begun intercourse about the time it was economically possible for them to marry and only delayed the actual marriage until a pregnancy occurred (Laslett 1971). This kind of "normal" premarital pregnancy appears to have become less common in the 1800s and early 1900s, although it was still practiced by a significant minority of the population. By the late 1970s and early 1980s, about one-quarter of all new brides were either pregnant or had already given birth. This was particularly true of teenage brides, about one-third of whom had their first child within eight months of the wedding.

The old pattern of pregnant brides diminished because the context for marriage has been changing. The results of premarital pregnancy are not as dire as they once were, and when marriage follows, it is less of a forced choice and more of a conscious decision. Couples living together, for instance, used to be likely to get married when the woman got pregnant or when they decided to have children. This still happens often, but as we discuss next, it is no longer the most common pattern.

The Nonmarital Childbirth Explosion

The proportion of births to unmarried women of all ages has gone up continuously during the last half of the twentieth century, so that now there are about 1.3 million nonmarital births each year in the United States (Figure 10.2). In 1960, nonmarital births accounted for 5 percent of all births, but by 1998, the share had increased to about one of every three children born. It used to be that most nonmarital births happened to teenagers, but the recent dramatic increase in the birthrate outside marriage is driven primarily by women age twenty and older, who now account for about 70 percent of all nonmarital births (see figure 10.2). Looking just at first births (as opposed to second and later births) reveals similar tendencies. About one in two births in the 1990s were conceived outside of marriage, including over 80 percent of first births to African Americans, over 50 percent of births to Latinos, and over 40 percent of births to whites (Bachu 1998).

To understand this rapid rise in the number of children born outside of marriage, we need to consider several factors. In general, we have seen that being single and having a baby does not carry the stigma that it used to, particularly among African Americans but also for whites and Latinos. This suggests that social norms might be important. Do trends in nonmarital births signal a major breakdown in the morals of young people, as some politicians and religious leaders suggest, or are they related to major social developments common to modern

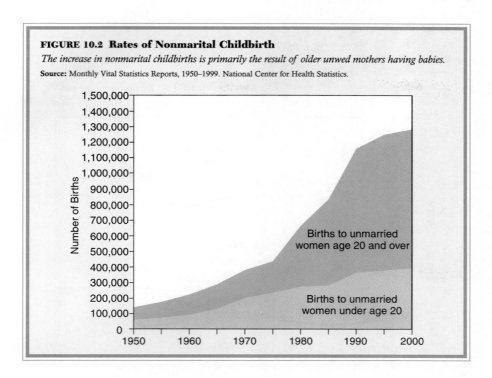

FIGURE 10.2 Rates of Nonmarital Childbirth

The increase in nonmarital childbirths is primarily the result of older unwed mothers having babies.

Source: Monthly Vital Statistics Reports, 1950–1999. National Center for Health Statistics.

industrial societies? We should also pay attention to the changing dynamics of marital relations in American society and consider especially the role of cohabitation. Other factors to consider include frequency of sex and women's chances of getting pregnant. Is there more nonmarital childbirth because young women are having more sex, or are they becoming more fertile? To answer such questions we need to take a look at how sexual activity, access to information about birth control, use of contraceptives, and availability of abortion might influence birthrates. Finally, to figure out what is going on, we need to make some distinctions between different ways of measuring pregnancy, conception, and birth.

Recent data from the *National Survey of Family Growth* (Abma et al 1997), *National Vital Statistics Surveys* (e.g., Ventura et al. 1999), and *Current Population Surveys* (Bachu 1998, 1999) allow us to examine these issues in some detail. Since 1950 the total number of babies born in the United States has fluctuated between about 3 and 4 million. In the late 1950s and early during the height of the baby boom, up to 4.3 million babies were born in a year. From the late 1960s to the early 1980s, the number hovered around 3.5 million, and since then approximately 4 million have been born each year.

Because the number of women of childbearing age has fluctuated over this period, it is more interesting to focus on the **birthrate** which refers to the number of births per 1,000 women between the ages of fifteen and forty-four. The birthrate is a measure of the "risk" a woman of childbearing age will give birth in any given year. The birthrate has fallen fairly steadily over the century, declining from about 25 births per 1,000 women during the baby boom era to under fifteen today. This simply means that women are less likely to have babies than they were earlier in the twentieth century. The decline in birthrates is common to all industrialized countries and is related to a wide range of economic, technological, nutritional, health, and social factors. The share of births

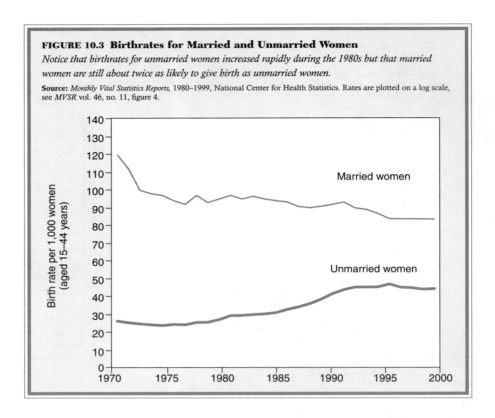

FIGURE 10.3 Birthrates for Married and Unmarried Women

Notice that birthrates for unmarried women increased rapidly during the 1980s but that married women are still about twice as likely to give birth as unmarried women.

Source: *Monthly Vital Statistics Reports,* 1980–1999, National Center for Health Statistics. Rates are plotted on a log scale, see *MVSR* vol. 46, no. 11, figure 4.

occurring outside of marriage in our country is similar to the share found in Canada, France, and England but only half the rate found in Sweden. In contrast, rates in Japan and some other countries are many times lower than in the United States (Ahlburg and DeVita 1992).

From the 1930s through the 1950s, about one in six first births were conceived outside of marriage, but over half of those mothers had married by the time of the birth. A major change occurred between the 1960s and the 1980s, a period of significant change among young people in the United States. Sexual practices became more open as we saw in the last chapter, but as we discuss later, birth control became more available and abortion laws were relaxed—factors hypothetically linked to the ability to avoid nonmarital birth. Nevertheless, rates of nonmarital birth rose substantially. By the early 1990s, over one-third of first births occurred before a first marriage, and by the late 1990s, both married and unmarried birthrates were declining (Bachu 1999). Figure 10.3 summarizes these trends by comparing the birthrates for married and unmarried women in the United States from 1970 to 1998. We can see that the rate for unmarried women climbed from the late 1970s to the early 1990's, when it stabilized, then declined slightly. The rate for married women, in contrast, has been declining pretty steadily since the early 1970s. Figure 10.3 also shows that the birthrate for married women is still about twice as high as that for unmarried women.

Looking at **pregnancy rates,** in addition to birthrates, can give us a better indication of what is going on. For 1996 (the last year with complete statistics for this sort of analysis), there were 105 pregnancies per 1,000 (women aged fifteen to forty-four). This pregnancy rate is the lowest it has been since 1976 (when it was 103). As for birthrates, pregnancy rates are highest for women in their twenties and higher for African Americans and Latinos than for whites.

In 1996, there were just over 6 million pregnancies but just short of 4 million live births in the United States. Obviously, some pregnancies do not result in births: 16 percent in 1996 ended in "fetal loss" (i.e., miscarriages, stillborns) and 22 percent in induced abortions (Ventura et al. 1999). These three pregnancy-related rates (fetal losses, induced abortions, and live births) have varied somewhat since the mid-1970s, but all have been declining since about 1990, with the rates for abortions falling the most. Pregnancy rates actually increased for women over thirty years old during this period and declined for women under twenty. The 1996 teen pregnancy rate of ninety-nine is the lowest recorded in the two decades for which this series of pregnancy rates is available. Although at a low point for the United States, it should be remembered that teen pregnancy rates in the United States are still many times higher than in most other industrialized countries.

Returning to the issue of marital versus nonmarital births, we can compare what happens to pregnancies depending on the marital status of the pregnant woman. From the 1980s through the 1990s, about three-quarters of pregnancies among married women ended in live births, whereas the proportion of pregnancies ending in live births for unmarried women rose from about one in three to almost one in two. The pregnancy rate for married women is still somewhat higher than the rate for unmarried women (by about 20 percent), but the differences in abortion rates are much larger; the birthrate for married women (84 per 1,000) is nearly ten times their abortion rate (9 per 1,000). In contrast, the birth and abortion rates for unmarried women are nearly equal (45 per 1,000 and 39 per 1,000, respectively) (Ventura et al. 1999). In other words, married women are only slightly more likely than unmarried women to get pregnant, but married women are considerably more likely to carry the baby to term.

What are the chances of the couple getting married as a result of the pregnancy? About one of four women who had premaritally conceived a child in the 1990s married before the child was born. Which women were most likely to get married? In general, patterns correspond to overall trends in marriage rates. African-American women were least likely to marry after they conceived a child outside of marriage, and women of all races and ethnicities with less than a high school education were also significantly less likely to marry. When compared with women in their early twenties, teenage women were least likely to marry before the birth. Women in their early thirties were the most likely to marry after conceiving a child premaritally, perhaps because they are more likely to be cohabiting in the first place (Bachu 1999).

By looking at rates of cohabitation, we can get a better idea of the context for decisions about birth and marriage. In the early 1980s, about 30 percent of births to unmarried women consisted of women who were cohabiting (presumably with the child's father). By the 1990s, that proportion had risen to about 40 percent. Thus, according to several researchers, almost all of the increase in nonmarital childbearing from the 1980s to the 1990s is accounted for by the increase in the number of cohabiting women who were getting pregnant and carrying the baby to term. The recent increases in nonmarital births can therefore be attributed to an increase in cohabitation and to the decline of the "shotgun marriage" (Bachu 1998; Manning 1993; Ventura et al. 1999). Findings from such studies also alert us to the fact that nonmarital birthrates are not the same as estimates of father absence: children born outside of marriage do not necessarily grow up in single-parent families. About two-thirds of children born to Latino women outside of marriage are born into two-parent cohabiting unions, compared with 57 percent for white women and 26 percent for black women (Bachu 1999).

To summarize, we call attention to the fact that most babies are still conceived and born within marriages but note that significant changes have occurred in the later part of the twentieth century. Although a majority of people continue to assume that childbirth should occur within marriage, the imperative to marry in order to have children has weakened considerably. It used to be the case that most people married to "legitimate" a birth, but only a minority of women who conceived premaritally now do so. As discussed later, an increasing number of single women, many in their thirties, are now deciding to have children on their own. Other women are cohabiting with the father of the child, delaying or sometimes avoiding a decision to marry. As we will see, more same-sex couples are also deciding to have children together. In short, there are a wider variety of household and family arrangements within which to bring children into the world.

Although increasing rates of nonmarital birth are sometimes interpreted as a problem of oversexed and amoral teenagers, we see little evidence to support this argument. In fact, the dramatic rise in nonmarital birth has little to do with recent changes in teenagers' sexual habits. Teen sex and teen births have been declining since about 1991, at the same time that overall levels of nonmarital birth have been increasing. Thus, as we will see in following chapters, we must look elsewhere to understand what is happening to marriage and child rearing in America. We also must point out that almost 1 million pregnancies still occur annually among American female teenagers, a level far higher than in similar industrialized countries. Because national differences in levels of sexual intercourse cannot account for the relatively high U.S. rate of teenage pregnancies, we need to focus on another important factor influencing birth—the underuse of contraception.

CONTRACEPTION

Historical Trends

Contraceptive methods of various types have been used for a long time, although they have not always been very reliable. Vaginal sponges soaked in lemon juice or wine (tannic acid and lemon juice are effective spermicides) were used in Europe, and cervical caps, penis sheaths, pessaries, and potions can be traced back to antiquity (Schneider and Schneider 1995). In 1844 the vulcanization of rubber was invented, and by the late 1800s, rubber condoms and diaphragms were in use (Himes 1963; Reed 1978). In the rather puritanical public atmosphere of the late nineteenth and early twentieth centuries, however, contraceptives were available only covertly. Margaret Sanger coined the term *birth control* and launched a campaign to bring information about it to the public, after visiting the deathbed of a young woman who died in 1909 while undergoing an illegal abortion. However, contraceptives remained illicit in many places, and some states followed Catholic doctrine in treating any form of contraception as murder. Federal customs agents used to confiscate women's diaphragms at the docks when they were returning from a trip abroad. Only in 1965 did the U.S. Supreme Court finally strike down laws prohibiting birth control, on the grounds that these laws unconstitutionally infringed on the individual's right to privacy.

In the early part of the century, even though contraception was still outlawed, social reformers convinced some state governments that so-called "inferior" people should be prohibited from making babies. To protect and "improve" the race,

Margaret Sanger, who coined the term birth control and promoted its practice, is confronted by a mother of seven in 1917.

eugenic sterilization was performed on individuals whose antisocial behavior allegedly gave evidence of hereditary defects. Elaine Tyler May (1995, 104) points out that sterilization was also performed to punish criminals who were believed to be sexually dangerous: "Given the racist assumptions of the day, it is not surprising that punitive eugenic policies were carried out with particular brutality toward black men." A small but energetic group of eugenics advocates in the midwest and far west were able to pass laws restricting immigration and requiring the forced sterilization of the "feebleminded" (*A Closer Look* 10.1). More recent attempts to sterilize women deemed undesirable have been directed at poor black, Hispanic, and Native American women who are on welfare (May 1995).

By the 1960s, voluntary forms of contraception, like the birth-control pill and the intrauterine device (IUD) had appeared. These constituted easier forms of birth control because they could be controlled by the woman (unlike the condom) and did not require preparing for every act of intercourse (unlike the diaphragm). As these became available and widely publicized, it was expected that unwanted pregnancies would virtually disappear. As it turned out, however, this was not the case, as we know from the increase in both nonmarital births and abortion.

In the last few decades there has been a revolution in the acceptance of birth control in most countries, including the United States. This is true even though the official doctrine of the Roman Catholic Church—America's single largest religious denomination—condemns all forms of birth control (except for abstinence or the rhythm method). By the 1980s, contraceptive use by Catholics was virtually identical to that of Protestants (Westoff 1986).

Contemporary Patterns

Because levels of sexual activity in the United States are comparable to levels in most Western countries, the high rates of teenage pregnancy in our country can

A CLOSER LOOK 10.1
Involuntary Sterilization

In the early part of the twentieth century, involuntary sterilization was a relatively common procedure in the United States. Modern Americans tend to assume that only in totalitarian dictatorships, like Hitler's Germany, did governments attempt to "improve the race" by forcing selected people to undergo surgical sterilization. In fact, countries as diverse as England, Spain, Denmark, and India adopted eugenics policies in this century with the specific intention of improving human heredity.

Derived from the Greek word meaning "well-born," the term *eugenics* was coined by Sir Francis Galton in 1883, who defined it as "the study of agencies under social control that may improve or impair . . . future generations either physically or mentally." Strongly influenced by social Darwinism, the pseudo-science of eugenics claimed as its goal "the betterment of the human race through the study and classification of genetic traits, establishing in the process a ranking of desirable and undesirable traits for human propagation" (Dugan 1993, 516). Eugenicists claimed that feeblemindedness was a remediable social evil that could be genetically isolated and effectively neutralized through the sterilization of those who were afflicted. "Despite the questionable scientific basis for this assertion, sterilization of the developmentally disabled captured the legislative imagination. This was partly due to the well-organized and politically

effective lobbying groups that espoused the cause to various state legislatures. Such legislation also became popular as a cost-saving device: these statutes typically provided for the discharge of the sterilized person from state custodial care, thus saving the state the expense of full-time care for that person, as well as for any possible offspring" (Dugan 1993, 516).

Indiana was the first state to pass an involuntary sterilization law in 1907, and twenty-five other states soon followed. In California, the new sterilization law led to twice as many sterilization procedures than in all other states combined. Between 1909 and 1929, 6,787 people (3,636 men and 3,151 women) diagnosed as insane or feebleminded underwent eugenic sterilization in California state institutions (Human Betterment Foundation, 1930). According to Paul Popenoe, a leading proponent of eugenic sterilization who conducted and published many studies for the Human Betterment Foundation, two-thirds of the persons sterilized in California were committed as insane, and one-third as feebleminded. Most of the men in the mental hospitals who underwent the involuntary operation were single and diagnosed as suffering from dementia praecox (schizophrenia), with an average age of thirty-three years. Most of the women labeled insane were married, with an average age of thirty, and diagnosed as manic-depressive. In the homes for the feebleminded, most of the operations were performed on people

who were between fifteen and twenty-five with an average IQ of 60, and three-fourths of the girls were labeled "sex delinquents." It was state policy to sterilize both men and women before they were released (Popenoe 1927a, 1927b, 1928a, 1928b, 1928c).

According to a 1938 report from Popenoe and Gosney of the Human Betterment Foundation, the "foreign-born" were over-represented among those sterilized, as were children whose fathers were "unskilled day-laborers," those from large families, and those from "broken homes." They suggest that "the records of public welfare departments and charitable organizations show that some of the least desirable citizens (beggars, vagrants, prostitutes, criminals) are chronic sufferers from dementia praecox in a mild form. . . . It is against this widespread and sinister menace, so little understood by the public, that the sterilization campaign is largely directed." Believing schizophrenia, manic depression, feeblemindedness, epilepsy, and related conditions were genetically inherited, eugenicists believed they were helping civilization by sterilizing people with these ailments. Worrying that modern civilizations were artificially protecting the weak and helpless, they decided they could not depend on nature to solve what they saw as the increasing problem of "the survival of the unfit." Popenoe and Gosney (1938, 21–22) state the eugenics view plainly:

[T]here are at least 5 or 6 million feebleminded persons in the United States. The cost to the community of these millions of defectives must be reckoned not merely in terms of dollars and cents, but still more in lowered standards of education, citizenship, industrial efficiency, and in human misery and suffering. There is little overlapping between the groups of mentally diseased and mentally deficient. Together they make up nearly 15 percent of the entire population that must be regarded as seriously handicapped mentally. Some of these persons have superior qualities in other respects, some are dull but harmless, perhaps useful workers. But they produce little leadership and much dependency. Also their multiplication furnishes a fertile soil for the production of dependents, delinquents, and criminals. Characterized by lack of foresight, self-control, and judgment, by ignorance, indifference, and frequently alcoholism, mentally defective persons do not limit their families as do the college and high school graduates. Hence investigations in many states agree with our researches in California which showed that educated families are dying out, while the families of the feebleminded and of the chronic paupers are not only holding their own but increasing.

Looking back on this period, historians such as Frank Dikotter (1998, 468) suggest that the eugenics movement was not so much a clear set of scientific principles as it was a new way to talk about social problems in "biologizing" terms: "Eugenics gave scientific authority to social fears and moral panics, lent respectability

to racial doctrines, and provided legitimacy to sterilization acts and immigration laws. Powered by the prestige of science, it allowed modernizing elites to represent their prescriptive claims about social order as objective statements irrevocably grounded in the laws of nature."

Before dismissing this movement as an irrelevant historical aside, consider the current Human Genome Project, a $3 billion program funded by the U.S. Government aimed at mapping and sequencing the genes that control our heredity. Many researchers see the rhetoric and implications of the Human Genome Project as reflecting similar tensions as the earlier eugenics movement. Supporters suggest that the project could isolate the genetic bases of schizophrenia and manic depression, among other diseases, and that expectant parents could screen their potential offspring for troubling anomalies. Detractors, on the other hand, worry that attempts to discover genetic bases for behavioral and sexual behaviors will lead to more discrimination against the less advantaged members of society:

More than a century ago, Francis Galton and other eugenicists dreamed of the day that the study of the wellborn would become part of the secular religion of developed countries. That time has come. In the Western world, the genetic revolution has brought new rhetorics of eugenics that celebrate the rise of new sciences like "genomics" and the development of new ways of assuring "perfect" babies. The media remind English and

American audiences of the power of genetics to explain everything from divorce to rising crime rates. The dense, discursive mythologies and narratives that once demanded the containment of racism and affirmed the inherent quality of all persons have been augmented with new lexicons that remind us of differences that are "hard-wired" (Hasian 1996, 156).

Sources:

Dikotter, Frank. 1998. "Race Culture: Recent Perspectives on the History of Eugenics." *American Historical Review* (April):467–78.

Dugan, James C. 1993. "The Conflict between 'Disabling' and 'Enabling' Paradigms in Law." *Cornell Law Review* 78:507–42.

Hasian, Marouf Arif, Jr. 1996. *The Rhetoric of Eugenics in Anglo-American Thought.* Athens, GA: University of Georgia Press.

Human Betterment Foundation. 1930. *Collected Papers on Eugenic Sterilization in California: A Critical Study of Results in 6000 Cases.* Pasadena, CA: Human Betterment Foundation.

Popenoe, Paul. 1927a. "Eugenic Sterilization in California: I. The Insane." *Journal of Social Hygiene* vol. 13, no. 5 (May):257–68. Reprinted in *Collected Papers.*

Popenoe, Paul. 1927b. "Eugenic Sterilization in California: I. The Feebleminded." *Journal of Social Hygiene* vol. 13, no. 6 (June): 321–30. Reprinted in *Collected Papers.*

Popenoe, Paul. 1928a. "Eugenic Sterilization in California: V. Economic and Social Status of the Sterilized Insane." *Journal of Social Hygiene* vol. 14, no. 1 (Jan.): 23–32. Reprinted in *Collected Papers.*

Popenoe, Paul. 1928b. "Eugenic Sterilization in California: VI. Marriage Rates of the Psychotic." *The Journal of Nervous and Mental Disease* vol. 68, no. 1 (July): 17–27. Reprinted in *Collected Papers.*

Popenoe, Paul. 1928c. "Eugenic Sterilization in California: XIV. The Number of Persons Needing Sterilization." *The Journal of Heredity* vol 19, no. 9 (Sept.): 405–11. Reprinted in *Collected Papers.*

Popenoe, Paul, and E. S. Gosney. 1938. *Twenty-Eight Years of Sterilization in California.* Pasadena, CA: Human Betterment Foundation.

be attributed to low levels of contraceptive use. Research has consistently shown that U.S. teens are relatively slow to adopt contraceptive techniques. Few teenagers report wanting to have a baby when they begin having sex, but some studies show that about 10 percent of whites and 20 percent of blacks report that they didn't think about pregnancy or didn't care about whether a pregnancy occurred when they first had sexual intercourse. (Miller and Moore 1991). About two-thirds of unmarried female teens who are having sex report using some form of birth control. Nevertheless, only 59 percent of females ages fifteen to nineteen report having used some method of contraception at first intercourse, with rates differing according to race/ethnicity. Almost two-thirds of white (64.8 percent), half of black (50.1 percent), and over a third of Hispanic (36.2 percent) teenage females who have ever had intercourse report that they used some form of birth control (Abma et al. 1997). Recent data suggest that condom use has increased substantially in the last decade primarily as a result of awareness among teenagers. Teenagers having sex in the 1990s reported significantly higher rates of condom use than other sexually active adults. In part because of educational programs, overall rates of contraceptive usage among teens who had first sexual intercourse in the 1990s climbed to over three of four (83 percent for whites, 72 percent for blacks, and 53 percent for Hispanics) (Abma et al. 1997, table 37).

The foregoing statement refers to teenagers; however, women in their twenties are most sexually active. About half of women aged twenty to twenty-nine have never married, and a majority of them have had intercourse at least once during the previous month. Nearly all these women have used a contraceptive at some time; they are not ignorant of contraceptive methods. Still, about one in ten report that they did not use a contraceptive at the time of their most recent intercourse. This represents more contraception than among teenagers, but it still reflects much unprotected intercourse. A recent comprehensive study shows that almost 33 percent of women using "coitus-dependent contraceptive methods" (condom, diaphragm, withdrawal, etc.) did not use them consistently; that is, during the past three months, they did not use any birth control at least once. An even higher proportion of teenagers (38 percent) reported some nonuse, but the highest levels of inconsistent contraceptive use (almost 42 percent) were reported by women twenty to twenty-four years of age (Abma et al. 1997). Not surprisingly, younger women are also more likely to become pregnant unintentionally.

The question then becomes: Why don't unmarried women use contraceptives more regularly? One response is that they do use them, especially if they have intercourse with more than one partner, have intercourse relatively frequently and over a long time, and discuss the contraception explicitly with their partner in advance. However, notice what this means: the women who are the most sexually active and sophisticated are the ones most likely to use contraceptives; the women who don't use them are the ones who have only occasional intercourse, with perhaps only one partner—and these are the women most likely to become pregnant. In short, the "good girls," who stay closest to the traditional pattern, are least apt to control unwanted pregnancies. To deliberately use contraceptives, and especially to talk about them in advance, is a way that a woman announces to herself (and perhaps to others) that she is sexually sophisticated. The young woman who regards herself as "virtuous" in the traditional sense is the one least likely to use contraceptives—not because she is ignorant of them but because doing so would contradict her moral self-image (Luker 1975, 1996). For example, religious young women are more likely to postpone sex but also less likely to use contraceptives when they have sex. One study found that Jewish women were the most likely to use contraception at the time of first intercourse and Fundamentalist Protestants, least likely (68 percent versus 39 percent). Women who wait until at least 19 and those with college-educated mothers were more likely to use contraceptives at first intercourse (Mosher and Bachrach 1996). As women become more sexually experienced, they become more consistent contraceptive users (Hofferth and Hayes 1992). Availability of family planning services and parental communication about sex also seem to increase contraceptive use (Miller and Moore 1991).

Which forms of birth control are used most frequently? Considering all U.S. women aged fifteen to forty-four, female sterilization is the most common contraceptive method (18 percent), followed by the oral contraceptive pill (17 percent), the male condom (13 percent), and male sterilization (7 percent) (Abma et al. 1997). Three new contraceptive methods introduced in the late 1980s and early 1990s, hormonal implants (1 percent), hormonal injectables (2 percent), and female condoms (under 1 percent), may become more popular. These new techniques are most often used by women under thirty, and future popularity will likely be shaped by legal and medical issues. Considering aggregate trends, use of birth control increases dramatically at about age twenty and remains relatively constant from age twenty-five to the end of a woman's reproductive years.

Sex Education

Contraception is still considered an improper topic for discussion by many Americans. The mass media rarely mention contraceptive use, family planning services are not readily available, many boys continue to view pregnancy as the young woman's problem, and teenage girls are reluctant to admit that they are sexually active.

In this context, sex education programs in the public schools play an important role in providing access to reliable information about contraception, pregnancy, and sexually transmitted diseases (STDs). Studies show that school-based sexuality education programs are effective in reducing rates of unprotected sex and unwanted pregnancies, particularly among teens. The educational programs themselves have been found to lower pregnancy rates, as do family planning services and clinics (Kahn, Brindis, and Glei 1999). According to national surveys, an overwhelming majority of U.S. adults support sex education in the public schools, but recent policy developments have politicized the design and implementation of programs offered by local school districts (Landry, Kaeser, and Richards 1999).

For several decades, self-described "profamily" groups in the United States have lobbied to remove sex education from the schools or to forbid schools from providing contraceptive information or services. Instead, they have argued, the government and public schools ought to address the issue of teenage sexual activity and pregnancy by promoting **abstinence** (not having sexual intercourse). Throughout the 1990s, debates about whether or how to teach sex education to teenagers garnered attention across the country, with one educational organization documenting more than 500 local controversies in 50 states, mostly involving the promotion of abstinence-only programs (Landry, Kaeser, and Richards 1999). In 1996, "profamily" lobbying efforts resulted in the passage of federal legislation to promote the teaching of abstinence in the schools. Through this legislation, Congress established a $250 million, five-year entitlement to states to support a variety of educational efforts that must have abstinence promotion outside of marriage as their exclusive purpose. As of 1999, all but two states had designed such programs (Landry, Kaeser, and Richards 1999).

Although more research needs to be conducted on such programs, a 1997 evaluation study concluded that there was no scientifically credible published research demonstrating that "abstinence-only" programs actually delayed the onset of sexual behaviors. The Consensus Panel on AIDS of the National Institutes of Health declared that the abstinence-only approach "ignores overwhelming evidence" that more comprehensive forms of sex education are effective in delaying the onset of sexual intercourse and increasing the chances that sexually active teens will use condoms (Landry, Kaeser, and Richards 1999). In the shadow of these funding restrictions and heated debates, local school boards have been reconsidering how sex education should be taught.

Recent studies show that 96 percent of teenagers received some formal sex education in the mid-1990s. The largest number reported that they had received instruction on "how to say no to sex," but most also reported learning about STDs and something about birth-control methods (Abma et al. 1997). A nationwide survey of school districts showed that about 69 percent had an explicit policy to teach sexuality education, with the remaining 31 percent leaving policy decisions to the individual schools or teachers (Landry, Kaeser, and Richards 1999). Only one in seven school districts with such policies called for **comprehensive sex education,** that is, treating abstinence as one option for adolescents in a broader sex education program considering various contraceptive methods to prevent pregnancy and avoid STDs. Half had an **abstinence-plus policy,** mandating the teaching of abstinence as the preferred option for adolescents, but also permitting discussion of contraception as an effective means of protecting against unintended pregnancy and disease. Finally, just over a third (thirty-five percent) of school districts adopted an explicit **abstinence-only policy,** mandating that teachers must portray abstinence as the only option outside of marriage. Abstinence-only policies prohibit discussion of contraception altogether or permit discussion of contraception only to emphasize its shortcomings.

As might be expected, regions of the country where women's opportunities are more tied to their status as wives and mothers are more likely to embrace school policies that limit young women's access to information about sexuality. School districts in the South are most likely to have abstinence-only policies, followed by districts in the Midwest and Mountain states. School districts in New England, the Mid-Atlantic, and the Pacific Coast are least likely to have abstinence-only programs and more likely than other states to have comprehensive sex education policies. In the wake of the 1996 federal legislation, the movement in the late 1990s was primarily from the more liberal comprehensive policies to the moderate abstinence-plus policies, with no net gain for abstinence-only policies (Landry, Kaeser, and Richards 1999). As a result, at the turn of the century, about one in three school districts continued to forbid the dissemination of positive information about contraception, regardless of whether students were sexually active or at risk for pregnancy or STDs. With little faith that telling students to "just say no" will lower rates of sex, family planning professionals have worried that unintended pregnancies will continue at high levels. Unintended pregnancies, in turn, force many women (and sometimes men) to face the difficult question of whether to consider having an abortion.

ABORTION

Forty-six million women around the world have **abortions** (medically induced expulsions of the embryo or fetus from the uterus) each year. The reasons women give for choosing to have an abortion are that they have had all the children they want, they want to delay their next birth, they are too young or too poor to raise a child, they are estranged from or on uneasy terms with their sexual partner, and they do not want a child while they are in school or working. In developing countries, where average desired family size is large, of the 182 million pregnancies occurring every year, an estimated 36 percent are unplanned, and 20 percent end in abortion. In developed countries, where average desired family size is small, of the 28 million pregnancies occurring every year, an estimated 49 percent are unplanned, and 36 percent end in abortion. Overall, women in developed and developing regions have strikingly similar abortion levels—39 procedures per 1,000 women and 34 per 1,000, respectively (Alan Guttmacher Institute 1999).

The abortion rate in the United States, although quite high compared with the number of pregnancies, is only average compared with those elsewhere. Some countries, such as the Netherlands, Britain, and France, have abortion rates that are half or even less than half of the American rate. At the other extreme, countries such as Hungary and Cuba have abortion rates that are two or even three times higher than ours. The United States is in the middle, along with Norway and Sweden, Germany, Denmark, and some Far Eastern countries.

The incidence of abortion in this country has risen sharply since abortion became legal in various states around 1970 (*A Closer Look* 10.2). Before that time, illegal abortions were occurring at a rate estimated at between 200,000 and 1.2 million per year. Illegal abortions decreased to low levels after legalization, whereas the number of legal abortions rose to a high point of 1.6 million per year in 1990. Since then the number of abortions has declined to levels comparable to the late 1970s (Figure 10.4). This is still a large number, considering that the total number of live births is almost 4 million annually. Since 1981, the abortion rate has been dropping and now sits at under 23 abortions per 1,000 women (*Statistical Abstract of the United States* 1999, table 123). This rate is influenced not only by the number of women choosing abortion but also by the pregnancy rate, which has been dropping. Not all pregnancies end in either birth or abortion; miscarriages and stillbirths (when the child is born dead) are also common. Thus, of 100 pregnancies, there are 63 live births, 23 abortions, and 14 miscarriages and stillbirths.

Back in the 1950s, the evidence seemed to indicate that most abortions were performed on married women who did not want more children (Calderone 1958, 60). By the mid-1970s, however, abortions were overwhelmingly being sought by unmarried women. Abortions are especially concentrated among teenagers and women in their early twenties. Teenagers have about one-fifth of all abortions, and another third are performed on women from twenty to twenty-four years of age (*Statistical Abstract of the United States* 1999). Although many people assume that abortions are associated with not using contraceptives, nearly half of all abortions are for women who are using some form of contraception during the month in which they conceive (Henshaw and Silverman 1988).

 Medical Abortion

New techniques of nonsurgical or **medical abortion** may change abortion practices and debates. French scientists developed a pill that can be taken within a few days of a missed menstrual period and that causes the uterus to expel the fertilized egg. Testing in Europe has found that RU-486 (mifepristone) is 96 percent effective in terminating a pregnancy and causes fewer side effects and

A CLOSER LOOK 10.2
Two Centuries of Abortion Battles

Before the early 1800s, the law in England and the United States allowed abortion in the first half of pregnancy. The crucial time was considered to be the "quickening": the point at which the fetus first began to move in the womb. This usually occurred late in the fourth month or in the fifth month, after which the fetus was considered to have begun a separate existence. After the 1840s, public controversy over abortion began to emerge in America. This was the height of the Victorian era, and abortion as evidence of sexual behavior was morally condemned. However, apparently large numbers of women, including married women, were resorting to abortion, mostly by medically unlicensed practitioners. Medical techniques in general were not very advanced, even in the regular medical profession, and the operation involved considerable pain and danger. By the 1860s, the American medical profession was attempting to reform itself, to get beyond its own harmful practices such as bleeding patients with leeches; antiseptics, for example, were not invented until 1865. As the best-educated doctors strove to upgrade their own practice, they also began to organize politically to drive out unlicensed practitioners. Among those whom they attacked were the abortionists, although the arguments that they chose were not so much ones of medical safety as moral arguments that abortion was murder.

The historian James Mohr (1978) has called this "the physician's crusade against abortion," and he argued that the political rhetoric that was most effective in getting antiabortion laws passed was an appeal to ethnic division. It was claimed that native-born white Protestant women were the ones who were having abortions, while the new European immigrants were breeding at a dangerous rate. Hence, outlawing abortion was claimed to be a way of keeping "pure" native American stock in the majority. The same sentiments were to lead, early in the twentieth century, to laws largely shutting off immigration. By 1900, virtually every state had passed a law making abortion of any kind a crime.

Throughout the first two-thirds of the twentieth century, abortion was illegal everywhere in the United States. But it did not go away; it only went underground. Kinsey (1953) found that 90 percent of the premarital pregnancies among the women in his sample had been terminated by abortion and that 22 percent of the women had an abortion during the time they were married. These illegal abortions were expensive, difficult to get, and sometimes dangerous, but they happened. In the 1960s, public opinion began to switch. The catalyst was the case of Mrs. Sherri Finkbine, a mother of four who had taken the drug thalidomide during the early months of her pregnancy. Thalidomide was a tranquilizer, but its effect on pregnancy was discovered to be that the child was born deformed, with "flippers" instead of arms and legs. Mrs. Finkbine applied for an abortion to her local medical board in Phoenix, Arizona, which denied her request after considerable national publicity and pressure from religious and political organizations. Mrs. Finkbine finally was able to have an abortion in a Swedish hospital, where she delivered a deformed fetus. As Mrs. and Mr. Finkbine desperately applied to one American medical board after another, the publicity dramatized the abortion issue, and 87 percent of American doctors favored a liberalization of abortion policy by 1967 (Mohr 1978, 256). In the same year, Colorado passed a law legalizing abortion, and seventeen other states had followed suit by 1972. In 1973, the U.S. Supreme Court, reviewing the case of *Roe v. Wade,* struck down all remaining antiabortion laws, dating from the late 1800s.

In the *Roe* decision, the Court held that three rights were involved: the constitutional right to privacy, the right to protect maternal health, and the right of the state to protect developing life. In reconciling these rights, the Court declared that in the first trimester (three months), pregnancy was a private matter involving a woman's right to decide her own future and struck down all laws interfering with that right. During the second trimester, as abortion became medically more dangerous to the mother, the state acquired an overriding right to regulate abortion by ensuring proper standards of medical procedure. During the last trimester of pregnancy, the Court held that the fetus had become "viable," in that it could likely survive outside the womb with artificial aids. During this time the right of the state to pro-

One issue that has polarized women is abortion rights. Abortion is a particularly explosive issue because childbearing, an activity traditionally identified as a woman's primary role, can interfere with her ability to take advantage of economic opportunities. Pro-choice rallies such as this one held on Boston Common support the belief that women should be able to control their reproduction.

tect potential life became paramount, and abortion could be legally prohibited except under extreme conditions to preserve the mother's health or life.

Since that time, there has been considerable legal and political controversy over abortion. An antiabortion, Right-to-Life movement sprang up in the 1970s in response to the *Roe* decision. The *Roe* decision was hailed by the women's movement, while opposition was led by the Catholic Church and fundamentalist Protestant denominations. There is evidence, though, that abortion is widely practiced in Catholic countries, as well as elsewhere, and that a large proportion of American Catholic women favor leaving abortion a matter of individual conscience rather than public law. Opinion polls have fluctuated somewhat since abortion

became a matter of controversy in the early 1970s. From the mid-1970s until the present, the majority of people agree on allowing abortion in at least certain cases. About 90 percent of Americans believe that abortions should be allowed if a woman's health is endangered, 80 percent if she became pregnant as the result of rape or if there is a strong chance of the baby being born with serious birth defects. There is much more controversy over the availability of abortion for other reasons, with the population about evenly split in their approval of a woman's right to choose to have an abortion if she is too poor to support the child or already has too many children. More people support a woman's right to an abortion under any circumstances than those who would deny women such a choice, but the majority of

Americans are in the middle, and most support putting some limits on the availability of abortions, especially for those under the age of eighteen. Catholics are about 6 percent less approving than Protestants, whereas Jews and persons with no religious affiliation are especially likely to approve abortion for any reason. Young people below the age of thirty are most strongly in favor of allowing abortion, but persons over age sixty-five were most conservative.

In 1979 the Supreme Court ruled that state laws may not prohibit women below age eighteen from obtaining an abortion even if they do not have their parents' consent. Since then, court decisions and state laws have challenged that right. Most states now have laws requiring minors to inform their parents or obtain their consent to get an abortion. Access to abortion, even for older married women, varies widely depending on where one lives. Eighty-six percent of all U.S. counties had no identified abortion provider in 1999. Because these were often in rural areas, 32 percent of women of reproductive age lived in those counties. Primarily because of sustained protests, the number of abortion providers declined during the 1990s (Henshaw 1998).

Recent debates over abortion have focused on late abortions, with legislation pending to outlaw such practices. New techniques of early medical abortion (as compared with the traditional surgical procedure—see section on Medical Abortion) promise to change the dynamics of abortion services and those who protest against them.

FIGURE 10.4 Legal and Illegal Abortions in the United States

Source: Cates 1982; *Statistical Abstract of the United States*, 1999.

complications than the traditional surgical abortion. In 1988, RU-486 was legalized for use as an abortifacient in France, where, about twelve-years later, junior high and high schools began dispensing "morning after" pills to schoolgirls in an effort to reduce unwanted pregnancies among teens (San Diego Union, February 8, 2000). The morning-after pill, or so-called abortion pill, commonly taken in conjunction with a second hormone (misoprostol) is now also used legally in China and several European countries.

How do medical abortion treatments work? In the **mifepristone-misoprostol regimen,** mifepristone (RU-486) blocks the production of the hormone progesterone, which is needed to maintain a pregnancy. Misoprostol, an already marketed prostaglandin causing uterine contractions, can be administered orally and is inexpensive and widely available. The combined regimen of mifepristone-misoprostol has demonstrated effectiveness roughly comparable to surgical abortion and can be used as soon as a pregnancy is detected up to seven weeks from a woman's last menstrual period. Barring congressional action to the contrary, and dependent on final Food and Drug Administration (FDA) approval, this regimen should soon become more available in the United States (Peyron et al. 1999). In an alternate **methotrexate-misoprostol regimen,** methotrexate blocks folic acid and prevents cell division. This drug is already marketed as a cancer-fighting drug, so its use as an abortifacient is legal (but technically "off-label") and also can be used as soon as a pregnancy is detected up to seven weeks into the pregnancy. Finally, so-called emergency contraception, which involves taking a high dose of oral contraceptive pills within 72 hours of unprotected intercourse, is also just becoming available on a widespread basis. Emergency contraception involves taking a short, strong course of ordinary birth control pills and has been declared safe and effective by both the American College of Obstetricians and Gynecologists and by the FDA.

Most Americans are still unaware of these new methods of ending an unwanted pregnancy. In 1997, fewer than half of U.S. women had even heard of either mifepristone or methotrexate (Saul 1999). In 1996, just 4,200 medical abortions were performed in the United States, but recent estimates suggest that the number has risen substantially. Support for medical abortion has been limited by lack of awareness and confusion over how the methods work. One myth is that the availability of medical abortion regimens would make unwanted pregnancy the medical equivalent of a headache: "pop a pill and it will go away." The implication is that because mifepristone or methotrexate might make the procedure "easier," their availability would diminish the seriousness with which women treat abortion (Saul 1999). According to the Alan Guttmacher Institute (a nonprofit organization that studies family planning), medical abortion is comparable to a miscarriage, except that medical abortion can take considerably longer, with the process spanning several days. Although the procedure has proven to be extremely safe, it also involves significant medical supervision. The Guttmacher Report notes, "in terms of what's involved in the actual procedures, medical abortions and surgical abortions each have relative advantages and disadvantages from the viewpoint of the women undergoing them; experience in Europe shows that some women prefer one and some the other. In countries where it has been available, however, there is no evidence to support the notion that the availability of mifepristone increases the overall incidence of abortion; in fact, the abortion rate in France has declined more or less steadily since mifepristone's advent in 1988" (Saul 1999).

The widespread availability of early medical abortion alternatives like the ones described will likely change the debate about abortions in this country. Although public opinion about abortion is radically divided (see *A Closer Look* 10.2), the American public has consistently supported abortion early in pregnancy. According to a *New York Times/CBS* poll conducted in early 1998 (at the time of the twenty-fifth anniversary of the Supreme Court's *Roe v. Wade* decision), almost two-thirds of Americans believe that abortion should be legal in the first three months of pregnancy, with support for legal abortion dropping sharply after the first trimester (Saul 1999). Abortion practices in the United States largely tend to parallel these public opinions. American women overwhelmingly have abortions early in pregnancy, with approximately half of all abortions occurring within the first eight weeks and about 9 in 10 by the end of the first trimester. The emergence and availability of medical abortions, along with advancements in early pregnancy detection and improved surgical techniques, will likely shift the timing of abortion in the United States even further toward the first half of the first trimester of pregnancy (Saul 1999).

In contrast to the development of these new early abortion techniques, most of the heated abortion debates in Congress in recent years have focused on rare late-term abortions, dubbed "partial birth abortions" by opponents. Although the graphic visual framing of abortion debates around these unpopular procedures has garnered some support for banning other forms of abortion, restrictions have been few and mostly focus on limiting federal funds or requiring approval from the parents of teenagers. In spite of such actions and continued violent attacks on clinics by antiabortion extremists, the increasing availability of new medical techniques should offer more women the option of early abortion. In addition, the availability of new forms of contraception that can be used after unprotected intercourse, but before a pregnancy actually begins, are also likely to shift abortion decisions to individual women and their doctors. Most experts agree, however, that the best way to reduce the prevalence of abortions is to make contraceptive information widely available through comprehensive sex education programs and to provide low-cost or free family planning services to a larger proportion of the population (Cohen 1998). As noted in our earlier discussion of federal support for abstinence-based sex education programs, we may be moving away from such a solution.

BIRTH TIMING

In the 1950s, most American women got married in their early twenties, and most had babies soon thereafter. If women worked before they were wed, they typically quit their jobs when they married or perhaps when they got pregnant. The ideal—and the practice among most women who could afford it—was to stay out of the labor force until after their children had grown up and left the home. Of course this was not possible for all women, but the majority followed this pattern. Today, most women are delaying having babies in the first place and then keeping their jobs after they have children (see figure 5.9). This leads some observers to claim that the most significant fertility trend during the last part of the century was the delayed timing of entry into parenthood (or at least delayed as compared with the early timing pattern of the baby boom era).

Over 80 percent of women in the United States eventually become mothers, but many are putting off childbearing until their later twenties or even into their thirties or beyond. The percentage of women waiting until they are past thirty to have their first child has more than doubled since the 1970s, with dramatic increases since about 1980. Figure 10.5 shows that the birthrates for women aged thirty to thirty-four and thirty-five to thirty nine have increased fairly steadily in the past two decades. The birthrates for women aged twenty to twenty-four and twenty-five to twenty-nine, while fluctuating, have remained at about the same level they were in 1980. The birthrate for women under 20 has remained at about the same level they were in 1980. The birthrate for women under twenty, in contrast, rose in the late 1980s but has been falling since the early 1990s. These trends are due to a variety of factors, but several stand out. First, as noted earlier, teenagers are more likely to use birth control or to avoid intercourse than they were in the recent past. Second, more women in their early twenties are attending college, getting and keeping full-time jobs, and avoiding or postponing marriage—all factors that reduce one's chances of bearing children. Third, women in their thirties are more likely than in the past to decide to have children, whether they are married or not. Finally, as discussed later in this chapter, advancements in reproductive technologies have enabled some women in their later childbearing years to become pregnant and give birth.

It is possible to track trends in the timing of birth by looking at the proportion of women who are childless at particular points in their lives. For example, figure 10.6 shows that the percentage of childless women is increasing in all age groups. Three of four women aged eighteen to twenty-four were childless in 1995, up slightly from the proportion childless two decades before. A more dramatic trend is evident during the prime childbearing years of the late twenties. Although only about a third of women aged twenty-five to twenty-nine were childless in the late 1970s, by the late 1990s, about half were still childless. Similarly, there has been a marked increase in the proportion of women who remain childless in the next three age groups: thirty to thirty-four, thirty-five to thirty-nine, and forty to forty-four. Although somewhat more women are remaining permanently childless, most women are just delaying having children longer than earlier cohorts. Delayed childbirth has been most common among employed and more highly educated women but has spread to all income and educational levels.

According to recent studies, the trend toward delayed childbearing is not fully explained by the tendency of people to wait longer to get married. Not only are people postponing marriage longer than was common during the 1950s and 1960s, but they are also postponing having children once they get married. The recent trend toward postponing having children has precedents from the mid-

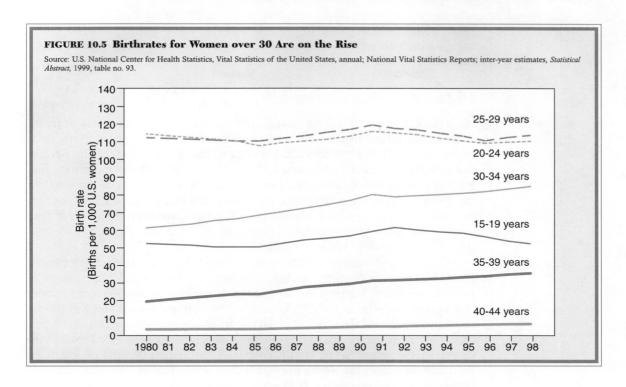

FIGURE 10.5 Birthrates for Women over 30 Are on the Rise

Source: U.S. National Center for Health Statistics, Vital Statistics of the United States, annual; National Vital Statistics Reports; inter-year estimates, *Statistical Abstract*, 1999, table no. 93.

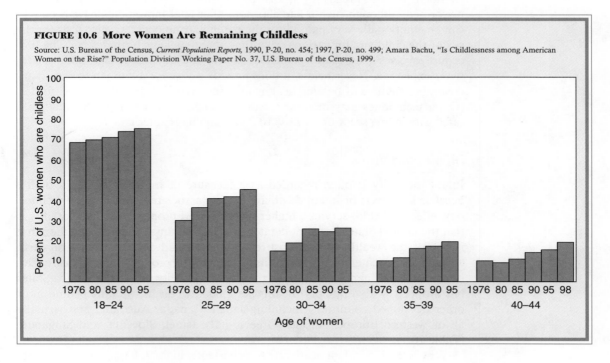

FIGURE 10.6 More Women Are Remaining Childless

Source: U.S. Bureau of the Census, *Current Population Reports*, 1990, P-20, no. 454; 1997, P-20, no. 499; Amara Bachu, "Is Childlessness among American Women on the Rise?" Population Division Working Paper No. 37, U.S. Bureau of the Census, 1999.

nineteenth century, but another demographic factor makes the recent trend even more consequential. Because of the postwar baby boom, there are significantly more people over thirty than there were ten or twenty years ago. Because there are more people in the later childbearing ages and because couples are now likely to delay childbearing, demographers predict that the number of children born to older women will continue to increase and that more women will end up remaining permanently childless. In short, the trend toward delaying having children will continue for some time in the future.

Medical Risks of Delayed Birth

How late can one wait to have a baby? Technically, women are capable of reproducing until menopause, in the late forties or early fifties. However, fecundity is highest in a woman's twenties and goes down sharply when she reaches her forties (Menken, Trussell, and Larsen 1986). Ovulation becomes more irregular, especially in women who already have irregular menstrual periods. So a couple who wait until their late thirties or forties may find that pregnancies do not necessarily follow unprotected intercourse. Of course, inability to conceive may occur at younger ages too; in fact, 8 percent of all women who wish to have children do not bear any (Westoff 1986).

The chances of genetic defects do go up somewhat as the parents grow older. **Down's syndrome** (formerly called "mongolism"), the most common form of mental retardation, happens in 1 of 1,500 pregnancies with mothers in their twenties; but the odds rise to 1 of 350 for mothers in their late thirties and 1 in 100 for those in their forties. The chances even for the oldest mothers are still only 1 percent but the risk can be dissuading. However, it is possible to test for Down's syndrome by the technique of **amniocentesis.** This method consists of removing fluid and cells from the womb through a thin needle between the sixteenth and twentieth week of pregnancy. If Down's syndrome is present, an abortion can be performed, although such a late abortion, after fetal movements have been felt, can be emotionally very painful for the parents. Amniocentesis also tests for the sex of the child and for various other (but not all) genetic birth defects. The procedure involves a less than 1 percent chance of causing miscarriage, but it is relatively expensive. Nevertheless, it does provide a way for a woman to greatly increase her chances of bearing healthy children even if she waits until her forties. (Other techniques may increase a woman's chances of bearing a healthy child safely late in life; see *A Closer Look* 10.3 later in this chapter.)

Infant Mortality

Infant mortality is often regarded as a measure of a society's quality of life because low levels of infant death are usually associated with modern technology, effective health services, higher incomes, and proper diets. The paradox is that the United States has a moderately high rate of infant death, even though it is one of the wealthiest and most technologically advanced societies on earth and spends a greater proportion of its gross national product on health care than other developed nations. Although the infant mortality rate in the United States has dropped, with a rate of 7 deaths per 1,000 live births, we rank behind most other developed countries—including Japan, Canada, Australia, and virtually all of western Europe, but also Greece, Italy, Israel, Slovenia, and Singapore (Table 10.1).

Why is the U.S. infant death rate so persistently high? A large part of the problem seems to be lack of adequate nutrition and prenatal care, for which minorities are especially at risk. Babies who are born with a low birth weight (under 2500 grams or about 5.5 pounds) have a smaller chance of surviving the first month of life, when most infant deaths occur. The major reason for low birth weights and infant death in this country is poverty, with its attendant poor nutri-

A boy with Down's syndrome displays a proud smile and rib-bons after a Special Olympics race. Special Olympics events help physically and mentally challenged individuals to realize their potential.

tion and lack of routine medical care. Rates for African Americans (14.2 per-cent) are more than double those of whites (6 percent), with Asians (5.2 percent) having the lowest rates. Hispanics had infant mortality rates only slightly higher than whites (6.1 percent), though subgroups varied from 5 percent for infants of South American descent to 8.6 percent for those of Puerto Rican descent (Fed-eral Interagency Forum 1999). In most other industrialized countries, pregnant women receive subsidized prenatal care and home visits after the baby is born. Ironically, our reluctance to spend money on poor mothers actually costs us more in the long run. It has been estimated that $400 worth of prenatal care could make the difference between a healthy baby and one requiring $400,000 of medical care and services throughout its life (Hale 1990). It remains to be seen whether recent attempts to reform the health care system and provide universal health insurance will help alleviate this problem.

Childless Families

Some married couples deliberately decide that they do not want children. At least 5 percent of all marriages, or one in twenty, are of this sort. The rate is even higher among college graduates, of whom as many as 15 percent of all couples are childless. The proportion of childless marriages is slowly rising (see Figure 10.6). They are most prevalent among people with career commitments in high-paying occupations. The apparent reason is that children are less appealing than the careers, and child-care responsibilities are considered a burden to be avoided (Campbell 1985).

Successful careers are a motivation for women as well as men to remain child-less, but noncareer motivations seem to be even more important. Childless couples

Table 10.1 Infant Mortality Rates in Comparative Perspective

The proportion of babies dying before they reach one year old is higher in the United States than in most other developed nations of the world but lower than in many less developed countries

Country	Infant Mortality Rate (per 1,000 births)	Country	Infant Mortality Rate (per 1,000 births)
Iceland	2.6	Cuba	7.2
Singapore	3.3	Malaysia	8.3
Sweden	3.6	Puerto Rico	10.5
Japan	3.7	South Korea	11.0
Norway	4.1	Chile	11.7
Finland	4.2	Kuwait	12.5
Austria	4.8	Costa Rica	14.2
Slovenia	4.8	United Arab Emirates	16.0
Switzerland	4.8	Sri Lanka	16.8
Germany	4.9	Thailand	25.0
France	5.0	Iran	26.0
Czech Republic	5.1	Colombia	28.0
Netherlands	5.1	Saudi Arabia	29.0
Denmark	5.2	China	31.4
New Zealand	5.3	Syria	34.6
Australia	5.3	Vietnam	34.8
Italy	5.5	Mexico	32.0
Spain	5.5	Turkey	42.7
Canada	5.6	Egypt	52.0
Israel	5.8	South Africa	52.0
United Kingdom	5.9	Zimbabwe	53.0
Belgium	6.0	India	72.3
Ireland	6.2	Nigeria	73.0
Greece	6.3	Bangladesh	82.2
Portugal	6.4	Pakistan	91.0
Taiwan	6.4	Iraq	127.0
United States	7.0	Afghanistan	149.8

Source: World Population Data Sheet, Population Reference Bureau, 1999, http://www/prb.org/pubs/wpds99/

seem to be somewhat different than others; studies indicate that they are more independent, live farther away from their own parents, and enjoy being alone more (Houseknecht 1987). The most common reason they give for not wanting to have children is to give more time to their spouse. In other words, their motivation might be broadly called erotic reasons. They wanted to spend their love on their adult partner, not on their children.

Being a childless family is not a once-and-for-all decision. Couples do not usually make up their minds to avoid childbearing until after they are married or cohabiting. In other words, the decision is not usually part of an originally thought-out plan. Also, many couples deliberately decide to delay childbearing. These couples seem to go through the same processes as the ones who forego parenting entirely. In both cases, it is usually the woman who has the stronger opinion; women also give as their main reason for delaying childbirth, not career considerations, but the desire to spend time with their husbands.

America is typically characterized as a strongly "pronatalist" society: that is, we tend to believe that all married couples should want children and that they should act on that desire. When cultural norms and values strongly encourage people to become parents, those who remain childless—through infertility or by choice—tend to be labeled as deviant. Those who are "childless by choice" are the most likely to be negatively stereotyped, and women who choose to forego motherhood are much more likely than men to be considered "unnatural" (Houseknecht 1987; Veevers 1980). Even women who are involuntarily childless often feel discredited. For instance, one woman felt that admitting to possible problems with reproduction was "an admission that you're not a whole person. . . . The ability to reproduce strikes at the very essence of one's being" (Miall 1987, 392). Another woman who tried unsuccessfully to have children commented that she and her husband felt like failures:

> We expected things to work out. We led the golden lives until this happened to us. It's the worst thing that ever happened to us. My husband and I are high achievers. We work together and it works out. Therefore, it's even harder for us to accept this because we don't have control over something that's so easy for other people. (Miall 1987, 391)

To avoid the negative reactions and judgments that followed when they told others of their inability to have children, many of the couples in Miall's study (1987) felt compelled to conceal or deceive others about their condition. These examples show how social pressure to have children can influence people's self-concepts, as well as their willingness to attempt to reproduce. Even though norms concerning childbearing are changing, most people, especially women, feel that they should have children in order to feel complete.

Compulsory Motherhood

The idea that women are required to have children and are fundamentally and essentially defined by being mothers is sometimes referred to as "compulsory motherhood." Although the average American woman spends only about one-seventh of her life either pregnant, nursing, or caring for preschool children, being a mother is seen as her ultimate reason for being. Motherhood is thus considered sacred but, according to conventional morality, only if it happens to a woman who is married. The stigma attached to women who give birth outside of marriage may not be as great as it once was, but can be quite intense. Labels of "unwed mother" and "illegitimate child" are still sometimes used to castigate women and children who are not connected to a man. Media images and political rhetoric also reveal an interesting double message about motherhood that is sent to white women and women of other races. Poor black women are sometimes threatened with sterilization or reduced welfare benefits in an effort to discourage their getting pregnant, but educated white women, in contrast, are bombarded with propregnancy propaganda that one commentator has called "maternity chic." The message seems to be that "Real women *should* have babies, successful women *are* having babies, and even feminists *want* to have babies" (Pogrebin 1983, 180).

What seems to be happening is a no-win situation for many women. On one hand, we still have the cultural ideal that fulfillment and true personhood come only with being a mother. On the other hand, we now have the "total woman" dictum that says being a mother is not enough, you must have a career as well. If women want only to be housewives and mothers, they are now made to feel guilty because that is not enough. If they try to "have it all" as suggested by countless articles in women's magazines, they end up perpetually exhausted. In the face of these conflicting pressures, it is no wonder that increasing numbers of women are delaying childbirth but maintaining their commitment to have children sometime in the future.

INFERTILITY

Although more couples are postponing childbirth or remaining childless, about 10 to 15 percent of couples want children but are unable to have them. *Infertility* is the label used to describe those who are unable to conceive a child. As a general rule, people are defined as infertile if they have not conceived after having had regular sexual relations for over a year without using any contraceptives. Infertility is also sometimes used to refer to what doctors call "impaired fecundity"—the inability to carry a pregnancy to live birth.

Infertility is about equally likely to be the result of problems attributed to the man as to the woman. Common causes of infertility in men include low sperm production; poor semen motility; effects of STDs such as chlamydia, gonorrhea, and syphilis; and enlargement of the prostate, limiting the passage of sperm. Common causes of infertility in women include blocked fallopian tubes, endocrine imbalances that prevent ovulation, dysfunctional ovaries, imbalances in cervical mucus, and effects of STDs (Knox and Schact 1994).

Of the estimated 6.2 million women in the United States with impaired fecundity, 44 percent had sought treatment for infertility (Abma et al. 1997; Stephen 1999). Although more people than ever before are seeking treatment for infertility, it is not the case that the actual incidence of infertility has increased. In fact, the National Center for Health Statistics estimates that the infertility rate declined from 1960 to 1990 (Mosher and Pratt 1990) and has remained steady since (Mosher and Bachrach 1996). The infertility rate is higher for those who delay childbearing, and because people are waiting longer to have children and older parents have more financial resources to treat the problem, more people are seeking medical help for infertility. As discussed next, the increase in help-seeking behavior is also linked to the development and availability of assisted reproductive technologies (ARTs) (see *A Closer Look* 10.3).

Doctors prescribe a variety of drugs to women with infertility problems. For example, fertility drugs can be used to treat hormonal imbalances, induce ovulation, and correct irregularities in the menstrual cycle. Two of the most commonly used drugs are Clomid and Pergonal. Taking fertility drugs drastically increases a woman's chances of giving birth to multiple children. In the 1980s, the rate of higher-order multiple births almost doubled, primarily because of the increase in the use of ovulation-inducing drugs for treating infertility (Kiely, Kleinmen, and Kiely 1992). As fertility treatments have increased, the ratio of triplets, quadruplets, quintuplets, and so on to normal births has climbed even faster. In 1986, the higher-order birth ratio was about 50, but by 1996, it exceeded 150. The use of fertility drugs accounts for about two-thirds of this increase, and the other third is attributed to the fact that more women are delaying childbearing until their late reproductive years (Stephen 1999). Greater incidence of multiple births affects the health of those infants and their chances for survival. Multiple-birth infants are at greater risk of preterm birth, low birth weight, developmental brain damage, and cerebral palsy (Stephen 1999). There is also evidence that use of some ARTs, such as intracytoplasmic sperm injection, may substantially increase the risks of congenital abnormalities (ISLAT 1998).

ASSISTED REPRODUCTIVE TECHNOLOGIES

The newest forms of reproductive technology involve procedures that bypass sexual intercourse. For instance, artificial insemination involves the introduction of sperm into the female reproductive system by means other than intercourse. (Children conceived in this fashion have been referred to by some as "turkey baster" babies.) Commercial and nonprofit "sperm banks" collect sperm from male donors (who are usually screened for genetic defects) and freeze it for future use. Artificial insemination is currently offered by the majority of infertility facilities in university clinics and is practiced by most of the larger medical centers in the country. The success rate for artificial insemination is lower than for natural methods but ranges from 40 to almost 90 percent depending on the woman's fertility potential, the condition of the sperm (fresh or frozen), and the timing and frequency of the insemination (Moghissi 1989, 124–5). It is estimated that over 60,000 children are conceived through artificial insemination in the United States every year, and another 15,000 per year are conceived by using in vitro fertilization *(A Closer Look 10.3)* (ISLAT 1998).

Some states have enacted new laws to handle this special kind of procreation. Such laws typically specify that a child conceived by means of artificial insemination carried out with the consent of the husband is the legal offspring of the couple. Some states further specify that the man who furnishes the sperm for the artificial insemination of someone other than his wife is not the child's legal father. In many states the child conceived through artificial insemination is still in legal limbo. Although the legal and ethical issues have yet to be resolved, some people suggest that we should freeze the sperm of especially talented men to be used for the artificial insemination of specially selected women. Others suggest that we should also freeze the ova from especially talented women. Proponents of eugenics (the science of controlled breeding) claim that we could produce superior children with these techniques and improve the human race (see *A Closer Look* 10.1 on the history of eugenics).

External fertilization is another technological possibility that is quickly gaining widespread acceptance. **In vitro fertilization** (literally "in a glass") is much more complicated than artificial insemination, but with rapid progress in medical technology, it is becoming a viable alternative.

Although the cloning of humans is a long way off, emerging fertility practices can now allow people to choose the sex of a child. Advances in ARTs are occurring so rapidly that it is impossible to predict what techniques might be available in the near future. Some envision a world in which couples can have "designer children" by choosing the genes for a child's hair color, height, and proclivity toward piano playing and where same-sex couples will be able to have a child who is biologically related to both of them (Silver 1997). Even today, newspaper advertisements across the country seek egg donors: one college newspaper caused an uproar because a couple was willing to pay $50,000 to an egg donor with SAT scores over 1400 who was at least 5'10" tall (Stephen 1999). Other legal issues include the question of the difficulty of establishing parenthood when there may be as many as five people involved: a sperm donor, an egg donor, a gestational mother, and the contracting mother and father (Eichler 1997). In one recent case, a doctor mistakenly mixed the embryos of two couples and one of the couples had twins—with one white and one African American baby. Another case involved a woman who gave birth using sperm frozen for 15 months that had been retrieved from her husband 30 hours after his death (Stephen 1999). Such developments raise important questions about what, who, and how new reproductive technologies should be regulated. Should genetic engineering companies be allowed to patent and market sperm and eggs? Should a woman who uses ART to conceive quintuplets be compelled to undergo "selective reduction" to minimize risks to the unborn?

A CLOSER LOOK 10.3
Assisted Reproductive Technologies

In the 1970s, scientists in England, Australia, and the United States developed a technique for fertilizing human eggs outside the mother's body called in vitro fertilization. The first baby born by this method was in Cambridge, England, in 1978. Since then, over 300,000 other children have been created worldwide by in vitro fertilization.

The method was developed to counteract sterility caused by blocked or damaged fallopian tubes, a condition shared by about 2 million American women. Blocked tubes can sometimes be opened with laser beam surgery or by inflating a tiny balloon in the constricted passage, but when this is not possible, test tube fertilization is an alternative. In this procedure, the ovaries are stimulated to produce multiple eggs, which are taken from the ovary through a small abdominal incision just before ovulation. In a glass petri dish in the laboratory, the eggs are mixed with sperm from the male donor—usually the woman's husband—and placed in a chemical incubator. Finally, one or more eggs are transferred back into the mother's uterus.

Success rates for in vitro fertilization vary, but it is estimated that only about one in ten couples give birth as a result of the procedure (U.S. Congress 1990; Health Facts 1999). Almost a fourth of those end up with two or more births because multiple eggs are placed in the uterus. In vitro fertilization is typically quite safe for the mother, though some research links abnormalities in children with the method, especially in cases of multiple births. The main drawback is that the method is expensive and is rarely covered by health insurance. Depending on the number of inseminations or special actions that are necessary, the cost of a pregnancy using in vitro techniques is typically over $10,000 and is often two or three times that much. In addition, the procedure entails hormone injections, blood samples, ultrasounds, physical exams, and periods of inactivity. Most couples thus report that in vitro techniques are time consuming and emotionally exhausting (Elmer-Dewitt 1991).

Several new techniques have been developed that are variations on these procedures:

In Vitro Fertilization (IVF)

IVF involves extracting a woman's eggs, fertilizing the eggs in the laboratory, and then transferring the resulting embryo(s) into the woman's uterus through the cervix. A variation on this method, popular with women ages forty and older, is to use **donated eggs.** In this procedure, a female egg donor takes ovulation-enhancing drugs to produce a large number of follicles. The donor's eggs are retrieved and fertilized with the woman's partner's sperm or that of a donor. The embryos are transferred to the woman's uterus forty-eight hours after fertilization.

Gemete Intrafallopian Transfer (GIFT)

Because only about one in five fertilized eggs implants on the wall of the uterus during in vitro procedures, doctors developed intrafallopian transfer. In this process, both egg and sperm are placed directly into the fallopian tube where they meet, fertilize, migrate down into the uterus, and hopefully successfully implant on the wall of the uterus. The GIFT procedure is reported to double or triple the chances of implantation over normal in vitro techniques (Knox and Schact 1994).

An estimated one of six couples who are infertile are now seeking help from ART clinics, and the industry now has an annual revenue of $2 billion. Although about 30,000 healthy infants are available for adoption each year in the United States, over 75,000 babies are born each year using ART. Every state has an elaborate regulatory mechanism for adoption, but legislation governing ART and agencies or boards overseeing it are still relatively rare. Many argue that problems like experimentation without appropriate review, use of embryos without consent, inadequate informed consent, conflicts over stored gametes and embryos, and failure to routinely screen donors for disease demand legislative and regulatory action. In the United Kingdom, where the first test-tube baby, Louise Brown, was

Zygote Intrafallopian Transfer (ZIFT)

This procedure combines in vitro techniques with GIFT techniques. The woman's eggs are extracted and fertilized with the man's sperm in a petri dish as in the normal in vitro procedure. The zygote is then transferred directly into the fallopian tube as in the GIFT procedure. In this way, the doctor ensures that fertilization has taken place, but the natural migration of the embryo enhances the chances of uterine implantation. Success rates for ZIFT procedures are similar to those for GIFT procedures (Knox and Schact 1994).

Partial Zona Drilling (PZD)

Eggs with shells that have been punctured have the highest chance of implanting on the walls of the uterus. The PZD procedure thus entails isolating eggs and drilling tiny holes in the protective shell that surrounds them. Although it is too early to determine the effectiveness of this technique, doctors hope that it will raise in vitro fertility rates.

Microinjection

In some cases, infertility results from sperm with low motility. To overcome this problem, doctors obtain sperm from the man and inject it directly into the egg using microinjection techniques.

Embryo Transplants or Ovum Transfer

In this technique, the sperm of the man is placed by a doctor in a surrogate woman. After about a week, her uterus is flushed and the contents are microscopically analyzed to detect the presence of a fertilized ovum. If one is found, it can be inserted into the uterus of an otherwise infertile woman, who may carry the baby to term. The embryo can be frozen and implanted later, but the highest implantation rates have been found for "fresh" zygotes (Levran et al. 1990). Embryo transplants allow the child to be the biological offspring of the father and afford the "mother" the experience of pregnancy and childbirth. As with other types of surrogacy, however, this technique raises some provocative moral and legal issues, and some suggest that it allows affluent women to exploit the reproductive capacity of less fortunate women.

Ovulation Drugs

Clomiphene citrate, Pergonal, or Metrodin enhance ovulation and may be used alone or in combination with one of the procedures listed next. The woman's cycle is closely monitored so that the insemination coincides with ovulation. Another shot may be administered to trigger ovulation.

Intrauterine Insemination

This method involves inserting prepared sperm into the woman's uterus.

Intracytoplasmic Sperm Injection (ICSI)

ICSI is used in dealing with male-related infertility. In ISCI, a single sperm is injected directly into an egg, and then the embryo is transferred into the woman's uterus using the standard IVF procedure.

Sorting Sperm

Sorting sperm before insemination allows doctors to select only those that produce girls (who do not carry any of the more than 300 known X-linked chromosomal diseases).

Preimplantation Genetic Diagnosis

This approach allows DNA-testing of embryonic cells. Using this technique, doctors of couples who fear passing on genetic diseases such as Tay-Sachs can confirm that only healthy embryos are transferred into the woman's uterus.

born on July 25, 1978, there are now fairly elaborate rules and administrative oversight for licensing and providing reproductive services. In the United States, a laissez-faire approach has resulted in some voluntary guidelines that are routinely ignored (ISLAT 1998). Most observers predict that the hard decisions and difficult ethical dilemmas posed by ARTs will require more direct attention by governments and medical overseers in the United States in the coming decades.

Surrogate Motherhood

One of the most controversial and troublesome outgrowths of the new reproductive technologies is the possibility of surrogate motherhood. Surrogacy has been described as a miracle cure for infertile couples but has also been called baby selling, womb renting, and reproductive exploitation. Commercial surrogacy involves a contractual agreement, often negotiated by an agency, between a couple and a fertile woman. The fertile woman is paid to be artificially inseminated with the husband's sperm, becomes impregnated, and carries the fetus to term, at which time this surrogate mother relinquishes the infant to its biological father and its social mother. At least, that's how it is supposed to work.

In the much publicized "Baby M" case in New Jersey, the surrogate/biological mother broke the signed contract and refused to surrender the baby to the biological father and his wife. After lengthy court battles, a judge awarded the baby to the father, but the surrogate/biological mother later won visitation rights. This case left many legal questions unanswered.

Two factors seem to make surrogate motherhood especially troublesome. First, the mother grows the fetus inside of her for nine months and during that time may change how she feels about keeping or giving up the baby. This is what happened in the case of Baby M. Second, there is usually an inequality between parties to the agreement. Some surrogate mothers say they do it because they enjoy being pregnant and want to make other, childless couples happy (Martin 1976; Ragoné 1994). Nevertheless, most surrogate mothers are young, poorly educated, have few job skills, and do it because they need the money. The other party to the contract, the infertile couple, is usually well-off financially and willing to pay all medical and nutritional costs, as well as to make a relatively large lump sum payment to the woman upon delivery of the baby. This sets up two classes of parents and leads some commentators to describe the relationship as one of exploitation. Dworkin (1987) likens surrogate motherhood to prostitution and suggests that poor women's reproductive capacity, like their sexuality, becomes a commodity that they are forced to sell to survive.

ADOPTION

A common solution to the problem of infertility has been adoption. Adequate adoption statistics do not exist because since 1975, no federal agency has collected such data. According to the best recent estimates, almost 120,000 adoptions take place each year in the United States (Flango and Flango 1993). About half of these are by relatives and half by nonrelatives. For nonrelatives, adoption by infertile couples are most common. Over a third of adoptions in the United States are through public agencies, about a third are through independent or private channels, and just under a third are through private agencies.

State agency adoptions are the least expensive because they require only legal fees, but the waiting times are long. It is not unusual to wait up to seven years to adopt a child through a state agency. State agency requirements for placement are also usually the most stringent. People over fifty are usually considered "too old," those under twenty-five are sometimes considered "too young," and those who violate traditional norms are given low priority. Unmarried heterosexual couples may sometimes adopt, but homosexual couples or individuals usually have great difficulty doing so. Private agency adoptions are usually faster, but they are more expensive, often costing $10,000 to $30,000 (Knox and Schact 1994). Some private agencies facilitate foreign adoptions, usually involving children from Third World countries. The waiting period for foreign children is usually less, often only one to two years. Over 15 percent of adoptions in the United States involve foreign-born children (Stolley 1993).

The number of international adoptions has grown steadily, from about 8,000 in 1980 to over 15,000 in 1998. This makes the United States the largest importer of adopted children in the world. The anthropologist Christine Ward Gailey (2000, 58) suggests that we cannot understand international adoptions today without an appreciation for the history of U.S. military operations:

> In the latter twentieth century, the patterns of U.S. adoption internationally are closely linked to the consequences of U.S. covert operations and Cold War activities. I realize that this is a strong statement, but the patterns are unmistakable. What is unusual about the U.S. as a conquering state is the aftermath: oftentimes the U.S. attempts to assimilate or incorporate the "enemy other." This is in all likelihood due to the religious orientation of the settling colonists and subsequent linkage of military conquest with notions of spreading "Christian civilization."

Until the Vietnam war, the pathway of American international adoptions directly followed military exploits. Korean and Filipino adoptions followed massive American military occupations. Latin American adoptions followed the Kennedy administration's "Alliance for Progress." In Brazil during the 1960s, Chile in the 1970s, and Guatemala in the 1980s, after the U.S. fostered military juntas and destabilized the governments, international adoptions followed within about five years. Gailey traces the political, economic, and cultural forces that shape who becomes available for adoption in these countries and debunks the myth that such children are "orphans." In general the term "orphan" does not mean the parents are dead but rather is expanded to include children whose parent(s) have relinquished legal rights by either giving them to the state or having them taken by the state. Gailey notes that "the legal fiction of orphan status" contributes to images of international adoptees that have important consequences for their eventual placement in U.S. families. Contrary to popular understandings, most international adoptees have been relinquished for reasons of poverty or because social conditions do not permit mothers to take care of them. For these and other reasons, grassroots political opposition to American adoptions is sometimes an impediment to the practices of international adoption agencies.

The third type of adoption—direct—entails contracts between a couple who want a baby and a pregnant woman. It is estimated that 80 percent of adoptions in California are of this type. Usually a pregnant teenager realizes that she does not have the resources to care for her baby, and her doctor or someone else helps arrange for the couple to legally adopt the child at birth (Knox and Schact 1994). The problems associated with direct adoptions can be similar to those encountered in surrogacy arrangements discussed previously.

CHILDBIRTH

The Medicalization of Childbirth

The reproductive technologies discussed earlier may seem like they belong in some far-out science fiction version of a brave new world. After all, they transform relatively "natural" processes—conception and childbirth—into medically controlled and legally governed contractual events. Nevertheless, new techniques like in vitro fertilization and amniocentesis are logical extensions of a process that has been occurring for several centuries: the medicalization of childbirth.

A CLOSER LOOK 10.4
Birth Practices of Some Nonindustrial Peoples

Éskimos

The Danish explorer Peter Freuchen first traveled to Greenland in 1906 and lived among the Hudson Bay Eskimos for many years. In his autobiography, *Book of the Eskimos,* he describes Eskimos' birthing practices:

> *Eskimo women used to talk about giving birth as being "inconvenient." This is not to say that it was any fun, but they had a remarkable short period of confinement. The women used to sit on their knees while giving birth. If the woman was in a tent or a house when her time came, she would most often dig a hole in the ground and place a box on either side of it to support her arms and then let the baby drop down into the hole. If she was in an igloo, the baby had to be content with the cold snow for its first resting place. If the birth seemed to take long, the husband would very often place himself behind his wife, thrust his arms around her, and help press the baby out. (Cited in Sorel 1985, 89).*

African Bushmen

In the early 1950s, Elizabeth Thomas went on three expeditions to the Kalahari Desert to study the African Bushmen. She describes how the women gave birth in *The Harmless People* (1958):

> *Day or night, whether or not the bush is dangerous with lions or with spirits of the dead, Bushman women give birth alone, crouching out in the veld somewhere. A woman will not tell anybody when she is going or ask anybody's help because it is the law of Bushmen never to do so unless a girl is bearing her first child, in which case her mother may help her, or unless the birth is extremely difficult, in which case a woman may ask the help of her mother or another woman. . . . When labor starts, the woman does not say what is happening, but lies down quietly in the werf, her face arranged to show nothing, and waits until the pains are very strong and very close together, though not so strong that she will be unable to walk, and then she goes by herself to the veld, to a place she may have chosen ahead of time and perhaps prepared with a bed of grass. If she has not prepared a place, she gathers what grass she can find and, making a little mound of it, crouches above it so that the baby is born into something soft. Unless the birth is very arduous and someone else is with the woman, the baby is not helped out or pulled, and when it comes the woman saws its cord off with a stick and wipes it clean with grass.*

Before modern times, pregnancy and childbirth were not considered medical problems, and women typically controlled the birthing process themselves, subject to the ritual practices and customs of their community. Although birthing practices have varied considerably throughout history and among cultures, the birth itself has almost always been regulated by the birthing woman herself, with emotional support and technical assistance offered by other women and sometimes by her husband (*A Closer Look* 10.4).

Our term for those attending women giving birth is *midwife,* which comes from the Anglo-Saxon "mid-wif," meaning literally, "with woman." Midwives have practiced their ancient art for thousands of years. Besides attending births, some midwives performed other functions: in ancient Rome, midwives often served as marriage brokers; in medieval England, midwives were called on to baptize babies; and in most societies, midwives have offered herbal remedies for various gynecological disorders and helped the new mother with domestic tasks (Sorel 1985; Sullivan and Weitz 1988).

Until the seventeenth century in Europe and America, midwifery was the exclusive province of women, as it was considered improper for men to be present during the delivery of a child (Merchant 1980). Midwifery represented women's control of a uniquely female process, though this was sometimes equated with witchcraft by powerful men. In seventeenth- and eighteenth-century Europe and America, it was not uncommon for midwives to be persecuted as witches (Merchant 1980; Ehrenreich and English 1973).

With the growth of man-midwifery, or obstetrics as it was called after 1828, childbirth began to be redefined as a pathological event that required monitoring and intervention by medical men. As part of a cultural shift toward scientific understanding of the world, the authority of physicians grew, and medical practices, such as forceps delivery and pain-killing medications, became prevalent during childbirth. During the era of the "Cult of True Womanhood" (Welter 1966, chapter 4), middle- and upper-class women were expected to be idle, even physically weak, as signs of their cultural delicacy and their husbands' wealth. These women entered the birthing chamber with muscles tensed by fear; rib cages and pelvises deformed by corsets; and natural strength weakened by lack of exercise, frequent pregnancies, and tuberculosis. They accepted pain relief, available only from physicians, as part of their lot (Wertz 1983; Wertz and Wertz 1977). In addition, the "true" woman in Victorian society was considered incapable of learning technical skills, and midwives were excluded from medical training (Sullivan and Weitz 1988, 7).

In the late nineteenth and early twentieth centuries, physicians effectively lobbied to better their position by restricting the practice of midwives, along with that of homeopaths, chiropractors, and others (Starr 1982). Medicalized childbirth techniques promised improved sanitary conditions, control of women's pain, and a reduction in the almost 1 percent of women and more than 4 percent of infants who died in childbirth at the turn of the century. As part of the campaign to convince the nation that obstetricians should replace midwives, physicians tended to exaggerate the dangers and difficulty of childbirth. By 1910, physician-attended births were typical among the middle class, but most American women still gave birth in their homes. At this time, the majority of all U.S. births were attended by midwives, primarily because poorer women continued to rely on midwives. In addition, the midwives themselves tended to be poor (large numbers were either immigrants or blacks), and they generally lacked the financial or political resources necessary to counter the physician's lobbying and claims of unsafe practices (Sullivan and Weitz 1988, 7). It did not take long for childbirth to become a medically controlled event. By 1935, the percentage of American infants delivered by midwives had dropped to just 11 percent. This included over half of all nonwhite infants but only about 5 percent of whites (Jacobson 1956).

In spite of real medical advances during this time, replacing midwives with physicians as birth attendants actually increased mortality and morbidity rates (Leavitt 1983; Wertz 1983). The White House Conference on Child Health and Protection in 1932 expressed concern that the United States had higher maternal mortality rates than European nations—where at least half of the births were attended by midwives. Studies at that time attributed birth-related deaths to unnecessary and inept medical intervention, such as forceps deliveries, cesarean sections, and manual removal of the placenta, made possible by the new anesthetics and analgesics and frequently performed by general practitioners (Sullivan and Weitz 1988, 17; White House Conference on Child Health and Protection 1933). Despite the absence of clear evidence indicating superiority of physician-attended

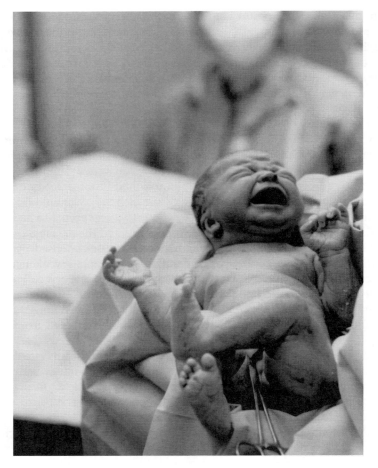

In the United States, about 92 percent of births occur in hospitals with physicians attending.

maternity care over traditional midwife care, the practice of midwifery was severely restricted and eventually was subsumed under "nurse-midwifery." The outcome is that licensing and training requirements for midwives today are similar to those for nurses, certified nurse-midwives receive additional graduate-level training, and in many instances nurse-midwives assume a subordinate role to the attending physician. Although many lay midwives continue to practice illegally, and some states allow home births with licensed midwives attending, all fifty states accept certified nurse-midwives in attendance at hospital births and free-standing birth centers, accounting for about 7 percent of all U.S. births. About 92 percent of births occur in hospitals with physicians attending.

QUESTIONING THE MEDICAL MODEL The organization of our medical delivery system and extensive media coverage of new medical discoveries encourage us to view medical technology as having almost unlimited power to better our lives. After all, since 1900, the average life expectancy of Americans has increased almost 30 years. The most common perception is that our increased life expectancy results from new developments in medical science such as vaccinations, X-rays, antibiotics, insulin, chemotherapy, open-heart surgery, and organ transplants. Although modern medicine has undoubtedly saved countless lives and bettered our chances for living longer, the primary reason for our increased life expectancy is improved living conditions and nutrition (Starr 1982; Sullivan and Weitz 1988).

Although modern medicine is practiced to save lives, our unquestioned faith in science and the authority of medical practitioners also act as a mechanism of social control and can lead to various negative consequences. Our tendency to believe that active medical intervention is superior to more passive or preventative practices has contributed to the premature adoption of some costly and questionable surgical interventions. Examples include the excessive number of tonsillectomies in the postwar years and the more recent proliferation of coronary bypass surgeries, radical mastectomies, and hysterectomies (Millman 1976). Women's consumer groups have consequently questioned whether the medicalization of childbirth has gone too far.

Because childbirth in the United States is typically considered to fall in the medical domain, decisions about the management of the birth are usually made by physicians who, drawing on medical technology, predict the course of labor. A reliance on technological solutions has encouraged birthing practices such as forceps delivery; amniocentesis; antiseptic and aseptic measures (enemas, shaved pubic areas, scrubbing, draping, and isolation); routine episiotomies (predelivery surgical incision of the perineum, the flesh between the vagina and anus); pain-control drugs (analgesics and anesthetics); and electronic fetal heart monitoring (electrodes attached to the fetus while inside the uterus). The proportion of births using electronic fetal monitoring increased from 68 percent of births in 1989 to over 83 percent in 1998 (Curtin and Park 1999). The increasing use of pain-control drugs during labor to lesson pain has been instrumental in promoting the use of other drugs such as oxytocin (Pitocin) to promote stronger uterine contractions. According to the National Center for Health Statistics, the proportion of births using drugs to induce labor more than doubled in the 1990s (from 9 to 18.4 percent) with the proportion of births using drugs to stimulate labor also increasing dramatically (from 10.9 percent in 1989 to 17.4 percent in 1997) (Curtin and Park 1999, table 7). The administration of these artificial labor-inducing medications and the use of mechanical fetal monitoring (which can produce false electronic heartbeat "reads" or varied ultrasound interpretations) have been blamed for increasing the likelihood of physicians deciding that labor is not progressing "normally." Such determinations often justify the use of cesarean section surgery, which has increased from just 4 percent of births in 1964 to more than 24 percent of all births in the late 1980s (A Closer Look 10.5) (Gleicher 1984; Rothman 1986; Sullivan and Weitz 1988). During the 1990s, the percent of U.S. babies born by cesarean section continued to remain above 20 percent of all births, a rate much higher than that found in most other developed countries (Curtin and Park 1999; Sakala 1993a) (see A Closer Look 10.5).

New categories of medical specialists, such as perinatologists (obstetricians trained intensively in high-risk maternity care), tend to encourage the use of the latest technological interventions. For example, the few elite intensive-care nurseries that existed in teaching hospitals in the 1950s have become the model for infant services in virtually every large hospital in the United States. Even some major medical research projects have concluded that intensive specialization in the delivery room can work against the needs of women with normal pregnancies, calling it "unwise and perhaps unsafe, for women with normal pregnancies to be cared for by obstetric specialists" (Eakin, Keirse, and Chalmers 1989; Sakala

A CLOSER LOOK 10.5
The Cesarean Explosion

Most births now take place in hospitals and under intense medical supervision. **Cesarean section** is the surgical procedure of removing the baby through the mother's abdominal wall instead of vaginally. The rate of cesarean deliveries—or "c-sections"—has increased from just 4 percent of births in 1964 to more than 20 percent of all births in the United States (Curtin and Park 1999). Even though rates of cesarean births in the United States have dipped slightly over the past few years, rates still exceed most other developed countries. One in five births in the U.S. are c-sections, whereas only about one in eight to ten births are c-sections in European countries (Sakala 1993a).

C-sections seem to have become a medical fad, despite the fact that the maternal mortality rate is far higher for c-sections than for normal vaginal births. Physicians seem to have changed the definition of "complications"; although not long ago well over 90 percent of all births were judged normal, that percentage has fallen drastically. In some cases in which a hospitalized woman in labor has refused a recommended cesarean delivery on religious or other grounds, the courts have upheld the right of the physician, as medical expert, to perform the operation against the woman's will (Jordan and Irwin 1989).

The threat of lawsuits, along with hospital policies, and the growth of specialists have encouraged increasingly interventionist medical strategies for childbirth, such as cesarean deliveries. There is also some suggestion that increasing rates for c-sections are related to doctors' convenience, insofar as c-sections are more likely to be performed at times of the day or week when physicians have competing professional and personal obligations (c-sections are much less time consuming than the natural process of childbirth).

Until recently, a woman who had a c-section had to have all subsequent births by cesarean delivery. Now, however, about one in four births to a woman who previously had a c-section is a vaginal birth, up significantly from rates in the 1980s (Curtin and Park 1999).

Most women who have a c-section do not typically plan it in advance with their doctors on the basis of some known problems such as a narrow pelvis. Rather, it is often something that is sprung on them in the delivery room. Frank discussion of this with the doctor beforehand may be the best way to avoid an unnecessary cesarean delivery.

1993b). The threat of lawsuits, along with rising insurance rates, hospital policies, and government regulations, have encouraged increasingly interventionist medical strategies. As a result, more babies are delivered via surgery, and newborn intensive-care units are populated by premature infants and babies with severe congenital abnormalities whose lives can be saved but whose conditions cannot be corrected by medical intervention. Taken together, the conditions surrounding medicalized maternity care raise a number of ethical, technical, legal, and emotional dilemmas that cannot be resolved easily.

Natural Childbirth

In response to what was perceived as "overmedicalized" maternal care in hospitals, a "natural" childbirth movement arose in the 1950s and grew during the following decades. Following the teachings of obstetricians like Grantly Dick-Read and Fernand Lamaze, childbirth education groups formed to prepare women and their "labor coaches" (usually their husbands), for a less medically controlled birth. Childbirth preparation classes help expectant couples understand the birthing process and teach women how to respond to uterine contractions with relaxation, structured breathing patterns, and muscular control (rather than with fear, tension, and pain). In general, research suggests that women trained in these techniques have shorter labors, lose less blood, require less anesthesia and analgesia, have fewer operative interventions, produce more alert and healthy babies, and experience greater personal satisfaction (Sullivan and Weitz 1988). In light of the observation that women who had taken child-birth classes and had a support person present were calmer and more compliant than other patients, even physicians who initially opposed "natural" childbirth became less hostile to it during the 1980s and 1990s.

Another aspect of the natural childbirth movement concerns husband-wife and parent-infant interaction. Following the treatment of birthing as a medical "problem," the standard hospital procedures in the 1950s and 1960s called for exclusion of the father from the delivery, isolation of the infant after birth, elaborate sanitary precautions, rigid feeding schedules, and long hospital stays. In a popular 1970s book on the ills of hospital birthing practices, one mother commented on how uncaring the rules seemed:

> I couldn't touch my baby for three days. He was in isolation in case there'd been an infection in utero. It was crazy! I had my baby and they took him away and I couldn't touch him. Then we went home. As I was leaving, they put me in a wheelchair, plopped the baby in my lap, and pushed me out of the hospital. (Arms 1975, 21)

Hospital practices were also criticized for discouraging breast-feeding by not allowing sufficient contact or flexibility to establish a mother-infant feeding routine. Breast-feeding, which was used by nine of ten mothers in 1922, was gradually replaced by bottle feeding during the era of "better living through science." By 1972, about eight mothers in ten fed their infants formula in a bottle instead of breast milk. In the 1960s and 1970s, researchers and proponents of breast-feeding began to document the advantages of breast-feeding over formula feeding. They generally found that breast-feeding provides better nutrition (particularly for preterm and low-birthweight babies), protects the infant against allergies and some diseases, and offers mothers some protection against possible breast cancer (Sullivan and Weitz 1988, 33). In conjunction with a growing concern for health issues of all sorts, breast-feeding began to regain popularity among the general population, so in the 1990s, the majority of birthing women advocated a preference for breast-feeding.

Admitting fathers to labor and delivery rooms was also prohibited by most hospitals in the 1950s and 1960s. The typical image is of the father smoking profusely and pacing back and forth in the waiting room until the doctor comes out to inform him "It's a boy!" In 1972, most hospitals still discouraged fathers from entering the delivery room, and only about one father in four was present at the birth of his child (May and Perrin 1985). By the late 1980s, almost all major hospitals had provisions for fathers being present during delivery, most encouraged the father's presence, and many provided special "homelike" birthing rooms or sleeping-in facilities for the father. By the 1990s, it was more usual for the father to be present at the birth than to be absent (Coltrane 1996).

The Birth Experience

> Oh—ah! Glorious! Fantastically excited! We were crying, my husband and I were crying and laughing and yelling, "It's a girl!" and I was kissing him and yelling—the feeling is—I don't think I've ever been happier. It's—you're so proud, you're so excited, you're—it's impossible to—put the emotion in words—none of the words I'm saying come anywhere near—the overwhelming emotion of it.

Pain—the pain just blocked everything. I was so scared.

Kind of excited and kind of unbelieving, you know.

Women's reactions to moment of birth

All beat to hell. I was stunned—it's very shocking. *It had a big blob of blood on its head and a bruise—two bruises on each side of its face—and milky-looking and yucky—it threw me. It really did.*

They tell you to expect the worst—and—I was surprised. He wasn't five minutes old and—um—he was just beautiful. In my eyes—he was really pretty. He had a white color, not reddish or anything.

Husbands' reactions to birth [Entwisle and Doering 1981, 102, 107]

The experience of childbirth can be a profound event, but parents vary a great deal in how they respond to it. Some parents find it a beautiful or ecstatic event. Others are neutral or rather negative about it. What causes these differences? In general, what is the course of pregnancy and childbearing like, as a social experience?

Doris Entwisle and Susan Doering (1981) followed a sample of 120 women, and about half of their husbands, through their pregnancies and the first six months after childbirth. The group included both middle-class and working-class couples, and for all of them it was their first birth. The patterns they displayed seem to be typical of many young couples in recent years.

For most women, the nine months of pregnancy are a time of good health. Those most likely to experience physical complaints were the ones who were most anxious about the childbirth. Most of them worried rather little, and their main concern was about how much weight they were gaining. Nevertheless, pregnancy did not emerge as a particularly pleasant time. The women's average opinion about their pregnancies was somewhat on the negative side, although many enjoyed the feeling "of something living inside you" (Entwisle and Doering 1981, 59). The researchers could not find any factor that seemed to predict which women would feel positive or negative about their pregnancy, though there seemed to be some connection to what the researchers called "sex-role ideology" (i.e., women who expected to breast-feed their babies felt better about their pregnancy).

One thing that women did often mention was how they felt about the fetal movements that began sometime around the fifth month. Some were very negative: "I . . . I'm shocked! I just didn't believe it would move that much; I really didn't. . . . It's very uncomfortable." More women, though, tended to like the sensation: "Oh, I love it! It's my favorite part of being pregnant. It feels like it's gonna come out dancing. It's wonderful feelings that there's life inside" (Entwisle and Doering 1981, 59).

Most of the women wanted to be able to go through their labor without drugs. Many had taken classes in natural childbirth and had practiced relaxation methods. Nevertheless, by the time the birth was over, the majority ended up taking at least some drugs. This was a disappointment; most women did not expect the delivery to be as painful as it was. They were simply not prepared for as much pain as actually occurred. The drugs given were fairly strong ones; over half the women had either Demerol (a narcotic), sedatives, or tranquilizers, whereas the rest had local anesthetics. The drugs cut off the chance of a "peak experience" for some of those who had been looking forward to it. A typical response was that of the woman who said she felt "very disappointed, just not *caring;* as if I had been dying." Most of those who were *less* drugged described the birth as a positive experience (Entwisle and Doering 1981, 96).

Almost all the husbands were present in the labor room and two-thirds were there during the delivery. By and large, the wives were pleased by this arrangement. The majority of them felt that their husband's presence was helpful, even though it was mostly a matter of holding their hands and giving emotional comfort. This was particularly so because the majority of women were left alone at some time during their labor or delivery, some for an hour, some for even as much as eight hours. Doctors were present in the room an average of 28 percent of the time, and nurses were there about half the time. Since these were first births, they tended to take longer than later births: about twelve to sixteen hours for the first stage of labor, and one to one and one-half hours for the delivery state. It was a long ordeal, and most of the women did not anticipate how physically arduous it would be.

The husbands in the Entwisle and Doering study tended to be anxious and worried during the birth, but their degree of worry did not seem to be associated with how much pain their wives experienced. The men in this study were not that closely attuned to what their wives were actually feeling. The pain tends to peak when the cervix is approaching full dilation, which is to say, just before the baby begins to move into the birth canal. The moment at which the head emerges, on the other hand, is least likely to be the most painful time. Some women, though a small minority (4 percent), did not experience any severe pain at any time. About 12 percent more said that the most painful thing was not the birth itself, but insertion of needles and other medical practices.

A key variable appears to be whether the women had learned to push correctly. Those who could simultaneously push while relaxing the muscles around the birth canal actually had a pleasurable moment of birth, whereas those who pushed while tightening their muscles reported excruciating pain. The required combination of relaxation of some muscles while pushing with others is difficult especially because the physical sensations are intense. Preparation before the experience, especially learning breathing techniques and defining childbirth discomfort as different from other types of pain can help most women. For instance, some natural-childbirth advocates even encourage couples to view the birth as a sensual experience *(A Closer Look 10.6)*.

The birth experience can be seen as a high-intensity interaction ritual, at least in the typical case when both parents are present and the mother is awake (if mothers receive medication, it is now usually a local or spinal anesthesia [epidural] which leaves the woman conscious). The essential components of a ritual are face-to-face interaction, attention focused on a common object or activity, a shared emotional tone that grows stronger, and an emotionally charged symbol (chapter 8). Most births have these elements because the father, particularly if he serves as "labor coach," is in close face-to-face contact with the mother; the couple is very focused on the breathing, and everyone there is focused on the birthing process; the intensity builds as the labor progresses, and the emotionally charged symbol is the newborn baby.

Fathers who are involved in childbirth in this fashion typically describe the experience as "incredible," "magical," "moving," "wonderful," and "exciting." One new father's comments highlight the ritual nature of childbirth:

I really participated in it. When I say we did the breathing together, I mean we really worked together and I really felt like I understood how she felt, in the sense that I was in the rhythm with her and what she was going through. . . . I cut the cord, which was symbolic, you know a kind of special thing to do, and the whole thing was very exciting and incredible, to hold the new baby at first (Coltrane 1996, 58).

A CLOSER LOOK 10.6
A Sensual View of Childbirth

Advocates of natural childbirth suggest that birth is a normal process that should occur at home, though most acknowledge that some pain is involved in giving birth. One unique group of midwives in Tennessee published a book, *Spiritual Midwifery* (1977), in which they attempted to redefine the childbirth process in positive, sensual terms:

> *I began having beautiful, rushing contractions that started low, built up to a peak, and then left me floating about two feet off the bed. [My husband] was lying beside me and going through the rushes too. I saw that I could breathe very deep and fast and rush higher with the contraction. The contraction would carry me and I would breathe harder and harder and then we would peak—it would slip off and leave us floating. It felt wonderful, and we were having a beautiful time. As the contractions got stronger, it felt like I was making love to the rushes and I could wiggle my body and push into them and it was really fine.*

Woman giving birth to her first child after two miscarriages (Gaskin 1977, 53)

> *Over and over again, I've seen that the best way to get a baby out is by cuddling and smooching with your husband. That loving, sexy vibe is what puts the baby in there, and it's what gets it out, too.*

Practicing Lay Midwife (Gaskin 1977, 52)

THE TRANSITION TO PARENTHOOD

The month or so following the birth is called the postpartum period. Most parents report that this is a time of significant emotional upheaval and adjustment to the demands of infant care. Although many parents experience profound joy during this period, most struggle to adjust to lack of sleep and the strain of being "on duty" all the time. Some researchers have reported that many mothers experience a period of "postpartum blues" characterized by depression, mood shifts, irritability, and fatigue. Although this used to be attributed to physiological factors such as changes in maternal hormones, research shows that fathers, as well as mothers, experience these symptoms (Zaslow 1981). What is probably happening is that the emotional intensity of the birth, coupled with lack of preparation for parenting duties and overwhelming time demands placed on infant caretakers, is stressful for both parents. Because the mother is more often the primary infant caretaker, she tends to be the one who is most subject to the blues. The emotional peaks and valleys of the postpartum period are quite common, but not everyone reports that they occur, and each person's postpartum experience is different.

There tends to be a sort of honeymoon atmosphere right after the mother and baby come home. The father is usually attentive, and the couple are emotionally close. However, after a few weeks, some strains and disappointments tend to set in. Most couples are overoptimistic about how easy it is to take care of an infant, and the pressures on their time and energy are larger than anticipated. In a recent study on the transition to parenthood, Susan Walzer quotes a woman who tried to stay positive in the face of the sometimes overwhelming demands of new motherhood:

> *This is probably the toughest thing that I have ever done. It's all the time and the minute you have the baby you just realize you would do anything, you would go to any extreme to make this baby happy, and you don't put any limits on yourself. You don't say, "I'm*

*going to work on this until 2:00 and if it doesn't work then forget it." You are just con-
stantly working at it and trying to make that baby happy. And I think by the end of the
day when that baby finally goes to sleep you think, "This is harder than any eight-hour
job I've ever had" (Walzer 1998, 30).*

Becoming parents is both joyful and extremely stressful. Earlier studies
showed that the birth of a first child tended to precipitate a crisis for the couple,
with the majority of new parents reporting major strains in their lives during the
family-formation period. Perhaps the most consistent finding across those studies is that both husband and wives reported declining satisfaction with marriage
and life in general, as they adjusted to the enormous time demands of newborn
care and realized the reality of widely differing expectations for mothers and fathers (Belsky, Spanier, and Rovine 1983; Hoffman 1979; LaRossa and LaRossa
1981; LeMasters 1957; Rossi 1968.)

More recent studies on the transition to parenthood have followed couples for
longer periods and focused on understanding variability in couples' adjustment
to parenthood. In general, these studies reveal how the transition to parenthood
is a complicated experience for both men and women, and how men sometimes
have more difficulty with the changes than women (Arendell 2000; Crnic and
Booth 1991). New parents continue to be surprised by how much work it takes
to care for an infant, and both women and men tend to report a decline in marital satisfaction during the first year after the birth (Demo and Cox 2000). Usually
in contrast to their expectations, mothers do most of the work, and the overall division of labor becomes more conventional (Cowan and Cowan 2000; Walzer
1998). During the second year, depressive symptoms decline, and marital satisfaction comes back up a little, suggesting that most couples weather the storm of
becoming parents and work out new ways to appreciate their family situations
(Cox et al. 1999; Gable, Belsky, and Crnic 1995). When new fathers are more
involved in infant care, marital satisfaction tends to be higher, but only when
the couple believes he should share responsibility for baby care (Coltrane 1996;
MacDermid et al. 1990).

Rossi (1968) first suggested how unprepared most parents are for assuming
parental duties. She noted that motherhood has been defined as essential to a
woman attaining full adult status in our society, but that unlike most other societies, we isolate mothers in individual households where they are unlikely to receive the help or training that they need. Unlike most roles that adults assume in
our society, we receive little preparation for being parents. Jobs typically allow us
to get some training, to serve as an apprentice, or to gradually assume full duties.
Parenthood, on the other hand, comes all at once, is irrevocable, and its training
period—pregnancy—teaches us almost nothing about infant care. Not only are
we unprepared for all the details of caring for a newborn baby, but after the baby
is born, someone has to be "on-duty" every day, twenty-four hours a day. It's no
wonder that new parents experience stress!

As noted earlier, things have changed somewhat since Rossi studied the transition to parenthood in the 1960s. For one thing, most women had babies at an
earlier age then. A more dramatic shift, however, has to do with the new
mother's labor force participation. When the Census department first began collecting data on working mothers in 1975, they found that 31 percent of married
women with a child under the age of one were in the paid labor force. That figure rose each year, so by 1985, 49 percent of wives with infants under a year were
employed. By 1998, 62 percent of mothers with children under one year old were
in the labor force (*Statistical Abstract* 1999, table 107).

The increase in maternal labor force participation has been especially dramatic during pregnancy. For those who gave birth to their first child in the early 1960s, fewer than half were employed at any point during their pregnancy. In contrast, two-thirds of those giving birth for the first time in the 1990s were employed during their pregnancy, and 80 to 90 percent of those were working full time. In the early 1960s, first-time mothers who worked during pregnancy were much more likely to be poor—teenagers, part-time workers, and high school dropouts. Employment patterns had changed substantially by the 1990s, however. Not only were many more pregnant women working, but the characteristics of those working had also changed. By the 1990s, women employed during pregnancy were more likely to be over age twenty-five and college educated. These older and more highly educated women were also more likely to work longer into their pregnancies than did their younger and unmarried counterparts from the 1960s. What we had, then, was a shift among middle-class women toward patterns that had once been common only for mothers at the lower end of the economic spectrum.

The economic necessity of having two workers in every family drove many new mothers to hold onto jobs that they would have abandoned in an earlier era. In the early 1960s, when less than one-half of women worked during their pregnancy, almost two-thirds quit their jobs before their first birth. Things had changed by the 1990s, when less than a quarter of women reported quitting their jobs altogether as the result of having a baby. Thus, in the space of twenty years, the relative percentages had switched: in the 1960s, women typically quit their jobs when they got pregnant, but by the 1990s, most kept their jobs.

Because most new parents need jobs to support themselves, their children, and if married or cohabiting, their partner, pressure has mounted for legislation that would encourage employers to offer maternity and paternity leave. Proposals for federal leave provisions have appeared since World War II, when the Women's Bureau of the Department of Labor recommended a six-week prenatal period and a two-month postbirth leave for women (Gerstel and McGonagle 1999). It was not until 1993, however, that the president signed legislation passed by the Congress mandating such leaves. The Family and Medical Leave Act (FMLA) stipulates that people employed for more than twelve months in companies with at least fifty employees within 75 miles of their work site can take up to twelve weeks of unpaid leave to care for newborn or newly adopted children (or seriously ill spouses, children, or parents) or to recover from their own serious health conditions, including pregnancy. Each of these provisions was the result of protracted political debates with an extensive coalition joining together to negotiate each part and each compromise. Although it represents a victory for family advocates, insofar as it begins to deal with some family needs previously ignored by federal policy, the act has been criticized on several grounds.

The law's limitations include the lack of provision for pay during the leave, a relatively short time of leave (twelve weeks), and a narrow definition of covered firms and types of families that are eligible (Gerstel and McGonagle 1999). According to analysts, the FMLA's provisions were targeted to cover those within the primary sector of the labor market who are more likely to have access to financial and family resources and to be white, middle-class, and married (Elison 1997; Marks 1997). Because it applies only to workplaces with fifty or more employees, the FMLA excludes 95 percent of employers and 50 percent of employees. In addition, it is of little assistance to part-time, seasonal, or temporary workers and is unavailable to same-sex couples, in-laws, or extended kin (Perry-Jenkins, Repetti, and Crouter 2000). Because the leave time is unpaid, low-income families often cannot take advantage of it. About one-fifth of workers expressed a need for such a leave in the years following passage of the act, and of these, about four-fifths took leaves. African Americans needed more leaves, but they were unable to take them as often as whites. A large number of men took leaves, but they were typically taken when they themselves got sick rather than to care for someone else. Women were more likely than men to express the need to take leaves and took longer leaves. Women were significantly more likely than men to take leaves to care for others. Motherhood increased the length of leaves for wives, but fatherhood reduced the amount of time husbands took for leave (Gerstel and McGonagle 1999). We can thus see that although the FMLA is putatively gender-, class-, and race-neutral, a disproportionate share of those in a position to take advantage of the unpaid leaves are white middle-class married women.

Other countries, including virtually all northern European nations, have made a much larger commitment to helping new parents. Over 100 countries throughout the world have legislated coverage that allows a mother to leave work for at least six weeks at the time of childbirth, guarantees that her job will be there when she returns, and provides a cash benefit to replace all or most of her earnings (Kamerman and Kahn 1995). Canadian workers, for example, receive fifteen weeks of family leave at 60 percent pay after the birth of a child. Swedish workers receive 90 percent pay for thirty-six weeks and prorated paid leave for the next eighteen months (Haas 1992).

Without leave policies that provide some form of income and without job protections for the majority of workers that ensure that they can return to their old jobs after they take time off to have or care for a new baby, Americans experience high levels of stress and economic uncertainty. Mothers carry an inordinate amount of the burden. Because of our expectations that mothers should do it all and because women's wages are usually much lower than men's, a new birth often means the creation of a sharp division of family labor based on gender.

Does Childbirth Promote Gender Stratification?

As noted previously, one of the major findings of the transition to parenting studies is that household divisions of labor and marital relations become more conventional after the birth of a child. In other words, women end up doing more of the cleaning, cooking, and child care, with men "helping" the mother but not assuming major responsibility for parenting or housework (chapter 11). Competing explanations for this trend focus on four different theoretical perspectives: sociobiology, personality differences, exchange theory, and interaction/conflict.

1. The sociobiological argument suggests that lactation, hormone secretion, and other physiological factors predispose women to be more proficient infant caretakers than men (Rossi 1977). More extreme versions of this approach posit a strong "maternal instinct" in women but not in men. Although prominent in popular culture, social scientists find little support for this position. Virtually, all parenting behaviors (like other human behaviors) are learned rather than being preprogrammed by biology or evolution.

2. The personality differences argument suggests that the experience of being raised by a mother cause girls to develop different "relational potential" than boys. According to Chodorow (1978), women grow up with fluid ego boundaries and experience the world through relations with others, predisposing them to want to mother their own children. Men, on the other hand, form their identities in opposition to women, form rigid ego boundaries, and consequently have less interest in and capacity for nurturing parental behaviors (see *A Closer Look* 12.3). (It should be noted that Chodorow's theory does suggest that some men are capable of nurturing and that, in fact, their assumption of responsibility for infant care could help promote gender equality by breaking the cycle of "oedipal asymmetries" [Chodorow 1976].) This view is also reflected in popular culture portrayals (e.g., *Men Are from Mars, Women Are from Venus*), but most social science evidence suggests that there is far more overlap between genders than differences. Experimental evidence suggests that rewards and expectations are stronger predictors of behavior (Howard and Hollander 1997).

3. The exchange theory explanation for marital roles becoming more traditional after childbirth rests on the assumption that women's bargaining power decreases after the birth (Nye 1978). New mothers typically make much less money than their husbands, often must cut back on employment after the birth, have few outside resources, and are perceived as lacking the stereotypical prebirth attractiveness that can give them leverage in the relationship. These factors make the woman dependent on her husband for economic and emotional support and reduce the likelihood that she will be able to bargain for increased domestic labor on his part. Although there is social science evidence in support of this position, the preponderance of evidence suggests that family patterns do not respond to simple rational choice trade-offs representing material power (Thompson and Walker 1989).

4. The interaction/conflict perspective on postbirth marital roles, like the exchange theory, assumes that spouses have competing needs but focuses on the ways in which they construct images of themselves as capable of various domestic tasks. In the face of extensive time demands, the new parents "do gender" (West and Zimmerman 1987) as they allocate and perform various child-care and housework chores. Fathers' lack of skill in caring for babies and mothers' superior nurturing capacities are created and reaffirmed through interaction with each other and those around them. Men, who get more free time and tend to focus on fun activities, are defined as "not good at" certain tasks or "not aware of" certain details, and their contributions are considered "helping out." Women, in contrast, are expected to and do perform most of the infant care and thus become "natural" caregivers who almost incidentally assume the position of household manager (Coltrane 1996; LaRossa and LaRossa 1989). This view receives the most support from recent research, though elements of the second and third theories are sometimes also supported.

Fathers' Participation in Infant Care

One interesting finding from recent research using the interaction/conflict perspective is that early assumption of infant care by the father seems to encourage later sharing of child care. For instance, fathers of six-month-old babies born by cesarean section have sometimes been found to be more involved in routine infant care than fathers of those born vaginally. This may be because the mother's incapacity requires a more active early caretaking role for the father and allows him to develop competence and confidence in his ability to perform infant care on his own. Interview and observational data also suggest that men develop parenting skills and construct nurturing self-images as a result of performing routine infant care (Coltrane 1996).

Early assumption of infant care by the father seems to encourage later sharing of child-care responsibilities.

I felt I needed to start from the beginning. Then I learned how to walk them at night and not be totally p.o.'ed at them and not feel that it was an infringement. It was something I got to do in some sense, along with changing diapers and all these things. It was certainly not repulsive and in some ways I really liked it a lot. It was not something innate, it was something to be learned. I managed to start at the beginning; if you don't start at the beginning, then you're sort of left behind. (Coltrane 1989)

Most of the fathers in this study, all of whom shared significant amounts of child care, talked about having to learn how to nurture and care for infants. Because most of the mothers breast-fed the babies and because everyone else assumed that the mother would be the major caretaker, these couples had to consciously encourage the father's participation. For example, one father commented:

She nursed both of them completely, for at least five or six months. So, my role was—we agreed on this—my role was the other direct intervention, like changing, and getting them up and walking them, and putting them back to sleep. For instance, she would nurse them but I would bring them to the bed afterward and change them if necessary, and get them back to sleep. . . . I really initiated those other kinds of care aspects so that I could be involved. I continued that on through infant and toddler and preschool classes that we would go to, even though I would usually be the only father there. (Coltrane 1996)

Other fathers talked about "nursing" their babies themselves, bottle feeding them with formula or with mother's milk that their wives had expressed using a simple hand pump. These examples illustrate how prescriptions about gender roles in baby care can be altered by those who try to change them. Sharing infant care, however, does not eliminate the stress that new parents experience.

Coping with the Stress

One of the reasons that the transition to parenthood is so stressful for most parents is that they feel isolated and alone. With all the diapers, sleepless nights, and financial worries of parenthood, most new parents sharply curtail their social lives. The isolation that many parents feel is eventually reduced by forming friendships with other parents, but that process takes time. Those who attend childbirth preparation classes have an opportunity to share hopes and fears with their peers and generally say they feel less isolated than others, report a more positive birth experience, and have a less anxious adjustment to newborn care than others (Cowan and Cowan 2000). Nevertheless, "prepared" mothers and fathers leave the childbirth class just before the baby is due, and while they may be ready for the birth itself, they continue to need support and social contact after the baby arrives. In more traditional societies and in some subcultures in modern American society, that support is provided by extended kin networks or by other people in the immediate community.

To break the isolation and find needed support during the difficult transition to parenthood, parents have recently turned to parents' support groups. Some are organized and run by mental health professionals, some are organized by schools and led by child development educators, and some are run by the parents themselves. All have become increasingly popular in the United States in the last two decades. These groups provide new parents with an opportunity to compare experiences, share advice, work on problems, and give and receive mutual support and encouragement. One of the primary benefits of such groups is that they help new parents make realistic appraisals of their expectations, their experiences, and the prospects for dealing with inevitable stresses and strains. Early evidence suggests that such groups help parents to feel less isolated and lead to higher levels of marital satisfaction and stability (Cowan and Cowan 2000). Extended family networks also continue to provide similar benefits to some new parents, though kin support tends to encourage conventional gender roles (Bott 1957; Riley 1990).

The best predictor of marital happiness and personal well-being after the birth of a first child is marital happiness and well-being *before* the baby is born (Belsky, Spanier, and Rovine 1983; Cowan and Cowan 2000; Wallace and Gotlib 1990). If a couple has a strong relationship with good communication and problem-solving patterns before conceiving a child, they are most likely to continue those patterns after the birth and are likely to report lower levels of stress. If, however, a couple decides to save an already troubled marriage by having a baby, the added stresses and strains of pregnancy and infant care are likely to make matters worse.

SUMMARY

1. In past times, high infant mortality rates served to limit population growth. In addition, most societies practiced some form of birth control, contraception, abortion, or infanticide. Cross-cultural studies show that when given the choice, women tend to have only a few children.

2. Rates of birth are determined especially by the economic value of children's labor for the family and whether the male or female has the power to decide on the size of the family. The U.S. birthrate has declined to near the rate of zero population growth, because children are no longer economically valuable and the power of employed women to decide on births has increased.

3. In the 1950s and 1960s, unmarried couples conceiving a child were likely to get married to "legitimate" the birth. Today only a minority of women who conceive a child outside of marriage marry the father. Because of the increasing prevalence of cohabitation, more births occur in the context of relatively steady nonmarital relationships than in past decades.

4. Rates of nonmarital childbirth have gone up dramatically in the past few decades. About half of all births in the 1990s were conceived outside of marriage, including over 80 percent of first births to African Americans, over 50 percent of first births to Latinos, and over 40 percent of first births to whites. The biggest increase in nonmarital births has been to women in their twenties and thirties. Rates of teenage births fell during the 1990s.

5. Although contraceptives are now widely available, a third of sexually active young people do not use them regularly. In recent years, the most common forms of birth control have become sterilization, birth-control pills, and condoms.

6. Although contraception education has been found to reduce unintended pregnancies among teenagers, one in three U.S. school districts in 1999 forbade the dissemination of any positive information about contraception, regardless of whether students were sexually active or at risk for pregnancy or STDs. Almost 1 million pregnancies still occur annually among American female teenagers. Unintended pregnancies force many U.S. teens to face the difficult question of whether to consider having an abortion.

7. Only about two-thirds of all pregnancies result in birth. Of the rest, 22 percent are terminated by abortion and another 16 percent by spontaneous miscarriages or stillbirths. The majority of abortions are performed on unmarried women below the age of twenty-five.

8. An increasing proportion of women are bearing their first child in their late twenties or in their thirties. It is possible for women to delay childbearing until the late thirties or early forties. The chances of birth defects do increase at this age; however, medical tests can reliably predict the presence of certain defects.

9. Assisted reproductive technologies, such as artificial insemination and in vitro fertilization, are making it possible for more infertile couples to have children. Surrogate motherhood, in which a fertile woman has a baby for an infertile couple, is rare but increasing. These new forms of reproduction raise complicated legal and ethical issues that have yet to be resolved.

10. Rates of childlessness are on the rise. Most decisions to avoid having children result from a series of decisions to delay childbearing. Voluntary childlessness is still stigmatized, though norms appear to be weakening.

11. The birth experience involves more discomfort than some women anticipate, and without preparation, women are likely to request pain medication. Women who receive less medication generally consider the birth to be a positive experience. Unlike practices in the 1950s and 1960s, most hospitals now invite fathers into the delivery room.

12. Most mothers of infants are now in the paid labor force. In the early 1960s, first-time mothers who worked were poor. Now mothers from all social classes are in the labor force, especially if they are well educated. Because most families now need two workers, women are less likely to give up their jobs when they have children.

13. Most young couples are unprepared for the transition in their lives that comes with having a baby. Childbirth preparation classes focus on the birth but do not prepare new parents for the round-the-clock duties and interrupted sleep that infant care demands. The transition to parenthood is usually a stressful time during which marital satisfaction declines.

14. After the first birth, household divisions of labor tend to become more conservative. This is primarily because women are socialized to become mothers, earn less money in the labor market, and become more dependent on their husbands after giving birth. Women "do gender" by caring for infants, and men "do gender" by paying attention to jobs. Nevertheless, men who share infant care develop the necessary skills and report interactions and relationships that are similar to those of mothers.

Key Terms

abortion	cesarean section	methotrexate-misoprostol
abstinence	comprehensive sex education	regimen
abstinence-only policy	Down's syndrome	mifepristone-misoprostol
abstinence-plus policy	eugenics	regimen
amniocentesis	in vitro fertilization	pregnancy rates
birthrate	medical abortion	zero population growth

Sociology Web Site

See the Wadsworth Sociology Resource Center, "Virtual Society," for additional links, quizzes, and learning tools:

http://www.sociology.wadsworth.com

Also on this web site you'll find InfoTrac College Edition, an online library of journals. Here you can search for electronic articles about central topics in sociology.

11

Domestic Life: Housework, Power, and Marital Happiness

Introduction

Who Gets More Out of Marriage?

Americans' Opinions About Marriage

Marriage and the Law

Historical Precedents

Same-Sex Marriage

A CLOSER LOOK 11.1: INEQUITIES IN PAST MARRIAGE LAWS

Marriage as a Social Institution

Home as a Workplace

A CLOSER LOOK 11.2: THE PARADOX OF LABOR-SAVING APPLIANCES

The Study of Household Labor

Who Does What?

Feelings About Housework

A CLOSER LOOK 11.3: PAID DOMESTIC LABOR

Women as Household Managers

Men's Participation in Housework

Attempting to Share the Strain

Predictors of Household Labor Sharing

INTRODUCTION

Is marriage good for people? Research shows that men and women who are married tend to have better health, accumulate more wealth, and report being happier than those who are not married. Because healthier, wealthier, and happier people are also more likely to get married in the first place, researchers have puzzled over whether it is marriage, per se, that leads to these positive outcomes, or whether they are attributable to what researchers call **selection effects.** Because people with more positive attributes and general promise tend to self-select into marriage, it is not surprising that married people look better, on average, than unmarried people. Because some of those not marrying have special problems, the comparison between the two groups is somewhat suspect.

In general, it has been difficult to conduct controlled comparisons to determine what influence being married actually has (Goldman 1993). Researchers are beginning to figure out the individual, social, and economic factors that help people operate successfully in the world, but isolating the specific contributions of marriage has proven more difficult. According to the most recent estimates from mathematical models, selectivity accounts for about half of the marriage premium (Waite 1995). In other words, the fact that more healthy and successful people are most likely to marry does not explain all of the differences between married and unmarried individuals. To understand the other half of the difference, we need to consider marriage as an institution.

We do know, for instance, that marriage provides various benefits to both men and women because it is socially approved. Particularly if a proposed mate is held in high esteem, parents, friends, and relatives tend to get very excited when wedding plans are announced. Being accepted into a family brings with it a host of material and emotional benefits. In addition to money, companionship, and practical help, marriage provides individuals—especially men—with someone who monitors their health and health-related habits and who encourages self-regulation (Ross 1995; Waite 1995). Marriage is also associated with a better diet, particularly for men, and especially after children are born, but even for women, as they begin to feel obligated to cook for "the family." Marriage also tends to be associated with a reduction in negative health behaviors—again, principally for men. Activities such as heavy drinking, the consumption of illicit drugs, fast driving, and other reckless behaviors like getting into fights tend to decline after marriage. As Umberson (1987, 1992) notes, marriage tends to limit such risky behaviors and encourages spouses to maintain an "orderly lifestyle."

When married people are compared with never married, separated, divorced, or widowed people, they are also typically found to have more money and long-term economic security. In large part, this is because marriages are increasingly likely to include two earners. Income and wealth accumulation advantages of marriage can also be attributed to economies of scale. Although two people cannot live as cheaply as one, they can live for about what it costs to maintain one-and-one-half people. Thus, married couples (and cohabitors) benefit financially from coresidence and combining domestic affairs (Waite 1995). In addition, as we discuss later, marriage tends to pull men into the labor force, often encouraging them to work longer hours in paid employment and to seek career advancement (Nock 1998). Although marriage has sometimes been portrayed as something that women want and men avoid, the analysis of everyday married life presented in this chapter suggests a different story.

Sociologist Jesse Bernard (1972) suggested that every marital union actually contains two marriages: "his" and "hers." The extent to which "his" is different from "hers" depends on a lot of rather mundane matters, which add up to living in different worlds. Economic realities outside the household shape husbands' and wives' class positions. Inside the home, these same realities are translated into the mundane experiences of doing the housework, leaving it to one's spouse, or hiring someone else to do it. Practical decisions have to be made, and someone has the power to make them. Will it be the wife, the husband, or both equally? The answers will affect all sorts of things, including "his" and "her" marital happiness.

WHO GETS MORE OUT OF MARRIAGE?

Who are more attached to marriage, men or women? Conventional wisdom says that women are. We have the popular image of men being dragged to the altar by women and then chafing under marital restraints and doing everything possible to escape. Like a lot of folklore, this image reveals a bit of truth but also a great deal of falsity. What seems to happen is this: men start out being less attracted to marriage than women, but once they are in it, they end up being more dependent on it than their wives. Why? The answer will tell us a great deal about the everyday reality of marriage, and its advantages and disadvantages for men and women.

First, part of the popular image seems to be true: men seem to be more reluctant to marry than women. One piece of evidence concerns the ages at which they typically marry. Men have always married at an older age than women. In the last few decades, men have waited on the average of two years longer than women to marry. Back around 1890, they waited four years longer. A more striking piece of evidence comes from surveys that simply ask people about their marital status. For the year 1998, one finds about 58.6 million married males and about 59.3 million married females (*Statistical Abstract* 1999). In other words, 700,000 more women than men *say* they are married. Yet men and women marry each other, so the number of married men and women must be equal. What is going on here? One possibility is bigamy, although it is hard to believe that three-quarters of a million men in the United States have two wives. Or, a certain portion of the respondents of both sexes may be lying; some unwed women may count themselves as married, while some married men may pass themselves off publicly as single. In general, what appears to be the case is that women are more willing than men to identify themselves as married, especially when they are under thirty years old.

Although marriage is associated with more happiness and better health among the general population, there is some evidence that men benefit more from being married than do women. We do know, for instance, that marriage has sometimes had an adverse effect on women's mental health (see *A Closer Look* 11.6 later in this chapter). Married women have higher rates of mental illness than married men (Gove 1972). One of four women has at least one episode of major depression in her lifetime, as compared with one of ten men (Holden 1986). Even though married people of both sexes have lower rates of mental illness than unmarried people, it has traditionally been true that marriage is more of a strain for the woman than it is for the man.

In addition, wives are more likely than husbands to report difficulties in the marital relationship, frustration, and contemplation of separation or divorce. Not only are wives more likely than husbands to be upset about finances, religion, sex, in-laws, and a lack of companionship, but they are also more likely to feel anxious and depressed and blame themselves for their troubles. Married

TABLE 11.1 Men Say They Are Happier with Marriage than Women

National surveys taken from the 1970s through the 1990s asked men and women, "Taking all things together, how would you describe your marriage?"

| | *Percentage Who Say They are "Very Happy" with Their Marriage* | | |
	1970s	1980s	1990s
Married men	68.9	64.9	64.4
Married women	65.4	61.8	60.0

Note the consistent gender difference in reporting about marital happiness and the slight decline in levels of marital happiness reported by both men and women. Some scholars suggest that men get more out of marriage than women.

Source: General Social Surveys (GSS) 1972–1996, National Opinion Research Center (NORC), University of Chicago.

women, especially if they are unemployed or have young children, report more troubles than women in other circumstances. The opposite pattern is true for men. Married men have better physical health, better mental health, and report that they are happier than unmarried men.

Jessie Bernard's (1972) provocative thesis that marriage is better for men than it is for women has generated much research and debate since the 1970s when Gove and his colleagues found that married women had higher rates of mental illness than married men (Gove 1972, 1979). Some subsequent studies have found few if any significant gender differences in marital happiness or mental health between married men and women (e.g., Glenn 1975; Johnson et al. 1986). Others, however, continue to find that husbands are happier with marriage than wives and garner more benefits from it (Finkel and Hansen 1992; Fowers 1991; Schumm and Silliman 1996; Schumm, Webb, and Bollman 1998). Fowers (1991) suggests that the differences in findings result from different approaches to measuring marital happiness, with those finding few gender differences more likely to use just a single questionnaire item or a simple scale (later in the chapter we discuss how marital happiness is measured and what promotes it; see *A Closer Look* 11.5 later in this chapter). Those who find differences between married men and women tend to use more detailed and elaborate measures of marital relationships and to focus on the individual couple; that is, they compare husbands and wives in the same marriage rather than relying on the average reports of men and women who are not married to one another (Schumm, Webb, and Bollman 1998). Because studies use different samples and measurement techniques, researchers continue to debate whether research supports Bernard's hypothesis about gender difference in marriage (Fowers 1991; Glenn 1991; Karney and Bradbury 1995).

Americans' Opinions About Marriage

Recent opinion polls provide support for the idea that men get more out of marriage than women. For example, the General Social Survey (GSS), conducted every year by the National Opinion Research Center at the University of Chicago, uses random sampling to select people from across the country and asks them questions about a wide variety of topics. By combining the results for all the years of the survey in each decade, we can get a sense of how attitudes might be changing. Table 11.1 shows that during the 1970s, over two-thirds (68.9

TABLE 11.2 Are Married People Happier than Unmarried People?

National surveys taken in the 1980s and 1990s asked if married people, in general, are happier than unmarried people. About half of respondents agree that married people are happier, but unqualified support is declining, and more men than women endorse this view.

Are Married People Happier than Unmarried People?

| | 1980s | | 1990s | |
| | Percent of | | Percent of | |
	Men	Women	Men	Women
Strongly agree	17.4	17.3	11.9	9.9
Agree	38.8	34.0	39.4	33.2
Neither agree nor disagree	29.4	31.5	28.9	33.6
Disagree	12.0	15.0	16.6	18.5
Strongly disagree	2.4	2.1	3.2	4.8

Source: General Social Surveys (GSS) 1980–1996, National Opinion Research Center (NORC), University of Chicago.

percent) of married men reported that they were "very happy" with their marriage. In the 1980s, that percentage had dropped to 64.9 percent, and it dropped only a little more during the 1990s (to 64.4 percent). Most married women also reported that they were "very happy" with their marriages during this period, but the percentage dropped from 65.4 percent in the 1970s, to 61.8 percent in the 1980s, and down to 60 percent in the 1990s (see Table 11.1). Note that the percentage of women saying they are very happy was below that for men in every decade and that the gap widened slightly in the 1990s. We thus see that there is a small but persistent gender gap in marital satisfaction, with men consistently more likely than women to report that they are very happy with their own marriage.

When GSS researchers asked American men and women whether married people (in general) were happier than unmarried people, they found similar patterns. This question focuses on other people and so reflects more global attitudes toward marriage, rather than evaluations of one's personal situation. Table 11.2 shows that most people agreed that married people were happier than unmarried people, though they were not as enthusiastic as they were about their own marriages. Similar to the self-ratings, we also see declining support for the idea that marriage automatically makes people happier. In addition, there are significant gender differences, with the gap between men and women largest in the more recent period. More men than women in the 1980s answered that they agreed or strongly agreed that married people were happier (56.2 versus 51.3 percent), with fewer answering that way in the 1990s and the gender gap increasing (51.3 versus 43.1 percent). At the other end of the spectrum, more women than men in the 1980s disagreed or strongly disagreed that married people were happier than unmarried people (17.1 versus 14.4 percent), with the percentages increasing in the 1990s and the gender gap increasing slightly (23.3 versus 19.8; see Table 11.2). Thus, although large-scale opinion polls may not be the best way to assess what is actually happening with American couples, they do tell us two important things about the way that men and women view happiness in marriage. First, although most people report that they are happy with their own marriage, there is less support for the idea that marriage is always good. Second, there is a consistent statistically significant difference in the views of men and women about marriage, with the gender gap widening slightly in the 1990s.

Why is it that marriage seems to benefit men more than women? To answer to that question, we first turn to legal images of marriage and then move on to consider the potential influence of unequal distributions of housework and marital power.

MARRIAGE AND THE LAW

Although most people do not usually think of marriage as a legal and financial contract, the law defines what marriage is, how spouses should relate, and what obligations husbands and wives have to each other and their children. Many people become aware of these laws only when they get divorced. While marriage laws are currently undergoing change, it is instructive to review the historical legal bases of the current institution of marriage.

Historical Precedents

Western law is based on traditions embodied in ancient Roman statutes, including the definition of family as the property of the male head of household (see *A Closer Look* 2.5). The laws of most Western societies have traditionally defined women as inferior beings who must be protected by a man. The ceremonial custom of a father "giving" his daughter away in marriage reflects the legal inferiority of women. In most cases, women have lost legal rights when they marry, and women's obligations in marriage have always been rather severe. As late as 1850, almost all states recognized a husband's right to beat his wife if she did not fulfill her wifely duties.

Traditional marriage law merged the identities of husband and wife. According to the feudal doctrine of coverture, the husband and wife became a unity at the time of marriage, and that unity was the husband. Symbolic loss of identity is still evident today in the custom of a married woman legally adopting her husband's name when she marries: Miss Jane Smith marries Mr. John Jones and becomes Jane Jones or Mrs. John Jones, whereas the man's legal identity remains exactly the same as it was before marriage. Some alternative customs are more common, but none is as yet standard. A married woman may keep her own name, or she may join her last name with that of her husband with a hyphen (e.g., Smith-Jones). Occasionally, a husband uses a hyphenated name too, but the rarity of this practice suggests that we still have a predominantly patrilineal marriage system.

In the past, marriage laws also required a wife to take her husband's legal domicile (place of residence) and obligated the wife to provide various services for her husband. Lenore Weitzman (1981) discusses four essential provisions of the marriage contract that were incorporated into our laws:

1. The husband is the head of the household.
2. The husband is responsible for financial support.
3. The wife is responsible for domestic services.
4. The wife is responsible for child care.

Although we might think of these provisions as outdated notions of the past, our courts were still relying on them in the 1970s, and their social ramifications

Ancient Greek and Roman ideals shaped our marriage laws.

remain with us today. As Weitzman shows in her review of court cases from the 1960s and 1970s, even liberal states tended to treat wives and their earnings as the property of their husbands. For example, the courts refused to enforce contracts in which the husband agreed to pay his wife for housekeeping, entertaining, child care, or other "wifely" duties. Even if the wife performed services considered to be "extra," such as working in the husband's business, the courts voided the contract that obligated the husband to pay her (Weitzman 1981, 1189) (*A Closer Look* 11.1). Today, most states have marriage laws that treat husbands and wives the same, though in some states married women still lose some legal rights to control property or enter into legal contracts on the same basis as men or unmarried women.

Same-Sex Marriage

Recent changes in marriage laws have focused on the legal rights of same-sex couples. There are cross-cultural and historical precedents for the recognition of same-sex marriages, and gay and lesbian couples in the United States are now seeking the right to marry. A major motivation is to obtain the spousal rights that marriages affords, such as health, retirement, social security, family leave, inheritance, and related benefits. Three Hawaiian same-sex couples applied for

A CLOSER LOOK 11.1
Inequities in Past Marriage Laws

In *The Marriage Contract* (1981), Lenore Weitzman presents some examples of the ways that laws regulating marriage gave a husband control over his wife's property. These summaries come from actual legal cases.

Husband's Rights After Wife's Death

Florence and Samuel Jorgenson, like most people married in the forties, chose a traditional relationship: Samuel brought home the paycheck, while Florence remained in the home to care for their sons, Bill and Ed. Each week, Samuel turned his paycheck over to Florence; she did the family's bookkeeping, paid the bills, and rationed out the money for household and personal expenses.

The Jorgensons didn't have much money. For most of their married life, they lived with Mrs. Jorgenson's mother, not even able to afford a home of their own. However, Florence believed that her boys should go to col-

lege, and she managed to save a little here and there for their education. She economized in all the little ways in which a housewife can economize. Occasionally, her mother made a donation to the college funds. Slowly but surely, two savings accounts grew.

When Florence Jorgenson died in 1963, she left her teenage sons two savings accounts: $2,000 in trust for Ed and another $3,000 in trust for Bill.

Samuel Jorgenson remarried and moved out of his mother-in-law's house. $5,000 seemed like a lot of money to him. He wanted it, and he went to court to get it. "I earned that money," he argued, "and I never gave it away. I merely entrusted it to Florence to pay the bills. The surplus belongs to me." The court agreed—despite the fact that there would have been no surplus without Florence's careful economizing, despite the fact that Florence's mother had contributed some of the funds. "The general rule in separate property

states," said the court, ". . . is that the excess left after paying the joint expenses of . . . the family remains the property of the husband." The boys' college money went to Samuel and his new wife. Florence's long years of scrimping and saving were an exercise in futility.

Husband as "Head and Master"

Married at eighteen, Helen Tarbel worked double shifts as a nurse to support her husband through four years of college. Her salary paid for all household expenses, and she saved enough to buy a small home. Her husband was unemployed for most of the year after he graduated from college, and Helen continued to support the family. When her husband proposed that they mortgage the house to borrow $5,000, Helen objected. But under the Louisiana "head and master" rule her husband had the full power to mortgage the home

marriage licenses in 1990, and when their applications were denied, they sued the state, claiming they were treated unfairly. The Hawaii Supreme Court sent the case back to the Circuit Court (*Baehr v. Lewin,* 74 Haw. 530, 1993) where the judge, hearing evidence that gays and lesbians parent as well as heterosexuals, ruled that the state could not deny marriage licenses to same-sex couples (*Baehr v. Miike,* Haw. 91-1394, 1996). Because marriages performed in one state are typically recognized by the other forty-nine, the Hawaii case prompted many legislative proposals at the federal and state levels. In 1996, Congress passed the "Defense of Marriage Act," stipulating that no state could be forced to recognize another state's same-sex marriage and defining the marital union as between a man and a woman (104th Congress, HR 3396). As of 1999, thirty states had passed "defense of marriage" legislation. It is notable, however, that twenty-one states defeated or blocked similar legislation in 1996, as did twenty-two states in 1997 and ten states in 1998. Under defense-of-marriage laws, a legally performed opposite-sex marriage would be recognized in any state, whereas a legal same-sex marriage performed in one state would not be recognized in a state banning same-sex marriages. As a consequence, defense-of-marriage laws are currently

she bought—and to control all of their community property—without even asking her. So he ignored Helen's objection, and without her signature or consent, mortgaged the home to take out a loan. When he failed to repay the loan the credit company sought to foreclose the mortgage and take their home. Helen Tarbel objected and challenged the constitutionality of the head and master rule. It was unfair, she said, for her husband to have the power to mortgage the home without her knowledge or permission. It was also unfair that he should control every cent she earned and every item she bought. But the Louisiana Supreme Court denied her appeal, and the U.S. Supreme Court refused to hear the case, thus leaving her creditors to foreclose the mortgage and letting the head and master rule stand.

Husband's Control of Wive's Earnings

Fred and Betty Nelson were married in 1950. They lived on a farm in Illinois that Fred had inherited before they were married. At first Betty was a housewife, caring for their children, performing the usual household tasks—gardening, preserving large quantities of food, and cooking five or six daily meals for the hired hands who worked on the farm. Later, when the children were in school, Betty took on an outside job. She continued to do the traditional tasks and, in addition, contributed part of her income for family expenses.

After twenty-two years of marriage, Fred divorced Betty. The court awarded her no alimony. Her share of their marital property consisted of only her own clothing and personal effects, a few household items that she owned before the marriage, and an automobile that she had purchased in her own name with her own funds. The house and furnishings, the farm with its machinery and livestock, the savings—all went to Fred.

Betty appealed the decision. "Surely, after twenty-two years of hard work, I am entitled to at least a portion of the assets I helped to accumulate," she thought, but the legal system saw things differently. The appellate court upheld the lower court's division of the property.

A spouse seeking part of the other spouse's property, explained the court, must show that she or he made valuable contributions to the property's worth. The court defined a valuable contribution as "money or services other than those normally performed in the marriage relation"; Betty's years of cooking, cleaning, and child rearing did not meet the court's definition of "valuable."

The money Betty contributed for family expenses from her outside employment was "valuable"; would she not be entitled to some recompense for this financial contribution? No, said the court; she did not keep clear records of what property was acquired through her own effort, and consequently, all property was properly awarded to her husband.

being challenged as a violation of the Equal Protection Clause of the United States Constitution. Some states, such as Vermont, are considering **civil unions** as a type of partnership that includes some, but not all, of the benefits of marriage. Whether same-sex couples should be afforded legal protections akin to marriage is a controversial topic that will undoubtedly be debated for many years to come.

Marriage as a Social Institution

Some recent research on marriage as a social institution emphasizes how the old legal and normative rules continue to apply, albeit in somewhat modified form. For example, the sociologist Steven Nock (1998, 6) suggests that Americans generally agree with six ideas about marriage:

1. Marriage is a free personal choice, based on love.
2. Maturity is a presumed requirement for marriage.
3. Marriage is a heterosexual relationship.
4. The husband is the head, and principal earner, in a marriage.

5. Sexual fidelity and monogamy are expectations for marriage.

6. Marriage typically involves children.

According to Nock, these ideas about what marriage *should be* constitute a normative definition of marriage in America today. Obviously, not all couples conform to such ideals, and we've seen that significant numbers of Americans reject one or more of these prescriptions for marriage. Many more do not follow at least one in actual practice, and we will explore later how people have challenged specific ideals (especially points 3, 4, and 6). However, for researchers who champion the positive aspects of marriage (e.g., Popenoe, Elshtain, and Blankenhorn 1996; Nock 1998; Waite 1995), this list is a good starting point for beginning to understand how and why marriage has tended to be beneficial—especially for men.

Nock (1998) suggests that the ideal of the breadwinner husband as head-of-household is so entrenched in practice and custom that it helps form the basis of marriage as a stable institution. For support, he suggests looking at how the U.S. Census Bureau collects data about such things. Until 1980, the Census Bureau automatically defined the husband as the "head of household" or "head of family" in its *Current Population Surveys*. Since then, they have used the term "householder" and "family householder." When a married person is interviewed, he or she is asked to identify the person in whose name the house or apartment is owned or rented. If there is no such person, any adult may be listed. If the house is owned or rented jointly by a married couple, the householder may be either the husband or the wife, so either can be listed. What can be inferred by how people answer this question? Nock suggests that it gives some indication of who owns married couples' property and provides insight into which spouse is most commonly considered to be the main "reference" person. In 1980, 96 percent of married individuals living with a spouse named the husband as the householder, and by 1994, Nock reported that the figure had dropped to 91 percent, offering "little evidence of major restructuring of the male-as-head-of-household practice" (1998, 38).

According to the most recent figures from the Census Bureau, this conclusion seems a bit premature. In the 1998 *Current Population Survey,* 77 percent of married couples with both husband and wife present listed the husband as householder, with the other 23 percent listing the wife as householder (*Statistical Abstract* 1999, table 69). This indicates at least a doubling of the practice of listing the wife as householder in a very short time. In light of such evidence, we would suggest that the institutional, legal, and normative practice of considering the husband as head and master of the household has weakened considerably, though it has definitely not disappeared.

More direct evidence about the changing attitudes of Americans toward the roles of husbands and wives comes from GSS pollsters, who in 1996 asked the following question:

> *In many married couples, women take the main responsibility for the care of the home and children, while men take the main responsibility for supporting the family financially. Do men benefit from this?*

As predicted by Bernard's idea of his and her marriages, there was a big gender difference in answers to this question. Over 60 percent of women said yes, men benefit from this traditional family arrangement, whereas just 45 percent of men said yes. When asked if this arrangement hurts women, 42 percent of women but just 29 percent of men said yes.

When GSS pollsters asked a related question about what people wanted in a marriage, Americans provided more evidence that the notion of the husband as head and master was no longer the popular ideal it once was:

Next we have some questions about the kind of relationships you would like with a spouse or partner. It doesn't matter whether you are now married or living with someone. Which type of relationship would you prefer: A relationship where the man has the main responsibility for taking care of the home and family, or A relationship where the man and the woman equally share responsibility for providing the household income and taking care of the home and family?

Over two-thirds of all respondents indicated that they would prefer an equal sharing marriage in which both men and women take responsibility for earning income and taking care of home and family. Unmarried people were most likely to endorse the equal sharing position, with 84 percent saying men and women should do both. About three-quarters of divorced and separated people (74 percent) said they endorsed sharing. Even among those who were married, just under two-thirds (64 percent) said that both husbands and wives should share responsibility for providing the household income and taking care of the home and family. Next we explore whether most American couples actually realize these ideals.

HOME AS A WORKPLACE

People usually marry out of an emotional attraction: love, sex, romance, companionship, and other idealized experiences that a couple look forward to. Once into the marriage, though, they experience the hard material realities of family life. The home, after all, turns out to be an economic unit, a place where work has to be done every day just to keep things going. Every home is a combination of hotel, restaurant, laundry, and often a child-care and entertainment center (*A Closer Look* 11.2). Each of these activities takes work, work that is often invisible when one is merely the recipient of these services (as children and men often are). However, the work becomes all too real when the young married couple finds out that they must take care of it themselves, or it will not get done at all. Even the pleasant parts of family life take behind-the-scenes work. Having a party can be fun for the guests but a chore for the wife (or possibly husband) who has to prepare everything, provide service while it is going on, and clean up afterward.

It has often been noted that love and romance tend to decline after the wedding. The reason is not simply the disappearance of the ritual conditions that made premarital love an emotional experience but also the rather rude shock of confronting the plain hard work that makes up an ongoing household. Moreover, the work is not merely mundane and demanding, it is also stratified. It is one of the main arenas in which men and women have different powers and payoffs and one of the activities around which latent struggle, and sometimes overt conflict, is likely to happen.

According to popular belief, housework is so trivial that it is hardly worthy of serious discussion. According to recent scholarship from sociology, economics, and women's studies, however, examining the allocation of housework may tell us more about marital power and gender relations than almost any other subject. As noted earlier, families cannot function unless someone does the routine shopping, cooking, and cleaning that it takes to run a household.

A CLOSER LOOK 11.2
The Paradox of Labor-Saving Appliances

One hundred years ago, the work of running a household was much more arduous than it is now. Many houses lacked hot and cold running water; water had to be hauled in from a well or street source and heated on the stove. Carrying water was traditional women's work and had been for thousands of years. To tend stoves and fireplaces, wood or coal needed to be brought in and ashes hauled away. Cooking could be very time consuming under these circumstances, and laundry was a major chore. Even as simple a matter as keeping a house lit at night required filling oil lamps and regularly cleaning off the accumulated smut.

A series of inventions transformed all that. Electricity came early in this century, although many rural homes did not have it, or running water, until the 1940s. Refrigerators meant that food did not have to be bought or prepared every day. Gas and electric stoves, washing machines, dryers, vacuum cleaners, and dishwashers have revolutionized housework. Or so it might seem.

A very surprising finding, though, is that *the total number of hours that American housewives spend on housework did not decline from 1926 to 1968.* Joann Vanek (1974) examined studies done on home economics since 1924 and found that full-time working housewives were working about fifty-two hours a week in 1926, a figure that *rose* to about fifty-five hours a week in the 1960s (Figure 11.1).

How can this be so? Vanek suggests that women's standards of home comfort have risen as new devices have become available. For instance, before automatic washers existed, doing laundry was very arduous. Usually, it was done only once a week. Now laundry is done much more frequently; it takes less time to do each load, but housewives now run a load almost every day. A result is that expectations for cleanliness have increased; clothing is changed more frequently.

In addition, women have shifted their housework time from certain tasks to others. Less time now is spent on preparing food and cleaning up after meals, due to all the advantages of labor-saving devices in the kitchen, although these continue to be the most time-consuming housework tasks. Cleaning is easier now, but housewives seem to have escalated their standards of how clean their homes should be. Whatever additional time has been made available has been taken up by other tasks. *Longer hours are spent on caring for children*—even though today's families have fewer of them than families of fifty years ago. More time is spent on shopping and general household management: in the 1920s women averaged two hours a week going to and from stores, whereas today

In fact, recent studies suggest that the total amount of time spent in unpaid family work is about equal to the time spent in paid labor (Robinson and Godbey 1997). In theoretical terms, family work—or social reproductive labor—is just as important to the maintenance of society as productive labor that occurs in the formal market economy. Nevertheless, housework is rarely afforded the attention that paid labor receives, primarily because marriage laws and customs have assumed that wives will provide domestic services free-of-charge to their husbands.

Modern history shows that it has been women who have done most of the routine family housework, including shopping, cooking, and cleaning. The cultural ideal of separate spheres, which suggests that women belong in the private/home sphere and men belong in the public/work sphere, has encouraged this unbalanced distribution of household labor. More recently, however, as wives have moved into the paid labor force and started to share breadwinning with their husbands, social scientists have generally expected men to begin to share housework responsibilities. Unfortunately for working women, this has not been the case, as most women have instead been expected to shoulder the

FIGURE 11.1 Distribution of Time Among Various Kinds of Household Work, 1926–1968

The data relate only to nonemployed women, meaning women who did not have full-time jobs outside the household. Top curve includes cleaning up after meals.

Source: From "Time Spent in Housework" by Joann Vanek, 1974. *Scientific American,* 231 (Nov.), pp. 116–20. Copyright © 1974 by Scientific American, Inc. All rights reserved.

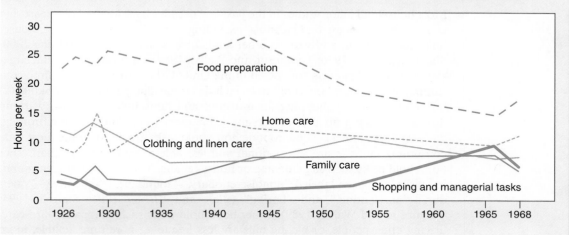

they spend a full day shopping every week.

Newer labor-saving devices have appeared in recent years, and still more are touted for the future. One can buy an electronic system that automatically turns on and off the sprinkler system or the lights, and home computers and robots are apparently about to become very common. If the lesson of the past tells us anything, though, it is that we can expect housework hours to decrease very little, if at all. Other studies (Cowan 1983) have confirmed that the amount of housework women do has not fallen since colonial days. There seems to be a status competition over home lifestyles, which will go on raising standards endlessly.

burden of a "second shift" of housework on top of their employment responsibilities (Hochschild with Machung 1989).

In the modern context, identifying who avoids doing everyday household tasks can be an excellent indicator of who has the highest status in the couple or family. Who feels entitled to household services, who is obligated to perform them, and how couples evaluate the fairness of divisions of labor can tell us something important about the subtle exercise of power in intimate relationships. Such seemingly trivial household matters can also help us understand how and why gender inequity is perpetuated in the society-at-large (Coltrane and Adams 2000).

In recent years, there have been some shifts in who is doing the housework in American families. Generally speaking, the amount of time per week that men spend doing housework has increased slightly. At the same time, women are now doing less housework then they used to, and therefore, the percent of total housework that is being performed by men has increased. Because the overall changes in men's contributions have been small, housework continues to be unequally distributed, with women doing roughly two or three times as much housework as men. In this section of the chapter, we look at how housework is studied, how it is distributed within households, how it came to be that way, how it is changing, and how we think these changes might affect the balance of power between men and women in the future.

The Study of Household Labor

Most household labor studies in the past few decades have focused on measuring tasks such as cooking and cleaning, excluding things like child minding, household management, and various other kinds of emotional labor (Coltrane 2000; Shelton and John 1996; Thompson and Walker 1989). According to several large-sample national surveys conducted in the United States, the five most time consuming of the major household tasks include (1) meal preparation or cooking, (2) house cleaning, (3) shopping for groceries and household goods, (4) washing dishes or cleaning up after meals; and (5) laundry (including washing, ironing, and mending clothes). These routine household tasks are not only the most time consuming but also are less optional and less able to be postponed than other household tasks such as gardening or house repairs. These seemingly never-ending tasks have been labeled "nondiscretionary," "mundane," "repetitive," "onerous," "unrelenting," and "boring" (Blair and Lichter 1991; Starrels 1994; Thompson and Walker 1989). Other household tasks such as repairs, yard care, driving other people, or paying bills are less frequent, more time flexible, more discretionary, and generally considered more enjoyable than the more time-consuming tasks noted previously. Some researchers refer to the routine housework chores of cooking, cleaning, and shopping as "female" or "feminine" and to tasks such as household repairs, mowing the lawn, and taking care of cars as "male" or "masculine." We prefer to use the terms *routine housework* and *other household labor,* calling attention to the character of the tasks themselves rather than to cultural beliefs about the suitability of one gender to perform them.

Information about household labor is usually collected using time diaries or surveys. In time diary studies, individuals are asked to complete logs accounting for time spent on various activities, usually for a twenty-four hour period. Time diaries generate the most accurate estimates of time spent on specific housework activities, though simultaneous activities are sometimes ignored or underestimated (Robinson and Godbey 1997). Phone, mail, and in-person interview surveys are also used to assess time spent on household labor. Respondents are typically asked how much time they "usually" spend per week on specific household activities or how much time they spent "yesterday" on selected activities. Both men and women tend to overestimate their own contributions in surveys and to double-count time spent in simultaneous activities. In general, researchers have moved away from asking simple proportionate questions (who does more housework?) and toward collecting hourly estimates of performance because questions about more narrowly defined tasks produce more accurate estimates (Shelton and John 1996). Past studies often collected information about various household members' task performance from wives only, whereas more recent studies have collected data from both men and women.

Who Does What?

National surveys and time-diary studies show that American household members spend about three hours on routine housework for every hour they spend on other household labor. According to the National Survey of Families and

Households (NSFH), the average married woman does about three times as much routine housework as the average married man (thirty-two versus ten hours per week), and the average married man does a little less than twice as much occasional household labor as the average married woman (ten versus six hours per week). Over the years, studies have found that wives perform over 96 percent of the cooking, 92 percent of the dishwashing, 90 percent of the vacuuming, and 94 percent of bed making (Blair and Lichter 1991; see also Berk and Berk 1979; and Berk 1985). Men, in contrast, do over 86 percent of household repairs, 75 percent of lawn mowing, and 77 percent of snow shoveling (Blair and Lichter 1991; see also Schooler et al. 1984). This division of labor is so influenced by gender that the average man would have to reallocate over 60 percent of his family work to other chores before gender equality would be achieved in the distribution of labor time across all domestic tasks (Blair and Lichter 1991).

Looking at study results over time, we can see some changes in who does what. Women, especially if they are employed, are doing less housework than they used to, and men are doing slightly more. Based on national time-diary studies, Robinson and Godbey (1997) reported that American women's time spent on housework declined from twenty-four hours per week in 1965 to sixteen hours in 1985, a decline of one-third. Employed women cut back on the time they devoted to housework the most and shifted many chores to the weekends, so they were doing about one-third less family work than nonemployed women by 1985. During that same two-decade period, men's contributions to routine housework increased from about two hours per week to about four hours per week. Questionnaire studies show that women's housework contributions continued to decline well into the 1990s and that men's contributions continued to increase slowly (Coltrane 2000). These trends have produced a shift toward a greater proportionate sharing of housework in couple households, even though most women still do about three times more housework than the men they live with.

Because gender is a major organizing feature of household labor, research has explored how men's and women's task performance differs and how their experience and evaluation of housework tend to diverge. In general, women have felt obligated to perform housework, and men have assumed that domestic work is primarily the responsibility of mothers, wives, daughters, and low-paid female housekeepers (*A Closer Look* 11.3). In contrast, men's participation in housework has appeared optional, with most couples—even those sharing substantial amounts of family work—characterizing men's contributions as "helping" their wives or partners (Coltrane 1996).

Feelings About Housework

Most women experience repetitive domestic chores as boring but important work that is performed for people they love. Even if they do not find the activities themselves enjoyable, most wives and mothers enjoy feeding and taking care of their families and derive considerable self-worth from meeting family members' needs. Because household labor is tied up with what it means to love and care for others, women often have ambivalent and contradictory feelings about it and report mixed reactions to being responsible for so many of the routine household chores (DeVault 1987; Ferree 1987; Thompson and Walker 1989).

Both women and men experience boredom, fatigue, and tension when they do household work alone (Baruch and Barnett 1986), but women are much more

A CLOSER LOOK 11.3
Paid Domestic Labor

At one time, most households had servants. A large household might have a butler, valets, footmen, and other male servants, but the majority of house servants were usually female. The number of servants has been on the decline in Western countries like Great Britain and the United States since the 1800s. As late as 1900, household servant was still the largest category of workers in Great Britain (Laslett 1977, 35). In 1850, every middle- and upper-class household in the United States had at least one servant (*Historical Statistics of the United States* 1965), usually a maid or housekeeper.

The growth of the middle class during the late nineteenth- and early twentieth centuries meant that many more "mistresses" demanded the help of maids and nannies to perform the work required in cooking, cleaning, and raising children. Black women were most likely to be servants and laundresses, especially in the South but in the North as well. In the Southwest, Chicanas were disproportionately concentrated in domestic service, and in the Far West (especially in California and Hawaii), Asian men were most often household servants (Glenn 1992). As the "hiring class" expanded, middle-class homemakers came to think of themselves as supervisors whose superior knowledge allowed them to manage and oversee the manual labor of the servants they employed (Dill 1988; Romero 1992). The rise of scientific mothering and home economics helped turn the middle-class home into the woman's domain (Cott 1977; Skolnick 1991). Treating homemaking and motherhood as a revered feminine profession gave middle-class wives managerial control over day-to-day domestic activities and thus solidified their power over the low-paid workers who had few other options for employment.

Because most working-class husbands could not earn enough to support the whole family, their wives had to work for wages or figure other ways to pool resources and make ends meet. Unlike middle-class women, poor women were forced to take in boarders, grow their own food, barter, take in laundry, or perform other kinds of domestic work for middle- and upper-class households. For immigrants and women of color who were typically hired as domestics, household work was thus both paid (because they cleaned wealthier families' houses, cooked for them, and cared for their children) and unpaid (because they still performed these tasks for their own families) (Dill 1994; Rollins 1985). Although the romantic ideal of separate spheres that developed during this time held that women

likely than men to perform household tasks in isolation. Although men tend to do much of their family work on weekends, women tend to do housework each day. Women also report doing an average of three household tasks at one time, which may help explain why they find household labor to be less relaxing and more stressful than men do (Thompson and Walker 1989).

Which tasks are most liked and disliked? According to some studies, the hierarchy from most favorite to least favorite work runs as follows: cooking, shopping, washing, cleaning house, and ironing. Cooking is considered the most enjoyable task, and ironing is the least enjoyable. When men contribute, they tend to avoid the unpleasant tasks (such as cleaning the toilets). When they pitch in, they are more likely to do child care than housework and more likely to watch or play with the children than to feed them or clean up after them. When they do housework, they are most likely to cook or wash dishes, followed by vacuuming or tidying up. Men are least likely to mop, clean bathrooms, and do laundry.

The exasperation experienced by working women whose husbands contribute little to routine housework is captured in books such as Francine M. Deutsch's *Halving It All* (1999). One woman she interviewed, employed fifty-two hours a week, still did virtually all of the housework:

should be sensitive and pure keepers of the home on a full-time basis, the reality was that women in less advantaged households had no choice but to be workers and mothers at the same time. As sociologist Mary Romero (1992) points out, even though middle-class and working-class women's lives increasingly came to be seen as revolving around the care of homes and children, patterns of paid domestic labor tended to accentuate class differences and reinforce racial discrimination.

During the middle part of the twentieth century, the use of paid domestic labor decreased, in-home residence declined, and trends in hiring began to shift. Old-fashioned live-in servants became much more rare, so by the 1980s, only about one in over a thousand households had a live-in domestic worker. As middle-class women's labor force participation levels rose, however, so did their use of paid housecleaners. According to the U.S. Bureau of Labor Statistics, in 1983 there

were 512,000 paid housecleaners working in private households (although we should keep in mind that many such arrangements are not reported to the federal government because the employers want to avoid paying income taxes and the employees want to avoid immigration investigations). Of these half a million domestic housecleaners in 1983, 96 percent were women, 42 percent were African American, and 12 percent were Hispanic. By 1998, there were 549,000 such workers, with 94 percent women, 15 percent African American, and 37 percent Hispanic (*Statistical Abstract* 1999). We thus see a dramatic increase in the number of Latinas who are cleaning others' houses for pay. In contrast, the number of child-care workers employed in private homes (sometimes called nannies, au pairs, mothers' helpers, governesses, etc.) has declined. In 1983, there were 408,000 private household child-care workers, 97 percent of whom were women, 8 percent African American, and 4

percent Hispanic. By 1998, there were 278,000 private household child-care workers, with 97 percent women, 9 percent African American, and 20 percent Hispanic (*Statistical Abstract* 1999). Although the number of nannies is declining, we can see that they are somewhat more likely to be Latinas than in the past.

Sources:

Cott, Nancy. 1977. *The Bonds of Womanhood.* New Haven: Yale University Press.

Dill, Bonnie Thorton. 1994. *Across the Boundaries of Race and Class.* New York: Garland.

Dill, Bonnie Thorton. 1988. "Our Mother's Grief: Racial Ethnic Women and the Maintenance of Families." *Journal of Family History* 13: 415–31.

Glenn, Evelyn Nakano. 1992. "From Servitude to Service Work: Historical Continuities in the Racial Division of Women's Work." *Signs* 18 1–43.

Rollins, Judith. 1985. *Between Women: Domestics and Their Employers.* Philadelphia: Temple University Press.

Romero, Mary. 1992. *Maid in the USA.* New York: Routledge.

Skolnick, Arlene. 1991. *Embattled Paradise: The American Family in an Age of Uncertainty* New York: Basic Books.

Statistical Abstract of the United States, 1999, Washington DC, table 675.

Although tired and stressed, working a double day, Carol doesn't expect [her husband] to do much: "I just want him to pick up after himself. I don't particularly expect that he is going to vacuum . . . My husband doesn't even know how dishes go in the dishwasher . . . All I would really like him to do is pick up behind himself" (Deutsch 1999, 67).

Another woman employed full-time singled out laundry as a particularly troublesome task: "I get so sick of doing laundry. I do laundry constantly . . . he won't lay a finger on laundry" (Deutsch 1999, 67). Vacuuming, washing dishes, and doing laundry, which these women's husbands refused to do, are household chores that women are typically expected to perform. Auto repairs and lawn maintenance, on the other hand, are chores that men are expected to do. How are these gendered expectations related? One husband explained to Deutsch that he "rides" on taking care of the cars when his wife is insistent that he clean up the house: "If the cars have needed a lot of work lately, I can ride on that for a bit because then I can have something quantifiable I point to that I have been doing" (Deutsch 1999, 71).

Gender-segregated patterns of household labor allocation tend to generate feelings of entitlement among men. When Deutsch asked Carol's husband how he responds to his more-than-full-time employed wife's pleas for help around the house,

There is some evidence that the gender gap in housework is beginning to narrow, especially for parents of young children.

he said "I just chuckle" (Deutsch 1999, 67). The husband in Deutsch's study who refused to do laundry said he felt entitled to relax after work and on the weekends:

> *She probably won't sit still on a Sunday . . . Sundays I usually relax . . . She's not happy unless she's doing something. That's the difference between her and I. She's not happy unless she's making a cake, making supper, doing laundry. She very rarely can sit down and watch television, take a break . . . She's not happy unless she's doing something. I'm different. I can relax (Deutsch 1999, 68).*

Women as Household Managers

Besides doing more of the actual chores, women also typically act as household managers; that is, they assume responsibility for planning and initiating the majority of household chores. Even in families where much of the child care and housework is shared, wives are much more likely than husbands to notice when a chore needs doing and to make sure that someone does it correctly. In most families, husbands notice less about what needs to be done, wait to be asked to do various chores, and require explicit directions if they are to complete the tasks successfully. In line with this division of the responsibility for management of household affairs, most couples characterize husbands' contributions to housework or child care as "helping" their wives (Coltrane 1996).

In general, then, women are more likely to carry the burden of managing the household, as well as doing most of the tasks, and tend to worry more about the planning and scheduling of various activities. Women usually plan meals so that the family can eat together, and they typically try to orchestrate the evening meal so that it is a calm and pleasant event (DeVault 1987; Thompson and Walker 1989). Women generally try to keep conversations going and ensure that everyone gets a chance to talk (Fishman 1978). For child-related tasks, women also usually take responsibility for most of the less pleasant activities and attempt to facilitate positive father-child interactions. Women in two-parent families almost invariably take responsibility for organizing, delegating, and scheduling children's activities (Ehrensaft 1987; Berk 1985).

Men's Participation in Housework

Relatively few cultural images exist suggesting that men could possibly be "as good" at housework as women are. With few exceptions, comic strips, television shows, and movies show us how inept men are when they attempt to perform "women's work." From Dagwood Bumstead in *Blondie* to Tim Allen on *Home Improvement,* men are shown as comic buffoons when it comes to doing the housework. Even films that celebrate fathers' efforts to care for their children, like Michael Keaton in *Mr. Mom,* Dustin Hoffman in *Kramer vs. Kramer,* and Robin Williams in *Mrs. Doubtfire,* illustrate how ill-prepared men are for performing mundane domestic work. Thus, there are few cultural role models for boys or men to follow in assuming greater responsibility for household tasks. Scholars argue about whether media images mimic everyday life and promote or inhibit social change, but in the area of housework, it is clear that popular culture is not pushing most men to do more (Coltrane and Allan 1994; Coltrane and Adams 1997).

In real life, men's self-proclaimed ineptitude at doing housework often becomes an excuse for them to avoid it. One wife's exasperation at her husband's (feigned) incompetence is expressed by Deutsch:

> *He plays, you know, "How do you do this kind of thing?" and asks me fifteen questions so it would almost be easier for me to do it myself than to sit there and answer all his questions. That makes me angry because I feel like he's just playing stupid because he doesn't want to do it (Deutsch 1999, 77).*

Deutsch notes that there are men who use praise of their wife as the "flip side" of their self-professed incompetence, as does the following father:

> *She's wonderful (as a mother) . . . Some women, like I say, are geared to be business-women; Florence is geared to be a mother. She loves it. She's good at it. I feel real lucky to have her as a partner because it takes a lot of the burden off me (Deutsch 1999, 77).*

By extolling Florence's virtues as a mother, her husband symbolically removes her from the public realm of business while at the same time praising her for allowing him unfettered access to it.

The Burden of housework falls disproportionately on women.

Why does men's participation in housework remain so far behind their wives' despite women's paid employment? A number of reasons have been suggested for men's "houswork lag." Some researchers believe that women may act as "gatekeepers" to men's involvement in family work, using their traditional role as household managers to restrict men's opportunities to learn how to do housework. By setting rigid standards for housecleaning, which men are unable (or unwilling) to maintain, it is argued, women may be able to sustain their (conceptual) dominance in the private sphere (see Allen and Hawkins 1999; Greenstein 1996). For example, Deutsch notes the benefits of "doing it all" for a school psychologist who works forty-five hours a week: "Embarrassed, Peg [said]: 'Another thing I can't ignore is I'm in control. That sounds terrible. That's not how I mean it, but I mean I'm able to structure things . . . I feel like I want to be in control'" (Deutsch 1999, 55). Although Peg's admission that she wants to be in control may represent an example of maternal gatekeeping, it is also likely a response to her husband's refusal to participate in family work in the first place (see Allen and Hawkins 1999). Ethan, Peg's husband, observes:

> *It's hard for me to do anything during the week. If I come home at six-thirty, seven, I'm tired, basically fatigued . . . so when six A.M. rolls around and the kids are getting up, the last thing I think I really want to volunteer for is extra duties. I shouldn't say extra duties, but [I'm] certainly not going out of my way (Deutsch 1999, 53).*

Most theories examining why men's housework continues to lag behind women's involve the issue of power in families and focus on the relative resources of husband and wife, time availability, economic dependency, and gender ideology (Coltrane 2000). Theories using **relative resource models** suggest that the person with more income (generally the husband) will do less housework, and **time availability theories** imply that when people spend more time in paid work they will spend less time doing housework. The **economic dependency model** of housework suggests that women make a contract with their husbands to exchange household labor in return for economic support from a main breadwinner (Brines 1994). Finally, theories drawing on gender ideology suggest that people brought up to believe in the gender segregation of work will conform to those beliefs when they later marry.

The difference in power between men and women in families has important implications for each of these theories and suggests that the partner with the most power in the relationship will be the one who can tacitly set the terms of the division of labor in the family. Because most people (men and women alike) do not find housework to be particularly enjoyable and most men have relatively higher earnings and social power, it is usually men who "opt out" of doing routine domestic chores. Moreover, power in marriage is not just about conflict or other overt behavior but also includes "invisible" or "hidden" power that relies on what Gramsci (1971) called "ideological hegemony." This notion suggests that both men and women tend to accept the idea that what is in the husband's best interest is in the wife's best interest as well. The invisible power that promotes an acceptance of men's interests as primary also encourages both husbands and wives to "buy into" men's reduced participation in housework (Komter 1989) and their excuses for doing little (Pyke and Coltrane 1996). Therefore, wives are often inadvertently complicit in permitting their husbands to avoid housework, protecting them from tasks that they find onerous. In this sense, it is not unusual for a woman to intentionally avoid asking her husband to do chores that she anticipates he will hate (Braverman 1991).

Attempting to Share the Strain

Studies focused on isolating the conditions under which men and women share housework show that it is usually a practical response to paid work and childcare demands rather than an idealistic attempt to change gender roles. Although most men and women in couple relationships now share the earning of income, substantial sharing of housework tends to occur only after wives actively bargain for it. Some studies find that if couples deliberately divide household tasks early in the relationship, a pattern of sharing develops and becomes self-perpetuating. If, however, couples make the assumption that sharing will happen on its own, the tendency is for women to end up doing virtually all of the routine domestic work (Coltrane 1996).

Arlie Hochschild (1989) interviewed and observed fifty-two couples over an eight-year period to see how they managed the complex relationships between paid and unpaid work. She reports that women are "far more deeply torn" between the demands of work and family than their husbands are. She describes how most women work one shift for pay and a **second shift** when they get off work and come home to take care of house and family.

Because they remain responsible for managing the household and performing the more onerous household tasks, Hochschild found that women tended to talk more intensely about being overtired, sick, and emotionally drained. She labeled the common situation of men's favored position in the household division of labor as the "leisure gap" because most men had more free time than their wives. In general, men did not consider the second shift to be "their issue."

Nevertheless, about 20 percent of the men that Hochschild interviewed shared housework equally with their wives. She reports that the men who did half of the domestic chores seemed to be just as pressed for time as their wives (Hochschild 1989). Berk (1985) reports that about 10 percent of the men in her sample also spent as much time as their wives on household chores. Coltrane (1996) selected couples on the basis of self-reported sharing of child care and discovered that highly involved fathers assumed significant responsibility for housework, though few did as much as their wives.

Predictors of Household Labor Sharing

A large number of studies were conducted in the 1990s to try to understand the factors associated with men doing more housework. We summarize research from such studies and refer interested students to the decade review from which these conclusions are drawn (Coltrane 2000).

WOMEN'S EMPLOYMENT Studies now find that women routinely spend less time on housework when they are employed longer hours. The relationship between women's employment hours and men's housework is more varied, principally because a fair number of men do so little. Some studies find that women's employment hours are related both to men's absolute hours and proportional contributions to housework, whereas others find that women's employment hours are significantly related only to men's share of the total. When women are involved in shift-work or flex-time employment, men contribute more to housework, especially if there is nonoverlap between spouses' employment hours, though such arrangements raise other issues for couples. Other aspects of women's employment may also influence household labor allocation. Some small sample studies suggest that women in professional jobs do more housework because they compensate for gender-atypical breadwinning patterns, but other studies using representative samples find that women with higher occupational prestige, or more workplace authority, tend to share more of the housework with their husbands.

MEN'S EMPLOYMENT As for women, less paid work generally means more family work for men, but low levels of housework and greater variation among men produces some mixed results. Using national samples, researchers typically find that men who are employed fewer hours do a greater share of the housework, as do men whose employment hours do not overlap with their wives'. Small sample studies from the 1990s continued to show that most men identify themselves as primary breadwinners and that both men and women are reluctant to accept wives as equal providers. In several studies, accepting wives as coproviders was identified as the key factor in reallocating family work.

EARNINGS In general, wives who make more money enjoy more equal divisions of labor. Although results were mixed in past studies, research in the 1990s suggests that when relative earnings between husbands and wives are more equal, the relative distribution of household tasks is more balanced. Some find that when women's absolute level of earnings go up, their absolute levels of time spent on housework go down. Smaller absolute income differences between husbands and wives are associated with more housework sharing, and wives' proportionate share of earnings is consistently associated with more equal divisions of housework.

A simple economic or power interpretation of these results does not hold across the full range of incomes. As noted earlier, when men are unemployed, they sometimes do less housework. For example, Brines (1994) finds that dependent husbands do less housework the more they are dependent on their wives for income, noting that this dynamic is particularly evident among (though not limited to) married men in low-income households. At the opposite end of the income pyramid, different patterns emerge. Wealthier men do little housework, but the amount done by their wives varies significantly. Women's higher occupational status and income (but not men's) are strongly associated with the purchase of domestic services. Results from sample surveys using quantitative data and results from historical and ethnographics studies using qualitative data thus converge on a general finding: women's economic resources allow them to reduce their own housework contributions and "buy out" of gendered domestic obligations. Upwardly mobile and well-educated women are the most likely to purchase domestic services, whether performed in their own homes or embedded in the food and products they purchase for the family from outside the home (Oropesa 1993). It is predominantly white middle-class women who consume these services and products, and immigrant, ethnic minority, and working-class women who produce and provide them (see *A Closer Look* 11.3).

EDUCATION Education is often used as a control variable in multivariate models predicting household divisions of labor. Interpretation of findings is complicated by conceptual confusion about whether years of education should be considered a measure of human capital accumulation, a relative resource, a component of social class, an indicator of ideology or attitudes, or a life course transitional experience. In general, studies suggest that women with more education do less housework, purchase more domestic services, and have children who do less housework. In contrast, men with more education generally do more housework.

GENDER ATTITUDES Studies from the 1990s show that women's egalitarian gender ideology is a consistent predictor of household labor sharing. When wives feel more strongly that both paid work and family work should be shared, and when they agree more fully with statements about equality between women and men, they are more likely to share housework with husbands. Some studies also show that more egalitarian men share more housework or child care. The fit between spouses' attitudes is also important: spouses with similar views are likely to put those ideals in practice (i.e., more congruent egalitarians share more housework, more congruent traditionals share less).

AGE AND LIFE COURSE ISSUES Because the meaning of housework varies between generations, some studies focus on cohort effects in its distribution. In general, younger women do less housework and share more than older women. Others find that when ideology and other variables are entered into multivariate models, cohort effects become nonsignificant (Presser, 1994). Another finding related to age is that the larger the age gap between spouses, the less the couple shares housework. Some studies find that men increase their contributions to household labor after retirement, though they remain in a helper role. Some suggest that retirement does not change the gender division of labor significantly, though many women expect that it should.

MARITAL STATUS Being married means more housework for women and less for men. Single and cohabiting women perform less housework than married women, but single and cohabiting men perform more housework than married men. Because single mothers perform about as much housework as married mothers, married fathers may do about as much household work as their presence creates. When single-mother and single-father households are compared, women do more housework than men, suggesting that even without a spouse, housework is still gendered. Nevertheless, single fathers do more housework than married fathers, so the difference between men's and women's housework in single-parent families is less than it is in two-parent families. The first marriage may be the most likely to produce gendered divisions of labor because remarried households share more than first-married ones.

PRESENCE OF CHILDREN As noted in the last chapter and explored more fully in the next, studies show that the transition to parenthood is associated with movement toward less sharing of family work between men and women. Women tend to feel more obligation to perform household labor when they have children, just as they do when they get married. When couples have children, men tend to work more hours at paid jobs but do not necessarily put in more hours of housework. Women, in contrast, tend to work fewer hours on the job and begin to put in significantly more hours of domestic work. Other studies show that more preschool children are associated with more hours of household labor for both men and women. Nevertheless, because women increase their hours more than men, they end up doing a larger proportionate share of family work as the number of children increases.

RACE/ETHNICITY Household labor studies in the 1990s began to take race seriously. Most studies found that black men do more housework than white men, net of other predictors, but that black women still do almost twice as much housework as black men. Some find that common predictor variables work somewhat differently for blacks, in part because of more egalitarian attitudes and greater employment/earnings equality between spouses. For example, employed black women do fewer hours of housework than other women, but black men do more hours of housework if they are employed. Some find unique patterns of labor allocation in black families when extended kin are included, with black adult children living at home contributing more than whites. Findings are contradictory concerning the sharing of family work in Latino families, with some suggesting there is slightly more sharing than among white families, and some suggesting there is less. Most studies show similar patterns of association between variables, whether the couples are Latino or Anglo. Research on other ethnic minorities in North America is still rare, though Johnson (1998) finds some cultural norms promoting sharing among Vietnamese and Laotians in the United States, and Brayfield (1992) finds that French Canadians share more than English Canadians.

Fairness and Entitlement in Housework

A recent public opinion poll showed that fully 88 percent of women and 78 percent of men believe that women do more of the housework chores in their family (To the Contrary Poll 1997). Interestingly, however, 60 percent of the women and 71 percent of the men answered "yes" to the question, "Is the way that you and your spouse share household chores fair to both of you?" This poll confirms what previous studies have suggested: Despite the fact that the burden of housework falls disproportionately on women, a majority of both men and women consider it to be fair.

Fairness in household labor does not automatically mean sharing tasks equally or putting in the same amount of time. To evaluate just how much housework is considered fair, sociologists Mary Clare Lennon and Sarah Rosenfield (1994) examined the amount of household labor that men and women are willing to do before they see the task division as unfair to themselves. According to Lennon and Rosenfield, women are willing to do roughly two-thirds of the household labor (about 66 percent) before they start to see it as unfair to themselves. Men, on the other hand, will do approximately 36 percent of the household labor before they begin to see it as unfair to themselves. Thus, there appears to be general acceptance of highly unbalanced divisions of household labor, with both sexes expecting women to put in many more hours than men on domestic tasks. The relevant question then becomes: Why are women willing to accept such an unequal division of housework?

Social theories suggest that wives label unbalanced divisions of household labor as fair because they have less power (including invisible power) in the marriage. One reason wives continue to perceive lopsided housework distribution as fair is because of differences in their sense of entitlement relative to that of their spouse (see Ferree 1990; Hochschild 1989; Major 1993; Thompson 1991). As noted earlier, men often see themselves as entitled to household services, a fact that sometimes makes women hesitant to ask for the help they need.

Hochschild (1989) described a **marital economy of gratitude** within which husbands' and wives' images of themselves as masculine or feminine encouraged them to see some actions in their marriage as gifts and others as burdens; for example, a man may see his wife's employment as either a gift or a burden, depending on how he views himself as a man. This "emotional economy of marriage" includes feelings about entitlements, as well (Pyke and Coltrane 1996). Although marital economies of gratitude are constantly negotiated in all couples, the more powerful spouse usually sets the terms for such negotiations. Thus, when one spouse is grateful for the actions of the other, he or she feels indebted and obligated to reciprocate. When, however, that spouse is displeased with the others' actions, he or she expects a spouse to compensate for the displeasing acts by doing more.

We know that it is generally the husband who is the more powerful spouse, resulting in large part from his greater financial and social leverage in the relationship. Therefore, the husband's pleasure or displeasure usually drives feelings of entitlement. These feelings of entitlement then reflect "invisible power" that leads both women and men to see fairness where fairness does not objectively exist. Men's culturally and economically driven feelings of entitlement to women's domestic services, therefore, may be important sources of women's acceptance of unbalanced distributions of housework (Coltrane and Adams 2000).

Other factors may contribute to a perception of fairness in the distribution of housework as well, such as the outcomes that are desired (more time, certain standards of cleanliness, care, or "keeping the peace," for example) (Major 1993; Thompson 1991). Also, whether individuals compare their household contributions with their partners (cross-gender comparison), or to other women's or men's (within-gender comparison) may significantly impact their fairness evaluations. Men's contributions to housework tend to be noticed and applauded, whereas women's are generally taken for granted (Coltrane 1996; Robinson and Spitze 1992; Thompson 1991). Finally, whether the procedures that created the existing unbalanced distribution of housework are considered appropriate also contributes to perceptions of fairness, including, for example, forgiving a partner's lack of housework because of supposed ineptitude or because a prior joint decision was made about who should do the work.

The more hours women work in the paid labor force, the less fair they see the division of labor in the home (see Greenstein 1996; Sanchez 1994; Sanchez and Kane 1996). A number of studies also show that both men and women with more egalitarian gender attitudes tend to see the existing division of household labor as more unfair to the wife (Blair and Johnson 1992; DeMaris and Longmore 1996; Sanchez and Kane 1996). Interestingly, when measuring gender ideology by asking if employed spouses should share housework, egalitarian men rate the existing division of household labor as more fair to their wives, but egalitarian wives rate the existing division of labor as less fair to themselves (DeMaris and Longmore 1996). Apparently, egalitarian husbands are hesitant to admit to an unbalanced distribution of housework, which contradicts their belief in equality. For egalitarian wives, on the other hand, an unequal division of tasks is likely to be particularly salient because of their belief in sharing, and their attitudes, therefore, are more likely to generate criticism of the present situation.

Women (and sometimes men) often see both their own and their spouse's housework as carrying emotional messages, such as love, caring, or appreciation. Symbolically equating housework with care can lead to demands for more task performance on the one hand, but it can also encourage women to consider men's expressions of affection or intent to do housework as sufficient (without actually *doing* the work), thereby encouraging these women to assess current unbalanced labor arrangements as fair. One woman who gave up a promising career to stay home with her children, asked only that her husband show some appreciation for her sacrifice and recognition of her domestic contributions: "You are at work with all these bigwigs and I'm home with children playing blocks. I've had a hard day too" (Deutsch 1999, 69).

Perceptions of fairness may also intervene between the division of household labor and personal or marital well-being. When housework is believed to be fair, wives display fewer symptoms of depression, but when it is perceived as unfair, women are more depressed (Glass and Fujimoto 1994; Lennon and Rosenfield 1994). Being satisfied with one's husband's contribution to housework is also related to better marital interaction and more marital closeness, less marital conflict, and fewer thoughts of divorce (Piña and Bengtson 1993). Moreover, although perceived unfairness predicts both unhappiness and distress for women, it predicts neither for men (Robinson and Spitze 1992).

Men are almost universally satisfied with the division of housework, whereas women, particularly egalitarian women who are content with their paid work, are typically less satisfied (Baxter and Western 1998). In the end, the single most important predictor of how fair a wife sees the distribution of housework to be is what portion of the routine housework (cooking, cleaning, and laundry) her husband contributes. Although husbands are making some limited progress in these areas, these also seem to be the areas in which they are most resistant to change.

What Difference Does Sharing Make?

Divisions of household labor are directly and indirectly linked to depression, which might help explain why many men seem to benefit from marriage more than women. Though detailed outcome studies are still rare, research indicates that performing larger amounts of routine, repetitive housework is associated with more depression in women and sometimes in men. It appears that it is primarily men's participation in the routine repetitive chores of cooking, cleaning, and washing that relieves women's burden, contributes to their sense of fairness, and hence lowers their chances of being depressed (Coltrane 2000). For their part, men often report some difficulty assuming more responsibility for family work, though initial frustration is typically short lived (Coltrane 1996; Cowan and Cowan 2000; Hawkins et al. 1994).

Several studies also find that marital satisfaction increases in relation to the amount of routine housework that is shared by spouses. Most studies find that the fit between husbands' and wives' gender attitudes is extremely important to marital satisfaction, as is the congruence between spouses' attitudes and actions. In general, if spouses align their attitudes and divisions of household labor, then their marital happiness is higher. Because men continue to do substantially less housework than women, however, a gender-bifurcated pattern emerges: women who believe in sharing housework tend to have lower marital satisfaction than others, and men who believe in sharing tend to have higher marital satisfaction than others. Similarly, when men are more egalitarian than wives, marital disagreements are fewer, but when wives are more egalitarian than husbands (the more typical case), then marital disagreements are more common (Lye and Biblarz 1993). Because housework is typically perceived as optional for men and required of women, it is generally up to women to bring about change. Only when women perceive the division of labor to be unfair does the level of marital conflict go up (Blair 1993; Perry-Jenkins and Folk 1994; Wilkie, Ferree, and Radcliff 1998). Marital conflict, in turn, is related to lower marital satisfaction and higher rates of depression. Women are thus faced with a double bind: They can push for change, threatening the relationship, or they can accept an unbalanced division of labor, labeling it "fair" (Hochschild 1989).

We can understand some of these findings on the division of household labor and couples' willingness to accept what look like extremely unequal arrangements in light of the concept of "doing gender" mentioned in the last chapter. West and Zimmerman (1987) point out that to be classified as a man or woman and to be judged a competent member of society, everyone must "do gender." Doing gender consists of interacting with others in such a way that people will perceive one's actions as expressions of an underlying masculine or feminine "nature." Thus, one is not automatically classified as a man or a woman on the basis of biological sex but on the basis of appearance and behavior in everyday social interaction.

Household labor offers people a prime opportunity to "do gender" because of our cultural prescriptions about the appropriateness of men and women performing certain chores. Doing household chores allows people to reaffirm their gendered relation to the work and to the world (Berk 1985). Thus, women can create and sustain their identities as women through cooking and cleaning house and men can sustain their identities as men by *not* cooking and *not* cleaning house. With peoples' sense of self so tied up with doing separate gender-linked activities, it is no wonder that people perceive their household arrangements as fair.

FAMILY POWER

There is a good deal of evidence that men usually have more power in marriages than women do. Let us be careful of what this means. The term *power* might seem to imply that there is a struggle over who gets to dominate whom and that men generally give the orders while their wives meekly obey. That, in fact, is a description of what traditional marriages once looked like, but it does not describe most people's experience today. On the surface, most marriages now seem to have very little to do with power. No one gives orders (except, sometimes, parents to their children). If interviewers ask, "Who has the power in this marriage?" typically, husbands and wives say that it is equally shared.

Nevertheless, this kind of answer is usually superficial. We can tell because more detailed studies, done when the researcher gets to know the respondents better, find there are real differences in who gets what inside a marriage. For example, power has often been studied in terms of who has how much influence over family decisions.

In one famous study that has been repeated many times, Blood and Wolfe (1960) asked Detroit area housewives about decision making for buying a car, purchasing life insurance, taking a new job, or going on a vacation (*A Closer Look* 11.4). Based on answers to these questions, they concluded that "the American family has changed its authority pattern from one of patriarchal male dominance to one of equalitarian sharing" (Blood and Wolfe 1960, 47). This optimistic conclusion has been challenged by many family scholars. Not only do people give superficial answers to such questions, but the scoring procedure treats all decisions as equal and assumes that people's answers always express power relations.

For major decisions, such as whether to move to a new house or a new city or to buy a car or appliance, men typically dominate, making the decision themselves or at least exercising veto power over a wife's decision (Rubin 1976, 110; Ostrander 1984, 51). The wife of a top executive, for example, recalled: "He wanted to move to the country, and I didn't. So we moved to the country." Another woman in this social class said: "When he's worked hard all day and calls up and says let's get a tennis game together, I do it. He'd get upset if I didn't. . . . If he calls me in the middle of the morning and wants to have a party the next day, I've got to do it" (Ostrander 1984, 37, 45). These women, however, have particularly powerful husbands, and their social lives are an adjunct to business. These couples do not fight over power, but they are maintaining a certain lifestyle in which the husband takes the initiative in what they do.

Life is less formal in the middle and working classes, and correspondingly, power differences may be less blatant. Moreover, there is the strong ideal of love between married partners, which in today's world, at least, has replaced the old-fashioned deference to the family patriarch. Especially in the minor decisions of the daily running of the household, women typically have great autonomy. However, there are subtle aspects of power that affect even the events of everyday life. For instance, as we've already seen, women do a good deal more housework than men. Even counting the different tasks that men do around the home (especially outside), women put in many more hours than men, and they do it even when they have jobs. Even when children are given household chores, girls work twice as many hours as boys—as if they were being groomed to fit the same gender pattern of housework in the next generation. Usually, not very much thought is given to these arrangements. They are just taken for granted, but they are a subtle form of power inside the household. Husbands and wives may love and

A CLOSER LOOK 11.4
Who Has the Power?

Think about a relationship you have been in, or about your parents, or some other couple you know. Who usually makes the final decision about the following items; the husband or the wife?

1. What car to buy?
2. Where to go on vacation?
3. What house or apartment to take?
4. Whether or not to buy some life insurance?
5. What job the husband should take?
6. Whether or not the wife should go to work or quit work?
7. What doctor to have when someone is sick?
8. How much money the family can afford to spend per week on food?

In the 1950s, researchers asked housewives these questions about family decision making. How well do you think these questions measure "family power"?

Source: Adapted from Robert O. Blood and Donald M. Wolfe, *Husbands and Wives* (New York: Free Press, 1960).

respect each other, but at the same time, there is another set of forces operating beneath the surface of their awareness—and these determine how much power they have at home, whether they think about it or not.

Occupation

It is generally true that the higher the husband's occupational level, the more likely he is to dominate at home. There is an especially sharp increase in power at the white-collar level, with many middle-class husbands having much more power than husbands who are skilled manual workers. The pattern is somewhat surprising, because middle-class husbands are more likely to *say* that they believe in equality between husbands and wives. Nonetheless, resources are what count, not sentiments, and these men's occupations give them the resources to get their own way. Although new career opportunities for women that have developed out of the feminist movement have challenged this state of affairs, women's occupational levels, and therefore power, are still generally lower than those of men.

What seems to happen is that middle-class men are less likely to demand that their wives defer to their wishes *as men,* but instead they ask them to defer *because of the importance of their jobs.* The ideology is different, but the outcome is even more biased against women than with an outright sexist claim to male domination. Working-class men, on the other hand, are more likely to speak overtly about male prerogatives, but their occupation brings them less prestige and other resources, so they are less able to translate their demands into realities. This is especially so when their resources do not outweigh their wives'.

In the lower part of the working class, the trend reverses again. Unskilled workers have relatively more power over their wives than the skilled workers of the upper working class. Because these men have even fewer resources, their power must have some other basis; there are some indications that it involves a greater willingness to use force (Rubin 1976; Szinovacz 1987).

Income

Income is even more important for marital power than occupation. The more income a man brings in, the more he gets his way at home. Again, this must be balanced against the wife's income. Ironically, this tends to make the gap between the husband's and wife's powers even wider in the upper-middle-class families than in working-class families. Because working women of all social classes tend to be in clerical and other low-paying jobs, upper-middle-class men tend to earn a much higher multiple of their wives' income than working-class men. This tends to make upper-middle-class men especially powerful at home.

In recent years, the proportion of married women who work outside the home has increased to about 60 percent and to over 70 percent for those aged twenty-five to forty-four, whereas the proportion of men has actually fallen slightly. The traditional barriers to women in high-paying jobs, such as business executives and the more lucrative professions, have come down, at least to the extent of allowing some women into these occupations. These trends, one might expect, would bring a shift in the domestic balance of power. Women's greater incomes should be translating into at least greater equality of power at home. We have no good data available on whether power in the household has really become more egalitarian since the 1950s. However, the figures on the incomes of men and women reveal an unpleasant surprise: there has been little change in the average amount that women get paid in relation to men. To be exact, in 1960, women were making 60 percent of what men were being paid in terms of hourly incomes; in 1997, women's hourly pay had risen to 72 percent (see Figure 1.3). However, these figures are for full-time year-round work and overstate women's gains relative to men, because women are much more likely than men to be unemployed, to work part-time, and to be employed seasonally.

A number of reasons have been advanced for the failure of women to achieve economic parity with men. One prominent argument has been that women are less committed to their careers because of their family responsibilities, moving to follow their husbands' jobs, or quitting to take care of the children. Nevertheless, these explanations do not hold up well against evidence that women with children do hold down jobs as consistently as men and that family responsibilities do not affect the pattern of job movement (Reskin 1984). The major explanation of women's lower wages remains that the job market is heavily segregated by gender (England and Browne 1992; Reskin and Padavic 1994). Even though some women have crossed the barrier into high-paying "men's jobs," they are not enough to offset the general pattern, and the lack of promotion to higher level positions—or "glass ceiling"—limits women's upward mobility.

Changing income and earnings potential set the scene for the shifts in the positions of men and women that have taken place in the last few decades—at exactly the time of the rise of the women's movement, the declining birthrate, and the rising divorce rate. Women are still disadvantaged in the marketplace, but they have left the old pattern in which only a minority of women worked for a new pattern in which work is the most typical female lifestyle. Women still have a long way to go to overcome economic discrimination; however, now a majority of women are confronting it directly in their own careers, whereas previously it was more typical for them not to make the attempt. In this sense, women have become economically more mobilized. Though they are discriminated against, they have more money simply because of doing more work; this has probably given them at least some leverage in subtle family maneuvers over power. Judging by the figures on how much housework wives and husbands do, women's economic mobilization hasn't had much effect yet on daily life, but it is probably implicated in the fact that the divorce rate has risen so sharply during this time. Being unable to change the power within marriages, women instead have tended to break up their marriages.

Recent data show that income is still a main source of domestic power. This is true both for married couples and for cohabiting couples who are not married. In both cases, whichever partner had more income had relatively more domestic power. Because men generally earned more than their wives or female partners, they generally had more domestic power. Blumstein and Schwartz (1983) compared homosexual couples on this issue. This is a useful comparison in that it enables us to see if income is really having the effect on power, independently of gender. The result is rather striking: income is the source of power, regardless of the couple's gender. That is, in couples composed of gay males, the person with the greater income played the role of the "husband," whereas the one with less income acted as the "wife," deferring to the other's job demands, yielding in major decisions, and generally taking care of the house. This suggests that the roles of "husband" and "wife" actually have little to do with sex. It is only because men have generally had more income (along with occupational power and other resources) that they have dominated their marriages. Traditional expectations built up along the assumption that the major income earner will be a man have made this into a cultural role.

Blumstein and Schwartz found one interesting exception to this pattern of income determining domestic power. Among lesbian couples, there was no relationship between income and power. Neither partner had more domestic power; both shared equally in doing the housework and in making decisions. Why this anomaly? It appears that males, even if they are homosexuals, carry on the dominant practice in which income translates into power. Females, when they are married to or living with a male, also follow this pattern to some extent. Lesbian couples are special. They are a relatively unusual type of household and may be operating with different ideas of what is fair and how it should be determined. For instance, England (1989, 26) suggests that we must consider the extent to which partners embrace a self-interested and competitive stance versus an empathetic and cooperative one. Various feminist theories describe how and why men and women develop different personalities, moralities, and ways of thinking. In summary, these theories suggest that men come to experience the world as distinctly separate individuals. England (1989) comments that economic, exchange, and rational-choice theories minimize emotional connection and altruism and inadvertently assume that all people feel and act like the stereotypical man in our society. To understand how marital power is exercised, experienced, and evaluated, we need to complicate our theoretical models to include consideration of when and why both genders would be empathetic, as well as selfish, and how preferences could change over time. Thus, although the spouse who has more income may exercise marital power by making more decisions or doing less housework, power may not be the primary concern of the people involved.

Education

Education also has an effect on domestic power. Generally, the spouse with more education tends to dominate. To some extent, this has helped women because women have tended to complete slightly more schooling on the average than men. Skilled workers, in particular, have been likely to marry high school graduates, which is one reason why they have tended to have relatively less power compared with their wives. Among the higher social classes, though, men tend to have a sizable advantage in education over their wives—another hidden reason why their egalitarian sentiments do not translate into reality. Where the wife in a professional-level family does have more education than her husband,

Feminist theories suggest that men tend to experience the world as distinctly separate individuals—not in connection with others.

that sometimes translates into greater outside contacts and prestige and hence into greater domestic power. Given the very recent tendency for women to attend college at a higher rate than men, we may eventually see another shift in domestic power. For the most part, it has not happened, despite the spread of feminist beliefs in the higher social classes, primarily because men still control the upper-level professions.

Again, among the lower part of the working class, the resource of relative educational levels does not count for much. Neither husband nor wife has a lot of education, but the lower-working-class husband is likely to be more in touch with the world outside his family. He thus takes on the patriarchal role of "representing the family to the world," while his wife is especially likely to be confined to a domestic routine.

Social Participation Versus Isolation

The more one participates in outside organizations, the more power one has at home. Some women thus acquire power because they are active in civic organizations, clubs, churches, or other areas. This gives them outside prestige that translates into some degree of deference at home. However, men are much more likely to participate in outside organizations than their wives; thus, this resource usually just reinforces the male advantage. Lower-working-class men in particular benefit from this resource; they usually don't belong to many *formal* organizations, but they are much more likely to have networks of acquaintances and to appear in the public arena of bars, street corners, and other "hangouts," unlike their wives, who are relatively isolated.

It has been noted that when families move from the city to the suburbs, the domination by the male tends to go *up.* The popular image of the suburbs is a place dominated by domestic and neighborly activities run by women. In fact, the suburbs tend to isolate women, especially working-class women who have neither the cultural resources nor the self-confidence to make widespread social contacts. Hence, their husbands, who are less isolated because they commute away from the suburbs to work, end up with even greater relative power.

Children

It might be thought that once a woman bears children her status would go up because she is fulfilling the important role of mother. However, *exactly the opposite tends to happen.* A woman's domestic power *declines* when the first child arrives, and it tends to decline further the more children she has (Szinovacz 1987). It reaches a low point during the time she has small children at home, before they go to school. Again, real resources count for far more than ideologies. Women with small children are maximally confined to the home, with the greatest number of pressures on them. They simply do not have the time to acquire any of the other resources—income, outside sources of prestige—that would give them power.

Outside Sources of Power

When women have more of these resources than their husbands, their power over domestic decisions tends to go up. Women who work at paying jobs have more power than housewives, and their power tends to go up the longer they hold their jobs. This is one reason why upper-working-class women have relatively better power positions via-à-vis their husbands than wives of middle-class men; they are more likely to be employed. In dual-career families, the wife's influence seems to be proportional to the amount she contributes to the family's total income. As noted, having relatively more outside contacts in formal organizations and being better educated than her husband tend to give her a relatively better power position. Except in the lower working class, it seems; there, if a wife earns more money than her husband, she may provide jealousy and anger, with resulting quarrels and violence.

HAPPINESS AND UNHAPPINESS

Survey researchers studying marital and relationship happiness use questionnaires like the Dyadic Adjustment Scale (Spanier 1976) (*A Closer Look* 11.5).

A CLOSER LOOK 11.5
Measuring Marital Adjustment

How "adjusted" are you and your partner? Even if you are not married, answer these questions about a relationship you are in or answer about your parents. As you will see, most of the questions focus on whether two people agree about various things.

I. How often do you and your partner agree or disagree about the following subjects? (Always agree, Almost always agree, Occasionally disagree, Frequently disagree, Almost always disagree, Always disagree)

1. Handling family finances
2. Matters of recreation
3. Religious matters
4. Demonstrations of affection
5. Friends
6. Sex relations
7. Conventionality (correct or proper behavior)
8. Philosophy of life
9. Ways of dealing with parents or in-laws
10. Aims, goals, and things believed important
11. Amount of time spent together
12. Making major decisions
13. Household tasks
14. Leisure time interests and activities
15. Career decisions

II. How often do you and your partner do the following? (All of the time, Most of the time, More often than not, Occasionally, Rarely, Never)

16. Discuss or consider divorce, separation, or terminating your relationship?
17. Leave the house after a fight?
18. Regret that you married? (*or lived together*)
19. Quarrel?
20. "Get on each other's nerves?"
21. Think that things between you and your partner are going well?
22. Confide in your mate?

III. How often would you say the following events occur between you and your partner? (Never, Less than once a month, Once or twice a month, Once or twice a week, Once a day, More often)

23. Have a stimulating exchange of ideas
24. Laugh together
25. Calmly discuss something
26. Work together on a project

IV. These are things about which couples sometimes agree and sometimes disagree. Indicate if either item below caused differences of opinions or were problems in your relationship in recent weeks.

27. Being too tired for sex
28. Not showing love

V. All things considered, how would you describe the degree of happiness of your relationship?

| Extremely Unhappy | Fairly Unhappy | A Little Unhappy | Happy | Very Happy | Extremely Happy | Perfect |

Source: Graham Spanier, February 1976. "Measuring Dyadic Adjustment: New Scales for Assessing the Quality of Marriage and Similar Dyads." *Journal of Marriage and the Family* 38:1, pp 15–28.

Using such scales to predict which couples will stay together and which will divorce has not been very successful. Starting in the 1930s, sociologists tried to isolate individual differences and social characteristics that might account for why some couples were happy and some were not (Burgess and Cottrell 1939). Despite an immense number of studies over many years, marriage and family scholars have expressed disappointment over their failure to identify the determinants of satisfaction and stability and to elucidate the relationship between marital quality and longevity (Adams 1988; Berscheid and Reis 1998; Nye 1988). Norval Glenn (1990; 818), for one, described the marital-quality literature as incohesive and atheoretical and as having produced "only a modest increment in understanding the causes and consequences of marital success," where marital success is defined as satisfaction in an intact marriage (see Berscheid and Reis 1998).

One of the reasons that scholars have had so much difficulty drawing firm conclusions about marital success is that the concept of marital quality can be measured many different ways. The most widely used scales purporting to measure **marital happiness** (or marital adjustment, marital satisfaction, etc.) tend to combine conceptually distinct components (Johnson et al. 1986). The often used Locke-Wallace scale (1959) combines satisfaction, disagreement, and interaction, and the popular Dyadic Adjustment Scale (see *A Closer Look* 11.5) combines interaction, disagreement, satisfaction, instability, and a variety of marital problems into a summated scale. As scholars develop more precise measures of these different components of marriage, predictions about marital success or failure should improve (Johnson et al. 1986).

A related issue has to do with how success is defined. Is success always staying in the relationship or always reporting that one is "very happy?" Some researchers criticize the concept of adjustment (Trost 1985) or suggest that people will provide answers that make them look "too good to be true." For instance, researchers find that many people agree with unrealistic statements such as, "If my mate has any faults, I am not aware of them," or "Everything I have learned about my mate has pleased me." People are influenced by what survey researchers label **social desirability** and tend to answer questions in a conventional way that makes them appear to be happier than they probably are. Even if we assume that some form of marital happiness or satisfaction is being measured with these questionnaires, there is a bias toward consensus and disapproval of conflict. Spouses who say they have different interests, admit to having disagreements, or acknowledge conflicts end up being rated as less "adjusted." In fact, however, many marriage counselors point out that recognition of differences and the ability to talk about them are signs of emotional maturity and can enhance the marriage. In general, there is no single type or style of marriage that makes the partners happy. Researchers have described a wide range of communication patterns and interaction styles that satisfy the partners and contribute to stability in the marriage (Berscheid and Reis 1998).

We have seen that married people are generally more likely to be happy than people who have never been married or those who are now separated, divorced, or widowed. Marital happiness is related to how much companionship, sociability, and other mutual enjoyments people derive from the relationship and is negatively correlated with the number of tensions. But the good parts and the bad parts of marriage seem to be independent of each other, and marriages can have a lot or a little of each. It is when there are especially few rewards to balance off the tensions that marriages are most unhappy.

What Makes a Successful Marriage?

Some of the answers that people gave when asked this question are listed in Table 11.3, ranked in order from most important to least important. College students' rankings of these factors were fairly similar in 1967 and 1994 (Barich and Bielby 1996). Respondents overwhelming identify love and affection as the most important element of marriage in both years. The ranking of the three least important expectations for a successful marriage also remain stable across the twenty-seven-year period: Moral and religious unity is consistently ninth, maintenance

TABLE 11.3 What Is Important for a Successful Marriage?

	All Respondents		Men		Women	
	1967	**1994**	**1967**	**1994**	**1967**	**1994**
Amount of love and affection	1	1	1	1	1	1
Healthy and happy children	2	4	3	3	2	4
Companionship	3	2	2	2	3	2
Emotional security	8	5	5	4	4	3
Satisfactory sexual relations	5	8	4	6	5	8
Common interests and activities	6	7	6	7	7	7
Personality development	7	6	8	8	6	6
Economic security	8	5	7	5	9	5
Moral and religious unity	9	9	9	9	8	9
Maintenance of a home	10	10	10	10	10	10
A respected place in the community	11	11	11	11	11	11

Source: Rachel Roseman Barich and Denise D. Bielby. 1996. "Rethinking Marriage: Change and Stability in Expectations, 1967–1994." *Journal of Family Issues* 17: 139–69.

of a home is tenth, and a respected place in the community is eleventh. Three other factors stayed fairly consistent across the two surveys: Common interests and activities ranked sixth in 1967 and seventh in 1994, and personality development ranked seventh in 1967 and sixth in 1994. Companionship was near the top in both years, ranking third in 1967 and second in 1994. Three factors showed the most movement: Emotional security climbed from eighth to fifth, and satisfactory sexual relations fell from fifth to eighth. Finally, having healthy and happy children fell from the second highest ranking in 1967 to fourth place in 1994. The increasing emphasis placed on companionship and emotional security suggests that personal fulfillment is now considered to be more important in marriage than it was about three decades ago. The lower ranking for children suggests that although many people still find that having kids is important in a marriage, child-bearing is no longer a foremost expectation of marriage. To explain the drop in sex as an important factor in the success of marriage, Barich and Bielby (1996) argue that because premarital sex and cohabitation are more prevalent than in the 1960s, the need for marriage to legitimate sexual activity is reduced.

When results are analyzed for women and men separately, Barich and Bielby (1996), report significant changes only for women (see Table 11.3). The most notable changes are an increase in women's ranking of the expectation that marriage should bring economic security, which rose from ninth place to fifth place, and a decrease in women's expectations for healthy and happy children, which dropped from second to fourth. These shifts suggest that instead of taking the economic security of marriage for granted as women in the 1960s did, women now must worry more about finances. Also, women used to consider marriage and children a "package deal," whereas they now have some autonomy in such decisions. Considering these and other findings, Barich and Bielby (1996, 160) conclude that relationship and marriage scripts for men and women were quite different in 1967 but that by 1994 "they were essentially the same."

Although asking college students what makes a successful marriage tells us something important about attitude changes, we should be cautious about interpreting the responses. For example, economic security ranks in the middle (fifth to eighth), and maintenance of a home ranks near the bottom (tenth); however money is the most common source of arguments (as we will see in the section on

Economic Success), and economic strains are often involved in spousal abuse and other relationship problems. We don't like to think our happiness depends on money, but lack of money, at least compared with what we want, is a source of trouble. It probably has its major effects because of the way it impacts the relative power of husbands and wives in the family.

People also seem to believe, in general, that having the same interests and activities is not very important in a successful marriage. However, if we look at the data on which marriages last and which are most likely to end in divorce we see again that "opposites do not attract," and in fact, the more similar the backgrounds of the couple, the more stable the marriage. That is what one would expect from the point of view that having a similar culture is what makes it possible for people to communicate well. People recognize that companionship is important but not the ingredients that make it happen.

The Pressure of Children

One major factor affecting happiness in marriage is children. The prevalent view of children is a rather sentimental one, and one might expect the presence of children to increase marital happiness. In fact, just the contrary is the case. Most studies have shown that marital happiness is highest at the beginning of the marriage, before children arrive. With the first birth, the happiness level begins to drop, and it continues to go down, through the children's preschool and school-age years, until it hits a low in the teenage years, just before the children are ready to leave home. Finally, as the children leave home, the curve abruptly goes up again (Glenn 1991; Spanier and Lewis 1980).

The conclusion is hard to escape: children are a strain on a marriage (McLanahan and Adams 1987). They tend to disrupt the emotional and sexual relationships of a husband and wife, give them more duties, and cause quarrels and bad moods. The strain seems to be worst in the teen years; perhaps because children are expensive then and perhaps also because the tensions between them and their parents are at their peaks. Children also contribute to marital strain because of the special pressure they put on women to carry out conventional responsibilities as mothers. This kind of parental "role strain" may be even more important than the life stage of the children.

Children can have both positive and negative effects on happiness. They are a source of companionship, love, and fun. They provide the center of attention around which family rituals are focused (birthdays, Christmas, Halloween, and the like). In this way they help hold families together, creating emotional bonds and the symbols that represent these ties. On the other hand, taking care of children is also a matter of practical work—work that largely falls to the woman. This is an unacknowledged source of strain in the relationship between husband and wife and probably has an indirect effect on the couple's happiness.

One of the best ways to measure whether the presence of children leads to declines in marital satisfaction is to follow families over time. These longitudinal studies sometimes call into question the conclusion noted earlier about a U-shaped curve of marital satisfaction that starts high at the honeymoon, drops precipitously with the arrival of kids, and then rises again after the kids leave home. Vaillant and Vaillant (1993) studied a group of married couples for forty years and found that when respondents were asked to report their marital happiness retrospectively, the usual U-shaped curve was found. However, actual measurement of satisfaction at several points during the four decades revealed a continuous decline in satisfaction, particularly for wives.

These results are similar to others reporting a gradual decline in the proportion of people who report that their marriages are "very happy" over the duration of the marriage. The steepest drop in satisfaction for both women and men occurs shortly after marriage and has been found to be related more to a decline in the frequency of positive behaviors than to an increase in the number of negative behaviors (Berscheid and Reis 1998). Although most research continues to show that drops in marital satisfaction are associated with the advent of children, some studies show that at least a part of the decline can be explained by how long the couple has been together. With more people cohabiting, it is difficult to make equivalent comparisons among groups. For example, the marital satisfaction of a couple who cohabited for two years before marrying might be lower at the end of their first year of marriage than that of a couple who did not cohabit but had been married for one year. The appropriate way to make such comparisons might be to focus on the length of the live-in relationship rather than the length of the official marriage.

Other Predictors

Many other predictors have been considered causes of marital happiness or success. For example, in one longitudinal study, nearly 200 variables were examined with the result that, in general, "positively valued variables—such as education, positive behavior, and employment predict positive marital outcomes, whereas negatively valued variables—such as neuroticism, negative behavior, and an unhappy childhood—predict negative marital outcomes" (Karney and Bradbury 1995, 18). Whether one considers behaviors and characteristics at the time of the marriage, or from the premarital period, no single factor has proved to be an especially potent predictor of satisfaction, and even groups of variables often account for a relatively small portion of the variance in marital quality (Berscheid and Reis 1998, 234). In addition, variables theoretically associated with relationship quality and stability, like love, trust, closeness, satisfaction, commitment, positive emotion, and so on, are all highly correlated with each other. Researchers can describe how such traits cluster, especially the negative ones. Compared with happy partners, distressed ones tend to escalate marital conflicts, locate causes of negative relationship events in the other person, and see the causes of those events as stable and global (Fincham and Bradbury 1993; Gottman 1998).

Several interesting new findings have emerged since scholars began conducting longitudinal studies. One is that children of parents with higher marital satisfaction tend to report higher relationship satisfaction when they form adult relationships (Larson and Holman 1994). Such findings support a life course perspective suggesting that children carry forward into their adult lives a set of attitudes, social skills, and interpersonal orientations learned in the family of origin and that these traits influence how intimate ties are formed and maintained. Some studies in this tradition find that low marital quality in parents' relationships is associated with earlier relationship formation (cohabitation and/or marriage) by their children but lower levels of eventual happiness or marital stability (Amato and Booth 1997).

Another finding in this general area concerns a topic we introduced at the beginning of this chapter: potential gender differences in the causes and consequences of marital quality, based on differences in the ways that men and women experience and describe relationships. As we saw in the analysis of college students' rankings of marital success factors, there are some differences in the expectations of men and women when they enter marriage. Most of these differences can be attributed to the assumptions we make about gender in marriage and to different patterns of socialization for boys and girls (chapter 12). Women are rated more favorably than men on helpfulness, kindness, compassion, expression of emotional support, and ability to devote oneself to another (Eagly 1987). Women are perceived as being more relationship oriented than men, are more willing to self-disclose than men, and often feel more responsible for the resolution of relationship difficulties (Fincham et al. 1997). These gender differences mean that his and her marriages, though related, can be experienced quite differently. For example, some recent longitudinal research suggests that men's and women's depression levels may be related to marital satisfaction in different ways. The typical pattern for women is to focus on relationship difficulties, to attempt to discuss them, and to blame themselves for any marital problems. Thus, researchers find that declines in marital satisfaction for women lead to higher levels of depression for them at a later time. Men, in contrast, have been socialized to express anger, but when confronted with relationship difficulties, they typically minimize the seriousness of their partners' concerns, take less responsibility for problems, and end up withdrawing emotionally from the conflict and sometimes the relationship (Gottman 1994). Researchers thus find a different causal path for men: increases in men's depression levels lead to later drops in marital satisfaction (Fincham et al. 1997).

Economic Success

People are satisfied with their marriages for different reasons, ranging from the intrinsically satisfying experience of warm, intimate relationships to cold-blooded reasoning that the arrangement is the most advantageous they can get. How many people are there of each?

Purely **utilitarian marriages** seem to be especially common in the higher social classes. John Cuber and Peggy Harroff (1965; see also Blumstein and Schwartz 1983, 345–60) interviewed a sample of very successful Americans: bank presidents, top corporation executives, high government officials, wealthy physicians, college presidents, and the like. They took only those with long, stable marriages who had never considered divorce. Their study produced the following findings. Eighty percent of the marriages were *utilitarian:* the couple stayed together because of economic, sexual, and status advantages, without any important emotional tie. They were either habituated to conflict, one partner (usually the wife) was completely passive, or the relationship was devitalized and empty. In the last type, the couple went through the motions of a marriage, occasionally conscious that the relationship once had more life and more love. Fifteen percent were *vital:* the couple meshed warmly and closely in at least a few areas of personal interest of emotional need. Five percent were *total:* the marriage satisfied virtually all of its partners' interests and needs, forming a little island against the world.

The majority of these marriages were empty, yet the group as a whole had never thought of divorce. In fact, the divorce rate in the upper and middle classes generally is lower than in other social classes (chapter 14). What seems to be happening is that success itself holds a marriage together. Working-class and middle-class couples have less reason not to separate, and in fact, they do separate more often, the less money there is coming in. The affluent upper classes, by definition, are enjoying a higher standard of living, and to a considerable extent, that is what holds them together. The wife of a successful executive, doctor, or

Money management is the main source of conflict in most marriages.

businessman has a strong economic incentive to stay married and even to try to keep her husband satisfied by staying attractive, cooking gourmet meals, and the like. Such wives share their husband's prestige and get to travel in his elevated social milieu, which they might have to give up if they lived on their own. The husbands have their careers to consider. Having a wife is not only essential for maintaining their professional images but for fulfilling their social obligations as well. What is more, their wives provide them with certain basic services (cooking, cleaning, child rearing), so they are free to pursue their careers aggressively. Therefore, the couples stay together at the cost of personal feelings.

There may also be quite a few empty marriages lower down in the class structure. The subject has never been well explored, although, as we know, more overt conflict (as well as separation and divorce) is seen as one proceeds downward into the middle and then the working and lower classes. The reason may be straightforward: money. Having lots of money motivates people to keep their marriages together; having little of it tends to create difficulties and arguments. Blood and Wolfe (1960, 241), for instance, found that money was far and away the most common answer (24 percent) to the question, "What do you most often argue about?" Following were children (16 percent), recreation (16 percent), personality conflicts (14 percent), in-laws (6 percent), and gender roles (4 percent). Fifteen percent of respondents reported no arguments. Fifteen years later, Scanzoni (1975) found that married couples tended to argue about the same things. Major conflict areas were money (33 percent), children (19 percent), recreation (8 percent), and gender roles (3 percent), with 17 percent reporting no conflicts (Scanzoni and Scanzoni 1988, 401). A more recent study by Blumstein and Schwartz (1983, 52, 67–93) confirms this finding; in their sample, the most peaceful relationships were those in which both partners felt satisfied with their income level and had equal control over spending it.

Studies in the 1990s continue to show the importance of income and wealth to marital happiness. Of course, some people at all locations in the occupational and educational hierarchy report that they are happy, but the average reports are higher as one moves up the income ladder. Having more education and a more prestigious occupation, earning more money, and having better health and vacation benefits all tend to make a marriage easier and thus "happier" (at least on the attitude surveys we use to measure such things). In general, researchers find that husbands with these characteristics report higher levels of marital quality (Larson and Holman 1994). In contrast, the stressful changes more common in the lives of those who are young, poorly educated, unemployed, or poorly paid tend to depress their levels of marital satisfaction (Kurdek 1993). Such effects tend to build up over generations; for example, Amato and Booth (1997) found that parents' socioeconomic resources increase the happiness and stability of children's marriages. Parents' material resources benefit children's marriages partly because they improve children's own educational and occupational attainment but also because parents provide direct economic support to their married children and grandchildren.

Single- and Dual-Career Marriages

We come now to a crucial point about the basis of family happiness. We have seen that money is frequently the cause of unhappiness in the family. It is often involved in family quarrels and in more extreme cases of spousal abuse. It probably has this effect because the power of husbands and wives is connected to their incomes. Quarreling about money may well be a surface manifestation of something more basic. A man angry with his wife over how she spends money is asserting, in a way, his claim to control her behavior because of his superior income. Conversely, a wife who criticizes her husband for his failure to make enough money is attacking his traditional source of dominance. Suppose he has the traditional mentality that symbolizes a man's standing by his economic success and by his dominance in his household—a combination that is particularly likely to be found in working-class families or other localistic, traditional communities. If this is the case, he may likely respond angrily to what he feels is an attack on the basic symbols of his position.

Even without overt quarrels and fights, money affects happiness. Supposing the husband makes all or most of the income, while the wife does most of the housework and defers on family decisions. The woman is, in effect, in a working-class position *inside the family,* although she would probably not define it that way. The idealization of marriage and the sexual and emotional ties of love cover up this practical aspect of the situation. However, being in a working-class type of position has its structural effects in the family, as it does in conventional jobs. The person who has the power is more attached to the role, whereas the person who is subordinate tends to be alienated from the role and tries to withdraw from it psychologically. Such a woman is usually cross-pressured: some factors cause her to identify with her role as wife (including love, sex, children, local community status); other factors cause her to be alienated (notably her power disadvantage caused by her economic position). Often, this is an uneasy balance. Sometimes the effects are unconscious, coming out in the greater tendency of women to have chronic depression than men (Holden 1986; Weissman 1987). Sometimes the result surfaces in overt dissatisfaction with marriage, resulting in quarrels or divorce.

A considerable amount of research over the years has focused on the importance of the "housewife role" in explaining why married women have higher rates of psychological distress than men. Bernard (1972) argued that housekeeping has a "pathogenic" effect on wives, and Gove (1972) contended that keeping house is so unstructured but so value-laden for women that housewives often judge themselves harshly. In general, housewives have fewer sources of personal gratification than husbands. Married men have traditionally had multiple roles inside and outside the home, but housewives have had to rely solely on their domestic role to bolster their self-esteem. Gove and others argued that this makes housewives more vulnerable to the problems associated with family life (*A Closer Look* 11.6).

This is not an irremediable situation, as social changes and subsequent research reveal. If women are being alienated, subtly or overtly, by the power imbalance in a marriage, then correcting the power imbalance should improve the happiness of the marriage. This is in fact what we do find. Where women improve the resources that give them more power in the marriage, there is a tendency for marital happiness to be greater (McLanahan and Adams 1987).

The emotional cost of a traditional, successful marriage seems to be paid especially by the wife. Women try hard to keep the marriage together, at the price of making themselves subservient to their husbands. This seems to happen most when the wife is a full-time housewife. When she has no job of her own, her self-esteem has been shown to go *down*, the more her husband's success goes *up*. At the same time, her husband's success makes her feel that the marriage itself is a success. However, wives who have professions of their own are more detached from their husband's success; it does not lower their own self-esteem, nor does it have much effect on how they feel about their marriages.

Thus, despite the conservative ideology often expressed abut the importance of maintaining the old-fashioned family, the fact seems to be that **dual-career marriages** are more successful. The traditional employed husband and full-time housewife combination often produces fights when the man is unsuccessful and empty marriages when he prospers. Working women in general are more likely to report happy marriages than nonemployed women, or at least not to report significantly more unhappiness in the marriage (Spitze 1991). Dual-career marriages often seem to have more energy and excitement, as well as more genuine equality and respect; hence, they don't detract from love but actually enhance it. Recent research suggests that marital satisfaction is greater in marriages in which both husband and wife share resources and share the traits of nurturing, caring, being affectionate, and devoting oneself to the other (Ickes 1993). The so-called traditional marriage with an emotional, feminine, stay-at-home wife and a masculine, take-charge, breadwinner husband is far from optimal (Larson and Holman 1994).

Recently, many couples have moved toward marriages of shared responsibility, in which both partners are employed and involved in housework and care work. As the number of these marriages has increased, the rate of depression among wives has decreased. If an employed wife thinks that men in general should do some housework, and if her husband makes significant contributions around the house, then she experiences better mental health (Ross, Mirowsky, and Huber 1983). If the husband shares the child-care responsibilities with his wife, she has lower-than-average depression levels (Ross and Mirowsky 1987). Such findings bode well for both wives and husbands in the future.

This does not mean that dual-career marriages are idyllic. They create new strains of their own. Both husband and wife may now be subject to pressures for success in the corporate world or in their professions (Hertz 1986). Work pressures can result in a shortage of time for one's partner. Women in these careers do not necessarily give up having children, although they go back to work

A CLOSER LOOK 11.6
Housewives in the Mental Hospital

Carol Warren (1987) studied the records of a group of housewives committed to mental hospitals in California in the late 1950s and early 1960s. The women were diagnosed as schizophrenics: psychotics who have lost all contact with reality. But their own complaints were those of typical housewives: they felt lonely, depressed, burdened by their housework, cut off from other adults, and overly dependent on their husbands. They complained of a lack of communication in their marriages and felt that their husbands had little interest in them except for the domestic and sexual services they provided. None of these women worked outside the home except for two who worked for their husbands—a situation providing no relief from the family circle. A number of these women said that they would like to work but their husbands would not let them. This was during the height of the wave of adulation of traditional home life during the 1950s. Some women in the sample who very strongly wanted a career were regarded as crazy for having that ambition; it was taken to be a rejection of femininity and a symptom of mental illness.

It was not their feelings of depression and loneliness that got them in the hospital but the fact that their behavior created troubles for their husbands. Typical troubles involved being sloppy and inadequate at housekeeping or child care, or money—spending too much, frivolously or "crazily" buying things the husband felt he could not afford. Often troubles involved sex as well, although this did not always lead to the husband defining his wife as psychiatrically disturbed. If the wife reacted by extreme withholding of sex, the husband was often angry but tended to regard the behavior as part of the typical female pattern. If, on the other hand, a wife was overly demanding sexually, this more likely raised suspicion of psychiatric problems.

Husbands, however, did not necessarily jump to the conclusion that their wives were mentally ill and ought to be committed, even though under California law at the time either spouse had the power to have the other committed, with the concurrence of one admitting physician. Most men resisted this course of action for some time because they did not want to lose their wife's services as housewife or her availability as a sex object. When troubles with spending, housework, and/or sex made the rewards no longer worth the cost, though, they would finally have their wives committed. This situation could be defined as a family emergency, and husbands usually got female relatives to help them out by taking over the housework.

In the hospital, the women often felt some respite from their housework, but it was hard for them to get away from their husbands and families psychologically. Many felt guilty about "abandoning" them, and several actually developed the delusion that their husbands and children were there in the hospital with them. Within a year, though, most of the women were out of the hospital, many of them having experienced electroshock treatment. There were numerous readmissions over the next few years, but by and large their social adjustment improved. The mental hospital did seem effective in getting the women in line. It was not always used merely for home troubles. One woman decided to leave her husband and had an affair with a fellow patient while on leave from the hospital, whereupon her permanent release was blocked by her irate husband, who procured papers to have her permanently committed.

Most of the women returned to their household responsibilities calmer and wiser. Many of them said in interviews that they were determined to behave and do what they were asked and that they did not want to be sent back to the hospital again.

quickly after childbirth and rely heavily on hiring other women to care for their children. As we have seen in chapter 6, jobs in the upper-middle class tend to absorb most of one's identity and one's energies. Although this previously applied mainly to men, it now can affect both spouses in a family. Thus, there are additional strains on the marriage, but these are compensated for by the fact that the wives are personally happier. Having two incomes also helps the overall family situation. On the whole, the result is probably an improvement in family satisfaction, though as we will see in chapter 13, such equality in marriage decreases women's dependence on men and gives them more options outside of marriage.

SUMMARY

1. Every marital union contains two marriages: his and hers. According to legal tradition, "her" marriage includes providing domestic services and deferring to her husband's wishes, whereas "his" marriage entails supporting the family and enjoying the benefits of his wife's household labor. Although laws have changed, wives still tend to do much more housework and child care than husbands and to pay much more attention to communicating with spouse and relatives.

2. Men and women who are married tend to have better health, accumulate more wealth, and report being happier than those who are not married. The fact that more healthy and successful people are more likely to marry in the first place does not explain all of the differences between married and unmarried individuals. Socially approved marriages provide the couple with money, companionship, and practical help. Marriage also provides individuals (especially men) with someone who monitors their diet and health and discourages risk taking and reckless behavior.

3. Major reasons that married people have more money and long-term economic security than never married, separated, divorced, or widowed people include economies of scale (it is cheaper for people to live together than alone) and higher earnings (marriages are increasingly likely to include two earners, and men tend to work longer hours when married). Intergenerational support also tends to be higher for married couples than unmarried individuals because parents are most likely to provide emotional and monetary support to adult children who are married.

4. Household and family labor tend to be divided on the basis of gender, with women doing most of the cooking, cleaning, laundry, and child care and men doing most of the household repairs and outdoor work. Husbands have increased the number of hours they spend each week doing household tasks, and wives have reduced their number of hours; however, wives still do two or three times more of the housework than their husbands. When husbands do more, wives are less depressed.

5. Wives usually act as household managers, planning and allocating various domestic tasks. When husbands care for children or clean house, they often perform the more pleasant tasks, and their contributions are typically described as "helping" the wife. Even though wives do much more than husbands, domestic arrangements are usually considered to be "fair." This is probably because unequal divisions of household labor reaffirm people's self-image as men and women.

6. Men tend to have greater power in families, especially based on their usually superior resources in occupational prestige, income, education, and participation in outside organizations. Where women have more of these resources, their relative power in the family generally goes up. A woman's domestic power declines when she has small children because her child care usually deprives her of the resources that bring power.

7. Happiness is generally highest at the beginning of the marriage, begins to go down with the arrival of children, and reaches its low point when children are in the teen years. Happiness goes up again after children leave home. The negative effect of children on happiness seems to operate especially through the strain it places on the wife, by increasing her responsibilities and reducing her sources of power and independence.

8. Among the higher social classes, traditionally there has been a high proportion of utilitarian marriages, held together not by personal feelings but by habituation and economic advantage. Lower down in the class structure, scarcity of money is the most common source of family troubles.

9. Inequality in marriages is a major source of strain, especially for wives. Dual-career marriages, in which power resources are more equal, are those in which a woman's self-esteem is highest and personal relationships seem to be best, provided this is what both spouses desire.

Key Terms

civil unions
dual-career marriages
economic dependency model
entitlement
family power
his and her marriages
household managers

marital happiness
marital satisfaction
marital economy of gratitude
other household labor
relative resource models
routine housework

second shift
selection effects
social desirability
time-diary studies
time availability theories
utilitarian marriage

Sociology Web Site

See the Wadsworth Sociology Resource Center, "Virtual Society," for additional links, quizzes, and learning tools:

http://www.sociology.wadsworth.com

Also on this web site you'll find InfoTrac College Edition, an online library of journals. Here you can search for electronic articles about central topics in sociology.

12 Raising Children: Motherhood, Fatherhood, and Socialization

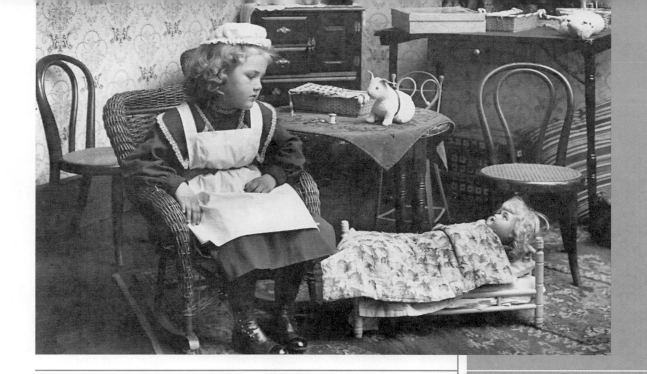

INTRODUCTION

Parenting may be the most important job we perform as adults, but as we saw in chapter 10, we are not always well-prepared for it. In this chapter we review what it means to be a parent or a child in our own society, with reference to how it was in past times. We look at recent changes in the organization of parenting that have resulted from the rapid influx of mothers into the paid labor force, including fathers' increased participation and trends in the use of outside child care. We also explore the various ways that children are socialized both inside and outside of families and highlight the ways in which child rearing contributes to stratification along the lines of gender and social class.

CHILDHOOD IN HISTORICAL PERSPECTIVE

The innocence and purity of childhood and the love that parents have for their children exemplify what we think is most noble about being human. Nevertheless, the close emotional ties between parents and children that we take so much for granted are a relatively recent historical invention. Before the seventeenth century, childhood was not considered a special time of life, and children were not sentimentalized as they are today. Very little distinction was made between children and adults, and the main value that children held was in their ability to contribute to the subsistence of the family (Ariès 1962; *A Closer Look* 12.1).

Although we know little about the day-to-day lives of children in preindustrial times, we can piece together a picture of how adults in Europe and America thought about them from existing historical records. The prevailing Christian view was that children were born in sin. In 1535, John Calvin remarked that children's "whole nature is a certain seed of Sin, therefore it cannot but be hateful and abominable to God" (Muir and Brett 1980, 3). A hundred years later, the American Puritans preached that God-fearing people must break the will of their children if they were to be saved from the devil: "Let a child from a year old be taught to fear the rod and cry softly. . . . Break his will now and his soul will live" (Muir and Brett 1980, 101). Throughout the eighteenth century and into the nineteenth century, Puritan and Methodist views on child rearing continued to stress the "corrupt nature" and "evil dispositions" of children. Parents were admonished to demand strict obedience and to use swift physical punishment to curtail the child's inherent evil (Synnott 1983). Duty and obligation were the hallmarks of parent-child relationships, and love, although probably present, was considered secondary. Much like the husband-wife relationships of the time, parent-child relationships were primarily instrumental, with children expected to do their part to keep the household running. (See chapter 3 for a discussion of the patrimonial household.)

The most significant changes in children's lives from medieval times to the modern era were ushered in by the rise of formal education and by the growth of what Stone (1977) calls "affective individualism." Both of these trends, in turn, were promoted by the growth of capitalism and the gradual shift from home-based production to waged labor, industrialization, and a market economy. Before the twentieth century, children were primarily valued for their economic contributions. They usually worked on farms or labored in their parents' trade, and by maintaining such pursuits, they provided a kind of insurance policy for their parents in old age. This does not mean that parents didn't love their chil-

A CLOSER LOOK 12.1
The Discovery of Childhood

According to many historians, the close ties between parents and children that we take for granted in the modern family were invented comparatively recently (Ariès 1962; for a contrasting view, see Pollock 1984). In the patrimonial household, children were just another part of the household political or business enterprise. They were valued for the fighting they could do, work they could provide, or marriage alliances they could bring; often they were apprenticed out at an early age. Many small children who died did not even have their names recorded. There was very little distinction made between children and adults. Children were dressed like little adults, and they were brought into adult activities as soon as they were big enough to take part. In medieval schools, children of ten might be reciting their lessons alongside young men in their twenties; in the army or navy, a young soldier would begin as an ensign or cabin boy in his early teens. Given the cramped living quarters at home and the lack of privacy, even small children were exposed to sexual activities and adult talk, unhampered by the modern attitude that such things ought to be kept from children.

When the private household began to emerge in the seventeenth and eighteenth, and especially in the nineteenth, centuries, the family took on a new shape. Affectionate relationships were supposed to hold, not only between husband and wife, but also between parent and child. Edward Shorter (1975) has referred to this as the "sentimental revolution," which he believed peaked around the years 1780–1830. It was during this period, especially in the United States, that the paramount ideals of the modern family were articulated (Degler 1980; Ryan 1981). Marriage was founded on romantic attraction between husband and wife, each of whom was to take care of his or her own separate sphere. The husband was to be the provider who left the home and returned with its economic support; the wife was to devote herself to the care of her children. Unlike the medieval family, with its practical, work-centered attitudes, children were now to be cared for, not merely

According to many historians, in the patrimonial household, there was little distinction made between children and adults, and children were often dressed like little adults.

physically but emotionally and morally. Motherhood, for the first time, became the object of a cult, conceived of as a moral ideal. Family ritual began to take over from community ritual. The medieval family took part in church ceremonies and community festivals; the modern family may have continued these practices, but it also created its own rituals: celebrating birthdays (part of the cult of childhood) and creating new family festivals like Thanksgiving. The family had become private, personalized, and sentimentalized.

dren, but they were valued less for their intrinsic worth than for the material contribution they could make to the household economy.

Death was much more visible in those days, which must have had some impact on the emotional bonds between parents and children (Uhlenberg 1980).

With fertility and mortality rates both much higher than they are today, love was "spread around" to more children who were more likely to die. Emotional involvement was likely to have been less, out of self-protection, though such things are difficult to measure (Synnott 1983, 29). Ariès (1962, 38) quotes a seventeenth-century woman as saying: "Before they are old enough to bother you, you will have lost half of them, or perhaps all of them." In the elite upper classes, where death was a less constant threat, children had more sentimental value, though still much less than among parents in the twentieth century.

From Economic Asset to Sentimental Object

Viviana Zelizer traces the changing social value of children in *Pricing the Priceless Child* (1985). She documents how the economically useful child of previous times was transformed into an "economically worthless, but emotionally priceless child" in the sixty years between 1870 and 1930. During this time, traditional forms of child labor came to be seen as harmful and inappropriate for those of "tender years." In 1870, if a child died in an accident and the courts concluded that another party was negligent, the parents were compensated for the value of the child's labor. By 1930, however, in cases of "wrongful death," the parents were compensated for incalculable emotional pain (Zelizer 1985). The emotional value of the youngest children rose the most, as the middle-class family came to be idealized as the only place where innocent and pure children could and should be protected. No longer considered evil creatures whose will had to be broken by their fathers, children had become precious creatures who needed nurturing and support from their mothers.

As in the case of past images of romantic love and of the Victorian family, this historical ideal of childhood innocence was attainable by only a minority of the population in Europe and America. In non-Western countries, other cultural ideals of children predominated and were also transformed as various social and economic changes unfolded. Among subgroups of the U.S. population, the sentimental ideal of innocent leisurely childhood often clashed with the need to make ends meet. For example, scholars have begun to debate how much white middle-class family norms and child-rearing ideals actually affected the lives of African Americans over the past two hundred years (Burgess 1995; McDaniel 1990). Images of children among the rich, white Southern gentry in the nineteenth century were certainly much different than those held by the slaves who were forced to work on their plantations (Griswold 1993). In addition, immigrant children in New York at the turn of the century were granted few opportunities to seem as playful and innocent as their wealthier contemporaries. In addition, important historical events, like war or economic depression, periodically change the way that we conceive of children, encouraging us to view them as potential soldiers or additional breadwinners (e.g., see Elder 1974; Elder, Modell, and Parke 1993; Hareven 1982).

The changing value of children is also reflected in the "market" for babies. As we saw in chapter 10, countless numbers of European infants were sent to foundling homes, where most of them died. Between the eighteenth and nineteenth centuries, there was a baby "glut," and infants were seen as relatively expendable. In the 1870s, a woman with an unwanted infant had to pay a "baby farm" to take the child off her hands because legal adoption was rare, and there was no market for babies. Many families were willing, however, to take older children (especially boys) into their homes because they could contribute to the household economy. By the mid-twentieth century, the situation had dramati-

cally reversed itself. By that time, it had become extremely difficult to find homes for older children—especially if they were boys. At the same time, adoption had become relatively common, and the sentimental value of younger children had blossomed. Couples who could not adopt an infant from a licensed agency in the 1950s were willing to pay as much as $10,000 for a baby on the black market (Zelizer 1985; Skolnick 1987, 307). By the 1990s, hopeful parents were willing to pay even larger sums of money to women who would agree to act as surrogate mothers and provide them with a "priceless" baby (see chapter 10).

Individual Versus Collective Patterns of Child Care

Our current child-rearing patterns are much more individualized than those of the past. Although women have been the primary caretakers of children in all known cultures, the assumption that children should be raised by their biological mother alone is a relatively recent invention. In most preindustrial societies, for instance, young children spent less than half of their waking hours in the care of their mothers. In preliterate tribes and in traditional agrarian societies, fathers, grandparents, siblings, and other women had important roles in taking care of the community's children (*A Closer Look* 12.2). Similarly, in Europe from the Middle Ages to the eighteenth century, the community was very involved in caring for children and shaping the individual's fate. In this older and more collective pattern, parent-child relations were regulated and monitored by relatives and other community members, and what happened inside the family was relatively public. A microcommunity of nearby adults and older children acted as surrogate parents, and there were always plenty of people around to offer advice on what to do in specific situations. It was not until the nineteenth century that the industrial revolution encouraged a withdrawal of the family from the outer world and created a relatively private family domain (Ariès 1979). According to the "cult of domesticity" that emerged around this time, women were defined as innocent, pure, and sensitive. Their virtue was reflected in motherhood, and they were increasingly confined to the home (see chapter 4).

In the twentieth century, the suburbs multiplied, and the family came to be seen more and more as a self-sufficient unit with a monopoly on emotions, raising children, and filling leisure time. By this time, the father's image was one of the **good provider,** who set a good table, provided a decent home, paid the mortgage, bought the shoes, and kept his children warmly clothed (Bernard 1981b). The woman, on the other hand, was expected to be consumed and fulfilled by her "natural" wifely and motherly duties. Isolated in her suburban home, she had almost sole responsibility for nurturing the children, aided by occasional reference to Dr. Spock or some other child-rearing expert. This was the era of the June Cleaver image of the happy household—suburban housewife, breadwinner husband, station wagon, dog Spot, and 2.4 children. This **individualistic model of parenting** was never a reality for most working-class women, but the ideal enjoyed unparalleled popularity during the 1950s.

Isolated and exclusive parenting by biological mothers is a sort of "all your eggs in one basket" approach to child rearing that produces unique stresses and strains for both parents and children. Because the mother does all of the child care, her feelings become overwhelmingly important to the children, and she has little opportunity for taking a break or deriving a sense of self-worth from alternate activities. This can set up a kind of "hothouse" environment in which mothers and children feel responsible for each others' feelings, and ego boundaries become confused. Exclusive and isolated mothering can create "double-binds" in

A CLOSER LOOK 12.2
Child Rearing in Nonindustrial Societies

The anthropologist Ronald Rohner (1975) studied child care in over one hundred nonindustrial societies throughout the world and came up with four central conclusions.

1. Parents who respond to infant distress by nurturing them and trying to reduce their discomfort tend to do the same for older children as well.

2. When children are rejected by their parents (denied love, esteem, and other forms of positive response), they grow up having difficulty managing hostility and aggression, have troubles with dependency, and are often emotionally unresponsive.

3. When children are accepted and given a warm and supportive upbringing, they tend to form less hostile relations.

4. Mothers who are unable to escape the intensity of continuous interaction with their children are more likely to reject them than mothers who can get away from time to time.

In other studies of nonindustrial societies (those relying on hunting and gathering, horticulture, herding, or agriculture; see chapters 2 and 3), Coltrane (1988, 1992) found that child-rearing patterns are linked to the status of women. If fathers have frequent contact with children, and especially if they participate in routine child care and are physically and emotionally supportive, then women in those societies are more likely to hold influential positions and participate in community decision making. Fathers' relationships to their children are also found to be related to ritual deference and "hypermasculine" behaviors. In societies with distant father-child relationships, men tend to fear and belittle women and are preoccupied with affirming their manliness through acts of bravery or physical prowess. In these distant-father societies, women are often excluded from ritual gatherings and are more likely to be required to bow to the men, serve them first, and stand in their presence. In contrast, if fathers are more involved with children, women are thought of as more equal and do not have to continually defer to their husbands. The other crucial factor is that in societies where women control property, they are less likely to be subjected to men's derogation, and men tend to be less aggressive and less concerned with proving their manhood. What we see then is that patterns of child rearing and property control are both important to the relative status of men and women. If men share child care with women, and if women share control of property with men, then there is a fair amount of sexual egalitarianism in the society. If, on the other hand, men control all of the property and women do all of the child rearing, there tends to be a fair amount of fear and antagonism between the sexes. Both patterns tend to be reflected in the mythical symbolism of the culture, with distant-father cultures having male-dominated creation myths and shared-parenting cultures having relatively gender-balanced stories of the way the world began (Sanday 1981).

which love and hostility are merged and children experience difficulty establishing independent identities. Nancy Chodorow (1978; Chodorow and Contratto 1992) shows how exclusive mothering contributes to personality differences between men and women and perpetuates itself by encouraging women to seek fulfillment through mothering (*A Closer Look* 12.3).

Our individualistic model of parenting also sets up a two-party conflict and a certain amount of anxiety that was not present in societies with more collective child-rearing practices. Because modern parents make their own rules, breaking or resisting the rules can easily be perceived as a personal threat, and personal battles between parents and children can ensue. In the modern case, parents have individualistic parental power but generally lack a larger sense of institutional legitimacy because each parenting decision is an individual choice. In the microcommunities of earlier times, children could disobey or fail to do their assigned tasks, but because it was not up the parents to establish the rules, the conflict was seen as outside of their control. They were simply enforcing a community standard. Not only that, but if the parent did not enforce the standard, someone else in the microcommunity would, providing validation that the parents were not individually responsible for setting standards. The earlier pattern could also be conflictual, and parents had little capacity to change parental practices; but they had much more institutional legitimacy, and the standards themselves were rarely challenged (Skolnick 1987, 1991). One of the important things about the more individualistic modern pattern of child rearing is that it allows for social change. What types of changes might result, and whether they are considered to be good or bad, are the subjects of many current debates over the future of families and children.

Current Idealized Images of Parents and Children

We continue to idealize and sentimentalize most aspects of parent-child relationships. In our advertisements and television shows, children are usually cute, clean, good-humored, and playful. Their parents are beautiful, contented adults, happily playing with them, caring for their little needs, amused by their pranks, and proudly watching their accomplishments as they grow up. Unfortunately, reality is not quite like this.

Small children just as often have runny noses and smelly bottoms; when and if they are clean and beautiful, it does not happen automatically but because the parents have put a good deal of effort, every day, into cleaning them up and feeding them wholesome food. Children sometimes do play happily, but one characteristic of young personalities is that they are very volatile and can switch from a happy mood to crying and whining almost instantaneously. Children can be very affectionate, and parents derive great pleasure from holding them in their arms; but children also tend to have few inhibitions about throwing things or hitting people. At certain ages, they are very possessive and selfish about their toys, and indeed about everything in their environment they call "Mine!" In short, children act childishly.

PARENTING TODAY

Most new parents have little idea of the difficulties involved in their new role. They have assimilated the ideals, but the realities often catch them unawares. Thus, for many couples, especially those who have their first child while they themselves are still young, its arrival can bring about a marital crisis. This is particularly likely for working-class families (chapter 6), who have fewer economic resources to ease the burden. If the baby arrives soon after the marriage, its parents face the problem of a double transition: having to adjust to living with another adult at the same time that the pressures of child care begin. Middle-class couples, and others who put off marriage and childbearing to a somewhat later age, tend to have certain advantages: more money, a better-settled routine, and more maturity for dealing with the children. Nevertheless, even here struggle is inevitable.

Not only that, but the worst may be yet to come. We have already seen in the last chapter that the family happiness of a married couple tends to hit its low point during the years that their children are teenagers. This may be partly due to the fact that parents are now in their forties and hence may be undergoing a midlife crisis of their own. However, dealing with children has its own costs from the time they are born; for that reason, family happiness starts creeping downward when the first child arrives, hops back up temporarily when the children are out of the house and beginning school, then keeps falling until they leave home for good, whereupon marital happiness jumps back up again. It is hard to avoid the conclusion that children are a strain on their parents.

A CLOSER LOOK 12.3
Why Do Women Mother?

Why do women do the mothering in our society rather than men? Mothering is not the same thing as childbearing or breast-feeding; it is taking care of a child, physically and emotionally. After a child is born, its own biological mother does not have to be the person who looks after it. Since bottle feeding was invented, there is no longer even the period of nursing that required a female, if only as a wet nurse. Mothering is a social, not a biological, role, and it is as much an emotional activity as a physical one: mothering means caring for one's children in both senses of the word.

The sociologist Nancy Chodorow (1978) has produced a theory to explain why this role is nearly always adopted by women. It is not instinctual because studies show that both men and women react similarly to infants' cries and smiles. In many animal species, males will care for an infant if left alone with one. There is nothing resembling a "maternal instinct" in women who are separated from their babies (for medical or other reasons) immediately after childbirth. The mothering role is learned.

Freud had pointed out that initially both the male and the female infant were cared for by a female. The mother is the first erotic love object for both sexes. This love is the prototype for all later love relations. It is a merging of the infant's self with that of the mother; in fact, according to Freud's theory, initially the infant had no sense of an individual self. An autonomous self emerges only as the child separates itself from this state of primal merging.

For the child to mature, the erotic attachment to the mother must be broken and directed outward. For the boy, this takes place during the classic Oedipus complex. Because of fear of his father's jealousy, the boy gives up his mother and instead displaces the erotic energy (libido) into a fantasy; this fantasy takes the form of an imaginary father inside his own head, which now serves as his conscience or superego. The boy's erotic renunciation of his mother, then, is a crucial step toward developing the adult psyche.

But what about little girls? A girl's problem is both to renounce her mother erotically (libidinally), while at the same time creating her own fantasy object of identification, her own superego. Because this would also be her adult sexual (and gender) identity, this identification has to be with her own mother. How this happens was never very clear in Freud's writings, and he never developed an adequate theory on this point.

This is where Chodorow offers her own theory. There is less pressure, she points out, for girls than for boys to break their deep primary attachment to their mothers. As a result, Chodorow says, the break is not so sharp. A girl is given more permission to be close and affectionate with her mother, even extending to physical caresses. This has more than a superficial significance because the girl is not required as much as the boy to separate herself sharply from her mother. She retains more of the original sense of merging with the world that the infant experienced in the original love relationship with the mother. This has a powerful effect on the development of the girl's personality.

What is the difference between feminine and masculine personalities? According to Chodorow, the feminine personality is less separated from other people and less sharply individuated. The boy, whose separation from the mother was made more sharply through the Oedipus complex, develops a sharper separation between himself and the world. Men have firmer ego boundaries; they tend to be more distant, domineering, and instrumental. They prefer the world of objects—action, machinery, science—to the world of people and of the self. Women prefer intimacy and warmth in personal relationships, and their

Who Takes Care of the Children?

How much of a strain children pose and to whom has very recently become a subject for debate. In our society and in virtually all other Western societies, it has traditionally been taken for granted that women take care of the children. In fact, the mother today still tends to put in most of the hours required to feed, clean, and dress them. Fathers tend to help more with the older children, especially in outdoor recreational matters: which is to say, in more of the

identities come more from how the group receives them than from their accomplishments. Their personalities are what Chodorow calls "relational," with more flexible ego boundaries between one's self and others.

The reason that women do the mothering, then, is because the maternal personality is simply a typically female one. A woman's personality needs are to be close to other people and submerge herself in the group. She surrounds herself with her husband and children because she herself remains underseparated from her own mother. Because she never broke her unconscious erotic ties with her mother, she continues to need this kind of close and nurturant relation with others. Women become mothers because their experience with their own mother has given them the kind of personality that needs to mother. Mothering thus reproduces itself in a chain across generations.

As a feminist, Chodorow asks: How can the chain be broken? If men begin to mother children, assuming more of the emotional care and the physical caressing as well as the physical work of looking after their needs, then the next generation of children will grow up with a different psychic structure. Will both boys and girls come to acquire combination male/female personalities as a result? Will boys become more

According to Freud and Chodorow, boys renounce their intimate connections with their mothers to develop a masculine self. Girls, on the other hand, can maintain connections with their mothers, developing relational capacities and preferences encouraging them to become mothers themselves.

relational and female-like, and girls more object-oriented and male-like? Chodorow does not say, but her theory remains a thoughtful challenge to our understanding.

pleasant tasks and fewer of the demanding and dirty ones. However, there has been a shift toward somewhat greater egalitarianism in caring for smaller children, and fathers do join in changing diapers or giving an infant a bottle in the middle of the night. The evidence seems to be that fathers are quite capable of carrying out all the child-care tasks, emotionally as well as physically (Coltrane, 1996; Parke 1996), but traditional gender roles still determine who does *most* of this work.

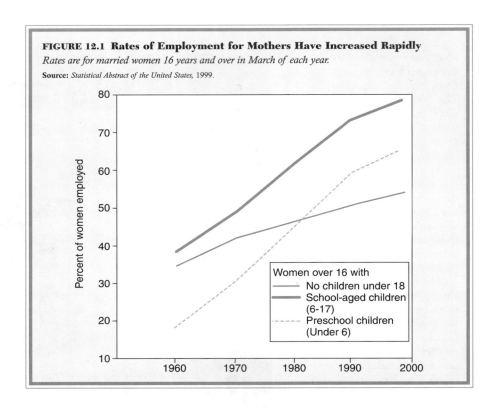

FIGURE 12.1 Rates of Employment for Mothers Have Increased Rapidly

Rates are for married women 16 years and over in March of each year.

Source: *Statistical Abstract of the United States, 1999.*

Maternal Employment and the Need for Child Care

Figure 12.1 shows that in the United States in 1998, over three-quarters of married women with school-age children, and over 63 percent of wives with preschool-age children were in the paid labor force. The change since the 1950s has been dramatic. In 1960, fewer than 19 percent of married mothers with children under six were employed. By 1998, that rate had more than tripled (*Statistical Abstract* 1999). Rates are highest for mothers of older children, but even mothers with the youngest children are likely to be working: 61 percent of mothers with a child under one year old were in the paid labor force in 1998. African American mothers are the most likely to be employed, with 79 percent of those with children employed in 1998 (*Statistical Abstract* 1999, table 660). Because single mothers are even more likely than married mothers to be employed, and because divorced mothers usually receive physical custody (see chapter 14), about three-quarters of all children under eighteen years are now living with employed mothers. What's more, over two-thirds of employed mothers now work full time.

Because most mothers must leave home to go to work, other caretakers have to be at least intermittently available. In two-parent families, fathers are a primary source of child care when mothers are working. One of every five preschool children were cared for by their fathers while their mothers worked outside the home in the 1990s. For school-age children, the primary source of care while mothers are at work is the formal school system, but fathers are the next most likely source of care (O'Connell 1993). Fathers are even more likely than mothers to be employed full time, so most families must turn elsewhere for child-care help.

In many working-class families, grandmothers regularly care for children, a practice that is more common among African Americans than any other

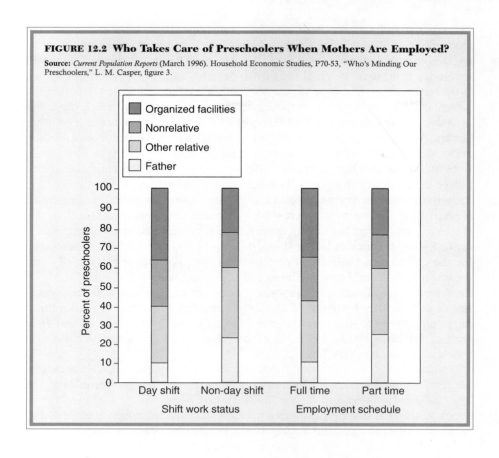

FIGURE 12.2 Who Takes Care of Preschoolers When Mothers Are Employed?

Source: *Current Population Reports* (March 1996). Household Economic Studies, P70-53, "Who's Minding Our Preschoolers," L. M. Casper, figure 3.

race/ethnic group. In many larger households and especially in Mexican American families, older children also play a major role in caring for younger siblings and cousins. Many children are cared for in family day-care homes, usually run by a working-class mother who watches several children in addition to her own. Although many of these homes are licensed by local government agencies to ensure adequate health and safety standards, others operate outside of government regulation. **Child-care centers,** in contrast, are routinely licensed and regulated and must meet government standards for adult-child ratios, indoor and outdoor space requirements, nutrition programs, health maintenance, and building safety. Some states also have requirements for teacher licensing and have additional curricular requirements for institutions that are designated as "child development centers." A few child-care centers are private for-profit enterprises, but most are run by nonprofit organizations such as schools, churches, or community service agencies. Almost a third of preschoolers of working mothers are cared for in organized and licensed child-care facilities, including day-care centers, child development centers, and preschools. This form of paid care is the fastest growing care arrangement for children when mothers are on the job. A smaller, but still significant, number of children whose parents are well-off continue to be cared for by nannies or in-home babysitters. As in the past, these caregivers tend to be immigrant or younger working-class women.

Figure 12.2 shows the child-care arrangements used for preschool-aged children when their mothers are working. You can see that the type of care used is shaped, in part, by the mother's job-scheduling requirements. For example, if the mother has a full-time job or works a day shift, her child is most likely to be in organized care or watched by relatives. If the mother has a part-time job or works a non-day shift, then the child is likely to be cared for by relatives or the father. Who watches the child depends on whether the mother is married, whether the father or other relatives live in the home or are working, how much money is available to pay for care, the availability of care facilities, and a number of other factors. As children get older, many end up being on their own until their mother, father, or other relative gets home from work.

Some businesses have begun to offer on-site child care in response to employee demands. Such centers have been found to increase worker productivity, reduce absenteeism, and minimize turnover and subsequent training costs. Nevertheless, only about 3 percent of medium and large businesses in the United States (defined as having more than 100 employees) and less than 1 percent of small businesses (fewer than 100 employees) provided on-site day care for the children of their employees in the 1990s. Another 1 percent of businesses managed child-care facilities away from the work site for the benefit of their employees. In addition, another 3 percent of medium and large businesses and another 1 percent of small businesses provided full or partial funding to employees so that they could pay for child-care expenses elsewhere (Bureau of Labor Statistics 1998). In all of private industry, rates of child-care provision have been very low.

The incidence of employer-provided child-care benefits was barely measurable for full-time workers in medium and large private companies in 1985. In 1990, when the Bureau of Labor Statistics first began tabulating such things, about 5,000 of 6 million businesses nationwide provided some form of child-care assistance (Morgan and Tucker 1991). By the late 1990s, child-care benefits had increased, but only slightly. About 2 percent of employees in small firms currently receive some form of child-care benefits, and about 10 percent of employees in larger firms receive some form of child-care benefits. Those working in blue collar and service occupations are least likely to obtain such benefits (7 percent in larger firms), and those working in professional and technical occupations are most likely to receive them (14 percent in larger firms) (*Statistical Abstract* 1999, tables 709 and 710).

With such low levels of workplace child care, one might suspect that Americans do not think that businesses or the government should provide such services to parents. According to recent opinion polls, however, there is substantial support for child-care subsidies, especially for lower income parents. Over half of parents surveyed in 1998 indicated that their community was doing only a "fair" or "poor" job providing quality child-care services for working families (California Center for Health Improvement 1998), and another poll found that nine of ten American adults reported that finding affordable quality child care was difficult (Harris Poll 1998). Almost three-fourths of respondents agree that the federal government should provide financial assistance to help pay for child care because they already provide families with financial assistance to help pay for college tuition. Eighty percent of Americans believe that if Congress provides new tax breaks to middle-income families, it should also provide child-care assistance for low-income working families (Children's Defense Fund 2000). A majority of Americans see child-care support as essential to helping working families move from welfare to work, with 86 percent agreeing that child care should be available to all low-income families so that parents can work (Kellog Foundation 1999).

Employer-supported child care enables both parents of preschool children to be part of the paid labor force.

Although most people say they support "quality" day care, the people who care for children in the United States can barely make a living doing it. Nine of ten child-care workers are women, and on average, they earn less than adults who tend our cars and parking lots, take care of our pets, or haul our garbage. In addition, nine of ten child-care providers who work in private homes earn below poverty-level wages. More than half of market-based child-care workers also earn wages below the federal poverty line (Phillips 1989; *Statistical Abstract* 1999). Because of low wages, poor benefits, and stressful working conditions, many child-care workers leave the child-care field for less demanding, less isolating, and better paying employment. The average day-care center in the United States has an employee turnover rate of over 40 percent per year (Etzioni 1993).

A look at the use of various types of child-care arrangements by working women since the 1970s shows some clear trends (Figure 12.3). Care by relatives accounts for under half of all the care provided to preschoolers with working mothers. Care by grandparents used to be the most common type of relative care, but it has declined somewhat since the 1970s. Care by other relatives has also declined. Father care has increased somewhat, though it dropped a little after the recession of the early 1990s. For nonrelative care, the use of sitters who come into the parent's home to stay with the children has declined a little since 1977 but dropped much more before that (from about 15 percent in 1965 down to 7 percent in 1977). The use of family day-care homes also dropped from 1985 to 1994. Only the use of child-care centers increased significantly, more than doubling between the mid-1970s and the mid-1990s (Casper 1996, 1998; O'Connell 1993; Phillips 1989).

Child-care arrangements change quite often because parents' work and family arrangements change, because of parental dissatisfaction, because the children grow older, and because of changing availability of caregivers (Leibowitz, Waite,

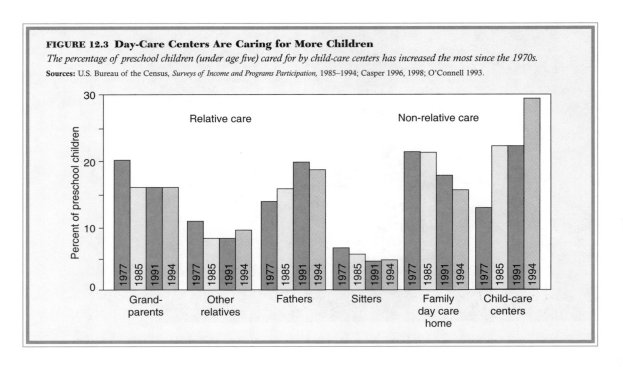

FIGURE 12.3 Day-Care Centers Are Caring for More Children

The percentage of preschool children (under age five) cared for by child-care centers has increased the most since the 1970s.

Sources: U.S. Bureau of the Census, *Surveys of Income and Programs Participation,* 1985–1994; Casper 1996, 1998; O'Connell 1993.

and Witsberger 1988). Shifts from relative care to some form of group care are the most common. To keep working on a continuous basis, most parents use multiple child-care arrangements and make adjustments when one form fails or when they cannot afford the higher quality options (Spitze 1988). Even though most child-care workers do not get paid much, the costs of child care are difficult for many working parents to meet. Among families with children who pay for child-care services, annual costs amounted to over $4,000 per year (Casper 1995). Families with incomes below the poverty line who pay for child care spend about one-fifth of their monthly incomes on child care. Families above the poverty line spent about 7 percent of their family incomes on child care (Casper 1995; O'Connell 1993).

With the strong emphasis now being placed on women having their own careers, the problem of child care has given rise to a number of new efforts at reaching a solution. Some women bring their children to work. For instance, about 6 percent of employed women were able to care for their preschool children at their place of work (Casper 1996). This arrangement tends to work best during the first six months of infancy, when the child is still rather immobile and quiet (and usually even this is possible only in some office environments). Some of these women are also able to work at home and thereby attempt to combine job and child care without disturbing other workers, but this solution is typically only available to more highly trained and better paid white collar workers. For most working-class women, the demands of the job are incompatible with child-care responsibilities (Uttal 1999).

Another fairly typical solution to child-care availability is shift-work (i.e., employment hours other than the typical 8:00 A.M. to 5:00 P.M. daytime schedule) or part-time work (less than thirty-five hours per week). A growing number of dual-earner families are juggling their work days and work schedules to meet both employment and child-care needs.

Over a third of preschoolers have mothers (or fathers) who work non–day shifts. More than one in ten dual-earner couples have no overlap between their hours, allowing for alternating child care between them. Shift work has increased significantly among both mothers and fathers in the past decade. In shift-work families, fathers are more likely to be the primary source of child care (apart from the mother) (O'Connell 1993).

Alternative work scheduling such as **flex-time** is also on the increase. Under such arrangements, workers can vary their schedules, depending on their work and family obligations, and in some cases perform duties at home as well as at the job site. Research suggests that alternative work schedules are associated with more routine child care by fathers (Pleck and Staines 1985; Presser 1988), though such arrangements can also place strains on families, in part because husbands and wives spend little time together (Presser 2000). Mothers are much more likely than fathers to be employed part-time, with over a third of employed mothers working less than thirty-five hours per week. Since the 1980s, part-time work by parents of preschoolers is also on the rise.

The Need for Child Care Is Increasing

Given the continued growth in the number of employed mothers (see figure 12.1) and considering that most work full-time, it is clear that the demand for child care has increased. Maternal employment is neither a fad that will reverse itself nor a phenomenon that is limited to mothers of older children, minority families, or single mothers, as many had previously claimed. Recent trends will undoubtedly put pressure on fathers to take on more of the routine parenting, on businesses to provide flexible scheduling and on-site day care, and on governments and social service agencies to provide decent day-care facilities.

The Family and Medical Leave Act (FMLA) of 1993 is the sort of family support reform that might be expected in the United States in the near future. This bill, originally vetoed by President Bush, was repassed by Congress and signed into law by President Clinton in early 1993. It mandates just twelve weeks of *unpaid leave* to care for a new baby or a seriously ill family member, and it only applies to companies with more than fifty employees. This legislation was considered a milestone in the area of family policy in the United States because it was the first major law that acknowledged work-family conflict and guaranteed that parents would not lose their jobs if they took time off to have a baby or care for a sick family member. Nevertheless, the FMLA excludes 95 percent of employers and 50 percent of employees; is of little assistance to part-time, seasonal, or temporary workers; and is unavailable to same-sex couples, in-laws, or extended kin (Perry-Jenkins, Repetti, and Crouter 2000). Because the leave time is unpaid, low-income families often cannot take advantage of it (Gerstel and McGonagle 1999). That it took a decade to pass this law, with its minimal benefits for parents, does not bode well for future significant reforms in this area.

Most other modern industrial nations make much larger contributions to the welfare of their families and children. In contrast to the minimal unpaid family leave that some U.S. parents now enjoy, Canadian workers receive fifteen weeks of family leave at 60 percent pay (Etzioni 1993, 60). Virtually every European nation provides at least that much support to individual parents, and most provide significantly more. For example, Swedish workers receive 90 percent pay for thirty-six weeks and prorated paid leave for the next eighteen months (Haas 1992). In addition, many European nations provide direct cash payments to parents when a child is born, and most also fund the construction and operation of group child-care facilities. The difference is that in Europe, child care is viewed as a public responsibility, and social welfare programs have a long history. Some European nations emphasize the benefits of state-supported child care in terms of women's employment and enhanced national productivity, whereas others emphasize the importance of such programs for optimum child development. Either way, these countries invest heavily in the future by subsidizing their children's care. In the United States, child care has received relatively little attention from politicians and lawmakers. As discussed in chapter 6, this has resulted in a growing portion of the nation's children who lack access to basic necessities like secure housing, adequate nutrition, minimal health care, and appropriate daily care while their parents are employed. One of the reasons that U.S. lawmakers have avoided subsidizing child care is that they have held on to the mistaken ideal that children will suffer unless they are raised by a full-time stay-at-home mom.

Working Mothers: Does a Child Need a Full-Time Mother?

Only two or three decades ago, the question "Who takes care of the children?" would hardly have arisen in a book of this kind. At that time, the "experts" were practically unanimous in the belief that if a mother worked the development of the child would be jeopardized. By the early part of the century, Sigmund Freud had already stressed the importance of libidinal/erotic interaction between infant and mother as a normal stage of psychological development. Then during World War II in Britain, many infants and young children were separated from their mothers because of bombing raids and other wartime disruptions. These youngsters, brought up in overcrowded and understaffed orphanages, were more likely than usual to succumb to physical illnesses, to lack normal energy, and to be psychologically withdrawn (Spitz 1945). Belief in the negative effects of "maternal deprivation" was further reinforced by the findings of the experimental psychologist Harry Harlow, whose infant rhesus monkeys were raised with a terry-cloth covered wire "mother" and grew up to be very disturbed psychologically.

The strongest advocate of this position was the British psychoanalyst John Bowlby. He was much impressed by the wartime damage to children deprived of their mothers and treated a number of other children for psychotic withdrawal that he attributed to the same cause. In *Child Care and the Growth of Love* (1953), Bowlby argued that a small child needs to have a continuous, unbroken, intimate relationship with the mother or some *one* person who is substituting for her. Even partial deprivation of such contact results in psychological disorders and instability, Bowlby said. He went so far as to declare that a mother should not leave her child for any reason except major emergencies for the first three years of life.

Nevertheless, it has turned out there is reason to doubt these extreme pronouncements. For one thing, the evidence of Spitz, Harlow, and Bowlby all concerns unusual cases. Spitz's orphans were not only deprived of their mothers but went through wartime shock and lived in understaffed institutions. Harlow's monkeys were deprived of *all* living contact. Bowlby examined psychotic children but did not compare them with a group of normal ones to see if the latter also had had mothers who were often away.

More recently, many opportunities have arisen to study children whose mothers are away at work for hours every day. The rising rate of female employment has made this a common situation (see Figure 12.1). Studies on children with employed mothers have shown no negative effects on their psychological development (Spitz 1991; Hayes and Kamerman 1983; Bianchi and Spain 1986). Using large representative samples and longitudinal data, studies show that neither maternal employment status, nor the timing and continuity of maternal employment, are consistently related to child outcomes (Harvey 1999). Research from the middle part of the century tended to seek and find negative impacts of mothers' employment on children, whereas research in the 1980s and 1990s tended to show positive and negative impacts of maternal labor force participation (Perry-Jenkins, Repetti, and Crouter 2000).

There is evidence that women who work outside the home provide better role models for their children and are especially likely to have daughters who develop a stronger sense of competence (Hoffman 1979). Daughters of employed mothers are likely to be independent and to plan employment for themselves (Moore, Spain, and Bianchi 1984). Both sons and daughters are also likely to hold more accepting attitudes about women having careers and men doing family work. Children of working mothers tend to hold less rigid gender stereotypes than others and view women in general (as well as their own mothers) as more competent (Bloom-Feshbach, Bloom-Feshbach, and Heller 1982; Wilkie 1987).

As the sociologist Terry Arendell (2000) suggests, maternal employment is conducive to mothers' mental health and parenting gratification, but it also produces some unique stresses and strains. Many mothers (and some fathers) experience a time bind and pay a high personal price trying to balance work and family demands (*A Closer Look* 12.4) (see Daly 1996; Hochschild 1997; Robinson and Godbey 1997; Schor 1991). Many experience loss of sleep, curtailed leisure time, and an ongoing feeling of overload and stress resulting from attempts to juggle work and child care (Arendell 2000; Coltrane 1996). Although employment is generally found to improve women's psychological well-being, employed mothers of young children tend to worry more than employed women without children (McLanahan and Adams 1987). This coincides with the more general tendency of parents with young children to report higher levels of depression and anxiety than nonparents. Higher levels of depression can also be the result of money problems. Ross and Huber (1985) found that having children at home leads to economic strain that promotes depression, but when economic strain is held constant, children actually improve women's psychological well-being. These findings suggest that mental health is more dependent on having enough money to live on than on having children per se.

Mental health is also dependent on having enough time to do everything that one feels is necessary. For instance, some studies have shown that the key factor in whether employment improves married women's mental health is whether husbands contribute to routine child care and housework (Kessler and McRae 1982; Ross, Mirowsky, and Huber 1983; Ross and Mirowsky 1987). Related findings show that employed single mothers experience the most distress and that employed married mothers who receive little help from their husbands are most likely to emphasize the direct costs of children: being tied down, being subject to burdensome demands, and having difficulty organizing one's time (McLanahan and Adams 1987). Studies of married women find that the psychological benefits of working and having a husband who does some of the domestic tasks depends on a couple's feeling about whether the wife should have a job (Coltrane 2000). Similarly, there is evidence that maternal employment can have adverse effects on children if family members hold unfavorable attitudes toward the mother's dual role. In other words, children and adults can suffer if they feel that the mother should be performing only the traditional duties of wife and mother.

A CLOSER LOOK 12.4
Families and Time

The meaning of family time has been shaped by a variety of historical forces. Capitalism, religious beliefs, technology, and changing patterns of labor force participation have both given rise to family time as a highly boundaried experience and contributed to its demise in the postmodern era. Although our current interpretations of family time tend to coalesce around notions of family togetherness and solidarity, this has not always been the case. In the nineteenth century, for example, community togetherness during key celebrations was viewed as a much more important source of fulfillment than spending time alone in families. However, in response to the fast pace of technology and the dramatic increase of women in the paid labor force, family time has been idealized as the private still point in an otherwise frenzied life pattern. In some ways, this hegemonic view of family time is a modernist belief because it emphasizes the distinction between the public world of productive work and the private world of personal feeling where family time is held out as the reward for successful productivity. Whereas a modernist view neatly dichotomizes work and family time into distinct and uniform domains, a postmodern perspective calls out for an analysis of the diverse experiences of family time. This includes portrayals of the negative aspects of outcomes of family time together and an analysis of the underlying conflicts and inequalities with respect to the experience of family time. In this regard, adults, children, women, and men make different contributions to family time, play different roles, and as a result, have different social constructions of the experience of family time.

Although the idea of private family time continues to be salient as a cultural ideal, a variety of forces make the realization of this goal difficult. One of these forces is the expansion of commerce into all days of the week, which has resulted in the deterioration of the distinction between work time and sanctioned days of rest. For example, with the growing pluralization of cultural practices within North America, what were once hegemonic Christian ideas about Sundays as a day of rest have begun to break down. The effect is that many families no longer have culturally prescribed days for togetherness. Similarly, on a day-to-day basis, family members typically disperse into their own temporal routines that have varying degrees of overlap. For these families, it is difficult to have fixed and shared meal times, which have traditionally been one of the focal points for the experience of family time.

Technology plays a major role in reshaping the meaning of family time. Devices such as personal phones and home computers have made the boundary between work time and family time much more permeable than it ever was. By opening a "hole in the fence" of private family time, technology keeps families in a state of interruption. At the same time, the availability of technology within the home also appears to be playing a role in keeping the family closer to home. The increasing availability of VCRs, compact discs, home computers, and home exercise machines means an increase in the opportunities for leisure at home. Similarly, electronic mail, phone mail, fax machines, and computer networks create an opportunity for family members to do their paid work at home. Television, too, with its expanding band of cable channels, continues to draw families into the home for time together. Although stereotyped notions suggest that families sit passively gazing at the tube, research indicates that a good deal of family interaction does occur as they sit around the electronic hearth. With the proliferation of technological devices within the home, it appears that families are at the crossroads of two very different tracks: one that provides them with many new opportunities for shared activity within the home and another that easily distracts them into the solitary world of technology that demands their undivided attention.

Source: Excerpted from Kerry J. Daly. 1996. *Families and Time: Keeping Pace in a Hurried Culture.* Thousand Oaks, Calif: Sage, Pp. 82–83. Copyright © 1996 by Sage Publications. Reprinted by permission of the publisher.

In that case they may end up resenting her for working outside the home, even if she is forced to do so for economic reasons.

Also available are comparative studies of other societies, including some traditional and tribal ones, in which the household arrangement is different from our nuclear pattern. These show that women are best able to fulfill their roles *as mothers* when there are other people around who regularly are able to help with caring for the children (Whiting 1963; Lambert, Hamers, and Frasure-Smith 1980; also see *A Closer Look* 12.2). These studies resonate with more recent ones showing that single mothers, among others, rely on a wide network of helpers that transcends narrow conceptions of mothers as sole caregivers (Arendell 2000).

There is no evidence, then, that upholds Bowlby's stern warning of the disastrous consequences of not having a mother continuously in contact with a small child. Many societies and social arrangements manage to do otherwise. An infant does need stability in the world immediately around it, and a special bond does tend to grow with whoever is the child's regular caretaker. However, there is no evidence that this person has to be the mother; it can just as well be an affectionate substitute caretaker, such as a relative, regular babysitter or child-care worker, or the child's own father. Also, having several persons who regularly care for the child seems to be a favorable experience for the child, not a negative one.

There has been less debate over psychic harm done to school-age children whose mothers hold jobs. Even these children, however, were called "latchkey kids" because they sometimes appeared at school with a house key hanging from a string around their neck. They were pitied in early journalistic accounts because they returned home to an empty house rather than to cookies and milk with Mom. Because of their economic situations, more and more parents have felt forced to leave school-aged children unattended, or in the care of their siblings, until the adults can get home from work. Today, few researchers or other people think school-age children are generally damaged when their mothers work, but some still worry that leaving children alone will have negative consequences. Primarily because there is no one around to check up on them, unattended children are more at risk for delinquency, drug use, and early sexual activity. Self-care may have some positive benefits as well, however, and some children report increased independence and maturity as the result of having to take responsibility for themselves.

The Special Strains of Single-Parent Families

Single parents are not much different from parents in general, but their life situations tend to present some unique challenges and some special problems. Single-parent families have always been a part of the social landscape in America. About one in four children born around the turn of the nineteenth century experienced the death of a parent before they reached the age of fifteen, and another 7 or 8 percent experienced parental separation or divorce (Amato 2000; Furstenberg and Cherlin 1991; Uhlenberg 1980). With parents living longer in the twentieth century, the percentage of children growing up in single-parent families declined and was at an all-time low in the 1950s. Increases in divorce (see chapter 14) and non marital births (see chapter 10) have increased the numbers of single parents dramatically in the past three decades (Bianchi 1995). In 1970, single parents represented just one in ten family groups in the United States, but by 1998, they were over one in four families with children (*Statistical Abstract* 1999). With almost 10 million single-parent households, some of the stigma of being in a single-parent family has been removed, but many problems remain. For one thing, the relative economic position of such families has not improved. Families headed by women are six times more likely than two-parent families to have incomes below the poverty line, and female-headed households now constitute the majority of poverty households in the United States. About half of single-mother households are living below the poverty level, with higher rates for African American and Latino single mothers (Council of Economic Advisors 1998; see chapter 6). This "feminization of poverty" creates hardship for women and children and increases the likelihood that the household will have to rely on extended kin networks for material and emotional support. Single-father households are actually increasing at a faster rate than single-mother households, but most single parents are still women. In 1990, women maintained 85 percent of the one-parent family groups with children and made up 81 percent in 1998 (*Statistical Abstract* 1999).

Many single mothers work at low-status jobs and are subjected to some subtle (and not so subtle) forms of discrimination. Over the past decades, as single-parent households have become more numerous, some myths about single mothers have emerged. One myth is that a low-income, separated, or divorced woman raising children alone is single by choice. In fact, many single mothers terminated bad marriages, some were deserted, and most are forced to support their children with little financial help from the absent father. A second myth is that single mothers could pull themselves up out of poverty if only they would get a job and work hard. The fact is that most single mothers are employed, but that the wages they earn are insufficient to keep them out of poverty. A third and related myth is that most single mothers are lazy and unworthy recipients of welfare who have babies just to receive government checks. Again, the fact is that welfare payments (formerly Aid to Families with Dependent Children [AFDC] and more recently Temporary Assistance to Needy Families [TANF]) are rarely sufficient to make ends meet, let alone generous enough to allow the mother and children to live comfortably. These myths are based on stereotypes that ignore diversity within the single-parent family population and ignore the realities experienced by these families (Arendell 2000; Mulroy 1988). Very young single mothers often encounter extraordinary and persistent economic hardship (Brooks-Gunn and Chase-Lansdale 1995). Although many single mothers have low-status jobs, over a third have attended college, and most of these have higher paying employment. Because of inequities in the labor market, single fathers earn 50 percent more income than single mothers (Amato 2000).

Because of the stigma of being in what was once labeled a "broken home," many parents in single-parent families feel pressure to be perfect. Many worry over the children's well-being, and some compensate by being overprotective. Most, however, develop especially close relationships with their children. Parent-child relationships in single-parent families tend to be more democratic than in two-parent families. Out of necessity, there tends to be more task sharing and more joint decisions. There is typically more negotiation and bargaining in these families, and the lack of resources and outside help often make it difficult for the parent to be authoritarian. Sometimes single-parent families are characterized as being more "permissive," but what this really means is that the parent tends to give more weight to the child's wishes and the child is not required to be as deferential to the parent. Children in single-parent families are also sometimes called on to provide emotional support to single parents, which reverses the traditional image of parent-child dependence (Arendell 1986, 2000; Weiss 1979). With the added responsibilities of being in a single-parent family, children can be said to "grow up a little faster" than children in more affluent two-parent families.

DISCOVERING FATHERS

We didn't just forget fathers by accident; we ignored them on purpose because of our assumption that they were less important than mothers in influencing the developing child (Parke 1981, 4).

Before the 1960s, social scientists were relatively unconcerned with interactions between fathers and their children. It was generally assumed that if the husband performed the instrumental tasks of his provider role, and the wife her expressive tasks as homemaker and mother, all would be well with the children, the nuclear family, and the larger society (Parsons and Bales 1955). In the 1960s, studies began to include the father but focused on "father-absent" families, suggesting that children without fathers had considerable difficulty in forming an appropriate gender identity and lagged behind others in academic achievement and moral development. These studies have been criticized for confusing father-absence with other social conditions such as living in a rough neighborhood or being poor. The **father-absence studies** have also been faulted because they used dubious measures that portrayed "normal" boys as tough unemotional loners who were emulating a narrow vision of instrumental masculinity. One of the assumptions of this research seems to have been that teenage boys needed fathers around to discipline them or else they might turn into juvenile delinquents. Another sexist assumption underlying some of the father-absence research was that boys from single-mother households lacked masculine role models and therefore might turn out to be effeminate "mama's boys."

Since the 1980s increasing attention has been paid to fathers as actual caretakers of young children. Research has focused on men's interest in caring for infants and toddlers, their capacity for child care, and differences between male and female styles of interacting with children. Most of this research has focused on father-child interaction when the children are babies or preschoolers. Studies have documented that many men want to become involved in routine infant care and that men can be competent caretakers of newborns, as well as of older children (Lamb 1997; Parke 1996). This is not to say that most fathers treat babies the same way that mothers do. Researchers have typically found that fathers are more likely than mothers to engage in rough-and-tumble play, to treat sons and daughters differently (favoring sons), and to be more directive in their interactions with younger, as well as older, children (Parke 1996). Mothers, in contrast, are typically found to be less directive, to more frequently offer verbal encouragement, and in home settings, to be involved in multiple household tasks while simultaneously tending to the children. The traditional pattern, then, is one in which child care is an ongoing and taken-for-granted task for the mother but a novel and fun distraction for the father. For example, time-use studies show that many fathers will contribute to household "labor' by playing with the children while the mother cooks or cleans (Berk 1985; Coltrane 2000).

Although some biologically based theories have suggested inherent limitations in men's abilities to nurture children (e.g., Rossi 1977), most researchers have assumed that sex differences in responsiveness to children are socially constructed. According to current social theories, men and women act differently toward children because (1) an unequal distribution of power and status requires women to do most of the mundane family labor; (2) different socialization practices for girls and boys create nurturant women and insensitive men; and (3) women exhibit their competence as gendered members of society

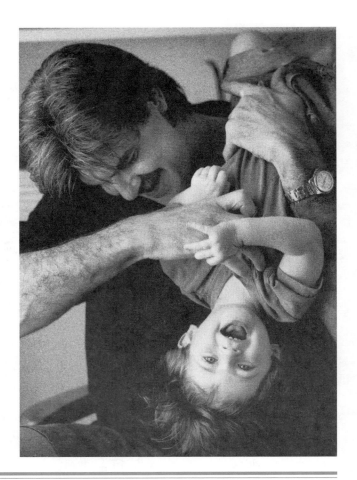

Fathers are more likely than mothers to engage in rough-and-tumble play.

through mothering, and men demonstrate their gendered selves through attention to work and inattention to children.

Although results are sometimes incomplete or contradictory, most child development studies find that fathers have positive effects on their children. For the most part, fathers who are both caring and demanding foster high self-esteem and promote academic achievement in their children. Most of this research, however, focuses on traditional families in which the father assumes few day-to-day caretaking tasks. Studies have found that fathers of preschool-age children average about two to four hours per week on direct child-care tasks (Coltrane 1996). With most fathers spending so little time interacting with young children, it is difficult to assess the potential impacts of fathers who play a more significant role in their children's lives. As Lamb (1981, 459) noted, "There is no reason to assume that 'nontraditional' fathers influence their children in the same way that traditional fathers do."

New Patterns of Fathering

If we are to believe prime-time television, fathers are acting more and more like mothers. They use the latest housecleaning products, change the baby's diaper, and even telephone young adult children who have gone off to college. Although most fathers do not live up to these media images, today's fathers participate more actively in their children's lives than fathers did a few decades ago. What's more, a small but growing number of fathers are assuming major responsibility for raising their children either as single parents (Greif 1985) or more typically as shared-parenting fathers (Coltrane 1996). Fathers involved in the routine details of daily child care are found to have even more beneficial impacts on children than kind, but distant, traditional fathers. In general, children with highly involved fathers are characterized by increased cognitive competence, increased empathy, less sex-stereotyped beliefs, and a more internal locus of control (Lamb 1997; Parke 1996; Radin and Russell 1983).

Because it is rare for fathers to share much of the child care in most families, it has been difficult to study the long-term impacts of active fathering. It may be that the positive child outcomes noted earlier are as much a result of the children being raised by two active caregivers, or in a relatively affluent two-income family, than with anything special about the father's contribution to the child. For instance, Lamb (1997) suggests that the positive family context for parents who are committed to sharing child care is probably what is contributing to beneficial child development: What matters is not so much who is at home, but how that person feels about being at home.

Some researchers focus on the potential benefits of shared parenting for children, whereas some focus more directly on the parents, claiming that further improvement in the status of women depends on the increased involvement of men in the family. Others focus on the gains that might accrue to both men and women if fathers were able to develop their emotional capacities through parenting. The idea is that active assumption of parenting duties could help men shed confining masculine stereotypes and relieve women of the sole burden of providing emotional support to family members. Because it is primarily men who physically and sexually abuse both women and children, it would be a mistake to assume that increased participation by fathers would be a good thing for all families. However, many families could benefit from fathers being more involved, especially if wives and children desire it.

Recent programs, classes, and research have suggested specific ways to encourage fathers' involvement in family life (Hawkins and Roberts 1992; Klinman and Kohl 1984; Levine and Pitt 1995; Levine and Pittinsky 1997; McBride 1990; Pleck 1997). Lack of skill and self-confidence discourage many men from attempting to assume a greater share of child care, but such tentativeness can usually be overcome with encouragement and practice. Mothers often act as **gatekeepers** to men's involvement with their children, and their actions can have substantial influence on a new father's developing sense of competence. If she helps him feel like he is doing a good job (which may not be apparent to either father or child), he is more likely to continue assuming responsibility for child care. If, on the other hand, she continually corrects him and tells him that he is inept, he may give up prematurely (Allen and Hawkins 1999).

For the most part, fathers, like mothers, learn by doing. The routine experience of caring for infants has been found to facilitate parental responsiveness in men and women (Coltrane 1996). Even inexperienced fathers who assume primary caregiving responsibility or share child care with their wives quickly develop the necessary parenting skills (Pruett 1987). If fathers assume major (or equal) responsibility for child care, their interactions with children come to resemble those of traditional mothers more than traditional fathers (Coltrane 1996). Unlike traditional fathers who tend to overstimulate infants and engage in rough-and-tumble play, primary-care fathers tend to interact verbally and let

A CLOSER LOOK 12.5
Can Men Mother?

In an interview and observational study of dual-earner couples with school-age children, Coltrane (1996) found that some of the fathers acted like mothers. One woman commented on how most people didn't believe her husband was as involved with their children as she was.

> *I think that nobody really understood that Jennifer had two mothers. The burden of proof was always on me that he was literally being a mother. He wasn't nursing, but he was getting up in the night to bring her to me, to change her poop, which is a lot more energy than nursing in the middle of the night. You have to get up and do all that, I mean get awake. So his sleep was interrupted, and yet within a week or two, at his work situation, it was expected that he was back to normal, and he never went back to normal. He was part of the same family that I was.*

For those families that shared most of the child care, the distinction between mothering and fathering became blurred. Parents reported that the children were as emotionally close to the father as to the mother and that children would seek comfort from either parent when upset or injured. Selection of one parent by the child typically followed being in the care of that person for the preceding period of time. For instance, if a child had been cared for by the father for the past several days and happened to awaken in the night from a dream, the child would call out for "daddy." If, on the other hand, the mother had more recently been "on-duty," the child would call for "mommy." In addition, many of the children whose parents shared child care roughly equally would confuse parental terms by calling the father "mommy" or the mother "daddy" without realizing that they had done so. In effect, the children were using the gendered form to signify "parent," and in so doing, indicated how mothering and fathering became merged if both men and women share "mothering" duties.

Other studies have come to similar conclusions about the social construction of parenting roles. Risman (1989) studied 141 single fathers and discovered that four of five had no outside help with housekeeping tasks such as grocery shopping, food preparation, house cleaning, or yard work. Virtually all of the single fathers worked hard to develop and maintain child-centered homes, and all had close relationships with their children. Risman concluded that when men "mother" they are very hard to distinguish from women.

> *Gender differences in our society are based as much on the differential expectations and role requirements males and females face throughout their lives as on internalized personality characteristics. . . . When males take full responsibility for child care, when they meet expectations usually confined to females, they develop intimate and affectionate relationships with their children. Despite male sex role training, fathers respond to the nontraditional role of single parent with strategies stereotypically considered feminine. (Risman 1989, 163)*

---

infants and toddlers direct the play. Unlike traditional fathers who tend to sex-type their children and give more attention to their sons, shared-parenting fathers appear to treat sons and daughters similarly (Parke 1996; Pruett 1987). Thus, shared-parenting families may provide children with the equivalent of two mothers, even though the total amount of time children spend with parents might be about the same as in other families (*A Closer Look* 12.5).

Why Do Some Parents Share Child Care?

The primary motivation for parents to share child care is that they both need to be employed. Whether they are near the bottom or the middle of the income pyramid, couples usually feel that two incomes are needed to maintain their standard of living. This means they need some form of daytime child care. Because full-day child care is costly, and because many believe that children should be with their parents, many are attempting to share the routine care of their children. Most must still use some form of paid child care, but an increas-

In families where child-care responsibilities are shared, children are as emotionally close to the father as to the mother, but fathers still tend to enjoy more leisure time than mothers.

ing number are trying to schedule employment so that they can alternate taking care of the children.

In addition, some adult men (and women) express regret that their own fathers were not involved in the emotional life of the family (Osherson 1986). When they contemplate having children, some vow to "do it differently" than their own parents did and attempt to set up schedules that ensure that both parents will have time to be with the children (Coltrane 1996). Other families, of course, believe that only mothers should take care of children, and if the husband earns enough money, the wife is able to stay home with the kids. This traditional pattern is increasingly rare among younger parents, however, who find that they both must work to make ends meet. As we saw in chapter 11, most employed mothers continue to shoulder the responsibility for domestic labor and thus work a "second shift" (Hochschild 1989). The high stress and "burn out" associated with the second shift are encouraging many mothers to demand help from their husbands. As a result, more and more couples are sharing the easier parts of child care—things like looking after children at home or taking them to the park. Whether large numbers of couples will share the more mundane aspects of child rearing—particularly the indirect child-care tasks of feeding and cleaning up after them—remains to be seen.

The handful of studies that have been conducted on shared parenting suggest some of the conditions that might promote such arrangements. There is some suggestion that sharing of child care occurs more frequently when children are younger and the total household labor burden is heaviest, but others find that fathers are more likely to take on or continue family work if child-care demands are low, that is, when there are fewer or older children. Some studies report that fathers' attendance at the birth or involvement in early infancy enhances chances for later shared responsibility (Coltrane 1996; Kimball 1983; Russell 1982; Russell and Radin 1983).

Some also find that couples who wait until their late twenties or early thirties to have children are more likely to share both child care and housework (Coltrane 1996; Daniels and Weingarten 1982; Deutsch 1999). This is probably because women are able to receive more education, establish careers, and set up patterns of task-sharing with husbands that persist after they have children. It is probably also the case that women who delay childbearing are able to establish an identity apart from family roles and may be more willing and able to bargain for participation by the prospective father (Gerson 1985, 1993). In addition, men who delay having children are probably somewhat more likely than younger fathers to have established themselves in their occupations and to desire an avenue to emotional fulfillment that active fathering offers (Coltrane 1996; May 1982).

Most researchers also report some differences in family interaction between shared-parenting families and more traditional ones. For instance, some find more conflict in shared-parenting families, probably because couples cannot rely on standard roles and must negotiate new divisions of labor (Deutsch 1999; Kimball 1983; Russell 1983; Russell and Radin 1983). Although negotiations of who does what need not be explicit (Hood 1983; Coltrane 1996), many shared-parenting couples report more talk and more acceptance of voicing anger than in other families (Gilbert 1985; Kimball 1983; Risman 1998). In some families, this translates into higher "marital satisfaction," whereas in others, it translates into "lower satisfaction." Russell and Radin (1983, 150) suggest that high levels of conflict may tend to be limited to the time when the couple is making the transition to shared parenting and may decrease as the couple adjusts to the sharing of tasks. Others report that shared-parenting arrangements often revert back to more traditional divisions of labor if things don't work out at first (Radin and Goldsmith 1983).

Reporting on an intensive study of a small number of shared-parenting families, Ehrensaft (1987) found that mothers retained major responsibility for managing child care and that fathers typically remained in a "helper" role. Even in shared-parenting families, women tend to be responsible for things like buying children's clothes, planning music lessons, and arranging for baby-sitters. Ehrensaft also suggests that there was a "duel of intimacy" in these families. She found that the father's relationship to the children was more threatening to the woman than anticipated, primarily because it was easier for fathers to be intimate with their children than with their wives. Along the same lines, some studies report that when men are more involved with routine child rearing, women are less satisfied with the child-care arrangements (Pleck 1983; Baruch and Barnett 1981). What seems to happen is that issues that are hidden when mothers do all of the child care are suddenly thrust to the foreground when fathers begin to participate. What was taken for granted in the past now has to be negotiated as both parents go about making daily decisions with and for the children.

Mothers' ambivalence about the father's participation in child care is probably also the result of having to give up authority in the one area of life that women have traditionally controlled. For instance, mothers sometimes report that encouraging fathers to participate fully in family work entails accepting his "looser" standards (Allen and Hawkins 1999; Coltrane 1996). This can create problems for the mother, who is still judged negatively by others if the house is a mess or if the kids wear something weird. Fathers, too, are sometimes discouraged from spending too much time with their children by comments from their male friends and co-workers. The presence of female relatives in the neighborhood, especially mothers, also plays a role in discouraging parents from experimenting with new forms of shared child care (Riley 1990). Because the 1950s ideal of exclusive maternal child care is still with us, mothers often feel guilty for not spending more time with their children, and fathers feel that they are intruding in the mother's domain. As is true with many forms of social change, actual behaviors often precede shifts in ideology.

GROWING UP

The Struggle over Love and Control

Parent-child interaction is not one-sided. It is not merely a matter, as one might think from reading some guidebooks for parents, of the adults doing certain things to make their children behave and develop in certain ways. Both the parents and the children have wishes and aims of their own, which often may clash or run at cross-purposes. A child, no matter how small, is an active agent. Hence, there tends to be a two-way struggle for control: the child is trying to control the parent, while the parent is trying to control the child.

Parents have certain advantages in this struggle: they are bigger and stronger, and hence they can often physically move the child around and make him or her do what they want. Moreover, they were there first; they have set up their local world the way they want it (or at least they have made their own adjustment to the world around them), and the children face a preexisting situation that they must fit into. At least at first, parents have the tremendous hidden power of being able to define reality for their children: to give them their world view, to explain "the way things are," and hence to shape their behavior by shaping their beliefs.

What advantages do children have? Their main advantage is simply their own attractiveness. There is a reason why small children are usually cute and cuddly; a special bond can develop between the parent and small child that makes the latter seem especially attractive, even if outsiders don't think so. This means that the child's main resource is the parents' love. Much of the analysis of parent-child bonds has concentrated on the child's side of the relationship—whether the caretaking is adequate to make the child bonded to the mother or other caretaker. However, the other direction is probably even more important—whether the parent becomes bonded *to the child* enough so that the parent will have an emotional need to take care of the child.

Parental bonding does not necessarily happen automatically. It is especially likely in our society (for reasons we will examine shortly); but even here, it is all too often that the parent is not very strongly bonded to the child, with the resulting potential for child neglect or abuse if a sense of duty does not prevail.

Freud believed that human relationships are derived from certain basic drives, especially the erotic one (libido). We need not take this literally, but there does seem to be an important element of truth in the general conception. The parent's bond to the child is a form of love, not unlike adult sexual attachments; similarly, the child's attachment to the parents and desire for parental love are analogous to later sexual demands. There is evidence that women become sexually more responsive after they have borne a child. Nipples are erogenous zones, as well as dispensers of milk for breast feeding; hence there is a kind of overlap or mixture of maternal and sexual behavior in the fact that a mother's nipples become erect and she experiences the desire to lactate upon hearing cries of her newborn infant (Rossi 1984). Also, women nursing a baby often feel uterine contractions similar to those resulting from orgasm; these help shrink the uterus back to normal size after childbirth.

The child's major resource, then, is to arouse a feeling of love in the parent (especially the mother, although there appear to be analogous processes in fathers). A good deal of children's cries and behaviors are methods of getting the parents to focus on them, to give them attention. Many of the little struggles that go on between parents and children are of this sort. For example, parents want to talk with visiting friends, while their children run around more and more excitedly making noise. "Why do you have to behave like this just when we have company?" a parent may say in exasperation. But that is just the point: the children act like this at exactly this time because the guests usurp their parents' attention. Moreover, it is one of the characteristics of the "primitive" desire for love that *any* form of attention—even negative—provides some satisfaction of the desire. Children will run around and misbehave if that is the only way they can get attention, even if the attention consists of angry commands or even punishments.

PARENTS' CONTROL TECHNIQUES Even though a family's resources—time, energy, and emotional investment, as well as money and possessions—do have limits, they can be squandered or multiplied, depending on how they are directed. For example, if parents and children are continually at loggerheads, both sides use up energy with little to show for it. For parents to help children progress through their life stages, resources of time and energy are required, especially in the short run. Over the long term, however, conflict will not consume so many resources, and more will be available for everyone's enjoyment and further growth. This is perhaps the real meaning of the aphorism, "Nothing succeeds like success."

As parents seek to guide their children while also preserving their own well-being, various techniques may be employed. Which are chosen depends on many factors—what part of the world they live in, their social class, their personal family customs, the resources available to them, and what they may have learned in their efforts to become proficient parents. Common methods may be categorized as reward, punishment, shame, and love.

Reward. This is one of the most common forms of control, although parents are often not aware of when they are rewarding children's behavior. There are various kinds of **rewards.** *Material* rewards may consist of giving children candy, money, or toys as an incentive for doing what the parent wants them to do: a dollar for a good report card, a cookie if you clean your room, and so on. One drawback of this method is that the child comes to expect a reward for every accomplishment and will not perform without one. In addition, children will focus on the reward rather than the action. For instance, they will see no intrinsic value in reading a book but only do it in a perfunctory way to get the reward.

Another type of reward is *social:* the parent rewards the child with attention, such as play or talk. Here again one has to consider that the child will become more attached to the reward than to the behavior and demand sociability in return for performance. One might see this as a desirable outcome: the child will like the parent and become sociably oriented.

Control by material rewards does require that the parents have enough wealth. Hence, we would expect this technique to be used more in wealthier societies and higher social classes. The social rewards of paying attention to children and devoting time to playing with them are in a sense even more costly because they require the parent to spend time and energy. These kinds of rewards have rather good outcomes for the children's behavior, as far as parents are concerned; however, not all parents are able to use such techniques because they simply lack the leisure time. Hence, it is not surprising that social rewards are used most by the affluent middle classes in wealthy societies with plenty of leisure time, such as our own.

Punishment. This type of control consists of either physically spanking or hitting the child or depriving him or her of something desired. Punishment can consist of threats and angry tones of voice, as well as overt actions. The psychologist B. F. Skinner (1969), who experimented largely with animals, argued that punishment is not a very effective means of control. When punished, any creature's first reaction is usually to fight back; if that is impossible because of the opponent's superior strength, then to run away; and finally to comply, but dully and unenthusiastically. If an extreme amount of punishment is used, the subject is too beaten up to be able to comply.

Nevertheless, physical punishment is fairly popular in our society and sometimes is even escalated to the extremes of causing bodily, and emotional, harm. It is generally used more against boys than against girls. The results tend to be:

■ Boys who fight back against their parents and who are aggressive toward outsiders

■ Boys who strongly identify with their fathers and acquire very masculine or "macho" personalities, with authoritarian and ethnocentric (bigoted) attitudes

■ Beliefs about right and wrong that are based not on internalized moral standards or conscience, but simply on fear of punishment if one is caught

Despite all these drawbacks, physical punishment is a common method of discipline. It may often be used, for instance, in hard-pressed working-class families or in rural cultures because it is a relatively cheap form of control, and these people may have no time to spend on more effective, psychological methods.

Shame. Shaming or ridicule is a kind of control in which the child is held up as a negative example to the group. It is widely used in a number of tribal societies, which anthropologists refer to as "shame cultures" (as compared with our own society, which is more typically called a "guilt culture"). However, it is used in some families in our own society and elsewhere in the modern world. The anthropologist Lawrence Wylie (1964) gives a classic description of how this approach was used for school discipline in a small town in rural France: the child who broke some rule was paraded through the town square wearing a placard around his neck, on which was written the offense ("I threw erasers in the classroom," or whatever), while everyone in the school and the town was lined up to watch. Many Americans would consider this a mortifying experience because our culture has fairly strong feelings against singling out persons for public embarrassment.

Shaming is a form of social punishment that is hard to counterattack or escape. Hence, it does not have some of the negative consequences of sheer physical punishment. It tends to produce personalities who strongly emphasize self-control, especially over public demeanor. They typically become very careful of how they express emotions; not that they are necessarily emotionless, but their behavior is calculated to conform to what is expected in a given situation. In other words, shaming leads to a personality type that is most strongly concerned with meeting group expectations. However, this is external conformity, not an internalized sense of right and wrong; when the group's demands change, people of this personality type rapidly change their behavior in response.

Control by shaming happens most often in societies in which people live in dense settlements with little privacy. It does not work very well in modern urban societies, which do not provide much surveillance over the individual, and in such societies it is not often used.

Love. Control by love is mostly discussed in the child-development literature as a form of manipulation of the parent's love, as a reward for compliant behavior and a punishment for disobedience. The term **love-deprivation** is used to refer to such commands as "All right, if you won't behave, I won't love you any more." These kinds of extreme and overt instances are especially common in the psychiatric and psychoanalytic literature, including such variants as "You don't love your mother. If you did love her, you wouldn't have been so bad as to do what you did." Even without saying these kinds of things (which indeed many parents do say), it is possible to convey to children that the rewards they receive from their parents are highly contingent on how they behave.

Such control by deprivation produces a child who has strongly internalized the parent's point of view. Deprivation of love poses a devastating threat to a small child, and hence there is little he or she can do but try to comply. Often this is difficult, because of the demanding personalities of the parents who use this kind of control (since they themselves were usually brought up this way). The resulting personality tends to feel that moral standards are absolute obligations, regardless of the consequences in the external world. Children brought up in this way tend to be emotionally inhibited and sexually repressed, though their sexual attitudes may come through in a highly romanticized and unrealistic view of their possible lovers. Often this is combined with strong self-discipline and striving for achievement. Love-deprivation is the only technique that produces strong feelings of guilt for breaking some rule, even if there is no chance of getting caught. In short, this technique produces Freud's classical strong superego and many of the classical Freudian neuroses.

Control by threatening to withdraw love requires that the parent spend a great deal of time and emotional energy on each child. It seems to occur most often in small modern families in which the mother is a full-time housewife with continuous contact with her children. If this arrangement produces a highly moralistic personality in the child, it is not necessarily because we have a guilt culture. Rather, the family pattern that produces this type of personality seems to be tied to the structural situation in which middle-class women derive all their status from their family position as wives and mothers. In such a situation, a great deal of psychological intensity can go into child rearing. If this method seems to be on the decline, supplanted in recent middle-class families by more emphasis on social rewards, the cause may be in the shift that allows women to derive more status from roles outside the family, especially careers.

There are, however, other ways to love one's children. The social rewards discussed previously are a use of love as a selective reward for performing certain behaviors, just as love-deprivation uses love as a punishment. As Skinner's principles would predict, the reward method generally has more positive consequences than the punishment method.

Best of all, many psychologists argue, is simply to *love one's children unconditionally.* The parents who spontaneously show affection for their children during the normal course of the day thereby help maintain a bond with them. This happens apart from whether the child is doing anything good or bad at the moment. It helps maintain a good fundamental relationship, whichever specific control techniques the parent uses in regard to the child's specific behavior (Rohner 1986).

It should be borne in mind that parents are not the only influence on how children develop. Parents may use one or another type of control or a mixture of them; nevertheless, the outcome will depend also on the child's life away from home. Child-care arrangements and, later, schools add external influences. Playmates and other peer groups become an important social reference point, especially by the preadolescent period. Perhaps even more important is a factor that has received relatively little attention: how much time the child spends alone. We know, for example, that an adult's degree of personal autonomy and creativity is strongly influenced by the amount of time the person spent in solitary pursuits as a child. Not only the controls used by parents but also how much a child is free of external controls affects personality. There are also important influences on personality that can happen after one grows up. Because social class settings influence one's style of thinking and acting, moving into different occupations will tend to change one's personality. Individuals will change depending on the social reference groups they belong to and the significant others they interact with. Personality is not inalterably set during childhood, and children are not merely passively manipulated by their parents; however, the childhood situation is an important beginning for what individuals do later.

The Social Context of Development

Most models of childhood development assume that socialization is a one-way process: as children pass through various developmental stages, parents socialize them to conform to a known set of cultural standards. This premise, sometimes called the "social molding" perspective (Peterson and Rollins 1987), has been criticized by scholars who have come to realize that children socialize parents too and that what gets learned is rarely what one thinks is being taught. More recent research tries to take into account that socialization is at least bidirectional and attempts to account for influences of social context, as well as direct parent-child effects. Things like the parents' marital relationship, sibling relationships, extended kinship ties, the neighborhood, ethnic identification, social class, schools, churches, peer groups, and television are all strong socializing influences on children.

One of the biggest differences between children of today and those of the past has to do with the fact that modern American children spend so much time in school or in front of a television set or computer. Schools, for instance, teach children to obey orders, to conform to rules, and to be punctual, along with trying to instill the three Rs. With compulsory education, more children are subjected to similar socialization pressures, which may lead to less regional diversity. Although children spend much of their time in school, the average American child spends a greater amount of time in front of the television. The mass media exposes children to styles, values, and stereotypes that may not be important in any particular family and again tends to standardize all children's experience. Commercial television also promotes consumerism and teaches children to be passively entertained. All of these influences—some good and some bad—shape how children look at the world and feel about themselves. To understand how children develop, we must take these varied influences into account. As our models of child development become more complicated, we begin to get a better sense of what's important, but we also realize that there are no easy answers to the question of what children need.

Psychologist Jerome Kagan (1976, 1994) notes that we carry three prejudices regarding the development of the child. The first is that the child is seriously

A CLOSER LOOK 12.6
Some Nondestructive Ways of Controlling Children

No matter how permissive parents are, there are always times when they have to say no to their children. When kids are small, parents have to prevent them from doing things that are unsafe. When they are larger, parents have to be peacemakers, to check their aggression against each other or even against adults. There are also head-on-collisions between what the adults and children want to do. Typical in almost all families are the times children simply won't go to bed, though the parents don't want to be interrupted or the children need to sleep so that they can get up early in the morning. How can this be handled?

The problem with parents' usual methods to get children to stay in bed is that the methods themselves reinforce the behavior that keeps the children awake. They stay awake by demanding attention, and they think of excuses to get it. The request for a glass of water or to go to the bathroom or any of a thousand other "problems" all are success-

ful if they get the parent to come and pay attention. Getting mad does not help, nor does making threats, offering enticements, or anything else the parent says, because the fact that the parent is talking to the child and providing attention is what the child wants. One method that works is to (1) attend to the child's physical wants before he goes to bed (bathroom, glass of water); (2) tell the child that you are not going to talk to him anymore until morning and the child shouldn't talk either; (3) be firm and stick to it. Every time the child comes in and wants something, the parent should gently, firmly, and *silently* take the child back to bed. This may have to be done repeatedly at first, but by the second or third night, the child will probably go right to sleep.

Another method called **time out** can be used effectively in situations where punishment is necessary. For instance, when a child is attacking other people or destroying things (children do tear things up from time to time)

or throwing a tantrum, parents need to intervene. However, punishment often does not work, and of course, it has many bad side effects. Yelling angrily at the child is even worse because it creates the hostility and fear that are negative results of punishment and often does not even stop the behavior. The parent who is doing something else and half attentively keeps telling a child "Don't do that!" is actually teaching the child that the words don't mean anything. The child may even keep on performing the annoying act just to provoke the parent into repeating the phrase. Attempting to shame the child or make him or her feel guilty also has negative effects on personality.

"Time out" avoids these drawbacks. It consists of putting the child *alone* in a room. Often this can be a bathroom or bedroom. The child is told something like: 'I'm putting you on two minutes of time out. Please remember while you're in there that you're not supposed to hit

influenced by the actions of others. The second is that children develop in a series of discrete stages that must be mastered in sequential order. The third and most important is that there is an identifiable set of psychological traits that are necessary for a child to develop into a happy, well-adjusted adult. Contrary to this set of assumptions, Kagan says that children do not require any specific actions from adults to develop optimally. Why is this so? First of all, assumptions about what is appropriate or healthy behavior for children and adults can vary tremendously from one culture to the next. To specify what children need, one must first know what the specific demands of the community are. For instance, middle-class Americans tend to value academic success, autonomy, and independence and to have permissive attitudes toward hostility and sexuality. These traits reflect an individualistic ego-ideal that tends to favor self-interest over intimacy, competitiveness over cooperation, and narcissism over altruism. When we compare **child-rearing patterns** in other cultures with our own, we see that these are not traits that all parents want to instill in their offspring. In an effort to identify what all children need, Kagan (1976, 1994) specifies the following general psychological requirements:

One of the most effective ways of controlling children's negative behavior is to remove them from the troubling situation.

your brother with a baseball bat. You can come out when the time is up." For small children, it is useful to use a kitchen timer with a bell, so that they know when their "time out" is over.

Why does the method work? Because it removes the child from the situation in which the misbehavior occurred and provides an opportunity for the child to calm down. It breaks the social situation in which the trouble arose, and it does not substitute a new troubling situation in the form of a fight with a parent, which is what happens in conventional punishment. Two minutes (or even one minute for small children below age five) does not seem very long, but it is effective for children, as their emotions tend to be very volatile and can change completely within a short period. What happens, one may ask, if the child insists on coming out of the bathroom and resuming the forbidden behavior? The parent simply puts the child, gently but firmly, back in "time out" and adds a minute to the time. As in the going-to-sleep-peacefully method, the first time or two this approach is used, the parent should be prepared to devote some time to following the routine repeatedly. Within a few days, a child's negative behavior can be very successfully controlled (Patterson 1976; Hetherington and Parke 1999).

■ Infants: assimilable environmental variety; regularity of experience; predictable human caretaking; and a chance to practice developing skills

■ Preschoolers: exposure to language (talk); affirmation of self-worth; models to identify with; and consistent standards

■ School age: mastery of school's requirements; success in some peer-valued activity, and models to identify with

The specific actions that parents take with their children may be less important than most of our theories suggest. Contrary to the volumes of material written on the "correct" way to raise children, what children need depends most on their social context. In today's world, that social context is changing faster than ever before.

Social-Class Differences in Child-Rearing Styles

Parents' social-class position strongly affects the way they bring up their children. There is a good deal of evidence (Kohn 1977; Kohn and Schooler 1978,

1983; Kohn and Slomczynski 1990) that middle-class and working-class parents hold different ideals of children's behavior. Middle-class parents, both mothers and fathers, tend to want their children to grow up happy, curious, interested in the world, and (especially girls) considerate of others. Working-class parents, on the other hand, want their children to be obedient (especially boys), neat, and clean. Interestingly enough, the working-class parents stress more than middle-class parents that their children should get good grades in school (even though, in fact, middle-class children tend to get the higher grades). Similar class differences have been found in Canada, Europe, and Japan (Lambert, Hamers, and Frasure-Smith 1980); in all of these places the working-class parents were more likely to censure insolence and temper and to insist on good manners while restricting children's autonomy and comfort.

What is going on here? Melvin Kohn's interpretation is that the working-class parents are stressing behaviors that are vital to at least maintaining the status they have. They want their kids to be obedient, neat, polite, and so forth because failing to meet these standards might lead to trouble with the law and living like "low-class," "not-respectable" people. Their position on the social ladder is often tenuous enough that these feel like real threats. Success in school and a good appearance, on the other hand, are perceived as steps to upward social mobility.

Working-class parents also stress obedience and punish insolence because polite, obedient, conforming behavior is seen as necessary to job survival, at least in their own lines of work. Middle-class parents, in contrast, tend to have jobs where initiative and self-direction lead to success. This may be the reason why they are less apt than working-class parents to stress conformity and obedience in their child rearing and more apt to teach independent decision making. Kohn tested this theory by dividing up all occupations according to whether they involved *close supervision or autonomy,* work of *high or low complexity,* and work of *high or low routine.* In fact, the middle-class occupations tended to have less routine and more autonomous, more complex work calling for personal decisions; the working-class jobs were the opposite. Moreover, working-class people whose positions did involve more autonomy and complexity and less routine had goals for their children rather similar to those of the middle class. (Conversely, middle-class jobs that actually had little freedom and challenge produced a working-class style of values.) Kohn also found that rural occupations tended to be more like working-class occupations; hence, rural child-rearing styles also tended toward the authoritarian, conformity-emphasizing style. These differences in child-rearing styles tend to perpetuate class divisions to the extent that they prepare kids to function best in the class to which they were born.

Instilling Gender Differences in Children

The process of childhood gender socialization often begins before babies are born. After receiving the results of sonograms or amniocenteses, the parents-to-be typically announce their new baby to the world using gender-coded language. Most adults go to great lengths to make male and female infants appear different, although they are, in fact, very similar to one another (Coltrane and Adams 1997). As soon as the infant's sex is known, parents and other adults begin gender differential treatment toward the child. Color-coded blankets and identification bracelets are generally provided by the newborn nursery, with pink identifying girls and blue distinguishing boys (though in the early part of the century, girls tended to be dressed in blue, and boys in pink or red; Kimmel 1996). Today, gifts are selected for newborns depending on their sex; girls generally receive pastel outfits, often with ruffles, and boys are given tiny jeans and bright, bold-colored outfits (Fagot and Leinbach 1993). So that other people can readily identify their infant's sex-class, parents habitually dress them in sex-appropriate clothes, as well as style their hair in stereotyped ways, going to such lengths as taping bows to the heads of female infants (Shakin, Shakin, and Sternglanz 1985). Event the bedrooms of infants are decorated and arranged based on gender stereotypes, with girls' rooms typically painted pink and populated with dolls and boys' rooms painted blue, red, or white and containing an abundance of vehicles and sports gear (Pomerleau et al. 1990).

Research shows us that gender labeling is a ubiquitous process, even though it is largely unnecessary. In one type of labeling study, people are exposed to a baby and then asked questions about the baby's personality traits or behaviors. Dressed in gender-neutral clothes, the baby is labeled "male" for some people and "female' for others. Other than the sex of the babies, people are typically given very little, if any, information about the infants. Because the baby's sex is always the same, these studies can effectively isolate the impact of calling the baby a boy or girl. Roughly two dozen such studies have been conducted, and although their specific results vary, in general they show that the actual sex of the babies makes little difference because people rely on gender stereotypes to rate the infants. Interestingly, when the people doing the rating are children, this becomes especially true (Cowan and Hoffman 1986; Stern and Karraker 1989). For example, in studies using child raters, boy babies were typically seen as bigger, stronger, and noisier; often as faster, meaner, and harder; and sometimes as angrier and smarter than girl babies. In several studies looking at how adults interacted with infants, babies labeled "girls" were given more verbalization, interpersonal stimulation, and nurturance play. Conversely, more encouragement of activity and more whole-body stimulation were given to those labeled "boys." This pattern has also been noted for parents (particularly fathers) with their own children (Fagot and Leinbach 1993; Stern and Karraker 1989).

Parent's attitudes and behaviors can have a substantial impact on the gender development of infants and toddlers, who develop gender schemata and gender scripts based on what they are exposed to in their immediate environment. Infants are actively engaged in processing information from their earliest days, and they are exposed to gender-relevant messages from their birth. By the time they are seven months old, infants can discriminate between men's and women's voices and generalize this to strangers. Infants under one year can also differentiate individual male and female faces. Even before they are verbal, although they have not yet developed gender schemata, young children are developing gender categories and making generalizations about people and objects in their environments (Fagot and Leinbach 1993).

Although they cannot always link gender to anatomical sex, preschool children between the ages of two and four are usually able to perceive gender labels for themselves and other children (Fagot and Leinbach 1993). Roughly 80 percent of American children can distinguish males from females on the basis of social cues like hairstyle and clothing by the time they are two years old, but only half of three-and four-year-olds can distinguish males from females if all they have to go on are biologically natural cues like genitalia and body physique (Bem 1989, 1993, 114). In other words, preschool children in the United States learn that the cultural accoutrements of gender are more significant than the underlying physical differences between boys and girls. In addition, children quickly incorporate new information into their developing gender schemata. Before they are five years old, American children have learned to allocate bears, fire, and something rough to boys and men, whereas they connect butterflies, hearts, and flowers with girls and women (Leinbach and Hort 1989). Although they are not directly taught to relate bears and men, by this age, children are able to categorize using a gender schema that connects qualities such as strength or dangerousness with males. In the same way, flowers and butterflies become associated with being female through a metaphorical cognitive process that identifies women with gentleness (Fagot and Leinbach 1993, 220).

Gender associations are constructed, in part, by providing unique toys to boys and girls and by creating distinct play environments for them. In general, parents give dolls to girl babies and footballs or hammers to baby boys, even when they are too young to play with them. During the preschool and kindergarten years, parents give toys that are increasingly gender stereotyped. Because different kinds of toys and furnishings promote different activities, they tend, in turn, to reinforce rigid gender schemata and scripts. "Masculine" toys such as trucks and balls foster independent or competitive activities, necessitating little verbal interaction, whereas "feminine" toys like dolls favor quiet, nurturing interaction with another playmate, encouraging physical closeness and verbal communication (Wood 1994). Not only do parents provide gender-stereotyped environments, but they also interact with preschool and school-aged children in sex-differential ways, often rewarding gender-typical play and punishing gender-atypical play. In this way, boys are deterred from playing house, and girls are dissuaded from engaging in vigorous competitive sports or games. In addition, essentially all studies looking at preschool and school-aged children find that parents engage boys in physical play more often than they do girls (Lytton and Romney 1991; Maccoby and Jacklin 1974).

Another area in which parents exhibit sex-differentiated treatment of their children is household tasks, for which studies show that parents consistently assign boys and girls gender-segregated chores. Household responsibilities such as cooking and cleaning are usually allotted to girls, whereas more active duties like lawn mowing are typically assigned to boys (Goodnow 1988; McHale et al. 1990). As with toys, performing different chores promotes specific perspectives on experiencing and understanding the world. Girls' chores usually take place inside the home, emphasizing nurturing activities and taking care of other people, whereas boys' chores usually take place outside the home, emphasizing the maintenance of things.

Research into gender socialization processes has shown that because boys and girls are treated differently and put into different learning environments, they develop different needs, wants, desires, skills, and temperaments. In short, boys and girls evolve into different kinds of adults—men and women—who barely question how or why they end up with such dissimilar attitudes and tendencies. Although the causal process is presently the subject of debate, the basic underlying model reflects the operation of a self-fulfilling prophecy (Bem 1993; Merton 1948; Rosenthal and Jacobsen 1968). Assuming that boys and girls are supposed to be different, people treat them differently and subsequently provide them with different developmental opportunities. This differential treatment promotes certain self-concepts and behaviors that then tend to recreate the preconceived cultural stereotypes about gender. Thus, a kind of social illusion is created, because as the process repeats itself across generations, gender stereotypes come to be seen as natural and impervious to change, even though such gender stereotypes are constantly recreated and modified in interaction.

Fathers tend to favor sons, engaging them in physical play more often than they do daughters, thus contributing to boys' socialization as unemotional, individualistic, and competitive.

Although the self-fulfilling prophecy concept is actually somewhat more complicated than this suggests, the basic idea is that if we treat boys and girls differently, they will develop in dissimilar ways. From a social constructionist framework, we can see that it is not just that boys and girls are fundamentally and unalterably different. Instead, they must learn how to fit in as appropriately gendered individuals to be considered competent members of society. This is a mandatory process, because of the importance of gender to the adults in our society, children are called on to conform to the gender standards currently in vogue. A large part of children's identity development involves forming a gender identity, and they must work very hard to make their actions and thoughts conform to the expectations of the people around them. In the process of interacting with adults and their peers, children literally "claim" a gender identity (Cahill 1986). By applying the concept of "doing gender" not only to adults but to children as well, we can see that gender is not something innate but rather something that is recreated on a continual basis (West and Zimmerman 1987; West and Fenstermaker 1993).

Intergenerational Transmission of Gender Stereotypes

In general, fathers enforce gender stereotypes more than mothers, especially in sons. This tendency extends across types of activities, including toy preferences,

A CLOSER LOOK 12.7
IQ: How Much Is Inherited? How Much Difference Does It Make?

The nature-versus-nurture debate has been going on for a long time in the field of IQ. By now, it is generally agreed that there is some biological, hereditary component in IQ. But how great is it, and how much is learned in response to social conditions?

The main research methods for answering this question attempt to compare the IQs of people who have varying degrees of kinship; *or* they measure changes in IQ over time and in different social circumstances. The classic kind of study is to compare identical twins who were reared apart. These have exactly the same genes (because identical twins are produced by the splitting of a single fertilized egg, whereas nonidentical twins are produced by separately fertilizing two different eggs). If they have been reared apart in different social environments, that should control the "nurture" effect, so whatever correlation there is between their IQs should be what is due to genetics.

As we can see from Table 12.1, studies of identical twins reared apart have found correlations between their IQs ranging from .60 to .90. These are the highest correlations in the table except for identical twins who are reared *together.* For many years, such figures were cited as

evidence of the importance of heredity. However, the sociologist Howard Taylor (1980) pointed out that, just because the twins were reared in different households, their environments were not necessarily different in important social ways. The psychologists who did these studies ignored patterns of social class or other features of the families and communities in which these twins were brought up. Taylor went back and looked at the primary data on these studies and separated them by degrees of real social difference. Many twins were actually brought up by foster parents whose occupational and educational levels were the same. The twins were usually separated because of some kind of family crisis, and often relatives adopted them. In many cases, the twins remained in the same town and went to the same school. Moreover, some twins were separated from each other at a relatively advanced age, which gave them a number of years in exactly the same environment.

Taylor's reanalysis had strikingly different results. Those twins who were reared in similar environments had IQ correlations ranging from .66 to .89, with an average of .87. Those who were reared in socially dif-

ferent environments had IQ correlations ranging from .36 to .51, with an average of .43—not particularly striking results, compared with many other figures in the table. Taylor's reanalysis thus showed that *the separated twins studies actually demonstrated greater importance of environment as compared with genetics.*

There is plenty of evidence for the same conclusion. For example (see table 12.1), just ordinary siblings who are reared together have higher IQ correlations than those who are reared apart. Nonidentical twins of the same sex have higher IQ correlations than those of different sex, almost certainly because two girls or two boys are treated more similarly than are a boy and girl. Nonidentical twins have higher IQ correlations than nontwin siblings, even though they are no more closely related genetically, again, because the twins are exposed to more common treatment.

How much of IQ, then, is hereditary? The statistical rule of thumb is that a correlation should be multiplied by itself to arrive at the percentage of variance that is explained; thus, taking the average r of .43 for identical twins who lived in genuinely different environments, we get an r^2 of .18. In other

---

play styles, chores, discipline, interaction, and personality assessments (Caldera, Huston, and O'Brien 1989; Fagot and Leinbach 1993; Lytton and Romney 1991). Despite the fact that both boys and girls receive gender messages from their parents, boys are, nevertheless, encouraged to conform to culturally valued masculine ideals more than girls are encouraged to conform to lower-status feminine ideals. Boys also receive more rewards for gender conformity (Wood 1994). Because society places greater emphasis on men's gender identity than on women's, there is a tendency for more attention to be paid to boys, reflecting an androcentric cultural bias that values masculine traits over feminine (Bem 1993;

TABLE 12.1 IQ and Kinship

Degree of Kinship	Correlation
Unrelated persons	
Reared apart	.0
Reared together	.18–.30
Foster parent/foster child	.20–.40
Parent/child	.20–.80
Siblings	
Reared together	.35–.75
Reared apart	.35–.45
Twins	
Nonidentical, opposite sex	.30–.75
Nonidentical, same sex	.40–.90
Identical, reared apart	.60–.90
Identical, reared together	.70–.95

Source: Taylor, 1980, 28–30.

words, something like one-fifth of all differences in IQ seem to be due to heredity.

What Difference Does It Make?

IQ scores thus seem to be more strongly influenced by the family environment and the way the children are treated than by heredity. But whatever its source, how much does IQ affect one's life? IQ is quite strongly related to school achievement and thus has an effect on one's placement in a career. Nevertheless:

■ IQ operates mainly within the school system and has little effect in predicting subsequent achievement once one has left school (Bajema 1968; Eckland 1979).

■ An individual's IQ score is at its peak during the teen years and declines as one grows older. Hence, children's IQs are much closer to their parents' IQ (average .69 correlation) than adults are to their own parents (average correlation .30) (Taylor 1980).

■ There is a wide range of IQs within the same occupation (Thomas 1956). IQ is not a *sine qua non* for doing any particular job. Even retarded persons having tested IQs of 60 (below the "normal" 100) may turn out later in life to do relatively well in jobs, especially if they have more middle-class patterns of dress and behavior (Baller, Charles, and Miller 1967).

The IQ test thus seems to be a ritual created by our bureaucratic school system, which measures and guides students in placement through the school. It is important in our society only because the educational credentials for procuring jobs have been driven up as a mass, and the competitive school system has expanded (see Collins 1979 for an analysis of this process). However, IQ is not intrinsically related to how well one can do in an actual career.

How Much Can IQ Scores Be Raised?

Major effects on IQ seem to be produced by the home. There have been some efforts to raise the IQ scores of poverty-level children by training them in preschool programs. These were able to raise IQs by seven points, although this difference tended to go away by the time the child had been in the regular school program through the sixth grade (Consortium for Longitudinal Studies 1983). The most notable long-term change took place in Japan after World War II. Japanese born during the 1940s have an average IQ of 104, whereas those born in 1960 have an IQ of 115—far above the U.S. average of 100 (Lynn 1982). The rapid change shows the variability of IQ scores, in this case probably responding to better nutrition and other environmental factors (such as stress on achievement at school).

Broverman et al. 1970; Lorber 1994). Masculine gender identity is also considered more fragile than feminine gender identity and takes more psychic effort because it requires suppressing human feelings of vulnerability and denying emotional connection (Chodorow 1978). Boys, therefore, are given less gender latitude than girls, and fathers are more intent than mothers on making sure that their sons do not become sissies. Later, as a result, these boys-turned-to-men will tend to spend considerable amounts of time and energy maintaining gender boundaries and denigrating women and gays (Connell 1995; Kimmel and Messner 1994).

Importantly, differences in the ways that parents act toward boys and girls perpetuate separate spheres for men and women. Parents promote nurturing behaviors for girls and autonomy for boys by creating different social environments and holding different expectations for them. In this way, parents promote the formation of gender schemata, gendered personalities, and taken-for-granted gender scripts that make gender differences seem natural and inescapable (Crouter, McHale, and Bartko 1993; Thompson and Walker 1989). Most girls, for example, grow up with an interest in babies. They are also more responsive to infants and take more responsibility for them than do boys (Ullian 1984). Experiencing more contact with young children, young girls also have more opportunities to develop nurturing capacities, and by the time they reach childbearing age, they are predisposed to want to bear children and to take primary responsibility for their care (Bem 1993; Chodorow 1978). Boys, on the other hand, are likely to be unemotional, competitive, and individualistic, developing into young adult men who are relatively uninterested in babies and unprepared to care for the emotional needs of others. By raising children to have the mental, psychological, and social faculties appropriate to children of their specific gender, parents prepare the next generation of boys and girls to occupy unequal positions in a hierarchy of gender relations (Coltrane and Adams 1997).

Placing Children in the Social Mobility Game

The development of gender-linked mental, psychological, and social faculties is one factor that shapes a person's adult life. Another factor is the social level of the family into which one is born. Children usually end up relatively near their parents' occupational level. This is not to say that they usually have identical positions (though there is a tendency, for instance, for doctors to be sons of doctors, and professors the sons of professors—both professions with a historically strong gender bias). The further away an occupation is on the prestige scale relative to parents' occupations, the less likely a child is to attain it. This is no surprise, even though a fairly small number of people today directly inherit an occupation (such as a family business). The family's main effect on social mobility (or, more likely, social immobility) is its influence on how well children do in school. This is ironic because the school system, at least officially, claims to be a "meritocracy" that operates to overcome the advantages and disadvantages that children acquire from their families. Instead, family influence on school achievement turns out to be one of the most important factors researchers have uncovered so far. (There is also IQ; but see *A Closer Look 12.7.*) Children from different social backgrounds come into school with different **cultural capital** and get tracked into different programs that support their achievements or stigmatize them and lower their levels of aspiration (DiMaggio and Mohr 1985; Rosenbaum 1980). In addition, schools in wealthier areas spend more on each pupil, further increasing the educational gap between social classes.

However, school only partly determines one's occupational achievement. What else is involved? Many factors are out of an individual's control, such as the state of the economy when one enters the workforce (there are more opportunities during economic booms than during depressions, for instance). Another factor is population patterns: there are fewer opportunities for an age group that enters the workforce right behind one that has filled most of the available vacancies. There are also social patterns of bias, such as those that have favored males over females in getting higher-level managerial and professional jobs and that have discriminated in favor of some ethnic and racial groups and against others. The family and the school are only some of the factors that determine how individuals move through their careers. Structural opportunities and lack of opportunities can thus cut across personality patterns acquired in childhood; they can also inflate or deflate the value of educational credentials that one has acquired at a particular time (Collins 1979).

SUMMARY

1.   We tend to think of children as pure, innocent, and deserving of special attention, but this view is a relatively recent historical development. In Western cultures before the seventeenth century, childhood was not distinct from other stages of life, and often children were assumed to be evil. Before the rise of capitalism and industrialization, children were valued primarily for the contributions they could make to the household economy. It was not until the nineteenth and twentieth centuries that children became emotionally "priceless."

2.   Compared with modern American child-care practices, child rearing in other times and in other cultures was much more collective. Many members of the community shared in watching after children, who spent the majority of their waking hours apart from their mothers. By the twentieth century, production had moved out of the home, the nuclear family had come to be seen as a private domain, and "true womanhood' had come to be associated with being a wife and mother. The growth of the suburbs in the 1950s further isolated families and made mothers individually responsible for their children's welfare.

3.   One of the most significant changes in the last half of the twentieth century is that most mothers are now employed. The trend toward increased maternal labor force participation is projected to continue for some time. When mothers are employed outside the home, young children are cared for by fathers, siblings, relatives, sitters, or in some form of group care. Family day-care homes are typically run by working-class women with their own small children. The fastest growing form of child care is day-care centers, but the demand for affordable day care far exceeds its availability. Most day-care workers earn wages below the poverty level.

4.   There is no evidence of disastrous psychological consequences for children who are not continuously attended by their mother during the first three years of life. Regular, predictable care by any caretaker or set of caretakers is all that is necessary for their psychological well-being. Sharing of caretaking of small children among several persons is favorable for both mother and child, and properly organized day-care centers are psychologically supportive and cognitively enriching.

5.   Having young children is associated with higher than average levels of stress and depression. Nevertheless, most of this stress results from lack of money and time, so mothers with sufficient resources and those living with husbands who contribute to housework and child care enjoy better mental health.

6.   Most of the poverty level households in the United States are headed by women. Low-income single-mother families tend to rely on kinship and friendship networks for emotional and material support. Out of necessity, parent-child relationships in single-parent families are often more democratic than those in traditional two-parent families.

7.   Because women have traditionally performed most of the child care, researchers have only recently begun to study fathers. Although most fathers still play a secondary role and interact with children differently than mothers, a growing number of men are sharing responsibility for parenting. When men perform the routine tasks of child care, their actions and feelings resemble those of mothers more than those of traditional fathers.

8.   Parents who control their children by material rewards tend to make the children oriented toward being rewarded rather than concerned with the intrinsic value of the action they did to get rewarded. If social rewards (parent's attention) are used, this results in making the child socially attached to the parent and perhaps to people in general. Control by physical punishment tends to produce children who fight back against parents and others, identify with an authoritarian father, and have a sense of right and wrong based on fear of punishment rather than internalized values. Control by shaming produces external conformity to the current standards of the group and control of one's emotional expressions. Control by threat of deprivation of love produces the "strong superego" personality, with a strongly internalized sense of guilt for trespasses, strong self-discipline, and emotional and sexual inhibition. When parents give unconditional love, children become attached to them personally and to social rewards generally.

9.   Very early on, parents begin socializing their daughters to be feminine and their sons to be masculine (young girls are often dressed in pink, given dolls to play with, and encouraged to be nurturing, whereas young boys are generally dressed in blue, given footballs or trucks to play with, and encouraged to be active and competitive). Although gender development seems natural and taken-for-granted, children learn to conform to culturally specific gender standards and expectations. Parenting practices encourage girls to grow into adult women predisposed to have babies and care for children and boys to grow into adult men who are disinterested in babies and not prepared to deal with others' emotional needs, thereby recreating the gender hierarchy.

10.   Our models of child development assume that parents' actions are extremely important to children's development but often ignore the fact that children socialize parents as well. We also tend to assume that there is a correct formula for raising optimum adults, but this varies with the individual, the family, the culture, and the historical period. In modern American society, children spend more time in front of the television than they do in school or interacting with parents.

11.   Middle-class parents are more likely to value bringing up their children with stress on personal autonomy, happiness, and curiosity; working-class parents stress cleanliness, good manners, obedience, and conformity. These values reflect the occupational structure of the parents' lives, especially the extent to which their work is autonomous or closely supervised, complex and challenging, or simple and routine. Parents thus provide children with cultural capital and reproduce the class structure through social psychological and structural processes.

Key Terms

child-care benefits
child-care centers
child-rearing patterns
control techniques
cultural capital
Family and Medical Leave Act
father-absence studies
flex-time

gatekeepers
gender socialization
good provider
ideal of childhood innocence
individualistic model
  of parenting
love-deprivation
maternal deprivation

maternal labor force
  participation
parents' social-class position
rewards
shared parenting
single-parent households
time out

Sociology Web Site

See the Wadsworth Sociology Resource Center, "Virtual Society," for additional
links, quizzes, and learning tools:

http://www.sociology.wadsworth.com

Also on this web site you'll find InfoTrac College Edition, an online library
of journals. Here you can search for electronic articles about central topics in
sociology.

13 Family Violence: Spouse Abuse, Child Abuse, and Elder Abuse

INTRODUCTION

Because the ideal family provides unique forms of love and support, we tend to think of families as places where people are nurtured and protected. In general, family members do care for each other, provide for each other, and help each other survive in a world that often seems hostile and uncaring. However, families also have a dark side. Families are one of the most common contexts of violence in American society. The likelihood that a man will be assaulted by someone who is a family member is more than twenty times greater than the odds he will be assaulted by someone outside his family. For women the situation is even worse. Women are 200 times more likely to be assaulted by a family member than by an outsider (Straus 1991, 18). Police receive more requests for help with "domestic disturbances" than with any other problem, and more police personnel are killed trying to settle family fights than in any other line-of-duty affairs. What's more, an individual is far more likely to be murdered by a family member than by anyone else (Straus 1991).

Because we like to think of families as protective and caring, it is sometimes difficult to accept the fact that so many people intentionally inflict pain and suffering on other family members. Although it is important to acknowledge that support and gentleness are typical of most families, it is also important to examine violence in families and to begin to understand why it exists. Only then will we be able to work toward reducing violence in American families.

In this chapter, we look at some recent research on three major types of family violence: spouse abuse, child abuse, and elder abuse. Besides investigating the incidence of each of these, we explore some of the myths about family violence and review some of the most common explanations as to why the different types of violence occur. Finally, we look at some attempts to reduce family violence through law enforcement, education, and treatment.

VIOLENCE IN SOCIETY

Before investigating the various forms of family violence, it is useful to turn our attention to the social context in which family violence occurs. Is violence a rarity in our society, or is it widespread? Although we might wish it were otherwise, we must conclude that violence is woven into the very fabric of American society. The United States was founded by revolutionaries who fought a violent war to end what they considered to be injustices perpetrated by the English monarchy. White settlers also waged a protracted and violent war against Native Americans, driving them from their homelands and imprisoning them on reservations. White American slaveholders used violence to enslave and control black Africans against their will. Violence was also a common form of justice in the American "Wild West," and contemporary movies continue to idealize rugged individualists who "take the law into their own hands." From John Wayne westerns to *Rambo*- and *Terminator*-type movies, our popular films tend to feature violent men as heroes.

Not only is violent behavior an integral part of our history and a continuing source of romantic cultural imagery, it is hauntingly present in our everyday lives. Although we fear violence, Americans accept and even condone threatening and abusive behavior that is forbidden in many other cultures. Official crime statistics, which tend to drastically *under*estiamte the amount of violent crime,

Ownership of handguns is more common in the United States than in any other country in the world.

show that an aggravated assault occurs in the United States about once every thirty seconds, a forcible rape every ten minutes, and a murder every twenty minutes. Our murder and rape rates are higher than any other nation that collects such crime statistics. For example, Americans are ten times more likely to die by homicide than the residents of most European countries and Japan (Currie 1985). One reason for our unusually high murder rate is that the ownership of firearms is more common in the United States than in any other country. This is especially so for handguns or pistols, which tend to be used in domestic homicides and school or workplace shootings.

The United States has the highest teen homicide rate in the world (Rollin 1999). In the 1990s, about half of young African American males reported that they had a pistol at home, and about nine in ten said they knew someone who had been shot (Price, Desmond, and Smith 1991). One study of teens in a detention center in 1996 showed that 84 percent of those who owned guns acquired them before they were fifteen. These teens felt safer with a gun and generally bought them, or were given them, on the street, where they were easily available (Ash et al. 1996; Rollin 1999). In national opinion polls, two of three adult Americans say that they can imagine a situation in which they approve of a man punching an adult male stranger. Over half of adults in the United States report that they have been punched or beaten by another person, and almost a quarter report that they have been threatened with a gun or shot at (General Social Survey 1999).

Although Americans are not alone in **condoning violence,** we seem especially concerned that men be tough, competitive, and aggressive. National surveys find that over 70 percent of adults think it is a good idea for boys to get in a few fistfights while growing up. In a never-ending barrage of American "action" movies and television shows, male heroes never let other men push them around. What's more, the hero usually overwhelms a beautiful, scantily clad woman, who feigns reluctance while yielding to his aggressive and persistent sexual advances. This one-sided image of male-dominated sexuality is also common in romance novels, popular magazines, video games, and daytime television. One study found that sexual aggression was the second most frequently shown type of sexual interaction on American television soap operas (Lowry, Love, and Kirby 1981). These stereotypes promote the violent rape of women.

The rape rate is ten times higher in the United States than in England or Japan, four times higher than in Australia, and nearly twice as high as in South Africa (Malamuth and Donnerstein 1984). According to the Federal Bureau of

Investigation (FBI), there were seventy forcible rapes per 100,000 females reported to United States law enforcement agencies in 1997 (Federal Bureau of Investigation 1997). Though this is a high rate relative to other nations, it grossly underestimates the actual number of rapes that occur annually. Data from the National Women's Study, a longitudinal telephone survey of a national household probability sample of women at least 18 years of age, show that almost 700,000 women are forcibly raped each year and that 84 percent of victims do not report the offense to the police (Kilpatrick, Edmunds, and Seymour 1992). In one study of a random sample of women in San Francisco, one in every four women reported that she had been the victim of at least one rape, and almost one of every three reported that someone had attempted to rape her (Russell 1984). A 1987 study of 6,000 college students found that 15 percent of women had been the victims of rape, 12 percent of women had been the victims of attempted rape, and one man of four reported engaging in some type of sexual aggression against women (Koss, Gidycz, and Wisniewski 1987). Women of all ages are raped, but the highest percentage of rape victims are between the ages of fifteen and twenty-five. Although we are just beginning to document the extent of rape, we do know that it most often occurs between people who know each other, rather than between strangers.

The National Crime Victimization Survey indicates that 92 percent of rapes are committed by known assailants. About half of all rapes and sexual assaults against women are committed by friends and acquaintances, and about one-fourth are committed by intimate partners (Bachman and Saltzman 1995). Myths that women provoke rape or somehow "ask for it" are still widespread, despite evidence to the contrary (*A Closer Look* 13.1).

Nonsexual violence is even more likely to be pictured in the media than sexual aggression. Prime-time television averages over thirteen acts of violence per hour, with even higher rates of occurrence in cartoons. In countless portrayals, frustrated people hit, shoot, and bomb others. By the age of sixteen, the average American child has witnessed a half million violent acts and over 50,000 murders on television (Signorelli 1991). Viewing so much violence on the screen tends to desensitize us to the real violence that is all around us and models aggressive behavior. After studying the effects of television violence on young viewers, the National Institute of Mental Health (1982) concluded that watching violent programs increases the incidence of violent behavior in children and teenagers. Although violent media images may serve as a catharsis for some viewers, they send a subtle message that it is all right to use physical force to solve one's problems.

Violence is not just perpetuated by the media, however. It is more likely that movies, television, and magazines are just reflecting some idealized image of what is socially acceptable. If we go back in history, we can see how certain forms of violence have been tolerated and condoned by our religions, schools, and governments. For instance, the Bible encourages stern discipline and admonishes parents who don't punish their children: "He who spares the rod hates his son, but he who loves him disciplines him diligently" (Proverbs 13:24). Even recent opinion polls show that three of four Americans approve of disciplining children with "a good, hard spanking" (General Social Survey 1999). The U.S. Supreme Court upheld the rights of school teachers to use corporal punishment, and surveys show that three-quarters of American parents agree with the ruling and want schools to use strict discipline practices (Currie 1985). State governments also tend to legitimate violence by supporting capital punishment for severe crimes, though usually only poor people are actually executed.

A CLOSER LOOK 13.1
Rape Myth Acceptance Scale

The following scale has been used to assess the extent to which people accept cultural myths about rape. How many do you agree with? A higher acceptance of **rape myths** has been found to correlate with sexual aggressiveness, traditional attitudes about gender roles, and low educational and occupational achievement.

1. A woman who goes to the home or apartment of a man on their first date implies that she is willing to have sex.

2. Any female can get raped.

3. One reason that women falsely report a rape is that they frequently have a need to call attention to themselves.

4. Any healthy woman can successfully resist a rapist if she really wants to.

5. When women go around braless or wearing short skirts and tight tops, they are just asking for trouble.

6. In the majority of rapes, the victim is promiscuous or has a bad reputation.

7. If a woman engages in necking or petting and she lets things get out of hand, it is her own fault if her partner forces sex on her.

8. Women who get raped while hitchhiking get what they deserve.

9. A woman who is stuck up and thinks she is too good to talk to guys on the street deserves to be taught a lesson.

10. Many women have an unconscious wish to be raped and may then unconsciously set up a situation in which they are likely to be attacked.

11. If a woman gets drunk at a party and has intercourse with a man she's just met there, she should be considered "fair game" to other males at the party who want to have sex with her too, whether she wants to or not.

12. What percentage of women who report a rape would you say are lying because they are angry and want to get back at the men they accuse?

13. What percentage of reported rapes would you guess were merely invented by women who discovered they were pregnant and wanted to protect their own reputations?

14. A person comes to you and claims he or she was raped. How likely would you be to believe his/her statement if the person were:

   a. Your best friend
   b. An Indian woman
   c. A neighborhood woman
   d. A young boy
   e. A black woman
   f. A white woman

For recent results of research based on college students who took this survey, see Johnson, Kuck, and Schander (1997) and Kopper (1996) articles in the journal *Sex Roles*.

Source: M. R. Burt, "Cultural Myths and Support for Rape," *Journal of Personality and Social Psychology* 38 (1980): 217–30.

INTIMATE PARTNER VIOLENCE

Violence has definitely been institutionalized within the family. As we saw in chapter 8, marriage laws have traditionally given husbands power over their wives. From the days of the Roman Empire, men's right to use physical force against wives has been lawful and even expected. English common law gave husbands this right, which was modified by the nineteenth-century "rule of thumb" allowing a husband to beat his wife with a rod no thicker than his thumb (Davidson 1977). Until the 1870s, wife beating was legal in most of the United States and remained quite common thereafter (Stets 1988). Aided by governmental neglect and protected by the privacy of their homes, men continued to "keep women in their place" with the threat and use of physical force. In the late 1960s, one in four Americans felt it was acceptable for a husband to hit his wife under certain conditions (Stark and McEvoy 1970). Until it was changed in 1977, the training manual for domestic-disturbance calls published by the

International Association of Chiefs of Police essentially recommended that hitting a spouse be treated as a "private matter" and that arrests should be avoided (Straus 1991, 27). It wasn't until the late 1970s that marital violence reached public awareness as a serious social problem that demanded some form of social or governmental action. In the 1990s, the O. J. Simpson trials further focused media attention on domestic violence. Although public consciousness about spouse abuse rose during this time, some misinformation was also spread, including Simpson's repeated references to himself as a battered husband (Jacobson and Gottman 1998).

How widespread is spouse abuse? Because this form of violence so often goes unreported to the police or other authorities, it is difficult to estimate its true incidence. Asking a random sample of husbands how often they beat their wives may yield some general indication of the extent of family violence, but we can safely assume that the answers to such a survey question would understate the extent of actual marital violence. Although wife beating has been condoned in the past, it carries a social stigma today, so most men would be reluctant to admit the extent of violence they inflict on their wives. It is also likely that many wives who have been assaulted by their husbands would tend to minimize such instances in an effort to focus on the more positive aspects of the marriage. When the husband is the victim, the social stigma attached to being battered by a woman may reduce men's reports of severe beatings by their wives. However, husbands who beat their wives might be eager to admit that their wives had also assaulted them because this would serve to justify their own violent actions. These kinds of conflicting social pressures undoubtedly influence people's responses to survey questions about family violence; hence, researchers estimate that the true rates might be twice as high as the rates reported in most studies (Straus and Gelles 1988, 19). Nevertheless, a surprising number of people admit to committing a wide range of violent acts against other family members.

Researchers at the University of New Hampshire have developed a **Conflict Tactics Scale** that measures violence between spouses, between parents and children, and between children. (See *A Closer Look* 13.2 for an example of the form used to measure husband-wife violence.) The Conflict Tactics Scale measures three different aspects of conflictual family interaction: (1) reasoning, (2) verbal aggression, and (3) violence or physical aggression. Investigators have used this scale to measure conflictual family interaction, with relatively consistent results for similar populations sampled in similar ways (Straus and Gelles 1988). The violent items in the most recent version of the Conflict Tactics Scale include the following: threw something at the other; pushed, grabbed, or shoved; slapped or spanked; kicked, bit, or hit with a fist; hit or tried to hit with something; beat up the other; burned or scalded (for children) or choked (for spouses); threatened with knife or gun; and used a knife or gun. As you can see, the items are arranged in order of severity, so if someone has used one of the harsher forms of violence, it is also likely that he or she has used one of the milder forms as well.

Using this scale and other social indicators, researchers have attempted to estimate just how often violence occurs between spouses. At the high end, some estimate that over one-third of all wives will be beaten by their husbands during their lives (Walker 1984). At the low end, Suzanne Steinmetz (1977) reported evidence that 7 percent of all wives and 0.5 percent of all husbands were severely beaten by their spouses at some time during their marriages. Less severe kinds of violence, such as slapping and shoving, tend to be even more common. Evidence from the late 1980s shows a decrease in the reported levels of spouse abuse, but 16 percent of all homes had some kind of spousal violence during the

A CLOSER LOOK 13.2
Measuring Couple Violence: The Conflict Tactics Scale

No matter how well a couple gets along, there are times when they disagree on major decisions, get annoyed about something the other person does, or just have spats or fights because they're in a bad mood or tired or for some other reason. They also use many different ways of trying to settle their differences. I'm going to read a list of some things that you and your (wife/husband/partner) might have done when you had a dispute, and would first like you to tell me for each one how often you did it in the past year and whether you have *ever* had it happen.

- Discussed the issue calmly
- Got information to back up (you/his/her) side of things
- Brought in or tried to bring in someone to help settle things
- Insulted or swore at the other one
- Sulked and/or refused to talk about it

- Stomped out of the room or house (or yard)
- Cried
- Did or said something to spite the other one
- Threatened to hit or throw something at the other one
- Threw or smashed or hit or kicked something
- Threw something at the other one
- Pushed, grabbed, or shoved the other one
- Slapped the other one
- Kicked, bit, or hit with a fist
- Hit or tried to hit with something
- Beat up the other one
- Threatened with a knife or gun
- Used a knife or gun

Source: Murray A. Straus, "Measuring Intrafamily Conflict and Violence," *Journal of Marriage and the Family* 41, 1 (1979): 75–88.

current year, and 28 percent at some time during the entire marriage. Applying these percentages to the approximately 55 million married couples in the United States, one can estimate that about 9 million couples experience at least one assault during a single year. Moreover, at least 15 million couples are estimated to have experienced some form of violence during the entire marriage, and in almost 4 million of these households, the violence has a relatively high risk of causing injury. These forms of domestic violence are labeled "battering" or "wife beating" and are the focus of the greatest concern.

The Myth of Battered Husbands

Using national survey data, researchers have found that in homes with couple violence, about one-fourth of the respondents report that men were victims but not offenders; another fourth allege that women were victims but not offenders; and one-half of the respondents report that both husbands and wives were violent (Straus, Gelles and Steinmetz 1980). Such surveys cannot detect whether violence by wives was retaliatory or not, and the reporting of simple estimates of incidence has touched off heated controversies. What appears to be consistent across studies using the Conflict Tactics Scale is that reported wife-to-husband assault is about as common as husband-to-wife assault. One set of questions about these patterns concerns the reasons for, and the consequence of, these different types of assault.

One important difference between men and women is physical size and strength. If a husband assaults his wife, he is much more likely to inflict pain and injury than if she assaults him. In addition, men tend to be more aggressive, so the same act, such as hitting with a fist, tends to be more violent coming from a man than from a woman. This does not imply that some women aren't bigger, stronger, and more aggressive than some men, but *on average* a husband punching a wife is much more apt to inflict injury than a wife punching a husband. When studies focus on the differences between husbands' and wives' violence, they discover that it is women, not men, who more frequently receive bruises, cuts, broken bones, and internal injuries from the assault. Official statistics show that of those treated in a hospital emergency department for injuries inflicted by an intimate partner, 84 percent are women (Rand 1997). Usually only the woman is injured, but in the relatively rare cases in which both spouses are injured, the wife's injuries tend to be much more severe than the husband's (Berk et al. 1983). In light of the general pattern of greater harm to wives, it is inappropriate to label examples of women hitting men as "husband abuse" or to consider marital violence simply a case of "mutual combat."

To understand the high rate of marital violence by wives, it is useful to compare women's family violence with women's nonfamily violence. Outside of the family, women rarely assault others, and they are victims of assault by outsiders far less often than they are victims within the family. In 1996, 30 percent of all female murders were perpetrated by husbands, ex-husbands, or boyfriends. Three percent of all male murder victims were killed by wives, ex-wives, or girlfriends (Federal Bureau of Investigation 1997). The major reason that women are violent within the family is that, for a typical American woman, home is the place where she is most at risk of being seriously assaulted (Straus and Gelles 1988).

Many of the assaults by women against their husbands should be considered as retaliation or self-defense (Johnson 2000; Straus 1980). In fact, the most frequent motive for violence reported by battered women is "fighting back" (Saunders 1988). Rarely do women report initiating an attack of severe violence, and their assaults are attempted in self-defense. As to those women who do initiate violence, some researchers speculate that they often sense impending violence from their husbands and initiate the violence themselves to stop the overwhelming buildup of tension (Walker 1984). One author, Kelly (1988, 120), describes a typical situation in which a woman experienced a continual threat of violence:

> *What he did wasn't exactly battering but it was* the threat. *I remember one night I spent the whole night in a state of terror, nothing less than terror* all night. . . . *And that was* worse *to me than getting whacked. . . . That waiting without confrontation is just so frightening. (Emphasis in original)*

Saunders (1988, 107) notes that battered women are often convinced that they or their children are in imminent danger of death or great bodily harm, even at times when the husband is not currently attacking or threatening. In these cases, the courts have sometimes held that the women were legally justified in using physical force in a kind of "preemptive strike."

Other concerns surrounding the reporting of supposed equal levels of violence between men and women pertain to the samples used in national surveys, the types of violence measured, and various unintended biases that influence results. Michael Johnson and Kathleen Ferraro (2000) note that surveys like the Conflict Tactics Scale are criticized because they merely count acts of violence, making no distinctions in terms of motives or consequences. They suggest that there are at least two types of couple violence. The first type, which they call **common couple violence** is roughly gender symmetric and involves occasional outbursts of violence that arise in response to escalating arguments. Much of the

Evidence from courts and police records suggests that women constitute roughly 95 percent of the victims in reported cases of domestic violence.

"mutuality" in these violent altercations should also be understood as self-defense by women, and Johnson (2000) also notes that women readily admit hitting men because both they and their partners see it as relatively harmless. The second type of intimate partner violence, which Johnson and Ferraro call **patriarchal terrorism** is decidedly male and involves violence as one tactic in the implementation of a general pattern of power and control. Among violent women living with patriarchal terrorists, 85 percent report that their violence was self-defensive (Johnson 2000). Contrary to popular myths, most of these women eventually leave their abusive partners. Johnson suggests that evidence about the nature of patriarchal terrorism is more accessible through samples obtained from battered women's shelters, whereas estimates of common couple violence are more readily accessible through general sample surveys like the Conflict Tactics Scale (CTS).

Evidence from courts and police records suggests that women constitute roughly 95 percent of the victims in reported cases of domestic violence (Dobash et al. 1992). Criminal victimization surveys using national probability studies similarly indicate that wives are much more often victimized than husbands, as do national crime surveys. Even in national probability surveys like the CTS, women are much more likely to report that they are injured than are men, with men often downplaying the women's injuries (Brush 1990). After analyzing the results of the United States National Crime Surveys, Schwartz (1987, 67) concluded, "There are still more than thirteen times as many women seeking medical care from a private physician for injuries received in a spousal assault." Many researchers thus reject the notion that there is sexual symmetry in domestic violence, even though sample surveys sometimes erroneously draw this conclusion. Yllö (1998, 613) points out that CTS-like surveys ask parallel questions to men and women about violent acts as if they were simply at the end of a normal continuum of items, including "discussed an issue calmly," "cried," and "stomped out." Such surveys do not assess the meanings, contexts, or consequences of these individual acts and do not include consideration of economic deprivation, sexual abuse, intimidation, isolation, stalking, and terrorizing—all common elements of wife battering and all rarely perpetrated by women (Yllö 1998). On the basis of putative equal reports of violence from phone surveys, it is absurd to consider "husband battering" to be the equivalent of "wife battering." To illustrate, Dobash and colleagues (1992, 82) call attention to the "enormous differences in meaning and consequence [which] exist between a woman pummelling her laughing husband in an attempt to convey strong feelings and a man pummeling his weeping wife in an attempt to punish her for coming home late."

Sexual Coercion and Marital Rape

According to recent studies, sexual coercion and sexual assault often accompany battering (Jacobson and Gottman 1998; Shields and Hanneke 1983). Rape is a violent crime against women that is about power rather than sex. Nevertheless, when it comes to rape within marriage, we have tended to ignore it. A high percentage of batterers rape their wives, but until recently, **marital rape** was considered an oxymoron (Jacobson and Gottman 1998). As noted in chapter 11, most states had a marital rape exemption, which stated that a husband could not rape his wife because he was entitled to sex as part of the marriage contract. As Jacobson and Gottman (1998, 150) suggest, this barbaric legal exemption has been abolished in most states, but many citizens and law enforcement officials continue to assume it still exists:

> Prosecutors don't spend much time either arresting people or prosecuting them for marital rape. Unfortunately, women still often feel guilty enough to submit to sex when they don't want it, even under coercive conditions. Even more unfortunately, many wives don't consider the sexual coercion as rape. Jacobson recently asked a female client if her husband had ever forced her to have sex, and she said no. A week later, she said: "I've thought about that question you asked me last week, you know, the one about forced sex. Would choking me until I give him a blow job count?"

This woman minimized this extremely violent act because she subscribed to the belief that her husband was entitled to have sex with her when and how he wanted it.

Several studies have looked at the incidence of marital rape. Russell (1982) found that 14 percent of the married women in a San Francisco sample had been raped at least once by their husbands. Gelles and Cornell (1990) found that at some time 10 percent of spouses or live-in partners used force or the threat of force to have sex. In a study of 323 Boston area women, Finkelhor and Yllö (1985) reported that one in ten women had been forced to have sex with their husbands or partners. Violence accompanied the rape in about half of the instances, and Finkelhor and Yllö identified three basic types of rape in marriage. The first, "battering rape," is found in marriages with a high overall level of violence. Wives were routinely beaten, and rape was an additional element of abuse and humiliation. The second type, "force-only rape," was found in marriages that were otherwise nonviolent. The researchers concluded that in this type of rape, the husband's desire for control led him to use sexual coercion, and he used just enough force to get his wife to comply. The third type, "obsessive rape," was the least common but most openly sadistic, as the man was obsessed with violent sex and inflicted pain to become sexually aroused. Although the women's experiences varied greatly in terms of physical pain and injury, all of the women reported being greatly traumatized (Finkelhor and Yllö 1985).

As noted previously, rape by intimate partners is much more common than rape by strangers. Although the impacts of marital rape are often different from the impacts of stranger rape, they are not less serious and are often more frightening (Russell 1982). The most profound psychological consequences of marital rape stem from the extreme violation of trust that the sexual violence signifies. Victims of marital rape are eleven times more likely to be clinically depressed and six times more likely to experience social phobia than nonvictims (Kilpatrick et al. 1988). Yllö (1988) reports that victims of marital rape experienced great anguish and most struggled to escape their marriages. For some, it took several years before they would let themselves trust men and develop intimate heterosexual relationships, and others felt that trusting men was now impossible. In other studies, psychological problems stemming from marital or date rape can still be evident as long as fifteen years after the assault (Kilpatrick et al. 1988).

When Do Battering Incidents Happen?

It has sometimes been asserted that marital violence is essentially a problem of the lower class. This is not true, because violence is found in all social classes. Among couples filing for divorce, for instance, more working-class couples than middle-class couples complain of physical abuse as a major cause of the breakup, but even among the middle class, about one-fourth complain of physical abuse. The kinds of violence that are most common are pushing, throwing things, or slapping. These actions often occur in the context of an argument that escalates beyond shouting or name-calling, but spouse battering is not usually an isolated incident. There is a continuum from the normal amount of quarreling through mild violence up to serious battering. Frequently couples treat even rather extreme violence as if it were nothing special, or else dismiss it as if it were an unpredictable aberration. Nevertheless, abused women report repeated attacks, with estimates ranging from three to eight times a year.

Gelles and Cornell (1990) summarized some factors that are typically associated with wife beating: (1) The husband is unemployed or employed only part time; the family income is low; or the husband's occupation is in the manual working class. (2) Both husband and wife are very worried about economic security, and/or the wife is strongly dissatisfied with the family's standard of living. (3) The wife is a full-time housewife; there are two or more children; and disagreements over the children are common. All factors are not always present, but in general, they make up a picture that is more common in the lower classes. Similarly, a portrait of men who were woman batterers (Fagan, Stewart, and Hansen 1983) found that 74 percent did not have more than a high school education and 30 percent were unemployed. Most were in their early thirties or younger, and the length of the relationship with their wives was less than five years. This looks like the pattern of early marriage strain found in many working-class marriages (chapter 6). Of course, when these conditions are found in other social classes, they also increase the chances of marital violence. In middle- and upper-class marriages, there are more incentives for wives to hide the fact that they are being physically abused, and there is less chance that others will report them to the police or to some other government agency. Consequently, we know more about spouse abuse in poor families than in others.

What situations tend to precipitate violent outbreaks? In one study of seriously battered women (Giles-Sim 1983, 141), the first incident of battering was likely to be preceded by some fairly major change in life patterns: 39 percent connected to moving to a new house or apartment; 32 percent connected to pregnancy or birth; and 19 percent associated with recent separation or divorce from another partner. Except for the separations and divorces, these do not seem like necessarily traumatic events, but these kinds of shifts in the surface of people's lives seem to unleash violent impulses that are present for underlying reasons.

Although it has not been possible to specify exactly when battering will occur, researchers have identified some of the situations under which batterers feel threatened and attempt to assert their authority. The most common theme is the claim for loyalty. Many battering men demand that wives continually show loyalty by talking in a certain way, cooking certain foods, showing affection in certain ways, and so forth. If an insecure man's wife or partner spends time with other people, even other women friends or relatives, this is sometimes taken to be a sign of rejection. Violent men are typically found to be very possessive (Dobash and Dobash 1979). These men tend to become sullen and irritable about their wives' involvement with others, which tends to exasperate the wife and cause her to withdraw. The husband interprets this as rejection, becomes more possessive, makes more demands that she show loyalty, and eventually becomes violent (Ferraro 1988).

Even children are sometimes viewed as threats to a wife's loyalty to her husband. Several researchers have discovered that the wife's becoming pregnant often precipitates beatings by the husband (Dobash and Dobash 1979; Ferraro 1988; Pagelow 1981). More recent estimates of the proportion of pregnant women who are subjected to domestic violence range from about 10 to 30 percent (Gazmararian 1996). One of Walker's informants illustrates this pattern:

> *He didn't really start to beat me until I started showing how pregnant I was, until my belly started swelling. Then it was like he was jealous of that baby, before he was even born. He was jealous of my babies even when he went in the Navy. He would come home and beat them and beat me. (Walker 1979, 119)*

Sexual jealousy is another form of insecurity experienced by many violent husbands. For many, the perceived threats to loyalty are extremely far-fetched, but the consequences are still tragic:

> *Sara had a three-year-old child from a prior relationship and was eight months pregnant with the child of her lover. He was so concerned about her sexual fidelity that he locked her in their small trailer each morning before leaving for work. The locked door and the small windows made her escape impossible even if she had been interested in pursuing a sexual relationship. He nevertheless returned each day to accuse her of being unfaithful, and to beat her for her transgressions. (Ferraro 1988, 133)*

As this example illustrates, violent men's consuming preoccupation with sexual fidelity is often irrational. In several studies, battering men were reported to fantasize lovers and develop paranoid perceptions of their wives' intentions, leading to excessively controlling behaviors (Ferraro 1988). For some men, the slightest indication of disloyalty on the part of the wife is taken to be a vicious attack on the man's sense of self. In part, this attitude is supported by attitudes in the general population. About two-thirds of Americans agree that acting violently in a situation involving a spouse's extramarital affairs is justified.

Another common theme in studies of wife beating is that violence often occurs around issues of control. Deciding how income is to be made or spent, how children are to be cared for, when and how sexual relations should occur, how meals are to be prepared, how leisure should be spent, or how deferential the wife should be all provide opportunities for exerting control. Violent men often make excessive demands for compliance with their wishes in these areas, and when wives question them or comply in a manner deemed unsuitable, the men feel threatened and fly into a rage. Often the issues that constitute threats to the men's control appear trivial to others but are interpreted by the men as "failure to fulfill the obligations of a good wife" (Bograd 1984). For example, Ptacek (1998, 627) reports how one batterer threatened his wife by telling her, "I should just smack you for the lousy wife you've been."

Sometimes men justify beating their wives because of things that their wives say to them, implying that wives should not challenge husbands. In one study, the most common explanation men offered for beating their wives was that they had been verbally provoked:

> She was trying to tell me, you know, I'm no fucking good and this and that . . . and she just kept at me, you know. And I couldn't believe it. And finally, I just got real pissed and I said wow, you know. I used to think, you're going to treat me like this? You're going to show me that I'm the scum bag? Whack. Take that. (Ptacek 1998, 625)

Some of the men equated women's verbal aggressiveness with their own physical violence:

> Women can verbally abuse you. They can rip your clothes off, without even touching you, the way women know how to talk, converse. But men don't. Well, they weren't brought up to talk as much as women do, converse as well as women do. So it was a resort to violence, if I couldn't get through to her by words. (Ptacek 1998, 625)

Although batterers often try to blame their wives for provoking their anger, researchers find that battered women rarely provoke abuse intentionally. One study found that battered women report relatively minor "misdeeds" as triggering abuse, such as preparing a casserole instead of fresh meat for dinner; wearing her hair in a ponytail; or commenting that she didn't like the pattern on the wallpaper (Martin 1976, 49). Sometimes a marriage can be nonviolent for a long time, and then some relatively minor event, such as cooking the wrong food, becomes a major threat to the man's control and triggers a violent response:

> Until we were married ten years or so there was no violence or anything. But then after a while, it just became, it just became too much. . . . I don't know if I demanded respect as a person or a husband or anything like that, but I certainly, you know, I didn't think it was wrong in asking not to be filled up with fatty foods. (Ptacek 1998, 627)

Once battering has happened, it seems to become more likely, unless something is done to head it off. Battering can become a routine. Hence, the factors that set off later incidents, including very severe ones, have no particular pattern but can include all the factors mentioned previously, as well as the man losing his job or the woman wanting to leave the relationship. Many such incidents happen for no apparent reason at all. For example:

> He hit me, and he said, "Now you are going to hit me." I said, "No, I'm not going to hit you. I don't want to hit you." I said, "Please leave me alone." I started crying. He did it again, and I still didn't hit back. Then he just chuckled and walked off. He never slapped me across the face, though.

I wanted to kill him, but I didn't. I just said, "Why, why is this happening to me?"
It was like he wanted to punish me.

There was another time when he kicked me. He just kept kicking me. I felt angry, but
I couldn't show it. I was too afraid. I don't think I felt. . . . I never felt like I could strike
him. I don't know if I ever could. I just didn't know how to give it back. Just couldn't
do it. I'm afraid of what I would do if I did. (Giles-Sim 1983, 1)

These incidents had become so normal that many couples had a history of violence with each other long before they were married. Although many couples experience their first incident of violence while they are dating or living together, they still go on to marry. Usually they interpret these early events as isolated incidents, as something to forgive and forget rather than the beginning of a pattern. Studies find that violence during courtship is common (see chapter 8). Many couples experience pushing, slapping, or shoving during dates, with the perpetrator of violence usually being the male. Why are these incidents not taken as warning signs? The answer is probably that violence, in a sexual context, may often be interpreted in our society as a sign of passion or of jealousy. Thus, Henton and colleagues (1983) found that in the cases of dating violence, about 25 percent of the victims and 30 percent of the offenders interpreted the violence as a sign of affection. The reinterpretation of violence as either normal or as affectionate is one of the main processes that keeps violent couples together and hence allows battering to be repeated.

Types of Batterers

Another common false assumption about domestic violence is that all batterers are alike. Earlier research and popular views tended to assume that men who abused their wives had similar pathological profiles. In fact, several different types of abusers have been profiled in the clinical and social science literature. One review of research identified three types: "family-only," "generally violent-antisocial," and "dysphoric-borderline" (Holtzworth-Munroe and Stuart 1994). Family-only abusers tend to engage in the least severe marital violence and are the least likely to engage in psychological and sexual abuse. This group is sometimes captured under the label "common couple violence" (Johnson 2000).

The generally violent-antisocial batterer (Holtzworth-Munroe and Stuart 1994) is similar to the **Cobra** discussed by psychologists Neil Jacobson and John Gottman (1998). They recruited a sample of couples who identified themselves as involved in violent relationships, interviewed them, videotaped their interactions, and monitored their physiological states as they were arguing. The Cobras were especially emotionally abusive and exhibited a "cold" physiology, even in the heat of vicious verbal attacks on their female partners, with heart rate and other physiological indicators suggesting a chilling internal calmness:

Cobras appear to be criminal types who have engaged in antisocial behavior since adolescence. They are hedonistic and impulsive. They beat their wives and abuse them emotionally to stop them from interfering with the Cobra's need to get what they want when they want it. Although they may say that they are sorry after a beating, and beg their wives' forgiveness, they are usually not sorry. They feel entitled to whatever they want whenever they want it and try to get it by whatever means necessary. (Jacobson and Gottman 1998, 37)

Some of the Cobras described by Jacobson and Gottman have been defined as psychopaths, which means they lack a conscience and are incapable of feeling remorse. They have diminished capacity for experiencing emotions or understanding the emotions of others. Psychopaths have difficulty experiencing sadness or remorse and rarely empathize with others, unless it is an act to get what they want. According to Jacobson and Gottman (1998), however, not all Cobras are psychopaths. Others are merely antisocial, but like the psychopaths, they are incapable of forming truly intimate relationships with others, and to the extent that they marry, they do so on their terms:

> *Their wives are convenient stepping-stones to gratification: sex, social status, economic benefits, for example. But their commitments are superficial, and their stance in the relationship is a "withdrawing" one. They attempt to keep intimacy to a minimum, and are most likely to be dangerous when their wives attempt to get more from them. They do not fear abandonment, but they will not be controlled. Their own family histories are often chaotic, with neither parent providing love or security, and they were often abused themselves as children. As adults, they can be recognized by their history of antisocial behavior, their high likelihood of drug and alcohol abuse, and the severity of their physical and emotional abuse. (Jacobson and Gottman 1998, 37)*

Cobras instill fear in those around them but particularly in their wives, who are often chronically depressed. At the same time, they exude a kind of "macabre charisma" (Jacobson and Gottman 1998, 38).

The third type of batterer, the "dysphoric-borderline" (Holtzworth-Munroe and Stuart 1994) has also been called "the cyclical batterer" (Dutton 1995). This type roughly corresponds to Jacobson and Gottman's **Pit Bulls.** Pit Bulls can be explosive, but they are more likely than Cobras to confine their violence to family members, especially their wives. They are less likely to have criminal records, but they commonly had fathers who beat their wives and thus learned that battering is an acceptable way to treat "their" women. Although they emotionally and physically abuse their wives, Pit Bulls are also quite emotionally dependent on them:

> *What they fear most is abandonment. Their fear of abandonment and the desperate need they have not to be abandoned produce jealous rages and attempts to deprive their partners of an independent life. They can be jealous to the point of paranoia, imagining that their wives are having affairs based on clues that most of us would find ridiculous. . . . The Pit Bulls, although somewhat less violent in general than the Cobras, are also capable of severe assault and murder, just as the Cobras are. (Jacobson and Gottman 1998, 38)*

Pit Bulls are relatively "hot-headed" and "hot-blooded." Jacobson and Gottman (1998) found that their physiological indicators matched their outward expressions; they are quick to anger, and their internal states tended to match their explosive outbursts. Most wives of both Pit Bulls and Cobras eventually leave the abusive relationship, but there is a tendency for partners of Pit Bulls to leave more quickly than partners of Cobras (Jacobson and Gottman 1998, 146).

Causes of Marital Violence

The traditional way of accounting for marital violence was either to ascribe it to lower-class culture or else to describe it as psychologically pathological and deviant. As we have seen, violence occurs at all class levels, even though incidence rates are higher among those who are economically marginalized. Although many batterers are indeed clinically "mentally ill," ascribing domestic violence merely to pathological individuals evades the question of explaining it.

A show of anger may precede family violence. "You're asking for it!" is a typical verbal prelude to physical abuse.

One particularly pernicious psychological account is the argument that women are masochistic and hence "ask for it." Actually, violent husbands often use this as an excuse for a description of what happened. What it implies, sociologically, is that there is a widespread cultural belief that women should behave in certain ways; if they do not, it is legitimate to use violence against them. In American society until the twentieth century, laws actually gave a husband the right to physically chastise his wife for nagging or other offenses against her "proper place" (Straus 1991). This suggests that we cannot understand domestic violence without taking gender and power into account (Yllö 1998).

Four main sociological explanations for why violence occurs in intimate relationships today are (1) **sexual property rights,** (2) **economic strain,** (3) **intergernational transmission of violence,** and (4) **social control.**

SEXUAL PROPERTY RIGHTS The traditional patriarchal attitude is that married women are essentially the property of their husbands, economically, sexually, and emotionally. Women who violate the male prerogative are felt to have broken a social tradition and hence to have brought upon themselves a legitimate angry response. This is, in fact, what one would expect given the theory of love and sex as rituals that establish emotional property (discussed in chapters 8 and 9). Dobash and Dobash (1979, 1992) show that men who beat their wives tend to regard their action as proper and to justify it as a defense of their traditional rights. This is particularly likely to happen when there is sexual infidelity or even minor flirtations causing jealousy; it also happens when the husband feels that his own sexual demands are being evaded. The "woman as property" attitude also condones and promotes violence in response to all kinds of actions on women's part that show their lack of respect for male prerogatives. As the theory of rituals shows, these become symbolic violations that call into question the whole structure of the relationship.

Another incident shows how male feelings about women being their sexual property result in wife battering:

> He went down in the cellar to fix the stove and I went down to try to talk to him. He wanted me to bend over so he could have sex downstairs. I said, "I don't drop my pants every time you turn around and want sex." He said, "You do what I tell you to do."
>
> When we went back upstairs to eat, we were sitting at the table and he was saying something— "You're gonna. . . ." do something. I said, "No, I don't want to," and he kicked me under the table. I said, "Don't kick me." He said, "I'll kick you if I want to. I'll do anything I want to you." Then he hit me with a stick and said he'd do anything he wanted to me because I was his property. (Giles-Sim 1983, 2)

As we have seen in chapter 6, married life is most intensely "ritualized," in the sense of creating strongly held traditional beliefs, in the typical living conditions of the working class. It is there that the highly confining social density of the local community creates the most strongly taken-for-granted attitudes. When traditional prerogatives are violated, the result is unthinking anger. Of course, these patriarchal prerogatives can be upheld in the higher social classes as well, although there the culture usually is more ideological and abstract and men tend to find more indirect and subtle ways of getting what they want.

The fact that the traditional structure of the entire community is involved in generating these beliefs, especially in the working class, means that the male's violence is often regarded as legitimate by persons outside his own family. The police, who tend to share the working-class culture, have traditionally been unwilling to intervene in what they regard as "domestic squabbles" (Pagelow 1981). Hence, battered women have frequently had no outside protection or redress.

ECONOMIC STRAIN Another explanation for family violence draws on the fact that it tends to increase during times of economic strain. This is another reason why lower-class families are especially prone to violence. Such families usually have a very traditional conception of gender roles, in which the husband is supposed to be the provider and head of household. When he fails to provide, he loses his standing both in his wife's eyes and in his own. Sometimes she is able to make up for some of their economic loss by her own work, but often this challenges the husband even more. A good deal of lower-class violence may have traditionally been an effort on the part of the male to overcome his feeling of social inferiority and reassert control (O'Brien 1971). More recent studies continue to

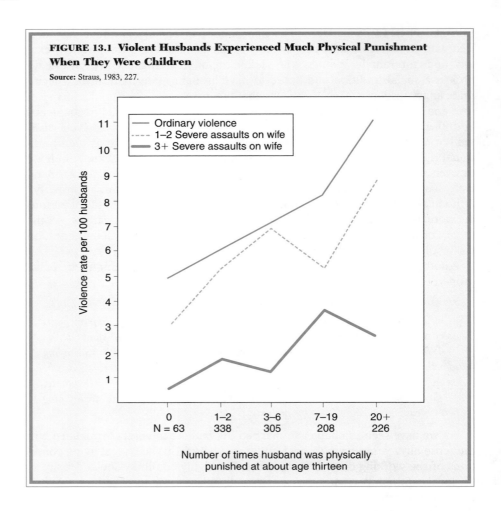

FIGURE 13.1 Violent Husbands Experienced Much Physical Punishment When They Were Children

Source: Straus, 1983, 227.

Number of times husband was physically
punished at about age thirteen

show that low relative income resources are associated with men perpetuating domestic violence (Anderson 1997).

Economic strains also play a role in keeping some battered women trapped in their marriages. Although most battered women leave their husbands, many stay because they can't afford to leave. Others express fear that if they leave, their husbands will come after them and punish them even more. Part of the reason that women with small children sometimes stay is that they are isolated and do not have enough economic resources to support themselves and their children.

INTERGENERATIONAL TRANSMISSION OF VIOLENCE There is also evidence that violence toward wives (and toward husbands, too) is passed along from the previous generation through parents' child-rearing practices. Straus (1983, 1994) shows that violence is contagious in a number of ways. A woman who was abused by her own parents is more likely to stay in a violent relationship with her husband or boyfriend. This is because she tends to perceive violence as normal or because she has low self-esteem and little sense that she could improve the situation. Even more important, husbands who were subjected to considerable physical punishment when they were children are especially likely to assault their wives. As Figure 13.1 shows, there is a direct correlation between the amount of physical punishment a boy had when he was about thirteen years old and the likelihood that he will later use violence on his wife. Not only does childhood punishment predict which men will frequently and severely attack their own wives (marked "3+" in Figure 13.1), but it also predicts which men will occasionally beat their wives (marked "1–2" in Figure 13.1) and even which ones will engage in "ordinary" levels of mild physical abuse.

Straus also shows that the more violent a husband is to his wife, the more likely she is to use violent punishment on her children. Violent husbands are also more violent to their children. This closes the circle and sets off the likelihood of children growing up to become spouse abusers in the next generation.

SOCIAL CONTROL One more pattern has to do with how many social pressures and controls there are from the surrounding society. That is to say, many of the factors that lead toward violence (sexual property feelings, economic strains, family backgrounds) can be neutralized or allowed to come forth, depending on how much control is present. In the past, our customs and laws have considered the family to be "private" and not subject to the same controls that operate in more public realms. The patriarchal ideal of "a man's home as his castle" has led to a hands-off attitude on the part of lawmakers and police. This idea of the family as a private "haven" has isolated women and children within families and allowed men to violently abuse them.

Family quarrels do not necessarily escalate into severe violence. This is most likely to happen when the couple has lived in their neighborhood less than two years and when they do not participate in an organized religion or other social groups (Gelles and Cornell 1990). In other words, the family is more socially isolated, with less pressure to restrain what goes on within it. Probably for the same reasons, spouse abuse is most common among couples who are young and who have not been together many years. Women who are full-time housewives are especially likely to be victims of battering because they are less likely to have social contacts and outside sources of social support. Isolated women are typically unable to find alternatives that would get them out of the victimized situation. This can be especially true for immigrant women. Eng (1995) reports high levels of battering among recent Asian immigrant communities, based on reports from a shelter in New York City. Acknowledgment of battering is highly shameful in some ethnic enclaves because women have been socialized to believe that marital failure is always the fault of the wife (Eng 1995; Song 1996).

A major choice point for battered women usually happens with some particular incident. Often, a woman decides to leave when she fears her children will be hurt. Some women leave when they become resentful toward their husbands for letting the children see their mother beaten or for exposing their violence to people outside the family (Giles-Sim 1983). Such dramatic incidents can sometimes rearrange an entire worldview. Jacobson and Gottman (1998, 138) describe how one frequently abusive husband exploded over a minor incident, beat up his wife Vicky, and stomped out of the house. As he left, she yelled "and don't you ever come back." He replied, "You try to keep me out of my house and I'll kill you." According to Jacobson and Gottman (1998, 138), "a lightbulb went on inside of her and she at last realized her dream was just a fantasy." These researchers report hearing similar stories from other battered women who remembered an incident when, instantaneously, they decided "I don't care what happens—eventually, I am going to leave him!"

Coping with Partner Violence

Most research on coping with domestic violence has focused on the more extreme forms that Johnson and Ferraro (2000) call patriarchal terrorism, exemplified by the Cobras and Pit Bulls described earlier. Growing out of the **battered women's shelter movement,** early research often focused on women in abusive relationships as victims and asked questions such as, "Why do they stay?" In the 1990s, the dominant view shifted to defining women who have experienced violence as "survivors," focusing on the decisions they make to escape, to end the violence, or to cope with it in some other manner (Ferraro 1997). One set of longitudinal studies found that over a two-and-a-half-year period, women showed great resourcefulness in their attempts to resist the pattern of violent control and intimidation they were facing (Campbell 1994, 1998). Women used strategies including conscious decisions to "make do" in a relationship, to subordinate the self, to respond to identifiable pivotal events, and to negotiate directly or indirectly with their partner. At the end of the two-and-a-half-year period, three-quarters of the battered women were no longer in a violent relationship, with 43 percent having left and 32 percent having successfully negotiated an end to the violence. Researchers note that strategies and outcomes can vary widely, depending on the type of violence being experienced and the social context of the relationship. For example, patriarchal terrorism and common couple violence may call for very different forms of negotiation and intervention, and dynamics can vary according to whether the relationship is one of dating, cohabiting, same-sex partnership, or marriage (Johnson 2000).

To correct for an earlier emphasis on women as victims, many researchers highlight the processes that abusive men use to entrap their partners and the processes that abused women use to engineer their escape. Kirkwood (1993) relies on the metaphors of a web of entrapment and a spiral of escape to capture the harrowing details of intimidation, conflict, and cyclical resistance in domestic partner relations. Men use a wide range of tactics of power and control not only to control the intact relationship but also to try to ensure that their partner will never be able to leave them. Johnson's (1998) analysis of the shelter movement addresses this process by focusing on how the abuser manipulates the personal, moral, and structural commitments to the relationship to entrap his partner. Temporary safe housing, support groups, empowerment counseling, networking with social support services, legal advocacy, coordinated community response, and other aspects of the women's shelter movement can neutralize the commitments that keep the women in violent relationships (Johnson 2000). When women can take advantage of these supports, they often go through a process of leaving and returning, each time gaining more personal and social resources, eventually gaining enough inertia to spiral outward until they escape from the web (Kirkwood 1993).

Intimate partner violence is not limited to married couples. As noted in chapter 8, dating and courtship violence is surprisingly common. Summarizing the results of many studies, the average prevalence for nonsexual dating violence is 22 percent among male and female high school students and 32 percent among college students (Centers for Disease Control 2000; Murphy 1988). Although distinctions among types of violence are rarely reported, Johnson (2000) suggests that common couple violence, rather than patriarchal terrorism, is the most common type of violence in dating relationships. Most findings suggest that dating violence, like marital violence, revolves around issues of power, with individuals becoming more violent and controlling when they feel threatened or want to assert their authority (Stets 1993). Levels of common couple violence in cohabiting relationships are sometimes reported to be higher than in marital relationships (Stets and Straus 1990), but the more extreme Cobra and Pit Bull behaviors are highest in marriages. Johnson and Ferraro (2000) suggest that marriage may not be a license to hit, but for some people, it is considered a license to terrorize.

Intimate partner violence is not limited to heterosexual partners. We know little about interpersonal violence in gay male relationships, even though it is likely to be higher than in lesbian relationships. We know more about violence among lesbians, in part because of the important role of the women's movement in generating research on domestic violence (Johnson and Ferraro 2000; Renzetti and Miley 1996). Claire Renzetti (1992) interviewed 100 women in same-sex couples who responded to fliers and newspaper advertisements. She found both similarities and differences between violence in lesbian couples and violence in heterosexual couples. She identified jealousy and emotional abuse in both types of relationships, concluding that power and control were central issues. Two of three women in her sample indicated that their partner's dependency was a source of conflict (similar to the Pit Bulls in Jacobson and Gottman's study). There are some special issues in lesbian couples, including psychological abuse related to threats of "outing," the failure of some in the lesbian community to acknowledge the abuse, and the reluctance of shelters and service providers to provide care for someone abused by a woman (Elliot 1990; Merrill 1996). Johnson (2000) suggests that violence motivated by a need for general power and control is not exclusively the province of men nor entirely a product of patriarchy, so we should not be surprised to find it when other precipitating factors are in place.

Intervention and Treatment

Although many women do not call the police to report battering, when they do so, the effects are usually positive. A U.S. Justice Department study found that when battered wives did not call the police, 41 percent were beaten again by their husbands within six months; but of those who did call the police, the proportion who were beaten again dropped to 15 percent. Typically, women fear reprisal by their husbands if they call the police, and sometimes the police do not take action. However, even the threat of police intervention tends to have a deterrent effect, at least on some men.

Sherman and Berk (1984) were among the first to conduct a field experiment to assess the impact of police intervention in cases of wife abuse. Three types of police response—immediate arrest, removing the husband from the home, and trying to cool down the situation through talk—were used according to a random assignment procedure. They discovered that if the man was arrested, there were fewer subsequent calls for help from the wife and fewer instances of repeated violent behavior. Later studies have shown that arrest is an effective deterrent and that the risk of retaliation by angry husbands can be reduced in many cases. As a result of research like this, and pressure by feminist activists, police are becoming more likely to arrest husbands for beating their wives. In addition, it appears that abused wives may now be more likely to call police for help. One reason women do not call the police, however, is that they are attached to their husbands; they love them and do not want to take an action that they feel will make the breach irreparable. These women are caught between opposing forces.

Official responses to social problems are not inevitable consequences of the problems themselves. In fact, most social problems exist long before they are recognized as conditions that demand some form of collective response. Various potential issues are identified, legitimated, and responded to as social problems in response to claims by advocacy groups to redress specific grievances (Spector and Kitsuse 1973). In the case of marital violence, the woman's movement provided the impetus for women to define husbands' use of force against wives as "abuse" and to identify victims of repeated violence as "battered women."

Battered women's shelters first emerged in the 1970s as local efforts by women to provide a supportive and safe environment for victims of abuse. Through volunteer efforts, including consciousness-raising groups, public speaking, lobbying, grant writing, and use of the media, feminists were able to conceptualize husbands' violence against wives as a community responsibility rather than an interpersonal problem limited to a few pathological families (Schechter 1982). Various local groups such as YWCAs, women's clubs, junior leagues, and church organizations solicited funds and lobbied for the establishment of shelters. Eventually, community service agencies, health organizations, child and family service agencies, and local governments began to fund programs for battered women. The shift from grassroots support to government funding professionalized battered women's shelters and initiated a shift from advocating women's rights to "treating clients." Some argue that the traditional social service agency approach to the problem focuses too much on the personality defects of battered wives and too little on the problem of men's responsibility for violence (Schechter 1988).

Some of the most recent attempts to address the problem of wife abuse *do* focus on the men. Pro-arrest policies and court-mandated rehabilitation programs are having some success, although treatment approaches vary widely (Lyon and Mace 1991). Probation and mandatory group counseling for batterers generally focus on inducing violent men to accept responsibility for their violence, develop communication skills, and limit their alcohol and drug abuse. Here too, there are debates between activists and professionals over how the problem should be defined and which clinical model to adopt. The following excerpt from an intake session at a Boston-area profeminist antibattering program illustrates how different counseling strategies can define the problem differently:

Counselor: What do you think causes you to hit your wife?

Client: Insecurity. I guess it goes way back. . . . My father was a drinker too. He wasn't just a drinker; he was a mean drunk. Once during Thanksgiving he got mad and threw the whole damned bird on the floor. . . . He'd whack my mother for no reason really. . . . I left home when I was seventeen, got married but it only lasted two months. She was pregnant of course. I've always been insecure with women.

Counselor: This is helping me to understand why you're insecure but not why you hit your wife.

Client: Sometimes I take things the wrong way. . . . I overreact I guess you could say, because of my insecurity. My shrink said I was like a time bomb waiting to go off. She [client's wife] might say something and I don't react at the time, but then the next day or maybe a few hours later I get to really thinking about that and I get really bull shit.

Counselor: A lot of people feel insecure but they are not violent. What I'm interested in finding out is how do you make the decision to hit your wife—and to break the law—even if you are feeling insecure?

Client: I never really thought of it that way, as a decision.

Counselor: But you were talking just now as if your violence is the direct result of your insecurity, or of something she says or does.

Client: Yeah, you're right, I do. But I'm still thinking about what you said about the decision. I honestly have to say that I never thought of it that way before. I mean, I'm really dumbfounded! I'm going to have to really think about that.

Counselor: What are you waiting for?

Client: What do you mean?

Counselor: I mean are you waiting to stop feeling insecure before you stop being violent?

Client: Yeah, I guess that's what I've been waiting for. (Adams 1988, 176–77)

This client had been through three years of individual and couple counseling that focused on his violent and abusive behaviors as an outgrowth of his drinking, insecurity, and poor communication. Although he had made some progress, the client continued to have "temper tantrums" during which he would grab, shake, slap, or kick his wife. The intent of the new counselor's questioning was to focus on violence as the treatment problem and to induce the client to stop beating his wife. Newer treatment techniques, such as persuading men to take "time out" when they sense they might get violent, avoid the problem of making nonviolence dependent on solving the man's other problems. First and foremost, these techniques teach men how to identify their violent tendencies and make themselves accountable for their violence (Adams 1988). As other researchers have noted, men hit their wives because they can get away with it. If we want wife abuse to stop, we must make such violence unacceptable.

CHILD ABUSE

When we got there this baby was crying and we could see his leg was twisted kind of funny to one side. He had bruises on his face. He looked pretty bad.

I ran because I didn't think they cared about me. The night before, my mom told me that they never liked me. She says, "Go live with your friend." And then she goes, "I don't give a damn about you. Just get the hell out of here. I never want to see your face in this house again." (Garbarino and Gilliam 1980, 10, 13)

Although it is difficult to obtain precise figures on how many children in the United States are abused, it is estimated that every year between 1 and 3 million children are abused, mostly by their parents. As Figure 13.2 indicates, the number of children reported as abused or neglected is on the rise, with over 3 million cases reported to Child Protective Service Agencies in 1997, and about a third of these were substantiated by investigative procedures (Wang 1999). Of these children, nearly 20 percent were infants or toddlers (Children's Defense Fund 1997). Children are often abused verbally, as well as being exposed to a wide range of physical and neglectful practices; they are starved, beaten, burned, cut, tied, chained, isolated, left unbathed and to lie in their own excrement, or sexually molested, and not a small number are murdered (Hetherington and Parke 1999). In 1997, more than 1,000 children died as a result of child abuse—about three children a day (Wang 1999). A small child has a greater chance of being

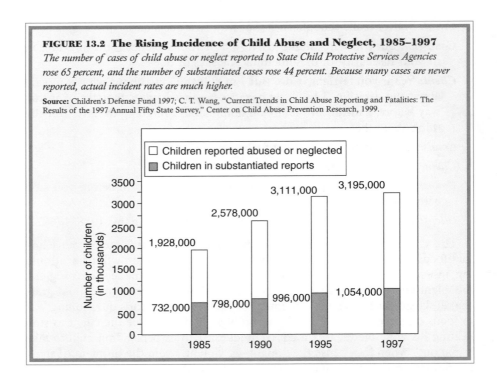

FIGURE 13.2 The Rising Incidence of Child Abuse and Neglect, 1985–1997
The number of cases of child abuse or neglect reported to State Child Protective Services Agencies rose 65 percent, and the number of substantiated cases rose 44 percent. Because many cases are never reported, actual incident rates are much higher.
Source: Children's Defense Fund 1997; C. T. Wang, "Current Trends in Child Abuse Reporting and Fatalities: The Results of the 1997 Annual Fifty State Survey," Center on Child Abuse Prevention Research, 1999.

severely injured or killed by parents than by anyone else. For children, the home can be a very dangerous place.

Assessing Child Abuse

Most cases of family violence are never made public because they do not come to the attention of police or child protection authorities. However, national probability surveys that asked persons how much violence they actually experienced or condoned in their own families found a considerable amount (Gelles and Straus 1987; Straus 1994). Most parents admitted to using some kind of violence on their children during the past year. The most common form of violence was fairly mild: over 50 percent said they had spanked or slapped a child, and over 30 percent had pushed, grabbed, or shoved one. However, 10 to 13 percent said they had hit a child with something, 3 to 5 percent had thrown something at a child, and 3 percent admitted kicking, biting, or hitting with a fist (Table 13.1). These sound like adults acting more like children. Moreover, when the time range was expanded beyond the past year to ask parents if they had *ever* done certain violent actions, 4 percent admitted beating up a child, and 3 percent had threatened to use or had actually used a knife or a gun on one.

Comparing the rates for 1985 to the rates for 1975 shows that there were few changes in the amount of overall violence against young children (at least as reported on national Conflict Tactics Scale surveys; see Table 13.1). There does appear to be a significant decline of severe violence against children over the decade. This finding may have more to do with public awareness than actual trends in behavior. National media campaigns, new child abuse and neglect laws, reporting hotlines, and almost daily media attention have transformed a problem that was ignored for centuries into a major social issue (Straus 1994).

Violence against children includes fairly mild forms, such as spanking, slapping, pushing, grabbing, or shoving. More severe forms include hitting a child with something, kicking, biting, or hitting with a fist.

TABLE 13.1 Kinds of Family Violence Against Children

| | *Families who reported violence the past year* | |
Type of Violence	1975	1985
Slapping or spanking child	58%	55%
Pushing, grabbing, or shoving child	32	31
Hitting child with something	13	10
Throwing something at child	5	3
Kicking, biting, or hitting child with fist	3.2	1.3
Beating up child	1.3	0.6
Burning or scalding child	NA	0.5
Threatening to use or using knife or gun	0.1	0.2
Overall violence	63	62
Severe violence	14	11

Source: Gelles and Straus 1987, 84; Straus 1994.

Opinion polls conducted in the mid-1970s showed that only about one American in ten considered child abuse a serious problem, but by the 1990s, nine of ten did (Magnuson 1983; Straus 1994). As a consequence, people responding to questions about violence against children may now be less likely to admit punching, biting, kicking, or beating children than they would have been ten years earlier. Alternately, one could assume that all this attention to child abuse could have altered violent parents' disciplinary practices and angry outbursts, resulting in lower actual rates of child abuse. The numbers from government agencies who investigate such things suggest otherwise, with more reports of abuse and more cases of substantiated abuse since systematic data began to be collected in the 1980s (see Figure 13.2).

Whether the likelihood of reporting child abuse is increasing or decreasing, we can assume that reported rates for severe violence against children are well below actual rates. Child abuse is typically hidden in the privacy of the home, and most people would be reluctant to admit the full extent of their abuse or neglect. On the surveys, nine of ten parents of three-year-olds admitted at least minor forms of physical punishment, such as spanking or slapping, and about one-third of parents of fifteen- to seventeen-year-olds also admitted using such tactics. When children of any age were more severely beaten, the episodes were reported as being repeated an average of once every two months (Gelles and Straus 1987).

About half of the reported cases of parental physical violence against children involve women, and about half involve men. When we consider the amount of time that children spend with women relative to men in two-parent families, and the fact that most single parents are women, we can see that the rate of violence is much higher for men than for women. Studies find that most men who abuse their wives also use violence against their children (McKibben, DeVos, and Newberger 1989). Most researchers also find that women who are abused by their husbands are most likely to use severe violence against their children, although these women account for only a portion of such cases. Although some early researchers speculated that mothers employed outside the home might be more violent toward children than housewives (Fontana 1973), just the opposite seems to be the case. Gelles and Hargreaves (1987) found that women who worked full time had the lowest rate of overall violence toward their children. Fathers with employed wives also had the lowest rates of violence toward children. The rate of child abuse among fathers with wives who did not work outside the home was 2.5 times greater than the rate for fathers married or living with women who worked full-time. Gelles and Hargreaves (1987, 101) speculate that men whose wives do not work maintain traditional beliefs about the man being the head of the household and believe it should be the father who takes responsibility for physically punishing the children ("Wait until your father gets home").

Violence against children is not the only form of violence that takes place in families. We have already seen that spouses often attack one another. Moreover, children often attack their parents when they get old enough to fight back. Almost 2.5 million (or about one of ten) teenagers every year commit violent acts against their parents, with almost 900,000 adolescents (3 percent) engaging in more severe attacks on their parents—such as punching, kicking, biting, or using a weapon. Mothers are much more likely than fathers to be the victims of both overall and severe acts of violence (Gelles 1987). About 2,000 parents are killed by their children each year, the most publicized example in the past decade being the Menendez brothers. It has also been estimated that people over age sixty-five who live with younger family members are also abused, at a rate of about 500,000 persons per year. Violence in the family may start with spouses fighting and parents abusing their children, but it spills over into reciprocal violence later on.

The most common type of family violence, in fact, is scarcely noticed because it is taken for granted. This is violence among siblings. Gelles (1977) found that 80 percent of the children between age three and eighteen who had siblings admitted trying to hurt their brother or sister during the preceding year. Boys were more violent than girls, but the difference was only a matter of degree. Nor was the violence minor. Almost half the children had kicked, punched, or bitten a sibling; 40 percent had hit one with an object; and about a sixth had beaten one up. More than half had committed a violent act that would be grounds for legal prosecution if it had been directed against someone outside the family.

Straus (1994, 103) examined whether children involved in sibling violence were assaulted by their parents and came up with some revealing findings. Children whose parents avoided using physical abuse or any form of corporal punishment (spanking, paddling, etc.) were the least likely to be involved in sibling violence. In contrast, over 40 percent of children whose parents used corporal punishment were found to have repeatedly and severely attacked a sibling. The rate of severe attacks went up even more for children who were physically abused. Over 75 percent of children whose parents abused them repeatedly and severely were also reported to have repeatedly or severely attacked a sibling (see Figure 13.4, later in the chapter). Such findings suggest that interpersonal violence is self-perpetuating. Straus (1994, 103) concludes, "Of course, much of the corporal punishment occurred precisely because the child hit a sibling. Granted that, these findings suggest that hitting children to get them to stop hitting a brother or sister does not work."

Folk Beliefs About Corporal Punishment

Why does this violence occur? One reason is that American culture supports the family use of violence. In national surveys, 90 percent of people agree that it is often desirable to spank or slap children. Many claim that spanking is for the child's own good and say they are glad their own parents used physical punishment on them (Straus 1994). Most parents admit that they hit children who are three or four years old, and over 20 percent also say they spank or slap infants (Straus 1994). This is an astounding finding, considering that infants have few cognitive or motor skills that enable them to respond to their environment or control their actions.

Many parents subscribe to the erroneous folk belief that children need to be physically punished to become moral human beings. In *Beating the Devil out of Them* (1994), Murray Straus documents how many parents' commitment to **corporal punishment** is embedded in religious traditions. He notes that many religious leaders who advocate corporal punishment believe that the traditional American family is declining and that society is being ruined because of overly permissive child rearing. According to Straus (1994, 15–16), authors promoting such beliefs tend to equate the putative decline in American society with NOT using corporal punishment:

> *Like just about everyone else, these authors [Guarendi and Eich 1990; Dobson 1970, 1988] are against child abuse, but, like more than 80 percent of the population, they believe that hitting a "willfully disobedient" child is an act of love and concern, not abuse. . . . They believe hitting a child is an obligation imposed by God, just as God expects parents to love and nurture children. Indeed, the two must go hand in hand. The child must come to understand that he or she is being hit as an act of love. Fundamentalist Protestants typically cite scriptures to show that children are inherently evil. Parents are responsible for shaping the will of the child, they believe. Like the early Puritans, these latter-day authoritarians generally describe the child as strong-willed and in a constant battle with parents. (Greven 1991)*

A belief in hitting children "for their own good" is not limited to Protestant fundamentalists or other religious denominations. The majority of Americans were spanked as children, so when they become parents, they tend to hold the attitude that using corporal punishment will actually help their children.

Moreover, these have tended to be official attitudes, as well as private and religious ones. Two-thirds of police and educators answering a poll were in favor of spanking, and the Texas legislature, as just one among many, enacted a law that stated: "The use of force, but not deadly force, against a child younger than 18 years is justified (1) if the actor is the child's parent or stepparent . . . (2) when and to the degree the actor believes the force is necessary to discipline the child" (Garbarino and Gilliam 1980). Ironically, much of the legislation passed in the last few decades to protect children from assault simultaneously reaffirmed the parents' right to hit their children "when necessary" (Straus 1991, 27). If we add to this the strong belief in the privacy of the household and the right of families to settle their own affairs, we find a strong set of viewpoints that condone and even favor the use of force inside the home.

"Normal" Violence

Most violence against children is ordinary physical punishment carried out by loving and concerned parents. To illustrate, Murray Straus (1991) uses the example of a ten-month-old child who picks up a stick and puts it in his mouth. The parent takes the stick away and says, "No, no, don't do that. You'll get sick. Don't put dirty things in your mouth." Unfortunately, children crawling on the ground are virtually assured of finding another stick, or rock, or something equally dirty and promptly put it back in their mouths. Eventually the parent notices, comes over to the child, and gently slaps the child's hand, saying, "No, no don't do that." This scene seems quite normal. As Straus points out (1991, 29), however, the problem is that these actions also teach the child the principle "those who love you are those who hit you." This lesson begins in infancy and continues for most American children until they leave home. Remarkably, one of three parents report that they continue to hit their children even when they are teenagers (Straus 1991, 28).

The fact that the hitting is done by someone who loves and cares for the child probably makes it worse, insofar as the child is more likely to obtain a deep moral sense that it is right to both give and receive physical punishment. The idea that people who love you hit you is also easily generalized to "those you love are those you can hit" (Straus 1991, 29). Because these lessons are learned early in life and ritualized in repeated encounters (*A Closer Look* 13.3), they tend to carry deep meaning for most people. As with spouse abuse, those who repeatedly or severely beat their children often feel a moral righteousness about it and justify it to themselves, and to others, on the grounds that the victim deserved it. As we will see, this tends to have devastating effects on children, who feel responsible for their mistreatment and who are at risk for treating their own children in similar fashion.

The Causes of Child Battering

In over 90 percent of all cases where children are abused, the aggressor is a member of the immediate family. However, recurrent child abuse cannot go on without the compliance or at least passive acquiescence of other persons besides the abusing parent and the child. Others who know about it must allow it to happen or at least refrain from reporting it. Because there are strong pressures in our society supporting the right of parents to physically punish their child, this is

A CLOSER LOOK 13.3
Corporal Punishment as a Ritual of Family Solidarity

Corporal punishment is the background upon which child abuse arises. Many parents regard physical punishment not only as legitimate but desirable, especially for boys (Straus, Gelles, and Steinmetz 1980). For these parents, punishment is not just a necessary evil but morally proper. We can explain this by using once again the sociological theory of rituals introduced in chapters 8 and 9 to explain the social ties of love and sex. Although it may seem strange that punishment operates in a fashion similar to love, in both cases there can be mixtures of solidarity and conflict; when sexual arousal is mixed in, the result may be clinically a sadomasochistic attachment.

A ritual is an interaction that creates or reinforces feelings of social membership, drawing boundaries between insiders and outsiders and giving off symbols by which members show their loyalty to the group. The best-known rituals are formal ones, such as flag-salutes, weddings, graduations, and holiday gift-exchanges. There is also a type of ritual in everyday life that represents the degree of intimacy in personal interactions, such as greeting and departure ceremonies, kisses, and shared drinks and meals. Because rituals convey membership, the flaunting of an expected ritual carries the implication that the relationship is being violated. This is the reason why a typical minor family quarrel breaks out when family members do not come to dinner on time; the sitting together at dinner is a ritual, and failing to do so is not just a practical issue but conveys a little break in family solidarity.

Rituals are not necessarily consciously recognized. They can occur as "natural rituals" (Collins 1988, 197–99), whenever the basic ingredients are present: the group assembled face-to-face; awareness of a common focus of attention, excluding outsiders; and a shared emotional mood. If these ingredients are present, the emotion grows more intense among all participants, resulting in an additional emotion—the feeling of attunement, which manifests itself in a feeling of membership. It is through this process that actions—such as spankings—can become symbols of relationships. The physical punishment situation brings the punisher and punishee together face-to-face (or at least face to rump), creates a strong focus of attention, and causes an emotional buildup. The conscious contents of the emotion need not be the same on both sides, but there is always some contagion of

nervous excitement, anger, and fear. "This is going to hurt me as much as it will hurt you" is not a strictly accurate statement, but it does convey an emotional truth that something is shared.

The result is that physical punishment tends to create a ritual bond between parent and child. The bond can be full of anger, but it is typical of rituals that there are two levels of emotion simultaneously. (In a funeral, there is grief but also solidarity; in a political rally against an enemy, anger but also solidarity; in a spanking, there is anger and fear but also a close focus of attention between parent and child.) When rituals have done their work, the surface emotion dissipates as time passes, whereas the solidarity feeling continues as long as the ritual is periodically repeated. This is how punishment rituals come to carry a moral connotation in families, especially where punishment is the most intense form of interaction. Men who were brought up by physical punishment will often speak of how their fathers would beat them ("Boy would I get a whipping!") with an affectionate and nostalgic tone. Children are often punished for ritualistic pranks; there are deliberate violations of conventional order, as if daring and provoking punishment. Many families and communities in the more traditional sector of society go through a reciprocating cycle of pranks and physical punishment, not unlike the conflictual cycles of tribal vendettas.

The fact that physical punishments so often carry the unconscious message of family solidarity is one reason why the more blatant domestic assaults are often ignored or excused. After all, if a certain degree of violence on children is felt to be a good thing, it is easy to overlook the points at which the violence gets really severe and a child is emotionally or physically hurt. Furthermore, the very success of a family tradition of punishment rituals helps perpetuate it. When children who have been routinely subject to physical punishment go on to use the same methods when they become parents, they are not necessarily displacing their pent-up aggression onto a weaker victim. The former punishee is not merely taking it out on someone else. This is the intergenerational transmission of a tradition, a feeling of carrying on family solidarity from generation to generation. It is because of these kinds of traditions that so many people are resistant to doing anything about controlling corporal punishment or seeing how it slides over into child abuse.

frequently what happens. However, often the root causes of the abuse itself come from outside the immediate perpetrator and the victim.

Some parents who abuse their children do so because they themselves are psychotic. They account for a relatively small percentage of all cases of child abuse, although they do tend to be involved in a greater proportion of the cases with severe injuries (Garbarino and Gilliam 1980). More commonly, abuse is perpetrated by "normal" parents, using the "normal" pattern of physical punishment and family authoritarianism, which then gets out of hand. Often the situation starts out with some mild control problem, which is made worse by the parent's use of some extreme form of punishment. A vicious circle is set up, which provokes yet further punishment, until injury occurs. The small child who cries too much (which at an early age may well be due to physiological causes) can provoke a sleepy or harassed mother, father, or mother's boyfriend into spanking, which at that age has little effect except to make the child cry more. As this goes on night after night, the adult may become extremely abusive and even produce physical injuries.

What are abusive families like? The key seems to be a situational match between a child who is a problem and parents who have the wrong skills and motivations to be caretakers (Cicchetti and Toth 1998; Garbarino and Gilliam 1980; Pianta, Egeland, and Erickson 1989), such as the following:

■ Children who are especially demanding are particularly likely to be abused. Thus, children who are ill or handicapped, hyperactive, or suffer emotional problems are especially likely to be victims.

■ Parents who have unrealistic expectations about their role are especially likely to become abusive. Becoming a parent requires reordering one's priorities away from the self-centeredness of the courtship stage to caring for the needs of a small child. Many couples who marry especially young, who have children early in their marriage, or who have unwanted children as a result of carelessness with birth control are likely to be in this category. For them, the arrival of the baby is a shock. Although much attention has been focused on young fathers or boyfriends who abuse children, it is even more often the mother in this situation who finds her life suddenly made frustrating and confining by her new child and who reacts with child abuse.

■ **Abusive parents** also tend to hold unrealistic beliefs about children's abilities and often respond inappropriately to their children's behaviors. They often expect their children to perform in an impossibly developmentally advanced way or to exhibit levels of independence and self-control that are unlikely in children of their age (Trickett et al. 1991).

■ For example, some parents spank children younger than twelve months or even six months—ages at which the spanking clearly represents the parents' emotions rather than any rational policy for controlling the child.

■ Parents who have had little opportunity to learn the role of caretaker often run into difficulty. Some in addition were badly treated or abused during their own childhood or grew up without regular caretakers with whom they had affectionate bonds. Child abuse thus tends to be perpetuated across the generations. Although parenting practices tend to be transmitted across generations, this does not mean that young parents are locked into their own parents' style of parenting. Only about a third of parents who were abused when they were young abuse their own children (Cicchetti and Toth 1998). Mothers who break this intergenerational cycle are more likely than others to have had a warm, caring

Abusive parents tend to hold unrealistic beliefs about children's abilities and often respond unrealistically to their children's behaviors.

adult in their background, to have established a close marital relationship, and to have received therapy (Hetherington and Parke 1999).

■ Many abusive parents are socially isolated (Belsky 1993). They seem to have relatively few friends, close kin relations, or neighbors, whom they can turn to for help. This isolation may contribute to the fact that these parents frequently do not seem to recognize the seriousness of their behavior and blame the child rather than themselves for what is occurring. They may even justify their behavior by saying that they are doing it for the child's good or that harsh discipline is necessary if children are to be taught what is right (Hetherington and Parke 1999, 502).

■ Stress in the parents' lives also acts to precipitate child abuse, just as it affects spousal violence. Particularly fathers who are under economic stress, because of unemployment or career setbacks, tend to increase their violence at home. This seems to occur in tandem with changes in the male-female power situation at home. The husband's relative power vis-à-vis his wife goes down as his contribution to the family income is reduced; hence, he may fall back on his remaining resource, domination by sheer force. This results not only in wife abuse but in child abuse as well.

It has been estimated that close to one-quarter of all American families are in danger of abusing their children because of a combination of these factors. This seems to be revealed by surveys of how mothers interact with their infants, in which 25 percent of the mothers are "at risk because they have child-rearing attitudes or experiences characteristic of abusers" (Garbarino and Gilliam 1980, 31). Intervention programs have been successful in lowering rates of child abuse, particularly in families with young children. These programs focus on educating parents about child development, explaining practical child-rearing techniques (see *A Closer Look 12.6),* raising parents' self-esteem, and providing supportive social networks (Garbarino 1989, 1995).

Reports of child abuse or maltreatment involving day-care centers and foster care homes tend to attract considerable media attention, but children are not at greater risk in such settings. Public fears and publicity have created the impression that abuse is common in these out-of-home settings, but this perception is out of line with reality (Finkelhor, Williams, and Burns 1988). According to Wang's (1999) review of states reporting this statistic for 1997, about 3 percent of confirmed abuse cases occurred in day-care centers, foster care, or other institutionalized care settings. These figures have been consistent for over a decade.

Abuse of Young Children and Teenagers

> *My dad started grounding me for finding dirty spots on the dishes. The last time he grounded me like that, it went on for six months and it got so bad I had to start asking for everything: if I could get up, if I could go to the bathroom, if I could sit down and eat with him, if I could get ready to go to bed, if I could take a bath. You know, everything you take for granted. And his answer to me always was, "Do you deserve it, do you think you deserve it?" Well, of course I deserved to eat. I have to eat to live, you know. And it just got really bad. (Garbarino and Gilliam 1980, 12)*

There seem to be two rather different patterns of child abuse. There are parents who start abusing their children rather early and who continue it until the children leave home. Garbarino and Gilliam (1980, 127–33) call these "long-term abusers." On the other hand, in some families, violence and overstrict discipline against children do not begin until the teenage years. Mothers are more likely to be the abusers of young children, whereas in the teen years, authoritarian fathers are the most likely perpetrators.

Most of the factors that have been reviewed previously are especially relevant to the families of long-term abusers. The general pattern is that people who abuse their small children are particularly likely to be suffering from the problems of poverty and to reflect the cultural patterns of being economically marginalized and socially isolated. Child abuse occurs in all social classes and in all religious, racial, and ethnic groups. At the same time, there is evidence that chronic (long-term) maltreatment is most likely to occur in economically deprived, poorly educated families who tend to experience more of the risk factors mentioned earlier (Hetherington and Parke 1999). Most research shows poor people to be greatly overrepresented among abusers of small children (Gelles and Cornell 1990). Leroy Pelton (1978) has referred to this as "the myth of classlessness." Hitting small children—as compared with hitting adolescents—is much less likely to be deliberately provoked by the children and more likely to result from environmental stresses. This is not to say that the danger of child abuse cannot exist to some degree in all classes, especially given our prevailing cultural belief that corporal punishment is desirable. It is true that injured children from wealthier families are much less likely than poor children to be labeled as "abused" by attending physicians. However, it is misleading to ignore

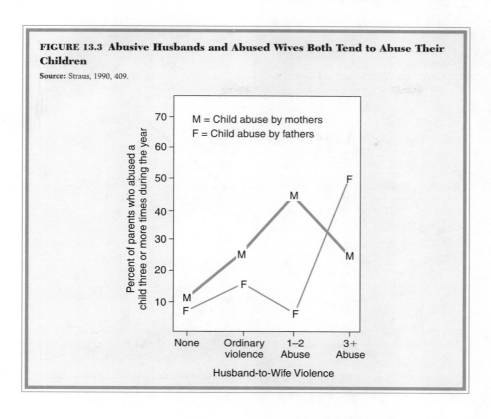

FIGURE 13.3 Abusive Husbands and Abused Wives Both Tend to Abuse Their Children

Source: Straus, 1990, 409.

poverty as a major cause of stress. In general, children of the poor are more likely to be exposed to violence in the home, just as they are more likely to be exposed to violence outside the home.

Abusers of teenage children (in contrast) have higher incomes and more stable family structures. Less is known about them and about the kind of interaction with their children that leads to the abuse. Teenage rebellion (see chapter 15) may interact with parental authoritarianism in an escalating spiral. Straus (1994) reports the highest rates of parents hitting teenagers in middle-class families, with lower rates in both lower- and higher-class families. He provides detailed analyses showing that lower-class youths are hit more frequently by their parents but that a greater share of middle-class parents use hitting as an occasional attempt to control their rebellious teenagers. Perhaps this pattern reflects tensions surrounding the declining economic security of the middle class and teens' growing sense of entitlement and awareness of their rights as individuals.

As noted earlier, researchers have also documented a pattern of the intergenerational transmission of violence. Parents who abuse one another are especially likely to have undergone severe physical punishment (which might now be classified as abuse) when they were children, and these parents are especially likely to abuse their own children. Straus (1990) found the amount of violence husbands used against their wives was correlated with the amount of abuse these fathers perpetrated against their children (Figure 13.3). Wife beating by husbands also correlated with the amount these mothers beat their children. In other words, the battered women tended to pass along the abuse to their kids. Straus also found that children who were beaten by their parents were highly likely to severely attack their siblings (Figure 13.4). All the way around, violence is contagious.

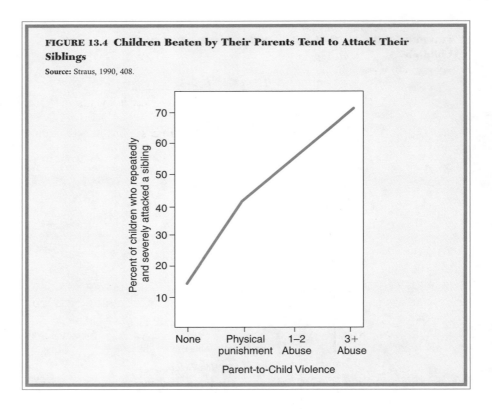

FIGURE 13.4 Children Beaten by Their Parents Tend to Attack Their Siblings

Source: Straus, 1990, 408.

Incest

If I was alone with my dad, he would touch and kiss me. I would try to please him in little ways because it was confusing to me. Then he came into my room one night. I was real scared, but he told me to relax and he wouldn't hurt me. It did hurt, but there were moments I remember enjoying. I mean, I was the ugly one and this was the closest thing I had ever experienced to love. The only affection I can remember were those times with my father. He told me he would kill me if I told, so I never told my mom. I still don't know if she knew. (Garbarino and Gilliam 1980, 151)

Most sexual abuse of children, like violent abuse, is perpetrated by their own families. Because of the extremely scandalous nature of incest, there are great pressures to keep it hidden, and it is difficult to estimate its actual rate of occurrence from the few cases that do come to light. National studies of cases reported to child protective services grossly underestimate the incidence of child sexual abuse, but even they report about 100,000 cases per year (*Statistical Abstract* 1999, table 378). In the mid-to-late 1990s, the number of child sexual abuse cases declined to about 7 percent of all reported cases of abuse or neglect (Wang 1999). Smaller studies have attempted to estimate the risk of being sexually abused. Russell (1986) found that one in forty-three girls (i.e., 2.3 percent) who had lived with their biological fathers (at least through age fourteen) reported they had been sexually abused; for girls who lived with stepfathers, however, the rate was one in six (16.7 percent). What researchers find out, though, depends on just what questions they ask. Questions that ask women generally about sexual abuse in childhood can get a prevalence rate of 6 to 14 percent, whereas more specific questions about sexual contacts with fathers or stepfathers will produce a prevalence rate of from 11 up to 62 percent (Finkelhor

1986, 43). For instance, the latter, very high figure (62 percent) is produced by asking women if their fathers ever touched them in ways that made them feel sexually uncomfortable. On a more restricted definition of sexual abuse, however, it still remains a sizable problem.

There are of course many possible types of **incest** (grandparent-grandchild, siblings, etc.), but of all these types, father-daughter incest is by far the most common. In a collection of various studies of incest cases, 424 involved incest between parents and their children. Of these, 399 were cases of father-daughter incest, 11 of father-son incest, 12 of mother-son, and 2 of mother-daughter incest.

As we can see, there is a small percent of homosexual incest cases; again, these largely involve the father rather than the mother. Mother-son incest cases were once believed to be so rare as to be nonexistent; in eight of the twelve cases reported earlier, the son forced his mother to have intercourse, so the incident could also be reported as rape. In most of these cases, the boy was psychotic or mentally retarded.

Most of the fathers who commit incest, on the other hand, are not psychotic (although psychosis has been described as responsible for 10 to 20 percent of cases [Garbarino and Gilliam 1980, 155]). Nor are these men socially "abnormal" according to the conventional culture. Incestuous fathers are often outstanding citizens to the public, with above-average levels of education and income. In fact, the fathers might be described as ultraconventional; not only are they likely to be churchgoing and "respectable," but they adhere strongly to traditional gender roles. They run authoritarian, male-dominated households, and their habitual violence is at a higher level than the average pattern (Herman 1981). Child sexual abusers themselves tend to have previously been the victims of childhood sexual abuse (32 percent, versus 3 percent of a control group of nonabusers; Finkelhor 1986, 102–3).

There is also sometimes a situational factor. In many cases the mother is ill, withdrawn, or actually separated from her children. Typically, the family roles have been rearranged, and a daughter has taken over the mother's role, either in helping to run the household or at least in being emotionally close to the father. Typically, there are poor relations between the parents and often between the mother and her daughter. Several studies found that victimized daughters were highly likely to be hostile or to dislike their mothers (79 percent in one study, 60 percent in another) much more than they were to feel this way toward their fathers (52 percent and 40 percent, respectively), even after the incest took place (Finkelhor 1986, 157). Incestuous affairs tend to go on for quite some time, typically beginning around age nine (though sometimes later) and continuing for three or four years. Most incestuous affairs are never reported; they end simply because the daughter grows old enough to move away, often by running away or early marriage. Although daughters may indeed feel considerable emotional attachment to their fathers (at least initially), they almost always have negative feelings about the incest. In about half the cases, the girls physically resisted or gave in under threat of force (Garbarino and Gilliam 1980, 157–58). Moreover, the long-term effects on the victims are often quite severe (Finkelhor 1986). Their self-images tend to be negative. Large numbers suffer from depression in their adult lives, almost half to the point of contemplating suicide. Drug abuse, alcoholism, and psychosomatic symptoms tend to be much more common than usual among these women, and they also turn out to be much more likely to become victims of spousal beatings (Herman 1981, 93, 99). The chain of victimization thus seems to repeat itself.

Is there any preventative for incest? Several courses of action are feasible. One is for families to become alerted to the conditions that tend to promote it (mother ill, withdrawn, or absent; a conventionally rigid, domineering father who is prone to violence). Another action is to explicitly warn children against keeping secret sexual pacts with adults. It is particularly important for not only the mother but the father to take part in these discussions. "If the father tells the child in a calm, guiding way, 'Never let an adult put his hands down your pants, and please don't keep secrets about that sort of thing from us,' the father has eliminated any chance he ever had of successfully victimizing his children. . . . If the mother alone warns the children, the warning might not clearly extend to adults who are part of the family" (Sanford 1980, 233–34).

ELDER ABUSE

In the 1960s and 1970s, child abuse became a pressing social issue as activists, researchers, service providers, and legislators turned their attention to violence against children. The 1970s and 1980s witnessed a similar concern over spouse abuse. In the 1980s and 1990s violence toward older persons burst onto the scene as a pressing social issue. Until then, **elder abuse** had gone relatively unnoticed. In the past twenty years, individual states became increasingly concerned with the problem of elder abuse both in private families and in institutions such as rest homes. As a consequence, new laws were introduced, and all fifty states now have statutes addressing elder abuse (National Center on Elder Abuse 1998). Although it is certainly not new, violence against the elderly is now beginning to achieve the status of a full-fledged social problem.

Research and policy on other forms of family violence laid the groundwork for focusing on the problem of elder abuse. In addition, because people are living longer and because a greater percentage of the American population than ever before is over sixty-five, more adult children are caring for their parents. The "sandwich generation" of baby boomers who delayed childbirth are now faced with caring for children and aging parents simultaneously (Carp 2000). Because there is more divorce than there used to be, as well as higher levels of single parenthood, and many more dual-earner couples than ever before, there is widespread anxiety over who will care for aging parents. The "crisis" in elder care seems particularly acute because health and other community services to assist stressed caregivers is inadequate, fragmented, and costly (Bruce 1994). As public concern over the care of the elderly has increased, more attention has been paid to the levels of abuse they endure.

As with other forms of violence, older people are more likely to be abused by family members than by strangers. We do not yet have accurate estimates of the extent of elder abuse. Like other forms of family violence, it tends to be underreported, and few surveys have focused extensively on the older population. Current estimates of the number of abused elders and frequency of abuse are based on reports to social service agencies. The National Elder Abuse Incidence Study found that approximately 450,000 elderly persons in domestic settings were abused and/or neglected during 1996. When elderly persons who experienced self-neglect are added, that number increases to approximately 551,000 (National Center on Elder Abuse 1998).

Elder abuse is a general term that can include physical, psychological, and material mistreatment or neglect. Although many older people are independent and in good health, the chances of poor health increase with age. Persons over the age of seventy-five are at increased risk of physical, mental, and financial dependency, which, in turn, make them susceptible to the following forms of abuse (Block and Sinnott 1979; Carp 2000; Gelles and Cornell 1990; Lau and Kosberg 1979; National Center on Elder Abuse 1998):

1. Physical violence, such as hitting, slapping, shoving, and other tactics common to spouse or child abuse

2. Psychological abuse, such as verbal assaults, threats, intimidation, and isolation

3. Physical maltreatment, such as physical restraint, excessive medication, or the withholding of personal care, food, medication

4. Financial or material exploitation, such a theft or misuse of money or other personal property

5. Violation of personal rights, such as forcing the elderly out of their dwellings or into other settings, such as nursing homes

6. Neglect, including the withholding of care or ignorance of the needs of the elderly

In summary, the American Medical Association (AMA) defines elder abuse and neglect as "actions or the omission of actions that result in harm or threatened harm to the health or welfare of the elderly" (Jones 1994, 846).

The most frequently cited estimates of elder abuse range between 450,000 and 2.5 million cases per year (Gelles and Cornell 1990; National Center on Elder Abuse 1998; Pillemer and Finkelhor 1988). These figures indicate considerably more cases than are actually reported to authorities (Carp 2000; Tatara 1989). Pillemer and Finkelhor (1988) suggest that only about one of fourteen cases of elder maltreatment come to the attention of the appropriate authorities, with at least 3 percent of all elderly being the victims of physical violence, verbal aggression, or neglect. For all these forms of abuse, the victim's spouse or adult children are most likely to be reported as the abuser.

Factors Associated with Elder Abuse

Women of very advanced age are the most likely victims of elder abuse (National Center on Elder Abuse 1998). Individuals with physical or mental impairments run the greatest risk of being abused, though major handicaps are not necessary precursors to abuse. For example:

> *An elderly husband and wife, aged 81 and 79, respectively, had by their own reports always had a "difficult relationship." The wife reported that her husband had struck her previously, but never repeatedly or severely. The situation became much worse when she suffered a severe stroke. She could not accept her new physical limitations and complained about and insulted her husband. During arguments, she would throw food and objects at her husband, which infuriated him. He responded by pushing and striking her more severely than ever before. (University of Massachusetts Center on Aging, 1984; cited in Pillemer and Finkelhor 1988)*

Other than spouses, middle-aged daughters caring for aged parents are the most numerous abusers. The fact that women are overwhelmingly responsible for elder care, however, masks the fact that men tend to be more violent, regardless of the age or gender of the victim. When researchers disaggregate elder abuse according to type, they find that men are more likely than women to use physical force and violence (Carp 2000; Filinson 1989; Sengstock 1991).

Elder abuse typically occurs under conditions of high stress. First and foremost on the list of stressors is the enormous financial burden of caring for the elderly; medical, food, housekeeping, and transportation costs can be very large. Second, the physical requirements of tending to an elderly dependent are substantial. The personal care, feeding, cleaning, and transporting tasks associated with elder care can be exhausting. Third, the emotional strain of caring for an aged dependent can be overwhelming. Watching and caring for a spouse who is deteriorating physically and mentally can be devastating. Similarly, caring for a dependent parent can entail special emotional burdens, particularly if there are unresolved issues from the past. Parents may continue to treat their adult caretakers as incompetent children or, conversely, may reverse the traditional roles and treat them as omnipotent parents. In either case, the emotional strains associated with power and dependency issues are often quite intense. Finally, multiple competing obligations often add significant stress to family elder care situations. Simultaneously meeting the needs of spouse, children, ailing parent, and job can prove to be difficult, if not impossible (Block and Sinnott 1979; Carp 2000; Gelles and Cornell 1990).

Another factor that makes the elderly vulnerable to abuse is their generally devalued status in society. Older people are subject to what has been labeled **ageism.** They are often forced to retire from productive occupational roles and can be isolated socially and geographically. Older people also tend to be discriminated against because of their age and are often judged on the basis of unfavorable stereotypes. Insofar as older Americans accept such stereotypes, they can be more willing to accept abuse. Insofar as adult children and other people define senior citizens as "useless," they may be more likely to perpetuate or condone elder abuse.

More and more individuals are surviving to old age. In fact, the most rapidly increasing portion of the U.S. population consists of individuals over the age of eighty-five (Dortch 1995). In 2000, there are twenty-six times as many Americans 85 and older as there were in 1990 (Carp 2000). The number of people over 100 years old is expected to increase at least ten-fold by 2030 (Carp 2000). With the aging of the American nation, we will undoubtedly see increased attention paid to the needs of older Americans and the emerging problem of elder abuse.

COMMON FEATURES OF FAMILY ABUSE

David Finkelhor (1983; Pillemer and Finkelhor 1988) has summarized some of the commonalities among the different forms of family abuse. Briefly, he suggests that spouse abuse, child abuse, and elder abuse are similar in the following ways:

1. *Abuse of power.* In general, it is the most powerful family members who abuse the least powerful. Thus, men tend to sexually abuse young girls; parents (especially men) tend to physically abuse young children; husbands tend to physically abuse wives who have few contacts or resources; and healthy caretakers tend to abuse frail elderly dependents.

2. *Response to perceived powerlessness.* Although family abuse is perpetrated by the strong against the weak, the violent acts tend to be acts carried out by abusers to compensate for the perceived lack or loss of power. Thus, men tend to beat their wives when they feel insecure or powerless and to abuse their children when they are unemployed or failing financially. Parents also tend to use physical punishment against children when they feel they have lost control of their children and their own lives.

3. *Shared effects on victims.* Not only are victims of family abuse exploited and physically injured, but they tend to be "brainwashed" into believing it was their own fault. Thus, abused children believe they are bad and unlovable; abused wives are persuaded that they are incompetent, hysterical, and frigid; sexually abused children are made to believe that their father's actions are normal and express love; and abused elders come to see themselves as a worthless burden. Victims of all types of family abuse tend to be intimidated by their abusers but often remain loyal to them, at least until they can leave. Long-term effects of all types of family abuse include self-blame, lowering of self-esteem, sense of stigma, despair, depression, suicidal feelings, self-contempt, and an inability to trust and develop intimate relationships.

4. *Shared characteristics of abusing families.* Many forms of family abuse tend to be more common in the lower socioeconomic strata; under conditions of unemployment or chronic poverty; in families with a male-dominated patriarchal structure; in families with drug or alcohol problems; and in families that are socially isolated. In addition, the various forms of family violence tend to co-occur, and children raised in abusive families are at risk of repeating those patterns in their adult lives.

WHEN CONFLICT TURNS INTO FAMILY VIOLENCE

All families have some conditions that cause conflict. When do these conflicts turn into violence? When do they remain relatively peaceful? The extreme forms of violence and abuse that we have reviewed in this chapter are related to the family realities that we have been analyzing elsewhere in this book. As we have seen, an intimate relationship or marriage is created by a process of bargaining and negotiation, even if it takes place at an emotional and symbolic level. Within the relationship, partners have greater or lesser power, depending on resources such as their occupations, networks, and sexual bargaining position. Power always has the potential for causing conflict, as does stress, which arises in any situation of change or difficulty. On top of this, traditional authoritarian roles that are found in some families exacerbate conflicts and cause them to turn more readily into family violence.

Shifts in Family Power

Power has the potential for creating conflicts, but these conflicts are likely to stay beneath the surface as long as the power situation is stable. No one really likes being controlled by others, and this situation generates at least a latent sense of resentment. However, powerless people tend to react passively if they can do nothing about the situation. It is when the power situation is changing that conflict is likely to come to the surface.

Until recently, men have held most of the resources that give individuals power in families. As we have seen in chapter 11, they have generally had more income, more powerful occupations, more education, and wider social networks. These conditions have been changing in the last few decades. It is certainly true that women have not yet caught up with men in income and occupational level, but women have been shifting into the full-time labor force. This has moved their power position, at least some. Women have increased their educational level much more rapidly, and it now exceeds that of men. These changes in education and in employment have no doubt expanded women's social networks, which is an important source of power in the home.

All these changes have been shifting the balance of power in marriages. It should not be surprising that there is an upsurge of family conflict at this time. This may turn out to be a temporary situation. In the long run, if women's occupational position continues to improve, families may stabilize at a point where power resources of men and women are more equal. When power resources stop shifting, we might expect that this source of conflict will settle down.

Stress Surrounding Family Life

Another source of conflict and potential violence within the family is the amount of stress that people are undergoing. Research on stress has turned up some points that are somewhat surprising. As we would expect, negative events are stressful, such as becoming unemployed, having one's car repossessed, or having problems with money. It is not surprising that families in these situations, especially working-class and poverty-level families, tend to have outbreaks of family violence. There are other sources of stress as well, including any major change in life circumstances. Taking a new job is stressful; so is the arrival of a baby. Many events that we might regard as happy events are also stressful and can be associated with an increase in conflict and violence.

For this reason, conflict tends to break out around any major family transition. The very process of getting married is a life change that can be quite stressful, particularly when the partners don't know what they are getting into. Similarly, the transition from childlessness to the pressures and responsibilities of being a parent is another stress. Some of these factors interact; for example, a woman's family power tends to go down when she has a baby because pregnancy and early motherhood usually reduce her economic power and her social networks. Putting these various factors together, we can see that family conflicts can build up when a new baby arrives and that the shifting power situation may lead to an outburst of violence that may encompass both wife battering and child abuse.

Of course, not every family that has a baby responds to the new stress by an outburst of violence. In fact, most families do not. How often these stresses and problems lead to open conflicts depends on how many resources the family members have for dealing with the situation. If they are economically strapped, lack social support networks, or are young and inexperienced at coping with household problems, then the rate of violence is likely to be much higher. Where various stressing conditions intersect, family violence tends to occur.

Traditional Authoritarian Gender Roles

Finally, we come to a social pattern that not only fosters violence but also makes it hard to eradicate. When families are organized around localized, encapsulated networks, there tends to be a sharp segregation of men's and women's roles. This traditional structure produces a high degree of pressure for conformity. People take their positions as rigidly fixed and immutable. They tend to see the world moralistically, with traditional behavior clearly marked as "right" and any other kind of behavior as "wrong."

Traditional authoritarian gender roles tend to promote violence in the family.

These families have a world view in which "men should be men, and women should be women," and the roles and powers of the two sexes should not be altered. They also draw a rigid line between the positions of parents and children and believe the power of the parents should be strongly enforced. *Violence in this type of family often has a symbolic significance.* It is a way of ritually acting out the traditional authority relationships. This is why violence can be set off by incidents that seem trivial, such as a husband getting furious about the way dinner was prepared or hitting the children for making noise at the wrong time. To the authority figure, the incident is taken as a provocation, a message that authority is being challenged and that the traditional relationship is threatened. Violence is a ritual for putting tradition back in order.

This is in keeping with ritual interaction theory because structures of enclosed, high-density interaction fill everyday life with rituals that people are supposed to conform to. (See Collins 1981 for an application of this theory to the patterns of violence throughout world history.) The ritual of spanking children is a way of reinforcing the authority of the parents, but it also reaffirms that the children are a part of the enclosed structure. Both parents and children become firmly attached to these rituals; children even grow up with a sentimental admiration for how tough their father was. In the same way, these traditional families tend to believe that men have the right to control their wives with force, that beating one's wife "is good for her." Because this is a ritual that acts out the marriage relationship, many wives who are deeply embedded in the structure end up interpreting the beating as a sign of affection.

This kind of traditional structure is one that takes sexual property rights seriously. Sex is highly ritualized, and any sign of disloyalty is regarded as a threat to the whole system of family authority. Husbands regard themselves as entirely within their rights to violently beat their wives for adultery or sometimes even for any action provoking jealousy. This is one area in which violence is likely to be more reciprocal; for sexual possession is supposed to go both ways, and wives also become intensely jealous of a husband's sexual behavior. A woman who is otherwise quite traditional and passive in her marriage may attempt to kill her husband when she feels she has been wronged in sexual matters.

This highly ritualized family structure reinforces violence because it creates a belief that the violence is morally justified. The violent husband—and sometimes the violent wife—is righteously angry and morally outraged at what he or she believes a spouse has done to violate "the way things should be." For this reason, the police are reluctant to intervene in family disputes because they are often confronted by a spouse who believes that (usually) he is acting within his rights and that no one outside the family has a right to interfere.

The Combination of Violence-Producing Conditions

When the several kinds of conditions we have reviewed here overlap, family violence is most likely to be severe; however, much of the time they do not overlap. Not all traditional family structures are abusive; as long as the power relationships remain stable and no one challenges them, as long as there are no important family stresses, and as long as there are plenty of economic and social resources for dealing with stress, things may be peaceful. Traditionalists like to point to this as the ideal family situation—the old-fashioned family in which everyone knew his or her place, husbands and parents were tough but loving, and everyone felt secure as a member of the family. The ideal combination of circumstances probably did not happen often, even in earlier historical periods. When conflict did break out in the traditional family structure, life could be extremely cruel.

Most sectors of modern society are moving away from that structure, but elements of it remain. It is most often found in working-class families but also lingers in middle-class families from traditional backgrounds and communities. The conditions of living in the poverty class may also re-create some of the same kinds of localized, encapsulated, and gender-segregated structures. On the whole, though, the trend is toward a different kind of structure. The newer structure is one in which the family is more of an open network, more connected to the outside society, with more equal power relations between husbands and wives and much less authoritarian control over children. The very fact of transition, of shifting power relationships in the family, tends to increase the level of conflict. In the long run, we can expect that family relationships will settle down at a new balance of power. As this comes about, we can look forward to a reduction in the level of family violence.

In the meantime, there are things that educators and social service providers can do to reduce family violence. Although a combination of stressors tends to increase the chances of family abuse occurring, a majority of people who are exposed to several risk conditions do not resort to violence. This makes intervention and education especially important. Social programs can break down the sense of fear and isolation that often accompanies family abuse, and with legal sanctions, treatment programs can help both perpetrators and victims redefine taken-for-granted violence as inappropriate and unacceptable. Interventions aimed at helping those undergoing stress associated with job loss, new babies, or the infirmity of a family member can also help to break the cycle of violence by encouraging people to develop realistic expectations about all aspects of parenting and family life. Just talking about the strains and hassles of everyday life and knowing that others face similar difficulties can bring a great sense of relief to family members who might otherwise resort to violence. Learning some simple self-help techniques to use in crisis situations has helped many people avoid re-creating ritualized family abuse.

We cannot expect family violence to disappear if the underlying conditions promoting it remain. As long as our culture idolizes violence, our schools and churches teach that physical punishment is moral, our marriages reflect **traditional authoritarian gender roles,** and our economy keeps a substantial portion of our workers in poverty, then family violence will continue at today's high levels. With change in at least some of these underlying conditions, private and government efforts at intervention and education are likely to meet with some success.

SUMMARY

1. Violence is widespread in American society, as evidenced by homicide and rape rates that are much higher than in most other industrialized countries. We idolize tough guys in the movies, portray aggressive male-dominated sexuality on television, and tend to believe that children should be physically punished for wrongdoing.

2. The family is the most common context for violence in American society. Although family violence often goes unreported, police still receive more requests for help with domestic disturbances than any other problem. Battering of wives and children is an extension of "normal" patterns of discipline that derive from patriarchal models of family authority. Child abuse, wife abuse, and elder abuse have recently been recognized as important social problems, but efforts to reduce the amount of family violence have been hampered by the assumption that the family is a private domain.

3. Although reporting problems make estimation difficult, at least one-quarter of American families will experience some violence between family members, and a substantial fraction will experience fairly serious violence. Although women push, shove, and hit men about as often as the reverse, women's violence is often in retaliation or self-defense, and men's violence against women is much more likely to result in serious injury. Common couple violence can be gender symmetrical, whereas men use patriarchal terrorism to exercise control over "their" women. Wife beating happens most frequently when the husband is poorly paid or unemployed, when there is great worry and dissatisfaction over economic resources, when the woman is a full-time housewife, and when the couple is socially isolated.

4. Men, and often the surrounding community and the police, justify violence against wives as just punishment for violation of male prerogatives of power and especially of sexual property. Economic strains also set off domestic violence, especially when the man is unable to support his family and feels downgraded by his wife's economic contributions. Violence is passed along intergenerationally; men who beat their wives tend to have experienced much physical punishment as children, and battered wives tend to use more violence on their children.

5. Social control does reduce domestic violence. Socially isolated families have more violence, and socially isolated women are less likely to leave an abusive relationship. Calling the police has some deterrent effect, especially when the man is arrested. In general, men hit their wives because they can get away with it.

6. Violence against children includes a very extensive use of mild physical punishment (spanking, slapping), but also injury-causing or even lethal beatings and use of weapons by parents on children, children on parents, and children on each other. Most child battering is not done by psychotic parents but by "normal" ones in particular circumstances. Children who are especially demanding, including handicapped children, are especially likely to be abused. Abusive parents tend to be those who have little commitment to the role of parent; those who themselves were abused in childhood or otherwise did not have an opportunity to learn the role of a good parent; and those who are under special stress, such as economic problems. Nearly 25 percent of all American children are judged to be at risk because of some combination of these factors.

7. Mothers are more likely to be the abusers of small children, and fathers of teenage children. Abuse of small children is most likely in the lower social classes; teenage abuse is more common in the middle class. Women who are abused by their husbands are most likely to use severe violence against their children.

8. Most incest is of the father-daughter kind. It is most likely to occur when the mother is withdrawn from the family and especially from her children, as a result of illness, absence, or family strain. Incestuous fathers tend to be ultraconventional persons in public and highly traditional about gender roles, who run authoritarian households and use above-average levels of violence. In about half the cases, the incest is forced on the daughter. Most cases are never reported but terminate when the daughter runs away or marries to leave home. Long-term effects on the daughters are quite negative, including high tendencies to alcoholism, suicidal depression, and becoming a victim of spousal beatings.

9. Elder abuse, including physical, psychological, and material mistreatment or neglect, is less studied than wife abuse or child abuse. Women of very advanced age run the greatest risk of abuse, which tends to be perpetrated by spouses or middle-aged daughters. Elder abuse, like other forms of family violence, tends to occur under conditions of high stress and social isolation. Because the elderly population is growing rapidly, we will undoubtedly see more attention focused on this issue in the future.

10. Violence often occurs within families that believe in traditional authoritarian gender and parent roles. A relatively insignificant event frequently triggers the violence in these kinds of families, and the physical outburst symbolically reaffirms relations of dominance and hierarchy.

11. The common features of family violence include the following: (1) it is an abuse of the powerless by the more powerful; (2) it is often a response to a perceived lack or loss of power; (3) its victims tend to suffer low self-esteem and blame themselves; and (4) it occurs under conditions of poverty, stress, male dominance, and social isolation.

Key Terms

abusive parents
ageism
battered women's shelter
    movement
child abuse
cobras
common couple violence
condoning violence
conflict tactics scale

corporal punishment
economic strains
elder abuse
incest
intergenerational transmission
    of violence
intimate partner violence
marital rape
myth of battered husbands

patriarchal terrorism
pitbulls
rape myths
sexual property rights
sibling violence
social control
social isolation
traditional authoritarian gender
    roles

Sociology Web Site

See the Wadsworth Sociology Resource Center, "Virtual Society," for additional
links, quizzes, and learning tools:

http://www.sociology.wadsworth.com

also on this web site you'll find InfoTrac College Edition, an online library of
journals. Here you can search for electronic articles about central topics in
sociology.

14

Uncoupling-Recoupling: Divorce, Remarriage, and Stepfamilies

INTRODUCTION

It is not surprising that there is such a thing as divorce, given what we have seen in previous chapters. Marriage is an arrangement of power, privilege, and conflict, as well as of love and solidarity. Often it is all of these at the same time. When the balance of emotional, economic, and sexual resources changes, the marriage too has to change its arrangement between the partners or else break up. Some marriages dissolve because they were not too well negotiated in the first place, whereas others come undone after a longer period of struggle. Many marriages do not come apart, of course, although some couples perhaps would be better off if they did separate instead of carrying a de facto emotional divorce inside a hollow frame. Others probably separate too soon, turning to divorce instead of working through marital problems by learning how to communicate more effectively. For all this, there is also remarriage, which happens to a large number of divorced people. It appears that we can't quite stand marriage, but we can't stand being without it either. In this chapter, we attempt to see how and why this is so.

HOW MUCH DIVORCE?

In an average year, about 1.2 million divorces are carried out in the United States. Is this a high or low number for a country as large as this one, with a population of some 270 million? Like most things, it depends on one's perspective. About 2.3 million marriages take place every year, so one can say that there are more marriages than divorces and that divorces are half the number of marriages (*Statistical Abstract* 1999). Most of the people who divorce in any one year, however, did not get married the same year but throughout a period that can stretch forty years or more into the past. Hence, statisticians seek alternative ways of calculating the divorce rate. One way is to divide the number of divorces by 1000 population. Because a marriage requires two people, one can count either the married men or the married women; statisticians for some reason usually count by the number of women, although (as already mentioned in chapter 11) there is a small tendency for more women to say they are married than for men to say so. At any rate, these figures are relatively easy to come by from the U.S. Bureau of the Census, and the result is a chart like Figure 5.3.

This shows us that the **divorce rate** is about 4 per 1,000 population. It also indicates that the divorce rate rose during most of the twentieth century, with dips during the 1930s, 1940s-50s, and 1980s-90s. The first noticeable dip was during the depression. Then, during World War II and immediately after, the divorce rate skyrocketed and then fell again almost as abruptly. War-related problems seemed to play havoc with the American family. It was the largest war ever fought by the United States, mobilizing over 16 million troops. It is conjectured that many marriages were contracted in haste, as the groom faced possible death in combat. In addition, the temporary separation of families at this time broke many bonds of affection. When couples were reunited, many were not compatible. Hence, the peak of the divorce rate was in the two years immediately after the war (see Figure 5.3). Aside from this episode, the curve has generally been upward: what seemed at the time to be an unprecedentedly high divorce rate in the early 1940s became the normal rate twenty years later and was surpassed by still higher rates in the 1970s. Only in the late 1980s did the divorce

When the war began, many couples married in haste. After the war, however, the divorce rate peaked.

rate dip down again, although it is standing at a level that is essentially at its all-time high.

Just how high it is can be seen by looking back a century. In 1860 the divorce rate was about 1 per 1,000 marriages. It grew steadily to about 8 per 1,000 in 1920 (the "Jazz Age"), more slowly to about 10 per 1,000 in 1960 (ignoring the sudden up-and-down around World War II), and then more than doubled in the next generation. There is no doubt that we have gone through the single largest and sharpest rise in divorce in American history and that the family system has been transformed by divorce as never before. Back in the 1800s, marriages were normally ended by the death of one of the partners: 97 percent were dissolved by death and only 3 percent by divorce (Thwing and Thwing 1887). Today, marriages end more or less equally by death or by divorce.

Another way to calculate divorce rates is to estimate the proportion of all marriages in any given year that will end in divorce (Figure 14.1). In 1890 the chances that a marriage would eventually end in divorce were only about one in ten. By 1930, the chances were about one in four, and by the 1970s, the chances were one in two. This is the rate where we have more or less stabilized today; hence, one would have to say that for marriages made today—the marriages of

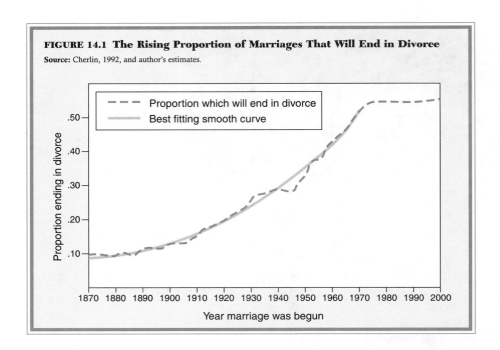

FIGURE 14.1 The Rising Proportion of Marriages That Will End in Divorce

Source: Cherlin, 1992, and author's estimates.

students who are reading this book—the statistical likelihood of divorce is about 50/50.

Divorce rates are higher in the United States than in Canada, Japan, and most of the western European nations. In fact, our rates have been about double the rates of these other countries since 1960 (Sorrentino 1990). However, divorce rates have gone up as fast or faster in other industrialized countries than they have here. Whatever has been going on, a unique breakdown in American cultural values is not the culprit.

When Does Divorce Happen?

The average divorce happens about six years after the marriage (U.S. Bureau of the Census 1992b). About 5 percent of all divorces happen within the first year, and the rate peaks at about 9 percent in both the second and third years. Then it starts falling again; until by the tenth year of marriage, two-thirds of those who are going to divorce have done so, and by the twentieth year of marriage, 90 percent of the divorces that are going to happen have already occurred (London and Wilson 1988; U.S. Bureau of the Census 1992b). The first few years of marriage are thus the main danger point.

Social Class and Divorce

If one reads the tabloid press, one would get the impression that rich people are the specialists in divorce. Much publicity is given to movie stars, but in general, the picture is just the opposite. The lower the social class, the more likely the marriage is to break down, whereas the higher the social class, generally speaking, the more likely the marriage is to survive. This is particularly true if we count desertions as well as formal divorces. Especially in the lower social classes, many people break up a marriage by simply leaving, and they never file for a formal divorce decree. This happens most often in hard economic times, when the expense of going through the legal arrangements is not felt to be worth it.

Research has consistently found that poverty is associated with a greater risk of separation and divorce, as well as higher rates of family violence (Rank 2000). As discussed in previous chapters, unemployment or chronic low-wage work puts great strain on a marriage or couple relationship. Levels of marital happiness and general well-being go down when families are under economic stress (Conger and Elder 1994). In addition, financial difficulties interfere with good parenting and other supportive family practices, adversely affecting children's well-being, which in turn creates more stress for the parents. It is no wonder that chronic poverty leads to dissatisfaction with marriage and to a greater likelihood of divorce or separation. Unfortunately, breaking up does little to solve the financial difficulties that are at the root of the problem.

If we use education as a measure of social class (this information is more easily standardized than information on people's occupations, which it closely parallels), we find that generally, as people's educational level goes up, the level of divorce goes down. This is true for both men and women, but educational level has tended to favor marital stability for men more than it has for women. For men, higher education generally brings a higher occupational level. These more affluent men are not only more likely to marry than poorer men but also more likely to keep their marriages together (Sassler and Schoen 1999). Apparently being married and staying married are part of the traditional respectability for men in the higher ranks of the class system. Moreover, until recently, men were obtaining most of the graduate education and hence usually married women with less education than themselves. Even now, men tend to monopolize most of the higher paying and prestigious jobs. Thus, high-ranking men have a lot of pull on the marriage market, and once they are married, they tend to have the resources to dominate their wives. These men's wives often provide considerable support for their careers and their egos and maintain a high level of domestic comfort. As far as affluent men are concerned, marriage is a good deal, and they don't usually want to lose it (see *A Closer Look* 6.5). From their wives' point of view, things may not be as rosy as they seem. However, in terms of hard economic realities, these women tend not to have the job opportunities to keep up their standard of living if they divorce. Thus, the economic situation tends to keep the couples together.

Unlike men, highly educated women who are lawyers, physicians, professors, and other highly skilled professionals and administrators have not been more likely to get married or stay married. Even given economic gender discrimination (i.e., female professors don't make as much as male professors, and so forth), highly educated women are comfortable in terms of their own incomes and careers. Such women are among the least likely to marry at all, and if they do, they are among the most likely to divorce. It also turns out that they are least likely to remarry after a divorce. This is probably because they are least willing to put up with traditional male domination of the marriage. They have the most resources to back up their own point of view, and if they cannot reach an egalitarian settlement, they are most willing to go off on their own. Also, highly educated women tend to be older when first married and older when divorced, so the pool of eligible mates for remarriage is smaller for them than for other women.

Age and Divorce

Another pattern is that divorces happen more often the younger the couple were when they married. Among women who marry at age sixteen or seventeen, over

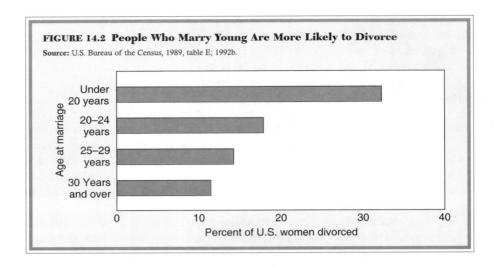

FIGURE 14.2 People Who Marry Young Are More Likely to Divorce

Source: U.S. Bureau of the Census, 1989, table E; 1992b.

a fifth are divorced within the next ten years. In fact, the rate is high for all teenage marriages. For example, a study by the U.S. Census Bureau (1992b) found that 32 percent of the women who had married when they were teenagers had divorced, but women who had waited until age thirty or more before they first married included only 12 percent divorced (Figure 14.2).

The effect of age, however, is mixed up with that of social class. Because persons from the lower social classes tend to marry earlier than those in the higher social classes, it may be that the age effect is secondary. The teenage marriages that show up on the chart may be mainly working-class and lower-class marriages, and that is the reason why the rate is so high. This seems to be true, but it probably accounts for only part of the divorce rate at young ages. Persons from any social class seem somewhat more likely to divorce if they marry young, and the effects of early marriage on divorce seem to be independent of early childbearing (White 1991).

Race and Religion

For a variety of reasons, there tends to be a higher divorce rate among African Americans than among whites, Asian Americans, or Hispanics in the United States. We have examined some of the causes in chapter 7: greater economic strains among the portion of the black population that is poor but also a particularly independent style among black women that makes them less willing to put up with male-dominated marriages.

As noted earlier, black women are less likely than others to marry in the first place and also more likely to divorce. According to the last Census Bureau report on divorce and remarriage, over nine of ten white and Hispanic women and about eight of ten black women had been married. Of these, over 35 percent of whites had been divorced, compared with about 45 percent of blacks and 27 percent of Hispanics (*Statistical Abstract* 1999, table 161). These figures tend to underestimate the extent of divorce (because they miss women who divorce after age forty-four and those with multiple divorces), but the percentages are useful for comparing race/ethnic groups and looking at changes over time. From the 1970s to the 1990s, the proportion of all women who had divorced was lowest for Hispanic women, with white women in the middle and black women most likely to divorce. About 10 percent more of each group was divorced in the 1990s compared with the 1970s, though the rate of increase was actually higher for whites and Hispanics than for blacks (Teachman 2000). Patterns for remarriage followed a slightly different pattern, with two-thirds of the white women but only half of the black and Hispanic women remarrying.

Because race/ethnicity and social class are correlated, these differences in divorce and remarriage are partially the result of economic factors. Groups and individuals with fewer resources can be expected to be less likely to form or stay in long-term relationships, but cultural factors come into play as well. For example, Latinos earn less, on average, than African Americans, yet they have higher marriage rates and lower divorce rates. Other influences, such as religion (discussed later) might account for these differences. Within the general category of "Hispanic" (the overarching term used by the Census Bureau, see chapter 7) there are differences between ethnic groups. Cubans are least likely to divorce, Mexicans are in the middle, and Puerto Ricans have divorce rates that approach those of African Americans. These patterns also appear to be driven by a combination of economic and cultural forces, because Cubans are somewhat more likely to be middle class and Puerto Ricans are somewhat more likely to live in poverty than Mexican Americans.

Religion has traditionally been very important in people's understanding of divorce, just as it plays an important role in shaping understandings of marriage. The sociologist Karla Hackstaff (1999, 40) describes how "marriage culture" has been promoted by Western Judeo-Christian religions in at least four interrelated ways: "(1) historically, religious institutions have regulated marital practices, sanctioning marriage and censuring divorce; (2) the marital practices of the religiously committed suggest that divorce is experienced as a last resort; (3) theoretically, religion has given meaning and purpose to family practices, especially marriage; and (4) religious institutions have generally reinforced the link between male dominance and marriage culture." Most Judeo-Christian teachings traditionally held that marriage is forever, so divorce was denounced and remarriage discouraged. In the twentieth century, however, most U.S. congregations have come to believe in divorce "as a last resort" (Bellah et al 1985; Hackstaff 1999). Even the traditionally divorce-resistant Catholic Church became more liberal between the 1950s and 1970s, with an increasing number of annulments and divorces among Catholics (McCarthy 1979).

In the 1980s and 1990s the influence of religion on divorce has continued to decline. It used to be that Protestants had higher divorce rates than Catholics. This is not surprising, given the Catholic Church's doctrine prohibiting divorce, but the rate of Catholic divorces has nevertheless gone up, parallel with Protestant divorces. By the late 1980s, several researchers reported that Catholics were not less likely to divorce than were non-Catholics (Bumpass 1990). Interreligious marriages, like interracial marriages, have also been subject to slightly higher divorce rates (Laumann et al. 1994), but in general, the influence of religion on marriage dissolution seems to be declining.

CAUSES OF DIVORCE

Why do divorces happen? Obviously, because people are dissatisfied with marriage and feel they would be better off apart. However, there is more to it than that. Divorces are somewhat like revolutions: the regime may be oppressive, but it doesn't come tumbling down just because people don't like it; something more specific has to happen that breaks it apart.

In our folklore of soap operas and gossip, the event that breaks a marriage apart is usually an illicit sexual affair. As we have already seen (chapter 9), most extramarital affairs are not discovered, and when they are, about half of the time, the aggrieved spouse forgives it. Sexual infidelity, although dramatic, is not really the most important cause of divorce.

Sexual relations of a different sort, though, may be quite important. Chapter 9 indicated that sexual dissatisfaction in marriage is fairly widespread and that it is correlated with a generally low level of marital happiness. The general unhappiness seems to prevent good sexual relations, which means that if individuals want to improve their sex lives, divorce may be the only route they know how to take. In fact, it does turn out that second marriages are sexually more satisfying than first marriages. Not only do people report that their sex lives are improved in quality, but even the frequency of sexual intercourse goes up. This happens even though people are older in their second marriages, and the rate of intercourse tends to decline with age. So it looks as if some people leave their first marriages in search of a better sexual relationship.

Economics and Divorce

Even though sex and other interpersonal relationships may be a motivation for getting divorced, they aren't the whole story. We can see this from the general economic trend. In times of prosperity, the divorce rate trends to go up; in times of economic depression, it drops (see Figure 5.3). During the Great Depression of the 1930s, the divorce rate dipped. This happened for all social classes. To some extent, the reason is that during hard times people are less able to afford a divorce, so instead, the couples simply separate, or one spouse (usually the husband) deserts the other. There is also some indication that families feel the need to stick together for economic support when they are in financial straits. To put it another way, when job opportunities are bad and the level of unemployment is high, spouses feel less confident about going off on their own. It doesn't necessarily mean that they feel more warmly about clinging together and supporting one another in time of need. More likely, wives who would otherwise leave their husbands realize that they just can't afford to do so; likewise, men who might desert their wives in times of prosperity find they are out of work and hang around so that they can live off their wife's salary or other income.

Economics is a very basic motivator in a marriage, and an economic downturn pulls realities closer to that bottom line. Marriage is a kind of trade-off of various resources: income, love and affection, domestic labor, and sex. When the trade-offs are unequal, one person or the other dominates, and this can give rise to considerable dissatisfaction. As long as the resources are too unequal, the situation may not change. If divorce is like a revolution, an unequal marriage is often like a stable dictatorship. It is only when one partner believes that he or she (very often she) has better opportunities outside the marriage that it falls apart.

Of all these resources, the economic ones are probably most crucial. A marriage is a way of providing a living, a communal economic unit, and that aspect can go on even if the love, common interests, and sexual ties all have dwindled away. The fact that marriages are more apt to stay together during an economic depression is not necessarily a good thing; one might say it is just one more toll that a depression takes on people's lives.

From the mid-1950s through the early 1970s, following the post-World War II deviation in the pattern, the rate of divorce went up in the United States. This was also a period of sustained economic growth. Some economic historians are now saying that it was an economic miracle that no society ever saw before and that we may never see again. We became a society in which the majority of the population—the working class included—could reasonably aspire to owning their own home, as well as a car (or several), television sets, household appliances, stereo system, and many other consumer goods. Near the end of this period, married women began to join the labor force in unprecedented numbers. It was also during this time that the divorce rate rose to a historic peak.

Since the late 1970s, the divorce rate has stabilized and is even inching downward. This is probably not because of some newly awakened surge of traditional morality; all other indicators of sexual behavior tend to indicate otherwise. What we are seeing instead is probably the mirror image of the earlier trend: divorce increases during prosperity and declines during depression. The American economy (like those of other industrial nations) has experienced irregular growth punctuated by downturns since the 1970s. We still remain a very wealthy society; the roller coaster has stalled nearer the top than the bottom. However, one result is apparently the end of the long trend of continuously rising rates of divorce, although the level at which the rise has topped off is historically an unprecedentedly high one.

Morality and Individualism

Toward the end of the 1980s and into the 1990s, much attention was paid to the decline of community and the rise of individualism (e.g., Bellah et al. 1985). The basic idea was that Americans were too self-absorbed and that they were no longer able to make the sacrifices that were necessary for their communities and their society to thrive. Religious leaders, politicians, and some family scholars picked up on this theme, as they have periodically for over a hundred years, and lamented the decline of The American Family (e.g., Popenoe 1993). The primary measure of this increasing hedonism, narcissism, and self-absorption was taken to be the rising divorce rate (White 1991). Also buttressing the view was the purported tendency of Americans to talk about families and express opinions in ways that characterized marriage as a personal, tentative, and fleeting commitment.

Drawing on this critique of individualism, in the late 1990s a coalition of Fundamentalist Christian organizations, father's rights groups, promorality political moderates, communitarians, and family scholars began mounting a national campaign to reexamine the proper ground rules for divorce and to promote lifelong marriage (Hackstaff 1999; Stacey 1996; Waite 1995; see *A Closer Look* 1.1). Through conferences, media campaigns, and political lobbying, they attempted to convince the public that marriage is good (which, if we can believe the opinion polls, is what most Americans have believed all along). Proposals to reinstitute a fault basis in legal divorce proceedings and impose extended waiting periods before granting divorce have been introduced in many state legislatures, and both Louisiana and Arizona enacted statutes that authorized a special **Convenant Marriage** option. It is too early to tell if this coalition will be successful in its efforts to limit no-fault divorce and bring religious codes concerning Covenant marriage into civil law.

As noted in earlier chapters, there are historical precedents for arguments that divorce is causing the decline of society. In *Framing American Divorce* (1999, 188) Norma Basch argues that antidivorce sentiment has been exploited for political purposes throughout American history. She notes that seventeenth-century royalist defenders of Charles I equated Parliament's rebellion with the anarchy of

A CLOSER LOOK 14.1
Dealing with the Legal Entanglements of Divorce

Divorce, like marriage, is a legal condition. It involves one in the complexities of courts and lawyers. Although some divorces are simple, others have considerable legal complications. Recent legal developments have begun to extend some of the liabilities of conventional marriage and divorce to the breaking up of cohabitation arrangements.

Grounds for Divorce

Until the early 1970s, divorce was legally very hard to get—at last under the strict letter of **fault-based divorce** law. At one time, couples would head for Las Vegas or Mexico to take advantage of places where one could acquire a quick divorce. In most states, the only grounds for divorce that were legally admissible were adultery, desertion, or extreme cruelty. In the case of well-known people filing for divorce, this led to quite a few lurid headlines. However, as the number of divorces began to increase, most people simply made up fictional complaints. Most common was the pat formula "extreme mental cruelty" as the ground for divorce.

In 1970, California enacted a **no-fault divorce** law, and more than half the states followed soon after. It was no longer necessary under these laws for one person to be the aggrieved party. Now the marriage could be ended by mutual consent, and it was no longer necessary to lie about the reasons for divorce. Opponents of these liberalized laws argued that they would make divorce all too easy; nevertheless, the rate of divorce in no-fault states has turned out to be little higher than the trend in states that kept the traditional laws (Cherlin 1992).

Who Needs a Lawyer?

With no-fault laws, it has become easier for couples to obtain a divorce without hiring lawyers to represent each side. In many states, it is possible for one spouse simply to obtain papers from the courthouse and file a petition for divorce. If the petition is not contested by the other spouse, the divorce is granted within a certain period of time (usually six months or one year from preliminary decree to final decree). In many large cities, there are "divorce clinics" where a professional will help persons who desire a divorce to file the necessary papers. However, these procedures are feasible only if there is no significant property involved or if there is no disagreement about custody or support of children. If divorce is complicated by either factor, both parties are well advised to retain their own lawyers to protect their interests. If no lawyers are involved, the judge will decide the property and child-related issues under conventional precedents. If one spouse is represented by a lawyer and the other is not, the latter's chance of getting an equitable settlement, from his or her own point of view, is rather slim. Lawyers usually charge by the hour for their legal services, and a minimum of about $750 for a divorce is typical. If the case is very complicated (i.e., if there is a great deal of property or a difficult custody fight), the cost may be considerably higher.

Property Settlements and Alimony

Different states have somewhat different laws regarding the set-

divorce. In the eighteenth century, American aristocrats like Benjamin Trumbull railed against easy divorce in Connecticut, calling statutory failure to provide punishment for the guilty spouse an "effectual opiate" that would undermine deeply rooted religious convictions (Basch 1999, 40). In nineteenth-century America, Horace Greeley rhetorically blamed the decline of Rome on its lax divorce code and attributed the economic supremacy of Europe over Asia to its adherence to Christian marriage. Greeley's message was clear: "Easy divorce generated political and economic decline; lifelong monogamy provided the foundation for capitalist enterprise and augured both a prosperous and orderly America" (Basch 1999, 93). Similarly, in the novels of T. S. Arthur, Harriet Beecher Stowe, and William Dean Howells, divorce was used as a symbol of the decline of the nation.

According to Basch (1999, 188), this combination of political imagery and evangelical Christianity was especially powerful in the hands of Victorian Amer-

tlement of marital property. In California and several other states, all property acquired during a marriage is what is legally known as **community property.** This includes all the salaries and other income to either spouse during the time they have been married, with the specific exception of inheritances, which are considered their separate property. Other separate property may consist of a house or a business that was owned before marriage, although these may become converted into community property by gift to the community, by being "mingled" with community property (e.g., mingling old funds with new contributions), or by making continued mortgage payments out of current income on a previously owned house. Because there are many complicated issues here, it is essential to have a lawyer if there is much property involved in a marriage. All community property is divided 50/50 between the divorcing partners, unless the couple has specifically made an agreement otherwise in the form of a marital contract.

In some other states without community property laws, there continues to be the traditional English common law, which vested all property in the husband, unless explicitly put jointly or singly into the name of the wife. Here divorced women might come away with little or no property at all, although in fact judges have tended to use their discretion to make an equitable distribution. In New York and elsewhere, "equitable distribution" is now the law, and the trend is for this form of law to become more widespread.

Alimony (sometimes called *spousal support*) consists of a fixed income to be paid by one partner to the other after the divorce. Its basis is the inability of one ex-spouse (traditionally the wife) to support herself after having given up her career for the domestic responsibilities of marriage. Although one hears about very large alimony payments to the ex-wives of some wealthy individuals, alimony is no longer typical in most divorces. In most states (such as Ohio), alimony is not awarded at all. In others, it is awarded only if the spouse did not work at an outside job at all during the marriage. It has now become common for a court to order spousal support payments for a limited period, based on an estimate of how many years of education and retraining it will take for a former housewife to support herself with a paying job.

Marital Contracts

Sophisticated lawyers now point out that one can avoid some of the legal difficulties of a divorce by planning ahead. No one goes into marriage expecting to get divorced, but because the probability is rather high (nearly 50 percent for first marriages, even higher for second marriages), it is rational for a couple to specify what will happen in that event. A **marital contract** can specify what property is to be communal and what is to be held separately; what spousal support, if any, is to be paid in the event of divorce; and many other matters relating to the rights and duties during the marriage, as well as in its dissolution. The only thing a marital contract cannot specify is the question of child custody and support. Because these affect the welfare of the children, the court legally retains jurisdiction to decide these matters after taking input from the contending parties themselves.

ican moralists who used it to make what they believed was a last-ditch effort to stem the tide of divorce:

> *When they exposed the perils of subjecting marriage to perennial rather than onetime consent, an outcome they anticipated with uncanny prescience, they advanced their argument by using marriage as a signifier of law and order and by equating divorce with political chaos. And when they championed the self-sacrificing communitarianism of marriage against the selfish individualism of divorce, they defined their campaign as nothing less than a contest between Christians and infidels. Their polarizing definition of themselves and their opponents exemplifies how they translated the divorce question into a symbolic focal point for competing worldviews. That the opposition between Christians and infidels, between order and anarchy could play so compelling a role in the acrimonious debates of the 1850s and 1860s suggests divorce was*

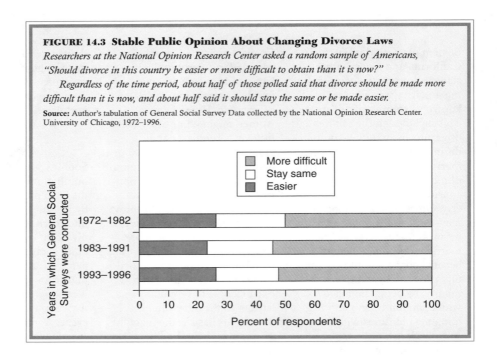

FIGURE 14.3 Stable Public Opinion About Changing Divorce Laws

Researchers at the National Opinion Research Center asked a random sample of Americans, "Should divorce in this country be easier or more difficult to obtain than it is now?"

Regardless of the time period, about half of those polled said that divorce should be made more difficult than it is now, and about half said it should stay the same or be made easier.

Source: Author's tabulation of General Social Survey Data collected by the National Opinion Research Center. University of Chicago, 1972–1996.

a lightning rod for deep-seated political anxieties that revolved around the positive and negative implications of freedom.

What should we make of more recent arguments that individualism and divorce are undermining society? It is true that marriage is less obligatory than it once was and that people are now able to end bad marriages fairly easily. Nevertheless, there is a problem with the argument that individualism, selfishness, or secular humanism caused the divorce rate to go up. The divorce rate has been stable for about 30 years (see Figure 14.1), and people's attitudes about divorce became slightly more negative during the 1980s but have remained fairly stable since the 1970s (Figure 14.3). In addition, when longitudinal studies are conducted, disapproval of divorce does not seem to affect the divorce rate, and attitudes toward divorce seem to change only after someone goes through it (Thornton 1989). Even though most Americans continue to believe in marriage, over two-thirds think that married couples should divorce if they don't get along, even if they have children (Figure 14.4). Apparently, attitudes toward divorce, or changing views toward marriage, are not the primary forces behind changes in divorce rates. It is more likely that both attitudes and divorce rates are responding to the same underlying economic and social pressures.

BREAKING UP

Most people who divorce only do so for strong and pressing reasons: they want out of their marriage. Even with all the incentives to leave, the experience of divorce is often a traumatic one. There tends to be a great deal of anger and resentment, as well as anxiety about the future. A psychological support has broken away, and a ritual tie that had been taken for granted has snapped.

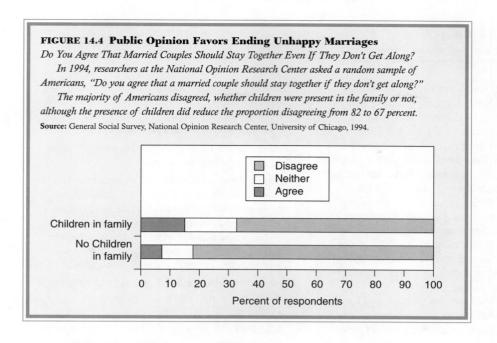

FIGURE 14.4 Public Opinion Favors Ending Unhappy Marriages

Do You Agree That Married Couples Should Stay Together Even If They Don't Get Along?

In 1994, researchers at the National Opinion Research Center asked a random sample of Americans, "Do you agree that a married couple should stay together if they don't get along?"

The majority of Americans disagreed, whether children were present in the family or not, although the presence of children did reduce the proportion disagreeing from 82 to 67 percent.

Source: General Social Survey, National Opinion Research Center, University of Chicago, 1994.

Although women are more economically dependent on marriage than men, they are most likely to initiate divorce (Braver with O'Connell 1998; Kitson and Holmes 1992). Sometimes they are driven to breaking up the marriage or long-term relationship because they are suffering emotional or physical abuse. As described by Demie Kurz, a surprisingly large number of divorces or separations initiated by women follow a history of abuse or neglect by husbands (*A Closer Look* 14.2). In other cases, women tend to initiate divorce because they are more focused on intimate relationships. According to several researchers, because women are more likely than men to define themselves in relation to others, they are more likely to sense and respond to deteriorating relationships (Chodorow 1989; Hackstaff 1999). Ironically, women's tendency to value intimacy often motivates them to question their relationships.

The greater likelihood that women will be the **initiators of divorce** reflects the divergent "his" and "hers" versions of marriage described by Jessie Bernard (1982; see chapter 11). The tendency for women to initiate divorce is also in line with findings that marriage serves to protect men more than women from poor health and even death (Hu and Goldman 1990). Because wives have traditionally been held responsible for the well-being of family members, they tend to be more attuned to the stresses and strains associated with emotion work, kin work, and care work (Hackstaff 1999; Hochschild 1989). When there are low levels of reciprocity in the relationship, women tend to notice it and initiate private talks or therapy to improve the situation. Men, in contrast, are more likely to be satisfied with communication in the marriage and often refuse, resist, or withdraw when wives want to talk about problems in the marriage (Blaisure and Allen 1995; Gottman 1994; Hackstaff 1999). If such talks fail to improve the relationship, women then become more likely than men to begin thinking about breaking up.

Although women, on average, begin thinking about a breakup before men, both men and women become unhappy with relationships, and anyone can initiate the actual breakup. In both heterosexual and homosexual relationships,

A CLOSER LOOK 14.2
Divorce and Marital Violence

In her 1995 book, *For Richer, For Poorer: Mothers Confront Divorce*, Demie Kurz reported on a study of a random sample of divorced mothers in a major metropolitan area of the United States. These mothers left their marriages for a number of reasons. However, they particularly identified gender issues as reasons for leaving. Some left because of their dissatisfaction with traditional gender roles; they found their husbands too controlling or unsupportive emotionally. In a second category of cases, women left their marriage because of their husbands' violence. Kurz found that 70% of the women had experienced marital violence, with one of five women reporting that they initiated divorce to end the violence.

Among the several reasons women gave for the violence, the most common reason was that violence occurred when they attempted to act independently:

> He was violent when I would go out and do things on my own. He didn't like that. For example, I went out and found myself a job. He didn't want that. He wanted me to always be home. My father gave me a car and I would take the kids places. He didn't like that. [28-year-old black poverty-level woman, high school education, seeking employment, married seven years, three children]

A working-class white woman reported that the violence started during the separation when she "started to change," that is, when she began to voice her own opinions:

> He wasn't violent during the marriage, but during the separation. When I started to change, that's when it started. During this time he was violent about once a month, including once in front of the kids. He usually took the phone off the wall when it happened. [27-year-old white working-class waitress, married seven years]

Another woman reported that her husband attempted to control arguments:

> He wasn't that violent a person. But if we're disagreeing he'll come and get me by the wrists or the shoulders. He knows karate. He broke my wrist. [36-year-old poverty-level black woman, married two years]

Another reported that her ex-husband was violent when things did not go his way:

> He would look in the cabinet and the glasses wouldn't be standing straight, or the kids ate the last piece of cake, or I'd be five minutes late and he'd accuse me of being in a motel. The children wouldn't go on a merry-go-round, it makes them sick, and he ended up beating me up in the parking lot. He used to go for my throat. [44-year-old, white secretary, married eight years, three children]

These women also spoke of generally controlling behavior on the part of their ex-husbands:

> I have friends now. Before I couldn't 'cause he would insult them and give me a hard time. He didn't want me to have anybody. I had to be totally dependent on him. [29-year-old poverty-level white woman, unemployed, married eight years, three children]

> I had no friends. I wasn't allowed to have friends. I stayed in. I couldn't say hi to anyone on the street. That's why I wasn't allowed to work. When I was thirty I started talking back. He said I watched too much Phil Donahue and talked to my friends too much. I couldn't go to the movies with my sister. [33-year-old poverty-level white woman, married fourteen years]

> He was very insecure. He wanted me at home. . . . He didn't mind if I worked when he was working, but if he got laid off, which he did sometimes, then he wouldn't let me work. One time he said that the worst thing that ever happened was that I went to school. . . . He also wanted to control everything about my life. He wanted to control my friends, my time. He wouldn't let my son see my brother, who is a successful businessman. [36-year-old black nurse, married fifteen years, two children]

Source: Demie Kurz. 1995. "For Richer, For Poorer: Mothers Confront Divorce." Pp. 67–68, 220. Copyright © 1995. Used with permission of Routledge, Inc.

however, the person who initiates the breakup is better prepared to weather the separation. According to Diane Vaughan (1986, 127), who studied the process of uncoupling, "regardless of sex, age, occupation, or income—regardless of social class, duration of relationship, social networks, or other factors known to be related to the disruptiveness of this experience—the initiator has the advantage over the partner." She suggests that the person who initiates the breakup has more alternatives and thus can more easily forge new identities independent of the partner relationship. In addition, according to Vaughan (1986) the initiator has the advantage of time, having begun the process of uncoupling earlier than the partner. Whoever initiates the breakup, uncoupling entails a restructuring of one's life and a redefinition of self. Through this process, many people gain new insights into who they are and what they want out of life. At the same time, the process of redefining oneself can be very disheartening, especially for the person who did not initiate the breakup (*A Closer Look* 14.3).

After a breakup, the formerly married person settles into a new life. For all the emotional flare-ups and the social readjustments that must be made, the most important problem is economic. A practical living arrangement has been broken up, and two new ones must be created. Historically, this has been a more severe problem for the woman. If she was a housewife before the divorce, she had the problem of getting a job or securing some other source of income. There used to be alimony decreed by the court to be paid by her ex-husband and, if there were children, child-support payments. However, alimony (where it existed) was rarely adequate to live on, and **child-support payments** were often lower than actual child-related expenditures.

More recently, there has been some improvement in **child-support awards** and collections, but alimony is rarely awarded. Of the 11.5 million custodial mothers of children in the United States in 1996 (including both divorced and unmarried mothers), over 7 million, or 61 percent, had child-support court orders or agreements requiring the absent father to pay. Of the women who were due payments in 1995, most received at least a portion of the amount they were owed, with the average amount received totaling $3,767. Of the $28 billion owed in child support in 1995, just under $18 billion was paid, leaving over $10 billion unpaid (Scoon-Rogers 1999). Compared with past decades, this represents an improvement in the proportion of children with awards and an improvement in the proportion of child-support payments made. Nevertheless, because more children are due payments than before, and because a large number still receive little or nothing, more children with an absent parent live in poverty (chapter 6).

Children who see their absent parent are more likely to receive child support. Over 10 million of the parents who were not living with their children in 1995 had joint custody provisions or visitation provisions for contact with their children. Having such provisions for continued involvement with children was associated with more regular payment of support, especially for fathers, who make up the majority of absent parents. Three-fourths of noncustodial parents with joint custody or visitation provisions made child-support payments in 1995, compared with only about a third of those who had no such provisions. In addition, over 3 million noncustodial parents provided health insurance for children living with their custodial parent (Scoon-Rogers 1999).

About one-third of custodial mothers with children living in the home have incomes below the federal poverty level. As one would expect, there are differences between groups of mothers in the likelihood of having a child-support award and in the actual receipt of child-support payments. Three of four divorced mothers with children are supposed to receive support payments, though less than half of these receive the full amount. Two-thirds of divorced mothers with a child-support award get something, with the amount received averaging almost $4,000 per year. Only about half of never-married mothers, in contrast, are supposed to receive child support, and about half of these receive any money, with the average amount received just $2,271 (Scoon-Rogers 1999). In general, women with more education, better jobs, and higher income are more likely to

A CLOSER LOOK 14.3
Uncoupling

Uncoupling is a railroader's term. When a relationship ends, it's like when a locomotive uncouples from a car or a car uncouples from another car. They're hooked together with knuckle couplers. They interlock, like your knuckles do when you clasp your hands together. When a locomotive uncouples, you pull a coupling pin on one side and one lets go, or you pull a coupling pin on the other side and the other lets go, or you pull both pins and they both let go. I know. I was a brakeman. I used to do that. Get the mechanical aspect of it? It's like a relationship. One can let go, the other can let go, or they can let go at the same time. But it's also a mechanical letting go because they no longer live together. They live separately. They do different things. They're no longer hooked up in mechanical ways. [Mechanic, age 39, divorced after 12 years]

Although now apart, each partner witnesses the other person's transition. Formerly participants in each other's life, they are now observers. They see the home transformed, or with new occupants. They see physical changes. They see the other person master new skills and demonstrate unexpected ones. They see the other person with someone else. For both initiator and partner, what they see brings with it some sense of exclusion and loss. Equally as important is what they don't see. Being physically excluded from the routine of life with the partner does not diminish one's awareness of it. Knowing it's the other person's birthday and not being present, hearing they are having a personal crisis and not giving a hand, realizing it's Sunday and they are at the ballgame—all contribute to each partner's redefining self and other as separate.

Both people also witness changes in themselves, which feed into this redefinition process. As they negotiate life on their own, they begin learning who they are without the other person. They gain insight about the relationship as they remove themselves from the patterns developed with the former partner. People describe this experience as one of discovery:

I lived all my life with 18th century antiques. I went out to buy furniture and discovered that I liked Swedish modern. [Lawyer, age 50, separated after 26 years of marriage]

Alex left, he said, because I didn't give him good enough sex. He said there wouldn't be anyone who would want me, as soon as they found out. Well, what I discovered was that Alex was a lousy lover. [Office manager, age 27, separated after living together 5 years]

I doubted that anyone would really come to visit just me. I discovered that they would. [Housewife, age 60, separated after 39 years of marriage]

She used to always want me to come watch TV with her after supper. I always had work to do. After we separated, I discovered myself watching TV in the evenings. I seldom worked. It must have just been my way of avoiding being with her. [Potter, age 32, separated after living together 9 years]

Some of the discoveries are pleasant; some are not. In the other person's absence, we have no ready scapegoat for the ills that befall us. When the same things keep happening over and over, but under different circumstances, we eventually have to confront the painful fact that we must be contributing to them.

I always thought Jamie was holding me back in my career. I discovered after she was gone that it wasn't her. [Lawyer, age 32, divorced after 9 years]

I always complained about the house. When I left, I moved in with a friend and discovered piles of newspapers, books, pencils, scraps of paper everywhere. I went home to visit and it was neat and clean. I was shocked to discover it was me all that time. [Teacher, technical school writer, age 39, divorced after 18 years]

I never had fun when we went out with other people. I thought it was because of who I was with. Since I've been seeing other people, I discovered I'm just not very good in groups. I get feeling inadequate, and that depresses me. [Leather craftsman, age 42, separated after living together 12 years]

Source: Diane Vaughan, *Uncoupling: Turning Points in Intimate Relationships*, pp 171–72. Copyrighted 1986 by Oxford University Press, Inc. Reprinted by permission of Oxford University Press, Inc.

get child-support awards or agreements in the first place, they are more likely to eventually receive payments, and they receive larger sums of money with which to support their children. As discussed in previous chapters, because U.S. government efforts to support children follow a privatized model of family care, policies and programs tend to benefit children who are better off in the first place. Because fathers at the bottom of the income hierarchy can barely support themselves, when they break up with the mother of their children, they are unlikely to pay child support on a regular basis.

Of the 13.6 million custodial parents in the United States in 1996, 2.1 million were men. As one might expect from the gender gap in wages, custodial mothers are more likely to receive child-support awards than custodial fathers. Whereas 61 percent of custodial mothers had child-support awards in 1996, 40 percent of custodial fathers did. When men receive child-support payments, the amount is not significantly different from what women receive. Nevertheless, men who received child-support payments had higher total individual incomes ($30,000) than women who received child-support payments ($22,000). Because of such income inequities, the poverty rate for custodial mothers was more than twice as high as for custodial fathers. Because women are far more likely to have child custody, this resulted in 3.9 million custodial mothers living in poverty in 1995 but only 0.3 million custodial fathers living below the poverty level (Scoon-Rogers 1999).

In 1996 the median family income of divorced mothers was about $20,000 compared with over $50,000 for married mothers (Lugaila 1998). As noted earlier, lack of child-support payments is one reason that divorced mothers have lower incomes than married mothers or divorced fathers. Other reasons include gender inequities in the labor market, the logistics of single parenting, and the postdivorce need to maintain two households. A large number of studies have documented how women's standard of living goes down after getting divorced, especially when compared with men who divorce. Using different samples and research methods, studies consistently show that women from a variety of different backgrounds experience relatively steep declines in most measures of economic status and well-being after they dissolve marriages or long-term relationships (Bianchi, Subaiya, and Kahn 1999; Duncan and Hoffman 1985; Holden and Smock 1991; Smock, Manning and Gupta 1999). This decline in income results from losing the man's income—which tends to be higher than the woman's—and sometimes from increased time demands from being the sole custodial parent. The latter is important because it limits the number of hours one can work for pay.

One's standard of living, including what one can buy and where one lives, typically goes down immediately following a divorce for everyone involved. One reason is that the same income now has to be shared across two households. Often the family home (if there is one) is sold to divide the proceeds so that both ex-spouses can find new accommodations. When there are children, the wife usually gets physical custody, so the standard of living needed by the ex-husband and the ex-wife tend to diverge. When studies take into account the number of people in the household that must be supported, most find that five to ten years after the divorce, the economic position of the woman is still below what it was when she was married (Morgan 1991). For the man, in contrast, the standard of living eventually improves, and as we will see, he is also more likely to remarry. For divorced women, the surest path to economic recovery is remarriage, though the odds of remarrying when they have young children are much lower than for men (Duncan and Hoffman 1985; Holden and Smock 1991; Smock, Manning, and Gupta 1999). Immediately following the divorce, both the husband and the wife experience a decline in their personal standard of living as a result of the loss of economies of scale. Usually at least one spouse is displaced from the previous family dwelling, and any savings realized by having two adults live together (such as appliances, utilities, and food buying) are lost when the couple moves apart.

The relative economic well-being of women and men after divorce differs by social class. In families at the poorest income levels, divorce or separation can actually improve women's per capita income. When men bring in little money, they can be a drain on the household economy, especially if they do not contribute much to the housework or child care (chapters 11 and 12). Although men at all class levels identify as breadwinners and tend to share their incomes with their families, employed women are even more likely to spend earnings on other family members. This is especially true in impoverished households, where women's incomes have always been essential to the survival of the family. At the margins, then, there can be more money to spend on the children (or on oneself) *after* the divorce. As we move up the income hierarchy, different patterns emerge. For women in the middle and upper classes, there is a greater disparity in earnings between men and women. Some of these women are not employed (see Figure 6.3), but even for those employed in professional careers, their husbands tend to make more money than they do. Under these conditions, divorce usually means more relative downward mobility for women, especially if they have been stay-at-home mothers for several years (*A Closer Look* 14.4).

Given these economic problems, one wonders how people can afford to get divorced in the first place; but they do, and in large numbers. There are several reasons why it is possible. For one thing, the divorce is usually an emotional blowup, and people do not think much (or, for that matter, know very much) about the economic consequences of what they are doing.

Second, economic factors do enter in covertly and unconsciously, so people are more likely to divorce if it is economically feasible for them. We can see this in the greater tendency for divorce in dual-income families (which makes the divorce settlement easier on the husband, surely, and also gives the wife some financial means—though generally not enough, realistically—to maintain the previous standard of living). Also, the tendency is for divorces to increase during good economic times and decrease in hard times; divorce is one of the luxuries that an economic boom makes possible.

Finally, there is the mitigating fact that divorces tend to occur relatively early in the marriage, when the couple has a lower standard of living. For a young couple, especially if they have not yet bought a home, the drop in standard of living may not be great if they separate. It may well be that a major reason why the divorce rate trails off to a minimum after about ten years of marriage is that couples at this age become too affluent. They have accumulated too many possessions to be able to afford a divorce, no matter what they feel about each other.

Because married women have higher family incomes than divorced women, some scholars and policy makers have argued that it is in women's best interests to get and stay married. As a remedy to child poverty and other social ills, they suggest that politicians and judges should consider making divorce more difficult to obtain (e.g., Gallagher 1996; Galston 1996; Whitehead 1996). Such proposals are not new (Basch 1999), but they gained momentum in the late 1990s, as politicians and religious leaders rushed to defend heterosexual marriage and fashion a symbolic crusade to revalue fathers as moral family leaders. The nostalgic family-values rhetoric behind such arguments suggests that if women would just commit to marriage instead of being so self-centered and career-focused, they would have more babies, be happier, get divorced less, and enjoy a better standard of living. This flawed reasoning assumes that changing one's attitudes toward marriage is enough to make it more stable. As noted earlier, there is little

A CLOSER LOOK 14.4
Divorce and Downward Mobility

For her book, *Mothers and Divorce*, Terry Arendell (1986) interviewed sixty middle-class women with children in Northern California who had been divorced. She found that even if they were employed before and after the divorce, the women experienced financial troubles and "downward mobility" after the divorce. Most could meet the most essential monthly expenses with caution and careful spending, but few had any money for emergencies or unanticipated demands. Some fell farther and farther behind, unable to establish any material security. One single mother, divorced nearly eight years, described her precarious existence:

> I've been living hand to mouth all these years, ever since the divorce. I have no savings account. The notion of having one is as foreign to me as insurance—there's no way I can afford insurance. I have an old pickup that I don't drive very often. In the summertime I don't wear nylons to work because I can cut costs there. Together the kids and I have had to struggle and struggle. Supposedly struggle builds character. Well, some things simply aren't character building. There have been times when we've scoured the shag rug to see if we could find a coin to come up with enough to buy milk so we could have cold cereal for dinner. That's not character building. (p. 37)

Most middle-class wives who divorce expect to recover financially and return to a middle-class standard of living. As Arendell discovered, most cannot realize that ideal. Forced to take low-paying jobs and often unable to keep up with the mortgage payments, many divorced women are forced to sell their houses and move to working-class neighborhoods. Giving up their previous standard of living also means not being able to provide one's children with the middle-class amenities they have come to expect:

> My husband liked really good food and always bought lots and the best. So when he left, it was really hard to cut the kids back. They were used to all that good eating. Now there's no food in the house, and everybody gets really grouchy when there's no food around. . . . I think I've cut back mostly on activities. I don't go to movies anymore with friends. We've lost $150 a month now, because my husband reduced the support. It gets cut from activities—we've stopped doing everything that costs, and there's nowhere else to cut. My phone is shut off. I pay all the bills first and then see what there is for food. . . . I grew up playing the violin, and I'd wanted my kids to have music lessons—piano would be wonderful for them. And my older two kids are very artistic. But lessons are out of the question. (p. 43)

Although the women in Arendell's study suffered through economic hardship, they also reported feeling stronger because of it. One woman whose husband had left her to marry another woman after twenty years of marriage reported realizing that she was better off without him:

> The best thing that's happened out of this divorce experience is acquiring a sense of myself. I've learned that I have a lot of abilities and skills and that I can make it. It's still scary jobwise, but the kids and I will survive. I've done an awful lot of growing, and I feel really good about that. And I feel pretty good about who I am and where I am. . . . The last time I saw my ex-husband, just a few months ago, I said to myself, "Thank God I'm not married to that man." I'd been thinking that for quite a while, but that was the first time I could say it and really mean it. (p. 146)

---

research support for this proposition. In addition, such arguments assume that staying married would substantially improve the economic prospects of women at all class levels, another proposition that deserves scrutiny.

Careful studies suggest that simple aggregate comparisons between married and divorced women overstate the potential economic benefits of marriage. We know, for example, that on average, unmarried women have higher *individual* incomes than married women, even though their family incomes are significantly lower. Research using random samples and longitudinal data also shows that because of selection effects, divorced women would not fare as well economically as married women if they remained married instead of divorcing (Peters 1993; Smock, Manning, and Gupta 1999, 809). Studies show that if women who actually stayed married had divorced instead, they too would suffer major economic loss, though perhaps they would fare slightly better than those actually divorcing (Smock, Manning, and Gupta 1999; Waite 1995).

How should we interpret these findings? Because of a societal division of labor in which women are paid less than men and are more likely to interrupt their careers to care for home and children, women are economically vulnerable outside of marriage. It is thus important to remember that marriage, divorce, and family values are related to economic conditions in multiple ways. Marriage is not a simple fix for custodial mothers' economic woes. Because marriage is a property relationship in addition to a romantic and practical one, the likelihood of marrying, divorcing, or remarrying is shaped by men's and women's economic circumstances and employment opportunities. Earlier in the book, we saw that marriage tends to be less necessary, and hence more fragile, when the patriarchal property relations supporting it are weak. In contrast, marriage is strongest when women's economic opportunities are limited and they are financially dependent on men. In the modern context, making marriage laws more restrictive might keep more middle-class women in marriages and gain more long-term financial security for children, but at what cost? More restrictive divorce laws are also likely to increase the incidence of domestic violence, depress women's wages, and eventually lead to more economic exploitation of women. For these and other reasons, it is unlikely that changing attitudes about marriage or returning to a fault-based divorce system would significantly improve the living conditions of the majority of women and children.

A New Social Life

If postdivorce economic realities are usually a down experience, socially and sexually the new divorced person, if childless, often has a more exciting time. At least some of the old friendship networks are broken up, and there is an incentive to make new acquaintances. Often one moves into different kinds of social circles. Married couples tend to socialize with their relatives and with other married couples. Especially in the middle class, women tend to be friends with the wives of their husbands' business or professional friends and vice versa; in other words, the couple system creates the friendship links for each of the persons within it. This arrangement tends to break up with the marriage. Although other people will try to remain friends, they typically find there is too much strain in keeping equal contact with both of the separated individuals. Other married couples find they have to choose which person they will continue to be friends with, and the other ex-partner gradually drifts away. Also, a single tends to have an awkward time of it in a social environment based on couples. Many divorced women report that married friends became jealous or suspicious of them around their husbands.

The same is true of friendships based on where one lives. The divorce means one person or the other (and sometimes both) moves out of the house or apartment, which breaks up yet another network of friendships. (One indirect way that we know about this, incidentally, is through the fact that the more close friends one has when married, the less likely one is to divorce. The inference is that people may stay together, not so much because they like each other, but because they don't want to break up their network of friends.)

For the divorced person, then, old friendship networks tend to break up, at least to a certain extent. He or she is less likely to be socializing with other married people. The way our housing is arranged is one reason. Married people (especially those with children) tend to live in certain neighborhoods, the ones with single-family dwellings. These include the typical "bedroom community" suburbs, with their yards, swings, tricycles and bicycles in the driveways, and other signs of conventional family life. Unmarried people usually live in apartments, more likely nearer the central city. These are the same places where one tends to find young married couples without children, and this is often where divorced people move, under the pressure of postdivorce economics, to find a smaller and cheaper dwelling.

Divorced people, then, end up back in the world of singles, only this time with greater sexual experience. It is not surprising that formerly married persons tend to have fairly active sex lives. Divorced people have equally active sex lives as unmarried people, generally, even though they are biologically older (Laumann et al. 1994). This is not to say that the postmarried state is an unending orgy of sexual bliss. There are many readjustments to be made, including a lot of strain in ending old relationships with friends and finding new ones. Loneliness and anxiety often accompany this transition, and financial strains in the background intensify the problems. For these reasons, the lives of the formerly married, especially for the first year or so, tend to fluctuate from one emotional extreme to the other.

Even though most women suffer economic hardship, the majority report substantial improvements in the quality of their social lives and sexual relationships after divorce. As noted previously this may be because women are more likely to initiate the divorce in the first place (Kelly 1986, 309). According to national surveys, women are over twice as likely as men to report having wanted a divorce, and most women report that they were much happier after the divorce (McLanahan 1989; Sweet, Bumpass, and Call 1988). Another long-term study of divorce (Wallerstein and Blakeslee 1989) found that five years after the divorce, two-thirds of the women and half of the men were more content with the quality of their lives.

Few divorces are impulsive. Usually the decision to divorce is preceded by months or years of accumulated grievances and unhappiness. Some studies find a "last straw" phenomenon in which some specific event, such as sexual infidelity, triggers the final decision to divorce. When it happens, few people report that it is a mutual decision, and as noted earlier, those who initiate the divorce fare much better in an emotional sense. Although spouses who initiate divorce often do so with sadness, guilt, and apprehension, they usually maintain a greater sense of control and do not experience as much humiliation and rejection as those who oppose the divorce (Kelly 1986; Vaughan 1986).

CHILDREN AND DIVORCE

Historically, a significant portion of American children experienced living with a single parent or a stepparent (Amato 2000). Although more children experience these arrangements than ever before, what has changed most over the years is the cause of these living conditions. Whereas the cause of single parenting used to be the death of a parent, it is now divorce or nonmarriage. As we see later in the chapter, although single parenting used to be followed by remarriage, it now often includes a period of cohabitation as well.

At the turn of the nineteenth century, one in four children experienced the death of a parent before they reached age fifteen (Uhlenberg 1980), and another 7 or 8 percent experienced parental separation or divorce (Furstenberg and Cherlin 1991). During the twentieth century, because of improvements in nutrition, living conditions, and medicine, the percentage of single families caused by the death of a parent declined dramatically. At the same time, the number of children experiencing the divorce of a parent increased. As we saw in chapter 5, dramatic increases in divorce during the 1960s and 1970s led to record levels of children living with a single parent, though most of them remarried. It was not until the 1970s that divorce displaced bereavement as the leading precursor to remarriage (Cherlin 1992). More recently, an increase in the number and percentage of children born outside of marriage has led to more children living in single-parent households, as well as to more children living in other household arrangements (Amato 2000; Seltzer 2000).

Although single-parent households and remarriage are thus not new, the proportion of women and children likely to experience them has increased substantially over the years. The likelihood that a child's family structure will change before leaving home is much greater than it was a few decades ago. Estimates are that more than half of children born in the 1990s will spend time in single-parent households (Bianchi 1995; Furstenberg and Cherlin 1991, 11). As we will see, remarriage is still common, with about one-third of children likely to live in a remarried or cohabiting stepfamily household before they reach adulthood (Bumpass, Raley, and Sweet 1994; Seltzer 1994). About 40 percent of adult women will likely reside in a remarried or cohabiting stepfamily household as a parent or stepparent at some time (Bumpass, Raley, and Sweet 1995).

We should remember, however, that many families continue to be relatively stable. About two-thirds of children under age eighteen in the United States live with two parents (including step and adoptive parents). More than a third of first marriages are likely to remain intact for life, and more than half of the children who begin life in a two-parent family will have intact families throughout their childhood (Bumpass 1990). Thus, although the likelihood of family change continues to increase for children, most still live with two parents. The difference is that they are more likely to have lived through a divorce and to be living with a stepfather than ever before.

Although race and ethnic differences are evident in the likelihood of children living in single-parent, stepfamily, or nonmarital households (chapter 7) these differences became even more pronounced at the end of the twentieth century. Although about 36 percent of children born to white parents can expect to live in a single-parent household, approximately 80 percent of African American children can expect this (Bianchi 1995). As noted in chapter 10, the racial gap is especially large for nonmarital births. About one-quarter of births to whites occur out of marriage, as compared with about two-thirds of births among African Americans (Amato 2000). Some commentators suggest that the increased rate of nonmarital births among African American women is due to a decline in the economic prospects and marriage potential of African American men (Wilson 1987), whereas others see recent trends as a continuation of traditional cultural patterns (Taylor 1994; Zuberi 1998; see chapter 7). Both factors are undoubtedly involved, and we are unlikely to see a reversal in such patterns unless the economic prospects for people of color, and for young African American men in particular, improve substantially.

About half of all divorces involve children. The reason the figure is not higher is that divorces tend to happen within the first few years of marriage, before children are born (Shiono and Quinn 1994; see Figure 14.2). For those couples who are already parents when they divorce, the typical number of children is two. In one study, Morgan, Lye, and Condran (1988) found that couples with boys were less likely to divorce than couples with girls. Fathers tend to be more involved with sons, and this greater involvement may integrate fathers more tightly into the family. In the event of divorce, however, we have single parents with two (usually small) children on their hands. How do they cope with the situation? Usually it is a mother who does the coping, because in most cases the woman receives custody. The 10 to 15 percent of the fathers who get custody represent a fairly new trend, and they have been experiencing some untraditional problems in their role as single parent. For the women in the more familiar role of single parent, these problems are not so foreign (see *A Closer Look* 6.4 and *A Closer Look* 14.4).

Reacting to the Change

Before the 1970s, the prevailing view was that divorce was indicative of individual pathology and that children of divorce would be likely to have major psychological problems. Divorce was viewed as a traumatic event that disrupted "normal" family functioning and was therefore likely to have negative emotional and psychological impacts on children (Herzog and Sudia 1973). Later research focused on some of the strengths of single-parent families and found that children who experienced divorce were not much different from children in "intact" families. Comparing children whose parents were divorced with children whose parents were still together showed some differences in terms of personality traits, average school grades, test scores, and behavior problems, but when children were followed over time, many of the differences were found to exist before the divorce (Cherlin et al. 1991). Thus, the physical absence of the father appeared to be no more harmful than his emotional absence or any number of other problems typically found in "intact" families (Krantz 1988).

More recently, scholars have focused on some of the deficits experienced by children who grow up with single mothers. They have asked questions such as: (1) Are children who grow up with only one biological parent less successful in adulthood, on average, than children who grow up with both parents?; (2) Are children with an absent parent less successful than children from two-parent families with similar characteristics (such as race or parent's income and education)?; and (3) Would children who grow up with only one parent have done better if their parents had stayed together? (McLanahan and Sandefur 1994, 9; see also Amato 1995; Amato and Booth 1997). As discussed in the following text, researchers agree that children from divorced families experience some short- and long-term negative consequences when compared with children who spend their entire childhoods with two biological parents, but debates about the extent of such differences continue. Conclusions about harmful effects depend on whether researchers use absolute or relative measures of disadvantage, with those more alarmed by divorce focusing on relative measures. In one of the best reviews of research in this area, Judith Seltzer (1994b, 239) concludes that "absolute differences between the two groups of children are generally small across various outcome measures." Because the majority of children who grow up in single-parent homes turn out to be happy, healthy, and productive adults, there may be too much emphasis placed on such comparisons. Nevertheless, direct comparisons between the two groups have received significant attention from scholars and policy makers in the past two decades.

One reason why children of divorce may be little worse off than children in two-parent households is that conventionally intact marriages often have major strains within them that are kept beneath the surface, so breaking up one of these marriages can be an improvement. A parent's personal fulfillment can facilitate a more healthy emotional environment, and sometimes children are better off living with one happy parent than with two unhappy parents. Examining children in various family situations shows that it tends to be the experience of conflict in the home that contributes to the children's positive or negative self-concepts not whether there has been a divorce (Amato and Booth 1997; Demo and Acock 1991). A major national survey found that children living with two parents who persistently quarrel over important areas of family life show higher levels of distress and behavioral problems than do the general class of children from disrupted marriages (Furstenberg and Cherlin 1991). Still, it is misleading to assume that divorce will automatically reduce conflict or produce happiness. Often, parents who divorce continue to have conflict for years after the initial separation.

At least one parent often feels relieved after a divorce, but this is not usually the case for the children. Only about one child in ten experiences relief when the parents separate, and that one case most likely involves a threat of violence (Wallerstein and Blakeslee 1989). Most children expect a reconciliation, and tend to reject the parents' initial decision as final. Although all children are profoundly affected by the divorce, there is no way to predict the long-term effects for a particular child based on the child's behavior at the time of the divorce. We do know that young children, particularly boys, suffer temporary deleterious effects when their parents divorce, whereas adolescents tend to be less affected, especially if little conflict is involved (Amato and Booth 1997; Demo and Acock 1991, 182). Three factors tend to aid children's adjustment after the divorce: (1) the effectiveness of the custodial parent (usually the mother) in parenting the child; (2) a low level of conflict between the mother and father; and (3) maintenance of a continuing relationship with the noncustodial parent (usually the father) (Furstenberg and Cherlin 1991). Many experts have claimed that a continuing relationship with the father is important to the child's later adjustment, but the results of studies on this are still inconclusive. When divorce ends a pattern of ongoing conflict, then the deleterious effects of the breakup for children are minimized (Amato and Booth 1997).

Although it is difficult to predict the particular responses of any one child to divorce, having one's parents split up carries with it special stresses and problems. Children are often upset by the violent emotions expressed by their parents and miss the parent who moves away. They often feel rejected (consciously or unconsciously) because their father or mother left. The new social and sexual life of their parents brings in unfamiliar people, who may be unpleasant or threatening. The divorce often precipitates a residential move for the custodial parent and the noncustodial parent, so the child's social life may be disrupted. Divorce increases the likelihood that a child will live in a disadvantaged community where jobs are scarce and schools are of low quality. Because the benefits of finishing high school and deferring parenthood are lower in such communities, adolescents are less likely to stay in school and more likely to become teen parents (Amato and Booth 1997; McLanahan 1989; McLanahan and Sandefur 1994; Seltzer 1994b).

Long-Term Impacts

What are the major long-term risks to children of divorce? In general, children who live apart from one or both parents are more likely to drop out of high school and less likely to attend college, more likely to marry and have children in their teens, and more likely to be single parents themselves. These tendencies increase the risk of long-term poverty and economic dependence. Findings are not always consistent; however, at least half of the major differences between children who live with one parent and children who live with two parents can thus be traced to the differing economic situations of the two groups (McLanahan 1989; McLanahan and Sandefur 1994; Seltzer 1994b).

Single-parent families, especially female-headed households, have significantly less money than two-parent families (Amato 2000). Not surprisingly, poverty causes high levels of physical, emotional, and psychological stress, and lack of money limits children's potential for achieving future upward mobility. One of the most important changes that occurs after a divorce is a decline in the parents' economic investment in children. This is partly the result of having to maintain two households instead of one and partly the result of the noncustodial parent making smaller monetary contributions to the household. In addition, divorce alters the quantity and quality of the time parents spend with their children. The noncustodial parent (usually the father) tends to spend less time with the children for a variety of reasons, not the least of which are transportation costs and ongoing conflict with the ex-spouse (Braver with O'Connell 1998). The custodial parent (usually the mother) tends to have less time to spend with the children as a result of working more hours to make up for lost income. In addition, the quality of the interaction between parents and children often declines because the parents are under stress and in conflict, and parental authority tends to be undermined (McLanahan 1989). Some researchers find that single parents and stepparents are less likely to monitor their children's school work and social activities, and they tend to have lower educational expectations than parents in two-parent families (Astone and McLanahan 1989; McLanahan and Sandefur 1994).

There is another way that divorce affects children's behavior after they have grown up. It seems to give them a more negative attitude toward marriage. Children whose parents were divorced are themselves more likely to go through a divorce (Glenn and Shelton 1983). Not only that, but they are less likely to marry in the first place or at least to marry as early as everyone else (Amato and Booth 1997; Kobrin and Waite 1983). This trend shouldn't be exaggerated, though, because these children of divorce are only slightly less likely to marry in their twenties than other people. However, there is a definite tendency in this direction, especially for some groups of people. White women in particular seem to acquire a noticeable disinclination to marriage as a result of their parents' divorce. Amato and Booth (1997, 109) found that adult children of divorce were not less likely to cohabit but that they were more likely to postpone or perhaps avoid marriage. In addition, researchers find that adult children of divorce are themselves more likely to get divorced, although differences between people from divorced families and their counterparts from intact families have been declining in recent years (Amato and Booth 1997; Wolfinger 1999).

Interestingly enough, this antipathy to marriage seems to operate whether the parents eventually remarry or not. In general, remarriages do not seem to affect children's psychological and personality traits very much. Perhaps this is because opposite effects balance out in the statistics: some children thrive in the new situation, whereas others become involved in new pressures and conflicts. Remarriages do improve the family finances, which has an important background effect on the level of happiness.

Debates about the effects of divorce on children often reflect disagreement or confusion about whether divorce is the "cause" of observed outcomes. Some studies do not distinguish single-parent households formed by divorce from those formed by widowhood or nonmarital birth. Often, studies cannot isolate whether it is the divorce, per se, that causes problems or whether the problems stem from economic and other effects of living in a single-parent family. Many studies do not specify the type of single-parent household, the recency of the divorce, nor control for the potential effects of remarriage on either adults or children. Although more studies are now considering the important role of conflict in producing negative child outcomes, many studies still fail to consider how much conflict was present before the divorce or how much conflict is present in comparison groups of "intact" families. Finally, it is usually very difficult to determine whether long-term family problems that contribute to parents getting divorced are also the major cause of subsequent child problems (Seltzer 1994b).

The best way to answer questions about the potential impact of divorce on children is to focus on children's developmental needs and to ask how divorce might influence the chances of meeting their needs. Major explanations for why marital or relationship dissolution might negatively affect children focus on disruption and the loss of elements that provide for children's basic needs: economic factors, living conditions, safety, emotional security, enduring social bonds, socialization, supervision, encouragement, limit setting, and love. Although researchers disagree about the extent to which each of these elements accounts for negative effects of marital dissolution on a child's well-being, they are all crucial for the welfare of children (Parke and Hetherington 1999, Seltzer 1994b). Because these elements are important for all children despite where they live or who their parents are, a focus on how to meet children's needs is more fruitful than any simple comparison between one-parent and two-parent families.

Child Custody

There are three basic forms of custody: **maternal custody, paternal custody,** and **joint custody.** Traditionally in English common law, legal divorces were extremely rare, and the father retained custody of the children. In the twentieth century in the United States, as divorce became more common, the situation switched to the opposite extreme, and courts routinely gave custody to mothers as the "natural" child-rearer. The sexism in this assumption has become more apparent in recent years, but in about 90 percent of all cases, mothers still get custody. This is partly because of a continued traditionalism on the parts of courts and partly because women are more likely than men to ask for custody.

The courts of virtually all states have recently begun to make two custody determinations: legal custody and physical custody. **Physical custody** relates to which parent the child will live with on a day-to-day basis, and **legal custody** relates to the parental power to decide a child's religion, education, and medical treatment. In cases of divorce, the courts respond to requests by the parties involved and grant maternal, joint, or paternal *legal* custody and maternal, joint, or paternal *physical* custody in any combination. Maternal physical and legal custody is the most common arrangement, but maternal physical custody with joint legal custody is becoming increasingly prevalent, with recent national estimates at about 20 percent of all cases (Braver with O'Connell 1998; Nord and Zill 1997).

In California, there is evidence that the majority of both mothers and fathers now request joint legal custody when they divorce and that the courts normally award it. Although joint custody only became legally possible in California in 1979, over three-fourths of divorce decrees now provide for either joint legal or joint physical custody (Mason 1999; Mnookin et al. 1990). In most cases the courts continue to award physical custody to mothers (approximately 80 percent; see Maccoby and Mnookin 1992). Contrary to media portrayals, most divorcing

A CLOSER LOOK 14.5
Joint Custody: What Actually Happens?

Joint physical custody involves sharing the children between two households. There are many different ways in which this is done, ranging between splitting the week (weekdays with one parent, weekends with the other), splitting the day (some hours in one place, other hours in the other), splitting the year (usually summer and other vacations with one parent), or alternating years between the parents.

When joint physical custody is awarded, the actual living arrangements may not reflect this pattern. For instance, in the California study (Mnookin et al. 1990), 20 percent of the total sample were awarded joint physical custody, but about half of these reported a one-parent living arrangement during the interview. Three times as many joint physical custody cases gravitated toward actual residence with the mother (39 percent) as with the father (13 percent). The divorced parents with joint physical custody decrees reported higher levels of conflict than mother-custody families, but those families with a joint decree, and actual residence with the mother, reported the highest conflict of all. In all, a total of 16 percent of the sample were found to have actual joint living arrangements for the children, in part because other physical custody decrees sometimes gravitated toward joint living arrangements (Mnookin et al. 1990). We can see from the California figures

that people don't always file legal papers to get what they want, nor do they always do exactly what the court decided in the divorce decree.

Social welfare professionals are sharply divided on whether joint physical custody is a good idea or not. One camp vociferously claims that joint custody is merely a way of dragging out a divorce, which allows the parents to continue to use their children as pawns in an ongoing fight. Another camp just as vigorously argues that joint custody ought to be normally awarded, unless there are special circumstances dictating otherwise. It turns out that there are advantages and disadvantages. It is *not* true, however, that joint custody is more of a strain on children than other forms of custody. Deborah Luepnitz (1982, 50) concluded that "joint custody at its best is superior to single-parent custody at *its* best."

The main advantages of joint custody are: (1) there are fewer court battles and fewer cases of parental child snatching; (2) mothers are more likely to receive regular support payments; and (3) both parents have a built-in break from continuous parenting and can rely on each other for "baby-sitting" when they need it. The main disadvantages are: (1) each parent is tied to the ex-spouse and cannot easily leave town for another job or other opportunity, and (2) hassles tend to arise in shuffling children between two houses.

---

families with children have very little legal conflict concerning custody or visitation before judgment. Over two-thirds of the California cases had either mild or negligible conflict. In 38 percent of the families, the divorce was uncontested, and the parental interview information suggested that there was no basic disagreement concerning custodial arrangements (Mnookin et al. 1990). This is primarily because both parties agreed that the mother should have physical custody.

Most divorcing mothers request sole physical custody for themselves. In the California study, more than 80 percent of the mothers said they personally wanted their children to live with them after the divorce. About 15 percent of the mothers indicated a desire for joint physical custody, but less than 2 percent said they wanted the father to have sole physical custody (Mnookin et al. 1990). As might be expected, the fathers' preferences were very different. Roughly equal proportions of the fathers desired joint custody, paternal custody, and maternal custody. Even if all fathers who could not be contacted in this study had indicated a preference for maternal custody, about half the total sample of fathers would still be in favor of joint or paternal custody.

When it came to actual legal requests for custody, mothers were more likely than fathers to act on their preferences. Nearly 80 percent of mothers requested

In single-parent custody situations, courts usually order some kind of visitation rights for the noncustodial parent with the children.

sole custody. Only about half of those fathers who said they wanted joint or paternal custody actually filed the papers requesting such an arrangement. There was a much greater likelihood that the divorcing parties would report intense conflict if the father requested joint or sole physical custody. In about one of ten cases, a parent asked for more physical custody than indicated during the interview (Mnookin et al. 1990). This might be related to strategic bargaining concerning the child custody and child-support provisions of the divorce agreement. With over 80 percent of the states now allowing joint custody by statute, and the rest permitting it by case law, such negotiations are likely to become more widespread (Braver with O'Connell 1998).

Visitation

In single-parent custody, courts usually order some kind of **visitation** rights for the noncustodial parent with his or her children. This often works out in fact somewhat differently than the court has specified. The custodial parent may use her or his control over when is an appropriate time to visit (because the exact hours are rarely stated in court orders) to punish the ex-spouse (Braver and O'Connell 1998). Conversely, some parents rarely use their visitation rights. This is more common in the case of fathers: about a quarter never visit their children if they lose custody of them, and another quarter visit rarely (Luepnitz 1982, 34). Mothers are less likely to do this if they lose custody, but still a quarter of them rarely or never visit. About half of both mothers and fathers visit their children quite frequently.

Visits mean that the noncustodial parent either takes the children out to the movies and ice cream parlor or the like or has them for a visit at his (her) home. Some children enjoy these visits; others do not. Here's what one seven-year-old girl said:

Q: Would you like to see your dad more or less?

A: Never!

Q: Why?

A: Because he makes mean faces and says things about Mom, and you get sick when you go with him.

Q: You get sick?

A: Yeah, I used to like to go before he started asking questions.

Q: What questions?

A: About Mom. Like if she goes out with men, and if she works, and if she leaves us with sitters. (Luepnitz 1982, 35)

Few states have provisions for considering the wishes of the children in visitation or custody adjudication or enforcement, and even those often fail to consider the different developmental competencies of the children (Mason 1999).

Advocates for divorced mothers tend to focus on the ways that the courts support men's visitation rights even when they do not pay child support or when the mothers have reason to believe such contact could be detrimental to the children (Arendell 1995). Advocates for divorced fathers, in contrast, tend to focus on the ways that mothers limit fathers' participation with their children, even when they are current with their support payments. Braver (1998, 176–77) recounts a story illustrating how mothers sometimes resist divorcing fathers' attempts to participate in their children's lives:

> *When we spoke to Pam prior to her divorce, she was livid about Drake. "He wants to trade in his sports care for a minivan," she told me. "But I'm not going to let him. My lawyer says she thinks she can stop him." When asked why she did not want him to buy a minivan, Pam responded, "Because then he'll want to participate in taking the boys to their activities in the car pool. He's never done this before! I always drove car pool. I was the soccer mom. He was always too busy, too interested in his job. Couldn't be bothered. Now we're getting a divorce and he suddenly wants to show off that he's such a great dad. It's a naked power grab, and I'm not going to let him do it."*

Braver describes how Drake was prevented from buying the minivan for about six months, but after permission was granted and he got the van, Pam got sick and needed him to drive six kids to soccer practice. This was the breakthrough that helped her see that they could both be involved in their children's lives. She said, "What was I resisting for? His having a minivan means I have to drive less, and I can take it easier and get more of my own stuff done." Braver (1998) suggests that a cooperative model of postdivorce relations, in which he got to participate in his sons' activities and she got some occasional relief, is a win-win model that can empower both divorcing spouses. As many studies show, however, not all couples who split up can get beyond the anger and hurt feelings associated with the breakup.

Studies find that the conflict associated with many divorces makes cooperative visitation or shared custody difficult to achieve. When there is less conflict, shared arrangements tend to work better. Most large-scale studies show that child support and visitation are correlated, though it is difficult to know which one comes first (Braver with O'Connell 1998; Seltzer 1994b). Soon after the divorce, when emotions are raw and conflict is high, visitation and child support can become pawns in hostile negotiations between bitter ex-partners who try to "win" by making the other parent "lose." The real losers in such situations are the children.

Support Payments

Traditionally husbands had the obligation under the law to support their children. The sexist language here has come under attack, but the spouse with the larger income is still obligated to pay for the children's support, and in most cases this is the father. In the past, judges have exercised wide discretion as to the amount of child support awarded. Court records indicate that amounts have ranged from 0 to over 100 percent of the noncustodial father's income (Garfinkel 1988). Child-support awards have tended to be regressive, insofar as wealthier fathers are ordered to pay a lesser share of their incomes than poorer fathers. However, the level of child support is supposed to reflect the standard of living the family had before the divorce, not merely provide the bare necessities of life. Some states use a sliding percentage of the father's income as a basis for awarding support: for example, 17 percent for one child, 25 percent for two children, 29 percent for three children, 31 percent for four children, and 34 percent for five or more children (Garfinkel 1988, 333). In dollar amounts, that means a father of two children who earns $20,000 per year is supposed to pay $417 per month in child support, whereas a father of two who earns $45,000 is supposed to pay $938 per month. From the women's viewpoint, these dollar amounts are criticized as being too low. On the other hand, some men feel unjustly deprived when a court awards more than a third of their income, and historically, the courts have been reluctant to award high levels of child support because it may "remove a man's incentive to earn" (Weitzman 1981, 123; see also Braver with O'Connell 1998).

According to recent research, most divorced men could pay more child support without assuming an unfair burden. It is estimated that noncustodial fathers could pay about two and one-half times their current legal obligations and three times what they are actually paying (Garfinkel and Oellerich 1989; Seltzer 1994b). The biggest problem has been divorced men's failure to pay anything at all. As discussed earlier, less than half of all divorced women receive the full amount of child support they are due.

In 1984 and again in 1988, Congress passed legislation concerning awarding and collecting child-support payments. Most of the changes were designed to ensure that the custodial parent (usually the mother) gets an award that is adequate to support the child and that the absent parent (usually the father) can be found and made to pay. Because of high levels of failure to pay in the past, the federal government now monitors states' efforts to collect delinquent child-support payments, helps in establishing paternity (pays for blood tests), and locates missing parents who are supposed to pay. Because of the difficulty in bringing nonpaying absent parents into court and getting them to pay, all new or modified support orders include a provision that payments are to be withheld from absent parents' wages automatically without regard to whether they are in arrears. Thus, child-support payments are beginning to be handled like income taxes and regularly deducted from the absent parent's paycheck. Although this has resulted in lower levels of non-payment, it has not lifted a significantly larger share of children out of poverty.

Effects of Custody Arrangements

All types of custody arrangements, like marriages, place strains on children (and on parents). We do know that children tend to have more postdivorce contact with fathers in joint custody than in mother-custody arrangements

(Bowman and Ahrons 1985; Braver with O'Connell 1998). Nevertheless, there is no evidence of any difference among children in maternal, paternal, or joint custody in terms of their psychological adjustment or behavior problems (Derdeyn and Scott 1984; Furstenberg, Morgan, and Allison 1987; Luepnitz 1982, 149).

Children in general tend to report being satisfied with whatever custody arrangements happen to exist. There is some evidence, though, that children are more socially competent if they live with the same-sex parent than if they live with the opposite-sex parent. This seems to be particularly true of boys, who do more poorly in their mother's custody than girls do (Luepnitz 1982, 11). There are also strains on the parents in *not* having their children. Psychologists have long used the term *father absence* or *mother absence* to refer to feelings of deprivation experienced by children in the absence of one parent. Now the term *child absence* has been coined to refer to the feelings of deprivation experienced by some parents whose ex-spouse has sole custody of their children. However, we do not have enough evidence as yet to conclude that a particular form of child custody worsens or improves the adjustment of the mothers and fathers after the divorce (Coysh, Tschann, Wallerstein, and Kline 1989).

REMARRIAGE

Remarriages are now quite common in the United States and other industrialized countries, with approximately half of U.S. marriages representing a remarriage for one or both partners (Bumpuss, Sweet, and Castro Martin 1990; Coleman, Ganong, and Fine 2000). Historically, researchers have treated remarriages as a uniform category, but more recently, studies have attempted to measure remarriages and stepfamilies in more complex ways (Bumpuss, Raley, and Sweet 1995; Coleman, Ganong, and Fine 2000). The term **remarriage,** for example, encompasses several different types of relationships. To be counted as a remarriage, only one of the partners need be marrying for at least the second time, but remarriages also encompass unions in which both partners are in a second or higher-order remarriage (e.g., a third or fourth marriage). Approximately three of every four people who divorce also remarry, and serial remarriages are becoming increasingly common. One in ten remarriages in the United States represents at least the third marriage for one or both partners (Coleman, Ganong, and Fine 2000).

Like marriage rates, remarriage rates have been dropping in the United States, but this does not mean that people are recoupling less often than they used to. Instead of marrying after marital dissolution, more people are cohabiting as an alternative to remarriage (Bumpass, Raley, and Sweet 1995; Bumpass, Sweet, and Castro Martin 1991). In addition, most couples who remarry cohabit before legally forming a union (Cherlin and Furstenberg 1994). For reasons discussed next, the divorce rate is slightly higher for remarriage than for first marriage, and remarriages ending in divorce do so somewhat more quickly than first marriages (Coleman, Ganong, and Fine 2000). Over time, the divorce rates of remarriages and first marriages tend to converge, suggesting that once one finds a compatible partner, similar forces tend to keep the relationships together, or alternately, drive them apart (Clarke and Wilson 1994). Finally, remarriages contracted by people after they are over 40 years of age may be more stable than first marriages (Wu and Penning 1997).

Who Remarries Whom?

Men are more likely to remarry than women, and they tend to do it faster as well. More than three of four divorced men remarry, in contrast to only about two of three divorced women. Again, as we saw in chapter 11, men actually find marriage more to their advantage than women do. The average remarriage occurs less than four years after a divorce: a length of time in which people can adjust to a "second-time-around" singles lifestyle and then grow tired of it. It is also enough time for people to find a new partner, if that is what they want to do. The interval to remarriage lengthens with advancing age for both men and women (Wilson and Clarke 1992).

Younger women remarry more quickly than older women. Women divorced after the age of 40 have a low probability of remarriage, even though remarriage rates among older widows have been rising (Ahlburg and DeVita 1992, 17). Age has little impact on the remarriage of men, again illustrating their favored position in the remarriage market. In addition, men with higher incomes or more education are more likely to remarry than those who are less well-off economically. Just the opposite is true for women: those with more resources are less likely to remarry. Although both men and women with children are less likely to remarry than those without, children lower women's chances of marriage more than men's. Women with a child under age six are least likely to remarry quickly (Sweeney 1997).

In remarriages as in first marriages, people tend to match themselves up on the marriage market with someone similar in social class, education, and ethnicity. An additional source of in-marriages among similar people is that divorced persons are most likely to marry another divorced person, just as single people tend (even more overwhelmingly) to marry other singles, and widowed people tend to marry widows. The second choice for single people is to marry a divorced person, which is also the second choice for widows. Thus, the divorced seem to be in between two other marriage markets, overlapping somewhat with both singles and widows, although each group marries primarily within itself.

Patterns of marriage and remarriage partially derive from people choosing to marry others of similar age. For first marriages, the trend is for men to be about two years older than their brides. However, in remarriages the age gap tends to widen but differs depending on one's gender and the age at which one remarries (Coleman, Ganong, and Fine 2000; Wilson and Clarke 1992). Men who remarry before age thirty tend to be about a year older than their brides, whereas men who remarry at ages thirty to thirty-nine are, on average, four years older. Men who remarry at age forty or older are an average of eight years older than their new brides and are probably able to marry younger women because they command more economic resources at that stage of their lives. In contrast, women who remarry before age thirty are three and one-half years younger than their grooms, whereas women remarrying after their thirtieth birthday are about two years younger. More than half of the men and women remarrying before age thirty marry single people, whereas about three-fourths of those remarrying after age forty marry divorced people. These trends are strongly influenced by the number of eligible mates in particular age groups. Because there are more divorced or single women than men in each age group going up from the thirties, men's position in the remarriage market is strengthened even further. Men's higher occupational status translates into higher likelihood of remarrying at all age levels, whereas women's higher occupational status at younger ages is associated with delayed remarriage. At relatively older ages,

In remarriage, the age gap between husband and wife tends to widen. Men who remarry at age forty or older are an average of eight years older than their new brides.

women's greater economic resources hasten remarriage (Coleman, Ganong, and Fine 2000).

Life in Remarriages

Because researchers have only recently turned their attention to studying the complexities of remarried households and **stepfamilies,** most findings are still tentative and often difficult to generalize. To begin with, the households in these categories can look quite different from one another. Not all remarriages include children from prior relationships, nor do all stepfamilies incorporate a remarriage. The presence of children in the household is an important factor to consider and can strongly influence family dynamics, as are the ages of children and their biological or legal relationship to other household members. Some first marriages create stepfamilies and stepparent-stepchild relationships, as when a never-married mother marries a man who is not the child's father (Coleman, Ganong, and Fine 2000).

Adding new children to the stepfamily mix is also a relatively common occurrence. About half of all women in remarriages give birth to at least one child, most within two years of remarriage (Wineberg 1990). Although the arrival of jointly conceived children can solidify the couple's relationship through shared parenting, it can also create difficulties in the new relationship. The arrival of a new baby demands significant care and investment of time from the parents, may reduce the couple's economic security, often interrupts the woman's (or sometimes the man's) career, and can complicate the parents' relationships with other children in the family, not to mention their relationships with various in-laws, ex-spouses, and ex-in-laws.

In addition to legally remarried couples, cohabiting households are very likely to have children. For classification purposes, these are often treated as belonging in the general category of stepfamilies. Almost a million cohabiting households in the United States contain at least one adult who brought a child from a prior relationship, thereby creating a cohabiting stepfamily household (Bumpass, Sweet, and Cherlin 1991). In fact, cohabiting couples are more likely to enter a new union with children from previous relationships than are remarried couples (Coleman, Ganong, and Fine 2000; Wineberg and McCarthy 1998). Although about a third of all U.S. children will have lived in a remarried or cohabiting stepfamily before they reach adulthood, many of these will also experience several different family forms in succession. A wide range of marital and cohabiting relationships for the parents create complex family histories for the children. This creates problems for demographers and offends narrow-minded family moralists, but the fluid complexity of such individual and family trajectories is likely to be the hallmark of the twenty-first century (Eichler 1997).

The social norms and rules governing remarriages are not as fixed as those governing first marriages (Cherlin 1978). Although this can create ambiguities and uncertainties, it also provides these families with opportunities for doing things a bit differently. When compared with first marriages, women in remarriages are generally more assertive and exercise more power in the relationship (Pyke 1994). Decision making and household labor allocation in remarriage are also perceived to be more equally shared than in first marriages, though they also tend to follow some traditional male-dominated patterns (Coltrane 1996; Pyke and Coltrane 1996). Women tend to be more involved in financial decision making in their remarriages than in their prior marriages, but it is difficult to study because finances in remarried families are complex. For example, remarried spouses are more likely than first-married spouses to maintain some economic resources under individual control, in part because of financial responsibilities for children from multiple unions and in part because they want to retain some financial independence (Burgoyne and Morison 1997; Coleman, Ganong, and Fine 2000). Remarried spouses may be more likely to express negative feelings than those in first marriages. This may stem from tensions and disagreements relating to stepchildren, such as discipline, rules for children, and the distribution of resources to children (Pasley, Koch, and Ihinger-Tallman 1993). It is no surprise that problems arise, given the complexity of stepfamily relations and the many opportunities for conflict in the face of competing loyalties and allegiances. Nevertheless, Kurdek (1999) found that children born to first marriages actually lowered marital quality more than stepchildren lowered remarital quality.

Contact between the children and the nonresident biological father often diminishes after remarriage, and in about half of the cases, fathers have little or no contact with their children from a prior marriage. In other cases, nonresident families maintain contact with their children but tend to be more distant and detached than they were before. Although relations between noncustodial parents and stepparents are sometimes strained, they are usually polite, and there is no evidence that they are especially conflict ridden (Pasley and Ihinger-Tallman 1987).

Are stepfamilies different from other families? Negative cultural stereotypes of "wicked stepmothers" and "neglected stepchildren" tend to prejudice us against the potential for remarried families to be similar to "biological" families (Coleman and Ganong 1987). Nevertheless, most research shows that daily life in stepfamilies is more similar than dissimilar to that of first-marriage families

(Furstenberg 1987). One difference, however, is that the role of stepparent is different from that of biological parent. Because there is often confusion over what authority the new parent should have, the biological parents, stepparents, and children must negotiate new patterns of relating. In general, stepparents tend to be somewhat more disengaged and are more apt to use an authoritative style of parenting than the resident biological parent. Stepparent-stepchild relations are typically found to be important determinants of marital satisfaction between the new marriage partners as well.

CONCLUDING THOUGHTS ON MARRIAGE AND DIVORCE

Divorces are tied to many problems and a great deal of suffering. There is the strain of the divorce itself, the negative effects on many children, and the economic loss suffered—especially by women. Would society be better off, then, if divorces were more difficult to obtain? Social and legal pressure could be applied to keep families together. It is doubtful that this would be a real solution, however. It would amount to turning the clock back to previous times when divorces were virtually impossible to obtain. Also, family life was much more authoritarian, based more on economic considerations and social respectability than on personal affection and concern for individual happiness (see the historical material in chapters 3 and 4). The humanitarian movements of the modern period moved in the direction of loosening up those authoritarian family controls and allowing individuals to escape from unsatisfactory marriages. If we went back to restrictions on divorce, people might be better off economically, but the old sources of family conflict would be back again, too. One can predict that, in that event, the pendulum would soon swing the other way, with movements springing up again to allow divorce.

The situation appears to be a dilemma, a trade-off between opposing concerns. Is there any way that the problems can be mitigated, without turning back the clock? Here are several suggestions.

There are two rather different types of divorces: those in which the couple has no children and those in which there are children. Divorces in childless marriages seem to have relatively few negative consequences other than the temporary unpleasantness of breaking up the relationship itself. These divorces tend to happen relatively early in a marriage. They are cases in which the couple discovers they made a mistake in the marriage market. They just weren't matched as well as they believed. Under these circumstances, the best thing for everyone's happiness is the divorce. In fact, such couples ought to be encouraged to get divorced as soon as they recognize their marriage will not work instead of waiting until they have children.

Divorces in which there are children, on the other hand, are full of problems. It is not true that children of divorce necessarily become neurotic or delinquent, but some children of divorce do have serious problems. Even without these, arrangements for custody of the children and visitation by the ex-spouses are full of dilemmas. Nevertheless, married couples who stay together "for the sake of the children" even though they are warring between themselves do not create a good environment for their children either.

The problem would be eased if adults were much more careful about planning when to have children. In particular, they should have every opportunity to gauge whether their own relationship is likely to be permanent before they have children. Cohabitation between unmarried couples as a kind of "trial marriage" is thus a useful thing. The prevalence of contraception, as well as abortion, is directly related to the number of children who are born who are genuinely wanted by their parents. The strain of having children, especially if the parents are not really committed to each other, is one of the factors that makes marriages unhappy and hence more likely to end in divorce. We see, for example, that divorce is higher among women who conceived premaritally (*Family Planning Perspectives,* 1986). Hence, one can predict that moves to restrict abortion and otherwise make it more difficult for adults to control childbirth would result in more children of divorce. Conversely, improvements in population control would reduce this problem.

Another potential solution to some of the hardships of divorce would be to provide more practical and economic help to families. Although such solutions are far from easy to implement, many child and family advocates suggest that divorce would be less prevalent, and less devastating when it did occur, if we could provide full employment and decent wages for every American family. A range of child-related services, such as day care and health care, have also been proposed by various advocacy groups, councils, and commissions.

Ultimately, a major factor in unhappy marriages is economic and social inequality between men and women. It is men's greater economic resources that give them power within the home and that in the past has motivated women to take a somewhat calculating attitude toward the marriage market (chapter 8). More genuine love relationships are fostered by greater equality because there is less feeling of power on one side and subtle coercion on the other. Many of the issues of spouse abuse, too, are based on the traditions of male domination, which make many men feel justified in using force to control their wives (chapter 13). In the long run, this suggests a potential solution to the problem of divorce. If and when women achieve a breakthrough against economic discrimination in the occupational sphere, they will enter marriages on a level of economic equality with men. A major source of family conflicts, the struggle over power, will have disappeared. When and if that happens, we may expect that the high divorce rates that characterized the period of most intense gender conflict (the 1970s, 1980s, and 1990s) may be a thing of the past.

SUMMARY

1. The divorce rate in the United States climbed steadily for over 120 years. The major exception was during and just after World War II, when the divorce rate suddenly skyrocketed and then fell for a few years, before beginning its climb again to still higher levels. Since the late 1970s the divorce rate has remained at a high level.

2. According to current projections, about half of all marriages will end in divorce. Divorce is most likely to occur two or three years after the marriage, whereafter the rate begins to fall. Half of all divorces occur within about six years of marriage.

3. Members of the lower social classes tend to divorce more often than those in the higher social classes. There is one major exception: highly educated career women (but not men) have a higher rate of divorce than those of lower educational and occupational levels. The reason is probably their greater reluctance to stay in a male-dominated marriage and their greater ability to support themselves independently.

4. Teenage marriages are particularly likely to end in divorce. There is also a tendency for religious and racial intermarriages to have higher divorce rates.

5. The most important factor affecting divorce is economic. Divorces generally bring a decline in the standard of living for both partners but especially for women. Divorce rates tend to decline during economic depressions and to go up during prosperity.

6. Family-values rhetoric blames divorce for many of society's ills, but evidence to support such claims is mixed at best. Throughout American history, divorce has been used as a symbol of moral decline by politicians and religious leaders who have reacted against women's attempts to be treated as equals.

7. Americans' opinions about divorce have remained fairly stable for the past thirty years. Most agree that married couples should divorce if they do not get along, whether they have children or not.

8. Women are more likely to initiate divorce and are less likely to remarry than men. The person initiating the divorce has an easier time with the transition than the person who was left.

9. Child-support payments are awarded to over half of custodial parents after divorce. Custodial mothers have lower incomes than custodial fathers or married mothers, in part because of low levels of child support.

10. The experience of divorce is difficult for children, but the majority suffer no long-term psychological difficulties. Two factors associated with divorce cause problems for children: conflict and poverty. If the divorced parents continue to engage in frequent conflict, children are more likely to experience distress and have more difficulty maintaining a positive self-image. The lack of financial resources often accompanying divorce can cause high levels of physical, emotional, and psychological stress. Children of divorced parents are more likely to drop out of school or have children while in their teens. Compared with children raised by both parents, children whose parents divorced tend to be more reluctant to marry and are more likely to have divorces of their own after they grow up, but such differences are declining.

11. Most divorces are uncontested, and most custody arrangements are agreed to by the divorcing spouses. In about 90 percent of all divorces, the mother is granted physical custody of the children. Many states are now awarding joint legal custody, wherein both mother and father retain rights to make decisions about children's medical treatment or educational endeavors. In some states, like California, about one in five divorcing couples are awarded joint physical custody, wherein the children actually reside for part of the time with each parent. Both types of joint custody arrangements tend to increase the amount of contact fathers have with children after the divorce, but we do not know the long-term impacts of such arrangements.

12. Remarriages are now a typical part of the life course. More divorced men eventually remarry than divorced women, probably because being married is more advantageous for men. Women with children are less likely to remarry and so are older and more highly educated women. Divorced people are especially likely to remarry other divorced persons.

13. Life in remarried families is not much different from life in first-married families, although spouses, stepparents, stepchildren, and stepsiblings must adjust to each other and negotiate new patterns of relating. There is some indication that power is shared more equally between spouses in remarriages but also some indication that conflict is more frequent.

Key Terms

alimony
child-support awards
child-support payments
community property
covenant marriage
divorce rate

fault-based divorce
initiators of divorce
joint custody
legal custody
marital contract
maternal custody

no-fault divorce
paternal custody
physical custody
remarriage
stepfamilies
visitation

Sociology Web Site

See the Wadsworth Sociology Resource Center, "Virtual Society," for additional links, quizzes, and learning tools:

http://www.sociology.wadsworth.com

Also on this web site you'll find InfoTrac College Edition, an online library of journals. Here you can search for electronic articles about central topics in sociology.

15

Life Course Transitions:
From Adolescence
to the Aging Family

The Late Life Transition

Death and Family

Summary

INTRODUCTION

The study of adult life transitions is a relatively new field of research for sociologists and psychologists. The theories that have drawn the most attention and achieved the greatest agreement are those applying to childhood and especially its first few years (chapter 12). However, led by Erik Erikson (1959; 1982), students of human development are now realizing that adulthood is not one uniform plateau but tends to be marked by peaks, valleys, and turning points. For instance, Rhona Rapoport (1963) was one of the first to note that individuals encounter a patterned set of family "crises," which she termed "normal critical transitions." Such transitions include leaving home, getting married, having children, launching children, retiring, surviving a spouse, and increasingly, getting a divorce.

LIFE COURSE VARIATIONS

Researchers studying the "life cycle" first attempted to identify regular stages of adult development through which everyone passed. More recent research reveals that life transitions do not occur in the same way or at the same time for everyone (Elder 1998; Furstenberg 2000; Mattessich and Hill 1987). Not only do individuals in the same culture tend to differ somewhat in the timing of the onset of various stages of adult development, but the length, sequence, and composition of life transitions tend to vary according to the period of history and the particular cultural milieu in which one lives. For instance, pastoral (herding) societies tend to socialize children to assume full adult responsibilities sooner than hunting and gathering societies (chapter 3). This is primarily because the type of subsistence economy—herding large domestic animals—demands many hours of relatively low-intensity labor that children can master. Similarly, societies existing in ecological scarcity tend to use child labor in any way that will contribute to the survival of the community. Societies living in scarcity therefore demand that children assume adultlike roles sooner than children in societies that enjoy relatively secure subsistence.

More modern social and economic developments, such as industrialization, urbanization, and the spread of mass education, have similarly impacted the meaning and timing of transitions into and out of childhood and early adult roles. The works of Ariès (1962) and Zelizer (1985) (chapter 12) illustrate how historical trends in Europe and America influenced our expectations about how children should act and when they should assume adultlike duties. Specific historical events, such as war or economic downturn, can also influence people's family experiences and alter the timing of important life transitions. For example, a lack of money during the depression discouraged young people from marrying or forming their own households and discouraged the already married from having children or getting divorced (Elder 1974, 1998). Similarly, the advent of war can influence the nature and timing of the transition to adulthood through recruitment of young men into the armed services. Being in the military, and especially having to fight in a war, makes young men "grow up" faster and also affects decisions about staying in school, pursuing a career, getting married, having children, or getting a divorce (Elder 1986).

When we look at broad historical trends in the United States, we can see that we are returning to timing patterns for getting married and having children that are more like events at the turn of the century than like those of the mid-twentieth century. Between 1890 and the late 1960s, we witnessed a gradual quickening of the pace of life course transitions. Over this time, people tended to get married, become parents, have children, leave home, and become grandparents sooner than did their own parents (Neugarten and Datan 1973). Since the 1960s, the trend has reversed, with both men and women marrying later and having children later than did their own parents. Returning to patterns of later marriage and later birth has been encouraged by women's increasing chances for going to college and getting jobs (chapters 5, 10, and 12).

Two other demographic trends are helping to shape the course of adult life. Life expectancy has increased a whopping twenty-seven years since 1900. The average U.S. woman can now expect to live to be eighty years old (*Statistical Abstract* 1999). The birthrate, on the other hand, has decreased gradually in this century, with only minor reversals around 1908, 1950, and 1980 (see chapter 5 and Figure 5.5). The average number of children born per female dropped from 3.7 children in 1900 to just under 2.0 today (National Center for Health Statistics 1999, table 3). These demographic trends are having some major impacts on the processes of growing up and growing old. In this chapter we look at some of these life course processes, beginning with the transition from teenage years to adulthood. We also examine some midlife and later-life transitions and briefly discuss the final transition, death.

ADOLESCENCE

Adolescence existed before the twentieth century, but it was not until the middle decades of the twentieth century that **adolescence** emerged as a discrete life stage in modern societies. As full-time education replaced full-time employment for many young people, adolescence became defined as a special period of life distinct from both childhood and adulthood. Historical scholars describe how the transition from childhood to adulthood became more predictable, rapidly accomplished, and socially organized—even if temporarily so—during the middle part of the century (Furstenberg 2000). As teenagers were pushed out of the labor market and sheltered from the adult world, they began to develop age-segregated patterns of social life with unique cultural qualities (Flacks 1971; Zelizer 1985). Ironically, as they were becoming more marginal, adolescents were also granted more autonomy (Modell 1989). Because they were threatened by it and did not understand it, many members of the older generation came to view youth culture with increasing suspicion and even hostility.

As the sociologist Frank Furstenberg suggests, the problematic features of adolescence that the media endlessly catalogues are structurally created and maintained by social institutions that isolate youths from adults—ironically to prepare them for future adult roles. He notes that "culturally, youth are simultaneously indulged and castigated—allowed or even encouraged to seek their own company yet reproached for being self-centered, irresponsible, and occupied with self- or socially-destructive behaviors (Furstenberg 2000, 5; see also Farkas and Johnson 1997). The sociological point is that advanced industrial societies create adolescence and early adulthood as life stages in ways that inevitably render them problematic. This paradox does not stop academics, politicians, religious leaders, and journalists from repeatedly focusing on the shortcomings of youth culture and highlighting the problems that adolescents supposedly cause. Although it is likely that on reaching middle age, virtually every generation complains about the shortcomings of youth, folk beliefs about irresponsible teenagers are also perpetuated through academic research. In a review of scholarly literature on adolescence published during the 1990s, Frank Furstenberg (2000) notes that at least half of the articles were principally about youthful misbehavior and maladjustment: delinquency and violence, substance abuse, school problems, mental health, and the like. A far smaller share of the literature concerned normal conventional behaviors, positive self-expression, developing mastery, and successful transitions into adulthood (Crockett and Crouter 1995; Graber, Brooks-Gunn, and Peterson 1996).

What is unique about being a teenager or adolescent? The teenage years are the period from thirteen to nineteen, though more realistically we usually stop calling someone a "teenager" when he or she has turned eighteen and graduated from high school. Adult status starts at eighteen in certain legal respects: being able to vote and marking the end of parents' liability for child support (again, in some states). However, other legal transitions occur at different ages: sixteen for drivers' licenses or twenty-one—the old legal age and still the drinking age in most places. The fact that the dates float around like this illustrates that the teenage or adolescent years are not merely biologically fixed but are also social. Of course, there is a biological factor—arriving at adult maturity—but precisely when this happens varies from individual to individual. This is overlaid by our society, which defines the transition in various ways and affects how people experience the teenage transition. Adolescents' self-images and coping abilities are affected by both the coming of puberty and by the shock of moving from a more intimate elementary school to the impersonal organization of junior and senior high school (Graber, Brooks-Gunn, and Peterson 1996; Simmons and Blyth 1987).

There is, of course, a physical side to the teenage transition. Children go through a growth spurt during their early teens, and most attain their full height by age seventeen. Some mature earlier or later, which creates a temporary lineup of physical advantages and disadvantages. Girls often feel unattractive when they reach their adult height at an early age, sometimes by age twelve or thirteen, and tower over the boys. Boys who attain their full height and weight earlier may have some tendency to push the less-quickly maturing around or at least dominate in athletics. Hence, sheer physical growth has an effect on the teenager's social prestige, though in different ways; because relative positions change during these years, egos can have some special ups and downs. Most of the males' later growth is in weight and musculature, which tends not to come in until the mid-twenties. By that time the adolescent transition, in the fullest sense, is usually over, and these kinds of physical rivalries settle down.

The adolescent transition is a time of sexual maturation as well. Again, different kids go through this at different paces, and the leads and lags can give rise to a status system. Though temporary, it may have a tremendous psychological impact on young egos. As social life becomes organized around boyfriends and girlfriends, dating, parties, and sex, a person's sexual attractiveness and popularity can be very important for placement in the teenage social networks. Teenagers are not only entering into the adult sexual marketplace; that by itself, as we have seen, creates both emotional excitement and anxieties. At the same time, physical changes are going on in their own bodies, which create additional temporary ups and downs. Putting these factors together, it is no wonder that teenage years are a time of considerable strain and have a reputation for wildness.

Mental illness rates are highest for this age, particularly rates of schizophrenia, a psychosis that involves extreme delusions and withdrawal from social reality (and often commitment to a mental hospital). If a person is going to acquire a homosexual rather than a heterosexual identity (or some mixed, bisexual identity), the teenage years are when this tends to occur (as we've seen in chapter 9). The delinquency rate is virtually zero at about age ten and then shoots up rapidly, so the highest level of juvenile delinquency is at about age sixteen. Thereafter it begins to dip again, and by the time young people are in their twenties, the rate at which they commit (and get caught for) burglaries, assaults, car thefts, and other crimes has dropped to a modest level, where it remains for the rest of their lives. Teenage drivers are the most likely of all to have auto accidents and are most likely to die from them. Perhaps because of these dangers (as well as the social and psychological strains of the sexual and other transitions), teenagers also form the group most likely to undergo a period of intense religious devotion.

One of the reasons that adult anxieties are aroused by teenagers is that they are less under their parents' control than they were when they were children. As direct parental supervision declines, contact with peers increases, and parents worry that other youths will encourage their teenagers to do something foolish. Surveys indicate that most Americans believe that parents, in general, provide too little supervision of teenagers. At the same time, surveys show that both parents and children report fairly close monitoring, at least for younger teens (Furstenberg 2000). Levels and types of control typically change during the course of adolescence, however, with direct monitoring declining over time and with a general relaxation of rules as children mature (Larson and Richards 1994; Furstenberg et al. 1999). Adolescents often follow their parents' directives, but undoubtedly they do so less frequently than most parents believe. For example, research indicates that parents tend to underreport their own children's use of alcohol and drugs, school problems, and sexual behavior (Bogenschneider et al. 1998). Some even suggest that an important developmental task of adolescents is learning what *not* to tell parents and how not to tell them (Furstenberg 2000).

Attaining Adulthood

When one thinks about it, many so-called adolescent problem behaviors could be called *adult* behaviors, insofar as they are practices in which the parents themselves engage (Furstenberg 2000). Such behaviors occur episodically and experimentally during the transition to **adulthood,** as adolescents gradually learn to deal with potentially risky behaviors like partying, drinking, and having sex (Jessor 1993). Peer groups become increasingly important in managing such behaviors, and parents worry that their children will "fall in with the wrong crowd." What many parents fear is that the **peer group** will condone unwanted behavior or allow too much experimentation. One particularly effective way that parents can exercise some control over the situation is by manipulating the environment. Not surprisingly, parents with greater financial resources are better able to choose the schools, neighborhoods, communities, or social activities that provide what they consider to be better supervision or the type of peers they would prefer their children to associate with. Influencing the choice of college is one of the main ways that middle- and upper-class parents subtly influence their children's choices during the transition to adulthood.

The foregoing analysis suggests that transitions from adolescence to adulthood vary according to social class. Working-class people typically go through the transition from adolescence to adulthood sooner than middle-class people.

Working-class teenagers are more likely to drop out of school, but even when they do get a high school diploma, they are more likely to take a full-time job by the time they are eighteen years old. Entering the adult world of routine employment often signifies the assumption of adult status. In addition, working-class couples are more likely to get married and have children while still in their teens or during their early twenties, again marking the adoption of full adult status. Middle-class children, in contrast, typically delay the full assumption of adult responsibilities by going to college, traveling, taking temporary instead of "permanent" jobs, working part-time instead of full-time, and continuing to receive financial assistance from their parents. Because they typically attend college until about the mid-twenties and sometimes return home to live with their parents for awhile, middle-class children have a later transition to full adult status. This is not to say that they are not considered adults by their peers or by the legal system, but they delay assumption of those occupational and family roles that have traditionally been used to mark the transition from adolescence to adulthood.

Employment is one of the most important socialization experiences bridging adolescence and adulthood. The overall labor force participation of sixteen- to nineteen-year-old males has been fairly steady throughout the century, whereas the labor force participation of sixteen- to nineteen-year-old women has increased substantially. The type and timing of the work have changed, with a larger number of young people working part-time, especially while they are attending school. The costs and benefits of employment for adolescents have been hotly debated by family scholars. Some researchers point to the harmful effects of work on school performance and educational commitment, but others focus on the potential benefits of employment for gaining skills, maturity, and social contacts (Greenberger and Steinberg 1986; Mortimer et al. 1996). In general, long hours, dangerous settings, and the absence of adult supervision have been found to contribute to the negative effects of work, whereas more favorable working conditions have been found to improve adolescent's social capital and promote responsibility, self-respect, and goal setting (Furstenberg 2000).

Because more adolescents at the beginning of the twenty-first century are attending college than in previous generations, on average, they are delaying conventional entry into adulthood. The usual markers of adulthood are permanent full-time employment, marriage, and parenthood. Virtually all studies concur that contemporary adolescents are delaying all three. In part, this is because they face an uncertain job market that provides less security and fewer chances for upward mobility than enjoyed by previous generations. As we saw in chapter 14, the marital instability faced by this generation's parents may make them more reluctant than others to embark on the marital or childbearing commitments that served as the markers of adulthood for their parents or grandparents. Some are returning periodically to their parents' homes, and others are delaying launching careers or families, perhaps to gain more education or job market experience (Goldscheider and Waite 1991; Schnaiberg and Golden 1989). As a recent commentary on "Generation X" attests (Howe and Strauss 1993), few of the stereotypes applied to contemporary adolescents were created by themselves, and most of the dialogue concerning the state of youth culture has been produced by earlier generations who worry that they are losing control of a world that is changing faster than they are. We speculate, however, that the next generation will provide this one with even more unanticipated worries about the decadence of the next adolescent peer culture.

THE MIDLIFE TRANSITION

The so-called **midlife crisis** became an important concept in the 1980s. This is a period around age thirty-five to forty-five or fifty when an individual may go through several years of psychological and social turmoil. People may become depressed or feel the need to shake up their lives by getting a new job or a divorce. The danger of alcoholism seems to be especially great at this age (Tamir 1982; Farrell and Rosenberg 1981).

Structurally, what is happening at this age is that one's career has reached a plateau or turning point. One has settled into a particular kind of work; the long series of stages of acquiring an occupational self-identity are over by sometime in the twenties, and now the individual has been in a chosen (or compromisingly accepted) field for ten or fifteen years. By now one knows realistically what kind of work it is, and one knows what one's chances are for moving up or into other kinds of work. This is part of the midlife crisis: the realization that you know what you are going to be doing for the rest of your life.

Often there is a feeling of panic connected with this. As Staude (1982) points out in his general theory of life transitions, the midlife period carries with it an aspect of death. What dies during the midlife period is one's youthful view of the world and one's role in it. The youthful period, as we have seen, is unsettled and vague. Although one's exact role in the world is unclear, the range of possibilities seems limitless. The midlife period closes this off. One now has a realistic view of what life is going to be like, and earlier ideals, hopes, and unrealistic dreams have to be put aside. More exactly: you do not usually *decide* at some point to "grow up" and put aside your dreams: instead, you gradually realize that your life has already fallen into a routine and that life itself has made its decision for you. Psychologically, the midlife crisis involves the realization of the death of one's fantasy self. Connected with this is a real sense of one's mortality. Not that one is close to death (because most people still have another thirty or forty years to live); but for the first time one has the experience of something like death. An idealized part of oneself is gone; what is left, realistically, is a mere mortal being who will eventually face the real biological death. The midlife crisis marks the end of the feeling that there are no limits.

It is not surprising that the midlife crisis became a popular topic in the late twentieth century, rather than at some other time in history, in the United States and other wealthy industrial countries. The midlife crisis is not a biological phenomenon but a social one. It is mostly a phenomenon involving white-collar workers, and it occurs among people who live a long time. Even in our own society, working-class people are much less likely to have a midlife crisis. For them, their occupational plateau comes much earlier. Already by his early twenties, a working-class man is likely to know what kind of work he will do for the rest of his life. (The case for women is somewhat different, as we will see later.) His income level, too, usually reaches its peak quite early, whereas middle-class persons usually have more steps on their salary scale and do not reach their peak earnings until they are in their forties or fifties. Working-class men thus seem to have their crisis of "awareness of limits' and "death of youthful dreams" very early. Our culture has not dignified this crisis with a grand label like the "midlife transition," but the working-class equivalent seems to show up in the early marriage crisis that is typical of many working-class families.

The midlife crisis, then, is largely built around the midcareer stage for middle- and upper-middle-class occupations and has the psychological aspect of a growing awareness of where and how one is stuck; hence, the often rather frenzied or anxious efforts are made to change one's job, to move somewhere else, or to get divorced and start over. Family pressures tend to add to this crisis atmosphere. Parents often go through their midlife crisis at the same time that their children are going through a crisis of adolescence. It is very likely that each crisis tends to build up emotional pressures that exacerbate the other.

There is also a third transition or crisis that may be superimposed on the midlife crisis, for a person around forty or fifty is likely to have parents who are retiring or even dying. There are the strains of watching one's own parents go through the transitions of later life. For some people in their forties or fifties, there is the problem or caring for aging parents. All this adds further pressure to the midlife period.

The Empty Nest Syndrome

It should be obvious that the concept of the midlife crisis was built up around the experience of men. It is their occupational career plateau that is most often at issue, and the "death of hopes" and the sense of limits are modeled on middle-class masculine experiences. In fact, the studies that first studied midlife crises and developed concepts to explain this time of life were based entirely on interviews with middle-aged men (Levinson 1978; Vaillant 1977).

A comparable turning point for noncareer women happened when their children were grown up and left home—the **empty nest syndrome.** Lillian Rubin (1979) produced a sensitive study of what traditionally used to be called "women of a certain age." She captured the upheaval that occurs when one long-standing routine is over and a woman faces the question of what to do for the rest of her life.

> *It's unbelievable when I think of it now. I never really saw past about age forty-two, where I am now. I mean, I never thought about what happens to the rest of life. Pretty much the whole of adult life was supposed to be around helping your husband and raising the children. Dammit, what a betrayal! Nobody ever tells you that there's many years of life left after that. He doesn't need your help any more, and the children are raised. Now what? (p. 123)*

Such women often feel as if they are back in adolescence, going through an earlier transition once more. There are the same feelings of anxiety, the unsettled self-image, the vague trying out of various new alternatives. Often this leads to the decision to make a new career, perhaps by returning for more formal education. Psychologically, the separation from their children may bring back some of the strains and scars of early childhood, when they themselves had to move away from the protective warmth of their own mothers (and perhaps fathers).

> *After twenty-five years of raising children, it's like I'm back to being twenty again— maybe only fifteen—and I have to start all over again. Only I'm more scared now than I was then. When you're a kid, you still think you own the world. But I'm not fifteen, I'm forty-five, and I know better now. (p. 123)*

On the other hand, some researchers (Stevens-Long 1984) questioned whether the "empty nest syndrome" always or even usually happens. For many, the departure of grown-up children from home is an event with which people successfully cope and sometimes offers an opportunity to take up new activities or careers. Cross-sectional studies (surveys conducted at one point in time) generally show that families with children in the home report less happiness than couples whose children have moved out (chapter 11). Those studies that followed a group of families from when children were in the home until after they moved out (longitudinal panel studies) found that women (and sometimes men) were happier after the children were "launched" (Menaghan 1983; McLanahan, Adams, and Sorenson 1985). One large-scale national survey found that marital happiness goes up only when all children are launched, not as each child departs (White and Edwards 1990). Because young adults are also increasingly likely to move back in with their parents for a time, the transition to an "empty nest" is less abrupt and total than it was just a few decades ago. It also appears that the period immediately after all teenagers depart is the happiest for the married couple and that overall satisfaction with life improves to the extent that the parents continue to see or talk to their launched offspring.

The assumption that middle-age mothers would necessarily feel depressed and aimless after the "nest" was empty appears to be unfounded, particularly if parents and young adult children remain on good terms. This makes sense in light of findings that parenthood is stressful, that it is most stressful when children are infants or teenagers, and that women are most stressed because they do most of the parenting (chapter 12). It is also the case that mothers who have other pursuits—as most now do—are least likely to depend totally on mothering activities to bolster their self-worth. For most such women, having teenagers grow up and move out is a liberating experience that provides time for rekindling marital romance and pursuing new opportunities for self-realization.

There is also the process of biological change, which occurs at about this time or a little later. **Menopause** typically occurs at about age fifty; it typically closes off the possibility of having more children (although in fact customarily most of the children have long since been born) and also thus affects one's sexual identity. As studies have shown, even such biological processes are strongly influenced by cultural expectations about appropriate gendered behaviors (*A Closer Look* 15.1).

THE LATE LIFE TRANSITION

Retirement may be a crisis for several reasons. One of the most important is that it marks the end of one's active involvement in a career. In contrast, in many cultures there is no such thing as formal retirement. People work until they are no longer able to do so. In our own society, retirement is to a large extent a product of our economic structure. Companies find it expedient to move managers out at a specified age, so that their juniors—having ideas of their own and feeling impatient—can step into the top spots at a predictable time, without too much conflict or unpleasantness. At lower levels in the work force as well, there is pressure to "make room" for the coming generation. Whether this retirement structure can continue remains to be seen as the proportion of the population that is aging increases. Society may not be able to comfortably support more and more people who are not working. How to solve the problem is likely to be an issue in coming decades. There are some indications that intergenerational conflict around this point is already shaping up.

In any case, as matters now stand, some persons die soon after their retirement because they cannot adjust to doing anything else but working. For others, though, the late life period can be a challenge and an opportunity in its own right. Individuals who adjust to it can have a happy and productive twenty years or more. In fact, many couples report that retirement is good for the marriage,

A CLOSER LOOK 15.1
Menopause: Stereotyping Older Women's Biological Aging

Anne Fausto-Sterling (1985) traces the history of medical interpretations of menopause and comes up with some illuminating findings. In the 1960s, some medical doctors began treating the "disease of menopause" by prescribing pharmaceutically produced estrogen. In medical journals and popular women's magazines, postmenopausal women were described as "unstable estrogen-starved women" who walk the streets stiffly in a "vapid cowlike negative state, seeing little and observing less" (Dr. Robert Wilson and Thelma Wilson 1963, cited in Fausto-Sterling 1989, 298). Another physician painted a bleak picture of what women could expect after menopause: "The vagina begins to shrivel, the breasts atrophy, sexual desire disappears. . . . Increased facial hair, deepening voice, obesity . . . , coarsened features, enlargement of the clitoris, and gradual baldness complete the tragic picture. Not really a man but no longer a functioning woman, these individuals live in the world of intersex" (Dr. David Reuben, 1969, cited in Fausto-Sterling 1989, 299). These poor, aimless, sexless women could be helped, of course, through "estrogen replacement therapy," a procedure that made Premarin (a brand name for estro-gen) one of the most popular drugs in the United States, despite the fact that some studies show estrogen treatment to be associated with uterine cancer.

Fausto-Sterling notes that only about a quarter of women going through menopause experience symptoms like hot flashes, and the majority of postmenopausal women have almost none of the symptoms that the estrogen-therapy proponents insisted accompanied this "disease." She concludes that "there are no data that support the idea that menopause has any relationship to serious depression in women" (1989, 303). Nevertheless, because our culture tends to define women as mothers and to idealize youthful bodies, estrogen-replacement therapy quickly became a treatment of choice. Fausto-Sterling suggests that the negative stereotyping of postmenopausal women came from using men's aging as a model and assuming that there was something "abnormal" about the natural process of aging in women. She reminds us that the labeling of illness is a social process that does not always reflect "objective" biological conditions, and she calls for more and better research on the subject of menopause.

and the majority of older people rate their marriages as happy or very happy. This may result in part from the fact that the unhappy marriages have had ample opportunity to be terminated, but most researchers find that marital satisfaction in the later years is much higher than it is in the middle years of marriage. Economic stability contributes to older couples' happiness, as do factors such as viewing one's mate as a best friend, liking one's mate as a person, agreeing on life goals, and maintaining humor and playfulness in marriage (Bengtson, Rosenthal, and Burton 1990).

Gender Trends in Longevity

The structure of the aging population is very gender divided and is becoming more so over time (Figure 15.1). Women used to outlive men by only one year on the average; now they tend to outlive them by six years (*Statistical Abstract* 1999, table 127). The gender gap in life expectancy was widest in 1979, when women lived 7.8 years longer than men, on average. Life expectancy for men began to catch up to that of women in the 1980s (Treas 1995). By 1997, female life expectancy at birth was 79.2 years, whereas that for males was 73.6 years, with whites living about six years more than blacks, on average (*Statistical Abstract* 1999). The result is that women are much more likely to be widowed than men: at age sixty-five and older, almost half of women but fewer than one in five men are widowed (Treas 1995). The average age at **widowhood** for

Retirement is a plus for many married couples; the majority of older people rate their marriages as happy or very happy.

women is about sixty-nine, whereas for men it is about seventy-two years (Schoen and Weinick 1993). Women spend an average of over fifteen years as widows, whereas men spend an average of about eight years as widowers.

The explanation for the **age gap in mortality** is not quite clear. It may be partly due to the decline in the maternal death rate as childbirth techniques have improved, but this can only account for a minor proportion of recent changes. It is probably not biological because the difference for the most part has only appeared recently in this century. It is sometimes said that men undergo more strain because of their work, and hence they die more quickly. However, there are several reasons to doubt this explanation. For one thing, it is precisely during the period when *the proportion of employed women has increased* that women have built up such a long lead in life expectancy. The "strains" of being in the labor force certainly have not reduced their longevity but exactly the opposite (at least until the mid-1990s). Moreover, it is probably not true that nonworking women are more "protected" from strains; as we've seen in chapter 11, housewives (and married women in general) have higher rates of mental illness and depression than married men. Whatever it is that has caused women to live longer remains a mystery, although it almost certainly has something to do with the gender roles in our society and possibly even involves some subliminal emotional struggle between the sexes that goes on inside families.

One important result of this pattern is that *the aging population is overwhelmingly female.* In the over-sixty-five population, there are about three females for every two males, and the disproportion grows with age (see Figure 15.1). Among

FIGURE 15.1 Changes in the Age Structure of the U.S. Population

These population pyramids show the progression of the cohort known as "baby boomers" from 1964 to 2024. The babies who were part of this cohort (persons born between 1946 and 1964) are indicated by the larger proportion of people aged nineteen and younger in the bottom pyramid. The "pinch" for ages twenty-five to thirty-four years in the bottom pyramid reflects the low birthrates during the Depression years. By 1994 (middle pyramid), baby boomers were swelling the ranks of the thirty- to forty-nine year-olds. By 2024 (top pyramid), the cohort will have reached the sixty- to seventy-nine-year-old age bracket. Note that older women significantly outnumber older men.

Source: U.S. Census Bureau, *International Data Base,* updated March 21, 2000.

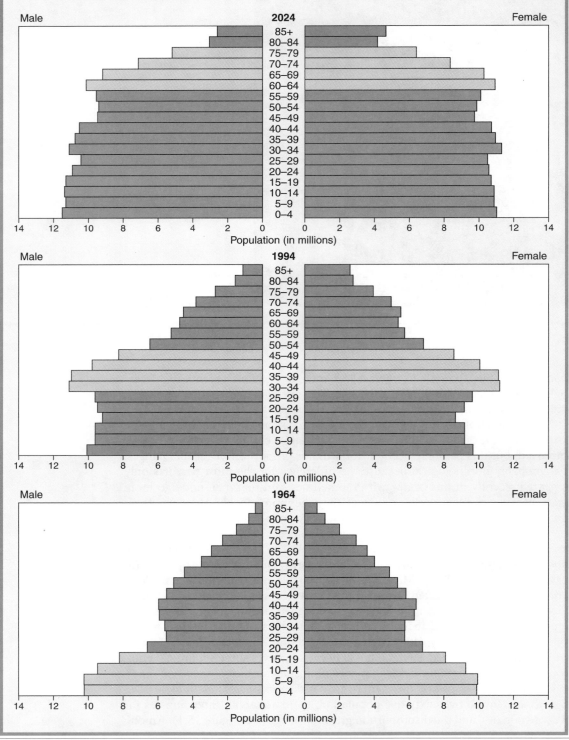

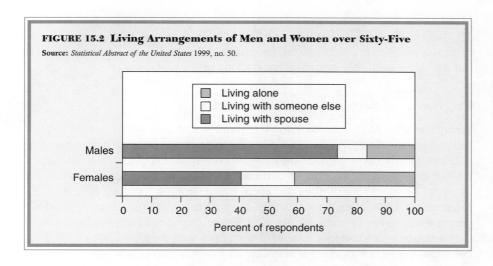

FIGURE 15.2 **Living Arrangements of Men and Women over Sixty-Five**

Source: *Statistical Abstract of the United States* 1999, no. 50.

people over eighty-five, the ratio of women to men is over two to one, and this group is the fastest growing of all segments of our population (*Statistical Abstract* 1999).

Moreover, these older men and women live in quite different household arrangements (Figure 15.2). Most men over sixty-five are married (about 75 percent of them), whereas less than half of the women are married (43 percent). As noted earlier, elderly women are more likely to be widowed and slightly more likely to be either single or divorced. As a result, the majority of older men are living with their spouses, whereas older women are just as likely to be living alone as with their husbands. Why is this? The cause is related to a combination of factors. Women tend to marry men who are on the average two years older in the first pace. In addition, women outlive men by about six years; the result is that wives outlive their husbands by many years. In addition, men over sixty-five are about seven times as likely to remarry (after widowhood or divorce) than are women at that age (Sweet and Bumpass 1987). Thus, we see the development of two predominant populations of the aged: old married men and old widowed women.

Compared with unmarried elderly, older married people are happier and have better health. Married elderly are more likely than their unmarried counterparts to report high levels of morale, life satisfaction, mental and physical health, economic resources, and social support. Widowhood, in contrast, initially tends to have negative effects on mental and physical health, even though these effects usually lessen over the long run (Ferraro 1984).

The major strains of old age, then, are especially experienced by women. There is the problem of living alone—at least for persons who may not like to live this way—although of course, some people prefer their privacy and independence. Over 40 percent of all elderly women live alone, whether by choice or necessity. Another 18 percent live with someone else, mostly relatives but also friends. Only a small proportion (about 5 percent) of either men or women lives in an old-age home or other institution; hence, an image of the elderly population as stuck away in formal institutions is not accurate. For the population as a whole, the biggest problem of old age is the economic strain that often goes along with it, especially for women living alone on a single means of support. Because most elderly women (except in the highest social classes) have relatively little income of their own and are not likely to receive much from their husband's estate (except small Social Security payments; see *A Closer Look* 15.2), they often can afford only mediocre living quarters. Add to this the cost of health problems, which becomes substantial in later life, and one can see that old age often produces its own, very realistic crises.

The Impact of Recent Demographic Trends

Today's elderly have been described as participants in a "quiet revolution" of intergenerational family life. People are growing old in different fashion and in different family contexts than they have in the past. People used to belong to two- or three-generation families in their later years, whereas today's elderly are likely to be members of four- or even five-generation families (Cherlin and Furstenberg 1986; Bengtson, Rosenthal, and Burton 1990). About half of all people over the age of sixty-five are members of four-generation families, that is, they are great-grandparents. One of five women who die after the age of eighty are great-great-grandmothers—members of five-generation families.

The amount of time that people spend in family roles has increased dramatically since the turn of the century. In the 1800s, parents and children in North America may have shared twenty to thirty years together, and grandparents and grandchildren may have spent only about ten years together (Juster and Vinovskis 1987). Aging parents today may be part of their children's lives for over fifty years. As grandparents, their ties to grandchildren often extend beyond twenty years (Barranti 1985; Preston 1984). Because women live much longer than before, and because they have fewer children, their later-life situations are markedly different from earlier patterns. For example, as compared with those in 1800, women spent four times the number of years as a daughter with both parents alive. The time they spend as an adult child of one or more parents over the age of sixty-five has increased from about seven years in 1800 to about eighteen years. The amount of time spent in a family with children under eighteen years of age has gone down, so the average number of years spent with parents over sixty-five now exceeds the average number of years spent with children under age eighteen (Watkins, Menken, and Bongaarts 1987). The median age at entry to grandparenthood is now estimated to be over forty-five years old. Because women can now expect to live to be about eighty years old means that most will spend close to half their lives as grandmothers.

Such projections are an oversimplification, of course, because not everyone is "average." In fact, there are two contrasting patterns that relate to the early versus late birth patterns (discussed in chapter 12). For those who give birth in their teens, mothers and grandmothers are quite close in age, and some researchers report that these younger women are not eager to become grandmothers (Elder, Caspi, and Burton 1985). In many instances, substitute childcare responsibilities are then shared with the great-grandmother, often herself about fifty years old. A contrasting pattern results from delayed marriage and birthing. If two consecutive generations delay first birth until age thirty-five, that means that the first parents will be seventy years old before they become grandparents. Clearly, the advanced age of delayed grandparents could limit the quality and quantity of interaction with their grandchildren (Parke 1988). Another important demographic trend that influences family interaction patterns in late life is divorce. When children of elderly parents divorce, the grandparents cannot as easily see their grandchildren, especially if their adult child does not receive custody. Moreover, the elderly parent is often forced to restrict relations with their former daughter-in-law or son-in-law. When

A CLOSER LOOK 15.2
Social Security

The **Social Security** Act of 1935 marked the beginning of a commitment to deal with the special problems of older people in the United States. Most older people rely on Social Security's Old Age and Survivors Insurance (OASI) program, operated by the Social Security Administration. The program provides monthly cash benefits to retired workers and their dependents or survivors. In the 1990s, an average of over 40 million Americans were OASI beneficiaries (Treas 1995). Social Security coverage is compulsory for most workers. To be eligible for benefits, workers must accumulate enough months of covered employment over their work history.

The average benefit is about $700 per month, but benefits are adjusted annually to take into account inflation. Starting in the year 2000 and continuing to 2022, the age for receiving full benefits will gradually rise from sixty-five to sixty-seven. Recently, Congress lifted the limits on earnings so that older citizens could receive the maximum amount of Social Security Benefits, regardless of their monthly income. Because of a combination of income from Social Security, Medicare, private pensions, and savings or investments, today's older Americans are enjoying a higher standard of living, on average, than any generation of elders before them. Social Security income is exempt (in whole or in part) from federal taxes, so almost half of the elderly in the United States owes no federal taxes.

Social Security payments account for about two of every five dollars that older Americans receive. If Social Security and other government payments were not counted, the poverty rate for the elderly would be four times higher than its current rate, and half of all persons age sixty-five and older would live in poverty. Nevertheless, because the amount of Social Security benefits is pegged to previous earnings, Social Security retirement income is no guarantee against poverty in old age for those who earned low wages throughout their working years (Treas 1995).

Social Security is financed mainly by payroll taxes paid by the employer and employee. The Social Security Trust Fund has been generating a surplus, but the federal government regularly borrows from it to offset the national debt. Because the ratio of workers to older people is declining, there are concerns that the Social Security System may be financially vulnerable in the future (see Figure 15.1, which tracks baby boomers over time).

---

divorce is followed by remarriage, multigenerational relationships must be reconstituted and renegotiated (Hilkevitch and Blieszner 2000).

How Much Contact with Relatives?

It is sometimes said that it is a shame that the old-fashioned family arrangement has disappeared, so that old people are no longer taken care of by their children. As we have seen, this is something of a myth about the past. Many people used to die by the time their children were fully adults. The ideal of the old people sitting by the fireplace surrounded by the younger generations was largely imaginary. Moreover, the myth is not even very accurate *today* as a picture of what people *want*. Surveys show that most parents and their grown-up, married children *do not want to live together*. Hence, the pattern of elderly people living alone or with roommates is largely their own choice.

That does not mean that parents become completely isolated from their children after they grow up. In fact, over half of adult children live within a one-hour drive of their parents, and two-thirds have weekly contact with their mothers (Lawton, Silverstein, and Bengtson 1994). Most national surveys show that close to 40 percent of adult children have face-to-face contact with their parents once a week or more (Lye 1996; Rossi and Rossi 1990). These are quite astounding figures, completely contrary to the image of old people as isolated and

People over age eighty may depend on "children" twenty to thirty years younger for various kinds of support.

ignored by the younger generation. Thus, few elderly parents live with their adult children, but most have frequent contact with them (Treas and Bengtson 1987; Treas 1995).

Some evidence shows, not unexpectedly, that elderly people's morale is higher the more family ties they have. However, this does not necessarily mean that elders become completely dependent emotionally on their children. This would be a change from the pattern when they were younger, and we might expect that old people would try to maintain their usual adult pattern. In fact, researchers who checked into the importance of nonkinship connections for older people found that *social contacts with friends are at least as important for morale* as contacts with relatives.

There is a social-class pattern to this. Working-class people are more likely to live near their relatives, whereas middle- and upper-middle-class persons have a greater tendency to move to another town or part of the country. Hence, working-class contacts are more likely to be face-to-face (this is at all ages, not just among old people). In fact, white-collar people are just as concerned to maintain interaction with their relatives as are blue-collar people; but living farther away, they are more likely to make contact by letter or telephone (Rossi and Rossi 1990). Somewhat surprisingly, children who are upwardly mobile, moving up and out of their parents' social class, do not thereby break off their family ties. On the contrary, the upwardly mobile are the most likely of all to believe in family ties and to keep up contacts with their aging parents and other relatives. Probably what is happening is that these are the "stars" and "success stories" of their families. Their families are proud of them and like to keep up contact with them, and the successful offspring enjoy the attention and deference they receive from their relatives. The opposite of this pattern is shown in the downwardly mobile—those who are the opposite of a family success story. These persons are most likely to say their relatives are unimportant to them and to know the fewest number of relatives. Even inside the family, then, one's financial and social success has a big effect on one's behavior.

Patterns of contact between elderly parents and adult children also vary by gender. Daughters tend to have more frequent interaction with parents than do sons. Middle-aged daughters most often act as **kinkeepers** regulating the amount and type of contact with parents (Hagestad 1988; Lye 1996; Rosenthal 1985). Thus, we can see that women act as emotional managers with aging parents, as well as with children. Marital status also influences the amount of contact between elderly parents and children, with unmarried children and widowed parents (usually women) the most likely to have frequent family contact. Ethnic and racial differences also account for some variation in contact between adult children and elderly parents, though results are inconsistent. Some studies show that Hispanics are most likely to have frequent contact, with whites and blacks having lower amounts of contact (Bengtson, Rosenthal, and Burton 1990). Other studies show that black adult children visit their parents more often than do whites, with Hispanic adult children visiting their parents less often than whites (Lye, Klepinger, Hyle, and Nelson 1995). Earlier studies showed that blacks and Hispanics tended to have the highest levels of material support exchanged, though more recent studies tend to contradict this finding (Lye 1996). Contact with siblings among the elderly also tends to vary by gender and family status, with women and those without children most likely to maintain frequent contact with sisters and brothers (Bengtson, Rosenthal, and Burton 1990).

People over the age of eighty are most at risk for physical and mental frailty and are thus more likely to be dependent on others for help. That help is typically provided by daughters, most of whom would then be in their fifties or sixties (Allen, Blieszner, and Roberto 2000). Because families are getting smaller, there are fewer siblings with whom to share the chores of elder care. Although much research has focused on the importance of adult children providing support to their elderly parents, recent studies reveal that one's spouse is often a preferred caregiver and that friendships with same-age peers continue to have beneficial impacts on elderly people's well-being, even if children are available. In addition, in many instances, elders provide more support to adult children than they receive, whether one considers such things as emotional support, financial contributions, provision of living quarters, or routine domestic tasks like cooking and cleaning (Mangen, Bengtson, and Landry 1988).

DEATH AND THE FAMILY

Death is the final transition; at least it is from the point of view of the individual. The deeper crisis of the old-age period is facing the inevitable coming of death. In our culture, death is something from which we try to insulate ourselves. Most people die alone, in bureaucratic medical settings, surrounded by nurses and attendants whose professional manner is to emphasize routine and minimize the disruptive or personal quality of what is going on. This response creates a superficial calmness because the overt thought of death is kept from people's consciousness, or at least from their talk. However, repressing the awareness of death, especially when the medical situation indicates that it is quite likely, only creates a split between people's feelings and their public consciousness. This emotional repression can do as much damage as any other aspect of the old-age crisis.

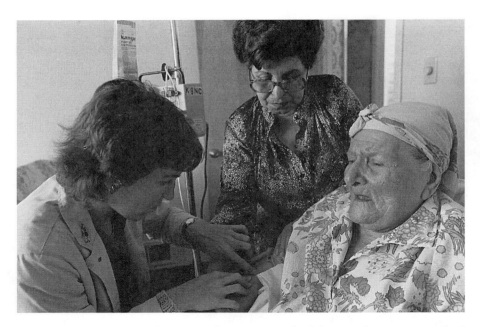

Home nursing provides an alternative to living out one's final days in a bureaucratic medical setting. Here, the patient's daughter is the primary caregiver, enabling this woman to remain at home in familiar surroundings.

In traditional India or China, the old person was given special honor because he had an institutionalized place in a religious tradition. (In such male-dominated societies, though, old women did not receive the same honor.) Old age was devoted to the cultivation of wisdom, to meditation, to religious study, or to philosophy and art. These roles do not exist in our society, and in fact, our public culture tends to be dominated by the extroverted action-oriented ideals of youth (and for that matter, the antiintellectual and antispiritual ideals of working-class and lower-middle-class youth). However, as our population becomes increasingly weighted toward the aged (see Figure 15.2), there may be some pressure to redevelop some of these kinds of social roles for developing inward, spiritual, and philosophical interests.

One emerging institution for care of the dying is hospice. The first hospice appeared in England in the 1960s and was brought to the United States in the 1970s. The term **hospice** refers to compassionate care that is given to terminally ill patients in a variety of settings. There are over 2,000 hospices across the country (Knox and Schact 1994). Most provide in-home services for cancer patients, though anyone who is dying is eligible. Because hospice care is given to terminally ill people, the goal is to make the person more comfortable rather than to cure the illness. Hospice stresses companionship and spiritual needs, as well as trying to make the patient comfortable and controlling the pain. Hospice staff members work with patients, family, and friends to help them discuss their feelings and expectations. Caring for dying persons at home is generally seen as more humane for the patient, though it can be quite stressful for those who provide care on a daily basis. Though hospice care is becoming more common, rituals associated with its provision have not been fully integrated into our culture.

There is, however, one area of life that is strongly ritualized. This is the **funeral.** Interestingly enough, the funeral is one of the strongest and most powerful rituals in our relatively deritualized society. Even among people who are not religious, funerals can be important and moving events. Why is this so? The key is to recognize that *a funeral is for those still living, not merely for the dead.* A funeral brings together the people who knew the dead person and assembles them face-to-face. Funerals are effective rituals because they have real work to do for their participants: fitting together the group after one of its members has left it. A death leaves a hole in the group; the purpose of the ritual is to bring everyone together so that they can explicitly recognize that the hole is there and then start to patch it over.

Often there is awkwardness over how to express one's condolences to the spouse or other close relatives and friends of someone recently deceased. One doesn't know quite what to say, for fear that saying anything at all will make the bereaved persons feel worse. In fact, mourning is a process that people need to go through. They must pass through the grief before they can recover emotionally. This is "working through" the loss. The funeral allows everyone to work this grief through together. At the funeral, condolences are natural; so are tears. Oddly enough, people at a funeral feel sadder than they thought they would; they can become emotionally wrought up over the death of someone they scarcely knew. This is because of the ritual situation: the focus of attention that strengthens the shared emotion in the group. Because of the shared emotion, the group feels itself stronger. The funeral does its work. By attaching a shared ritual onto the end of life, society can ensure that death itself ends up strengthening the bonds among those who carry on afterwards.

SUMMARY

1. Individuals encounter a patterned set of family transitions including such events as leaving home, getting married, having children, getting divorced, launching children, retiring, and surviving a spouse. The timing, length, sequence, and composition of life transitions vary according to the historical period and culture in which we live and are influenced by significant events such as war and major shifts in the economy.

2. The teenage or adolescent transition involves biological maturity but also a shifting place in the social structure. Adolescents experience considerable tension as they become ranked in sexual and social markets and in the beginning stages of the adult occupational system. Dangers of delinquency and accidental death are highest at this time.

3. Problematic features of adolescence are created and maintained by social institutions that isolate youths from adults. Advanced industrialized societies exclude young people from employment and isolate them socially, at the same time blaming them for a variety of social ills.

4. The midlife crisis in men tends to occur in upper-middle-class occupations and involves the recognition of limitations on careers. The result is the death of early ideals and of the sense of unlimited possibilities. This crisis often coincides with the teenage crisis of the man's children and possibly with the late life crisis of his own parents.

5.   Although early research assumed that the empty nest syndrome represented a traumatic transition for middle-aged wives, recent research shows that women's overall happiness and marital satisfaction improve after the children move out of the house. Earlier depictions of women's reactions to menopause were similarly negative but have given way to a more positive view of women's midlife biological changes.

6.   Because men tend to marry women younger than themselves and women tend to live six years longer than men, wives are especially likely to outlive their husbands. The aging population consists largely of married men living with their wives or widowed women living alone or with friends or relatives. The proportion of the aged population that lives in institutions is very small.

7.   A large proportion of elderly live near their children and see them quite often. Older people's morale depends on contact with their children, but social contacts with their friends are even more important. Upwardly mobile persons are most likely to maintain contact with a large number of relatives. Women maintain contact with relatives more than men.

8.   Because people live longer and have smaller families, intergenerational family roles have changed dramatically in this century. People spend less time raising small children and more time as adult children of older parents. We are much more likely to be members of families with great-grandparents or great-great-grandparents than ever before. Because U.S. women can now expect to live to be eighty, they are likely to spend close to half of their lives as grandmothers.

9.   People over the age of eighty are most at risk for physical and mental frailty and are thus more likely to depend on others for help. Wives and daughters are the most likely family members to provide that help. Although our stereotypes depict older people as helpless, elders often provide more financial and emotional support to their children than they receive.

10.   Death is largely repressed in our society, especially as a result of bureaucratized hospital procedures. This creates many submerged psychological stresses. The aged in our society lack socially esteemed roles for the development of wisdom and spirituality. Funerals, however, do provide powerful rituals for the final life transition, especially for reintegrating the society among those who are left.

Key Terms

adolescence	hospice	peer group
adulthood	kinkeepers	retirement
age gap in mortality	life course	social security
empty nest syndrome	menopause	widowhood
funeral	midlife crisis	

Sociology Web Site

See the Wadsworth Sociology Resource Center, "Virtual Society," for additional links, quizzes, and learning tools:

http://www.sociology.wadsworth.com

Also on this web site you'll find InfoTrac College Edition, an online library of journals. Here you can search for electronic articles about central topics in sociology.

16 The Future: Social Policy and Family Well-Being

INTRODUCTION

What can we expect for the future? We know what the current trends have been, and we can extend them through the next decade or even further. However, it is dangerous to assume that trends will just keep on flowing in the same direction. There are famous cases such as the predictions made in the 1930s, when the birthrate was going down and it looked like the U.S. population would level out by the 1960s. Instead, everyone was surprised by the postwar baby boom, which overwhelmed all projections.

We can be a little more confident if we are not just extrapolating a trend but using a theory that explains the causes of social changes. Let us take a look, then, at the main trends we have witnessed throughout this book. In connection with them, we will review the theories that we have used to explain these phenomena. We can estimate whether we should expect the causes of these trends to continue or whether we will turn in other directions.

THE MAIN TRENDS

■ *The marriage rate* has been dropping, although the decline has been leveling out in the last few years. The age at marriage has been going up sharply. Most women do not marry until their late twenties, and the trend may reach the point at which a majority will not marry until after thirty. Men have always married later than women, but their age at marriage has been rising more slowly than that of women (see Figures 5.2 and 5.8). How far can these trends go? Will the average woman marry a man her own age? Will people ordinarily be getting married in their thirties or forties? Will we reach the point at which most people never marry?

■ *Premarital sexual experience* has increased, so it now occurs for a large majority of both men and women. Women's sexual behavior has tended to converge with that of men, both in premarital and extramarital sex (see Figure 5.1 and the data in chapter 9). Does this mean that marriage as a sexual relationship is being displaced?

■ *The divorce rate* has risen to a level at which half of all marriages are expected to end in divorce. In the last two decades, however, the divorce rate has remained stable and even dropped a little (review Figure 5.3). Will it stay at this high level? Or is this just a temporary resting point in a curve whose shape has been uphill for more than a century? On the other hand, are there social conditions that could reverse its direction and bring the divorce rate back down again?

■ *The remarriage rate* has been running parallel to the marriage rate. Divorced people tend to get remarried at about the same rate as they married in the first place (chapter 8), so divorces aren't fatal to the family system. One could even say marriages are popular because people are having more of them. Nevertheless, as the marriage rate has been dropping, the remarriage rate has been falling off somewhat too. Has the pattern leveled out, with most people going through several marriages in succession? Or are we in transition to a situation in which most people are going to be either unmarried or permanently divorced, and the marrieds-plus-remarrieds will become a declining majority?

■ *Family violence,* including spouse abuse, child abuse, and elder abuse, has come into the spotlight (chapter 13). Here we lack good trend data. The amount of abuse that is *officially reported* has gone up sharply in recent years, but it is likely that previously a great deal of family violence was not reported. In any

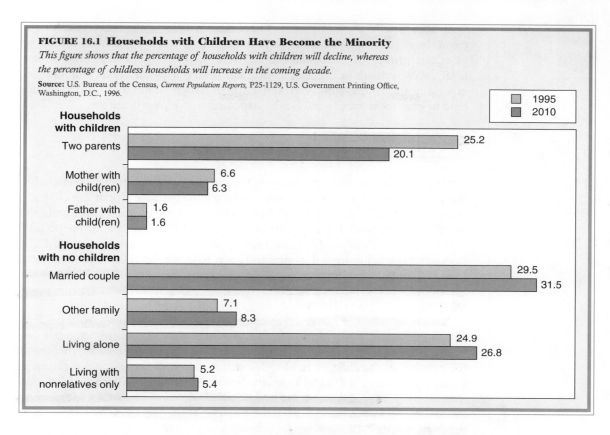

FIGURE 16.1 Households with Children Have Become the Minority

This figure shows that the percentage of households with children will decline, whereas the percentage of childless households will increase in the coming decade.

Source: U.S. Bureau of the Census, *Current Population Reports,* P25-1129, U.S. Government Printing Office, Washington, D.C., 1996.

Legend: 1995, 2010

Households with children

- Two parents: 25.2 (1995), 20.1 (2010)
- Mother with child(ren): 6.6 (1995), 6.3 (2010)
- Father with child(ren): 1.6 (1995), 1.6 (2010)

Households with no children

- Married couple: 29.5 (1995), 31.5 (2010)
- Other family: 7.1 (1995), 8.3 (2010)
- Living alone: 24.9 (1995), 26.8 (2010)
- Living with nonrelatives only: 5.2 (1995), 5.4 (2010)

case, is today's family a violent battleground, and will this situation continue into the future? Can it be that this family conflict is an indication of forces that are in the long run tearing the family apart? Are there any prospects of bringing down the amount of family violence?

■ *The birthrate* has been dropping over the long term but rising in the past few years. For decades it has hovered around replacement level, with two children born for every woman (see Figure 5.5). The *age of childbearing* has also been rising (see Figure 10.5), which parallels the later age of marriage. Women who put off marrying until their late twenties and early thirties are delaying having children for another few years after that. How far can this go? There has been a biological upper limit on women's childbearing by the late forties; as women get married later, their number of years of childbearing are narrowed. This may mean that birthrates will drop in the future. What is there to keep the birthrate from falling all the way to zero?

There are a number of side effects of the falling birthrate:

■ Nonmarital births are increasingly common. Most of the increase is due to unmarried women over the age of twenty giving birth (see Figures 10.2 and 10.3). In the black population, where the marriage rate is especially low, the proportion of births that are nonmarital has risen to well over half.

■ *The average household size* has gotten smaller. Few families have large numbers of children. On the other hand, there are many singles and childless couples; at the upper end of the age spectrum, we see an increasing number of senior citizens, with a preponderance of widows among the oldest. The mom-and-dad-and-kids family has become a minority (Figure 16.1).

■ *The ethnic composition of families* has been changing in the United States. The trend toward declining birthrates is strongest in the white population comprised of ethnic groups of European origin. With their birthrates near replacement level, recent growth in U.S. population has been due primarily to immigration, especially to groups from Asia and Latin America. Immigrant ethnic groups tend to have more traditional family structures and higher birthrates.

ETHNIC DIVERSITY

During most of the twentieth century the minority population was overwhelmingly African American, and it comprised a relatively stable proportion (13 to 15 percent) of the U.S. population. Between 1900 and 1960 the minority population doubled, but its rate of growth was only slightly higher than that of the population as a whole. Between 1960 and 1990 the situation changed dramatically. The minority population tripled in size and grew to comprise about a fourth of the total U.S. population. By the year 2010, minorities are projected to account for about a third of the nation's people, with Hispanics becoming the most numerous (*Statistical Abstract* 1999).

With these number increasing so rapidly, it will soon become a misnomer to label such groups "minorities." In fact, in some states, "minority" children are already in the majority. Over three-quarters of the children in Hawaii, two-thirds (67 percent) of the children in New Mexico, and half of those under eighteen in Texas and California are black, Asian, Native American, or Hispanic (*Statistical Abstract* 1999). By 2010 over half of the children in Florida, New York, and Louisiana will be "minority." By 2010, nineteen states will be composed of more than one-third minority children (Schwartz and Exter 1989).

Thus, we can project that early in the twenty-first century, a substantial proportion of the people in the United States will belong to these ethnic groups (Figure 16.2). What consequences will this have for the structure of the American family? We will return to this question later in this chapter.

Let us now briefly look at the points of sociological theory that explain the causes driving these trends. Using theory we should be able to guess what directions these trends will take in the future.

SEXUAL AND EMOTIONAL PROPERTY

The core of the marital relationship is a claim to permanent and exclusive sexual possession of one's partner. Now that there is so much sex outside of marriage, is this core going to fade away, and will the family disappear along with it?

The evidence suggests that this is not going to happen. For one thing, although there is more sex outside of marriage than there used to be, *there is still more sex inside of marriage than anywhere else.* Premarital sex is common, in the sense that most people these days have experienced it, but it is not nearly as regular and frequent as marital sex. Many people have premarital sex with the person they eventually marry; in fact, this may be the way in which they negotiate themselves into a relationship that culminates in marriage.

Extramarital sex is also fairly widespread now, but these are mostly rather short episodes. Although they may be dramatic and exciting, they make up a comparatively small part of a person's sex life. In fact, the excitement comes precisely from the fact they are illicit. If we compare arrangements that try to

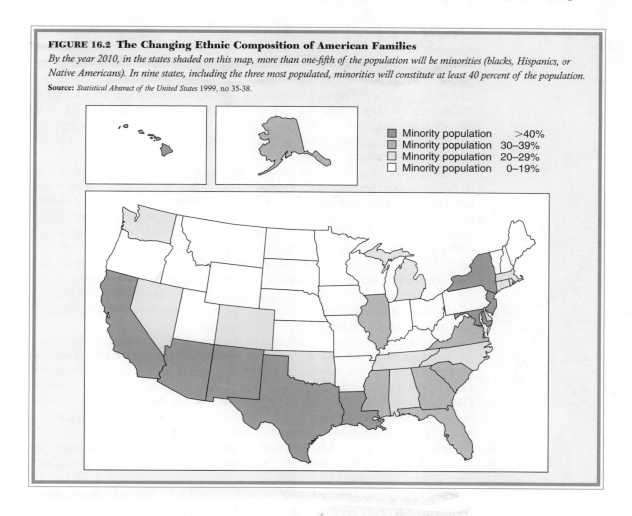

FIGURE 16.2 The Changing Ethnic Composition of American Families

By the year 2010, in the states shaded on this map, more than one-fifth of the population will be minorities (blacks, Hispanics, or Native Americans). In nine states, including the three most populated, minorities will constitute at least 40 percent of the population.

Source: *Statistical Abstract of the United States 1999*, no 35-38.

Minority population >40%
Minority population 30–39%
Minority population 20–29%
Minority population 0–19%

"normalize" extramarital sex, the evidence is that these experiments ultimately show that ordinary sexual possessiveness is more powerful. Communes, we recall (chapter 9), have found that sex is a volatile force and that sexual affairs between individuals tend to tear the group apart. Hence, most communes have attempted to institute total celibacy or some other rigid form of control over sex. Open marriages have been advocated and tried (chapter 9), but they seem to be extremely unstable. Open marriages usually terminate within months, either by separation or by a return to exclusive sexual possession.

Cohabitation

One form of nonmarital union that has become rather common is the cohabiting couple, living together but not married. According to the U.S. Census Bureau, half of all American women between the ages of twenty-five and forty have cohabited (*Statistical Abstract* 1999). The important point to focus on is that *cohabiting couples act much the same as married couples,* except for the fact that they are not legally married. About one-third of all cohabiting opposite-sex couples have children, which is not remarkably different from the proportion of married couples living with children. In fact, most of the increase in the number of nonmarital births can be accounted for by the increase in the number of cohabiting women who got pregnant and decided to carry the baby to term (chapter 10). Cohabiting couples have about the same patterns of sexual possessiveness as married couples, and they also have a joint economic relationship as in marriage. Researchers find relatively few differences between cohabiting couples and married couples. A cohabitation may be a little easier to break up because the legal formality of a divorce is not required. However, even here there is a tendency for the legal system to treat the partners as having property obligations to each other, if one of them brings the matter to court.

Perhaps the most important reason that sexual possession is not going away is that it is linked to emotional possession. Sex is linked with sexual love; a partner wants an exclusive place in his or her lover's heart, and sexual relationships operate as a symbol of this emotional property. As we have seen (chapter 4), historically the ideal of mutual love and marriage as a sexual possession binding on both partners came into existence with the rise of the individual marriage market. At first impression, the idea of a market seems rather heartless, but in fact, love and seeking one's own marriage partner go together. When persons started negotiating their own marriages, rather than serving as pawns in a system of kinship networks or property transactions of patriarchal households, they found themselves in a situation conducive to romance.

How this comes about is explained by the theory of interaction rituals (chapter 8). Having to negotiate one's own fate in an open competition for sexual partners generates the emotional ingredients that create a tie with the person one finds who matches up with oneself. Erving Goffman (1967), who developed the theory of interaction rituals in everyday life, points out that the face-to-face negotiations of modern life produce a cult of individualism. In modern private lives, the self becomes the main sacred object. A love relationship is this cult of the self-enhanced to the highest degree, a private ritual in which the lovers idealize each other. At the same time, their cult is a shared one, and sex becomes a symbol of the relationship.

Although there can be sex without love, emotional symbolism tends to grow out of a sexual relationship, and compatible partners make claims on each other for exclusiveness and permanence. Marriage simply ratifies this relationship. This is one important reason why it looks like marriages are not going to disappear. Also, if we expand the notion of "marriage" to include both legal marriage and cohabitation, we can see that there has been little decline in the institution of marriage.

ECONOMICS AND THE MARRIAGE MARKET

There is no doubt that the marriage market has been changing. Both men and women are waiting longer to get married. One might say they are warier about what kind of deal they are striking. This is spurred by the fact that women have been improving their positions in the marriage market.

Women have become committed to jobs as never before. Young women's career aspirations are higher than ever in the past, and women are implementing them by achieving an unprecedented level of education. Throughout most of the twentieth century, males attended college in much greater numbers than females. The popularity of education was increasing throughout this period, as the percentage of young adults who attended college went from 4 percent in 1900 to 53 percent in 1970 (Collins 1979, 4). After 1970 an important shift took

A son congratulates his mother at her college graduation with a kiss. As women achieve higher levels of education and experience more success in their careers, they alter their bargaining position in the marriage market.

place. College attendance by males leveled off; however, the rate of female college attendance has continued to rise, and in 1980 there were more women than men enrolled for the first time in history (*Statistical Abstract* 1987). Now females make up a substantial majority of all college students (55 percent in 1999).

The same thing has been happening in the professional schools, though at a slower rate. Women used to pursue professional training only in traditional areas such as nursing and primary and secondary education. In the late 1990s, women earned more than half of all master's degrees awarded nationwide. Still, they were less well represented at the highest levels. Only 38 percent of all professional degrees and 40 percent of doctoral degrees went to women in 1996 (*Statistical Abstract* 1999). Women have not yet made very large inroads into higher management and the more lucrative professions. Only 2 percent of top corporate managers are women, and fewer then one in four of all mangers and administrators are women. Nevertheless, it is clear that women's aspirations have changed.

Women have less incentive to enter into an unequal relationship and less incentive to stay in a marriage if they are unhappy with it. The result is that the marriage market is now more active than ever. If getting married means going off the market—becoming no longer available to negotiate for sexual and emotional partners—getting divorced puts one back on the market. Then, the tendency of bargaining over sex and love takes over again, and within two to three years, a large proportion of divorced persons have made a new exclusive relationship and remarried. Of those who haven't, many enter the pseudomarital relationship of cohabitation; in fact, divorced persons make up about a third of the cohabiters.

The marriage market is not disappearing. Men and women are bargaining more equally and are more careful about what they get from it. One indication is the increasing similarity in age between partners for first marriages. Men used to be four years older, on average, than the women they married. Now they are only about two years older, and we expect the male older pattern to decline even further as women bring more resources to the relationship. One might say that there is more marital bargaining than ever, with it being more prolonged and with more willingness to reopen negotiations.

DUAL-INCOME FAMILIES AND CLASS STRATIFICATION

Besides sexual and emotional ties, there is one more important incentive for people to get married. The marital relationship has become the most crucial element affecting one's social class position.

As we have seen in chapter 1 (see Figure 1.2) and in more detail in chapter 6, dual-income households have a huge advantage over most single-income households in their standard of living. To have an upper-middle-class lifestyle today, one usually needs to combine two professional or managerial incomes. Even a moderate middle-class standard of living depends more on the combination of two incomes—even from working-class occupations—than on any single source of income. This becomes not just a matter of keeping up with the Joneses, because these dual-income families have driven up the cost of housing in the United States. As prices go up, there is an incentive to put together two good incomes in a marriage, if only to be able to own a nice place to live.

Family events thus have become a prime determinant of the ups and downs of one's economic position. Whether one marries and who one marries can push income up considerably, and this is now true for men as well as women. Divorce or death of one's partner can send one spinning downward (chapters 6 and 14).

Not only does marriage affect the upper part of the income structure, but its absence also has an effect—a negative one—on the poverty sector. Both aging families and single-parent families are especially likely to find themselves in this sector. Again, the marriage market is involved. In the aging population of seventy and beyond, there are far more women than men and hence few opportunities to make combined households. In the case of black women who are single parents, there is a severe shortage of marriageable men (chapter 7). When these women have a better source of income than the men who are available, there is no economic advantage in putting together a joint household. This family structure is really the flip side of the dual-income couples who make up the new upper middle class.

We can say, then, that there are strong economic incentives and opportunities to get married, especially in the middle and upper-middle parts of the occupational distribution. The lower-income levels show the opposite pattern. Class stratification has become the bulwark of the modern family system.

SOCIAL CAUSES OF DIVORCE

Divorce rates go up when people can afford it. Overall economic prosperity has made divorce a reasonable alternative for people who are dissatisfied with their marriages. The increasing proportion of women in the labor force has been especially important because it gives women the resources to be independent. Women tend to suffer economically after divorce because their incomes are still considerably lower than men's incomes. However, compared with the housewives of a generation ago, who had no economic resources whatsoever to fall back on, today's employed women have more room for choice in leaving a marriage.

In the future, even more women will be employed full time, and the proportion who have higher-level jobs with good incomes should increase. The prognosis, then, is that the divorce rate probably will stay at a high level. As of now, women still have much to lose from a divorce. As their incomes improve, divorce may become even more common. One reason the divorce rate may remain stable or even decline, however, is that fewer people will marry in the first place. The same reasons that allow women to end marriages will also encourage them to avoid marriage in the first place. Women's increased economic opportunities will thus encourage more cohabitation and more nonmarital birth in the future. In addition, if women's economic prospects approach those of men, we can expect men to begin assuming more of the everyday cooking and cleaning that it takes to run households and raise children. This pressure on men, in turn, is likely to lead to more negotiation and conflict in marriage and thus will tend to keep the divorce rate near its present high level.

One of the reasons that couples get divorced is because of one partner's sexual affairs. Extramarital sex violates the bond of sexual possession; when the other partner reacts angrily and demands a divorce, it shows that the ritual of possessiveness is still taken seriously. Sexual affairs thus provide a "push" factor breaking up marriages. On the other hand, such affairs also can provide a "pull" factor, as a husband or wife finds a new sexual partner he or she prefers to the old one and leaves the marriage for the new lover. How often people have extramarital affairs depends partly on their opportunities. Men have traditionally had more extramarital affairs than women because they were more likely to work outside the home, where they could come into contact with potential sexual partners. As more women have moved into occupations that bring them into contact with men, their chances for affairs have increased. Married women in many jobs are beginning to have as many adulterous affairs as married men. As women obtain better jobs in the professions and management, we can expect more adulterous affairs on both sides, merely because there will be more opportunities. If gender equality increases in the future, so will the number of divorces.

FAMILY CONFLICT

Family conflict ranges from quarrels and feelings of dissatisfaction to the extremes of family violence. There are several different sources of conflict: shifts in power relations, stressful situations, and traditional authoritarian roles.

Power Shifts

Inequality in power sets the stage for potential conflicts. People who have power feel justified and confident in controlling others; people who are powerless tend to be resentful, but they behave passively if they can do nothing about it. It is when power resources shift that conflict usually breaks out.

Until the last few decades, men have had most of the power resources in families, primarily because of their economic control. One reason for an upsurge of conflict is that women have become mobilized to improve their power position in the home. Women's occupations have given them more economic resources; although these usually do not match the incomes of men, the shift has been

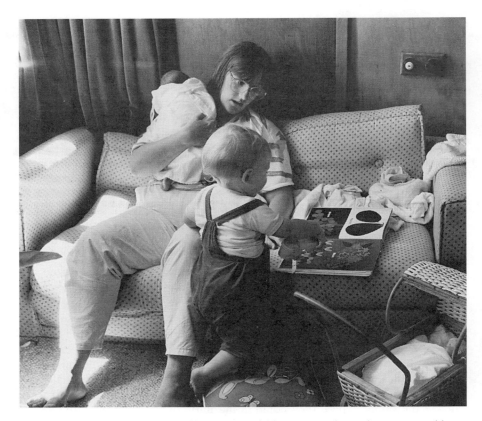

Working-class marriages are typically more stressful because couples tend to marry and have children earlier, when they have less money and are less equipped emotionally to deal with their situation's demands.

enough to at least set off battles over the balance of power. Women's education level has gone up even more rapidly, and this has increased both consciousness around issues of equality and women's aspirations. Changes in occupations and education have probably increased the size of women's networks, which is another important resource for domestic power (chapter 11). We see that the absence of these resources is what makes women most powerless; for instance, battered women who feel unable to leave their husbands are those who lack jobs and networks of their own (chapter 13).

In the long run, we might expect that this source of conflict will rise and fall in a bell-shaped pattern. When men had most of the power resources, conflict was relatively low because women accepted their position passively. As women's power resources increase, conflict goes up as women challenge men. Eventually, we might expect that the conflict level will fall again and that things will stabilize around a more egalitarian situation.

Stressful Situations

Family conflict goes up in stressful situations. Families react to these times at different levels of intensity: some merely by feeling unhappy, others by outbursts of violence. All kinds of negative events create stress, but so do events that on the surface may appear only to be change. Losing one's job is stressful, but so is taking a new job. The initial phase of marriage itself is a shock of adjustment, as is the arrival of children or moving to a new home. All of these situations are associated with increases in stress and at the extreme with incidents of violence.

How families react to stress depends on what other resources they have for dealing with stress. For this reason, working-class and poverty-level families are especially likely to have extreme forms of conflict around these incidents. Working-class persons tend to marry earlier, which makes the marital adjustment more of a shock. They also have children earlier, when they have less money and are less equipped emotionally to deal with their demands. These are some of the factors associated with spouse abuse and child abuse.

The long-run picture is one of fewer early marriages and a declining birthrate, both of which reduce these particular sources of stress. Economic stratification, on the other hand, is getting more severe, and the result will be a high level of stress-inducing transitions connected with careers. Here the picture for family conflict is likely to become worse.

Traditional Authoritarian Roles

Probably the most important factor that turns conflicts and stresses into violence is the extent to which men's and women's roles in the family are organized around a traditional pattern of segregated social networks. Here again we can see how ritual density and conformity (chapter 6) help explain class cultures. When families are organized into local, encapsulated networks, with sharp segregation of gender roles, this enforces the tradition of male domination and female passiveness. There are strong pressures for conformity, and violations even of minor matters are taken as symbolic affronts to authority and propriety. In this kind of family, as we saw in chapter 13, men tend to feel that they have the right or even the duty to keep their wives in line; similarly, parents believe that physically punishing their children is "good for them."

These families, then, tend to react to power shifts and to life stresses by a ritualized use of violence. This is also the situation in which ideals of sexual property are most strongly cherished. When there is adultery, or even just a hint of violating sexual possession, the result is likely to be intense jealousy. The jealous husband in particular feels justified in using violence against his wife because he traditionally has had the power to control her sexually. Wives, too, may react with violence when jealous of their husbands. Often, trivial incidents set off violence; we should note that these are symbolic of the whole relationship in high-ritual-density situations.

On this factor, we can be optimistic about future trends. Traditional encapsulated family structures are giving way to more cosmopolitan networks, as more jobs shift toward the middle-class occupations and more women have careers. Conflicts will still exist, but the extremely violent reactions to them that are produced by traditional authoritarian roles should be in decline.

MOTIVATIONS FOR CHILD REARING

The family has always been an economic as well as a sexual/emotional unit. With regard to children, economic payoffs that used to exist in other historical periods have disappeared. Children are simply an economic cost, and the cost has been rising. Housing and medical expenses have gone up drastically. Because children stay in school longer, parents have even higher expenses, continuing for more years. As more mothers enter the labor force, the old source of household labor and child-care is diminishing. It is no surprise that the birthrate declined. As women make even more inroads into professional and managerial careers with their high demands on time and energy, children may become even more of a practical problem. All factors point to the birthrate falling even further in the future.

What's to keep the birthrate from falling all the way to zero? From an economic viewpoint, it is surprising that the birthrate is as high as it is now, near the replacement level of two children per woman.

The countervailing force seems to be that the *emotional side of child rearing* has become more central. There is little reason to have children now except for love. There are fewer children per family, and the births are more carefully planned. This makes children more important to their parents, not less.

From the point of view of sociological theory, this is another instance of the focus on private, personal relationships in modern society. We have a cult of the individual, which reaches a particularly intense form in the search for sexual love between adults. These same conditions act to make children into sacred objects, centers of a great deal of symbolic attention. However, practical conditions in modern life are simultaneously undercutting the amount of care parents can give to their children. Children are spending more time in the care of teachers, day-care centers, and other forms of surrogate parenting. This is one reason why children are at the center of so many controversies: concerns over child abuse, sexual exposure, school achievement, drugs, athletics, and entertainment. Parents are concerned about their children, at the same time that they are less able to control them. Like love/sex relations among adults, the love relations between parents and children are heavily ritualized and symbolic, and they are undercut by practical realities. Even so, the forces for ritualizing and idealizing these relationships are strong and show every sign of continuing.

A major area of change right now is the pressure to expand the activities of *fathering* along with *mothering*. As women have moved into the labor force and the power balance in the household has shifted, women have adjusted primarily by cutting down the amount of time they spend in housework and child care. So far men have taken up only a small amount of the slack by increasing their own household work. However, there is a growing tendency of men to be involved in caring for children, from early infancy on. The main forces operating here are not so much an increasing ideological belief—that men ought to share equally in household and child-rearing responsibilities—but rather practical economic pressures. We have seen (chapter 11) that working-class men are less likely to subscribe to this ideology than are middle-class men, but practical pressures actually cause the former to pitch in more at home. In the upper middle class of intensely career-oriented dual-income families, both men and women are extremely busy. The pressure there is toward greater equality of responsibilities because their work lives don't allow either one of them to put in much time at home.

As we move from one generation to the next, the effects of greater gender equality in parenting should produce a cumulative effect. Evidence shows (chapter 12) that when men participate in child rearing, there is less overall conflict between the sexes and less gender stratification in the surrounding society. We can expect that as there is more active fathering, boys and girls will grow up into more gender-egalitarian adults. The shift toward increased fathering should

We will probably become more child centered symbolically as we become less child centered in fact.

prove to be the beginning of a pattern that will gain even more momentum in the next generation.

The overall picture of parents and children is a combination of increasing ideals but also numerous strains. We are inching toward parenting equality, and the dual-income pattern will strengthen it even more. Parents idealize the relationship with their children and devise many symbolic gestures to represent their love. At the same time, the practical strains intrinsic to child rearing are increasing. Paradoxically, these conditions may reinforce one another. For practical economic and career reasons, the number of children will probably decline still further. However, as children become more scarce, they become even more special, and this increases the tendency of their parents to make them centers of a ritual of parental love. We will probably become more child centered symbolically as we become less child centered in fact.

New technological advances will enable more affluent couples to bear children and will allow for more choices to be made surrounding the birth. For example, some lesbian couples are now implanting the fertilized embryo of one woman into the uterus of the other so that the child born from such a process will be the "biological" child of both women. This is done, in part, because the courts have sometimes denied parental rights to gay and lesbian people who act as parents but who do not have a genetic link to their children. Donor insemination, surrogacy arrangements, and other technologically assisted reproductive practices promise to make baby making and the laws surrounding it more complicated than ever (chapter 10).

These new developments highlight the fact that parent-child relationships, like other family relationships, are fundamentally social relationships. Regardless of any biological link between family members, the people involved and the institutional authorities in charge must define the nature of the relationships involved: who is in, who is out, and what rights and responsibilities accrue to each. In this sense, families may be becoming more intentional and fluid than they were in the past. Marriages and families have always been socially constructed, but with more elaborate ways of conceiving, bearing, and raising children (biological, step, surrogate, adoptive, joint versus full physical custody, joint versus full legal custody, visitation arrangements, co-parenting, grandparenting, etc.) and more diverse ways of being in committed relationships (married, cohabiting, same-sex, etc.), the possibilities for forming different types of families are likely to increase.

ETHNIC FAMILY PATTERNS

Ethnic changes are particularly visible in the United States at the beginning of the twenty-first century. Nevertheless, the most obvious things about ethnicity are the most superficial—that people have different skin color or hair or speak with a different language or accent. Such differences have little effect on social behavior, whether in the family or in economics or politics. If ethnic groups have different patterns, it is primarily because *sociological causes are operating in different proportions.*

One instance of this was noted in the discussion about class stratification and the marriage market. African American women have an extremely high nonmarital birthrate. This is not a direct consequence of being black; rather, it is due to a combination of several sociological conditions: the shortage of marriageable African American men, the class position of African American women, discrimination against blacks that keeps these women in this class position, and some intergenerational cultural influences.

Similarly, in chapter 7 we saw that the traditional male-dominated structure of many Latino families can be explained by their background in rural societies that are similar to the patriarchal households of earlier agrarian societies. The sociological principles that account for historical differences among societies (chapter 3) continue to explain many differences that we find in the twenty-first century, as families migrate from agrarian societies into industrial ones. In the United States, the trend for Latino women to move into the labor force like other American women is already shifting the power structure of the Latino family toward a more egalitarian pattern.

In the long run, it probably will not make much difference in the twenty-first century that the ethnic composition of American families will have changed. The family structure of ethnic groups is not static. There are two processes contributing to change. The first is that social conditions affecting all groups in American society will push family structures toward similar forms. This is not to say that all families will look the same regardless of ethnicity. Rather, we will see the retention of some cultural traits, the diminishment of others, and some

Intermarriage will continue to contribute to changes in the family. Asian-American women have particularly high tendencies to marry out of their ethnic group.

diffusion of customs and practices across groups. In addition, all groups will experience similar types of pressures for families, marriages, and households to respond to changing economic and social conditions. As discussed in earlier chapters, demographic trends tend to be moving in the same direction for all racial/ethnic groups, even though they start at higher or lower levels (e.g., cohabitation, marriage, fertility, nonmarital birth, divorce, single parenting, and household size).

Another major process contributing to change is *intermarriage*. As we noted in Figure 8.2, most ethnic groups have fairly high proportions of intermarriage with other groups. Asian Americans, especially those of Japanese and Chinese origin, have particularly high tendencies to marry out of their ethnic group. Hispanics also have a history of considerable intermarriage. The only group with a very low intermarriage rate is blacks. A realistic projection is that most ethnic groups will be widely absorbed into a multiethnic population through intermarriage within the next century. This implies that most ethnic differences in family structure are going to blur or rather that ethnicity may no longer be as noticeable as class differences. Unless trends change, blacks will be the exception. American families in the twenty-first century apparently will continue to be segregated along lines of black and white.

GENDER AND CLASS STRATIFICATION

The most important process driving almost all of these changes is the movement of women into full-time careers in the labor force. This trend began with the demand for large numbers of white-collar workers, especially secretaries and clerical workers. However, these are relatively low-paid jobs, so these workers really constitute a white-collar working class. Another phase of women's careers began later, when women caught up with men in higher education and began to cross over into the professional and managerial occupations. As we saw in earlier chapters, not only have women been entering the labor market in record numbers, but the pattern of their labor force participation is coming to resemble that of men. The most dramatic changes have occurred for married women with children. Pregnant women are less likely to quit their jobs and more likely to return to work shortly after giving birth than ever before.

What will happen with women's labor force participation in the future? As discussed next, that depends on developments in the larger economy, but there are strong indications that things will continue in the direction they have been heading (see Coltrane 1998). Additional information to address this question comes from young people's plans for the future. Considerable evidence suggests that the educational and career expectations of young American women and men have converged and that they will continue to do so. Boys used to have higher educational and occupational aspirations than girls, but now there are few if any gender differences. Some surveys even find that girls have higher occupational aspirations than boys. Over 90 percent of adolescent girls now expect to be employed after they marry, and virtually all of those expect to be working after they become mothers. Most adolescent boys similarly expect their future wives to hold paid jobs, though many express uncertainty about whether their wives will continue to hold jobs after having children (Dennehy and Mortimer 1992). The fact that teenage girls have expectations for more continuous labor force participation than teenage boys could portend future trouble. If young women's attitudes are changing faster than young men's, marital conflicts over jobs and family work are likely to ensue.

Young people's expectations about the future have a way of shifting to correspond to the practical choices they face in life. Most of those choices will be shaped by developments in the larger economy. Analysis of recent changes in the economy are many and varied, but there is a broad consensus on some of the most essential points. Analysts agree that there has been a global transformation of finance, business, and industry, with capital becoming more internationalized and competitive markets around the world becoming more closely linked than ever. Employers in the United States and throughout the world have attempted to maximize profits and increase efficiency through restructuring, technological innovation, and reduction of labor costs. They have often moved production from factory and office to private homes, from older urban areas to less expensive rural areas, and from industrialized countries to less developed countries (Acker 1992). These changes in the global division of labor have occurred as the general economy has shifted away from large-scale manufacturing, particularly in the male-dominated "smokestack industries." The movement of capital and employment has been toward service industries that are the bedrock of the information economy.

Because of continued global restructuring and growth in the service sector, the demand for lower-wage jobs is projected to increase. Most of the new jobs will be filled by women. Women are a majority of workers in the three occupational categories that are projected to have the most job openings in the early twenty-first century: service, administrative support, and professional specialty. Health and personal service occupations, both composed primarily of women, are also projected to have extremely high growth rates in the coming decades. By the year 2010, women will likely constitute half of the total labor force.

In fields where women are already significantly represented, female representation in higher-paying jobs (e.g., professional specialties, executive, administrative, and managerial jobs) will likely increase. Others predict that gains for women in management will level off as restructuring eliminates significant numbers of middle-level management positions and favors the incumbency of men (Acker 1992). Part-time positions (about 20 percent of all workers) are also projected to increase substantially. Part-time and flexible work situations typically allow integration of family work and paid work; however, such positions also tend to lack benefits (health insurance, vacation pay, parental leave, etc.), and few offer opportunities for career advancement. Women hold almost two-thirds of current part-time jobs, and growth in this area will mean employment for many more women in the near future. Nevertheless, because of financial necessity, job availability, and personal preference, most employed women will continue to work full time.

Even if the economy falters or overheats, we can expect more women in the labor force in the next decades. Some women will continue to enter higher-paying management and professional positions, but the majority will occupy lower-paying second-tier positions. Aggregate projections for minority women and men are relatively pessimistic, with continued overall decreases expected in their earning power. As manufacturing jobs decline, however, the relative positions of men and women will continue to converge. Men's real wages (adjusted for inflation) are likely to continue their downward slide, and in the more bleak projections, unemployment for men is expected to increase significantly. The major economic dividing line between families will become the number of earners in the household, with two-job couples faring better, on average, than single-breadwinner families or single-parent families. For subgroups with high rates of male unemployment or underemployment and/or increasing female earnings (i.e., African Americans and Latinos), women will have less incentive to stay married or to marry in the first place. Single mothers and their children, especially in minority communities, will continue to struggle economically, and most will be at risk for living in poverty.

Equality of employment is definitely not just around the corner, but according to the economic forces outlined earlier, we can expect a continued narrowing of the gap between men's and women's labor force participation rates and pay scales. This should lead to more parity in men's and women's bargaining positions on the marriage market and within the home. Nevertheless, increasing equality between men and women will occur in the context of increasing income inequality in the population at large. Real wages have fallen since the 1970s, and the gap between rich and poor has widened. Particularly for those with little education, wages have remained low and are expected to continue at low levels (Weinberg 1996). Despite the positive evaluation of the U.S. economy from most observers and increasing dividends for corporations and wealthy individuals, we see that restructuring and downsizing have not benefited most American workers. The income of the middle class is declining, and the middle class itself appears to be shrinking. The ranks of the poor are swelling, with people of color and single-mother households most likely to be mired in poverty (O'Hare 1996).

If income inequality in the overall population continues in the direction it is heading, there will be increasing pressure on both spouses in married couples to become and stay employed. Two-job families will be able to stay in the middle class, whereas one-earner families will have to fight to stay out of poverty. Workers at the bottom of the income pyramid, such as undereducated Latinos, will tend to live in poverty-level households, even if they are married. Because income inequality is increasing, we should guard against thinking that women's increasing labor force participation is simply a movement toward equity for women. Recent developments represent significant new pressures to work at undesirable jobs for both men and women and to spend more hours in paid employment per family member than at anytime in the past forty years.

In the long run, we may be seeing a renewed importance of class stratification. Gender equality for some women is not gender equality for all; nor is a society of gender equality going to be a society of economic equality across social classes. For the families of the future, social class is likely to shape the biggest differences.

THE NEED FOR FAMILY SUPPORTIVE PROGRAMS

Compared with other modern industrial democracies, the United States shows a marked reluctance to fund child-centered and (parent-centered) programs (such as paid parental leave, maternal health care, child-care centers, elder-care facilities, and universal health insurance). Other nations tend to support parents and encourage healthy child development by subsidizing programs that serve children directly. In the United States, in contrast, we assume that private families are individually responsible for the health and well-being of their children. Because we have so many poor mothers and fathers with limited access to jobs or insurance, this puts large numbers of children (and a disproportionate share of ethnic minority children) at risk. The assumption that individual parents are the only ones responsible for their children also puts pressure on families (mostly mothers) to work out their own child-care and health-care arrangements. If, on the other hand, we began to recognize the links between child poverty and other social problems and if we reoriented our thinking to be more community minded, then we might move away from scapegoating single mothers and their children. We could then move our social policies toward collective responsibility for all our children and develop "safety-net" programs such as those that exist in European-style democracies (see Table 6.1).

What changes in child-care practices can we expect in the future? As noted in chapter 12, child care by relatives has been the most common form of care for preschool children in the past, especially among the poor. Today, about half of all preschoolers with employed mothers are cared for by family or relatives (including fathers, grandparents, siblings, and other family members). Patterns vary by part-time versus full-time employment of the mother, but fathers and grandparents will continue to provide substantial amounts of care. Among the middle and upper classes, in-home sitters, nannies, and housekeepers are still common and will continue to be, especially if the income inequality trends discussed earlier continue. The use of family day-care homes, recently on a downward trend, will continue to be important in the future, especially in working-class neighborhoods where there are few other alternatives. Organized child-care facilities are likely to become even more important in the future and could serve at least half of all preschool children if adequate funding were provided. Nevertheless, because child-care centers are not lucrative profit-making enterprises, the current affordable child-care shortage is likely to continue.

Because maternal employment is not just a passing fad, there will be mounting pressure on businesses to provide job leaves, flexible scheduling, and on-site child care. There will also be more pressure on governments to fund community day-care facilities, to regulate private and nonprofit care centers, and to provide recreation facilities for school-aged children. In the face of economic restructuring and an aging population, however, it is unlikely that a European-style commitment to child care will emerge in the United Sates. We've seen that women's labor force participation is the engine driving most of the family changes reviewed here. For this reason, perhaps the easiest way to promote family-friendly programs in this country would be to standardize wage rates and equalize the occupational opportunity structure in the labor market. Promoting comparable worth, pay equity, and gender-fair promotions would probably do more for children and mothers than any other policy initiative currently being contemplated in the United States.

CONCLUDING THOUGHTS ON AMERICAN INDIVIDUALISM AND SEPARATE SPHERES

In his classic 1830s commentary on democracy in America, the French observer Alexis de Tocqueville (1969) worried about the tendency of Americans to isolate themselves from their responsibilities to the larger society by focusing exclusively on themselves and their families. American individualism, according to Tocqueville, could isolate people from their ancestors, their contemporaries, and even their descendants as they came to believe that they controlled their own destinies, owed nothing to others, and expected nothing in return (Bogenschneider 2000). The individualism and private family ideals observed by Tocqueville almost two centuries ago still pervade American culture today (Bellah et al. 1996). An emphasis on individual rights, a belief in the unfettered pursuit of self-interest, unbounded faith in free-market capitalism, and an idealized vision of the private family continue to dominate the American consciousness.

Tocqueville observed that American individualism could be tempered by civic and political participation that fostered a sense of public responsibility. Similarly, in the modern context, Wilson (1999) suggests that membership in civic associations can counter isolationism and build tolerance and social trust. He notes that declining civic engagement, along with deteriorating economic prospects for a large segment of the population, account for many contemporary social ills. What role might families play in building and maintaining a sense of community? Could a symbolic or legal revaluing of traditional marriage and family ideals (as proposed by those calling themselves "family values" advocates) counter the isolationism observed by Tocqueville and Wilson? Would changing how people think about marriage and divorce or returning to nineteenth-century laws regulating family life help Americans regain a sense of community? Probably not.

As we have discussed throughout this book, political debates about family topics like divorce, cohabitation, nonmarital birth, marital rape, abortion, or same-sex marriage tend to take on symbolic significance well beyond the specific issues at hand. In this sense, "The Family" carries much more emotional and political significance than the actual personal processes involved in families. In particular, such debates tend to reflect widespread anxiety about the changing roles of men and women in society, with particular worry about whether women will stay committed to children if they are not economically dependent on men. Scholars like Hochschild (1997) and Schor (1991) suggest that Americans, especially women, are investing increasingly more time and emotional energy into their jobs. Does this mean that Americans are abandoning their families? The answer, in our view, is decidedly no. Even with most women employed outside the home, marriage rates dropping, and divorce being common, most people still decide to enter committed relationships and become parents. Although Americans may have less time to spend with family members than the idealized and lampooned 1950s housewife, they are placing increasing emotional significance on family time and family relations. This symbolic valuing of family in the modern era both resurrects and transcends the historical ideal of separate spheres for men and women.

Tocqueville was also among the first to note that Americans are obsessed with separating the practical activities and symbolic worth of men and women: "In America, more than anywhere else in the world, care has been taken to constantly trace clearly the distinct spheres of action of the two sexes and both are required to keep in step, but along paths that are never the same" (Tocqueville [1832] 1969: 601). Despite significant changes in the past two centuries, this American preoccupation with separate gender spheres can still be seen in the family ideals and practices that organize everyday life. At the same time, Americans' preoccupation with personal freedom increasingly conflicts with traditional notions of home and family as the woman's domain and the man as "head and master" of the household. The false distinction between the private family and the public market that came to symbolize differences between women and men has increasingly been called into question. As our laws and social traditions have been transformed to treat women and men as equals, social anxiety about the mythical decline of The American Family has increased.

In the coming decades, we predict that symbolic and actual conflicts between individualism and separate spheres will repeatedly surface in political campaigns, family policy debates, and private family discussions. Idealized visions of marriages and families will continue to be used rhetorically to advance various causes, but the actual content of family and household relationships will continue to change in response to the larger social and economic forces that have always shaped them. We have no doubt that individuals will continue to fall in love, decide to have children, and make long-term commitments to each other and to their children. Because of Americans' simultaneous preoccupation with both individualism and separate spheres, we are less sanguine about the nation's commitment to furnish children and their parents with what they need for optimum growth and development. We are concerned with diminishing levels of government support and the limited opportunities available to the increasing numbers of children who live in poverty. Reversing this trend will require a commitment of financial resources that Americans have been reluctant to provide in the past.

In the face of increasing inequality between rich and poor, more ethnic diversity in the population, and expanding choices for those with sufficient resources, we expect to witness even more diversity in family forms and practices. Families of various structures and types have sustained people throughout human history and will continue to do so in the future. Because families have also served to regulate peoples' emotions, sexuality, personal power, and access to property,, they are also implicated in the maintenance of social hierarchies. We have seen that families have promoted gender, class, and ethnic inequalities, and also that they have been a major site of resistance against many forms of exploitation and oppression. Although families thus exercise considerable control over money, sex, and power, they also provide invaluable love, trust, and care to both adults and children. As we close this book, one of the few predictions we can be sure about is that families in the twenty-first century will continue to be a mix of all of these things.

Sociology Web Site

See the Wadsworth Sociology Resource Center, "Virtual Society," for additional links, quizzes, and learning tools:

http://www.sociology.wadsworth.com

Also on this web site you'll find InfoTrac College Edition, an online library of journals. Here you can search for electronic articles about central topics in sociology.

Glossary

Abortion Expelling a fetus or embryo from the uterus either spontaneously or via medical procedures.

Abstinence Deliberately refraining from sexual intercourse.

Abstinence-only policy Sex education policy that advocates not having sexual intercourse as the only acceptable means of contraception.

Abstinence-plus policy Sex education policy that advocates not having sexual intercourse as the preferable means of contraception, but also includes discussion of alternative means of contraception.

Adultery Extramarital sexual intercourse.

Advanced horticultural societies Tribal societies that practice simple agriculture and possess metal tools.

Aging population A population in which the proportions of adults and elderly increase, while the proportions of children and adolescents decrease.

Agrarian plow cultures Societies that utilize the technology of large-scale agriculture (plows or irrigation, draught animals, metalworking), and usually have high levels of literacy and urban centers.

Alimony Income paid to support a divorced spouse. Also called *spousal support*.

Alliance theories Theories that analyze kinship systems in tribal societies as forms of political alliance.

Amniocentesis Medical test performed during pregnancy to determine the presence of birth defects and the sex of the child.

Annulment A legal judgment that a marriage is void.

Artificial insemination Method of fertilizing a female egg with sperm from a male donor through laboratory implantation rather than sexual intercourse.

Baby boom The unexpected increase in the birthrate that took place from roughly 1946 to 1964.

Bilateral sexual possession Social arrangement in which each sexual partner has exclusive sexual access to the other.

Bilineal descent Family membership and inheritance traced through the lines of both male and female parents to their children. Compare *matrilineal descent; patrilineal descent*.

Birth control Any method used to avoid pregnancy.

Birth order Sibling position based on age.

Birth rate The number of births in a specified population per unit of time.

Birth ratio The number of non-marital births divided by the number of marital births.

Celibacy Abstention from sexual activity.

Cesarean section (c-section) Procedure by which the baby is surgically removed through the wall of the uterus at birth.

Child custody Legal award of responsibility for residence, control, and care of a child.

Child support Income paid to a divorced spouse for support of the dependent children from a marriage.

Cohabitation Residence of a couple in a shared household, with mutual sexual access, but without legal sanction; essentially an informal marriage.

Cohort Group of persons who are born at approximately the same time and who subsequently go through life stages together.

Coitus Sexual intercourse.

Communes Group residential arrangements in which many or all economic resources are shared.

Community property Wealth, income, or other assets legally belonging jointly to both marriage partners.

Compadrazgo A system of mutual exchange and support linking godparents to children in some Mexican and Mexican American families.

Comprehensive sex education Sex education that treats abstinence as one option in a broader sex education program that considers various contraceptive methods to prevent pregnancy and avoid sexually transmitted diseases.

Conflict theories Sociological theories that explain institutions, such as the family, in terms of the self-interest of their individual members and the conflicts arising from unequal distribution of power and resources.

Conjugal family *See* Nuclear family.

Contraception *See* Birth control.

Cosmopolitan Having exposure to a large number of diverse social contacts.

Courtly love A code of romantic behavior popular among the aristocracy of medieval Europe. Its tenets, which included idealizing one's lover, were carried to Europe's courts in the songs and poems of royal troubadours.

Courtship A process of acquaintance, selection, and attachment between potential mates that leads to the formation of strong sexual ties and possibly marriage.

Dependency ratio The ratio of persons in the ages defined as dependent to persons in the ages defined as economically productive in a population.

Descent rule Lineage through which family membership and inheritance is traced; can be matrilineal, patrilineal, or bilinear.

Down's syndrome A form of mental retardation due to birth defects.

Dual-career marriage A marriage in which both wife and husband are committed to full-time occupational careers.

Economic class Groups within a society distinguished and often stratified according to economic, occupational, and social factors.

Economic determinism The theory that social institutions are determined primarily by the economic system.

Economic theories of gender stratification Theories of the economic causes and effects of social inequalities between men and women.

Economic property rights Socially recognized rights of an individual over economic assets, including physical objects, land, currency and credit, and labor.

Educational attainment The highest level of school completed or the highest degree received.

Egalitarian Equality in social arrangements. A society based on the principle that all individuals are entitled to equal treatment and rights in the society.

Elderly Persons 65 years and older.

Emotional possession The belief that one's emotional attachments to another imply a personal right or property; typically, the feeling that someone's love is owed exclusively to oneself.

Emotion work Efforts to shape and control one's emotions by socially defining them through conversations or personal reflections on one's feelings.

Empty nest syndrome State of psychological depression or search for new roles that ostensibly occurs among full-time housewives whose children have grown up and left home.

Endogamy Marriage to a partner within the same group, clan, tribe, or ethnicity.

Episiotomy Cutting of the membrane between the anus and the vagina; often performed to facilitate childbirth.

Equity theory Theory that love is based on an equal exchange of social resources between partners.

Ethnic identity Identification with a social group that has a common cultural tradition, common history, and common sense of identity and exists as a subgroup in a larger society.

Exchange theory Theory that holds that human behavior is determined by the rewards individuals give to each other.

Exogamy Marriage to a partner outside one's group, clan, tribe, or ethnicity.

Extended family A family grouping that extends beyond the immediate relationships of husband, wife, and their children, and includes several generations.

Fallopian tubes Tubes connecting the ovaries with the uterus.

Familism The idea that family members should give priority to family (versus individual) interests, welfare, and survival, hence family members should cleave to one another and support one another under all circumstances.

Family Any unit in which there exists a sharing of economic property, mutually held, and relatively permanent rights of sexual access between sexual partners, a sense of commitment or identification with the other members, including any children born to or raised by members. The many forms of the family include: nuclear, extended, single-parent, communal, and homosexual. See chapter 2 for census and historical definitions of families.

Family property All economic assets held by a family.

Feminist movement Political movement advocating economic and social rights for women.

Feminist theories Theories that identify and criticize male domination in the family, economy, or other spheres of the social and cultural world.

Fictive kin A friend who is assigned a kin relationship based on his or her exceptional participation in sharing exchanges of emotional support, goods, and services, thus meeting the dual expectations of friends and kin.

Fishing societies Groups of people living in fishing communities, which formed in coastal areas like the Pacific Northwest.

Fortified households In medieval civilizations, patriarchal households were built like fortresses, with everyone who could owning weapons and each household having its own armed forces.

Functionalism Type of social theory that explains social institutions by the contribution they make to the functioning and survival of society.

Gay liberation Political movement advocating open and legal recognition of homosexuality.

Gender Socially defined "acceptable" behavior and characteristics for men and women.

Gender stratification, theory of Theory of the causes and effects of social inequalities between men and women at different periods of history.

Gestation Period of growth of an embryo in the womb.

Group marriage Arrangements in which more than two individuals consider themselves all equally married to each other, especially in regard to sexual access.

High-density situation A situation in which a person is in close physical proximity to a number of other people.

High ritual density The ongoing experience of people who live in circumstances where they are constantly in the presence of others, all doing the same thing, maintaining a high focus of attention on each other, and sharing common moods.

High social density Social density is the extent to which a person is physically in the presence of others. The higher the social density, the more individuals will conform to the group's customs and beliefs, and the more they will expect conformity from others.

Homogamy The tendency for people to marry those with similar social characteristics.

Homophily The tendency for people to associate with those having similar social characteristics and status.

Hospice A facility or program designed to provide a caring environment for supplying the physical and emotional needs of the terminally ill.

Household A residential unit in which co-residents typically share some resources. Households vary in membership and composition. Some households are families (parents and children, or husbands and wives), but many families are not households (some family members may not live in the same residence).

Household labor The labor involved in maintaining and operating a household, including cooking, laundry, child care, shopping, repairing items, and yard care.

Hunting-and-gathering societies Small tribal societies that utilize stone and wooden implements and survive by hunting animals and gathering wild plants.

Hypergamy The tendency of women to marry into a higher social class.

Immigration Denotes geographical movement of relatively settled people, especially those making a comparatively permanent change of residence.

Incest Sexual intercourse between family members who are socially defined as closely related.

Incest taboo Societal prohibition against intercourse between family members.

Industrial societies Large-scale, complex societies whose economic basis derives from the use of inanimate energy sources such as coal, oil, and electricity.

Infanticide Killing an infant.

Interaction ritual (theory) Any activity in everyday life that brings persons together, focuses their attention on a common activity or set of symbols, and builds up a shared emotion or mood, resulting in a feeling of group membership.

Intergenerational property rights Socially recognized power of parents to control their children's labor and sexual and other activities; also the rights of offspring to inherit and receive economic support from their parents.

In vitro fertilization A method of conception in which an egg is removed from the mother's ovaries and fertilized outside her body

In vitro fertilization—*Cont'd.* with sperm from a male donor. Often referred to as conception "in a test-tube."

IUD (intrauterine device) A small plastic or metal birth-control device placed within the uterus to prevent implantation of a fertilized egg.

Joint custody Legal award of responsibility for residence, care, and control of a child to both parents jointly after a separation or divorce. *Joint legal custody* refers to the divorced legal parents sharing decisions regarding such things as a child's education and medical treatment. *Joint physical custody* refers to a child living with each parent for alternating periods of time.

Kinship systems Any form of social organization relating individuals to each other by marriage or descent.

Labor force participation Paid employment or the active search for paid employment.

Longevity Length of time (years) that people live.

Love A strong emotional attachment to another person.

Love-deprivation A parental technique of controlling a child by threat of withholding love.

Love revolution The historical period during which traditional marriages arranged for economic or political purposes gave way to the modern ideal that marriages should be based primarily on love.

Low-density situation A low-density situation is one in which a person is not in close physical proximity to other people.

Low social density Social density is the extent to which a person is physically in the presence of others. The lower the social density, the less individuals will conform to the group's customs and beliefs, and the more autonomous they will be.

Machismo Exaggerated masculine values and behaviors enacted by men to maintain power and elicit respect and deference from others (i.e., hypermasculinity).

Marital contract Legal aspects of marriage, specifying rights and duties of the partners and especially the disposition of property in case of death or divorce.

Marriage market The total set of interactions among all potentially marriageable persons available to one another, which results in different degrees of attraction of any particular person to any other person, by comparison to the attraction of other potential partners.

Maternal custody Legal award to the mother of rights and responsibilities of a child's residence and care after separation or divorce.

Matrilineal descent Family membership and inheritance traced through the female line, from a mother to her children.

Matrilocal system Residence of married partners in the wife's family's home.

Medical abortion Termination of pregnancy by nonsurgical means, such as by taking RU-486.

Melting pot A place where racial amalgamation and social and cultural assimilation occur.

Menopause The cessation of ovulation and menstruation in an adult woman, usually taking place around age fifty.

Methotrexate-misoprostol regimen Medical abortion method by which methotrexate blocks folic acid and prevents cell division; misoprostol is a prostaglandin causing uterine contractions.

Midlife crisis The psychological difficulties associated with the role transition that can take place in one's thirties or forties, when growing recognition of one's career limitations and of the unrealistic expectations of one's youth causes questioning and a renewed search for meaning in life.

Mifepristone-misoprostol regimen Medical abortion method by which mifepristone (RU-486) blocks production of hormone progesterone, which is needed for pregnancy; misprostol is a prostaglandin causing uterine contractions.

Miscegenation Marriage or cohabitation between persons of different races.

Monogamy Marriage and sexuality confined to two persons.

No-fault divorce Law providing for divorce by mutual consent of the partners.

Nuclear family Family group consisting of a mother, a father, and their children. Also called "conjugal family."

Oldest old Persons 85 years and older.

Order givers Occupational class of people, usually high-ranking managers or professionals, who tell other people what to do.

Order takers Occupational class of people, generally the working class, who have little power in their jobs and take orders from others.

Out-marriage Marriage to someone of a different race or ethnicity (*see also* exogamy).

Pastoral societies Groups of people who live by herding cattle, sheep, goats, or other animals.

Paternal custody Legal award to the father of rights and responsibilities of a child's residence and care after separation or divorce.

Patriarchal household Traditional household that comprised a unit of economic production and military defense headed by an elder male who exercised control over all household members.

Patriarchy Male domination within the family, and within the economic and political spheres of society.

Patrilineal descent Family membership and inheritance traced through the male line, from father to his offspring.

Patrilocal system Residence of a married couple with the husband's family.

Physical custody Relates to which parent the child will live with on a day-to-day basis; can be maternal, paternal, or joint.

Political theory Theories of the political causes and effects of social inequalities between men and women.

Polyandry System of marriage involving one woman with multiple husbands.

Polygamy System of multiple marriage, including both polyandry and polygyny.

Polygynous Relating to polygyny, the system of marriage involving one husband with multiple wives.

Population pyramid A bar chart that shows the distribution of a population by age and sex. By showing numbers or proportions of males and females in each age group, a pyramid presents a "picture" of a population's characteristics.

Power The capacity to control another's behavior while escaping reciprocal influence.

Pregnancy rates The number of pregnancies per specified population of women (usually 1,000 women between the ages of fifteen and forty-four).

Primitive horticultural societies Relatively small tribal societies whose economic base consists of simple agriculture, carried out with hoes or sticks.

Principle of legitimacy Theory formulated by Malinowski that every child must have a socially recognized father.

Propinquity Nearness of physical location.

Racial identity Identification with a particular racial category (white, black, Asian/Pacific Islander, American Indian/Alaska Native are the four official U.S. Census categories for race); distinct from ethnic identity (Hispanic or non-Hispanic).

Residence pattern Where one lives; pattern can be matrilocal or patrilocal.

Rewards A parental technique for controlling a child's behavior by giving material or social rewards for good behavior.

Rights of sexual possession The socially sanctioned right of one individual to erotic access to another.

Romantic love Passionate attachment to another person, accompanied by a high degree of idealization and usually erotic desire.

Selection effects The bias introduced when groups are not equivalent before an event occurs (e.g., cohabitation, marriage, divorce, remarriage).

Sex (1) Biological characteristics of males and females; (2) erotic behavior, especially involving the genitals, such as intercourse.

Sexual division of labor The assignment of different work tasks to men and women.

Sexually transmitted disease (STD) Any contagious disease passed on specifically by erotic contacts with other persons. Also called venereal disease.

Social density The proportion of time that a person spends in the presence of other people, rather than alone.

Socialization The process of training a new member in the beliefs and practices of a social group.

Social-psychological theories Theories that focus on the face-to-face interaction of individuals or on their subjective experiences.

Social ritual Any activity that brings people together, focuses their attention on a common activity or set of symbols, and builds up a shared emotion or mood, resulting in a feeling of group membership.

Spousal support *See* alimony.

Spouse abuse Physical or psychological mistreatment of a marital partner.

Status The symbolic and social benefits or limitations accruing to one who occupies a specific position in a group or social structure.

Straight-line assimilation theory View suggesting that immigrants become increasingly similar to the host population as they spend more time in the country.

Stratification Division of society into a hierarchy of economic class, power, and status. The pattern of social equality or inequality.

Symbolic interactionism Theory stressing the interpretations that individuals make of social situations, other people, their own selves, and lines of action.

Time out A parental technique of controlling a child's misbehavior by isolating him or her for a short time.

Transition rituals Ceremonies that mark a change in a person's social role or standing.

Transition to parenthood The changes and adaptations that accompany the birth of the first child.

Unilateral sexual possession Exclusive rights of erotic access by one person over another, without equally exclusive rights of the second person over the first.

Utilitarian marriage A marriage with relatively few emotional ties, held together by practical concerns over social status and career.

Vasectomy A method of permanent sterilization performed by surgically cutting the tubes between the testicles and the seminal vesicles.

Victorian revolution Historical period during which women were highly idealized but also sharply segregated into domestic roles and away from men's more public activities.

Working-age population Persons in the ages defined as economically productive.

Zero population growth Occurs when birthrates and death rates are approximately equal so that the population does not grow but remains stable.

References

Abma, Joyce, Anjani Chandra, William Mosher, Linda Peterson, and Linda Piccinino. 1997. Fertility, Family Planning, and Women's Health: New Data From the 1995 National Survey of Family Growth. Hyattsville, MD: National Center for Health Statistics, U.S. Department of Health and Human Services.

Abrahamson, Mark. 1998. *Out-of-Wedlock Births: The United States in Comparative Perspective.* Westport, CT: Praeger.

Acker, Joan. 1992. "The Future of Women and Work: Ending the Twentieth Century." *Sociological Perspectives* 35(1):53–68.

Adams, Bert N. 1988. "Fifty Years of Family Research: What Does It Mean?" *Journal of Marriage and the Family* 50:5–17.

Adams, David. 1988. "Treatment Models of Men Who Batter." In *Feminist Perspectives on Wife Abuse,* edited by Kersti Yllö and Michele Bograd. Newbury Park, CA: Sage.

Agassi, Judith Buber. 1991. "Theories of Gender Equality: Lessons from the Israeli Kibbutz." In *The Social Construction of Gender,* edited by Judith Lorber and Susan Farrell. Newbury Park, CA: Sage.

Aguirre, Adalberto, and R. Martinez. 1984. "Hispanics and the U.S. Occupational Structure." *International Journal of Sociology and Social Policy* 4:50–59.

Ahlburg, Dennis, and Carol DeVita. 1992. "New Realities of the American Family." *Population Bulletin* Washington, DC: 47(2). Population Reference Bureau.

Alan Guttmacher Institute. 1999. *Sharing Responsibility: Women, Society and Abortion Worldwide.* New York: Alan Guttmacher Institute.

Aldous, Joan. 1982. *Two Paychecks: Life in Dual-Earner Families.* Beverly Hills: Sage.

Allen, Katherine R., Rosemary Blieszner, and Karen A. Roberto. 2000. "Families in the Middle and Later Years: A Review and Critique of Research in the 1990s." *Journal of Marriage and the Family* 62.

Allen, Sarah M., and Alan J. Hawkins. 1999. "Maternal Gatekeeping: Mothers Beliefs and Behaviors That Inhibit Father Involvement in Family Work." *Journal of Marriage and the Family* 61:199–212.

Allgeier, Elizabeth Rice, and Michael W. Wiederman. 1991. "Love and Mate Selection in the 1990s." *Free Inquiry* 11:25–27.

Amadiume, Ifi. 1987. *Male Daughters, Female Husbands: Gender and Sex in African Society.* London: Zed.

Amato, Paul R. 1995. "Single Parent Households As Settings for Children's Development, Well-Being, and Attainment: A Social Network/Resources Perspective." *Sociological Studies of Children* 7:19–47.

Amato, Paul R. 2000. "Diversity within Single-Parent Families." Pp. 149–72 in *Handbook of Family Diversity,* edited by D. H. Demo, K. R. Allen, and M. A. Fine. New York: Oxford University Press.

Amato, Paul R., and Alan Booth. 1997. *A Generation at Risk: Growing Up in an Era of Upheaval.* Cambridge, MA: Harvard University Press.

Anderson, K. L. 1997. "Gender, Status and Domestic Violence: An Integration of Feminist and Family Violence Approaches." *Journal of Marriage and the Family* 59(3):655–69.

Arendell, Terry. 1986. *Mothers and Divorce.* Berkeley: University of California Press.

Arendell, Terry. 1995. *Fathers and Divorce.* Thousand Oaks, CA: Sage.

Arendell, Terry. 2000. "Conceiving and Investigating Motherhood: The Decade's Scholarship." *Journal of Marriage and the Family* 62.

Ariès, Phillipe. 1962. *Centuries of Childhood.* New York: Random House.

Ariès, Phillipe. 1979. "The Family and the City in the Old World and the New." In *Changing Images of the Family,* edited by V. Tuft and B. Myerhoff. New Haven, CT: Yale University Press.

Arms, Suzanne, 1975. *Immaculate Deception.* New York: Bantam.

Ash, Peter, Arthur L. Kellerman, Dawna Fuqua-Whitley, and Amri Johnson. 1996. "Gun Acquisition and Use by Juvenile Offenders." *The Journal of the American Medical Association* 275(22):1754–58.

Astone, Nan M., and Sara S. McLanahan. 1989. "The Effect of Family Structure on School Completion." Paper presented at the Population Association of America annual meeting, Baltimore, MD.

Axinn, William G., and Jennifer S. Barber. 1997. "Living Arrangements and Family Formation Attitudes in Early Adulthood." *Journal of Marriage and the Family* 59:595–611.

Bacdayan, Albert S. 1977. "Mechanistic Cooperation and Sexual Equality among the Western Bontoc." In *Sexual Stratification,* edited by Alice Schlegel. New York: Columbia University Press.

Bachman, Ronet, and Linda E. Saltzman. 1995. *Violence against Women: Estimates from the Redesigned Survey.* Washington, DC: U.S. Department of Justice, Office of Justice Programs. (Bureau of Justice Statistics Special Report, No. NCJ-154348).

Bachu, Amara. 1999. "Is Childlessness among American Women on the Rise?" Population Division Working Paper No. 37. Washington, DC: U.S. Bureau of the Census.

Bachu, Amara. 1998. "Trends in Marital Status of U.S. Women at First Birth: 1930–1994." Population Division Working Paper No. 20. Washington, DC: U.S. Bureau of the Census.

Bachu, Amara. 1999. "Trends in Premarital Childbearing." *Current Population Reports* P23–197. Washington, DC: U.S. Bureau of the Census.

Bajema, C. J. 1968. "Interrelations among Intellectual Ability, Educational Attainment, and Occupational Achievement." *Sociology of Education* 41:317–19.

Baker, C. A., G. J. Gilson, M. D. Vill, and L. B. Curet. 1993. "Female Circumcision—Obstetric Issues." *American Journal of Obstetrics and Gynecology* 169(6):1116–18.

Baller, W. R., D. C. Charles, and E. L. Miller. 1967. "Midlife Attainment of the Mentally Retarded." *Genetic Psychology Monographs* 42:235–327.

Barich, Rachel Roseman, and Denise D. Bielby. 1996. "Rethinking Marriage: Change and Stability in Expectations, 1967–1994." *Journal of Family Issues* 17:139–69.

Barnett, Rosalind, and Caryl Rivers. 1998. *She Works/He Works.* Cambridge, MA: Harvard University Press.

Barranti, C.C.R. 1985. "The Grandparent/Grandchild Relationship: Family Resources in an Era of Voluntary Bonds." *Family Relations* 34:343–52.

Barrera, Mario. 1979. *Race and Class in the South.* Notre Dame, IN: University of Notre Dame Press.

Baruch, Grace, and Rosalind Barnett. 1981. "Fathers' Participation in the Care of Their Preschool Children." *Sex Roles* 7:1043–55.

Baruch, Grace, and Rosalind Barnett. 1986. "Consequences of Fathers' Participation in Family Work: Parents' Role Strain and Well-Being." *Journal of Personality and Social Psychology* 51:983–92.

Basch, Norma. 1999. *Framing American Divorce: From the Revolutionary Generation to the Victorian.* Berkeley, CA: University of California Press.

Baxter, Janeen and Mark Western. 1998. "Satisfaction with Housework: Examining the Paradox." *Sociology* 32:101–120.

Bean, Frank D., Russell L. Curtis, Jr., and John P. Marcum. 1977. "Familism and Marital Satisfaction among Mexican Americans: The Effects of Family Size, Wife's Labor Force Participation, and Conjugal Power." *Journal of Marriage and the Family:* 759–67.

Bean, Frank, and Marta Tienda. 1987. *The Hispanic Population of the United States.* New York: Russell Sage Foundation.

Beauvoir, Simone de. [1952] 1953. *The Second Sex.* Translated and edited by H. M. Parshley. [1st American ed.] New York: Knopf.

Bellah, Robert N., Richard Madsen, William M. Sullivan, Ann Swidler, and Steven M. Tipton. 1996. *Habits of the Heart: Individualism and Commitment in American Life.* Berkeley: University of California Press.

Bellah, Robert, Richard Madsen, William Sullivan. Ann Swidler, and Steven Tipton. 1985. *Habits of the Heart: Individualism and Commitment in American Life.* Berkeley: University of California Press.

Belsky, J., G. B. Spanier, and M. Rovine, 1983. "Stability and Change in Marriage across the Transition to Parenthood." *Journal of Marriage and the Family* 45:567–77.

Belsky, Jay. 1993. "Etiology of Child Maltreatment: A Developmental–Ecological Analysis." *Psychological Bulletin* 114:413–34.

Bem, Sandra L. 1993. *The Lenses of Gender.* New Haven, CT: Yale University Press.

Bem, Sandra Lipsitz. 1989. "Genital Knowledge and Gender Constancy in Preschool Children." *Child Development* 60(3):649–62.

Bengtson, Vern, Carolyn Rosenthal, and Linda Burton. 1990. "Families and Aging: Diversity and Heterogeneity." In *Handbook of Aging and the Social Sciences,* 3d ed., edited by R.H. Binstock and L. George. New York: Academic Press.

Bennett, Neil G., David E. Bloom, and Patricia H. Craig. 1989. "The Divergence of Black and White Marriage Patterns." *American Journal of Sociology* 3: 692–722.

Benokraitis, Nijole V. 1993. *Marriages and Families: Changes, Choices, and Constraints.* Englewood Cliffs, NJ: Prentice Hall.

Berger, Bennett. 1960. *Working-Class Suburb.* Berkeley: University of California Press.

Berger, Bennett. 1971. *Looking for America.* Englewood Cliffs, NJ: Prentice Hall.

Berger, Peter, and Hansfried Kellner. 1964. "Marriage and the Construction of Reality." Diogenes 46:1–23.

Berk, Richard A., and Sarah Fenstermaker Berk. 1979. *Labor and Leisure at Home: Content and Organization of the Household Day.* Beverly Hills, CA: Sage.

Berk, Richard A., Sarah F. Berk, Donileen R. Loeske, and David Rauma. 1983. "Mutual Combat and Other Family Violence Myths." In *The Dark Side of Families,* edited by David Finkelhor, Richard J. Gelles, Gerald T. Hotaling, and Murray A. Straus. Beverly Hills, CA: Sage.

Berk, Sarah F. 1985. *The Gender Factory. The Apportionment of Work in American Households.* New York: Plenum Press.

Berkner, Lutz K. 1972. "The Stem Family and the Developmental Cycle of the Peasant Household: An Eighteenth Century Austrian Example." *American Historical Review* 77:398–418.

Bernard, Jessie Shirley. 1982. *The Future of Marriage.* New Haven, CT: Yale University Press.

Bernard, Jessie. 1972. *The Future of Marriage.* New York: World.

Bernard, Jessie. 1981b. "The Good Provider Role: Its Rise and Fall." *American Psychologist* 36:1–12.

Bernstein, Basil. 1971–75. *Class, Codes, and Control.* 3 vols. London: Routledge and Kegan Paul.

Bernstein, Jared, and Lawrence Mishel. 1997. "Has Wage Inequality Stopped Growing?" *Monthly Labor Review* 120:3–17.

Berreman, Gerald Duane.1993. *Hindus of the Himalayas: Ethnography and Change.* 2d ed. New York: Oxford University Press.

Berscheid, E., K. K. Dion, E. Walster, and G. W. Walster. 1971. "Physical Attractiveness and Dating Choice: A Test of the Matching Hypothesis." *Journal of Experimental Social Psychology* 7:173–89.

Berscheid, Ellen, and Harry T. Reis. 1998. "Interaction and Close Relationships." Pp. 193–281 in *The Handbook of Social Psychology*, 4th ed., edited by D. T. Gilbert, S. T. Fiske, and G. Lindzey. Boston, MA: McGraw-Hill.

Berscheid, Ellen. 1985. "Interpersonal Attraction." In *The Handbook of Social Psychology*, edited by Gardner Lindzey and Elliot Aronson. New York: Random House.

Berscheid, Ellen. 1994. "Interpersonal Relationships." *Annual Review of Psychology* 45:79–129.

Bianchi, Suzanne M. 1995. "The Changing Demographic and Socioeconomic Characteristics of Single-Parent Families." *Marriage and Family Review* 20:71–97.

Bianchi, Suzanne M., and Daphne Spain. 1986. *American Women in Transition.* New York: Russell Sage Foundation.

Bianchi, Suzanne M., Lekha Subaiya, and Joan R. Kahn. 1999. "The Gender Gap in the Economic Well-Being of Nonresident Fathers and Custodial Mothers." *Demography* 36:195–203.

Bielby, Denise D. 1992. "Commitment to Work and Family." *Annual Review of Sociology* 18:281–302.

Bielby, Denise D., and William Bielby. 1988. "She Works Hard for the Money." *American Journal of Sociology* 93:1031–59.

Bielby, William, and Denise D. Bielby. 1989. "Family Ties: Balancing Commitments to Work and Family in Dual Earner Households." *American Sociological Review* 54(5):776–89.

Bielby, William, and Denise D. Bielby. 1992. "I Will Follow Him: Family Ties, Gender Role Beliefs, and Reluctance to Relocate for a Better Job." *American Journal of Sociology* 97:1241–67.

Billingsley, Andrew. 1968. *Black Families in White America.* Englewood Cliffs, NJ: Prentice Hall.

Blackstone, William, Sir. 1765. *Commentaries on the Laws of England.* Adapted by R. M. Kerr. Boston: Beacon Press.

Blair, Sampson Lee and Michael P. Johnson. 1992. "Wives' Perceptions of the Fairness of the Division of Household Labor: The Intersection of Housework and Ideology." *Journal of Marriage and the Family* 54:570–81.

Blair, Sampson Lee, and Daniel T. Lichter. 1991. "Measuring the Division of Household Labor: Gender Segregation of Housework among American Couples." *Journal of Family Issues* 12:91–113.

Blair, Sampson Lee. 1993. "Employment, Family, and Perceptions of Marital Quality among Husbands and Wives." *Journal of Family Issues* 14:189–212.

Blaisure, Karen R., and Katherine R. Allen. 1995. "Feminists and the Ideology and Practice of Marital Equality." *Journal of Marriage and the Family* 57:5–19.

Block, M. R., and J. D. Sinnott. 1979. *The Battered Elder Syndrome: An Exploratory Study.* College Park: University of Maryland, Center on Aging.

Blood, Robert O. Jr., and Donald M. Wolfe. 1960. *Husbands and Wives.* New York: Free Press.

Bloom-Feshbach, Sally, Jonathan Bloom-Feshbach, and Kirby A. Heller. 1982. "Work, Family, and Children's Perceptions of the World." In *Families That Work: Children in a Changing World,* edited by Sheila B. Kamerman and Cheryl S. Hayes. Washington, DC: National Academy Press.

Blumberg, Rae. 1976. "Kibbutz Women: From the Fields of Revolution to the Laundries of Discontent." In *Women in the World: A Comparative Study,* edited by Lynne Iglitzin and Ruth Ross. Oxford: ABC Clio.

Blumberg, Rae. 1978. *Stratification: Socioeconomic and Sexual Inequality.* Dubuque, Iowa: William C. Brown.

Blumberg, Rae. 1984. "A General Theory of Gender Stratification." In *Sociological Theory 1984,* edited by Randall Collins. San Francisco, CA: Jossey-Bass.

Blumstein, Philip, and Pepper Schwartz. 1976. "Bisexual Women." In *The Social Psychology of Sex,* edited by Josephine P. Wiseman. New York: Harper & Row.

Blumstein, Philip, and Pepper Schwartz. 1983. *American Couples: Money/Work/Sex.* New York: William Morrow.

Blumstein, Phillip, and Peter Kollock. 1988. "Personal Relationships." *Annual Review of Sociology* 14:467–90.

Bogenschneider, Karen, Ming-Yeh Wu, Marcela Raffaelli, and Jenner C. Tsay. 1998. "'Other Teens Drink, but Not My Kid': Does Parental awareness of Adolescent Alcohol Use Protect Adolescents from Risky Consequences?" *Journal of Marriage and the Family* 60:356–73.

Bogenschneider, Karen. 2000. "Has Family Policy Come of Age? A Decade Review of the State of U.S. Family Policy in the 1990s." *Journal of Marriage and the Family* 62.

Bograd, M. 1984. "Family Systems Approaches to Wife Battering: A Feminist Critique," *American Journal of Orthopsychiatry* 54:558–68.

Boserup, Ester. 1970. *The Role of Women in Economic Development.* New York: St. Martin's Press.

Bott, Elizabeth. 1957. *Family and Social Network.* London: Tavistock.

Boulding, Elise. 1976. *The Underside of History.* Boulder, CO: Westview Press.

Bourdieu, Pierre. 1984. *Distinction: A Social Critique of the Judgement of Taste.* Richard Nice, trans. Cambridge, MA: Harvard University Press.

Bowlby, John. 1953. *Child Care and the Growth of Love.* Abridged and edited by M. Fry. Baltimore, MD: Penguin Books.

Bowman, Madonna E., and Constance R. Ahrons. 1985. "Impact of Legal Custody Status on Fathers' Parenting Postdivorce. *Journal of Marriage and the Family* 47:481–88.

Braver, Sanford L., with Diane O'Connell. 1998. *Divorced Dads: Shattering the Myths.* New York: Jeremy P. Tarcher/Putnam.

Braverman, Lois. 1991. "The Dilemma of Housework: A Feminist Response to Gottman, Napier, and Pittman." *Journal of Marital and Family Therapy* 17(1): 25–28.

Brayfield, April A. 1992. "Employment Resources and Housework in Canada." *Journal of Marriage and the Family* 54:19–30.

Brines, Julie, and Kara Joyner. 1999. "The Ties That Bind: Principles of Cohesion in Cohabitation and Marriage." *American Sociological Review* 64:333–55.

Brines, Julie. 1994. "Economic Dependency, Gender, and the Division of Labor at Home." *American Journal of Sociology* 100:652–88.

Broman, Clifford L. 1988. "Household Work and Family Life Satisfaction of Blacks." *Journal of Marriage and the Family* 550:45–51.

Brooks-Gunn, Jeanne, and P. Lindsay Chase-Lansdale. 1995. "Adolescent Parenthood." Pp. 113–49 in *Handbook of Parenting*, vol. 3, edited by Marc H. Bornstein. Mahwah, NJ: Lawrence Erlbaum.

Broverman, I., D. Broverman, F. Clarkson, P. Rosenkrantz, and S. Vogel. 1970. "Sex-Role Stereotypes and Clinical Judgements of Mental Health." *Journal of Consulting and Clinical Psychology* 34:1–7.

Brown, J. 1975. "Iroquois Women: An Ethnohistorical Note." Pp. 235–51 in *Toward An Anthropology of Women*, edited by R. R. Reiter. New York: Monthly Review.

Bruce, C. H. 1994. "Elder Abuse." *Journal of the American Academy of Physician Assistants* 7:170–4.

Brush, Lisa. 1990. "Violent Acts and Injurious Outcomes in Married Couples: Methodological Issues in the National Survey of Families and Households." *Gender & Society* 4:56–67.

Bullough, Vern. 1974. *The Subordinate Sex. A History of Attitudes towards Women.* Baltimore, MD: Penguin.

Bumpass, Larry L. 1990. "What's Happening to the Family? Intersection between Demographic and Institutional Change." *Demography* 27:484–98.

Bumpass, Larry L., and James A. Sweet. 1972. "Differentials in Marital Instability." *American Sociological Review* 37:754–67.

Bumpass, Larry L., R. Kelly Raley, and James A. Sweet. 1995. "The Changing Character of Stepfamilies: Implications of Cohabitation and Nonmarital Childbearing." *Demography* 32:425–36.

Bumpass, Larry, James A. Sweet, and Andrew Cherlin. 1991. "The Role of Cohabitation in Declining Rates of Marriage." *Journal of Marriage and the Family* 53:913–27.

Bumpass, Larry, James A. Sweet, and Teresa Castro Martin. 1990. "Changing Patterns of Remarriage." *Journal of Marriage and the Family* 52:747–56.

Burgess, Ernest W., and Paul Wallin. 1953. *Engagement and Marriage.* Philadelphia, PA: Lippincott.

Burgess, Ernest, and Leonard Cottrell. 1939. *Predicting Success or Failure in Marriage.* New York: Prentice Hall.

Burgess, Norma J. 1995. "Female-Headed Households in Sociohistorical Context." In *African American Single Mothers: Understanding Their Lives and Families,* edited by B. J. Dikerson. Thousand Oaks, CA: Sage.

Burgoyne, Carole B., and Victoria Morison. 1997. "Money in Remarriage: Keeping Things Simple and Separate." *Sociological Review* 45:363–95.

Buriel, Raymond, and Terri De Ment. 1997. "Immigration and Sociocultural Change in Mexican, Chinese, and Vietnamese American Families." Pp. 165–200 in *Immigration and the Family: Research and Policy of U.S. Immigrants,* edited by A. Booth, A. C. Crouter, and N. Landale. Mahwah, NJ: Lawrence Erlbaum.

Burt, Martha R. 1980. "Cultural Myths and Supports for Rape." *Journal of Personality and Social Psychology* 38(2):217–30.

Buss, D. M. 1989. "Sex Differences in Human Mate Preference: Evolutionary Hypothesis Tested in 37 Cultures." *Behavioral and Brain Sciences* 12:1–49.

Buss, David M. 1998. "Sexual Strategies Theory: Historical Origins and Current Status." *The Journal of Sex Research* 35:19–31.

Cahill, Spencer. 1986. "Childhood Socialization As a Recruitment Process: Some Lessons from the Study of Gender Development." *Sociological Studies of Child Development* 1:163–86.

Cahill, Spencer. 1989. "Fashioning Males and Females: Appearance Management and the Social Reproduction of Gender." *Symbolic Interaction* 12(2):281–98.

Caldera, Y. M., A. C. Huston, and M. O'Brien. 1989. "Social Interactions and Play Patterns of Parents and Toddlers with Feminine, Masculine, and Neutral Toys." *Child Development* 60:70–76.

Calderone, Mary S. 1958. *Abortion in the United States.* New York: Harper and Row.

California Center for Health Improvement. 1998. *Children and Youth Survey: Early Infant/Child Development.* Conducted by the Field Institute (July).

Call, Vaughn, Susan Sprecher, and Pepper Schwartz. 1995. "The Incidence and Frequency of Marital Sex in a National Sample." *Journal of Marriage and the Family* 57:639–50.

Cameron, S., and A. Collins. 1998. "Sex Differences in Stipulated Preferences in Personal Advertisements." *Psychological Reports* 82:119–23.

Campbell, Elaine. 1985. *The Childless Marriage: An Exploratory Study of Couples Who Do Not Want Children.* London: Tavistock.

Campbell, Jacqueline C., Linda Rose, Joan Kub, and Daphne Nedd. 1998. "Voices of Strength and Resistance: A Contextual and Longitudinal Analysis of Women's Responses to Battering." *Journal of Interpersonal Violence* 13(6):743–62.

Campbell, Jacqueline C., Paul Miller, Mary M. Cardwell, and Ruth Ann Belknap. 1994. "Relationship Status of Battered Women over Time." *Journal of Family Violence* 9(2):99–111.

Cancian, Francesca, and Stacey Oliker. 2000. *Caring and Gender.* Thousand Oak, CA: Pine Forge Press.

Cancian, Francesca. 1993. "Gender Politics: Love and Power in the Private and Public Spheres." In *Family Patterns: Gender Relations,* edited by Bonnie J. Fox. Toronto: Oxford University Press.

Cancian, Francesca. 1985. "Marital Conflict over Intimacy." In *The Psychosocial Interior of the Family,* edited by G. Handel. New York: Aldine.

Cancian, Francesca. 1987. *Love in America: Gender and Self-Development.* New York: Cambridge University Press.

Carp, Frances Merchant. 2000. *Elder Abuse in the Family.* New York: Springer.

Carter, Hugh, and Paul C. Glick. 1976. *Marriage and Divorce: A Social and Economic Study.* Cambridge, MA: Harvard University Press.

Casper, Lynne M. 1995. "What Does It Cost to Mind Our Preschoolers?" *Current Population Reports:* Household Economic Studies, Series P-70, No. 52. Washington, DC: U.S. Bureau of the Census.

Casper, Lynne M. 1996. "Who's Minding Our Preschoolers?" *Current Population Reports:* Household Economic Studies, Series P-70, No. 53. Washington, DC: U.S. Bureau of the Census.

Casper, Lynne M. 1998. "Who's Minding Our Preschoolers?" Fall 1994 (Update). *Current Population Reports:* Household Economic Studies, Series P-70. Washington, DC: United States Bureau of the Census. http://www.census.gov.

Catalano, Ralph. 1991. "The Health Effects of Economic Insecurity." *American Journal of Public Health* 81: 1148–52.

Cates, Willard, Jr. 1982. "Legal Abortion: The Public Health Record." *Science* 215:1586–90.

Centers for Disease Control and Prevention. 2000. *Dating Violence.* National Center for Injury Prevention and Control. http://www.cdc.gov/ncipc/factsheets/datviol.htm

Chadwick, Bruce, and Tim Heaton. 1992. *Statistical Handbook on the American Family.* Phoenix, AZ: Oryx Press.

Chafetz, Janet S. 1984. *Sex and Advantage: A Comparative, Macro-Structural Theory of Sex Stratification.* Totowa, NJ: Rowman and Allanheld.

Chafetz, Janet S. 1990. *Gender Equity: An Integrated Theory of Stability and Change.* Newbury Park, CA: Sage.

Chafetz, Janet, and A. Gary Dworkin. 1986. *Female Revolt: Women's Movements in World and Historical Perspective.* Totowa, NJ: Rowman and Allanheld.

Chafetz, Janet, and A. Gary Dworkin. 1987. "In the Face of Threat: Organized Antifeminism in Comparative Perspective." *Gender and Society* 1:33–60.

Chafetz, Janet, and A. Gary Dworkin. 1989. "Action and Reaction: An Integrated, Comparative Perspective on Feminist and Antifeminist Movements." In *Cross-National Research in Sociology,* edited by Melvin Kohn. Newbury Park, CA: Sage.

Chao, Ruth K. 1994. "Beyond Parental Control and Authoritarian Parenting Style: Understanding Chinese Parenting Tthrough the Cultural Notion of Training." *Child Development* 65:1111–19.

Cherlin, Andrew J. 1978. "Remarriage As an Incomplete Institution." *American Journal of Sociology* 84:634–50.

Cherlin, Andrew J. 1992. *Marriage, Divorce, Remarriage.* 2d ed. Cambridge, MA: Harvard University Press.

Cherlin, Andrew J. 1997. "What's Most Important in a Family Textbook?" *Family Relations* 46:209–11.

Cherlin, Andrew J., and Frank F. Furstenberg. 1994. "Stepfamilies in the United States: A Reconsideration." *Annual Review of Sociology* 20:359–81.

Cherlin, Andrew J., and Frank F. Furstenburg. 1986. *The New American Grandparent.* New York: Basic Books.

Cherlin, Andrew, Frank Furstenberg, P. Lindsay Chase-Landsdale, Kathleen Kiernan, Philip Robins, Donna Morrison, and Julien Teitler. 1991. "Longitudinal Studies of Effects of Divorce on Children in Great Britain and the United States." *Science* 252:1386–89.

Chesney, Kellow. 1970. *The Victorian Underworld.* New York: Schocken Books.

Child, Meredith, Kathryn Graff, Dheryl McDonnell McCormick, and Andrew

Cocciarella. 1996. "Personal Advertisements of Male-to-Female Transsexuals, Homosexual Men, and Heterosexuals." *Sex Roles* 34:447–56.

Children's Defense Fund. 1994. *Wasting America's Future.* Boston, MA: Beacon Press.

Children's Defense Fund. 1997. *The State of America's Children: Yearbook 1997.* Washington, DC: Children's Defense Fund.

Children's Defense Fund. 2000. "Polls Indicate Support for Child Care." www.chilrensdefense.org.

Chilman, Catherine Street. 1993. "Hispanic Families in the United States." Pp. 141–64 in *Family Ethnicity: Strength in Diversity*, edited by H. P. McAdoo, Thousand Oaks, CA: Sage.

Chodorow, Nancy, and Susan Contratto. 1992. "The Fantasy of the Perfect Mother." Pp. 191–214 in *Rethinking the Family: Some Feminist Questions*, edited by B. Thorne with M. Yalom. Boston, MA: Northeastern University Press.

Chodorow, Nancy J. 1976. "Oedipal Asymmetries and Heterosexual Knots." *Social Problems* 23:454–67.

Chodorow, Nancy J. 1978. *The Reproduction of Mothering.* Berkeley: University of California Press.

Chodorow, Nancy. 1989. *Feminism and Psychoanalytic Theory.* New Haven, CT: Yale University Press.

Christopher, F. Scott, and Susan Sprecher. 2000. "Sexuality in Marriage, Dating, and Other Relationships: A Decade Review." *Journal of Marriage and the Family* 62.

Ciccheti, D., and S. L. Toth. 1998. "Perspectives on Research and Practice in Developmental Psychopathology." Pp. 479–583 in *Handbook of Child Psychology*, vol. 4, edited by W. Damon, I. E. Sigel, and K. Renninger. New York: John Wiley.

Cicerello, Antoinette, and Eugene P. Sheehan. 1995. "Personal Advertisements: A Content Analysis." *Journal of Social Behavior and Personality* 10:751–56.

Clark, M. 1959. *Health in the Mexican-American Culture.* Berkeley: University of California Press.

Clarke, Sally Cunningham, and Barbara Foley Wilson. 1994. "The Relative Stability of Remarriages: A Cohort Approach Using Vital Statistics." *Family Relations* 43:305–10.

Clignet, Remi. 1970. *Many Wives, Many Powers.* Evanston, IL: Northwestern University Press.

Cohen, Susan A. 1998. "Issues and Implications: 25 Years after Roe: New Technological Parameters for an Old Debate." *The Guttmacher Report* 1(1).

Coleman, Marilyn, and Lawrence H. Ganong. 1987. "The Cultural Stereotyping of Stepfamilies." In *Remarriage and Stepparenting*, edited by Kay Pasley and Marilyn Ihinger-Tallman. New York: Guilford.

Coleman, Marilyn, Lawrence Ganong, and Mark Fine. 2000. "Remarriage and Stepfamilies." *Journal of Marriage and the Family* 62.

Collins, Randall. 1971. "A Conflict Theory of Sexual Stratification." *Social Problems* 19(Summer):3–21.

Collins, Randall. 1979. *Conflict Sociology: Toward an Explanatory Science.* New York: Academic Press.

Collins, Randall. 1979. *The Credential Society. An Historical Sociology of Education and Stratification.* New York: Academic Press.

Collins, Randall. 1981. "Three Faces of Cruelty: Towards a Comparative Sociology of Violence." In *Sociology Since Midcentury*, edited by Randall Collins. New York: Academic Press.

Collins, Randall. 1986a. "Weber's Theory of the Family." In *Weberian Sociological Theory.* New York: Cambridge University Press.

Collins, Randall. 1986b. "Courtly Politics and the Status of Women." In *Weberian Sociological Theory.* New York: Cambridge University Press.

Collins, Randall. 1988. *Theoretical Sociology.* San Diego, CA: Harcourt-Brace Jovanovich.

Coltrane, Scott, and Elsa Valdez. 1993. "Reluctant Compliance: Work/Family Role Allocation in Dual-Earner Chicano Families." pp. 151–75 in *Men, Work and Family*, edited by Jane C. Hood. Newbury Park, CA: Sage.

Coltrane, Scott, and Kenneth Allan. 1994. "'New' Fathers and Old Stereotypes: Representations of Masculinity in 1980s Television Advertising." *Masculinities* 2(4):43–66.

Coltrane, Scott, and Michele Adams. 1997. "Children and Gender." Pp. 219–53 in *Contemporary Parenting: Challenges and Issues*, edited by T. Arendell. Thousand Oaks, CA: Sage.

Coltrane, Scott, and Michele Adams. 1997. "Work—Family Imagery and Gender Stereotypes: Television and the Reproduction of Difference." *Journal of Vocational Behavior* 50(2):323–47.

Coltrane, Scott, and Michele Adams. 2000. "Men, Women, and Housework" In *Gender Mosaics: Social Perspectives*, edited by D. Vannoy. Los Angeles, CA: Roxbury Press.

Coltrane, Scott. 1988. "Father-Child Relationships and the Status of Women." *American Journal of Sociology* 93:1060–95.

Coltrane, Scott. 1989. "Household Labor and the Routine Production of Gender." *Social Problems* 36:473–90.

Coltrane, Scott. 1992. "The Micro-Politics of Gender in Non-industrial Societies." *Gender & Society* 6:86–107.

Coltrane, Scott. 1996. *Family Man: Fatherhood, Housework, and Gender Equity.* New York: Oxford University Press.

Coltrane, Scott. 1997. "Scientific Half-Truths and Postmodern Parody in the Family Values Debate." *Contemporary Sociology* 27(1):7–10.

Coltrane, Scott. 1998. *Gender and Families.* Thousand Oaks, CA: Pine Forge Press.

Coltrane, Scott. 2000. "Modeling and Measuring the Social Embeddedness of Routine Family Labor." *Journal of Marriage and the Family* 62.

Conger, Rand D., and Glen H. Elder, Jr., Frederick O. Lorenz, Katherine J. Conger, Ronald L. Simons, Les B. Whitbeck, Shirley Huck, and Janet N. Melby. 1990. "Linking Economic Hardship to Marital Quality and Instability." *Journal of Marriage and the Family* 52:643–56.

Conger, Rand D., and Glen H. Elder, Jr. 1994. *Families in Troubled Times: Adapting to Change in Rural America.* New York: Aldine de Gruyter.

Conger, Rand D., Katherine J. Conger, Glen H. Elder, Jr., Frederick O. Lorenz, Ronald L. Simons, and Les B. Whitbeck. 1992. "A Family Process Model of Economic Hardship and Adjustment of Early Adolescent Boys." *Child Development* 63(3):526–41.

Connell, R.W. 1995. *Masculinities.* Berkeley: University of California Press.

Consortium for Longitudinal Studies. 1983. *As the Twig Is Bent … Lasting Effects of Preschool Programs.* Hillsdale, NJ: Erlbaum.

Constantine, Larry L., and Joan M. Constantine. 1973. *Group Marriage.* New York: Macmillan.

Coontz, Stephanie. 1992. *The Way We Never Were: American Families and the Nostalgia Trap.* New York: Basic Books.

Coontz, Stephanie. 1997. *The Way We Really Are: Coming to Terms with America's Changing Families.* New York: Basic Books.

Cooper, John M. 1946. "The Patagonian and Pampean Hunters." In *Handbook of South American Indians*, edited by Julian Steward. Washington, DC: Smithsonian Institution.

Coser, Rose L. 1964. "Authority and Structural Ambivalence in the Middle Class Family." In *The Family: Its Structure and Functions.* New York: St. Martin's Press.

Cott, Nancy F. 1978. "Passionlessness: An Interpretation of Victorian Sexual Ideology, 1790–1850." *Signs* 4:219–36.

Cott, Nancy. 1977. *The Bonds of Womanhood.* New Haven: Yale University Press.

Council of Economic Advisors. 1998. Changing America: Indicators of Social and Economic Well-Being by Race and Hispanic Origin. Washington,

DC: President of the United States (http://www.ehouse.gov /WH/EOP/ CEA/html/publications.html).

Cowan, Carolyn Pape, and Philip A. Cowan. 2000. *When Partners Become Parents: The Big Life Change for Couples.* Mahwah, NJ: Lawrence Erlbaum Associates.

Cowan, G., and C. Hoffman. 1986. "Gender Stereotyping in Young Children: Evidence to Support a Concept-Learning Model." *Sex Roles* 14:211–24.

Cowan, Philip A. 1993. "The Sky IS Falling, but Popenoe's Analysis Won't Help Us Do Anything About It." *Journal of Marriage and the Family* 55:548–53.

Cowan, Ruth S. 1983. *More Work for Mother. The Ironies of Household Technology from the Open Hearth to The Microwave.* New York: Basic Books.

Cox, Martha J., Blair Paley, Margaret Burchinal, and C. Chris Payne. 1999. "Marital Perceptions and Interactions across the Transition to Parenthood." *Journal of Marriage and the Family* 61(3):611.

Coysh, William S., Janet R. Johnston, Jeanne M. Tschann, Judith S. Wallerstein, and Marsha Kline. 1989. "Parental Postdivorce Adjustment in Joint and Sole Physical Custody Families." *Journal of Family Issues* 10(1):52–71.

Creighton-Zollar, Ann, and J. Sherwood Williams. 1992. "The Relative Educational Attainment and Occupational Prestige of Black Spouses and Life Satisfaction." *The Western Journal of Black Studies* 16:57–63.

Critelli, J., and L. R. Waid. 1980. "Physical Attractiveness, Romantic Love, and Equity Restoration in Dating Relationships." *Journal of Personality Assessment* 44:624–29.

Crnic, Keith A., and Cathryn L. Booth. 1991. "Mothers' and Fathers' Perceptions of Daily Hassles of Parenting across Early Childhood." *Journal of Marriage and the Family* 53(4):1042–50.

Crockett, Lisa J., and Ann C. Crouter, eds. 1995. *Pathways through Adolescence: Individual Development in Relation to Social Contexts.* Mahwah, NJ: Lawrence Erlbaum.

Crouter, Ann C., Susan M. McHale, and W. Todd Bartko. 1993. "Gender As an Organizing Feature in Parent–Child Relationships." *Journal of Social Issues* 49:161–74.

Cuber, John F., and Peggy B. Harroff. 1965. *The Significant Americans: A Study of Sexual Behavior among the Affluent.* Baltimore, MD: Penguin Books.

Currie, Elliott. 1985. *Confronting Crime: An American Challenge.* New York: Pantheon.

Curtin, Sally C., and Melissa M. Park. 1999. "Trends in the Attendant, Place, and Timing of Births, and in the Use of Obstetric Interventions: United States, 1989–97." *National Vital Statistics Reports* 47(27).

D'Emilio, John, and Estelle Freedman. 1988. *Intimate Matters: A History of Sexuality in America.* New York: Harper and Row.

Daley, Suzanne. 2000. "Morning-after Pills Will Be Free to French Schoolgirls | The Goal Is to Reduce Unwanted Pregnancies." *San Diego Union-Tribune,* February 8, p. A-11.

Daly, Kerry J. 1996. *Families and Time: Keeping Pace in a Hurried Culture.* Newbury Park: Sage.

Daniels, Arlene K. 1987. "The Hidden Work of Constructing Class and Community: Women Volunteer Leaders in Social Philanthropy." In *Families and Work,* edited by Naomi Gerstel and Harriet Gross. Philadelphia, PA: Temple University Press.

Daniels, Arlene K. 1988. *Invisible Careers.* Chicago: University of Chicago Press.

Daniels, Pamela, and Kathy Weingarten. 1982. *Sooner or Later: The Timing of Parenthood in Adult Lives.* New York: Norton.

Davidson, T. 1977. "Wifebeating: A Recurring Phenomenon Throughout History." In *Battered Women: A Psychological Study of Domestic Violence,* edited by M. Roy. New York: Van Nostrand Reinhold.

Davis, Kingsley. 1936. "Jealousy and Sexual Property." *Social Forces* 14:395–405.

Davis, Kingsley. 1949. *Human Society.* New York: Macmillan.

Davis, Simon. 1990. "Men As Success Objects and Women As Sex Objects: A Study of Personal Advertisements." *Sex Roles* 23:43–50.

Deaux, Kay and R. Hanna. 1984. "Courtship in the Personals Column: The Influence of Gender and Sexual Orientation." *Sex Roles* 19:131–42.

Degler, Carl N. 1980. *At Odds: Women and the Family in America from the Revolution to the Present.* New York: Oxford University Press.

DeMaris, Alfred, and Monica A. Longmore. 1996. "Ideology, Power, and Equity: Testing Competing Explanations for the Perception of Fairness in Household Labor." *Social Forces* 74:1043–71.

Demo, David H. 2000. "Families with Young Children: A Review of Research in the 1990s." *Journal of Marriage and the Family* 62.

Demo, David, and Alan Acock. 1991. "The Impact of Divorce on Children." pp. 162–91 in *Contemporary Families: Looking Forward, Looking Back,* edited by Alan Booth. Minneapolis, MN: National Council on Family Relations.

Dengler, Ian C. 1978. "Turkish Women in the Ottoman Empire: The Classical Age." In *Women in the Moslem World,* edited by Lois Beck and Nikki Keddie. Cambridge, MA: Harvard University Press.

Dennehy, K. and J. Mortimer. 1992. "Work and Family Orientations of Contemporary Adolescent Boys and Girls in a Context of Social Change." Paper presented at the 87th Annual Meeting of the American Sociological Association, Pittsburgh, PA.

Dennis, Wendy. 1992. *Hot and Bothered: Sex and Love in the Nineties.* New York: Viking/Penguin.

Derdeyn, A., and E. Scott. 1984. "Joint Custody: A Critical Analysis and Appraisal." *American Journal of Orthopsychiatry* 54(Apr.):199–209.

Deutsch, Francine. 1999. *Halving It All: How Equally Shared Parenting Works.* Cambridge, MA: Harvard University Press.

DeVault, Marjorie L. 1987. "Doing Housework: Feeding and Family Life." In *Families and Work,* edited by Naomi Gerstel and Harriet E. Gross. Philadelphia, PA: Temple University Press.

DeVita, Carol. 1996. "The United States at Mid-Decade." *Population Bulletin* 50:4. Washington, DC: Population Reference Bureau.

Diamond, Stanley. 1996. "Dahomey: The Development of a Proto-State." *Dialectical Anthropology* 21:121–216.

Dickens, A. G. 1977. *The Courts of Europe, Politics, Patronage and Royalty, 1400–1800.* New York: McGraw-Hill.

Dikotter, Frank. 1998. "Race Culture: Recent Perspectives on the History of Eugenics." *American Historical Review:* 467–78.

Dill, Bonnie Thornton. 1988. "Our Mother's Grief: Racial Ethnic Women and the Maintenance of Families." *Journal of Family History* 13:415–31.

Dill, Bonnie Thornton. 1994. *Across the Boundaries of Race and Class.* New York: Garland.

DiMaggio, Paul, and John Mohr. 1985. "Cultural Capital, Educational Attainment, and Marital Selection." *American Journal of Sociology* 90:1231–61.

Divale, William T. 1984. "Migration, External Warfare and Matrilocal Residence." *Behavior Science Research* 9:75–133.

Divale, William T., and Marvin Harris. 1976. "Population, Warfare, and the Male Supremacist Complex." *American Anthropologist* 78:521–38.

Dix, Dorothy (Elizabeth M. Gilmer). [1939] 1974. *How to Win and Hold a Husband.* New York: Arno/New York Times/ Doubleday.

Dobash, R. Emerson, and Russell Dobash. 1979. *Violence against Wives.* New York: Free Press.

Dobash, Russell, R. Emerson Dobash, Margo Wilson, and Martin Daly. 1992. "The Myth of Sexual Symmetry in Marital Violence." *Social Problems* 39:71–91.

Dobson, James C. 1970. *Dare to Discipline.* Wheaton, IL: Tyndale.

Dobson, James C. 1988. *The Strong-Willed Child*. Wheaton, IL: Tyndale.

Dobyns, Henry F. 1983. *Their Numbers Become Thinned: Native American Population Dynamics in Eastern North America*. Knoxville, TN: University of Tennessee Press.

Dodson, Jualynne Elizabeth. 1997. "Conceptualizations of African American Families." Pp. 323–32 in *Black Families*, 3d ed., edited by H. P. McAdoo. Thousand Oaks, CA: Sage.

Domhoff, G. William. 1990. *The Power Elite and the State: How Policy Is Made in America*. New York: Aldine de Gruyter.

Dorkenoo, Efua. 1994. *Cutting the Rose: Female Genital Mutilation: The Practice and its Prevention*. London, UK: Minority Rights Group.

Dortch, Shannon. 1995. "The Future of Kinship." *American Demographics*.

Dover, K. J. 1978. *Greek Homosexuality*. Cambridge, MA: Harvard University Press.

Drake, St. Clair, and Horace R. Cayton. 1945. *Black Metropolis: A Study of Negro Life in a Northern City*. New York: Harcourt and Brace.

Draper, Patricia. 1975. "Kung Women: Contrasts in Sexual Egalitarianism in Foraging and Sedentary Contexts." In *Toward an Anthropology of Women,* edited by R. Reiter. New York: Monthly Review Press.

Driver, Harold E. 1969. *Indians of North America*. 2nd ed. Chicago: University of Chicago Press.

DuBois, W. E. B. 1899 [1973]. *The Philadelphia Negro: A Social Study*. Millwood, NY: Kraus-Thomson.

Dugan, James C. 1993. "The Conflict between 'Disabling' and 'Enabling' Paradigms in Law." *Cornell Law Review* 78:507–42.

Duncan, Greg J. et al. 1984. *Years of Plenty: The Changing Economic Fortunes of American Workers and Families*. Ann Arbor: Institute for Social Research, The University of Michigan.

Duncan, Greg J., and Saul D. Hoffman. 1985. "A Reconsideration of the Economic Consequences of Divorce." *Demography* 22:485–97.

Duncan, Greg J., Martha S. Hill, and Saul D. Hoffman. 1988. "Welfare Dependence within and across Generations." *Science* 239(Jan. 29):467–71.

Durkheim, Emile. 1893/1947. *The Division of Labor in Society*. New York: Free Press.

Durkheim, Emile. 1912/1954. *The Elementary Forms of the Religious Life*. New York: Free Press.

Dutton, Donald G., with Susan K. Golant. 1995. *The Batterer: A Psychological Profile*. New York: Basic Books.

Dworkin, Andrea. 1987. *Intercourse*. New York: Free Press.

Eagly, Alice Hendrickson. 1987. *Sex Differences in Social Behavior: A Social-Role Interpretation*. Hillsdale, NJ: L. Erlbaum Associates.

Easterlin, Richard. 1980. *Birth and Fortune: The Impact of Numbers on Personal Welfare*. New York: Basic Books.

Eckland, B. K. 1979. "Genetic Variance in the SES-IQ Correlation." *Sociology of Education* 52:191–6.

Edelman, Marian Wright. 1997. "An Advocacy Agenda for Black Families and Children." Pp. 9–40 in *Black Families*, 3d ed., edited by H. P. McAdoo. Thousand Oaks, CA: Sage.

Edmonson, Brad. 1997. "Asian Americans in 2001." *American Demographics* 19(2).

Edwards, J. H., and Alan Booth. 1994. "Sexuality, Marriage, and Well-Being: The Middle Years." In *Sexuality across the Life Course*, edited by Alice S. Rossi. Chicago: University of Chicago Press.

Ehrenreich, Barbara, and Deidre English. 1978. *Witches, Midwives, and Nurses*. Old Westbury, NY: Feminist Press.

Ehrensaft, Diane. 1987. *Parenting Together*. New York: Free Press.

Eichler, Margrit. 1997. *Family Shifts: Families, Policies, and Gender Equality*. Toronto: Oxford University Press.

Eitzen, D. Stanley. 1983. *Social Problems*. 2d ed. Boston, MA: Allyn and Bacon.

Elder, Glen H., and Avshalom Caspi. 1988. "Economic Stress in Lives: Developmental Perspectives." *Journal of Social Issues* 44:25–45.

Elder, Glen H., Jr. 1974. *Children of the Great Depression*. Chicago: University of Chicago Press.

Elder, Glen H., Jr. 1986. "Military Times and Turning Points in Men's Lives." *Developmental Psychology* 22:233–45.

Elder, Glen H., Jr. 1998. "The Life Course As Developmental Theory." *Child Development* 69:1–12.

Elder, Glen H., Jr., A. Caspi, and L. M. Burton. 1985. "Adolescent Transitions in Developmental Perspective: Sociological and Historical Insights." In *Minnesota Symposium on Child Psychology,* edited by M. Gunnar. vol. 21. Hillsdale: NJ: Erlbaum.

Elder, Glen H., Jr., John Modell, and Ross D. Parke. 1993. *Children in Time and Place: Developmental and Historical Insights*. New York: Cambridge University Press.

Elison, Sonja Klueck. 1997. "Policy Innovation in a Cold Climate: the Family and Medical Leave Act of 1993." *Journal of Family Issues* 18(1):30–54.

Elliot, P., ed. 1990. *Confronting Lesbian Battering: A Manual for the Battered Women's Movement*. St. Paul, MN: Minnesota Coalition for Battered Women.

Ellwood, David, and Lawrence H. Summers. 1986. "Poverty in America: Is Welfare the Answer or the Problem?" In *Fighting Poverty: What Works and What Doesn't,* edited by Sheldon H. Danzinger and Daniel H. Weinberg. Cambridge, MA: Harvard University Press.

Ellwood, David, and Mary Jo Bane. 1985. "The Impact of AFDC on Family Structure and Living Arrangements." In *Research in Labor Economics* 7, edited by Ronald Ehrenberg. Greenwich, CT: JAI Press.

Elmer-Dewitt, P. 1991. "Making Babies." *Time* (September 30):56–63.

Ember, Melvin, and Carol Ember. 1971. "The Conditions Favoring Matrilocal vs. Patrilocal Residence." *American Anthropologist* 73:571–94.

Eng, P. 1995. "Domestic Violence in Asian/Pacific Island Communities." Pp. 78–88 in *Health Issues for Women of Color*, edited by D. L. Adams. Thousand Oaks, CA: Sage.

Engels, Friedrich, 1884/1972. *The Origin of the Family, Private Property and the State*. New York: International Publishers.

England, Paula, and Irene Browne. 1992. "Trends in Women's Economic Status." *Sociological Perspectives* 35:17–51.

England, Paula. 1989. "A Feminist Critique of Rational Choice Theories." *American Sociologist* 19.

Enkin, Murray, Marc J. N. C. Keirse, and Iain Chalmers, eds. 1989. *A Guide to Effective Care in Pregnancy and Childbirth*. Oxford: Oxford University Press.

Entwisle, Doris R., and Susan G. Doering. 1981. *The First Birth: A Family Turning Point*. Baltimore, MD: Johns Hopkins University Press.

Epstein, Cynthia F. 1988. *Deceptive Distinctions: Sex, Gender, and the Social Order*. New Haven: Yale University Press.

Eriksen, Julia A., with Sally Stefen. 1999. *Kiss and Tell*. Cambridge, MA: Harvard University Press.

Erikson, Erik H. 1959. "Identity and the Life Cycle." *Psychological Issues* 1(1).

Erikson, Erik H. 1982. *The Life Cycle Completed: A Revision*. New York: Norton.

Espiritu, Yen Le. 1992. *Asian American Panethnicity*. Philadelphia, PA: Temple University Press.

Espiritu, Yen Le. 1997. *Asian American Women and Men: Labor, Laws, and Love*. Thousand Oaks, CA: Pine Forge Press.

Etzioni, Amitai. 1993. "Children of the Universe." *Utne Reader*. Vol. 57.

Evans-Pritchard, E. E. 1951. *Kinship and Marriage among the Nuer*. London: Oxford University Press.

Fagan, Jeffrey A., Douglas K. Stewart, and Karen V. Hansen. 1983. "Violent Men or

Violent Husbands? Background Factors and Situational Correlates." In *The Dark Side of Families,* edited by David Finkelhor. Beverly Hills, CA: Sage.

Fagot, Beverly I., Mary D. Leinbach, and Cherie O'Boyle. 1992. "Gender Labeling, Gender Stereotyping, and Parenting Behaviors." Developmental Psychology 28:225–30.

Fagot, Beverly I., and Mary D. Leinbach. 1993. "Gender-Role Development in Young Children: From Discrimination to Labeling." *Developmental Review* 13:205–24.

Family Planning Perspectives. 1985. "Recently Wed Women More Likely to Have Had Premarital Sex." Vol. 17, Sept./Oct., p. 142.

Family Planning Perspectives. 1986. "Women in 30s Experiencing Record-High Divorce; Level Expected to Decline among Younger Women." 18(May/June):133–34.

Farkas, S., and J. Johnson, with A. Duffett and A. Bers. 1997. *Kids These Days: What Americans Really Think about the Next Generation.* New York: Public Agenda.

Farley, R., and A. J. Hermalin. 1971. "Family Stability: A Comparison of Trends between Blacks and Whites." *American Sociological Review* 36:207–22.

Farrell, Michael P., and Stanley D. Rosenberg. 1981. *Men at Mid-life.* Boston, MA: Auburn House.

Fausto-Sterling, Anne. 1985. *Myths of Gender: Biological Theories about Women and Men.* New York: Basic Books.

Fausto-Sterling, Anne. 1989. "Hormonal Hurricanes: Menstruation, Menopause, and Female Behavior." In *Feminist Frontiers II: Rethinking Sex, Gender, and Society,* edited by Laurel Richardson and Verta Taylor. New York: Random House.

Federal Bureau of Investigation. 1997. *Crime in the United States: 1996.* Washington, DC: U.S. Government Printing Office.

Feingold, A. 1982. "Physical Attractiveness and Romantic Involvement." *Psychological Reports* 50:802.

Feingold, Alan. 1992. "Good Looking People Are Not What We Think." *Psychological Bulletin* 111:304–41.

Ferraro, Kathleen J. 1984. "Widowhood and Social Participation in Later Life." *Research on Aging* 6:451–68.

Ferraro, Kathleen J. 1988. "An Existential Approach to Battering." In *Family Abuse and Its Consequences,* edited by Gerald T. Hotaling, David Finkelhor, John T. Kirkpatrick, and Murray A. Straus. Newbury Park, CA: Sage.

Ferraro, Kathleen J. 1997. "Battered Women: Strategies for Survival. Pp. 124–40 in *Violence among Intimate Partners: Patterns, Causes and Effects,* edited by A. Carderelli. New York: Macmillan.

Ferree, Myra M. 1987. "Family and Job for Working-Class Women: Gender and Class Systems Seen from Below." In *Families and Work,* edited by Naomi Gerstel and Harriet Gross. Philadelphia, PA: Temple University Press.

Ferree, Myra Marx. 1987. "She Works Hard for a Living." Pp. 322–47 in *Analyzing Gender,* edited by B. Hess and M. M. Ferree. Newbury Park, CA: Sage.

Ferree, Myra Marx. 1990. "Beyond Separate Spheres: Feminism and Family Research." *Journal of Marriage and the Family* 52:866–84.

Filinson, Rachel, 1989. "Introduction." In *Elder Abuse: Practice and Policy,* edited by Rachel Filinson and Stanley R. Ingman. New York: Human Sciences Press.

Fincham, Frank D., and Thomas N. Bradbury. 1993. "Marital Satisfaction, Depression, and Attributions: A Longitudinal Analysis." *Journal of Personality and Social Psychology* 64: 442–52.

Fincham, Frank D., Steven Beach, Gordon Harold, and Lori Osborne. 1997. "Marital Satisfaction and Depression: Different Causal Relationships for Men and Women?" *Psychological Science* 8:351–57.

Finck, Henry T. 1899. *Primitive Love and Love Stories.* New York: Charles Scribner's Sons.

Finkel, J. S., and F. J. Hansen. 1992. "Correlates of Retrospective Marital Satisfaction in Long-Lived Marriages: A Social Constructivist Perspective." *Family Therapy* 19:1–16.

Finkelhor, David, and Kersti Yllö. 1985. *License to Rape: Sexual Abuse of Wives.* New York: Holt, Rinehart, and Winston.

Finkelhor, David, Linda Williams, and Nancy Burns. 1988. *Nursery Crimes: Sexual Abuse in Day Care.* Newbury Park, CA: Sage Publications.

Finkelhor, David. 1983. "Common Features of Family Abuse." In *The Dark Side of Families,* edited by David Finkelhor, Richard J. Gelles, Gerald T. Hotaling, and Murray A. Strauss. Beverly Hills, CA: Sage.

Finkelhor, David. 1986. *A Sourcebook on Child Sexual Abuse.* Beverly Hills, CA: Sage.

Firestone, Shulamith. 1970. *The Dialectic of Sex.* New York: William Morrow.

Fischer, Claude S. 1982. *To Dwell among Friends: Personal Networks in Town and City.* Chicago, IL: University of Chicago Press.

Fischer, Claude S., Michael Hout, Martin Sanchez Jankowski, Ann Swidler (Contributor), Samuel R. Lucas (Contributor). 1996. *Inequality by Design: Cracking the Bell Curve Myth.* Princeton, NJ: Princton University Press, 1996.

Fisher, Wesley A. 1980. *The Soviet Marriage Market.* Cambridge, MA: Harvard University Press.

Fishman, P. 1978. "Interaction: The Work Women Do." *Social Problems* 25:398–406.

Fitzgerald, F. Scott. 1931/1956. "Echoes of the Jazz Age." In *The Crack-Up.* New York: New Directions.

Flacks, Richard. 1971. *Youth and Social Change.* Chicago: Markham.

Flandrin, Jean-Louis. 1979. *Families in Former Times: Kinship, Household and Sexuality.* Cambridge: Cambridge University Press.

Flango, Eugene, and Carol Flango. 1993. "Adoption Statistics by State." *Child Welfare* 72:311–19.

Flexner, Eleanor. 1959. *Century of Struggle: The Woman's Rights Movement in the United States.* Cambridge, MA: Harvard University Press.

Fontana, V. 1973. *Somewhere a Child is Crying: Maltreatment—Causes and Prevention.* New York: Macmillan.

Ford, Clellan S., and Frank A. Beach. 1951. *Patterns of Sexual Behavior.* New York: Harper & Row.

Foucault, Michel. 1978. *The History of Sexuality.* vol. 1. New York: Random House

Fowers, Blaine. 1991. "His and Her Marriage: A Multivariate Study of Gender and Marital Satisfaction." *Sex Roles* 24:209–21.

Frazier, E. Franklin. 1950. "Problems and Needs of Negro Children and Youth Resulting from Family Organization." *Journal of Negro Education* 19:276–77.

Frazier, Patricia A., and E. Esterly. 1990. "Correlates of Relationship Beliefs: Gender, Relationship Experience and Relationship Satisfaction." *Journal of Social and Personal Relationships* 7:331–52.

Freud, Sigmund. 1924. *A General Introduction to Psychoanalysis.* New York: Boni and Liveright.

Friedl, Ernestine. 1975. *Women and Men: An Anthropologist's View."* New York: Holt, Rinehart and Winston.

Furstenberg, Frank F. 2000. "The Sociology of Adolescence and Youth in the 1990s." *Journal of Marriage and the Family* 62.

Furstenberg, Frank F., Thomas D. Cook, Jacquelynne Eccles, Glen H. Elder, Jr., and Arnold J. Sameroff. 1999. *Managing to Make It: Urban Families and Adolescent Success.* Chicago: University of Chicago Press.

Furstenberg, Frank F, and Andrew Cherlin. 1991. *Divided Families.* Cambridge, MA: Harvard University Press.

Furstenberg, Frank F. 1987. "The New Extended Family: The Experience of Parents and Children after Remarriage." In *Remarriage and Stepparenting,* edited by Kay Pasley and Marilyn Ihinger-Tallman. New York: Guilford.

Furstenberg, Frank F., S. Philip Morgan, and Paul Allison. 1987. "Paternal Participation and Children's Well-Being after Marital Dissolution." *American Sociological Review* 52:695–701.

Fustel de Coulanges, Numa Denis. 1973. *The Ancient City.* Baltimore: Johns Hopkins University Press.

Gable, Sara, Jay Belsky, and Keith Crnic. 1995. "Coparenting during the Child's 2nd Year: A Descriptive Account." *Journal of Marriage and the Family* 57(3):609–16.

Gailey, Christine Ward. 2000. "Seeking Baby Right: Race, Class, and Gender in U.S. International Adoption." Pp. 52–81 in *Mine, Yours, Ours and Theirs: Adoption, Changing Kinship and Family Patterns*, edited by A. L. Rygvold, M. Dalen, and B. Saetersdal. Oslo, Norway: University of Oslo.

Gailey, Christine Ward. 1987. "Evolutionary Perspectives on Gender Hierarchy." Pp. 32–67 in *Analyzing Gender*, edited by B. Hess and M. M. Ferree. Newbury Park, CA: Sage.

Gallagher, Maggie. 1996. *The Abolition of Marriage: How We Destroy Lasting Love.* Washington, DC: Genery.

Gallup, George H. 1972. *The Gallup Poll: Public Opinion 1935–1971, Vols. 1 to 3.* New York: Random House.

Galston, William A. 1996. "Divorce American Style." *The Public Interest* 124:12–26.

Gans, Herbert. 1962. *The Urban Villagers.* New York: Free Press.

Gans, Herbert. 1967. *The Levittowners.* New York: Random House.

Gans, Herbert. 1974. *Popular Culture and High Culture.* New York: Basic Books.

Garbarino, James, and Gwen Gilliam. 1980. *Understanding Abusive Families.* Lexington, MA: DC Health.

Garbarino, James. 1989. *The Psychologically Battered Child.* San Francisco: Jossey-Bass.

Garbarino, James. 1995. *Raising Children in a Socially Toxic Environment.* San Francisco: Jossey-Bass.

Garcia, Mario T. 1980. "La Familia: The Mexican Immigrant Family, 1900–1930." In *Work, Family, Sex Roles, Language*, edited by Mario Barrera, Alberto Camarillo, and Francisco Hernandez. Berkeley, CA: Tonatiua-Quinto Sol International.

Garcia-Bahne, Betty. 1977. "La Chicana and the Chicano Family." In *Essays on La Mujer*, edited by R. Sanchez. Los Angeles, CA: UCLA Chicano Studies Center Publications.

Garfinkel, Irwin, and Donald Oellerich. 1989. "Noncustodial Fathers' Ability to Pay Child Support." *Demography* 26:219–34.

Garfinkel, Irwin, and Sara S. McLanahan. 1986. *Single Mothers and Their Children: A New American Dilemma.* Washington, DC: Urban Institute.

Garfinkel, Irwin. 1988. "Child Support Assurance: A New Tool for Achieving Social Security." In *Child Support: From Debt Collection to Social Policy,* edited by Alfred J. Kahn and Sheila B. Kamerman. Nesbury Park, CA: Sage.

Gaskin, Ina M. 1977. *Spiritual Midwifery.* Summertown, TN: Schocken Books.

Gazmararian, Julie A., Suzanne Lazorick, and Alison M. Spitz. 1996. "Prevalence of Violence Against Pregnant Women." *Journal of the American Medical Association* 275:1915–20.

Geiger, H. Kent. 1968. *The Family in Soviet Russia.* Cambridge, MA: Harvard University Press.

Gelles, Richard J. 1977. "Violence in the American Family." In *Violence and the Family,* edited by P. Martin. New York: Wiley.

Gelles, Richard J. 1987. *Family Violence.* Newbury Park, CA: Sage.

Gelles, Richard J., and Claire Pedrich Cornell. 1990. *Intimate Violence in Families.* 2d ed. Beverly Hills, CA: Sage.

Gelles, Richard J., and Eileen F. Hargreaves. 1987. "Maternal Employment and Violence toward Children." In *Family Violence,* edited by Richard J. Gelles. Newbury Park, CA: Sage.

Gelles, Richard J., and M. A. Straus. 1987. "Is Violence Toward Children Increasing? A Comparison of 1975–1985 National Survey Rates." *Journal of Interpersonal Violence* 2:212–22.

General Social Survey (GSS). 1999. University of Michigan, Inter-University Consortium for Political and Social Research. http//:www. icpsr.umich.edu.

Gerson, Kathleen. 1985. *Hard Choices.* Berkeley: University of California Press.

Gerson, Kathleen. 1993. *No Man's Land: Men's Changing Commitments to Family and Work.* New York: BasicBooks.

Gerstel, Naomi, and Katherine McGonagle. 1999. "Job Leaves and the Limits of the Family and Medical Leave Act." *Journal of Work and Occupations* 26(4):510–34.

Gilbert, Lucia. 1985. *Men in Dual-Career Marriages.* Hillsdale, NJ: Erlbaum.

Giles-Sim, Jean. 1983. *Wife Battering: A Systems Theory Approach.* New York: Guilford Press.

Gilligan, Carol. 1982. *In a Different Voice. Psychological Theory and Women's Development.* Cambridge, MA: Harvard University Press.

Girouard, Mark. 1980. *Life in the English Country House.* New Haven, CT: Yale University Press.

Gitlin, Todd. 1987. *The Sixties, Years of Hope, Days of Rage.* New York: Basic Books.

Glass, Jennifer, and Fujimoto, Tetsushi. (1994). "Housework, Paid Work, and Depression among Husbands and Wives." *Journal of Health and Social Behavior* 35:179–91.

Glass, Shirley P., and Thomas L. Wright. 1977. "The Relationship of Extramarital Sex, Length of Marriage, and Sex Differences on Marital Satisfaction and Romanticism." *Journal of Marriage and the Family* 39:691–703.

Gleicher, Norman. 1984. "Cesarean Section Rates in the United States: The Short-term Failure of the National Consensus Development Conference in 1980." *Journal of the American Medical Association* 252:3273–78.

Glenn, Evelyn N. 1983. "Split Household, Small Producer and Dual Wage Earner: An Analysis of Chinese-American Family Strategies." *Journal of Marriage and the Family* 45:35–46.

Glenn, Evelyn Nakano. 1992. "From Servitude to Service Work: Historical Continuities in the Racial Division of Women's Work." *Signs* 18:1–43.

Glenn, Norval D. 1975. "Psychological Well-Being in the Postparental Stage: Some Evidence from National Surveys." *Journal of Marriage and the Family* 37(Feb):105–10.

Glenn, Norval D. 1975. "The Contribution of Marriage to the Psychological Well-being of Males and Females." *Journal of Marriage and the Family* 37:594–600.

Glenn, Norval D. 1982. "Interreligious Marriage in the United States." *Journal of Marriage and the Family* 44:555–68.

Glenn, Norval D. 1990. "Quantitative Research on Marital Quality in the 1980s: A Critical Review." *Journal of Marriage and the Family* 52 (4):818–831.

Glenn, Norval D. 1991. "Quantitative Research on Marital Quality in the 1980s." Pp. 28–41 in *Contemporary Families,* edited by A. Booth. Minneapolis, MN: National Council on Family Relations.

Glenn, Norval D. 1997. "A Critique of Twenty Family and Marriage and the Family Textbooks." *Family Relations* 46(3):197–208.

Glenn, Norval D., and Beth Ann Shelton. 1983. "Pre-Adult Background Variables and Divorce." *Journal of Marriage and the Family* 39:405–10.

Glenn, Norval. 1997. "Marriage Is Not a Dirty Word: College Textbooks Teach Otherwise, Presenting a Dated, One-Sided Distortion." *Los Angeles Times*, September 16, p. B-7.

Glenn, Norval. 1997. *Closed Hearts, Closed Minds: The Textbook Story of Marriage.* New York: Institute for American Values.

Goffman, Erving. 1967. *Interaction Ritual.* New York: Doubleday.

Goldberg, Marilyn P. 1972. "Women in the Soviet Economy." *Review of Radical Political Economics* 4:1–15.

Golding, Jaqueline M. 1990. "Division of Household Labor, Strain, and DepressiveSymptoms among Mexican Americans and Non-Hispanic Whites." *Psychology of Women Quarterly* 14:103–17.

Goldman, Noreen, Charles F. Westoff, and Charles Hammerslough. 1984.

"Demography of the Marriage Market in the United States." *Population Index* 50(1):5–25.

Goldman, Noreen. 1993. "Marriage Selection and Mortality Patterns Inferences and Fallacies." *Demography* 30:189–98.

Goldscheider, Frances K. 1991. *New Families, No Families?: The Transformation of the American Home.* Berkeley: University of California Press.

Goldstone, Jack A. 1986. "The Demographic Revolution in England: A Re-examination." *Population Studies.* 49:5–33.

Gonzales, Marti Hope, and Sarah A. Meyers. 1993. "Your Mother Would Like Me: Self-Presentation in the Personal Ads of Heterosexual and Homosexual Men and Women." *Society for Personality and Social Psychology* 19:131–42.

Goode, Erich. 1996. "Gender and Courtship Entitlement: Responses to Personal Ads." *Sex Roles* 34:141–70.

Goodnow, Jacqueline J. 1988. "Children's Household Work: Its Nature and Functions." *Psychological Bulletin* 103(1):5–26.

Goody, Jack. 1983. *The Development of the Family and Marriage in Europe.* Cambridge: Cambridge University Press.

Gordon, Linda. 1982. "Why Nineteenth-Century Feminists Did Not Support Birth Control and Twentieth-Century Feminists Do." In *Rethinking the Family: Some Feminist Questions,* edited by Barrie Thorne. New York: Longman.

Gordon, Milton Myron. 1964. *Assimilation in American Life.* New York: Oxford University Press.

Gottman, John M. 1994. *What Predicts Divorce? The Relationship between Marital Processes and Marital Outcomes.* Hillsdale, NJ: Erlbaum.

Gottman, John M. 1998. "Psychology and the Study of Marital Processes." *Annual Review of Psychology* 49:169–97.

Gove, Walter R. 1972. "The Relation between Sex Roles, Marital Status, and Mental Illness." *Social Forces* 51:34–44.

Gove, Walter R. 1979. "Sex Differences in the Epidemiology of Mental Disorder: Evidence and Explanations." In *Gender and Disordered Behavior: Sex Differences in Psychopathology,* edited by E. S. Gomberg and V. Franks. New York: Brunner/Mazel.

Graber, Julia A., Jeanne Brooks-Gunn, and Anne C. Peterson, eds. 1996. *Transitions through Adolescence: Interpersonal Domains and Context.* Mahwah, NJ: Lawrence Erlbaum.

Gramsci, Antonio. 1971. *Selections from the Prison Notebooks,* edited and translated by Q. Hoare and G. Nowell-Smith. London: Lawrence and Wishart.

Gray, John. 1992. *Men Are from Mars, Women Are from Venus: A Practical Guide for Improving Communication and Getting What You Want in Your Relationships.* New York: Harper Collins.

Greeley, Andrew M. 1991. *Faithful Attraction: Discovering Intimacy, Love, and Fidelity in American Marriage.* New York: Doherty.

Greenberger, Ellen, and Laurence Steinberg. 1986. *When Teenagers Work: The Psychological Costs of Adolescents Employment.* New York: Basic Books.

Greenstein, Theodore N. 1996. "Husbands' Participation in Domestic Labor: Interactive Effects of Wives' and Husbands' Gender Ideologies." *Journal of Marriage and the Family* 58:585–95.

Greven, Philip. 1991. *Spare the Child: The Religious Roots of Physical Punishment and the Psychological Impact of Physical Abuse.* New York: Knopf.

Grief, G. L. 1985. *Single Fathers.* Lexington, MA: Lexington Books.

Griffith, Samuel B. 1963. "Introduction." In *Sun Tzu: The Art of War.* Oxford: Oxford University Press.

Grigsby, Jill S. 1992. "Women Change Places." *American Demographics* 14:46–50.

Grimsley, Kristin Downey. 1998. "MBA No Ticket to Top for Women." Washington, Post, March 24, p. A1.

Griswold del Castillo, Richard. 1984. *La Familia.* Notre Dame, IN: University of Notre Dame Press.

Griswold, Robert. 1993. *Fatherhood in America: A History.* New York: Basic Books.

Guarendi, Ray, and David Paul Eich. 1990. *Back to the Family: How to Encourage Traditional Values in Complicated Times.* New York: Villard Books.

Gusfield, Joseph R. 1963. *Symbolic Crusade. Status Politics and the American Temperance Movement,* Urbana: University of Illinois Press.

Gutman, Herbert G. 1976. *The Black Family in Slavery and Freedom,* New York: Pantheon.

Guttentag, Marcia, and Paul F. Secord. 1983. *Too Many Women? The Sex Ratio Question.* Beverly Hills, CA: Sage.

Haas, Linda. 1992. *Equal Parenthood and Social Policy.* Albany: State University of New York Press.

Hackstaff, Karla B. 1999. *Marriage in a Culture of Divorce.* Philadelphia: Temple University Press.

Hadas, Moses. 1950. *A History of Greek Literature.* New York: Columbia University Press.

Hajnal, J. 1965. "European Marriage Patterns in Perspective." pp. 101–43. In *Population in History,* edited by D. V. Glass and D. E. C. Eversley. Chicago, IL: Aldine Pub. Co.

Hale, Christine. 1990. "Infant Mortality: An American Tragedy." *Population Trends and Public Policy* 18:1–16. Washington, DC: Population Reference Bureau.

Halle, David. 1984. *America's Working Man, Work, Home, and Politics among Blue-Collar Property Owners.* Chicago, IL: University of Chicago Press.

Hareven, Tamara K. 1982. *Family Time and Industrial Time: The Relationship between the Family and Work in a New England Industrial Town.* Cambridge, England: Cambridge University Press.

Harris Poll. 1998. Cited in *Polls Indicate Support for Child Care.* Children's Defense Fund. www.childrensdefense.org. 2000.

Harriss, John, ed. 1991. *The Family: A Social History of the Twentieth Century.* New York: Oxford University Press.

Hartmann, Heidi. 1981. "The Family as the Locus of Gender, Class, and Political Struggle: The Example of Housework." *Signs* 6:366–94.

Hartzler, Kaye, and Juan N. Franco. 1985. "Ethnicity, Division of Household Tasks and Equity in Marital Roles: A Comparison of Anglo and Mexican American Couples." *Hispanic Journal of Behavioral Sciences* 7(4):333–44.

Harvey S. Marie, Linda J. Beckman, Christy Sherman, and Diana Petitti. 1999. "Women's Experience and Satisfaction with Emergency Contraception." *Family Planning Perspectives* 31(5):237–40 and 260.

Harvey, Elizabeth. 1999. "Short-Term and Long-Term Effects of Parental Employment on Children of the National Longitudinal Survey of Youth." *Developmental Psychology* 35:445–59.

Hasian, Marouf Arif, Jr. 1996. *The Rhetoric of Eugenics in Anglo-American Thought.* Athens, GA: University of Georgia Press.

Hatala, Mark Nicholas, and Jill Prehodka. 1996. "Content Analysis of Gay Male and Lesbian Personal Advertisements." *Psychological Reports* 78(2):371–74.

Hatcher, Robert A., Gary K. Stewart, Felicia Steward, Felicia Guest, David W. Schwartz, and Stephanie A. Jones. 1980. *Contraceptive Technology: 1980–1981,* 10th rev. ed. New York: Irvington Publishers, Inc.

Hatcher, Robert A., Gary K. Stewart, Felicia Steward, Felicia Guest, David W. Schwartz, and Stephanie A. Jones. 1998. *Contraceptive Technology.* 17th rev. ed. New York: Ardent Media.

Hatcher, Robert A., Gary K. Stewart, Felicia Steward, Felicia Guest, David W. Schwartz, and Stephanie A. Jones. 1997. *Essentials of Contraceptive Technology.* Baltimore: The Johns Hopkins School of Public Health, Population Information Program.

Hatchett, Shirley J., and James S. Jackson. 1993. "African American Extended Kin Systems." Pp. 90–108 in *Family Ethnicity: Strength in Diversity,* edited by H. P. McAdoo. Thousand Oaks, CA: Sage.

Hatfield, Elaine, and Richard L. Rapson. 1996. *Love and Sex: Cross-Cultural Perspectives.* Needham Heights, MA: Allyn and Bacon.

Hatfield, Elaine, and Susan Sprecher. 1986. *Mirror, Mirror: The Importance of Looks in Everyday Life.* Albany: State University of New York Press.

Hatto, A. T., trans. 1969. *The Nibelungenlied.* Baltimore, MD: Penguin Books.

Hauser, Arnold. 1951. *The Social History of Art.* New York: Knopf.

Hawkins, Alan, and T. Roberts. 1992. "Designing a Primary Intervention to Help Dual-earner Couples Share Housework and Childcare. *Family Relations* 41:169–77.

Hawkins, Alan. J., Tomi-Ann Roberts, Shawn L. Christiansen, and Christina M. Marshall. 1994. "An Evaluation of a Program to Help Dual-earner Couples Share the Second Shift." *Family Relations* 43:213–20.

Hayes, Cheryl D., and Sheila B. Kamerman, eds. 1983. *Children of Working Parents: Experiences and Outcomes.* Washington, DC: National Academy Press.

Hayghe, Howard. 1990. "Family Members in the Workforce." *Monthly Labor Review* 113:14–19.

Hays, Sharon. 1996. *The Cultural Contradictions of Motherhood.* New Haven, CT: Yale University Press.

Health Facts. 1999. "Intertility Treatments: Weighing the Risks and Benefits." *Health Facts.* Center for Medical Consumers.

Heller, Celia. 1966. *Mexican American Youth: Forgotten Youth at the Crossroads.* New York: Random House.

Hendrick, Clyde, and Susan Hendrick. 1996. "Gender and the Experience of Heterosexual Love." Pp. 131–48 in *Gendered Relationships,* edited by J. T. Wood. Mountain View, CA: Mayfield.

Hendrick, Clyde, and Susan Hendrick. 1989. "Research on Love: Does It Measure Up?" *Journal of Personality and Social Psychology* 56:784–94.

Hendrick, Susan, and Clyde Hendrick. 1992. *Romantic Love.* Newbury Park, CA: Sage.

Henshaw, S. K., and J. Silverman. 1988. "The Characteristics and Prior Contraceptive Use of U.S. Abortion Patients." *Family Planning Perspectives* 20:158–68.

Henshaw, Stanley K. 1998. *Abortion Incidence and Services in the United States.* New York: Alan Guttmacher Institute.

Henton, J. R., A. Cate, J. Koval, S. Lloyd, and S. Christopher. 1983. "Romance and Violence in Dating Relationships." *Journal of Family Issues* 4:467–82.

Herman, Judith L. 1981. *Father-Daughter Incest.* Cambridge, MA: Harvard University Press.

Herrera, Ruth S., and Robert L. Del Campo. 1995. "Beyond the Superwoman Syndrome: Work Satisfaction and Family Functioning among Working-Class, Mexican American Women." *Hispanic Journal of Behavioral Sciences* 17:49–60.

Hertz, Rosanna. 1986. *More Equal than Others: Women and Men in Dual-Career Marriages.* Berkeley: University of California Press.

Herzog, Elizabeth, and Celia E. Sudia. 1973. "Children in Fatherless Families." In *Child Development and Social Policy,* edited by Bette M. Caldwell. Chicago: University of Chicago Press.

Hetherington, Mavis, and Ross D. Parke. 1999. *Child Psychology: A Contemporary Viewpoint.* 5th ed. New York: McGraw-Hill

Hewlett, Barry S. 1991. Intimate Fathers: The Nature and Context of Aka Pygmy Paternal Infant Care. Ann Arbor: University of Michigan Press.

Hilkevitch, Victoria B., and Rosemary Blieszner. 2000. "Older Adults and Their Families," Pp. 216–31 in Handbook of Family Diversity, edited by D. H. Demo, K. R. Allen, and M. A. Fine. New York: Oxford University Press.

Hiller, Dana Vannoy, and William W. Philliber. 1986. "The Division of Labor in Contemporary Marriage: Expectations, Perceptions, and Performance." *Social Problems* 33:191–201.

Hills, Charles T., Zick Rubin, and Letitia A. Peplau. 1976. "Breakups before Marriage: The End of 103 Affairs." *Journal of Marriage and the Family"* 32:147–68.

Himes, Norman E. 1963. Medical History of Contraception. New York: Gamut Press.

Historical Statistics of the United States. Series A255–257. 1965. Washington, DC: U.S. Government Printing Office.

Hochschild, Arlie R. 1983. *The Managed Heart.* Berkeley: University of California Press.

Hochschild, Arlie R. 1989. *The Second Shift: Working Parents and the Revolution at Home.* New York: Viking.

Hochschild, Arlie. 1997. *The Time Bind: When Work Becomes Home and Home Becomes Work.* New York: Metropolitan Books.

Hodgson, Marshall. 1974. *The Venture of Islam.* Chicago: University of Chicago Press.

Hoebel, E. Adamson. 1954. *The Law of Primitive Man.* Cambridge, MA: Harvard University Press.

Hofferth, Sandra, and Cheryl Hayes, eds. 1992. *Risking the Future: Adolescent Sexuality, Pregnancy, and Childbearing.* Washington, DC: National Academy Press.

Hofferth, Sandra, and D. A. Phillips. 1987. "Child Care in the United States, 1970–1995." *Journal of Marriage and the Family* 49:559–71.

Hoffman, Saul D., and Greg J. Duncan. 1988. "What Are the Economic Consequences of Divorce?" *Demography* 25(4):641–45.

Holden, Constance. 1986. "Depression Research Advances, Treatment Lags." *Science* 233:723–26.

Holden, Karen, and Pamela J. Smock. 1991. "The Economc Costs of Marital Dissolution: Why Do Women Bear a Disproportionate Cost?" Annual Review of Sociology 17:51–78.

Holtzworth-Munroe, and Amy; Gregory L. Stuart. 1994. "Typologies of Male Batterers: Three Subtypes and the Differences among Them." *Psychological Bulletin* 116(3):476–97.

Hondagneu-Sotelo, Pierette. 1992. "Overcoming Patriarchal Constraints: The Reconstruction of Gender Relations among Mexican Immigrant Women and Men." *Gender & Society* 6:393–415.

Hood, Jane. 1983. *Becoming a Two-Job Family.* New York: Praeger.

Hooper, Linda M., and Claudette Bennett. 1998. "The Asian and Pacific Islander Population in the United States." *Current Population Reports* P20–512. Washington, DC: U.S. Census Bureau.

House-Midamba, Bessie, and Felix K. Ekechi, eds. 1995. *African Market Women and Economic Power.* Westport, CT: Greenwood Press.

Houseknecht, Sharon K. 1987. "Voluntary Childlessness." In *Handbook of Marriage and the Family,* edited by Marvin B. Sussman and Suzanne K. Steinmetz. New York: Plenum Press.

Houston, Jean W. 1985. *Beyond Manzanar: Views of Asian American Womanhood.* Santa Barbara, CA: Capra.

Howard, Judith A., and Jocelyn A. Hollander. 1997. *Gendered Situations, Gendered Selves: A Gender Lens on Social Psychology.* Thousand Oaks, CA: Sage.

Howe, Neil, and Bill Strauss. 1993. *13Th GEN: Abort, Retry, Ignore, Fail.* New York: Vintage Books.

Howery, Carla. 1998. "Sociologists Differ about Family Textbooks' Message." *Footnotes* 26(1):7.

Hu, Yuanreng, and Noreen Goldman. 1990. "Mortality Differentials by Marital Status: An International Comparison." *Demography* 27:233–50.

Huber, Joan, and Glenna Spitze. 1983. *Sex Stratification, Children, Housework, and Jobs.* New York: Academic Press.

Huber, Joan. 1991. *Macro-Micro Linkages in Sociology.* Newbury Park, CA: Sage.

Human Betterment Foundation. 1930. *Collected Papers on Eugenic Sterilization in California: A Critical Study of Results in 6000 Cases.* Pasadena, CA: Human Betterment Foundation.

Hunt, Janet, and Larry Hunt. 1982. "The Dualities of Careers and Families: New Integrations or New Polarizations?" *Social Problems* 29.

Hunt, Morton M. 1974. *Sexual Behavior in the 1970s.* New York: Dell.

Ickes, William 1993. "Tradiational Gender Roles: Do They Make, and Then Break, Our Relationships?" *Journal of Social Issues* 49:71–85.

Illouz, Eva. 1997. *Consuming the Romantic Utopia: Love and the Cultural Contradictions of Capitalism.* Berkeley: University of California Press.

ISLAT (Institute for Science, Law and Technology Working Group, Illinois Institute of Technology). 1998. "ART into Science: Regulation of Fertility Techniques." *Science* 281:651–52.

Jacobson, Neil S., and John M. Gottman. 1998. *When Men Batter Women: New Insights into Ending Abusive Relationships.* New York: Simon and Schuster.

Jacobson, P. H. 1956. "Hospital Care and the Vanishing Midwife." *Milbank Memorial Fund Quarterly* 34:253–61.

Jankowiak, William R., and Edward F. Fischer. 1992. "A Cross-Cultural Perspective on Romantic Love." *Ethnology* 31:149–55.

Jarvenpa, Robert. 1988. "The Political Economy and Political Ethnicity of American Indian Adaptations and Identities." In *Ethnicity and Race in the U.S.A.,* edited by Richard Alba. New York: Routledge.

Jaynes, Gerald D., and Robon M. Williams, Jr., eds. 1989. *A Common Destiny: Blacks and American Society.* Washington, DC: National Academy Press.

Jessor, Richard. 1993. " Successful Adolescent Development among Youth in High-Risk Settings." *American Psychologist* 48:117–26.

John, Daphne, Beth Anne Shelton, and Kristen Luschen. 1995. "Race, Ethnicity, Gender and Perceptions of Fairness." *Journal of Family Issues* 16:357–79.

John, Robert. 1988. "The Native American Family." pp 325–66 in *Ethnic families in America: Patterns and Variations,* 3d ed., edited by C. H. Mindel, R. W. Habenstein, and R. Wright. New York: Elsevier.

Johnson, Barbara E., Douglas L. Kuck, and Patricia R. Schander. 1997. "Rape Myth Acceptance and Sociodemographic Characteristics: A Multidimensional Analysis." *Sex Roles: A Journal of Research* 36(11–12):693–707.

Johnson, David R., Lynn K. White, John Edwards, and Alan Booth. 1986. "Dimensions of Marital Quality: Toward Methodological and Conceptual Refinement." *Journal of Family Issues* 7:31–49.

Johnson, Michael P. 2000. "Conflict and Control: Images of Symmetry and Asymmetry in Domestic Violence." In *Couples in Conflict,* edited by A. Booth, A. C. Crouter, and M. Clements. Hillsdale, NJ: Lawrence Erlbaum.

Johnson, Michael P., and Kathleen J. Ferraro. 2000. "Research on Domestic Violence in the 1990s: Making Distinctions." *Journal of Marriage and the Family* 62.

Johnson, Miriam M. 1988. *Strong Mothers, Weak Wives: The Search for Gender Equality.* Berkeley: University of California Press.

Johnson, Miriam M. 1997. "Review of Marriage and Family Textbooks." *Contemporary Sociology—A Journal of Reviews* 26(3):395–99.

Johnson, Phyllis J. 1998. "Performance of Household Tasks by Vietnamese and Laotian Refugees: Tradition and Change." *Journal of Family Issues* 19:245–73.

Jones, J. S. 1994. "Elder Abuse and Neglect: Responding to a National Problem. *Annals of Emergency Medicine,* 23:845–48.

Jordan, Brigitte, and Susan L. Irwin. 1989. "The Ultimate Failure: Court-Ordered Cesarean Section." In *New Approaches to Human Reproduction,* edited by Linda Whiteford and Marilyn Polan. Boulder, CO: Westview Press.

Joyce, James. 1934. *Ulysses.* New York: Random House.

Juster, S., and M. Vinovskis. 1987. "Changing Perspectives on the American Family in the Past." *Annual Review of Sociology* 13:193–216.

Kagan, Jerome. 1976. "The Psychological Requirements for Human Development." In *Raising Children in Modern America,* edited by Nathan Talbot. Little, Brown.

Kagan, Jerome. 1994. *The Nature of the Child.* New York: Basic Books

Kahn, Alfred J., and Sheila B. Kamerman, eds. 1988. *Child Support.* Newbury Park, CA: Sage.

Kahn, James, Claire Brindis, and Dana Glei. 1999. "Pregnancies Averted among U.S. Teenagers by the Use of Contraceptives." *Family Planning Perspectives* 31.

Kalmijn, Matthijs. 1998. " Intermarriage and Homogamy: Causes, Patterns, Trends." *Annual Review of Sociology* 24(1):395–421.

Kamerman, Sheila B., and Alfred J. Kahn. 1995. "Innovations in Toddler Day Care and Family Support Services: An International Overview." *Child Welfare* 74(6):1281–300.

Kanowitz, Leo. 1969. *Women and the Law.* Albuquerque: University of New Mexico Press.

Kanter, Rosabeth Moss. 1977. *Men and Women of the Corporation.* New York: Basic Books.

Karney, Benjamin R., and Thomas N. Bradbury. 1995. "The Longitudinal Course of Marital Quality and Satbility: A Review of Theory, Method, and Research." *Psychological Bulletin* 118:3–34.

Keefe, Susan E., and Amado M. Padilla. 1987. *Chicano Ethnicity.* Albuquerque: University of New Mexico Press.

Kellogg Foundation Poll. 1999. Cited in *Polls Indicate Support for Child Care.* Children's Defense Fund. http://www.chilrensdefense.org. 2000.

Kelly, Joan Berlin. 1986. "Divorce: The Adult Perspective." In *Family in Transition: Rethinking Marriage, Sexuality, Child Rearing, and Family Organization,* 5th ed., edited by Arlene S. Skolnick and Jerome H. Skolnick. Boston: Little, Brown.

Kelly, Liz. 1988. "How Women Define Their Experiences of Violence." In *Feminist Perspectives on Wife Abuse,* edited by Kersti Yllö and Michele Bograd. Newbury Park, CA: Sage.

Kephart, William M. 1967. "Some Correlates of Romantic Love." *Journal of Marriage and the Family* 29:470–74.

Kessler, Ronald, and James McRae. 1982. "The Effects of Wives' Employment on the Mental Health of Married Men and Women." *American Sociological Review* 47:216–27.

Kibria, Nazli. 1997. "The Construction of 'Asian American': Reflections on Intermarriage and Ethnic Identity among Second-Generation Chinese and Korean Americans." *Ethnic and Racial Studies* 20(3):523–44.

Kiecolt, K. Jill, and Mark A. Fossett. 1995. "Mate Availability and Marriage among African Americans." Pp. 121–42 in *The Decline in Marriage among African Americans: Causes, Consequences, and Policy Implications,* edited by B. Tucker and C. Mitchell-Kernan. New York: Russell Sage Foundation.

Kiely, J. L., J. Kleinmen, and M. Kiely. 1992. "Triplets and Higher-Order Multiple Births." *American Journal of Diseases in Children* 146:862–68.

Kikumura, A., and Harry Kitano. 1973. "Interracial Marriage: A Picture of the Japanese Americans." *Journal of Social Issues* 29(Spring):67–81.

Kilpatrick, D. G., C. N. Edmunds, and A. K. Seymour. 1992. *Rape in America: A Report to the Nation.* National Victim Center.

Kilpatrick, D. G., C. L. Best, B. E. Saunders, and L. J. Veronen. 1988. "Rape in Marriage and in Dating Relationships: How Bad Is it for Mental Health?" *Annals of New York Academy of Sciences* 528:335–44.

Kimball, Gayle. 1983. *The 50/50 Marriage.* Boston: Beacon.

Kimmel, Michael, and Martin P. Levine. 1992. "Men and AIDS." In *Men's Lives.* Edited by Michael Kimmel and Michael Messner. New York: Macmillan.

Kimmel, Michael, and Michael Messner, eds. 1994. *Men's Lives.* New York: MacMillan.

Kimmel, Michael. 1996. *Manhood in America: A Cultural History.* New York: Free Press.

Kinsey, Alfred C., Wardell B. Pomeroy, and Clyde D. Martin. 1948. *Sexual Behavior in the Human Male.* Philadelphia, PA: W. B. Saunders.

Kinsey, Alfred C., Wardell B. Pomeroy, Clyde E. Martin, and Paul H. Gebhard. 1953. *Sexual Behavior in the Human Female.* Philadelphia, PA: W. B. Saunders.

Kinsman, Gary. 1992. "Men Loving Men: The Challenge of Gay Liberation." In *Men's Lives.* Edited by Michael Kimmel and Michael Messner. New York: Macmillan.

Kirkwood, Catherine. 1993. *Leaving Abusive Partners: From the Scars of Survival to the Wisdom for Change.* Newbury Park, CA: Sage.

Kitano, Harry H. L. 1976. *Japanese Americans: The Evolution of a Subculture.* Englewood Cliffs, NJ: Prentice Hall.

Kitano, Harry H. L., and Roger Daniels. 1995. *Asian Americans: Emerging Minorities.* 2d ed. Englewood Cliffs, NJ: Prentice Hall.

Kitano, Harry, and Roger Daniels. 1988. *Asian Americans: Emerging Minorities.* Englewood Cliffs, NJ: Prentice Hall.

Kitson, Gay C., and William Holmes. 1992. *Portrait of Divorce: Adjustment to Marital Breakdown.* New York: Guilford Press.

Klatch, Rebecca. 1992. "The Two Worlds of Women of the New Right." Pp. 529–52 in *Women, Politics, and Change,* edited by L. Tilly and P. Gurin. New York: Russell Sage.

Kleinberg, Seymour. 1992. "The New Masculinity of Gay Men." In *Men's Lives.* Edited by Michael Kimmel and Michael Messner. New York: Macmillan.

Klinman, Debra G., and Rhiana Kohl. 1984. *Fatherhood U.S.A.* New York: Garland.

Knox, David, and Caroline Schact. 1994. *Choices in Relationships.* Minneapolis/St. Paul: West Publishing.

Koball, Heather. 1998. "Have African American Men Become Less Committed to Marriage? Explaining the Twentieth Century Racial Cross-Over in Men's Marriage Timing." *Demography* 35:251–58.

Kobrin, Frances E., and Linda J. Waite. 1983. "Effects of Family Stability and Nestleaving Patterns on the Transition to Marriage." Paper presented at the Annual Meeting of the American Sociological Association.

Kohlberg, Lawrence. 1966. "A Cognitive–Developmental Analysis of Children's Sex-role Concepts and Attitudes." In *The Development of Sex Differences,* edited by Eleanor Maccoby. Palo Alto, CA: Stanford University Press.

Kohn, Melvin L. 1977. *Class and Conformity.* Chicago, IL: University of Chicago Press.

Kohn, Melvin L. 1979. "The Effects of Social Class on Parental Values and Practices." In *The American Family: Dying or Development?*

edited by David Reiss and Howard A. Hoffman. New York: Plenum Press.

Kohn, Melvin L., and Kazimierz M. Slomczynski. 1990. *Social Structure and Self-direction: A Comparative Analysis of the United States and Poland.* Cambridge, MA: Blackwell.

Kohn, Melvin L., and Carmi Schooler, 1978. "The Reciprocal Effects of Substantive Complexity of Work and Intellectual Flexibility: A Longitudinal Assessment." *American Journal of Sociology* 84:24–52.

Kohn, Melvin L., and Carmi Schooler. 1983. *Work and Personality: An Inquiry into the Impact of Social Stratification.* Norwood, NJ: Ablex.

Kohn, Melvin L., Atsushi Naoi, Varrie Schoenbach, Carmi Schooler, and Kazimierz M. Slomczynski. 1990. "Position in the Class Structure and Psychological Functioning in the United States, Japan, and Poland." *American Journal of Sociology* 95(4):964–1009.

Komarovsky, Mirra. 1962. *Blue-Collar Marriage.* New York: Random House.

Komter, Aafke. 1989. "Hidden Power in Marriage." *Gender and Society* 3(2):187–216.

Kopper, Beverly A. 1996. "Gender, Gender Identity, Rape Myth Acceptance, and Time of Initial Resistance on the Perception of Acquaintance Rape Blame and Avoidability." *Sex Roles: A Journal of Research* 34(1–2):81–93.

Koss, M. P., C. A. Gidycz, and N. Wisniewski. 1987. "The Scope of Rape: Incidence and Prevalence of Sexual Aggression and Victimization in a National Sample of Higher Education Students." *Journal of Consulting and Clinical Psychology* 52:162–70.

Krantowitz, Barbara, and D. Witherspoon. 1987. "The December Dilemma: How to Reconcile Two Faiths in one Household." *Newsweek,* December 28, p. 56.

Krantz, Susan E. 1988. "The Impact of Divorce on Children." In *Feminism, Children, and the New Families,* edited by S. M. Dornbush and M. H. Strober. New York: Guilford.

Kurdek, Lawrence. 1993. "Predicting Marital Dissolution: A 5–year Prospective Longitudinal Study of Newlywed Couples." *Journal of Personality and Social Psychology* 64: 221–42.

Kurdek, Lawrence. 1999. "The Nature and Predictors of the Trajectory of Change of Marital Quality of Husbands and Wives over the First 10 Years of Marriage." *Developmental Psychology* 35:1283–96.

Kurz, Demie. 1995. *For Richer, for Poorer: Mothers Confront Divorce.* New York: Routledge.

Lamb, Michael E., ed. 1997. *The Role of the Father in Child Development.* 3d ed. New York: Wiley.

Lamb, Michael. 1981. *The Role of the Father in Child Development.* New York: Wiley.

Lamb, Michael. 1986. "The Changing Roles of Fathers." In *The Father's Role: Applied Perspectives,* edited by M. Lamb. New York: Wiley.

Lambert, Wallace E., Josiane F. Hamers, and Nancy Frasure-Smith. 1980. *Child-Rearing Values: A Cross National Study.* New York: Praeger.

Lamphere, Louise, Patricia Zavella, and F. Gonzales. 1993. *Sunbelt Working Mothers: Reconciling Family and Factory.* Ithaca, NY: Cornell University Press.

Landale, Nancy S. 1997. "Immigration and the Family: An Overview." Pp. 281–91 in *Immigration and the Family: Research and Policy of U.S. Immigrants,* edited by A. Booth, A. C. Crouter, and N. Landale (eds.), Mahwah, NJ: Lawrence Erlbaum.

Landry, David, Lisa Kaeser, and Cory Richards. 1999. "Abstinence Promotion and the Provision of Information about Contraception in Public School District Sexuality Education Policies." *Family Planning Perspectives* 31.

Langer, W. L. 1972. "Checks on Population Growth: 1750–1850." *Scientific American* 226:93–100.

Laqueur, Thomas. 1990. *Making Sex.* Cambridge, MA: Harvard University Press.

LaRossa, Ralph, and Maureen LaRossa. 1981. *Transition to Parenthood.* Beverly Hills, CA: Sage.

LaRossa, Ralph, and Maureen LaRossa. 1989. "Baby Care: Fathers vs. Mothers." In *Gender in Intimate Relationships,* edited by Barbara J. Risman and Pepper Schwartz. Belmont, CA: Wadsworth.

Larson, Jeffry H., and Thomas B. Holman. 1994. "Premarital Predictors of Marital Quality and Stability." *Family Relations* 43:228–37.

Larson, Reed W., and Maryse H. Richards. 1994. *Divergent Realities: The Emotional Lives of Mothers, Fathers, and Adolescents.* Princeton, NJ: Princeton University Press.

Laslett, Peter. 1971. *The World We Have Lost: England Before the Industrial Age.* New York: Scribners.

Laslett, Peter. 1977. *Family Life and Illicit Love in Earlier Generations.* Cambridge: Cambridge University Press.

Lau, E. E., and J. Kosberg. 1979. "Abuse of the Elderly by Informal Care Providers." *Aging* 299:10–15.

Laumann, Edward O., John H. Gagnon, Robert T. Michael, and Stuart Michaels. 1994. *The Social Organization of Sexuality: Sexual Practices in the United States.* Chicago, IL: University of Chicago Press.

Lawton, Leora, Merril Silverstein, and Bengston 1994: "Solidarity Between Generations in Families." Pp. 19-42 in *Intergenerational Linkages: Hidden Connections in American Society,* edited by V. L. Bengston and R. A. Harootyan. New York: Springer Publishing Co.

Leavitt, Judith W. 1983. "Science Enters the Birthing Room: Obstetrics in America since the Eighteenth Century." *Journal of American History.* 70:281–304.

Lee, Sharon, and Marilyn Fernandez. 1998. "Trends in Asian American Racial/Ethnic Intermarriage." *Sociological Perspectives* 41:323–42.

Lee, Sharon. 1998. "Asian Americans: Diverse and Growing." *Population Bulletin* 53(2). Washington, DC: Population Reference Bureau.

Leibowitz, Arleen, Linda J. Waite, and Christina J. Witsburger. 1988. "Child Care for Preschoolers: Difference by Child's Age." *Demography* 25:205–20.

Leinbach, Mary D., and Barbara Hort. 1989. "Bears Are for Boys: 'Metaphorical' Associations in the Young Child's Gender Schema." Paper presented at the Biennial Conference of the Society for Research in Child Development, Kansas City, MO.

LeMasters, E. E. 1957. "Parenthood as Crisis." *Marriage and Family Living* 19:352–5.

LeMasters, E. E. 1975. *Working Class Aristocrats: Life Styles at a Working-Class Tavern.* Madison: University of Wisconsin Press.

Lennon, Mary Clare, and Sarah Rosenfield. 1994. "Relative Fairness and the Division of Housework: The Importance of Opinions." *American Journal of Sociology* 100:506–31.

Lenski, Gerhard E., and Jean Lenski. 1974. *Human Societies.* New York: McGraw-Hill.

Lenski, Gerhard E., and Jean Lenski. 1991. *Human Societies.* 6th ed. New York: McGraw-Hill.

Lévi-Strauss, Claude. 1969. *The Elementary Structures of Kinship.* Boston, MA: Beacon Press.

Levine, James A., and Edward W. Pitt. 1995. *New Expectations: Community Strategies for Responsible Fatherhood.* New York: Families and Work Institute.

Levine, James A., and Todd L. Pittinsky. 1997. *Working Fathers: New Strategies for Balancing Work and Family.* New York: Harcourt Brace and Company.

Levine, Robert V. 1993. "Is Love a Luxury?" *American Demographics* 15(2):27–29.

Levinson, Daniel J. 1978. *The Seasons of a Man's Life.* New York: Knopf.

Levran, D., J. Dor, E. Rudak, L. Nebel, I Ben-Shlomo, Z. Ben-Rafael, and S. Mashiach. 1990. "Pregnancy Potential of Human Oocytes." *New England Journal of Medicine* 323:1153–56.

Lewin, Tamar. 1997. "Study Criticizes Textbooks on Marriage as Pessimistic." *New York Times,* September 17, p. A-21.

Lewis, W. H. 1957. *The Splendid Century: Life in the France of Louis XIV.* New York: Doubleday.

Lichter, Daniel T. 1995. "The Retreat from Marriage and the Rise in Nonmarital Fertility." Pp. 137–46 in *Report to Congress on Out-of-Wedlock Childbearing.* (PHS 95–1257). Washington, DC: U.S. Department of Health and Human Services.

Lichter, Daniel. 1997. "Poverty and Inequality among Children." *Annual Review of Sociology* 23:121–45.

Liebman, Robert C., and Robert Wuthnow. 1983. *The New Christian Right.* Chicago, IL: Aldine.

Linton, Ralph. 1936. *The Study of Man.* New York: Appleton.

Lips, Hilary M. 1993. *Sex and Gender: An Introduction.* 2d ed. Mountain View, CA: Mayfield Publishing.

Lloyd, Peter C. 1965. "The Yoruba of Nigeria." In *Peoples of Africa,* edited by James L. Gibbs. New York: Holt, Rinehart and Winston.

Locksley, Ann. 1982. "Social Class and Marital Attitudes and Behavior." *Journal of Marriage and the Family* 44:427–40.

London, Kathryn A., and Barbara Foley Wilson. 1988. "Divorce." *American Demographics* 10(10):23–26.

Lopata, Helena Z. 1971. *Occupation: Housewife.* New York: Oxford University Press.

Lorber, Judith. 1994. *Paradoxes of Gender.* New Haven, CT: Yale University Press.

Lothrop, Samuel Kirkland. 1928. *The Indians of Tierra del Fuego.* New York: Heye Foundation.

Lowry, D. T., G. Love, and M. Kirby. 1981. "Sex on the Soap Operas: Patterns of Intimacy." *Journal of Communication* 31:90–96.

Luepnitz, Deborah Anna. 1982. *Child Custody: A Study of Families after Divorce.* Lexington, MA: D. C. Heath.

Lugaila, Terry. 1998. *Marital Status and Living Arrangements: March 1997. Current Population Reports* P20–506. Washington, DC: U.S. Government Printing Office.

Luker, Kristin. 1975. *Taking Chances: Abortion and the Decision to Contracept.* Berkeley: University of California Press.

Luker, Kristin. 1984. *Abortion and the Politics of Motherhood.* Berkeley: University of California Press.

Luker, Kristin. 1996. *Dubious Conceptions: the Politics of Teenage Pregnancy.* Cambridge, MA: Harvard University Press.

Lye, Diane N. 1996. "Adult Child–Parent Relationships." *Annual Review of Sociology* 22:79–92.

Lye, Diane N., and Timothy J. Biblarz. 1993. "The Effects of Attitudes toward Family Life and Gender Roles on Marital Satisfaction." *Journal of Family Issues* 14:157–88.

Lye, Diane N., Daniel H. Klepinger, Patricia Davis Hyle, Anjanette Nelson. 1995. "Childhood Living Arrangements and Adult Children's Relations with Their Parents." *Demography* 32(2):261–80.

Lynd, Robert S., and Helen Merrell Lynd. 1929/1956. *Middletown: A Study in American Culture.* New York: Harcourt Brace Jovanovich.

Lynn, Richard. 1982. "IQ in Japan and the U.S. Shows a Growing Disparity." *Nature* 297(20 May): 222–23.

Lyon, Eleanor, and Patricia Goth Mace. 1991. "Family Violence and the Courts." pp. 167–79 in *Abused and Battered,* edited by Dean Knudsen and JoAnn Miller. New York: Aldine de Gruyter.

Lytton, Hugh, and David M. Romney. 1991. "Parents' Differential Socialization of Boys and Girls: A Meta-Analysis." *Psychological Bulletin* 109(2):267–96.

Maccoby, Eleanor E. 1992. "The Role of Parents in the Socialization of Children: An Historical Overview." *Developmental Psychology* 28:1006–17.

Maccoby, Eleanor E., and Carol Nagy Jacklin. 1974. *The Psychology of Sex Differences.* Stanford, CA: Stanford University Press.

Maccoby, Eleanor E., and Robert H. Mnookin. 1992. *Dividing the Child: Social and Legal Dilemmas of Custody.* Cambridge, MA: Harvard University Press.

MacDermid, Shelley M., Ted L. Huston, and Susan M. McHale. 1990. "Changes in Marriage Associated with the Transition to Parenthood: Individual Differences as a Function of Sex-role Attitudes and Changes in the Division of Household Labor." *Journal of Marriage and the Family* 52(2):475–56.

MacDorman, Marian F, and Jonnae Atkinson. 1999. "Infant Mortality Statistics from the 1997 Period Linked Birth/Infant Death Data Set." *National Vital Statistics Report* 47(23). Washington, DC: U.S. Department of Health and Human Services.

Macfarlane, Alan. 1979. *The Origins of English Individualism.* New York: Cambridge University Press.

Macfarlane, Alan. 1986. *Marriage and Love in England: Modes of Reproduction 1300–1840.* Oxford: Blackwell.

Madsen, William. 1973. *Mexican-Americans of South Texas.* 2d ed. New York: Holt, Rinehart and Winston.

Magnuson, E. 1983. "Child Abuse: The Ultimate Betrayal." *Time,* Sept. 5, pp. 20–22.

Major, Brenda. 1993. "Gender, Entitlement, and the Distribution of Family Labor." *Journal of Social Issues* 49:141–59.

Malamuth, M. M., and E. Donnerstein, eds. 1984. *Pornography and Sexual Aggression.* New York: Academic Press.

Malinowski, Bronislaw. 1929. *The Sexual Life of Savages in North-Western Melanesia.* New York: Harcourt.

Malinowski, Bronislaw. 1964. "Parenthood. The Basis of Social Structure." In *The Family: Its Structure and Functions,* edited by Rose Laub Coser. New York: St. Martin's Press.

Mangen, D. J., V. L. Bengston, and P. H. Landry, eds. 1988. *Measurement of Intergenerational Relations.* Beverly Hills, CA: Sage.

Mann, Susan A., Michael Grimes, Alice Abel Kemp, and Pamela Jenkins. 1997. "Paradigm Shifts in Family Sociology? Evidence from Three Decades of Family Textbooks." *Journal of Family Issues* 18(3):315–49.

Manning, Wendy, and Pamela Smock. 1995. "Why Marry? Race and the Transition to Marriage among Cohabitors." *Demography* 32:509–20.

Manning, Wendy. 1993. "Marriage and Cohabitation Following Premarital Conception." *Journal of Marriage and the Family* 55:839–50.

Marcus, Steven. 1964. *The Other Victorians. A Study of Sexuality and Pornography in Mid-Nineteenth Century England.* New York: Basic Books.

Marks, Michelle Rose. 1997. "Party Politics and Family Policy: the Case of the Family and Medical Leave Act." *Journal of Family Issues* 18(1):55–70.

Martin, Calvin. 1978. *Keepers of the Game.* Berkeley: University of California Press.

Martin, D. 1976. *Battered Wives.* New York: Pocket Books.

Martinez, Gladys M., and Andrea E. Curry. 1999. "School Enrollment—Social and Economic Characteristics of Students: October 1998 (Update)." *Current Population Reports,* P20–521. U.S. Bureau of the Census.

Mason, Mary Ann. 1999. *The Custody Wars.* New York: Basic Books.

Mason, Patrick. 1996. *Joblessness and Unemployment: A Review of the Literature* (LR-JU-96-03). Philadelphia, PA: National Center on Fathers and Families.

Massey, Douglas, and Audrey Singer. 1995. "New Estimates of Undocumented Mexican Migration to the United States and the Probability of Apprehension." *Demography* 32: 203–13.

Massey, Douglas, and Nancy Denton. 1993. *American Apartheid: Segregation and the Making of the Underclass.* Cambridge, MA: Harvard University Press.

Masters, William H., and Virginia E. Johnson. 1966. *Human Sexual Response.* Boston, MA: Little, Brown.

Masters, William H., and Virginia E. Johnson. 1975. *The Pleasure Bond.* New York: Bantam.

Mattessich, Paul, and Reuben Hill. 1987. "Life Cycle and Family Development." In *Handbook of Marriage and the Family,* edited by Marvin B. Sussman and Susan K. Steinmetz. New York: Plenum.

May, Elaine Tyler. 1995. *Barren in the Promised Land: Childless Americans and the Pursuit of Happiness.* New York: Basic Books.

May, K. 1982. "Factors Contributing to First-Time Fathers' Readiness for Fatherhood: An Exploratory Study." *Family Relations* 31:353–61.

May, K., and S. Perrin. 1985. "Prelude: Pregnancy and Birth." In *Dimensions of Fatherhood,* edited by S. Hanson and F. Bozett. Beverly Hills, CA: Sage.

McAdoo, Harriette Pipes. 1997. "Upward Mobility Across Generations in African American Families." Pp. 139–62 in *Family Ethnicity: Strength in Diversity,* edited by H. P. McAdoo. Thousand Oaks, CA: Sage.

McBride, B. A. 1990. "The Effects of a Parent Education/Play Group Program on Father Involvement in Child Rearing." *Family Relations* 39:250–56.

McCarthy, James. 1979. "Religious Commitment, Affiliation, and Marriage Dissolution." Pp. 179–97 in *The Religious Dimension,* edited by Robert Wuthnow. New York: Academic Press.

McDaniel, Antonio. 1990. "The Power of Culture: A Review of the Idea of Africa's Influence on Family Structure in Antebellum America.*" Journal of Family History* 15:225–38.

McDaniel, Antonio. 1994. "Historical Racial Differences in Living Arrangements of Children." *Journal of Family History* 19:57–77.

McHale, Susan M, W. Todd Bartko, Ann C. Crouter, and Maureen Perry-Jenkins. 1990. "Children's Housework and Psychosocial Functioning: The Mediating Effects of Parents' Sex-Role Behaviors and Attitudes." *Child Development* 61(5):1413–26.

McKibben, L., E. De Vos, and E. Newberger. 1989. "Victimization of Mothers and Abused Children: A Controlled Study." *Pediatrics* 84:531–35.

McLanahan, Sara, and Gary Sandefur. 1994. *Growing Up with a Single Parent: What Hurts, What Helps.* Cambridge, MA: Harvard University Press.

McLanahan, Sara, and Julia Adams. 1987. "Parenthood and Psychological Well-Being." *Annual Review of Immunology* 5:237–57.

McLanahan, Sara, and Lynn Casper. 1995. "Growing Diversity and Inequality in the American Family." Pp. 1–45 in *State of the Union: American in the 1990s,* vol. 2, edited by R. Farley. New York: Russell Sage Foundation.

McLanahan, Sara, Julia Adams, and A. B. Sorenson. 1985. "Life Events and Psychological Well-Being Over the Life Course." In *Life Course Dynamics,* edited by G. H. Elder. Ithaca, NY: Cornell University Press.

McLanahan, Sara. 1989. "The Two Faces of Divorce: Women's and Children's Interests." Paper presented at the American Sociological Association. San Francisco, August.

McLoyd, Vonnie C. 1997. "The Impact of Poverty and Low Socioeconomic Status on the Socioemotional Functioning of African-American Children and Adolescents." Pp. 7–34 in *Social and Emotional Adjustment and Family Relations in Ethnic Minority Families,* edited by R. C. Taylor and M. C. Wang . Mahwah, NJ: Lawrence Erlbaum.

Mead, George. 1934/1967. *Mind, Self, and Society.* Chicago, IL: University of Chicago Press.

"Men Control Female Sexuality with Circumcision." 1993. *AIDS Weekly,* March 29, p. 10.

Menaghan, Elizabeth, and Toby Parcel. 1990. "Parental Employment and Family Life." *Journal of Marriage and the Family* 52: 1079–98.

Menaghan, Elizabeth. 1983. "Marital Stress and Family Transitions: A Panel Analysis." *Journal of Marriage and the Family* 45:371–86.

Menken, Jane, James Trussell, and Ulla Larsen. 1986. "Age and Infertility." *Science* 1390:1389–94.

Merchant, Carolyn. 1980. *The Death of Nature.* San Francisco, CA: Harper & Row.

Merida, Kevin, and Barbara Vobejeda. 1998. "Couples in Conflict over Roles." *Washington Post,* March 25, p. A1.

Merrill, G. S. 1996. "Ruling the Exceptions: Same-Sex Battering and Domestic Violence Theory." Pp. 9–21 in *Violence in Gay and Lesbian Domestic Partnerships,* edited by C. M. Renzetti and C. H. Miley. New York: The Haworth Press.

Merstein, Bernard. 1972. "Physical Attractiveness and Marital Choice." *Journal of Personality and Social Psychology* 22:8–12.

Mertes, Kate. 1988. *The English Noble Household. 1250–1600.* Oxford: Blackwell.

Merton, Robert K. 1948. "The Self-Fulfilling Prophecy." *Antioch Review* 8:193–210.

Miall, Charlene E. 1987. "The Stigma of Involuntary Childlessness." In *Family in Transition,* edited by Arlene S. Skolnick and Jerome H. Skolnick. Glenview, IL: Scott, Foresman.

Miller, Brent, and Kristin Moore. 1991. "Adolescent Sexual Behavior, Pregnancy, and Parenting." pp. 307–26 in *Contemporary Families Looking Forward, Looking Back,* edited by Alan Booth. Minneapolis, MN: National Council on Family Relations.

Miller, Dorothy. 1979. "The Native American Family: The Urban Way." In *Families Today,* edited by Eunice Corfman, Washington, DC: U.S. Government Printing Office.

Miller, E. R., B. Shane, and Elaine Murphy. 1998. *Contraceptive Safety.* Washington, DC: Population Reference Bureau.

Miller, Joanne, and Howard H. Garrison. 1982. "Sex Roles: The Division of Labor at Home and in the Workplace." *Annual Review of Sociology* 8:237–62.

Miller-Loessi, Karen. 1992. "Toward Gender Integration in the Workplace." *Sociological Perspectives* 35:1–15.

Millett, Kate. 1970. *Sexual Politics.* New York: Doubleday.

Millman, Marcia. 1976. *The Unkindest Cut.* New York: Morrow.

Mills, Judson, and Margaret S. Clark. 1994. "Communal and Exchange Relationships." Pp. 29–42 in Theoretical Frameworks for Personal Relationships, edited by R. Erber and R. Gilmour. Hillsdale, NJ: Lawrence Erlbaum Associates.

Min, Pyong Gap. 1995. *Asian Americans: Contemporary Trends and Issues.* Thousand Oaks, CA: Sage.

Mirandé, Alfredo. 1988. "Chicano Fathers: Traditional Perceptions and Current Realities." In *Fatherhood Today: Men's Changing Role in the Family,* edited by P. Bronstein and C. P. Cowan. New York: Wiley.

Mirande, Alfredo. 1997. *Hombres y Machos: Masculinity and Latino Culture.* Boulder, CO: Westview.

Mishel, Lawrence. 1997. "Capital's Gain." *American Prospect* 33:71–4.

Mnookin, Robert H., Eleanor E. Maccoby, Charlene F. Depner, and Catherine R. Albiston. 1990. "Private Ordering Revisited: What Custodial Arrangements Are Parents Negotiating?" In *Divorce Reform at the Crossroads,* edited by S. Sugarman and H. Kay. New Haven, CT: Yale University Press.

Modell, John. 1989. *Into One's Own: From Youth to Adulthood in the United States, 1920–1975.* Berkeley: University of California Press.

Moen, Phyllis. 1985. "Continuities and Discontinuities in Women's Labor Force Activity." Pp.113–55 in *Life Course Dynamics,* edited by G. H. Elder, Jr. Ithaca, NY: Cornell University Press.

Moffitt, Robert A. 1995. "The Effect of the Welfare System on Nonmarital Childbearing." Pp. 167–76 in *Report to Congress on Out-of-Wedlock Childbearing.* (PHS 95–1257). Washington, DC: U.S. Department of Health and Human Services.

Moghissi, Kamran S. 1989. "The Technology of AID and Surrogacy." In *New Approaches to Human Reproduction,* edited by Linda Whiteford and Marilyn Polan. Boulder, CO: Westview Press.

Mohr, James C. 1978. *Abortion in America: The Origins and Evolution of National Policy, 1800–1900.* New York: Oxford University Press.

Monthly Vital Statistics Reports, 1980–1999. Hyattsville, MD: National Center for Health Statistics.

Moore, Kristin, Daphne Spain, and Suzanne M. Bianchi. 1984. "The Working Wife and Other." *Marriage and Family Review* 7:77–98.

Moore, Kristin. 1995. "Nonmarital Childbearing in the United States in Report to Congress on Out-of-Wedlock Childbearing." (PHS 95–1257). Washington, DC: U.S. Department of Health and Human Services.

Morgan, Hal, and Kerry Tucker. 1991. *Companies That Care.* New York: Simon & Schuster.

Morgan, Leslie A. 1991. *After Marriage Ends: Economic Consequences for Midlife Women.* Newbury Park, CA: Sage.

Morgan, S. Philip, Antonio McDaniel, Andrew T. Miller, and Samuel H. Preston. 1993. "Racial Differences in Household and Family Structure at the Turn of the Century." *American Journal of Sociology* 98:799–828.

Morgan, S. Philip, Diane Lye, and Gretchen Condran. 1988. "Sons, Daughters, and the Rise of Marital Disruption." *American Journal of Sociology* 94:110–29.

Morgan, S. Philip. 1991. "Late Nineteenth- and Early Twentieth-Century Childlessness." *American Journal of Sociology* 97:779–807.

Morin, Richard, and Megan Rosenfeld. 1998. "With More Equity, More Sweat." *Washington Post,* March 22, p. A1.

Mortimer, Jeylan T., Ellen Efron Pimental, Seongryeol Ryu, Katherine Nash, and C. Lee. 1996. "Part-Time Work and Occupational Value Formation in Adolescence." *Social Forces* 74:1405–18.

Mosher, W. D., and W. Pratt. 1990. "Fecundity and Infertility in the United States, 1965–1988." *Vital Health Statistics.* Hyattsville, MD: National Center for Health Statistics.

Mosher, William D., and Christine A. Bachrach. 1996. "Understanding U.S. Fertility: Continuity and Change in the National Survey of Family Growth, 1988–1995." *Family Planning Perspectives* 38.

Moynihan, Daniel Patrick, Paul Barton, and Ellen Broderick. 1965. *The Negro Family: The Case for National Action.* Washington, DC: U.S. Department of Labor.

Muir, Frank, and Simon Brett. 1980. *On Children.* London: Heinemann.

Muir, Grant, Kimberly Lonsway, and Diana L. Payne. 1996. "Rape Myth Acceptance among Scottish and American Students." *Journal of Social Psychology* 136(2):261–62.

Muller, V. 1985. "Origins of Class and Gender Stratification in Northwest Europe." *Dialectical Anthropology* 9:93–105.

Mulroy, Elizabeth A., ed. 1988. *Women As Single Parents.* Dover, MA: Auburn House.

Murdock, George P. 1967. *World Ethnographic Atlas.* Pittsburgh, PA: University of Pittsburgh Press.

Murphy, John E. 1988. "Date Abuse and Forced Intercourse among College Students." In *Family Abuse and Its Consequences,* edited by Gerald T. Hotaling, David Finkelhor, John T. Kirkpatrick, and Murray A. Straus. Newbury Park, CA: Sage.

Murphy, Robert F. 1957. "Intergroup Hostility and Social Cohesion." *American Anthropologist* 59:1018–35.

Murphy, Robert F. 1959. "Social Structure and Sex Antagonism." *Southwestern Journal of Anthropology* 15:89–98.

Murstein, Bernard I., and P. Christy. 1976. "Physical Attractiveness and Marital Adjustment in Middle Age Couples." *Journal of Personality and Social Psychology* 34:537–42.

Nye, Ivan F. 1978. "Is Choice and Exchange Theory the Key?" *Journal of Marriage and the Family"* 40:219–34.

Nance, John. 1975. *The Gentle Tasaday.* New York: Harcourt Brace Jovanovich.

National Center for Health Statistics. 1999. "Highlights of Trends in Pregnancies and Pregnancy Rates by Outcome: Estimates for the United States, 1976–1996." *National Vital Statistics Reports* 47:1–10.

National Center for Health Statistics. Home page. http://www.cdc.gov/nchs/. January, 2000.

National Center on Elder Abuse. 1998. *The National Elder Abuse Incidence Study: Final Report.* Washington, DC: Administration on Aging and U.S. Department of Health and Human Services.

National Institute of Mental Health. 1982. *Television and Behavior: Ten Years of Science Progress and Implications for the Eighties.* Rockville, MD: National Institute of Mental Health.

National Opinion Research Center, University of Michigan. General Social Surveys, 1980–1998. http://www.icpsr.umich.edu/gss. January, 2000.

Neugarten, B., and N. Datan. 1973. "Sociological Perspectives on the Life Cycle." In *Life-Span Developmental Psychology: Personality and Socialization,* edited by P. Baltes and K. Schaie. New York: Academic Press.

Neumann, Franz. 1944. *Behemoth: The Structure and Practice of National Socialism.* New York: Oxford University Press.

Nock, Steven. 1995. "A Comparison of Marriages and Cohabiting Relationships." *Journal of Family Issues* 16:53–76.

Nock, Steven. 1998. *Marriage in Men's Lives.* New York: Oxford University Press.

NORC-GSS 1972–1998 Cumulative Study. 1998. Storrs, CT: Roper Center for Public Opinion Research.

Nord, Christine, and Nicholas Zill. 1997. "Noncustodial Parents' Participation in Their Children's Lives." Child Support Report 19:1–2.

Nye, F. Ivan. 1988. "Fifty Years of Family Research, 1937–1987." *Journal of Marriage and the Family* 50:305–16.

O'Brien, John E. 1971. "Violence in Divorce-Prone Families." *Journal of Marriage and the Family* 33:692–98.

O'Connell, Martin. 1993. *Where's Pappa: Fathers' Role in Child Care.* No. 20. Washington, DC: Population Reference Bureau.

O'Connor, Anne-Marie. 1998. "New Lives for Women from Iran." *Los Angeles Times,* December 10, pp. A1, A16–18.

O'Hare, William P. 1992. "America's Minorities: The Demographics of Diversity." *Population Bulletin* Vol. 47, no 4. Washington, DC: Population Reference Bureau.

O'Hare, William P. 1996. "A New Look at Poverty in America." *Population Bulletin* 51(2):2–48.

O'Hare, William P., and Judy Felt. 1991. "Asian Americans: America's Fastest Growing Minority Group." *Population Trends and Public Policy.* No. 11. Washington, DC: Population Reference Bureau.

O'Hare, William, P., Kelvin M. Pollard, Taynia L. Mann, and Mary M. Kent. 1991. "African Americans in the 1990s." *Population Bulletin* 46(1). Washington, DC: Population Reference Bureau.

O'Neill, William L. 1970. *The Woman Movement: Feminism in the United States and England.* London: Allen and Unwin.

O'Sullivan, Lucia F., and M. E. Gaines. 1998. "Decision-Making in College Students' Heterosexual Dating Relationships: Ambivalence About Engaging in Sexual Activity." *Journal of Social and Personal Relationships* 15:347–63.

O'Sullivan, Lucia F., and E. Sandra Byers. 1992. "College Students' Incorporation of Initiator and Restrictor Roles in Sexual Dating Interactions." *The Journal of Sex Research* 29:435–46.

Ogbomo, Onaiwu W. 1995. "Esan Women Traders and Precolonial Economic Power." Pp. 1–22 in *African Market Women and Economic Power,* edited by B. House-Midamba and F. K. Ekechi. Westport, CT: Greenwood Press.

Olsen, M. E. 1970. "Social and Political Participation of Blacks." *American Sociological Review* 35:682–97.

Olson, J. S., and R. Wilson. 1984. *Native Americans in the Twentieth Century.* Provo. Utah: Brigham Young University Press.

Oropesa, R. S. 1993. "Using the Service Economy to Relieve the Double Burden: Female Labor Force Participation and Service Purchases." *Journal of Family Issues* 14:438–73.

Ortner, Sherry, and Harriet Whitehead, eds. 1981. *Sexual Meanings: The Cultural Construction of Gender and Sexuality.* Cambridge: Cambridge University Press.

Osherson, S. 1986. *Finding Our Fathers: The Unfinished Business of Manhood.* New York: Free Press.

Ostrander, Susan A. 1984. *Women of the Upper Class.* Philadelphia, PA: Temple University Press.

Pagelow, Mildred D. 1981. *Woman-Battering: Victims and Their Experiences.* Beverly Hills, CA: Sage.

Paige, Karen E., and Jeffery M. Paige. 1981. *The Politics of Reproductive Ritual.* Berkeley: University of California Press.

Parenti, Michael. 1983. *Democracy for the Few.* 4th ed. New York: St. Martin's Press.

Park, Robert E. 1950. *Race and Culture.* Glencoe, IL: The Free Press.

Parke, Ross D. 1981. *Fathers.* Cambridge, MA: Harvard University Press.

Parke, Ross D. 1988. "Families in Life-Span Perspective: A Multilevel Developmental Approach." In *Child Development in Life-Span Perspective,* edited by E. M. Hetherington, R. Lerner, and M. Perlmutter. Hillsdale, NJ: Erlbaum.

Parke, Ross D. 1996. *Fatherhood.* Cambridge, MA: Harvard University Press.

Parsons, Talcott, and Robert Bales. 1955. *Family Socialization and Interaction Process.* Glencoe, IL: Free Press.

Pasley, Kay, Mark G. Kock, and Marilyn Ihinger-Tallman. 1993. "Problems in Remarriage: An Exploratory Study of Intact and Terminated Remarriages." *Journal of Divorce and Remarriage* 20:63–83.

Pasley, Kay, and Marilyn Ihinger-Tallman, eds. 1987. *Remarriage and Stepparenting.* New York: Guilford Press.

Patterson, Gerald R. 1976. *Living with Children: New Methods for Parents' and*

Teachers. Champaign, IL: Research Press.

Patzer, Gordon L. 1985. *The Physical Attractiveness Phenomena.* New York: Plenum Press.

Pelton, Leroy. 1978. "The Myth of Classlessness in Child Abuse Cases." *American Journal of Orthopsychiatry* 48:569–79.

Pepitone-Rockwell, Fran. 1980. *Dual Career Couples.* Beverly Hills, CA: Sage.

Perry-Jenkins, Maureen, and Karen Folk. 1994. "Class, Couples, and Conflict: Effects of the Division of Labor on Assessments of Marriage in Dual-earner Families." *Journal of Marriage and the Family* 56:165–80.

Perry-Jenkins, Maureen, Rena L. Repetti, and Ann C. Crouter. 2000. "Work and Family in the 1990s." *Journal of Marriage and the Family* 62.

Peters, H. Elizabeth. 1993. "The Importance of Financial Considerations in Divorce Decisions." *Economic Inquiry* 31:71–86.

Peters, Joan K. 1997. *When Mothers Work: Loving Our Children without Sacrificing Our Selves.* Reading, MA: Addison-Wesley.

Petersen, James R., Arthur Kretchmer, Barbara Nellis, Janet Lever, and Rosanna Hertz. 1983. "The Playboy Readers' Sex Survey." *Playboy* 30(January):108, 241–50.

Peterson, Gary, and Boyd C. Rollins. 1987. "Parent–Child Socialization." In *Handbook of Marriage and the Family,* edited by Marvin B. Sussman and Suzanne K. Steinmetz. New York: Plenum Press.

Pettigrew, Thomas F. 1989. "The Changing—Not Declining—Significance of Race." In *Caste and Class Controversy on Race and Poverty,* edited by Charles Vert Willie. Dix Hills, NY: General Hall.

Peyron, Remi, Elisabeth Aubeny, Veronique Targosz, Louise Silvestre, et al. 1999. "Early Termination of Pregnancy with Mifepristone (RU 486) and the Orally Active Prostaglandin Misoprostol." *New England Journal of Medicine* 328(21):1509–13.

Phillips, Deborah. 1989. "Future Directions and Need for Child Care in the United States." In *Caring for the Children,* edited by Jeffery S. Lande, Sandra Starr, and Nina Gunzenhauser. Hillsdale, NJ: Erlbaum.

Pianta, Robert, Byron Egeland, and Martha Farrell Erickson. 1989. "The Antecedents of Maltreatment." Pp. 203–53 in *Child Maltreatment: Theory and Research on the Causes and Consequences of Child Abuse and Neglect,* edited by D. Cicchetti and V. Carlson. New York: Cambridge University Press.

Pillemer, Karl and David Finkelhor. 1988. "The Prevalence of Elder Abuse: A Random Sample Survey." *Gerontologist* 28:51–57.

Piña, Darlene L., and Vern L. Bengtson. 1993. "The Division of Household Labor and Wives' Happiness—Ideology, Employment, and Perceptions of Support." *Journal of Marriage and the Family* 55:901–12.

Pinal, Jorge del, and Audrey Singer. 1997. "Generations of Diversity: Latinos in the United States." *Population Bulletin* 52(3). Washington, DC: Population Reference Bureau.

Pleck, Joseph H. 1997. "Paternal Involvement: Levels, Sources, and Consequences." In *The Role of the Father in Child Development*, 3d ed., edited by M. E. Lamb. New York: John Wiley and Sons.

Pleck, Joseph H. 1983. "Husband's Paid Work and Family Roles: Current Research Issues." In *Research in the Interweave of Social Roles*, edited by H. Lopata and J. Pleck. Greenwich: Jai Press.

Pleck, Joseph H., and Graham L. Staines. 1985. "Work Schedules and Family Life in Two-Earner Couples." *Journal of Family Issues* 6:61–82.

Pogrebin, Letty C. 1983. *Family Politics*. New York: McGraw-Hill.

Pollock, Linda A. 1984. *Forgotten Children: Parent–Child Relations from 1500 to 1900*. Cambridge: Cambridge University Press.

Pomerleau, Andree, Daniel Bolduc, Gerard Malcuit, and Louise Cosette. 1990. "Pink or Blue: Environmental Stereotypes in the First Two Years of Life." *Sex Roles* 22:359–67.

Pomeroy, Sarah B. 1975. *Goddesses, Whores, Wives and Slaves. Women in Classical Antiquity*. New York: Schocken.

Popenoe, David, Jean Bethke Elshtain, and David Blankenhorn. 1996. *Promises to Keep: Decline and Renewal of Marriage in America*. Lanham, MD: Rowman & Littlefield.

Popenoe, David. 1993. "American Family Decline, 1960–1990." *Journal of Marriage and the Family* 55:527–44.

Popenoe, David. 1996. *Life without Father: Compelling New Evidence That Fatherhood and Marriage Are Indispensable for the Good of Children and Society*. New York: Martin Kessler Books.

Popenoe, Paul, and E. S. Gosney. 1938. *Twenty-Eight Years of Sterilization in California*. Pasadena, CA: Human Betterment Foundation.

Popenoe, Paul. 1927a. "Eugenic Sterilization in California: I. The Insane." *Journal of Social Hygiene* 13(5):257–68. Reprinted in *Collected Papers*.

Popenoe, Paul. 1927b. "Eugenic Sterilization in California: I. The Feebleminded." Journal of Social Hygiene13(6):321–30. Reprinted in *Collected Papers*.

Popenoe, Paul. 1928a. "Eugenic Sterilization in California: V. Economic and Social Status of the Sterilized Insane." *Journal of Social Hygiene* 14(1):23–32. Reprinted in *Collected Papers*.

Popenoe, Paul. 1928b. "Eugenic Sterilization in California: VI. Marriage Rates of the Psychotic." *The Journal of Nervous and Mental Disease* 68(1):17–27. Reprinted in *Collected Papers*.

Popenoe, Paul. 1928c. "Eugenic Sterilization in California: XIV. The Number of Persons Needing Sterilization." *The Journal of Heredity* 19(9):405–11. Reprinted in *Collected Papers*.

Poposil, Leopold. 1963. *The Kapauku Papuans of West New Guinea*. New York: Holt, Rhinehart and Winston.

Population Reference Bureau. 1999. World Population Data Sheet. http://www.prb.org/pubs/wpds99. January, 2000.

Portes, Alejandro, and Min Zhou. 1993. "The New Second Generation: Segmented Assimilation and Its Variants." *Annals, AAPSS* 530:74–96.

Portes, Alejandro, and Richard Schauffler. 1996. "Language and Second Generation," Pp. 8–29 in *The New Second Generation*, edited by A. Portes. New York: Russell Sage Foundation.

Power, Eileen. 1975. *Medieval Women*. Cambridge: Cambridge University Press.

Presser, Harriet B. 1988. "Shift Work and Child Care among Young Dual-earner American Parents." *Journal of Marriage and the Family* 50:133–48.

Presser, Harriet B. 1994. "Employment Schedules among Dual-earner Spouses and the Division of Household Labor by Gender." *American Sociological Review* 59:348–64.

Presser, Harriet. 2000. "Nonstandard Work Schedules and Marital Instability." *Journal of Marriage and the Family*. 62:93–110.

Preston, S. 1984. "Children and the Elderly: Divergent Paths for America's Dependents." *Demography* 21:435–57.

Price, James H., Sharon M. Desmond, and Daisy Smith. 1991. "A Preliminary Investigation of Inner City Adolescents' Perceptions of Guns." *Journal of School Health* 61:225–36.

Pruett, Kyle. 1987. *The Nurturing Father*. New York: Warner.

Ptacek, James. 1988. "The Clinical Literature on Men Who Batter: A Review and Critique." In *Family Abuse and Its Consequences*, edited by Gerald T. Hotaling, David Finkelhor, John T. Kirkpatrick, and Murray A. Straus. Newbury Park, CA: Sage.

Ptacek, James. 1998. "Why Do Men Batter Their Wives?" Pp. 619–33 in *Families in the U.S.: Kinship and Domestic Politics*, edited by K. Hansen and A. Garey, Philadelphia: Temple University Press.

"Puberty Rite for Girls Is Bitter Issue across Africa." 1990. *New York Times International*, January 15.

Pyke, Karen, and Scott Coltrane. 1996. "Entitlement, Obligation, and Gratitude in Family Work." *Journal of Family Issues* 17:60–82.

Pyke, Karen. 1994. "Women's Employment As a Gift or a Burden? Marital Power across Marriage, Divorce, and Remarriage." *Gender and Society* 8:73–91.

Queen, Stuart A., and Robert W. Habenstein. 1967. *The Family in Various Cultures*. Philadelphia, PA: Lippincott.

Radin, Norma, and Graeme Russell. 1983. "Increased Father Participation and Child Development Outcomes." In *Fatherhood and Family Policy*, edited by Michael Lamb and Abraham Sagi. Hillsdale, NJ: Erlbaum.

Radin, Norma, and R. Goldsmith. 1983. "Predictors of Father Involvement in Child Care." Paper presented at the meeting of the Society for Research in Child Development, Detroit, MI.

Ragoné, Helena. 1994. *Surrogate Motherhood*. Boulder, CO: Westview Press.

Rainwater, Lee, 1964. "Marital Sexuality in Four Cultures of Poverty." *Journal of Marriage and the Family* 26:457–66.

Rainwater, Lee, and Timothy M. Smeeding. 1995. "Doing Poorly: The Real Income of American Children in a Comparative Perspective." Working Paper No. 127. Luxembourg Income Study. Maxwell School of Citizenship and Public Affairs. Syracuse, NY: Syracuse University.

Raley, Kelly R. 1999. "Then Comes Marriage? Recent Changes in Women's Response to a Non-Marital Pregnancy." Paper presented at the 1999 annual meeting of the Population Association of America, New York, March 25.

Rand, M. R. 1997. "Violence-related Injuries Treated in Hospital Emergency Departments." Bureau of Justice Statistics, Special Report. Washington, DC: U.S. Department of Justice.

Rank, Mark R. 2000. "Poverty and Economic Hardship in Families." Pp. 293–315 in *Handbook of Family Diversity*, edited by David H. Demo, Katherine R. Allen, and Mark A. Fine. New York: Oxford University Press.

Ransford, H. Edward, and Jon Miller. 1983. "Race, Sex, and Feminist Outlooks." *American Sociological Review* 48:46–59.

Rapoport, R. 1963. "Normal Crises, Family Structure, and Mental Health." *Family Process* 2:68–80; 312–27.

Rapoport, Rhona, and Robert Rapoport. 1971. *Dual Career Families*. Baltimore, MD: Penguin.

Rapoport, Robert, and Rhona Rapoport. 1976. *Dual Career Families Re-examined: New Integrations of Work and Family*. London: Martin Robertson.

Rapp, Rayna. 1982. "Family and Class in Contemporary America: Notes Toward an Understanding of Ideology." In *Rethinking the Family,* edited by Barrie Thorne and Marilyn Yalom. New York: Longman.

Red Horse, John. 1980. "Family Structure and Value Orientation in American-Indians." *Social Casework* 61:462–67.

Reed, James. 1978. *From Private Vice to Public Virtue: The Birth Control Movement and American Society since 1830.* New York: Basic Books.

Reiger, D. A., W. E. Narrow, D. S. Rae, R. W. Manderscheid, B. Z. Locke, and F. K. Goodwin. 1993. "The De Facto U.S. Mental and Addictive Disorders Service System. Epidemiologic Catchment Area Prospective 1–Year Prevalence Rates of Disorders and Services." *Archives of General Psychiatry* 50:85–94.

Reiss, Ira L. 1960. *Premarital Sexual Standards in America.* New York: Free Press.

Reiss, Ira L. 1967. *The Social Context of Premarital Sexual Permissiveness.* New York: Holt, Rinehart, and Winston.

Remez, L. C. 1995. "Confronting the Reality of Abortion in Latin America." *International Family Planning Perspectives* 21:32–36.

Renzetti, Claire M. 1992. *Violent Betrayal: Partner Abuse in Lesbian Relationships.* Newbury Park, CA: Sage.

Renzetti, Claire M., and Charles Harvey Miley. 1996. *Violence in Gay and Lesbian Domestic Partnerships.* New York: The Haworth Press.

Reskin, Barbara, and Irene Padavic. 1994. *Women and Men at Work.* Thousand Oaks, CA: Pine Forge Press.

Reskin, Barbara. 1984. *Sex Segregation in the Workplace.* Washington, DC: National Academy Press.

Reyes, Olga, Kimberly Kobus, and Karen Gillock. 1999. "Career Aspirations of Urban, Mexican American Adolescent Females." *Hispanic Journal of Behavioral Sciences* 21(3):366–82.

Riesman, David. 1950. *The Lonely Crowd.* New Haven, CT: Yale University Press.

Riley, Dave. 1990. "Network Influences on Father Involvement in Childrearing." In *Extending Families,* edited by Cochran, Larner, Riley, Gunnarsson, Henderson, and Cross. Cambridge: Cambridge University Press.

Risman, Barbara and Donald Tomaskovic-Devey. 1998. "Sociologists Differ about Family Textbooks' Message." *Footnotes* 26(1):10.

Risman, Barbara J. 1998. *Gender Vertigo: American Families in Transition.* New Haven: Yale University Press.

Risman, Barbara. 1989. "Can Men 'Mother'? Life as a Single Father." In *Gender in Intimate Relationships,* edited by Barbara J. Risman, and Pepper Schwartz. Belmont, CA: Wadsworth.

Rivers, Rose Merry, and John Scanzoni. 1997. " Social Families among African

Americans." Pp. 333–48 in *Black Families,* 3d ed., edited by H. P. McAdoo. Thousand Oaks, CA: Sage.

Roberts, Elizabeth. 1986. *A Woman's Place: An Oral History of Working-Class Women, 1890–1940.* New York: Blackwell.

Robinson, John, and Geoffrey Godbey. 1997. *Time for Life.* University Park, PA: Pennsylvania State University Press.

Robinson, John, and Glenna Spitze. 1992. "Whistle While You Work? The Effect of Household Task Performance on Women's and Men's Well-Being." *Social Science Quarterly* 73:844–61.

Robinson, John. 1977. *How Americans Use Time.* New York: Praeger.

Robinson, John. 1988. "Who's Doing the Housework?" *American Demographics* 10:24–28, 63.

Rohner, Ronald. 1975. *They Love Me, They Love Me Not.* Human Relations Area Files, Inc.

Rohner, Ronald. 1986. *The Warmth Dimension. Foundations of Parental Acceptance-Rejection Theory.* Beverly Hills, CA: Sage.

Rollin, Lucy. 1999. *Twentieth-Century Teen Culture by the Decades: A Reference Guide.* Westport, CT: Greenwood Press.

Rollins, Judith. 1985. *Between Women: Domestics and Their Employers.* Philadelphia, PA: Temple University Press.

Romero, Mary. 1992. *Maid in the USA.* New York: Routledge.

Rosen, Ellen I. 1987. *Bitter Choices: Blue-Collar Women In and Out of Work.* Chicago, IL: University Press of Chicago.

Rosenbaum, James. 1980. "Track Misconceptions and Frustrated College Plans." *Sociology of Education* 53:74–88.

Rosenthal, C. J. 1985. "Kin-Keeping in the Familial Division of Labor." *Journal of Marriage and the Family* 45:509–21.

Rosenthal, Robert, and Lenore Jacobson. 1968. *Pygmalion in the Classroom: Teacher Expectations and Pupil's Intellectual Development.* New York: Holt.

Ross, Catherine E. 1995. "Reconceptualizing Marital Status As a Continuum of Social Attachment." *Journal of Marriage and the Family* 57:129–40.

Ross, Catherine E., and J. Huber. 1985. "Hardship and Depression." *Journal of Health and Social Behavior* 26:312–27.

Ross, Catherine E., and John Mirwosky. 1987. "Children, Child Care, and Parents' Psychological Well-Being." Paper presented at the annual meeting of the American Sociological Association, New York.

Ross, Catherine E., John Mirowsky, and Joan Huber. 1983. "Dividing Work, Sharing Work, and In-Between: Marriage Patterns and Depression." *American Sociological Review* 48:809–23.

Rossi, Alice S. 1968. "Transition to Parenthood." *Journal of Marriage and the Family* 30:26–39.

Rossi, Alice S. 1977. "A Biosocial Perspective on Parenting." *Daedalus* 106:1–31.

Rossi, Alice S. 1984. "Gender and Parenthood." *American Sociological Review* 49:1–19.

Rossi, Alice S., and Peter H. Rossi. 1990. *Of Human Bonding: Parent–Child Relations across the Life Course.* New York: Aldine de Gruyter.

Rothman, Barbara K. 1986. *The Tentative Pregnancy.* New York: Norton.

Rougemont, Denis de. 1956. *Love in the Western World.* New York: Pantheon.

Rourke, Mary. 1998. "A Woman's Place: What the Denominations Think." *Los Angeles Times,* June 16, p. E-2.

Rubel, A. 1966. *Across the Tracks: Mexican-Americans in a Texas City.* Austin: University of Texas Press.

Rubin, Gayle. 1975. "The Traffic in Women: Notes on the 'Political Economy' of Sex." In *Toward an Anthropology of Women,* edited by Rayna Reiter. New York: Monthly Review Press.

Rubin, J., R. Provenzano, and Z. Luria. 1974. " The Eye of the Beholder: Parents' Views on Sex of Newborns." *American Journal of Orthopsychiatry* 44:512–19.

Rubin, Lilian. 1983. *Intimate Strangers.* New York: Harper and Row.

Rubin, Lilian. 1994. *Families on the Fault Line: America's Working Class Speaks about the Family, the Economy, and Ethnicity.* New York: Harper, Collins.

Rubin, Lillian. 1976. *World of Pain. Life in the Working-Class Family.* New York: Basic Books.

Rubin, Lillian. 1979. *Women of a Certain Age: The Midlife Search for Self.* New York: Harper and Row.

Rubin, Zick, Letitia Peplau, and C. Hill. 1981. "Loving and Leaving: Sex Differences in Romantic Attachments. *Sex Roles* 7: 821–35.

Ruggles, Steven. 1994. "The Origins of African-American Family Structure." *American Sociological Review* 59:136–51.

Rumbaut, Ruben G. 1997. "Ties That Bind: Immigration and Immigrant Families in the United States." Pp. 3–46 in *Immigration and the Family: Research and Policy of U.S. Immigrants,* edited by A. Booth, A. C. Crouter, and N. Landale. Mahwah, New Jersey: Lawrence Erlbaum.

Rushwan, Hamid. 1995. "Female Circumcision." *World Health* 48(SPEISS):16–17.

Russell, Diana. 1982. *Rape in Marriage.* New York: Macmillan.

Russell, Diana. 1984. *Sexual Exploitation.* Newbury Park, CA: Sage.

Russell, Diana. 1986. *The Secret Trauma: Incest in the Lives of Girls and Women.* New York: Basic Books.

Russell, Graeme, and Norma Radin. 1983. "Increased Paternal Participation: The Father's Perspective." In *Fatherhood and Family Policy,* edited by M. Lamb and A. Sagi. Hillsdale, NJ: Erlbaum.

Russell, Graeme. 1982. "Shared Caregiving Families: An Australian Study." In *Nontraditional Families: Parenting and Child Development,* edited by M. Lamb. Hillsdale, NJ: Erlbaum.

Ryan, Mary P. 1981. *Cradle of the Middle Class: The Family in Oneida County, New York, 1790–1865.* New York: Cambridge University Press.

Sacks, Karen. 1979. *Sisters and Wives: The Past and Future of Sexual Equality.* Westport, CT: Greenwood.

Safilios-Rothschild, Constantina. 1977. *Love, Sex, and Sex Roles.* Englewood Cliffs, NJ: Prentice Hall.

Sakala, Carol. 1993a. "Medically Unnecessary Cesarean Section Births: Introduction to a Symposium." *Social Science and Medicine* 37(10):1177–98.

Sakala, Carol. 1993b. "Midwifery Care and Out-of-Hospital Birth Settings: How Do They Reduce Unnecessary Cesarean Section Births?" *Social Science and Medicine* 37(10):1233–50.

Salovey, Peter, ed. 1991. *The Psychology of Jealousy and Envy.* New York: Guilford Press.

Sanchez, Laura, and Emily W. Kane. 1996. "Women's and Men's Constructions of Perceptions of Housework Fairness." *Journal of Family Issues* 17:358–87.

Sanchez, Laura. 1994. "Gender, Labor Allocations, and the Psychology of Entitlement Within the Home." *Social Forces* 73:533–53.

Sanday, Peggy Reeves. 1981. *Female Power and Male Dominance: On the Origins of Sexual Inequality.* Cambridge, Eng.: Cambridge University Press.

Sanford, Linda. 1980. *The Silent Children: A Parent's Guide to the Prevention of Child Sexual Abuse.* New York: Doubleday.

Sassler, Sharon, and Robert Schoen. 1999. "The Effect of Attitudes and Economic Activity on Marriage." *Journal of Marriage and the Family* 61:147–59.

Sattel, Jack W. 1992. " The Inexpressive Male." Pp. 350–70 in *Men's Lives,* edited by M. S. Kimmel and M. A. Messner. New York: Macmillan.

Saul, Rebekah. 1999. "The Political Challenges and Educational Opportunities around Very Early Abortion." *The Guttmacher Report* 2(1).

Saunders, Daniel G. 1988. "Wife Abuse, Husband Abuse, or Mutual Combat?" In *Feminist Perspectives on Wife Abuse,* edited by

Kersti Yllö and Michele Bograd. Newbury Park, CA: Sage.

Savage, David G. 1990. "1 in 4 Young Blacks in Jail or in Court Control, Study Says." *Los Angeles Times,* Feb. 27.

Scanzoni, John H. 1975. *Sex Roles, Life Styles, and Childbearing: Changing Patterns in Marriage and Family.* New York: Free Press.

Scanzoni, John H. 1977. *The Black Family in Modern Society.* Chicago, IL: University of Chicago Press.

Scanzoni, John H. 1997. "Fashioning Families and Policies for the Future, Not the Past." *Family Relations* 46:213–17.

Scanzoni, Letha D., and John Scanzoni. 1988. *Men, Women, and Change.* New York: McGraw-Hill.

Schacter, Jim. 1989. "The Daddy Track." *Los Angeles Times Magazine,* Oct. 1:7–16.

Schechter, Susan. 1982. *Women and Male Violence: The Visions and Struggles of the Battered Women's Movement.* Boston, MA: South End.

Schechter, Susan. 1988. "Building Bridges between Activists, Professionals, and Researchers." In *Feminist Perspectives in Wife Abuse,* edited by Kersti Yllö and Michele Bograd. Newbury Park, CA: Sage.

Schlegel, Alice. 1972. *Male Dominance and Female Autonomy: Domestic Authority in Matrilineal Societies.* New Haven, CT: HRAF Press.

Schnaiberg, Alan, and Sheldon Goldenberg. 1989. "From Empty Nest to Crowded Nest: The Dynamics of Incompletely-Launched Young Adults." *Social Problems* 36:251–69.

Schneider, Peter, and Jane Schneider. 1995. "Coitus Interruptus and Family Repectability in Catholic Europe." Pp. 177–94 in *Conceiving the New World Order: The Global Politics of Reproduction,* edited by F. Ginsburg and R. Rapp. Berkeley: University of California Press.

Schoen, Robert, and Robin W. Weinick. 1993. "The Slowing Metabolism of Marriage." *Demography* 30:740–41.

Schoen, Robert, and Robin M. Weinick. 1993. "Partner Choice in Marriages and Cohabitations." *Journal of Marriage and the Family* 55:408–14.

Schooler, Carmi, Joanne Miller, Karen A. Miller, and Carol N. Richtand. 1984. "Work for the Household: Its Nature and Consequences for Husbands and Wives." *American Journal of Sociology* 90:97–124.

Schooler, Carmi. 1996. "Cultural and Social-Structural Explanations of Cross-National Psychological Differences." *Annual Review of Sociology* 22:323–49.

Schor, Juliet. 1991. *The Overworked American: The Unexpected Decline of Leisure.* New York: Basic Books.

Schuler, Sidney Ruth. 1987. *The Other Side of Polyandry.* Boulder, CO: Westview Press.

Schumm, Walter R., and Benjamin Silliman. 1996. "Gender and Marital Satisfaction: A Replication with a Sample of Spouses from the Christian Church (Disciples of God)." *Psychological Reports* 79:496–98.

Schumm, Walter R., Farrell Webb, and Stephan Bollman. 1998. "Gender and Marital Satisfaction: Data from the National Survey of Families and Households." *Psychological Reports* 83:319–27.

Schwartz, Joe, and Thomas Exter. 1989. "All Our Children." *American Demographics* 5(May):34–37.

Schwartz, Martin D. 1987. "Gender and Injury in Spousal Assault." *Sociological Focus* 20: 61–75.

Schwartz, Pepper, and Virginia Rutter. 1998. *The Gender of Sexuality.* Thousand Oaks: Pine Forge Press.

Scoon-Rogers, Lydia. 1999. "Child Support for Custodial Mothers and Fathers: 1995." *Current Population Reports* P60–196. U.S. Bureau of the Census.

Scott, Joan W., and Louise A. Tilly. 1975. "Women's Work and the Family in Nineteenth-Century Europe." *Comparative Studies in Society and History* 17:36–64.

Segura, Denise. 1984. "Labor Market Stratification: The Chicana Experience." *Berkeley Journal of Sociology* 29:57–91.

Seltzer, Judith. 1994a. "Intergenerational Ties in Adulthood and Childhood Experience." Pp. 153–63 in *Stepfamilies: Who Benefits? Who Does Not?,* edited by A. Booth and J. Dunn. Hillsdale, NJ: Erlbaum.

Seltzer, Judith. 1994b. "Consequences of Marital Dissolution for Children." *Annual Review of Sociology* 20:235–66.

Seltzer, Judith. 2000. "Families Formed Outside of Marriage." *Journal of Marriage and the Family* 62.

Sengstock, M. C. 1991. "Sex and Gender Implications in Cases of Elder Abuse." *Journal of Women and Aging* 3:25–43.

Sennett, Richard, and Jonathan Cobb. 1973. *The Hidden Injuries of Class.* New York: Random House.

Shakin, M., D. Shakin, and S. H. Sternglanz. 1985. " Infant Clothing: Sex Labeling for Strangers." *Sex Roles* 12:955–63.

Shehan, Constance L., E. W. Bock, and Gary R. Lee. 1990. "Religious Heterogamy, Religiosity, and Marital Happiness: The Case of Catholics." *Journal of Marriage and the Family* 52:73–79.

Shelton, Beth Anne, and Daphne John. 1993. " Ethnicity, Race, and Difference: a Comparison of White, Black, and Hispanic Men's Household Labor Time." Pp. 131–50 in *Men, Work, and Family,* edited by J. C. Hood . Newbury Park, CA: Sage.

Shelton, Beth Anne, and Daphne John. 1996. "The Division of Household Labor." *Annual Review of Sociology* 22:299–322.

Sherman, L. W., and R. A. Berk. 1984. "Deterrant Effects of Arrest for Domestic Violence." *American Sociological Review* 49:261–72.

Shields, Nancy M., and Christine R. Hanneke. 1983. "Wives' Reactions to Marital Rape." In *The Dark Side of Families,* edited by David Finkelhor, Richard J. Gelles, Gerald T. Hotaling, and Murray A. Straus. Beverly Hills, CA: Sage.

Shinagawa, Larry Hajime, and Gin Yong Pang. 1996. "Asian American Panethnicity and Intermarriage." *Amerasia Journal* 22:127–52.

Shiono, Paul, and Linda Quinn. 1994. "Epidemiology of Divorce." *Future of Children* 4:8.

Shorter, Edward. 1975. *The Making of the Modern Family.* New York: Basic Books.

Shotland, R. Lance, and Hunter, Barbara A. 1995. "Women's 'Token Resistant' and Compliant Sexual Behaviors Are Related to Uncertain Sexual Intentions and Rape." *Personality and Social Psychology Bulletin* 21:226–36.

Sidel, Ruth.1996. *Keeping Women and Children Last: America's War on the Poor.* New York: Penguin.

Signorelli, Nancy. 1991. *A Sourcebook on Children and Television.* New York: Greenwood Press.

Silver, Lee. 1997. *Remaking Eden: Cloning and Beyond in a Brave New World.* New York: Avon.

Simmons, Roberta G., and Dale A. Blyth. 1987. *Moving into Adolescence: The Impact of Pubertal Change and School Context.* New York: Aldine de Gruyter.

Simpson, J. A., B. Campbell, and Ellen Berscheid. 1986. " The Association between Romantic Love and Marriage: Kephart (1967) Twice Revisited." *Personality and Social Psychology Bulletin* 12:363–72.

Sinclair, Andrew. 1965. *The Emancipation of the American Woman.* New York: Harper and Row.

Skinner, B. F. 1969. *Contingencies of Reinforcement.* New York: Appleton.

Skolnick, Arlene S. 1987. *The Intimate Environment: Exploring Marriage and the Family.* Boston, MA: Little, Brown.

Skolnick, Arlene. 1991. *Embattled Paradise: The American Family in an Age of Uncertainty.* New York: Basic Books.

Skolnick, Arlene. 1997. "The Battle of the Textbooks: Bringing in the Culture War." *Family Relations* 46(3):219–22.

Slater, Philip E. 1963. "On Social Regression." *American Journal of Sociology* 28:339–64.

Smith, Jane Ellen., V. Ann Waldorf, and David L. Trembath. 1990. "Single White Male Looking for Thin, Very Attractive . . ." *Sex Roles* 23:675–85.

Smith, Tom W. 1991. "Adult Sexual Behavior in 1989: Number of Partners, Frequency of Intercourse and Risk of AIDS." *Family Planning Perspectives* 23:102–7.

Smith, Tom W. 1994. *The Demography of Sexual Behavior.* Menlo Park, CA: Kaiser Family Foundation.

Smock, Pamela J., Wendy D. Manning, and Sanjiv Gupta. 1999. "The Effect of Marriage and Divorce on Women's Economic Well-Being." *American Sociological Review* 64:794–12.

Snipp, C. Matthew. 1989. *American Indians: First of This Land.* New York: Russell Sage Foundation.

Snodgrass, Anthony. 1980. *Archaic Greece.* Berkeley: University of California Press.

Sonenstein, Freya L., Joseph Pleck, and Leighton Ku. 1991. "Levels of Sexual Activity, Condom Use, and AIDS Awareness among Adolescent Males in the United States." *Family Planning Perspectives* 21:152–58.

Song, Young I. 1996. *Battered Women in Korean Immigrant Families: The Silent Scream.* New York: Garland.

Sorel, Nancy C. 1985. *Ever Since Eve.* London: Michael Joseph.

Sørensen, Annemette. 1994. "Family and Class." *Annual Review of Sociology* 20:27–47.

Sorrentino, Constance, 1990. "The Changing Family in International Perspective." *Monthly Labor Review* 113:41–58.

Spanier, Graham B. 1976. "Measuring Dyadic Adjustment: New Scales for Assessing the Quality of Marriage and Similar Dyads." *Journal of Marriage and the Family* 38(1):15–28.

Spanier, Graham B., and Paul C. Glick. 1986. "Mate Selection Differentials between Whites and Blacks in the United States." *Social Forces* 58(3):707–25.

Spanier, Graham B., and Robert A. Lewis. 1980. "Marital Quality: A Review of the Seventies." *Journal of Marriage and the Family* 42(4):825–38.

Spector, Malcolm, and John I. Kituse. 1973. "Social Problems: A Re-Formulation." *Social Problems* 21(2):145–58.

Spiro, Melford E., 1956. *Kibbutz: Venture in Utopia.* Cambridge, MA: Harvard University Press.

Spitz, Rene. 1945. "Hospitalism: An Inquiry into the Genesis of Psychiatric Conditions in Early Childhood." *Psychoanalytic Studies of the Child* 1:53–74.

Spitze, Glenna. 1988. "Women's Employment and Family Relations: A Review." *Journal of Marriage and the Family* 50:595–618.

Spitze, Glenna. 1991. "Women's Employment and Family Relations." pp. 381–404 in *Contemporary Families: Looking Forward, Looking Back,* edited by Alan Booth. Minneapolis, MN: National Council on Family Relations.

Sprecher, Susan. 1989. " The Importance to Males and Females of Physical Attractiveness, Earning Potential and Expressiveness in Initial Attraction." *Sex Roles* 21:591–607.

Stacey, Judith. 1993. "Good Riddance to 'The Family': A Response to David Popenoe." *Journal of Marriage and the Family* 55:545–47.

Stacey, Judith. 1996. *In the Name of the Family.* Boston, MA: Beacon.

Stacey, Judith. 1998. "Sociologists Differ about Family Textbooks' Message." *Footnotes* 26(1):10.

Stack, Carol. 1974. *All Our Kin.* New York: Harper and Row.

Staples, Robert, and Alfredo Mirandé. 1980. "Racial and Cultural Variations among American Families: A Decennial Review of the Literature on Minority Families." *Journal of Marriage and the Family* 33:119–35.

Staples, Robert. 1985. "Changes in Black Family Structure: The Conflict between Family Ideology and Structural Conditions." *Journal of Marriage and the Family* 47:1005–13.

Staples, Robert. 1997. "An Overview of Race and Marital Status." Pp. 269–72 in *Black Families,* 3d ed., edited by H. P. McAdoo. Thousand Oaks, CA: Sage.

Stark, Rodney, and James McEvoy. 1970. "Middle Class Violence." *Psychology Today* 4:52–54, 110–12.

Starr, Paul. 1982. *The Social Transformation of American Medicine.* New York: Basic Books.

Starrels, M. E. 1994. "Husbands' Involvement in Female Gender-Typed Household Chores." *Sex Roles: A Journal of Research* 31:473–91.

Statistical Abstract of the United States 1971–1999. Washington, DC: U.S. Government Printing Office.

Staude, John. 1982. *The Adult Development of C. G. Jung.* London: Routledge and Kegan Paul.

Stein, L., and A. K. Baxter. 1974. "A Note about this Volume," In *How to Win and Hold a Husband.* New York: Arno Press/New York Times.

Steinmetz, Suzanne K. 1977. *The Cycle of Violence: Assertive, Aggressive, and Abusive Family Interaction.* New York: Praeger.

Steinmetz, Suzanne, Sylvia Clavan, and Karen F. Stein. 1990. *Marriage and Family Realities.* Grand Rapids, PA: Harper and Row.

Stendhal. 1967 (1826–42). *On Love.* New York: Grosset and Dunlap.

Stephen, Elizabeth Hervey. 1999. "Assisted Reproductive Technologies: Is the Price Too High?" *Population Today* 27:1–2, 7.

Stern, Marilyn, and Katherine Hildebrandt Karraker. 1989. " Sex Stereotyping of

Infants: A Review of Gender Labeling Studies." *Sex Roles* 20:501–22.

Stets, Jan E. 1988. *Domestic Violence and Control.* New York: Springer-Verlag.

Stets, Jan E. 1993. "Control in Dating Relationships." *Journal of Marriage and the Family* 55(3):673–85.

Stets, Jan E., and Murray A. Straus. 1990. "The Marriage License As Hitting License: a Comparison of Assaults in Dating, Cohabiting, and Married Couples." Pp. 227–44 in *Physical Violence in American Families: Risk Factors and Adaptations to Violence in 8,145 Families,* edited by M. A. Straus and R. J. Gelles. New Brunswick, NJ: Transaction.

Stevens-Long, Judith. 1984. *Adult Life: Developmental Processes.* Palo Alto, CA: Mayfield.

Stewart, S. 1996. "Changing Attitudes toward Violence against Women: the Musasa Project." Pp. 343–62 in *Learning about Sexuality: A Practical Beginning,* edited by S. Zeidenstein and K. Moore. New York: International Women's Health Coalition, The Population Council.

Stolley, Kathy S. 1993. "Statistics on Adoption in the United States." In *The Future of Children* 3(1). Los Altos, CA: The David and Lucille Packard Foundation.

Stone, Lawrence. 1977. *The Family, Sex, and Marriage in England, 1500–1800.* New York: Harper and Row.

Strathern, Marilyn, ed. 1987. *Dealing with Inequality: Analysing Gender Relations in Melanesia and Beyond: Essays by Members of the 1983/1984 Anthropological Research Group at the Research School of Pacific Studies, the Australian National University.* New York: Cambridge University Press.

Straus, Murray A. 1991. "Physical Violence in American Families: Incidence, Rates, Causes, and Trends." pp. 17–34 in *Abused and Battered,* edited by Dean Knudsen and JoAnn Miller. New York: Aldine de Gruyter.

Straus, Murray A. 1979. "Measuring Intrafamily Conflict and Violence: The Conflict Tactics (CT) Scales." *Journal of Marriage and the Family* 41(Feb.):75–88.

Straus, Murray A. 1983. "Ordinary Violence, Child Abuse, and Wife-Beating: What Do They Have in Common?" In *The Dark Side of Families. Current Family Violence Research,* edited by David Finkelhor, Richard J. Gelles, Gerald T Hotaling, and Murray A. Straus. Beverly Hills, CA: Sage.

Straus, Murray A. 1990. "Ordinary Violence, Child Abuse, and Wife Beating: What Do They Have in Common?" In *Physical Violence in American Families: Risk Factors and Adaptation to Violence in 8,145 Families,* edited by Murray A. Straus and Richard J. Gelles. New Brunswick, NJ: Transaction.

Straus, Murray A., and Richard J. Gelles. 1988. "How Violent Are American Families? Estimates from the National Family Violence Resurvey and Other Studies." In *Family Abuse and its Consequences,* edited by Gerald T. Hotaling, David Finkelhor, John T. Kirkpatrick, and Murray A. Straus. Newbury Park, CA: Sage.

Straus, Murray A., Richard J. Gelles, and Suzanne K. Steinmetz. 1980. *Behind Closed Doors: Violence in the American Family.* New York: Doubleday.

Straus, Murray A., with Denise Donnelly. 1994. *Beating the Devil out of Them: Corporal Punishment in American Families.* San Francisco: Jossey-Bass.

Suarez, Zulema E. 1993. "Cuban Americans: From Golden Exiles to Social Undesirables." Pp. 164–83 in *Family Ethnicity: Strength in Diversity,* edited by H. P. McAdoo. Thousand Oaks, CA: Sage.

Sudarkasa, Niara. 1993. "Female-Headed African American Households." Pp. 81–9 in *Family Ethnicity: Strength in Diversity,* edited by H. P. McAdoo. Thousand Oaks, CA: Sage.

Sudarkasa, Niara. 1997. "African American Families and Family Values." Pp. 9–40 in *Black Families,* 3d ed., edited by H. P. McAdoo. Thousand Oaks, CA: Sage.

Sue, S., and Harry Kitano. 1973. "Asian American Stereotypes." *Journal of Social Issues* 29(Spring):83–98.

Sugarman, D. B., and G. T. Hotaling. 1989. "Dating Violence: Prevalence, Context and Risk Markers." Pp. 3–32 in *Violence in Dating Relationships,* edited by M. Pirog-Good and J. Stets. New York: Praeger.

Sullivan, Deborah A., and Rose Weitz. 1988. *Labor Pains.* New Haven, CT: Yale University Press.

Surgeon General. 1999. *Mental Health: A Report of the Surgeon General.* http://www.surgeongeneral.gov/library/mentalhealth/toc/html

Surra, Catherine A. 1990. "Research and Theory on Mate Selection and Premarital Relationships in the 1980s." *Journal of Marriage and the Family* 52:844–65.

Surra, Catherine A., 1991. "Research and Theory on Mate Selection and Premarital Relationships in the 1980s." pp. 54–75 in *Contemporary Families: Looking Forward, Looking Back,* edited by Alan Booth. Minneapolis, MN: National Council on Family Relations.

Sweeney, M. 1997. "Remarriages of Women and Men After Divorce: The Role of Socioeconomic Prospects." *Journal of Family Issues* 18:479–502.

Sweet, James A., and Larry L. Bumpass. 1987. *American Families and Households.* New York: Russell Sage.

Sweet, James A., Larry L. Bumpass, and Vaughn Call. 1988. "The Design and Content of the National Survey of Families and Households." NSFH Working Paper No. 1, Center for Demography and Ecology, University of Wisconsin, Madison.

Synnott, Anthony. 1983. "Little Angels, Little Devils: A Sociology of Children." *Canadian Review of Sociology and Anthropology* 20:79–95.

Szinovacz, Maximiliane. 1987. "Family Power." In *Handbook of Marriage and the Family,* edited by Marvin B. Sussman and Susan K. Steinmetz. New York: Plenum.

Tamir, Lois M. 1982. *Men in Their Forties: The Transition to Middle Age.* New York: Springer.

Tatara, Toshio. 1989. "Toward the Development of Estimates of the National Incidence of Reports of Elder Abuse Based on Currently Available State Data." In *Elder Abuse: Practice and Policy,* edited by Rachel Filinson and Stanley R. Ingman. New York: Human Sciences Press.

Tavris, Carol, and Susan Sadd. 1977. *The Red Book Report on Female Sexuality.* New York: Delacorte.

Taylor, Howard F. 1980. *The IQ Game.* New Brunswick, NJ: Rutgers University Press.

Taylor, Robert J., Linda M. Chatters, M. Belinda Tucker, and Edith Lewis. 1991. "Developments in Research on Black Families: A Decade Review." pp. 275–96 in *Contemporary Families: Looking Forward, Looking Back,* edited by Alan Booth. Minneapolis, MN: National Council on Family Relations.

Taylor, Ronald L. 1994. "Minority Families and Social Change." In *Minority Families in the United States: A Multicultural Perspective,* edited by R.L Taylor. Englewood Cliffs, NJ: Prentice Hall.

Teachman, Jay D. 2000. "Diversity of Family Structure: Economic and Social Influences." Pp. 32–58 in *Handbook of Family Diversity,* edited by D. Demo, K. Allen, and M. Fine. New York: Oxford University Press.

Terman, Lewis M. 1938. *Psychological Factors in Marital Happiness.* New York: McGraw-Hill.

Thomas, Elizabeth M. 1958. *The Harmless People.* New York: Knopf.

Thomas, Lawrence. 1956. *The Occupational Structure and Education.* Englewood Cliffs, NJ: Prentice Hall.

Thompson, Linda. 1991. "Family Work: Women's Sense of Fairness." *Journal of Family Issues* 12:181–96.

Thompson, Linda. 1993. "Conceptualizing Gender in Marriage: The Case of Marital Care." *Journal of Marriage and the Family* 5:557–69.

Thompson, Linda., and Alexis J. Walker. 1989. "Gender in Families: Women and Men in Marriage, Work, and Parenthood." *Journal of Marriage and the Family* 51:845–71.

Thompson, Sharon. 1996. *Going All the Way: Teenage Girls' Tales of Sex, Romance, and Pregnancy*. New York: Hill and Wang.

Thornton, Arland. 1989. "Changing Attitudes toward Family Issues in the United States." *Journal of Marriage and the Family* 51:873–93.

Thwing, C. F., and C. F. B. Thwing. 1887. *The Family: An Historical and Social Study*. Boston, MA: Lee and Shepard.

Tilly, Louise A. 1981. "Women's Collective Action and Feminism in France, 1870–1914." In *Class Conflict and Collective Action*, edited by Louise A. Tilly and Charles Tilly. Beverly Hills, CA: Sage.

To the Contrary Poll. 1997. http://www.pbs.org/ttc/speakup/pollresults.html. 5 March, 1999.

Tocqueville, Alexis de. [1832] 1969. *Democracy in America*. New York: Anchor.

Treas, J., and V. L. Bengtson. 1987. "Family in Later Years." In *Handbook on Marriage and the Family*, edited by M. Sussman and S. Steinmetz. New York: Plenum.

Treas, Judith. 1995. "Older Americans in the 1990s and Beyond. " *Population Bulletin* 50(2). Washington, DC: Population Reference Bureau.

Trickett, Penelope K., J. Lawrence Aber, Vicki Carlson, and Dante Cicchetti. 1991. "The Relationship of Socioeconomic Status to the Etiology and Developmental Sequelae of Physical Child Abuse." *Developmental Psychology* 27:148–58.

Trost, Jan E. 1985. "Abandon Adjustment!" *Journal of Marriage and the Family* 47(4):1072–73.

Tucker, M. Belinda, and Claudia Mitchell-Kernan. 1995. "Trends in African American Family Formation: A Theoretical and Statistical Overview." Pp. 3–25 in *The Decline in Marriage Among African Americans: Causes, Consequences, and Policy Implications*, edited by B. Tucker and C. Mitchell-Kernan. New York: Russell Sage Foundation.

U.S. Bureau of Labor Statistics. 1998. "Employment and Earnings." Washington, DC: U.S. Government Printing Office.

U.S. Bureau of the Census. 1967-1999. *Current Population Survey* (March). Washington, DC: Government Printing Office.

U.S. Bureau of the Census. 1978a. "Perspectives on American Fertility." *Current Population Reports* (July) Series P23, no. 70. Washington, DC: Government Printing Office.

U.S. Bureau of the Census. 1985-1999. *Surveys of Income and Programs Participation*.

U.S. Bureau of the Census. 1989. "Studies in Marriage and the Family." *Current Population Reports* P23, no. 165. See also Series P20, no. 436, Washington, DC: U.S. Government Printing Office.

U.S. Bureau of the Census. 1990. *Current Population Reports* P20-454. Washington, DC: U.S. Government Printing Office.

U.S. Bureau of the Census. 1992a. "Population Projections of the United States, by Age, Sex, Race, and Hispanic Origin." *Current Population Reports* P25, no. 1092. Washington, DC: U.S. Government Printing Office.

U.S. Bureau of the Census. 1992b. "Marriage, Divorce, and Remarriage in the 1990s." *Current Population Reports* P23, no. 180. Washington, DC: U.S. Government Printing Office.

U.S. Bureau of the Census. 1992c. "Marital Status and Living Arrangements." *Current Population Reports* P20, no. 468. Washington, DC: U.S. Government Printing Office.

U.S. Bureau of the Census. 1992d. "Special Studies." *Current Population Reports* P23, no. 181. Washington, DC: U.S. Government Printing Office.

U.S. Bureau of the Census. 1996. "Projections of the Number of Households and Families in the United States." *Current Population Reports* P-25-1129. Washington, DC: U.S. Government Printing Office.

U.S. Bureau of the Census. 1998. "Marital Status and Living Arrangements." *Current Population Reports* P20-514. Washington, DC: U.S. Government Printing Office.

U.S. Bureau of the Census. 1998a. "Money Income in the United States." *Current Population Reports* P60-200. Washington, DC: Government Printing Office.

U.S. Bureau of the Census. 1998b. "Poverty in the United States." *Current Population Reports* P60-201. Washington, DC: Government Printing Office.

U.S. Bureau of the Census. 1998c. "Dynamics of Economic Well-Being, Poverty 1993-94: Trap Door? Revolving Door? or Both?" *Current Population Reports* P70-63. Washington, DC: Government Printing Office.

U.S. Bureau of the Census. 1999. "Poverty in the United States: 1998." *Current Population Reports* P60-207. Washington, DC: U.S. Government Printing Office.

U.S. Bureau of the Census. 1999. "Historical Income Tables." *Current Population Survey* (March), Series P-33. Washington, DC: Government Printing Office.

U.S. Bureau of the Census. 2000. *International Data Base*. Washington, DC.

U.S. Department of Education. 1997. "Fathers' Involvement in Their Children's Schools." NCES 98-091. Washington, DC: U.S. Government Printing Office.

U.S. Department of Labor. 1998. *Equal Pay: A Thirty-Five Year Perspective*. Washington, DC: Women's Bureau, U.S. Government Printing Office.

U.S. National Center for Health Statistics. *Vital Statistics of the United States, 1980-99*.

National Vital Statistics Reports. Washington, DC: U.S. Government Printing Office.

Uhlenberg, Peter. 1980. "Death and the Family." *Journal of Family History* 5(3):313–20.

Ullian, D. 1984. "Why Girls Are Good: A Constructivist View." *Sex Roles* 11:241–56.

Umberson, Deborah. 1992. "Gender, Marital Status, and the Social Control of Behavior." *Social Science and Medicine* 34:907–17.

Umberson, Deborah. 1987. "Family Status and Health Behaviors: Social control As a Dimension of Social Integration." *Journal of Health and Social Behavior* 28:306–19.

United Nationals Development Programme. 1998. *Human Development Report*. New York: Oxford University Press.

Uttal, Lynet. 1999. "Using Kin for Child Care: Embedment in the Socioeconomic Networks of Extended Families." *Journal of Marriage and the Family* 61(4):845.

Vaillant, Caroline O., and George E. Vaillant. 1993. "Is the U Curve of Marital Satisfaction an Illusion? A Forty Year Study of Marriage." *Journal of Marriage and the Family* 55:230–39.

Vaillant, G. 1977. Adaptation to Life. Boston, Mass.: Little, Brown.

Valdivieso, Rafael, and Cary Davis. 1988. "U.S. Hispanics: Changing Issues for the 1990s." *Population Trends and Public Policy*, no. 17. Washington, DC: Population Reference Bureau.

Vance, Carole S. 1989. *Pleasure and Danger: Exploring Female Sexuality*. London: Routledge.

Vanek, Joann. 1974. "Time Spent in Housework." *Scientific American* 231(Nov.):116–20.

Vaughan, Diane. 1986. *Uncoupling: Turning Points in Intimate Relationship*. New York: Oxford University Press.

Veevers, Jean E. 1980. *Childless by Choice*. Toronto: Butterworths.

Vega, William A. 1991. "Hispanic Families in the 1980s." pp. 297–306 in *Contemporary Families: Looking Forward, Looking Back*, edited by Alan Booth. Minneapolis, MN: National Council on Family Relations.

Ventura, Stephanie J. 1995. "Births to Unmarried Mothers: United States, 1980–92." National Center for Health Statistics. Vital Health Statistics 21(53).

Ventura, Stephanie, J., Joyce Martin, Sally Curtin, and T .J. Mathews. 1998. "Report of Final Natality Statistics, 1996." *Monthly Vital Statistics Report* 46(11). Washington, DC: U.S. Department of Health and Human Services.

Ventura, Stephanie, William Mosher, Sally Curtin, and Joyce Amba. 1999. *Highlights of Trends in Pregnancies and Pregnancy Rates by Outcome*. Hyattsville, MD: National Center for Health Statistics, U.S. Department of Health and Human Services.

Vieille, Paul. 1978. "Iranian Women in Family Alliance and Sexual Politics." In *Women in the Moslem World*, edited by Lois Beck and Nikki Keddie. Cambridge, MA: Harvard University Press.

Vinokur, Amiram D., Richard Price, and Robert Caplan. 1996. "Hard Times and Hurtful Partners: How Financial Strain Affects Depression and Relationship Satisfaction of Unemployed Persons and Their Spouses." *Journal of Personality and Social Psychology* 71:166–79.

Voydanoff, Patricia. 1987. *Work and Family Life*. Newbury Park, CA: Sage.

Voydanoff, Patricia. 1991. "Economic Distress and Family Relations: A Review of the Eighties." pp. 429–45 in *Contemporary Families: Looking Forward, Looking Back*, edited by Alan Booth. Minneapolis, MN: National Council on Family Relations.

Vygotsky, Lev S. [1934] 1962. *Language and Thought*, edited and translated by E. Hanfmann and G. Vakar. Cambridge, MA: MIT Press.

Waite, Linda. 1995. "Does Marriage Matter?" *Demography* 32:483–507.

Walker, L. E. 1979. *The Battered Woman*. New York: Harper and Row.

Walker, L. E. 1984. *The Battered Woman Syndrome*. New York: Springer.

Wallace, Pamela M., and Ian H. Gotlib. 1990. "Marital Adjustment during the Transition to Parenthood: Stability and Predictors of Change." *Journal of Marriage and the Family* 52(1):21–9.

Wallerstein, Judith S., and Sandra Blakeslee. 1989. *Second Chances: Men, Women, and Children. A Decade after Divorce*. New York: Trenor and Fields.

Walster, Elaine, and G. William Walster. 1978. *A New Look at Love*. Reading, MA: Addison-Wesley.

Walster, Elaine, G. William Walster, and Berschied. 1978. *Equity, Theory and Research*. Rockleigh, NJ: Allyn and Bacon.

Walzer, Susan. 1998. *Thinking about the Baby: Gender and Transitions into Parenthood*. Philadelphia, PA: Temple University Press.

Wang, Ching-Tung. 1999. *Current Trends in Child Abuse Reporting and Fatalities: The Results of the 1997 Annual Fifty State Survey*. Working Paper No. 808. Chicago: The Center on Child Abuse Prevention Research, Prevent Child Abuse America.

Warren, Carol A. B. 1987. *Madwives: Schizophrenic Women at Mid-Century*. New Brunswick, NJ: Rutgers University Press.

Washington Post/Kaiser Family Foundation/Harvard University Survey of Americans' Attitudes toward Gender, March 22–26, 1998.

Watkins, S. C., J. A. Menken, and J. Bongaarts. 1987. "Demographic Foundations of Family Change." *American Sociological Review* 52:346–58.

Watt, Ian. 1957. *The Rise of the Novel*. Berkeley: University of California Press.

Weber, Max. 1922/68. *Economy and Society*. New York: Bedminister Press.

Weber, Max. 1923/61. *General Economic History*. New York: Collier-Macmillan.

Weinberg, D. H. 1996. "A Brief Look at Postwar U.S. Income Inequality." *Current Population Reports*, Household Economic Studies, Series P60, no. 191. Washington, DC: U.S. Government Printing Office.

Weinberg, Martin S., and Colin J. Williams. 1980. "Sexual Embourgeoisment? Social Class and Sexual Activity: 1938–1970." *American Sociological Review* 45:33–48.

Weinberg, Martin, Colin Williams, and Douglas Pryor. 1993. *Dual Attraction: Understanding Bisexuality*. New York: Oxford University Press.

Weiss, Robert S. 1979. "Growing Up a Little Faster: The Experience of Growing Up in a Single-Parent Household." *Journal of Social Issues* 35(Fall):97–111.

Weissman, M. M. 1987. "Advances in Psychiatric Epidemeology: Rates and Risks for Major Depression." *American Journal of Public Health* 77:445–51.

Weitzman, Lenore J. 1981. *The Marriage Contract: Spouses, Lovers and the Law*. New York: Free Press.

Wellman, Barry, ed. 1999. *Networks in the Global Village: Life in Contemporary Communities*. Boulder, CO: Westview Press.

Welter, Barbara. 1966. "The Cult of True Womanhood: 1820–1860." *American Quarterly* 18:151–74.

Wertz, Dorothy C. 1983. "What Birth Has Done for Doctors: A Historical View." *Women and Health* (Spring):7–24.

Wertz, Richard W., and Dorothy C. Wertz. 1977. *Lying-In: A History of Childbirth in America*. New York: Free Press.

West, Candace, and Don H. Zimmerman. 1987. "Doing Gender." *Gender and Society* 1:125–51.

West, Candace, and Sarah Fenstermaker. 1993. "Power, Inequality, and the Accomplishment of Gender: An Ethnomethodological View." Pp. 151–74 in *Theory on Gender/Feminism on Theory*, edited by P. England. New York: Aldine de Gruyter.

Westoff, Charles F. 1974. "Coital Frequencies and Contraception." *Family Planning Perspectives* 3:136–41.

Westoff, Charles F. 1986. "Fertility in the United States." *Science* 234:554–59.

Weston, Kath. 1998. *Long Slow Burn: Sexuality and Social Science*. New York: Routledge.

White, Lynn, and John N. Edwards. 1990. "Emptying the Nest and Parental Well-Being: An Analysis of National Panel Data," *American Sociological Review* 55(Apr.):235–42.

White, Lynn. 1991. "Determinants of Divorce: A Review of Research in the Eighties." pp. 141–49 in *Contemporary Families: Looking Forward, Looking Back*, edited by Alan Booth. Minneapolis, MN: National Council on Family Relations.

Whitehead, Barbara Dafoe. 1996. *The Divorce Culture: Rethinking Commitments to Marriage and Family*. New York: Vintage.

Whiting, Beatrice. 1963. *Six Cultures: Studies in Child Rearing*. New York: Wiley.

Whyte, William H. 1956. *The Organization Man*. New York: Doubleday.

Wilkie, Jane R. 1987. "Marriage, Family Life, and Women's Employment." In *Working Women*, 2nd ed., edited by Ann Helton Stromberg and Shirley Harkess. Mountain View, CA: Mayfield.

Wilkie, Jane Riblett; Myra Marx Ferree, and Kathryn Strother Ratcliff. 1998. "Gender and Fairness: Marital Satisfaction in Two-Earner Couples." *Journal of Marriage and the Family* 60(3):577–94.

Williams, Christine L. 1993. "Psychoanalytic Theory and the Sociology of Gender." Pp. 131–49 in *Theory on Gender/Feminism on Theory*, edited by Paula England. New York: Aldine de Gruyter.

Williams, Florence. 1998. *Glamour* 286–7; 294–6.

Williams, Norma. 1988. "Role Making among Married Mexican American Women: Issues of Class and Ethnicity." *Journal of Applied Behavioral Science* 24(2):203–17.

Willie, Charles V. 1981. *A New Look at Black Families*. Bayside, NY: General Hall.

Willie, Charles V. 1985. *Black and White Families*. Bayside, NY: General Hall.

Wilson, Barbara Foley, and Sally Cunningham Clarke. 1992. "Remarriages: A Demographic Profile." *Journal of Family Issues* 13:123–41.

Wilson, William J. 1978. *The Declining Significance of Race*. Chicago, IL: University of Chicago Press.

Wilson, William J. 1987. *The Truly Disadvantaged: The Inner City, the Underclass, and Public Policy*. Chicago, IL: University of Chicago Press.

Wilson, William J., and Kathryn M. Neckerman. 1986. "Poverty and Family Structure: The Widening Gap between Evidence and Public Policy Issues." In *Fighting Poverty: What Works and What Doesn't*, edited by Sheldon H. Danzinger and Daniel H. Weinberg. Cambridge, MA: Harvard University Press.

Wilson, William Julius. 1999. *The Bridge over the Racial Divide: Rising Inequality and Coalition Politics*. Berkeley, CA: University of California Press.

Wineberg, Howard, and James McCarthy. 1998. "Living Arrangements after Divorce: Cohabitation versus Remarriage." *Journal of Divorce and Remarriage* 29:131–46.

Wineberg, Howard. 1990. "Childbearing after Remarriage." *Journal of Marriage and the Family* 52: 31–8.

Winikoff, Beverly. 1995. "Acceptability of Medical Abortion in Early Pregnancy." *Family Planning Perspectives* 27(4):142–48.

Wittgenstein, Ludwig. 1953. *Philosophical Investigations*. New York: Macmillan.

Wolfinger, Nicholas W. 1999. "Trends in the Intergenerational Transmission of Divorce." *Demography* 36(3):415–20.

Woll, Stanley B., and Peter Young. 1989. "Looking for Mr. or Ms. Right: Self-Presentation in Videodating." *Journal of Marriage and the Family* 51:483–88.

Women's International Network. 1993. "Progress Report: Campaign to Stop FGM." *WIN News* 19:29.

Wood, Julia T. 1994. *Gendered Lives: Communication, Gender, and Culture*. Belmont, CA: Wadsworth.

Wood, Julia T. 1996. "She Says/He Says: Communication, Caring, and Conflict in Heterosexual Relationships." Pp. 149–62 in *Gendered Relationships*, edited by J. T. Wood. Mountain View CA: Mayfield.

Wright, Erik O. 1985. *Classes*. London: Verso.

Wright, Erik Olin. 1997. *Class Counts: Comparative Studies in Class Analysis*. New York: Cambridge University Press.

Wu, Zheng, and Margaret J. Penning. 1997. "Marital Instability after Midlife." *Journal of Family Issues* 18:459–78.

Wu, Zheng, and T.R. Balakrishnan. 1995. Dissolution of Premarital Cohabitation in Canada. *Demography* 32:521–9.

Wylie, Lawrence. 1964. *Village in the Vaucluse*. Cambridge, MA: Harvard University Press.

Ybarra, Lea. 1982. "When Wives Work: The Impact on the Chicano Family." *Journal of Marriage and the Family*:169–78.

Yellowbird, Michael, and C. Matthew Snipp. 1994. "American Indian Families." In *Minority Families in the United States: A Multicultural Perspective*. Englewood Cliffs, NJ: Prentice Hall, edited by R. L. Taylor.

Yllö, Kersti A. 1998. "Through a Feminist Lens: Gender, Power, and Violence." Pp. 609–18 in *Families in the U.S.: Kinship and Domestic Politics*, edited by K. Hansen and A. Garey. Philadelphia: Temple University Press.

Yllö, Kersti. 1988. "Political and Methodological Debates in Wife Abuse Research." In *Feminist Perspectives on Wife Abuse*, edited by Kersti Yllö and Michele Bograd. Newbury Park, CA: Sage.

Young, Michael, and Peter Willmott. 1957. *Family and Kinship in East London*. London: Routledge and Kegan Paul.

Zablocki, Benjamin. 1980. *Alienation and Charisma. A Study of Contemporary American Communes*. New York: Free Press.

Zarsky, Lyuba, and Samuel Bowels, eds. 1986. *Economic Report of the People*. Boston, MA: South End Press.

Zaslow, Martha. 1981. "Depressed Mood in New Fathers." Paper presented at the Society for Research in Child Development, Boston (April).

Zavella, Patricia. 1987. *Women's Work and Chicano Families: Cannery Workers of the Santa Clara Valley*. Ithaca, NY: Cornell University Press.

Zelizer, Viviana. 1985. *Pricing the Priceless Child*. New York: Basic Books.

Zhou, Min. 1997. "Growing up American: The Challenge Confronting Immigrant Children and the Children of Immigrants." *Annual Review of Sociology* 23:63–95.

Zinn, Maxine B. 1980. "Employment and Education of Mexican-American Women: The Interplay of Modernity and Ethnicity in Eight Families." *Harvard Educational Review* 50(1):47–62.

Zinn, Maxine B. 1982. "Qualitative Methods in Family Research: A Look Inside the Chicano Families." *California Sociologist* (Summer):58–79.

Zinn, Maxine B., and D. Stanley Eitzen. 1987. *Diversity in American Families*. New York: Harper and Row.

Zuberi, Tukufu (Antonio McDaniel). 1998. *African American Men, Inequality and Family Structure*. WP-98-12C. Philadephia, PA: National Center on Fathers and Families.

Photo Credits

Name Index

Subject Index